METAL AND INORGANIC WASTE RECLAIMING ENCYCLOPEDIA

METAL AND INORGANIC WASTE RECLAIMING ENCYCLOPEDIA

by

Marshall Sittig

NOYES DATA CORPORATION
Park Ridge, New Jersey, U.S.A.
1980

Library of Congress Catalog Card Number: 80-21669
ISBN: 0-8155-0823-9
Printed in the United States

Published in the United States of America by
Noyes Data Corporation
Noyes Building, Park Ridge, New Jersey 07656

Library of Congress Cataloging in Publication Data

Sittig, Marshall.
Metal and inorganic waste reclaiming encyclopedia.

(Pollution technology review ; no. 70) (Chemical technology review ; no. 175)
Includes bibliographical references and index.
1. Recycling (Waste, etc.)--Patents. 2. Metals--Patents. 3. Materials--Patents. I. Title.
II. Series. III. Series: Chemical technology review ; no. 175.
TD794.5.S56 660.2'82 80-21669
ISBN 0-8155-0823-9

FOREWORD

This book describes 328 recent processes for reclaiming metal and inorganic wastes, and is arranged in encyclopedic form, alphabetically by the product recovered.

The recovery and recycling of these industrial wastes is assuming ever-increasing importance due to the need for ore and mineral conservation, and the requirements of pollution regulations. The technology of each specific recovery and recycling process is outlined, as well as background as to the origin of the waste material, some of the problems posed by the specific waste, and possible alternative processes for reclaiming, recovery, and recycling.

There is also an index listing the sources of waste feed materials, and each entry lists the product that may be reclaimed from the waste source.

Contents and Subject Index

Introduction

The recovery and recycling of industrial wastes are assuming ever-increasing importance in the 1980s due to a combination of the following factors.

(1) Increasing world scarcity of economically viable deposits of some ores.

(2) Increasing national and international efforts to conserve and control mineral prices along the lines of OPEC and petroleum.

(3) Increasing pressure in a somewhat unfriendly world for major nations in particular to be reasonably self-dependent as regards supplies of strategic commodities.

(4) Increased emphasis on pollution control and toxic materials control making dumping of waste slags and the like much less desirable.

(5) Rapidly increasing energy prices which accentuate in many cases the virtues of using recycled rather than virgin raw materials.

Due to the amount of material available, it was necessary to divide the material in order to make a volume of reasonable size and hence this book has a companion volume entitled *Organic and Polymer Waste Reclaiming Encyclopedia* (1).

These volumes will focus only on the technology available for the recycling of industrial wastes. Process economics will not be considered because of the following:

(a) Rapidly fluctuating prices of some ores and commodities.

(b) Rapidly increasing energy costs which make economics meaningless unless carefully dated and/or adjusted.

(c) Rapid monetary inflation in most countries which again makes costs very difficult to state accurately in terms of today's situation, or worse, of future projections.

In each case, an attempt will be made, not only to outline the technology of a specific recovery and recycling process, but to give some background as to:

- The origin of the waste material;
- Some of the problems posed by the waste; and
- Some possible alternative processes for recovery and recycling.

Emphasis is placed in this volume and in the companion volume on the recycling of industrial wastes, often within the factory fence, rather than on postconsumer wastes such as waste newspapers, discarded beverage bottles and the like. However, it will be appreciated that there is no clear line of demarcation between these areas. Thus, some industrial wastes may quickly end up as municipal wastes and must of necessity be recycled from that point.

U.S. Import reliance of selected minerals and metals are shown in Figure 1. These data emphasize the need for the U.S. to develop effective techniques for the recovery and recycling of many materials.

This volume is arranged like any dictionary or encyclopedia in alphabetical manner. Every effort is made to cross-index entries within this volume and frequent cross-references are also made to the companion volume, *Organic and Polymer Waste Reclaiming Encyclopedia* (1).

As the reader will observe, an index to the sources of waste feed for the various recycling processes is provided at the end of this volume. Together with the alphabetical arrangement by recovered material, this index of waste sources should make access to useful information both rapid and easy.

The majority of references in this volume are to U.S. Patents. A secondary source is the published work of the U.S. Bureau of Mines. A bibliography at the end of this volume lists the general publications cited in the text.

This volume, together with its companion volume, *Organic and Polymer Waste Reclaiming Encyclopedia*, represents an updating of an earlier work by the publisher of this volume entitled *Resource Recovery and Recycling Handbook of Industrial Wastes* (3) published in 1975.

The reader's attention is also called to a number of other published works including a summary volume by Barton (4), monthly materials surveys by the U.S. Bureau of Mines (2), a volume of forecasts for resource recovery to the year 1990, made in 1975 (5), a discussion of energy use patterns for metal recycling (6), an Office of Technology Assessment review of technical options for the conservation of metals (7) and a series of bibliographies on solid waste reclamation and recycling prepared by the National Technical Information Service covering the period 1964 to July 1979 (8)-(13) and dealing with:

Packaging and Containers (8)
Plastics (9)
Metals (10)

Glass (11)
Paper (12)
Tires (13)

Detailed discussions of various metal scrap recycling processes and their environmental impacts have been reviewed by Nack et al (15).

Figure 1: U.S. Net Import Reliance of Selected Minerals and Metals as a Percent of Consumption in 1978

MINERALS AND METAL	NET IMPORT RELIANCE* AS A PERCENT OF APPARENT CONSUMPTION** (0%–100%)	MAJOR FOREIGN SOURCES (1974-1977)
COLUMBIUM	100	BRAZIL, THAILAND, CANADA
MICA (sheet)	100	INDIA, BRAZIL, MALAGASY REPUBLIC
STRONTIUM	100	MEXICO, SPAIN
MANGANESE	98	GABON, BRAZIL, SOUTH AFRICA
TANTALUM	97	THAILAND, CANADA, MALAYSIA, BRAZIL
COBALT	97	ZAIRE, BELG.-LUX., ZAMBIA, FINLAND
BAUXITE & ALUMINA	93	JAMAICA, AUSTRALIA, SURINUM
CHROMIUM	92	SOUTH AFRICA, U.S.S.R., SOUTHERN RHODESIA, TURKEY
PLATINUM - GROUP METALS	91	SOUTH AFRICA, U.S.S.R., UNITED KINGDOM
ASBESTOS	84	CANADA, SOUTH AFRICA
FLUORINE	82	MEXICO, SPAIN, SOUTH AFRICA
TIN	81	MALAYSIA, BOLIVIA, THAILAND, INDONESIA
NICKEL	77	CANADA, NORWAY, NEW CALEDONIA, DOMIN. REP.
CADMIUM	66	CANADA, AUSTRALIA, BELG.-LUX., MEXICO
ZINC	62	CANADA, MEXICO, AUSTRALIA, BELG.-LUX.
POTASSIUM	61	CANADA, ISRAEL, W. GERMANY
SELENIUM	61	CANADA, JAPAN, YUGOSLAVIA, MEXICO
MERCURY	57	ALGERIA, CANADA, SPAIN, MEXICO, YUGOSLAVIA
GOLD	54	CANADA, SWITZERLAND, U.S.S.R.
TUNGSTEN	50	CANADA, BOLIVIA, PERU, THAILAND
ANTIMONY	48	SOUTH AFRICA, BOLIVIA, CHINA
SILVER	41	CANADA, MEXICO, PERU, UNITED KINGDOM
BARIUM	40	PERU, IRELAND, MEXICO
TITANIUM (ilmenite)	39	CANADA, AUSTRALIA
GYPSUM	34	CANADA, MEXICO, JAMAICA, DOMIN. REP.
IRON ORE	29	CANADA, VENEZUELA, BRAZIL, LIBERIA
VANADIUM	27	SOUTH AFRICA, CHILE, U.S.S.R.
COPPER	19	CANADA, CHILE, PERU, ZAMBIA
IRON & STEEL SCRAP	(17) NET EXPORTS	
IRON & STEEL PRODUCTS	13	JAPAN, EUROPE, CANADA
LEAD	11	CANADA, MEXICO, PERU, AUSTRALIA
ALUMINUM	10	CANADA
SULFUR	10	CANADA, MEXICO
SALT	9	CANADA, BAHAMAS, MEXICO
CEMENT	7	CANADA, NORWAY, BAHAMAS, MEXICO, UNITED KINGDOM
PUMICE & VOLCANIC CINDER	5	GREECE, ITALY

*Net Import Reliance = Imports - Exports + Adjustments for Government and Industry Stock Changes

**Apparent consumption = U.S. Primary + Secondary Production + Net Import Reliance

Source: Bureau of Mines, U.S. Department of the Interior (Import-Export Data from Bureau of the Census) Reference (2).

A

ABRASIVE MATERIAL FROM SCARFER SPITTINGS

Scarfer spittings are a steel mill waste material produced by scarfing steel ingots, blooms, slabs, billets and bars prior to or during processing to remove surface defects. Scarfing consists of rapidly removing the surface of steel by the use of a fuel containing oxygen applied to the surface by means of a torch or torches. The oxygen oxidizes a portion of the steel thereby generating heat and increasing the temperature to cause the steel to become molten. The partially oxidized generally globular product is referred to as scarfer spittings.

The scarfer spittings are cooled and flushed from the surface of the steel by high pressure water and are collected in a water bath. The scarfer spittings range in size from less than a No. 100 sieve size, that is, less than 0.149 mm to larger than 50.8 mm in diameter. The spittings are comprised of an outer shell of iron oxides surrounding an inner metallic core which has a chemical composition similar to that of the scarfed steel.

Scarfer spittings have no specific use. In recent years efforts have been directed to recycling the spittings in the steel mill to recover the iron which they contain. A portion of the spittings are mixed with mill scale, steel borings and the like and are used as part of the charge to a sinter strand to recover the iron. However, only the larger sized particles or spittings can be so used. Hence, most of the spittings must be removed from the mill and stored. In recent years, increased emphasis on the surface cleanliness of steel has resulted in an increase in the use of automatic scarfing machines to scarf the steel surfaces. As a result, the volume of scarfer spittings produced in a steel mill has increased, thereby increasing the time required to transport the spittings and the amount of storage space required for the spittings.

A process developed by *H.W. Hitzrot, Jr.; U.S. Patent 4,190,422; February 26, 1980; assigned to Bethlehem Steel Corporation* has as its object the use of scarfer spittings comprised of inner metallic cores and outer brittle shells of iron oxides for use in machine or manual blast cleaning equipment to blast clean metallic and nonmetallic surfaces. This process yields a product having a hardness of about R_C 20 to 35 and a lathlike martensitic matrix which can be used as a size-graded metallic abrasive having good toughness, extended service life and improved cleaning efficiency when compared with commercially available size-graded steel shot and grit abrasives.

The waste material is screened to simultaneously separate it from all the foreign matter collected therewith and also to separate it into a fraction containing particles larger than about 6.35 mm and a fraction containing particles smaller than about 6.35 mm. The fraction containing the larger particles is recycled in the steel mill. The fraction containing the smaller particles is charged into a grinding mill containing appropriate grinding media. The smaller fraction remains in the mill for a time so that the outer shells are broken into small pieces substantially all of which are removed from the surfaces of the inner cores. The pieces of the shells and the metallic cores are separated from each other by screening. The metallic cores are graded into a plurality of sizes by screening. The pieces of shells are recycled in the plant.

The metallic cores have a microstructure of untempered lathlike martensite substantially free from intergranular and intragranular cracking, a hardness of about R_c 20 to 35, and a grain size of between about 3 and 4 and are characterized by having good impact toughness and extended service life.

ABRASIVE MATERIAL FROM SPENT CATALYST

The last few decades have seen a steep rise in the use of heterogenous catalytic reactions. The ever increasing use of crude oil and its distillation products, which must be desulfurized to protect the environment, has led to a considerable increase in the use of catalysts useful in such reactions. In one widely used catalytic desulfurization process, the hydrocarbon starting materials are contacted with an MoCo catalyst on an alumina (Al_2O_3) carrier and the developing hydrogen sulfide is removed. In some instances, MoNi- and WNi-type catalysts or other metallic catalyst combinations are used.

During the catalytic reaction, the catalyst absorbs or adsorbs various chemical elements or compounds from the reactants and eventually becomes inactive. While the catalyst may be regenerated, it will eventually be spent and must be replaced. It is of considerable economic importance to utilize at least portions of such spent catalytic materials.

Various processes have been proposed for working up hydrodesulfurization catalytic materials of the indicated types. One process involves roasting the catalytic material with sodium chloride (NaCl) after calcining and then extracting vanadium, molybdenum, alumina, nickel and/or cobalt, the ammonium salts of molybdenum and vanadium as well as aluminum hydroxide ($Al(OH)_3$) being obtained after several process steps. Cobalt and nickel remaining in the extraction residue must be extracted in a further step if they are to be recovered.

Other processes work with soda (sodium carbonate) instead of sodium chloride. All the known processes are chemically complex and technically expensive, thus making them commercially unattractive. Furthermore, it is quite difficult to separate molybdenum from vanadium and cobalt from nickel.

It is the primary object of the following process to provide for preparing an abrasive material from spent catalytic materials of the indicated type, the alumina carrier component of the starting material providing the abrasive material and the metallic catalyst component providing valuable alloys which may be used directly or after refining in the steel and alloying industry, or which may be readily separated chemically into their chemical elements after having been stripped of the carrier.

This process has been described in some detail by *H. Zeiringer; U.S. Patent 4,142,871; March 6, 1979; assigned to Treibacher Chemische Werke AG, Austria.*

It is one in which an abrasive material is prepared from a starting material comprised of a spent metallic catalyst on an alumina carrier by melting the starting material with a reducing agent to obtain a melt consisting of a melt component including the alumina on an alloy residue, cooling the melt at a speed correlated with a desired crystallite size of the abrasive material to be obtained, and mechanically separating the melt component from the alloy residue before or after solidification, the melt component constituting the abrasive material.

The size of the corundum crystallites obtained in the process may be influenced within wide limits by the rate of cooling of the melt, the crystallite size of the abrasive determining its usefulness for various fields of utilization. Considerable differences in the crystallite sizes and the corresponding utility of the abrasive can be obtained by solidifying the melt very slowly in a block, on the one hand, and rapidly cooling by casting it over steel balls, on the other hand. The resultant sizes of the corundum crystallites may accordingly vary, depending on the method of cooling, from 1 mm to 0.001 mm.

The solid alumina product of the process is an excellent abrasive after comminution and classification into various grain sizes, with or without heat treatment. Depending on the type and amount of additives as well as the crystallite size, these abrasives may be used for polishing steel, precision polishing under very light pressures or polishing wood.

The catalyst component accumulating as an alloy at the bottom of the melting furnace vessel is either cast or tapped with the alumina melt component, or it is permitted to solidify therewith in a block. Depending on the type of catalyst, the alloy is comprised primarily of MoCo, WNi, MoCoVNi and various impurities, such as sulfur, carbon, silicon, iron, titanium or chromium. After the alloy has been mechanically separated from the abrasive component, it may be used directly in the steel or alloying industry. If impurities, such as silicon, sulfur or carbon, are present in undesirable amounts, the alloy may be refined in any conventional manner.

ALUMINA FROM ALUMINUM REDUCTION PLANT WASTES

In the production of metallic aluminum by electrolysis of reduction grade Al_2O_3, the electrolysis is generally carried out in reduction cells or pot lines which are lined with a carbonaceous material. During the life of the cells, this carbon lining is gradually destroyed by penetration of bath materials into the lining, for example, metallic aluminum, cryolite and alumina. Also, due to the high temperatures employed in the electrolytic reduction process, gradual aging of the carbonaceous lining takes place.

The combined result of penetration and aging can reach a stage where the further operation of the cell or cells reaches an economically prohibitive point and replacement of the carbonaceous lining becomes a must. The unusable or spent pot lining is then removed and in most instances stockpiled. In large aluminum reduction facilities, this lining replacement is a continuous process and, consequently, the quantity of spent lining stockpiled increases from day to day.

In aluminum reduction facilities, where metallic aluminum is produced by the electrolysis of Al_2O_3 in the presence of a fluoridic electrolyte, such as cryolite (Na_3AlF_6), the electrolysis results in off-gases of high fluoride content. In addition to the fluoride content, the off-gases generated in the reduction process contain gaseous and particulate impurities, for example, volatilized metallic compounds and carbon derivatives, together with solid matter and nonvolatile carbonaceous materials. The quantity of volatilized and solid carbon compounds in the off-gases vary within wide limits depending on the type of anode used in the reduction system. Soderberg carbon anodes generate far more of these materials than prebaked carbon anodes.

In order to protect the environment and to provide healthy operating conditions in the reduction facility, these off-gases must undergo a purification process for the removal of harmful constituents. A common process for cleaning the off-gases is to subject them to a dry scrubbing treatment which effectively removes essentially all of the environmentally harmful impurities from the off-gases. In the dry scrubbing treatment of reduction off-gases, alumina is usually employed as the scrubbing medium. The alumina readily absorbs the fluoridic components of the off-gases and also captures the particulate impurities. It further removes harmful high molecular weight carbon derivatives. Consequently, the dry scrubbing of reduction cell off-gases with alumina is an effective purification process resulting in purified off-gases containing only environmentally harmless components.

While scrubbing of the off-gases solves the environmental and health problems, it poses a serious disposal problem. The spent alumina from the scrubber system is heavily laden with impurities and cannot be directly employed as feed for reduction cells without introducing unacceptable alloying components in the metal to be produced and without seriously interfering with the efficient operation of the reduction cells. Since the alumina is spent, it cannot be used for further scrubbing without purification.

In the production of metallic aluminum by the electrolytic reduction of Al_2O_3 in a series of cells, a significant quantity of impure metal and contaminated aluminum oxide feed are also generated in the form of floor sweepings, channel and trench cleanings. These materials, due to their high impurity content, cannot be directly employed for making metallic aluminum of commercial purity and, in general, if not blended with pure feed materials, are considered as waste with no convenient way of disposal.

Thus, from the above, it becomes clear that the producers of aluminum by the electrolytic process have major problems relative to the disposal of spent pot linings, exhausted alumina from the dry scrubbers, floor sweepings, channel and trench cleanings. These problems have been acutely recognized by operators of aluminum reduction facilities all over the world and partial solutions have been offered to overcome one or more of the problems associated with the generation of these spent materials.

However, there is a need for an integrated system which is capable of dealing with all of these spent materials with simultaneous recovery of all of the valuable components from these by-products of the electrolytic aluminum reduction process.

Such an integrated system has been proposed by *N. Bell, J.N. Andersen, and H.-K.H. Lam; U.S. Patent 4,113,832; September 12,1978; assigned to Kaiser Aluminum & Chemical Corporation.* The waste materials, such as spent pot linings, channel and trench cleanings, floor sweepings and spent alumina from off-gas purifying dry scrubbers, are combined, then pyrohydrolyzed at elevated temperature. Fluoridic values, such as NaF and HF can be recovered from the off-gas generated by pyrohydrolysis, while alumina and Na_2O values, or if desired, sodium aluminate, are reclaimed from the solid residue of pyrohydrolysis.

The fluoridic values from the pyrohydrolysis off-gas can be used for the manufacture of both electrolytes for aluminum reduction cells and also for the production of anhydrous HF. The alumina from the pyrohydrolysis residue can be reclaimed by a Bayer process-type leach with a caustic solution and the recovered high purity alumina utilized, for example, as reduction cell feed and/or for scrubbing reduction cell off-gases. If the solid residue of pyrohydrolysis contains significant amounts of sodium aluminate, this material can either be directly used for dry scrubbing cell off-gases, or if desired, utilized for production of high purity alumina. The process is shown schematically in Figure 2.

Figure 2: Kaiser Aluminum & Chemical Corp. Process for the Utilization of Waste Materials from Electrolytic Aluminum Reduction Systems

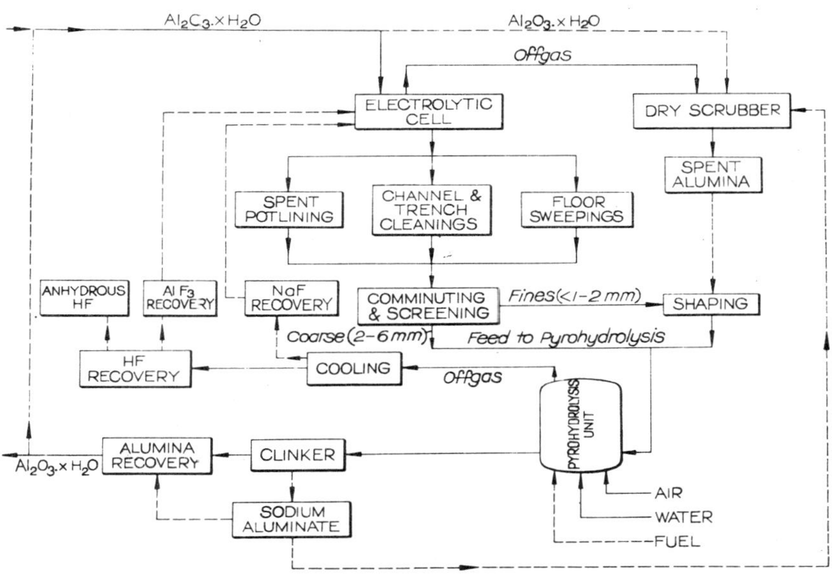

Source: U.S. Patent 4,113,832

See also "Aluminum for Aluminum-Glass Cloth Residues," "Aluminum from Furnace Skim or Dross," and "Cryolite from Potline Recovered Particulates," as well as "Hydrogen Fluoride from Furnace Pot Linings."

ALUMINA FROM RED MUD FROM BAYER PROCESS

Bauxite, a sedimentary oxide ore of variable composition, containing principally the oxides of aluminum, iron, silicon, and titanium, is a principal raw material for production of aluminum oxide.

Digestion of the bauxite with sodium hydroxide after the well-known Bayer process selectively dissolves out the amphoteric aluminum oxide as aluminate, which is subsequently further treated to produce aluminum oxide, which is, for example, a starting material for the production of aluminum by electrolysis of a melt (using molten cryolite as the solvent). The fine grained residue formed in the Bayer process contains, besides other oxides, also Fe_2O_3 and is therefore red in color, for which reason it is known as red mud. Red mud usually has the following composition, by weight of the dry materials: Al_2O_3, 22 to 39%; Fe_2O_3, 20 to 38%; TiO_2, 7 to 25%; SiO_2, 7 to 15%; Na_2O, 6 to 11%; CaO, 1 to 2%; and hot drying loss (H_2O, etc.), 8 to 12%.

Although the high aluminum, iron and titanium content suggests utilization of the red mud as a secondary ore, a suggestion that has not failed to lead to experiments to extract one or another element according to the raw materials situation, red mud is mainly disposed of as a waste product of aluminum production in large dumps.

In the usual Bayer process, the caustic aluminate solution is separated from the red mud and subjected to precipitation to enable recovery of the alumina values. The red mud, which also contains occluded or trapped caustic soluble alumina values is generally washed to remove most of these values from the mud in order to improve the yield of the Bayer process.

However, in many instances, the red mud also contains considerable quantities of alumina which, under conditions of the Bayer process, is unavailable to caustic extraction processes and constitutes unrecoverable values. This is particularly true when the ore which is subjected to the Bayer processing has an appreciable quantity of silica because the silica and the alumina react during the Bayer process to form an insoluble desilication product, thereby causing a loss in alumina and caustic values.

It has been heretofore proposed to subject low iron-containing aluminiferous ores, including the aforementioned red mud to the so-called lime-soda-sinter process. In this process, an alkaline earth compound such as lime, and an alkali metal compound, such as soda ash, is admixed with the red mud and sintered. The function of the alkaline earth compound is to react with the silica and to form an insoluble compound of calcium and silica. The function of the alkali metal compound is to react with the alumina and to form a soluble alkali metal aluminate. After the sintering has been complete, the sinter is then leached for the purpose of recovering therefrom soluble alumina and caustic values.

Although the lime-soda-sinter process has been previously known, there have been many problems associated therewith in an attempt to arrive at an economic balance between the cost involved in recovering the added values of alumina and alkali, and the cost of the treatment itself. There have also been technical problems in an attempt to recover the maximum amount of alumina and caustic values. Thus, for example, the sintering operation must be carried out in such a manner as to cause fritting or sintering of the particles without substantial fusion or melting of large portions of the mixture. Melting of substantial portions of the mass would interfere with the recovery of the values during subsequent leaching operations, thereby seriously affecting the overall efficiency.

Various techniques have existed for treating aluminiferous ores having a high iron content. However, the presence of high iron oxides content in the ore involved an irretrievable loss of limestone or soda ash reagents and usually alternate treatments were used in order to reduce the iron oxide content of the ore. Thus, for example, it is known to incorporate into the lime-soda-sinter treatment process a carbonaceous material which was stated to react with the iron oxides to form magnetic iron which could then be separated by other techniques.

Other proposed methods involved the use of an acid extraction step so as to produce crude alumina contaminated with iron, but which was relatively free of silica. This product was then subjected to the lime-soda-sinter process. Quite obviously, the combination of an acid and a caustic process can lead to additional operating expenses. However, there has not been much work relating to processing a Bayer red mud having a high iron content. Quite obviously, an acid process would destroy the soda values and the heretofore mentioned magnetic separation techniques might have been thought to be too costly.

A process developed by *K.G. Hrishikesan; U.S. Patent 4,045,537; August 30, 1977; assigned to Reynolds Metals Company* is one for the recovery of caustic and alumina values from a red mud resulting from the conventional Bayer process wherein the mud or ore contains a substantial quantity of iron and a substantial quantity of silica. This process is applicable to Bayer red muds which contain significant quantities of soda which had heretofore been wasted. It has been found that if the Fe_2O_3/SiO_2 weight ratio in the mud desired to be processed is above about 0.4, the recoveries of soda and alumina from the sinter utilizing conventional methods have drastically been reduced.

For reasons which are not completely understood, it has been found that as the iron to silica ratio in the mud desired to be treated increases, the amount of recoverable alumina and caustic decreases.

It has been found that excellent recoveries of both soda and alumina can be obtained by incorporating a carbonaceous substance into the high iron-containing materials and subjecting them to a lime-soda-sinter operation followed by extraction of the sinter.

In a process developed by *E. Zimmer, A. Nafissi, and G. Winkhaus; U.S. Patent 4,119,698; October 10, 1978; assigned to Kernforschungsanlage Jülich GmbH, and Vereinigte Aluminum-Werke AG, Germany* the red mud is digested with concentrated sulfuric acid or with sulfur trioxide gas to produce sulfates that can be leached out to the resulting mass with water. The solution is then heated at a pH of 1 to precipitate titanium oxide hydrate by hydrolysis. The remaining sulfates of the solution are then obtained in solid form by evaporation, or by precipitation with acetone, and the solid is then roasted to convert the aluminum and iron to the oxide. After leaching out the sodium sulfate with water, the aluminum and iron oxide are separated by the Bayer process.

The aluminum oxide remaining after the last leaching step, in which the sodium sulfate is removed from the mixture of aluminum and iron oxides, has surprisingly been found to be highly soluble in the application of the Bayer process to ore separation from the iron oxide. This result was not to be expected, because x-ray diffraction patterns showed that the aluminum oxide was present mainly as α-aluminum oxide which was not technically digestable by the known Bayer process without the use of extreme temperatures. The aluminum oxide resulting from this process, however, is amenable to successful treatment by the Bayer process without resort to such extreme temperatures.

ALUMINUM FROM ACETYLENE SCRUBBING SOLUTIONS

Bimetallic salt complexes that have the generic formula $M_I M_{II} X_n$·Aromatic are known to be useful in the separation from gas mixtures of such complexible ligands as olefins, acetylene, aromatics, and carbon monoxide. For example, U.S. Patent 3,651,159 discloses a process in which a liquid sorbent that was a solution of cuprous aluminum tetrahalide in toluene was used to separate ethylene, propylene, and other complexible ligands from a gas feedstream. The complexed ligands were recovered by ligand exchange with toluene. The resulting solution of cuprous aluminum tetrahalide-toluene in toluene was recycled and used to separate additional quantities of the complexible ligands from the feedstream. U.S. Patent 3,647,843 discloses a process in which a hydrocarbon pryolysis gas stream was contacted with a cuprous aluminum tetrachloride solution in toluene to separate acetylene from the gas stream as a solution of the complex $HC{\equiv}CH \cdot CuAlCl_4$ in toluene. Acetylene was stripped from this complex with toluene, and the cuprous aluminum tetrachloride-toluene complex was recycled.

In processes such as those specified above in which a liquid sorbent containing a bimetallic salt complex is recycled without purification and is used for long periods of time, there is a gradual increase in the amounts of reaction by-products and other impurities in it until sufficient impurities are present to interfere with the efficient operation of the process.

For example, when the liquid sorbent is contacted with a gas feedstream that contains an olefin having 2 to 4 carbon atoms, some of the olefin reacts with the aromatic hydrocarbon or halogenated aromatic hydrocarbon in the sorbent to form alkylated aromatic compounds, and some undergoes polymerization to form olefin oligomers. Any water that is present in the gas stream reacts with the bimetallic salt complex to form the complex $M_I M_{II} X_n \cdot M_{II}OX$·Aromatic, which has limited solubility in the liquid sorbent. When it contains amounts of these and other contaminants that are sufficient to coat heat exchangers, clog lines, and otherwise foul the equipment, the liquid sorbent must be purified, for example, by cooling to precipitate a sludge that comprises the slightly soluble bimetallic salt complexes and separating this sludge from it, or replaced by fresh liquid sorbent.

When a Group I-B metal halide is reacted with a Group III-A metal halide in the presence of an aromatic hydrocarbon solvent to from the liquid sorbent, there is usually also formed a small amount of sludge that contains a major amount of the bimetallic salt complex $M_I M_{II} X_n \cdot M_{II}OX$, resulting from the presence of contaminants such as $M_{II}OX$ and water in the reactants. This sludge, which is only slightly soluble in the liquid sorbent, is removed before the sorbent is used to separate complexible ligands from gas feedstreams.

Because of their high metal contents, spent liquid sorbent, sludge separated from liquid sorbent, and sludge formed during the preparation of the liquid sorbent cannot be discharged into sewers or waste ponds without causing serious pollution problems. In addition, it is economically desirable to recover the metals, which are usually copper and aluminum, from the waste materials.

Several processes have been proposed for the recovery of metals from sludge and spent liquid sorbents containing the bimetallic salt complex $M_IM_{II}X_n$·Aromatic. U.S. Patent 3,845,188 discloses a process in which the Group I-B metal in the spent liquid sorbent is recovered as its halide by contacting the liquid sorbent with anhydrous ammonia and separating from the sorbent the metal halide that precipitates. It has been proposed that the waste materials be burned, but since these materials contain about 30% of metal salts, this process leaves a substantial metal ash that must be disposed of in a manner consistent with environmental conservation. These processes are generally uneconomical and impractical to carry out on an industrial scale, and they may cause pollution problems.

An improved process for metal recovery from such complexes has been developed by *D.A. Keyworth and J.R. Sudduth; U.S. Patent 4,153,452; May 8, 1979; assigned to Tenneco Chemicals, Inc.* In this process, metals are recovered from waste materials that contain an organic component and a bimetallic salt complex component that comprises complexes having the formula $M_IM_{II}X_n$·Aromatic and/or the formula $M_IM_{II}X_n \cdot M_{II}OX$·Aromatic, wherein M_I is a Group I-B metal, M_{II} is a Group III-A metal, X is halogen, n is the sum of the valences of M_I and M_{II}, and Aromatic is a monocyclic aromatic hydrocarbon or halogenated aromatic hydrocarbon having 6 to 12 carbon atoms.

More specifically, the process is carried out by hydrolyzing the waste material with dilute hydrochloric acid to form a hydrolysis mixture that separates into an organic phase and an aqueous phase, and, after removing the organic phase, adding to the aqueous phase sufficient aluminum or another reducing metal to precipitate the Group I-B metal. After the precipitated metal has been removed from it, the aqueous solution is treated with sufficient base to precipitate the Group III-A metal hydroxide, which is recovered from the aqueous solution. This process can be used, for example, to recover copper and aluminum from waste materials that contain $CuAlCl_4$·toluene and/or $CuAlCl_4 \cdot AlOCl$·toluene.

ALUMINUM FROM ALUMINUM-GLASS CLOTH RESIDUES

In pouring molten aluminum into various solid shapes suitable for rolling or extrusion operations, the molten aluminum passes through a woven glass cloth filter. The purpose of this filtering is to minimize the number of hard particles, e.g., aluminum borides and aluminum oxides, contained in the molten metal from entering the solid shape. Such a procedure using a glass cloth filter is described in Canadian Patent 554,853, March 25, 1958.

This filtering technique is widely used throughout the aluminum industry and a significant amount of metal freezes on the glass cloth on completion of each pour of molten aluminum. The usual method for recovering the aluminum deposited on the glass cloth is by differential melting whereby the glass cloth with the deposited aluminum is heated to a temperature at which the aluminum melts but below the point at which the glass cloth melts. This combined with differences in specific gravity results in their separation, the glass cloth rising to the surface of the bath to be skimmed off as part of the dross. This has been done either by returning the aluminum-glass cloth residues to the melting furnace from which the aluminum has passed through the glass cloths or it is even sometimes done in totally different locations.

An improved and simpler mechanical technique for this separation has been developed by *J.P. Skelton; U.S. Patent 4,044,955; August 30, 1977*. The method comprises the steps of: breaking down the residue into small pieces, squeezing each piece of residue between hard surfaces on a pair of rotating cylindrical rolls with sufficient squeezing pressure to extrude the aluminum coating to destroy its integrity and expose the interior glass fibers along substantially the entire length of each piece and segregating the resulting glass fibers from the remaining aluminum pieces by air separation. The recovered aluminum pieces can be then reused in various industrial process. Such a process is shown in Figure 3.

Figure 3: Apparatus for Recovery of Aluminum from Aluminum-Glass Cloth Residues

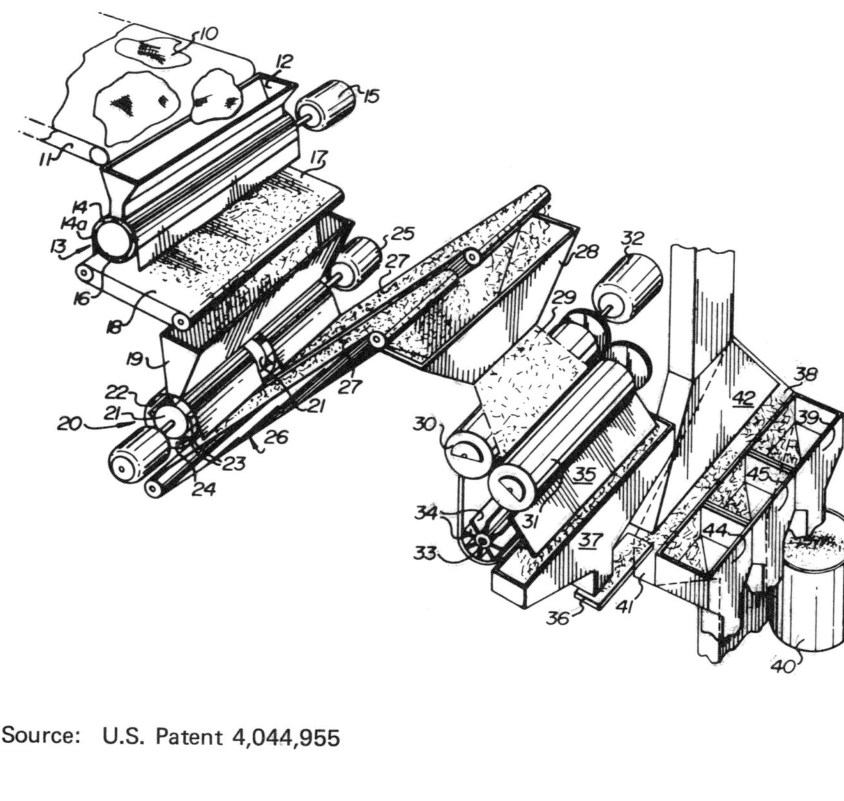

Source: U.S. Patent 4,044,955

As shown, used glass cloth filters **10** coated with aluminum are carried along by conveyor **11** and dumped into a hopper **12** from which they are delivered to a chopping station **13**. This chopping station includes a rotating drum **14** having bar shaped cutters **14a** rotatable past a fixed bed blade or block to chop the filters into pieces. The drum is driven by means of a motor **15**.

Surrounding the drum is a screen **16** in the form of a plate having holes of about 1" dimension through which broken down pieces of the glass fiber filter pass. Thus, the glass fiber filter pieces will be carried around by the blades **14a** and recut, perhaps many times, until they are small enough to pass through the holes in the screen **16**.

The pieces **17** that pass through the holes are conveyed by conveyor **18** into hopper **19**. From this they are delivered into a second chopping station **20** where they are broken down into still smaller pieces. Each cutting station **20** includes a drum **21** with cutting blades **23** passing a fixed bed blade **22** and driven by motor **25**. Once again, the cutter is surrounded by the screen **24** having openings of a size of about ⅜". Thus, the pieces of aluminum-glass cloth residue are chopped within the chopping station **20** until they are small enough to pass through the holes in screen **24** and are delivered onto conveyor **26** as small pieces **27**.

The small aluminum-glass fiber residue pieces are delivered into hopper **28** and onto a flow spreader **29** in the form of a fluted sheet so as to cause the pieces to fall relatively uniformly over a wide area. These pieces then fall between a pair of hard rolls **30** and **31** having hard, relatively unyielding surfaces.

The squeeze rolls **30** and **31** are preferably formed of hardened steel and have sufficiently large diameters as to provide surfaces which will frictionally grip the aluminum-glass fiber pieces and carry them into and through the nip while squeezing the aluminum surrounding the glass cloth fibers. The rolls are driven by an electric motor **32** and can conveniently be rotated at a speed of about 200 rpm.

Emerging from the bottom of rolls **30** and **31** are pieces of aluminum metal and shredded pieces of glass cloth fibers. In order to approach 100% efficiency, leaving practically no glass cloth fragments on the aluminum pieces, the squeezed pieces are preferably rapped sharply to knock apart any pieces that may be stuck together after squeezing. This can be done by a rotating paddle wheel **33** having blades **34** which strike the pieces sharply knocking them against plate **35**.

The aluminum pieces and the glass cloth pieces free from one another, are received by hopper **37** and then carried by a vibratory metering conveyor **36** to a segregating station in the form of a conventional segregator where the pieces are segregated by blowing, sucking and vibrating. This can take the form of a vibrating screen **38** inclined upward in the direction of flow and transversely of flow. It is vibrated in the direction of flow, causing the denser pieces of metal to move preferentially upward along the screen until they are discharged at the end through chute **39** into receptacle **40**. The lighter pieces of glass cloth fibers are preferentially levitated by a blast of air from a conduit **41** below the table and are sucked away through a conduit **42** above the table.

In the event that there are some aluminum pieces with some glass cloth fibers still adhering thereto, these can also be separated on the vibrating screen via intermediate chutes **44** and **45**.

ALUMINUM FROM ELECTRICAL CONDUCTOR SCRAP

It is estimated that about 1.5 billion pounds of insulated electrical wire of all types are manufactured each year in the United States. A portion of the insulated wire manufactured, e.g., about 4%, is deemed to be defective due to improper application of the insulation. However, the wire itself is in generally good condition and is capable of being reused provided the insulation can be removed without damaging the wire.

Various techniques have been proposed for stripping insulation from insulated wire. Among the suggestions made are the following:

(1) Mechanical stripping, as by abrasion or shearing;

(2) Thermal stripping, as by use of elevated temperature to destroy the insulation, or use of low temperature to weaken the insulation; and

(3) Chemical stripping of the insulation.

Although each of the aforementioned techniques may be useful for the particular application referred to, none of these conventional techniques, however, is suitable where the wire itself is to be preserved for immediate reuse. The problem associated with aforementioned prior art techniques is that in stripping off the insulation, the wire structure itself is altered and not in electrically undamaged condition. By damaged it is meant that the wire surface may be nicked or scratched by the sharp cutting blades, or its surface condition and metallurgical condition may be altered when high temperature is used to remove the insulation. Another disadvantage of the prior art techniques is the relatively high cost of carrying out the stripping operations.

ALUMINUM FROM FLY ASH

Large quantities of fly ash carried by the combustion products of power plants burning pulverized coal exist throughout the country and more is being created by operation of these plants. This accumulation creates a disposal problem and represents a waste of metal values, particularly aluminum, as a typical fly ash contains up to 14% aluminum by weight. Lesser amounts of iron, titanium and other useful metals are present in fly ash.

No satisfactory process exists for economically recovering aluminum from fly ash having the required purity for commercial sale because of the difficulty of separating it from other metals present in the fly ash, particularly, iron. Separation through the chlorination route to recover aluminum as aluminum chloride looks attractive, however, the process must produce an aluminum chloride of substantial purity. For example, purity requirements for aluminum chloride feed material to an Alcoa-type aluminum cell limit the Fe_2O_3 content of the feed to 0.03%. Furthermore, in the chlorination process, the chlorination of unwanted metals, such as silicon, must be suppressed to restrict the consumption of chlorine; otherwise, the process becomes prohibitively expensive.

A further problem involved in recovering the metal values from fly ash through the chlorination route, is the disposal of alkali and alkaline earth metal chlorides remaining in the final residue.

A process developed by *J.E. Reynolds and A.R. Williams; U.S. Patent 4,159,310; June 26, 1979; assigned to Public Service Company of New Mexico* for recovering aluminum from fly ash containing iron, silicon and titanium comprises:

(a) Chlorinating the fly ash in an oxidizing atmosphere to selectively chlorinate and vaporize iron chloride from the remaining chlorides;

(b) Chlorinating the residue from step (a) in a reducing atmosphere of carbon monoxide, in the presence of added silicon chloride to suppress the chlorination of silicon, and vaporizing the chlorides of aluminum, silicon, titanium, and the residual iron;

(c) Separating and recovering the vaporized chlorides by selective condensation, and treating the residue of step (b) with sulfuric acid to convert calcium chloride to gypsum, and to regenerate a chloridizing and binder solution for pelletizing fly ash feed.

Such a process is shown schematically in Figure 4. As shown there, the fly ash accompanying the combustion products of the boilers heated by burning of pulverized coal is collected in a fly ash hopper. The particular fly ash used was recovered from power plants using San Juan coal from the Four Corners area of the United States. Of course, the metal value content of the fly ash will vary depending upon the area in which the coal was mined.

Periodically the collected dry fly ash is transferred to a dry storage hopper.

The dry fly ash can be sent to a dry magnetic separation step. Optionally, in this step 50 to 60% of magnetite iron can be removed by magnetic separation.

The fly ash is next sent to the pelletizing step where a hydrochloric acid binder solution is added and it is pelletized into high density, high strength pellets in conventional equipment. Following pelletizing, the pellets are dried at about 300°C in a direct fired tunnel dryer. Dry pellets are inventoried for feed to the shaft chlorinator or furnace. The fly ash may be ground before pelletizing; however, it was found that this does not affect the recovery of the metal values. Pelletizing is mandatory for a shaft reactor. Sequential chlorination techniques are amenable to the plug-flow nature of the shaft chlorinator.

Both hydrochloric acid and sulfuric acid are suitable binders for carbon-free pellet compositions. Fly ash without any binder produces a weak pellet when sintered at 300°C. The presence of carbonaceous material also reduces the pellet strength.

The pellets are dried with fuel-air or by recuperation of heat from high temperature gases exiting the oxidation chlorinator and stored pursuant to chlorination.

Figure 4: Public Service Co. of New Mexico Process for Recovering Aluminum and Other Metal Values from Fly Ash

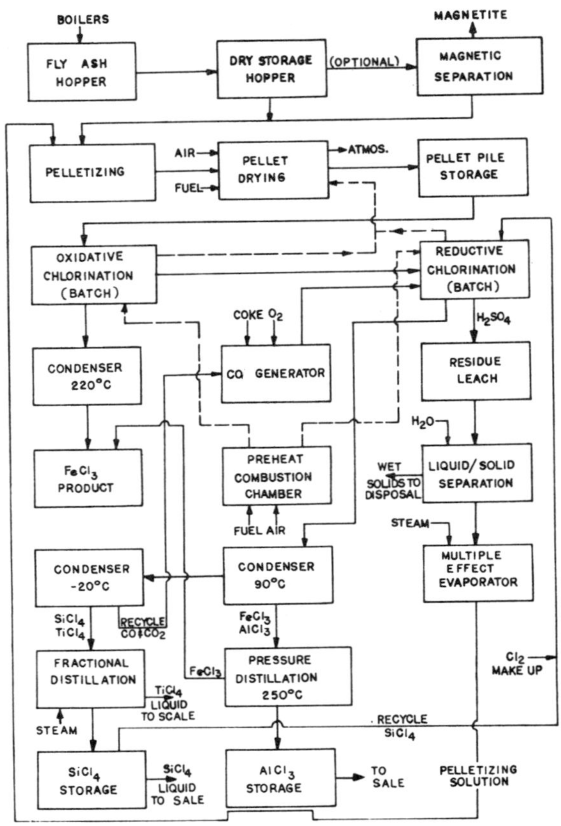

Source: U.S. Patent 4,159,310

As seen from the flow sheet, the oxidative chlorination step comes next followed by reductive chlorination. It was found that the most effective procedure was to first remove the iron by selective chlorination in an oxidative chlorination step followed by volatilization of the formed ferric chloride and its recovery by condensation. Up to 98% of the iron was volatilized with substantially no chlorination or volatilization of the other metal values. It is important, of course, that substantially no aluminum chloride be formed or volatilized at this stage. As one of the big economic factors involved with the process is the use of chlorine, it is also important to suppress the chlorination of the other metal values, particularly silicon, as the fly ash contains over 25% silicon.

It was found that the degree of silica chlorination in the reductive chlorination step can be greatly reduced by using only carbon monoxide as a reducing agent rather than a mixture of carbonaceous material and fly ash. Carbon monoxide improves the selectivity of alumina chlorination over that of silica. The injection of silicon tetrachloride into the reaction gas mixture of chlorine and carbon monoxide was found to be very effective in almost completely eliminating silica chlorination at 950°C, for example.

It was found that the overall chlorination procedure resulted in chlorination of alkali and alkaline earth metals present. Suppression of the chlorination of these elements which end up in the final residue as chlorides was not emphasized because a feasible way of disposing of the chlorides in the residue was found. It was found, however, that the best reaction conditions for minimizing chlorination of sodium and magnesium was the absence of carbon during the chlorination and chlorination at a temperature of about 950°C.

The oxidative chlorination for the selective removal of iron is preferably performed on fly ash pellets with a hydrochloric acid binder in a shaft chlorinator. Attempts to remove iron from the pellets by perchlorination under reducing or neutral conditions were not feasible because of cochlorination of excessive amounts of alumina.

A number of shaft furnace chlorinators used as batch chlorinators is preferred. These chlorinators are operated with staggered sequence of operation designed for optimum heat recuperation. For the oxidative chlorination, the charge is brought up to the proper temperature with hot, neutral combustion gases from a coal-fired furnace. A mixture of chlorine and oxygen gases is then circulated for about three hours through the charge to perchlorinate and volatilize about 90 to 95% of iron content. The volatilized ferric chloride is collected in an air-cooled scraped condenser. The next step is the reductive chlorination.

Carbon monoxide gas is added to the chlorinator. The reaction with carbon monoxide is sufficiently exothermic to be self-heating. The chlorinator is operated for about four to eight hours to collect a small amount of residual iron chloride in the first stage condenser and a high purity aluminum chloride in the second stage condenser.

About 3% silicon chloride by volume is injected during the reduction to suppress silica chlorination. A third stage condenser collects the chlorides of titanium and silicon. The on-stream chlorinator is then purged with ambient air to remove residual chlorine and cool the residue. The purged gas is routed to a chlorinator coming on line for heat up and to react with the residual chlorine and silicon chloride.

A preferred method of introducing the silicon chloride is to run the chlorine through the liquid silicon chloride before it enters the reactor. The cooled depleted pellets are conveyed to the leach circuit where water-soluble chlorides are removed and calcium is converted to gypsum with sulfuric acid. The residue solids are filtered, washed and sent to the disposal while the hydrochloric acid solution is evaporated as required for water balance control and recycled to the pelletization step for reuse as pellet binder and perchloridizer.

ALUMINUM FROM FURNACE SKIM OR DROSS

When aluminum metal is melted, there is formed on the surface of the molten metal a skim or dross comprising a mixture of the reaction products of aluminum with air. These reaction products include aluminum oxide, aluminum nitride and other nonmetallic impurities. Prior to tapping or pouring the molten aluminum, the dross is skimmed or raked off the surface of the molten metal.

During the raking, skimming and removal of the dross from the top of the metal melt, the dross becomes compressed into pasty mudlike masses. These masses of dross, when removed from the furnace, will vary from small lumps on the order of one inch in dimension and smaller to large lumps approaching one foot in dimension. The amount of free metal and/or alloy in the dross may vary from 30 to 95% by weight depending upon a number of factors, such as the composition of an alloy being melted, the melting procedure followed, and the care with which the dross is skimmed or raked from the melt. It has been estimated that as much as 1 to 2% of the total aluminum melted ends up in the dross fraction.

Because the dross contains a high percentage of aluminum metal and because the dross fraction represents a substantial processing loss, efforts are ordinarily made to reprocess the dross to recover as much of its aluminum content as possible. Typical processing approaches used commercially involve the collection, crushing and sizing of dross followed by smelting using a salt flux.

Fluxes used typically comprise a mixture of sodium chloride and potassium chloride with small amounts, ordinarily 1 to 5%, of cryolite. Flux is mixed with the dross and charged into a rotating furnace which is heated to a temperature above the melting point of aluminum and of the flux. The salt flux selectively wets the nonmetallic components of the dross and promotes the separation of metallic aluminum from the nonmetallic fraction. Molten metal is thereafter tapped from the furnace and the residue is then dumped from the furnace. It has been the practice to dispose of the residue in a landfill.

It has been estimated that primary and secondary aluminum plants in the United States generate over 200,000 tons of high salt residue annually from the processing of aluminum dross. Disposal of these residues by landfill techniques is coming under sharp criticism because ground water leaches the salts from the residue. Furthermore, use of salt fluxes creates corrosive fumes which must be recovered from the smelting operation.

Thus, a variety of processes have been developed for aluminum recovery from furnace skim or dross. These include: processes using fluxes; fluxless recovery processes; and processes in which the skim or dross is used per se without treatment in various metallurgical processes.

A process developed by *D. Montagna; U.S. Patent 3,999,980; December 28, 1976; assigned to U.S. Secretary of the Interior* is one is which aluminum metal is recovered from processing wastes, such as dross, beverage can scrap and the like, without use of a flux by heating the wastes to a temperature above the melting point of aluminum but below about 800°C while maintaining the wastes under a protective gaseous atmosphere which is substantially inert to molten aluminum. After melting of aluminum contained in the wastes is complete, the heated material is agitated gently to agglomerate the molten aluminum and to cause its settling and collection in the bottom of the containment vessel. Molten aluminum is thereafter removed from the vessel leaving a substantially inert residue. Argon is preferred as the protective atmosphere. Such a process is shown in Figure 5.

Figure 5: U.S. Dept. of Interior Process for Fluxless Recovery of Metallic Aluminum from Wastes

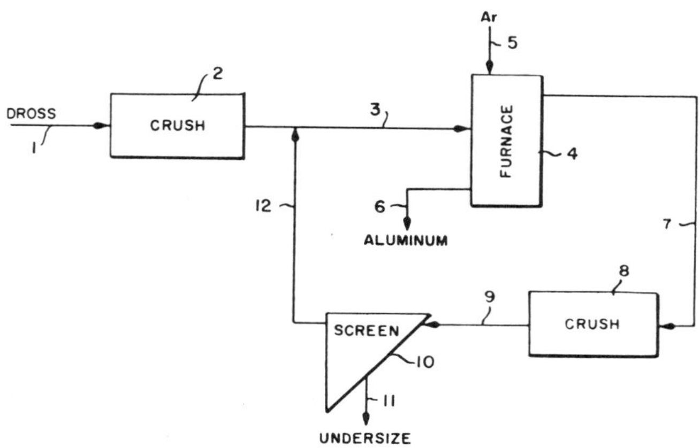

Source: U.S. Patent 3,999,980

Dross fraction **1**, produced by the working of molten aluminum, is subjected to a crushing step in means **2** which may be a roll crusher or like device. This step may be omitted if the dross is in a granulated form free of large chunks which would interfere with its convenient handling in subsequent processing steps. Crushed dross, conveniently having a maximum granule size of about 2", is fed via means **3** to furnace **4**. The furnace may comprise any indirect heated device capable of attaining temperatures of at least 750°C and having a sealing or cover sufficient to maintain a blanket of inert gas such as argon, introduced via means **5**, within the furnace during the heating cycle.

In addition, the furnace must have provisions for gently agitating the dross after the aluminum content contained in the dross has been melted. Agitation means associated with the furnace may comprise a simple propeller-type blade, constructed of materials such as silicon carbide coated steel, which is arranged to travel from the top to the bottom of a pot-type furnace. When this mode of agitation is used, rotational speed of the blade appropriately may be in the range of about 5 to 20 rpm.

Alternatively, the furnace may be of the rotary type in which agitation is accomplished by rotating the furnace. Agitation must be continued for a time sufficient to cause agglomeration of the melted aluminum globules and their accumulation into a molten pool at the bottom of the furnace. Generally, agitation times of 2 to 15 minutes are sufficient to accomplish that result.

After agitation has ceased, molten aluminum **6** is tapped or poured from the furnace into suitable molds. It is neither necessary nor advantageous to leave a quantity, or heel, of metallic aluminum within the furnace in preparation for the next dross charge as is conventional in the practice of the salt flux method. The solid residue remaining after tapping aluminum from the furnace may be characterized as a low metal content dross. It is dumped from the furnace and introduced, via means **7**, into crushing means **8**. It is desirable that means **8** imparts an abrading action to the residue fraction in order to free metallic aluminum particles from entrapment in aluminum oxides.

After being subjected to crushing and abrasion in means **8**, the residue is passed via means **9** to screening means **10** where it is sized into a fine and a coarse fraction. The fine fraction **11**, comprising mostly aluminum oxide, is removed from the process. This fine fraction may be discarded or may be used for other purposes such as for hot-topping steel ingots. Coarse fraction **12** comprises the metallic aluminum-rich portion of residue **7** and preferably is recycled back to the furnace for extraction of additional metallic aluminum. It is possible to operate this process on a strictly batch basis dispensing with the crushing, sizing and recycling steps but this decreases the ultimate aluminum recovery from the dross.

In a process developed by *P.N. Papafingos and R.T. Lance; U.S. Patent 4,030,914; June 21, 1977; assigned to Alumax Mill Products, Inc.* aluminum drosses, skims and slags are treated by melting under a cover flux comprising sodium chloride or potassium chloride or mixtures thereof in combination with calcium chloride, the calcium chloride comprising from about 1 to 50% of the flux composition.

Removal of aluminum dross or skimmings carries with it a large amount of molten aluminum, much of which is entrained in the dross and protected from further oxidation. Much of the metallic aluminum, however, is merely attached to the oxide clumps and is exposed to the air. When the aluminum dross, or skimming, is removed from the furnace, it is at a temperature of about 1200°F and any molten aluminum which is exposed to air is very vulnerable to rapid oxidation. In fact, since oxidation by itself generates heat, the molten or semimolten mixture catches fire as it oxidizes, a phenomena called thermitting or aluminothermic reaction.

Present methods of handling aluminum include processes which encourage thermitting to maintain the dross and aluminum at high temperatures to allow aluminum metal to drain out and separate from the dross, and methods in which aluminum dross is comminuted during cooling to permit aluminum metal to drain from its entrained condition in the dross.

In addition to the substantial loss of aluminum by the thermitting action, large amounts of "smoke," finely disbursed particles of aluminum oxide carried in the air, are generated presenting health and other problems.

Thus, a technique has been developed by *P.N. Papafingos and R.T.. Lance; U.S. Patent 4,039,173; August 2, 1977; assigned to Alumax Mill Products, Inc.* in which dross, principally oxides and nitrides of aluminum and entrained metallic aluminum, is transferred quickly from the melting furnace to a cooling drum and cooled in less than 3 minutes to below 400°F without substantial removal of metallic aluminum before recovery of metallic aluminum.

In a number of prior art processes for recovering aluminum metal value from aluminum dross, the dross is heated under a flux comprised of various salts or combinations of salts, usually comprising chloride salts of alkali or alkaline earth metals or mixtures of these salts. Fluoride salts are still used in some countries.

In the fluxing step aluminum metal, protected by the flux, is removed from the dross by draining or other gravity or mechanical separating techniques leaving a product sometimes referred to as salt cake which is composed of the fluxing salt mixture, aluminum dross which is primarily oxides of aluminum, entrained aluminum metal particles or chunks, and, in general, all impurities which are less dense than aluminum or which are formed into particles with sufficient porosity to be less dense than aluminum. Traditionally, the salt cake is allowed to cool in large rocks or chunks resembling mine slag. These chunks are extremely hard, quite large, and, in general, are extremely difficult to handle.

These chunks or rocks of salt cake usually are broken up by heavy, specially designed machinery, or by laborious manual or mechanical operations to recover the aluminum metal values which are entrained in the salt cake. The aluminum metal may appear as chunks ranging from very fine to pieces several inches in thickness. Following removal of the valuable aluminum metal, the salt cake has, in the past, usually been discarded since it has no value in the form in which it exists and it has been too expensive and difficult to recover the flux salt components therefrom.

Disposal of the salt cake presents particular problems and improper disposal often has resulted in serious ecological disruptions. Any disposal which presents the immediate or long-term potential of exposure to moisture presents the hazardous possibility that salts may be leached from the salt cake and carried into agricultural, industrial or potable water supplies. In the past, salt cake has been disposed of by simply depositing it in marble quarries, etc., in isolated locations. This type of disposal is unacceptable from the safety and ecological point of view since it presents the serious possibility that salts leached from the salt cake may find their way into underground or surface water streams or reservoirs.

A process developed by *P.N. Papafingos and R.T. Lance; U.S. Patent 4,073,644; February 14, 1978; assigned to Alumax Mill Products, Inc.* provides for recovering not only aluminum metal values from salt cake but for recovering and recycling or otherwise using the fluxing salt components of salt cake. This process is shown schematically in Figure 6.

A process has been developed by *P.J. Cromwell; U.S. Patent 4,126,673; Nov. 21, 1978; assigned to Cromwell Metals, Inc.* which is shown in Figure 7.

Dross which has been previously milled and screened from whole dross into particle sizes ranging for example from 1/10 to ¼" are placed in hopper **10** and conveyed upwardly by a bucket elevator **12** to a double deck screen **14**. It is to be understood that the instant method is more efficient when the dross charged into hopper **10** has been appropriately sized into selected size ranges. With respect to dross particles 2" in dimension and smaller, the size ranges of dross for charging into hopper **10** could include 1/40 to 1/20" in dimension, 1/20 to 1/10", 1/10 to ¼", ¼ to ½", ½ to 1", 1 to 1½", and 1½ to 2". Of course these ranges could be varied somewhat.

Figure 6: Alumax Mill Products, Inc. Recovery of Aluminum and Salt Cake from Fluxing Process for Furnace Skim or Dross Treatment

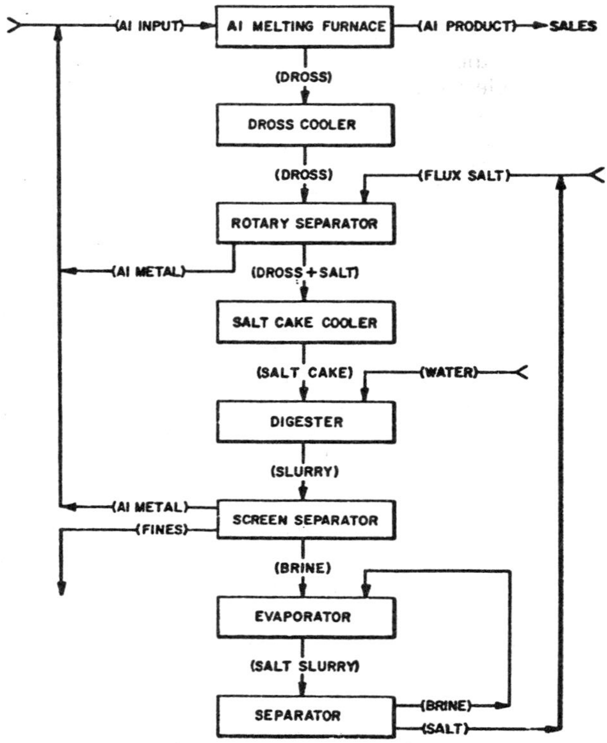

Source: U.S. Patent 4,073,644

Oxide and aluminum dust having particles smaller than $\frac{1}{20}$" for example pass downward through the double deck screen **14** into hopper **22** and are collected as an end product for sale as low percentage metallic content aluminum oxide dust. In referring to the separation of nonmetallic substances from metallic substances/concentrates, it is to be understood that minute particles of metal substantially in the form of dust will be commingled with the nonmetallic substances which also will be of minute particulate size.

Necessarily, the dust collected in hopper **22** will have a varying percentage of aluminum dust commingled therein. Container **22a** may simply be a removable barrel for receiving the aluminum oxide dust from hopper **22**. The dross material not falling through the double deck screen passes onto a magnetic separating means **24** which removes the loose ferrous substances, subsequent to which the dross is conveyed to hopper **26**. The dross in hopper **26** is in turn fed to a conveyor **28** which conveys the dross material to a vibrating feeder **30** which feeds the dross material to a first pair of spaced roller means **32**.

Figure 7: Cromwell Metals, Inc. Process for Fluxless Recovery of Aluminum from Furnace Skim or Dross

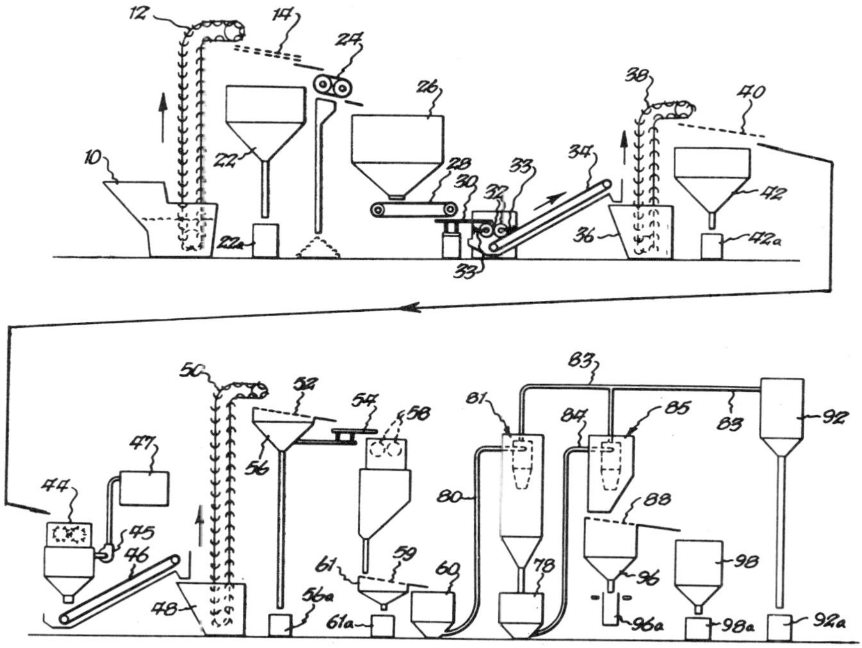

Source: U.S. Patent 4,126,673

The rollers **32** are preferably resiliently mounted with respect to one another as for example by means of springs **33**. The rollers have a predetermined spacing less than the selected size range of dross particles passed therebetween whereby the rollers may resiliently separate one from the other upon passage of the dross therebetween.

For example, for dross 1/10 to 1/4" fed to the rollers, the spacing therebetween could be on the order of 1/20". The spacing would be correspondingly increased or decreased for other dross size ranges so that it is less than the smallest size of particles passed therethrough. The rollers are adjusted to compress the dross passed therebetween to a limited degree without substantially crushing the dross so as to substantially break the bonds between the metallic and nonmetallic substances therein. The spring pressure on the rollers, which can be adjusted independently of the spacing, should be increased or decreased respectively for the larger or small dross sizes so as to impart the above effect on the dross.

The crushed dross flows from the rollers up an inclined conveyor means **34** to hopper **36** which feeds a bucket elevator **38**. The bucket elevator feeds the dross onto a single deck vibrating screen **40** which screens out nonaluminum and aluminum dust substances in a manner similar to the collection of such substances in hopper **22** described previously. The dross materials which do not fall through screen **40** into the underlying hopper **42** and container **42a** are passed to a hammer mill means **44**. The hammer mill is of a type well known to those skilled in the art and which impacts the dross, previously fractured to a limited degree, so as to knock off loose oxide materials. In this regard the prior compressing of the dross by rollers **32** is significant so as to enable the hammer mill to disassociate the oxides from the metal after the rollers have broken the bonds therebetween. However, the hammer mill does not necessarily deform the aluminum particles and chunks, but yields particles of dross having significantly higher percentage of metal content (concentrates) with oxides completely removed in fact from some of the surface area of the metal.

An air separator or separating means **45** is in fluid communication with the lower housing of mill **44** and pulls a certain amount of nonmetallic and metallic dust from the housing to a remotely located baghouse **47** for example.

The dross as processed by the hammer mill falls to the underlying conveyor **46** which passes the same to hopper **48** and the bucket elevator **50**. The bucket elevator transports the chunks of aluminum concentrates and disassociated oxide materials not pulled off by air separator **45** upward onto a single deck vibrating screen **52**. The resulting relatively high metallic concentrates may be passed to another magnetic separator (not shown) and onto vibrating feeder **54**.

The milled nonmetallic substances as well as minute aluminum particles which may be present fall through screen **52** into hopper **56** as an end product and then into container **56a**, in a manner similar to that described with respect to screens **40** and **14**. In actual practice, it has been found that the limited compression imparted by the rollers removes on the order of 7 to 8% of the original oxides present and at the further step of milling as at **44** for example removes a significant additional amount of nonmetallic substances (oxides) on the order of 11 to 12% of the original nonmetallic substances.

The dross or aluminum concentrates conveyed to vibrating feeder **54** are a marketable product, without the need for further processing. Further processing would increase the purity of the concentrates but at the cost, of course, of further processing operations. The concentrates provided at the feeder have been cleaned of oxides to such a degree that they may be charged into, for example, a swarf furnace for recovery of metal in ingot form. The charging of relatively clean concentrates into a swarf furnace is to be contrasted with the prior art methods of charging considerably less pure concentrates into a rotary furnace for the recovery of metal which involves all of the drawbacks referred to previously.

Alternatively, the concentrates on feeder **54**, instead of being taken off as an end product, may be passed to the spaced rollers **58** which deform the aluminum concentrates into flattened, flakelike pieces.

Rollers **58** preferably include fixed mountings whereby the concentrates from feeder **54** are compressed into flakes which may be on the order of 1⁄16" thick for example with respect to 1⁄10 to ¼" dross originating in hopper **10**. The spacing of rollers **58** would be set to a corresponding degree and could be increased or decreased, respectively for larger or small dross sizes originating in hopper **10**.

The flakes fall to the underlying screen **59** which passes some additional aluminum and oxide dust into hopper **61** and associated container **61a**.

At this point the flakes include a somewhat higher percentage metal content than the concentrates conveyed to feeder **54**. Consistent with this approach, the flakes residing on screen **59** may be pulled off as a marketable product not requiring further processing. With respect to conversion to ingot form, the flakes may be viewed as more desirable than the concentrates conveyed to feeder **54** since the flakes for example have a higher density and accordingly will sink below the surface of a swarf furnace more rapidly as a consequence. Of course, it is important that the charge to the swarf furnace (concentrates/flakes) sink below the furnace surface so as to avoid potential combustion with the atmosphere.

Should it be desired to convert the flakes on screen **59** to yet a more pure form, the flakes may be conveyed from screen **59** to a hammer mill **60**.

The flakes are processed by mill **60** and flow through passage **80** to a cyclone type of air separation means **81**. Cyclone **81** is a device well known to those skilled in the art wherein the milled flakes would enter the cyclone at a tangential angle so as to create a low pressure area in the central portion of the cyclone. The low pressure draws aluminum and oxide dust into the central portion which pass in turn through conduit **83** to a remotely located baghouse **92**. This baghouse empties into an underlying removable container **92a** which is similar to containers **61a**, **56a**, etc. The heavier metallic particles in cyclone separator **81** drop to a second mill **78** underlying the cyclone **81**. The output from mill **78** is conveyed through passage **84** to a second cyclone means **85** which is similar in all respects to cyclone **81** and which conveys aluminum and oxide dust to conduit **83** and the baghouse **92**.

Metallic concentrates, which may be in the form of pellets fall from cyclone **85** to the underlying screen means **88**. The aluminum pellets falling onto this screen are of a higher purity than the flakes charged into mill **60** in view of the oxides removed by cyclones **81** and **85**, and further in view of the oxide, as well as aluminum dust, which passes through screen **88** to the underlying hopper **96** and container **96a**. The pellets residing on screen **88** are in turn conveyed to hopper **98** and to the underlying container **98a** as an end product.

The pellets or concentrates collected in container **98a** are the most desirable of the various products produced since they are of the relatively greatest purity. Of course, the pellets in container **98a** necessitate more processing than the other concentrates which are of lesser purity in correspondence to their degree of processing. Accordingly, a user will be able to selectively process metallic concentrates to various degrees of purity based on operating expenses and market conditions, etc.

ALUMINUM FROM MUNICIPAL WASTE

See particularly U.S. Patent 3,945,575 under "Glass from Municipal Waste."

See also U.S. Patents 4,113,185 and 4,129,440 under "Ferrous Metals from Municipal Waste."

Aluminum constitutes less than 1% by weight of municipal solid waste but comprises the bulk of nonferrous metals. The estimated quantity of aluminum in municipal waste is 800,000 tons a year. Current recovery of aluminum is estimated to be 80,000 tons. Most of this is being recovered in industry programs of separate collection through recycling centers.

Aluminum, like glass, is a valuable and desirable material for recycling from the standpoint of the user, in this case secondary aluminum smelters. The market price for secondary aluminum is many times that of other materials in mixed municipal waste.

The major constraint to aluminum recycling is supply. Recovery through consumer separation efforts has been successful in some instances and after some years of such practice, it appears that these efforts are becoming institutionalized and will continue. Aluminum recovery has been considered in most large-scale recovery systems under development, but the actual separation process (heavy media separation in most instances) has not yet been demonstrated.

Aluminum recovery will be tested as a part of some of the resource recovery systems EPA is sponsoring. Aluminum recovery could also be significantly affected by beverage container measures eliminating the beverage can or the pull-tab top.

A study of *Aluminum as a Component of Solid Waste and a Recoverable Resource* has been prepared by the National Environmental Research Center, Cincinnati, Ohio and is available as PB 235 770, Springfield, VA, Nat. Tech. Information Service (September 1974).

Some aspects of aluminum recovery from mixed municipal waste, particularly from admixtures with glass, have been discussed by W. Herbert and W.A. Flower in Public Works 102, (8), 70, 110, 112 (August 1971).

ALUMINUM FROM NONFERROUS SLUDGES

See "Chromium from Nonferrous Sludges."

ALUMINUM FROM SCRAP ALUMINUM

Aluminum scrap as used herein refers to alloys containing at least 70% aluminum but which cannot be used commercially in large quantities because of their excessive content of one or more impurities such as iron, copper, manganese, titanium, calcium, chromium, phosphorus, zinc, nickel, and of other less common substances.

In general the composition of well segregated aluminum base scrap is 99+% Al, but scrap is often contaminated with other materials. Stainless steel is especially troublesome because it cannot be segregated magnetically, is difficult to detect visually, and dissolves more readily in molten aluminum than does common steel. Free zinc is present in some borings, in jar covers, and as die castings. Magnesium, whether free or alloyed, may be disadvantageous since the principal use for scrap is for permanent mold and die casting alloys which contain little magnesium. Nonmetallic contaminants in aluminum scrap such as paint, oil, plastic, insulation, and rubber are a major source of air pollution in melting scrap, and if such contamination is too great the recovery of the metal values is not profitable (14).

Aluminum base scrap is usually melted, sometimes with primary ingot metal, to recycle the aluminum. The only refining of aluminum base scrap is the removal of magnesium by bubbling chlorine through the molten metal or by treatment with various fluxes such as aluminum chloride or mixtures of sodium and potassium chlorides. To facilitate handling, a significant proportion of the old aluminum scrap and a small part of the new scrap are melted to form a solid ingot called "sweated pig," which must be further treated to produce specification-grade ingot. Aluminum is recovered from drosses and slags by leaching the soluble salts from the small particles of metal usually left in such material, or by a dry method wherein the material is crushed, screened to remove fines, and separated magnetically.

In the United States an estimated 10 to 20% of the aluminum that is put into use as an industrial or consumer end product is eventually recycled as old scrap. The recycle time for old scrap, virtually all of which is purchased, varies considerably depending on the form and end use, but is normally much longer than for new scrap. Some may be in use 10 to 30 years prior to its return to the production stream. Aluminum cans and certain types of aluminum foil are exceptions and may return to the production stream a few months after the aluminum industry has shipped metal for their manufacture. Major aluminum producers and some can and beverage manufacturers have extensive aluminum can reclamation programs.

A process developed by *H. Martin; U.S. Patent 3,900,313; August 19, 1975* provides an economical method to remove excess iron and other elements from aluminum scrap without introducing other undesirable elements, physical properties, or impurities. The basic raw material can be waste materials of any sort having aluminum-containing pieces capable of being separated by conventional means as crude aluminum scrap.

Impurities such as iron are undesirable because, if present in excess amounts, they form brittle intermetallic crystals in aluminum, resulting in reduced mechanical properties, poor corrosion resistance, etc. Die-casting aluminum alloys, however, generally require small amounts of iron, 0.60 to 1.20%, to minimize die wear. The detrimental effects of the brittle intermetallic constituents caused by this amount of iron are avoided due to the rapid solidification during die-casting and consequent formation of small, less harmful crystals.

An industrial society creates ever increasing amounts of aluminum scrap contaminated with undesirable elements. Motor pistons are an example, having steel rings and other nonaluminum parts attached to them, which, during remelting of the aluminum pistons contaminate the alloy excessively with iron and other elements. The situation is similar with aluminum recovered from household garbage. An economical scrap conversion process is, therefore, most desirable. Such a process has an additional ecological impact. To reduce aluminum newly from ores requires at least 6 kWh/lb while recovering aluminum from scrap requires less than 1 kWh/lb. Recycle processes such as the aluminum scrap conversion process conserve raw materials and energy.

Aluminum scrap with high iron content of more than 1.5% Fe is mostly used as a deoxidizer in the steel industry. Homeopathic dosages are also added to commercial aluminum alloys. The high iron content prevents its large-scale addition to the commercial alloys.

There has been no commercially attractive method known in the aluminum industry to preferentially remove excess iron from aluminum scrap.

The cited process produces die-casting alloys by dissolving silicon in the scrap melt cooling the melt to precipitate second-phase iron-rich particles, coarsening the precipitated second-phase particles and separating them from the liquid near the eutectic AlSi temperature.

The purpose of adding silicon is threefold. It suppresses the solubility of other elements in aluminum so that these come out of solution during cooling of the melt and form solid particles. Secondly, the binary eutectic AlSi solidifies at 580°C, while pure aluminum solidifies at 660°C. Multiphase eutectic compositions with AlSi have still slightly lower eutectic temperatures. At the lower temperature of 580°C the solubility of many undesirable elements in liquid aluminum is further reduced since any solubility depends on the composition of the dissolving medium and on its temperature. Thirdly, silicon is a most desirable element in virtually all aluminum die-casting alloys.

Existing fuel fired aluminum melting furnaces are generally direct fired with limited areas for heat absorption, such that the exhaust gases are emitted at relatively high temperatures. A conventional melting furnace may utilize hot gases formed by the combustion of carbonaceous materials, e.g., hydrocarbons, such as fuel oil, natural gas, powdered coal and the like, to produce temperatures in the range of 3000° to 3400°F, the exhaust gases delivered to the stack generally ranging in temperature from about 2000° to 2500°F.

When using conventional direct fired furnaces of the type utilizing high temperature gases referred to above, it is not uncommon for the fuel input to correspond to about 2,000 to 3,500 Btu/hr/lb of aluminum melted, the thermal efficiency of the operation being rather low, for example, less than 30%, the thermal efficiency in many instances ranging from about 10 to 20%.

Aluminum scrap, such as turnings, borings, grindings and other forms of machinings, generally contain lubricating oil or other forms of hydrocarbon contaminants which must be removed to avoid contamination of the environment. Generally, separate charge driers are employed together with a separate afterburner combustion chamber in order to incinerate the hydrocarbons emitted from the scrap.

One can increase the thermal efficiency to the range of about 40 to 50% by employing the integrated scrap melting system of a process developed by *W.E. Stephens; U.S. Patent 4,010,935; March 8, 1977; assigned to Alumax Inc.*

A high efficiency aluminum scrap melting system is provided in which a portion of the hot combustion gases employed to melt the aluminum in a hearth is recycled through scrap-receiving preheat compartments to preheat scrap and volatilize contaminants as the scrap is fed to the hearth, the remainder of the hot combustion gases leaving the hearth as hot exhaust gases being optionally employed through a recuperator to preheat air which is fed to a combustion burner system used to burn hydrocarbon fuel and provide heat for the overall system. Energy available in the contaminants also supplements the energy supply to the system.

Such a system is shown in Figure 8 where a combustion chamber **10** is shown having a burner section **11** and an afterburner section **12**, fuel **13** and air **14** being fed to the burner at predetermined ratios to provide together with recycle effluent hot gases fed to the afterburner section, hot gases in the afterburner section having a temperature of about 2000° to 2500°F. The final effluent hot gases are fed to scrap-melting hearth **15** to melt the aluminum scrap fed thereto from preheat compartment **17** or **17A**. Preheat compartment **17** comprises a rotary kiln used for chopped materials and preheat compartment **17A** for receiving compressed bales.

A portion of effluent hot gas **16** leaving the hearth is passed through scrap-receiving preheat compartment **17** to preheat the scrap therein prior to feeding the scrap into the hearth, the scrap being preheated to about 900°F, the gases being withdrawn via fan **18A** either through the preheat compartment **17** or preheat compartment or duct **17A** and recirculated to the afterburner section **12** as shown. The remainder of the effluent hot gases exhaust to recuperator **20** or via bypass valve or damper **19** directly to exhaust stack via line **23**. As molten aluminum builds up in the hearth, it is tapped at **15A**. A damper **18** permits bypassing preheat compartment **17** for temperature control.

Figure 8: Alumax, Inc. High Efficiency Aluminum Scrap Melter

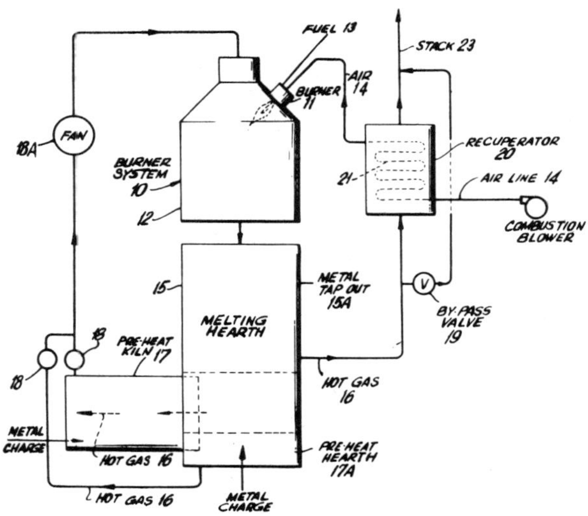

Source: U.S. Patent 4,010,935

It is preferred, in order to make use of as much residual heat as possible to conduct the remainder of the effluent hot gases from the hearth as hot exhaust gases directly to recuperator **20** to heat tubes **21** therein for preheating air **14** for feeding to burner **11** of the combustion chamber, the gas going to the stack via line **23** at a temperature of the order of about 1200° to 1500°F.

In the melting of aluminum, about 200 Btu/lb aluminum can be used to preheat the scrap to the 900°F range without excessive oxidation. About 300 Btu/lb aluminum will then be added to the charge by conduction through the heel charge of molten metal. This enables the reduction in temperature of the exhaust gases to a range of about 1200° to 1500°F, instead of 2000° to 2500°F common to standard direct fired melting furnaces.

Aluminum or aluminum alloy extrusions have been widely utilized in various fields as structural members, for example, in buildings due to their light weight and corrosion resistance property. Thus, the amount of manufacture of such aluminum or aluminum alloy extrusions has increased year by year. Such increases in production of extrusions cause, as a matter of fact, an increase in the amount of discard or waste produced in the extrusion process. Further, a remarkable amount of scrap is produced during fabrication and assembling of the extrusions. The total amount of such scrap reaches 20 to 30% of the billet weight. Thus, there is an increased need for development of a technique for effective recovery of such scrap and waste.

Known processes for recovering scrap include steps of melting the scrap and forming into billets, and then extruding the billets through an appropriately shaped mold or die to form extrusions. In the process, however, it is required to add an adequate amount of virgin metal in the melting step in order to ensure a satisfactory extrudability. Further, the process is disadvantageous with respect to high electrical power consumption in the melting step as well as heavy metal losses.

Therefore, it has been proposed to extrude aluminum or aluminum alloy scraps directly through a mold to form extruded structural profiles. For example, U.S. Patents 2,391,752 and 2,302,980 teach extrusion of aluminum or aluminum alloy scraps directly through the mold. As was recognized in the patents, one of the problems encountered in the direct extruding of aluminum or aluminum alloy scraps is the treatment of oxide skins or oxide films coating the scrap pieces. If the extruding pressure is not sufficient to disrupt these oxide skins or films, they will prevent metallic contact of individual scrap pieces resulting in products of less uniformity and less mechanical strength.

Thus, it was proposed to have aluminum and aluminum alloy scraps and wastes subjected to pressure of 3 to 8 ton/in^2 in an extruder housing under the temperature of about 300°C in order to loosen the oxide skins and then extrude the scraps and wastes under the pressure of 18 to 40 ton/in^2 at a temperature between about 350° and 450°C. This process has been found effective to solve the problems inherent to the oxide skins existing on the scrap pieces, however, it is not effective to eliminate problems caused by air entrapped in the extruder housing.

Such air entrapped in the extruder housing provides resistance to the movement of an extruding ram and forms voids in the extrusions. Thus, existence of air in the extruder housing has adverse effects on the quality and strength of the final products. Further, it is not easy in actual practice to charge the chip materials into the extruder and control the correct amount of such chip materials which is required for one extrusion stroke. The above charging process takes a lot of time and this causes the temperature of chip materials to drop below the temperature suitable for hot extrusion.

T. Takahashi, T. Nagano, K. Nakamura, and M. Kikuchi; U.S. Patent 4,028,795; June 14, 1977; assigned to Riken Keikinzoku Kogyo KK, Japan have developed an improved process in which aluminum or aluminum alloy scraps or wastes are broken into chips and converted into a cylindrical body to be extruded under a temperature suitable to produce extruded structural profiles. The method is characterized by the fact that the interior of the extruder is evacuated prior to or simultaneously with the extrusion step so as to exhaust air entrapped in the cylindrical body.

Similar process coverage is that by *T. Takahashi, T. Nagano, K. Nakamura, M. Kikuchi, and K. Suzuki; U.S. Patent 4,033,024; July 5, 1977; and T. Takahashi, T. Nagano, S. Iguchi, M. Kikuchi, K. Nakamura, and A. Mochizuki; U.S. Patents 4,050,142; September 27, 1977; and 4,117,703; October 3, 1978; all assigned to Riken Keikinzoku Kogyo KK, Japan.*

A drawback is encountered in the extrusion step performed under vacuum in that the flow rate of the exhausted air sucked by the vacuum pump per unit time is limited in practice. Consequently, if a vacuum must be created for each extrusion step, the working time ratio of the extrusion press is low.

Moreover, as is well known, aluminum is recognized as a metal having a relatively high consolidation rate. This property would seem to be advantageous in the extrusion of aluminum scrap. However, instead it poses a serious problem in obtaining a substantially void-free, dense and scab-free product in the extrusion step.

More specifically, when the individual scrap particles constituting the aluminum scrap are compressed, they are deformed. An increase in the compression pressure causes metallic contact of the scrap particles and their consolidation into metallic bodies. Since, in this instance, metallic union of the individual scrap particles is completed within an extremely short period of time because of the high consolidation rate of the scrap metal, exhausted air trapped between the scrap particles cannot escape out of the metallic bodies. This results in the formation of voids or perforations in the final product. These voids or perforations in the metal cause deterioration of mechanical properties such as strength and corrosion resistance.

In addition, because of these defects, bloated portions are formed near the surface of the finished product in the course of the extrusion thereby providing poor surface appearance of the extruded article. These bloated portions will hereinafter be referred to as scabs. When the finished articles contain scabs on their surfaces, the commercial value of the recycled products is significantly reduced.

In summary, it has been theoretically recognized that it is advantageous to perform the extrusion of aluminum scrap in a vacuum or after the compression of the aluminum scrap into a mass with a density close to the theoretical complete density of the aluminum metal. These techniques aid in eliminating the formation of voids or perforations during treatment of the aluminum scrap including compression, heating and extrusion. However, these expedients have disadvantages as discussed above in that an increase in energy consumption and design of a large-size apparatus are required.

Moreover, it is extremely difficult to produce void-free and scab-free finished articles using apparatus of limited size with a limited amount of energy consumption. Consequently, there is a need for a new process for treating scrap aluminum whereby the finished product will be free of these defects.

Such an improved process is claimed by *A. Asari and K. Tatsuno; U.S. Patent 4,059,896; November 29, 1977; assigned to Kobe Steel, Ltd., Japan*. It is a process for converting aluminum scrap into finished articles by compaction, heating and extrusion.

The process comprises the steps of compacting the aluminum scrap at ambient temperature into a billet having a density between 1.8 and 2.4 g/cm^3; heating the compacted billet to a temperature between 350° and 550°C; compacting the heated billet into a dense mass having a density about 2.4 g/cm^3 either in an extrusion container, while maintaining conditions under which the air pressure at the outer periphery of the billet is lower than that in the central portion of the billet, or externally of the extrusion container. Subsequently, after completion of the second compaction, the second compacted billet is consolidated by extrusion through an extrusion die fitted to the container while maintaining the above mentioned conditions, whereby void-free and scab-free finished articles of good quality are efficiently produced with a small apparatus using little energy.

Recycling systems for metal scrap are normally required to be at least competitive with established manufacturing systems for producing such metal. To be competitive, such recycling systems must be efficient which can often require that they be designed to reclaim certain or specific kinds of metal scrap. With respect to the kinds of metal scrap encountered in the light metals field, particularly aluminum, it should be observed that it can range from scrap generated very early in the manufacturing process to that resulting from used products, e.g., metal containers.

Scrap generated early in the manufacturing of aluminum, for example, are scalping chips. These chips result from preparing the surface of an aluminum ingot for rolling or cladding or some such purpose when often large irregularities are removed. Chips removed at this time can constitute as much as 6% of the total metal in the ingot and therefore obviously can amount to a considerable amount of metal which can be recycled. Scrap typical of used metal containers which can be beneficial to recycle are metal beverage cans and the like. Recycling such cans serves to conserve resources and also to provide a cleaner environment.

These two types of scrap are often considered to present different problems from the standpoint of reclamation. For example, container or can scrap normally comprises bits and pieces of metal which can range in size from relatively small particles to crushed cans. Often the cans have holes pierced in them for purposes of delacquering using solvents. The pierced holes, while aiding the flow of solvent through the mass of cans, result in jagged edges which, together with the generally nonuniform shapes and sizes of crushed cans, make this scrap relatively nonflowable. Furthermore, such scrap tends to be rather buoyant with respect to the melting media, requiring special equipment to forcibly submerge it. A particular type of equipment for handling this type of scrap is disclosed in U.S. Patent 3,873,305 where a rotating wedge forcibly submerges this type of scrap into the melting media. In the system described in this patent, additional equipment is required to recirculate the molten melting media.

With respect to the scalping chips, or scrap which may be characterized as flowable, for example, such as filings and cuttings or even large pieces, these often do not present great difficulty in submerging as does the abovementioned container scrap. Nevertheless, it is imperative that such flowable scrap be melted and reclaimed in the most efficient manner in order to provide an incentive to operate a remelting system as a source of metal, as compared to the established manufacturing system for producing such metal.

With highly oxidizable metals, such as aluminum, out of which the aforementioned containers and beverage cans are often made, it is important to provide a remelting system which provides maximum heat transfer for efficient melting. Also, it is important that the metal scrap be submerged with minimum turbulence at the melting media surface in order to minimize oxidation of the molten metal. Oxidation leads to skim formation, a problem inherent in most systems reclaiming highly oxidizable metals such as aluminum.

The skim referred to includes the metal oxide, e.g., Al_2O_3, and molten metal which becomes entrained in such oxide. Skim is undesirable and therefore should be minimized not only because of the molten metal lost to it, but also because it can cause problems by acting as an insulator in the heating bay. That is, when combustion units discharging their heat on the surface of the molten media in the heating bay are used, skim floating on the surface thereof makes it difficult to heat the molten media. Also, such floating skim by rejecting heat can seriously shorten the useful life of the equipment by causing it to overheat.

An improved form of apparatus which has been claimed to overcome these problems has been developed by *J.H.L. van Linden, R.J. Claxton, J.R. Herrick, and R.J. Ormesher; U.S. Patent 4,128,415; December 5, 1978; assigned to Aluminum Company of America.* This apparatus for melting metal scrap in a molten melting media includes a housing generally cylindrical in cross section and having upper and lower portions.

Metal scrap is introduced to a body of molten melting media contained in the upper portion of the housing. A supply of molten melting media is added to the upper portion of the housing through a volute located in the lower portion. The molten melting media is supplied or added by action of an impeller located in the lower portion and mounted on a drive shaft extending through the upper portion. Vanes are mounted on the drive shaft to control the flow motion of the body of molten melting media and metal scrap in the upper portion of the housing by creating a vortex in this body for purposes of mixing the melting media and metal scrap. An opening is provided in the upper portion to remove melted scrap and melting media at a rate substantially commensurate with their rate of introduction.

Such an apparatus is shown in general schematic plan and in more detailed sectional elevation in Figure 9.

Figure 9: Aluminum Co. of America Apparatus for Aluminum Scrap Reclamation by Molten Bath Treatment

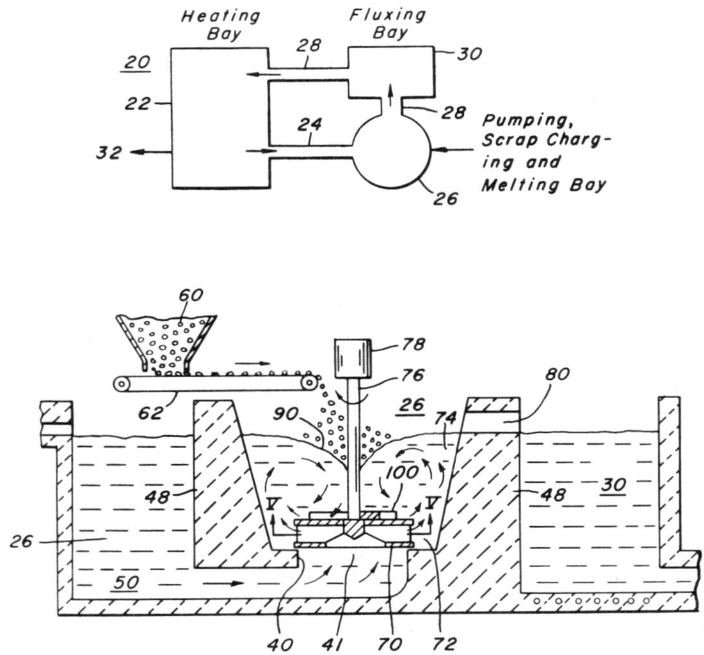

Source: U.S. Patent 4,128,415

There is shown a schematic of a circulatory melting system **20** incorporating the process. In the circulatory melting system, molten melting media is circulated from the heating bay **22** along line **24** to a bay **26** wherein pumping, scrap charging and melting are accomplished simultaneously. Thereafter, the melt is circulated back to the heating bay via line **28** which can pass through a fluxing bay **30**. The heating bay can have a discharge line **32** which in the circulatory system can discharge molten metal at a rate commensurate with the feed rate of scrap to bay **26**. This type of circulatory system is highly efficient in its use of heat input and in its freedom from skim generation, particularly when used in conjunction with the fluxing bay.

The melting media may be heated in the heating bay by combustion units discharging their heat upon its surface. Alternatively, electric induction or resistance heaters immersed in the molten media may be used.

The melting media can be molten metal of similar composition to the metal charge or it may be a molten salt. If a molten salt is used, a salt-metal separation would normally be necessary to facilitate the removal of the melted metal. When molten aluminum is the melting media, a typical temperature leaving the heating bay is about 1400°F although this temperature can range from 1325° to 1475°F but with a higher temperature there is a greater tendency to form skim. A typical temperature reentering the heating bay is about 1300°F. This provides about a 100°F temperature drop across the melting bay to provide heat to melt the charge.

In the major sectional view in the figure, there is shown bay **26** which allows for charging metal scrap **60**, for circulating molten melting media **50** and melting of metal scrap, all of which can occur simultaneously. As will be noted, bay **26** is preferably generally circular in cross section and can be generally cylindrical in shape or may have walls **48** which are downwardly, converging, as illustrated.

In operation, the molten melting media is circulated from the heating bay to bay **26** and back to the heating bay by means of a rotating impeller, generally referred to as **70**. The action of the impeller located in lower portion **72**, as well as providing circulation in the system, draws melting media up into upper portion **74** of bay **26**. That is, the impeller continuously replenishes the body of the melting media contained in bay **26**.

Scrap **60** is fed along an endless belt **62** into bay **26** where it is ingested by the molten melting media. The combination of melting media and molten metal from the melted scrap is removed from bay **26** through opening **80** located in upper portion **74** of bay or housing **26**. The removal is at a rate substantially commensurate with the rate of their introduction thereto. The combination may be introduced to a fluxing bay, for example, as shown for purposes of removing skim or dross.

The metal scrap is ingested and melted in bay **26** free from problems of metal scrap becoming wedged in the impeller or becoming wedged between the impeller and volute wall **40**. That is, it will be observed by inspection of bay **26** that in spite of the melting media being drawn up through the lower portion **72** of bay or housing **26**, the flow of molten melting media is controlled so as to provide vortex **90**. The vortex is effective in forcibly submerging or ingesting metal scrap into the melting media. The flow of molten media introduced by impeller **70** is controlled by vanes **100** projecting from drive shaft **76**. Vanes **100** are mounted on the drive shaft substantially adjacent to the impeller.

However, it should be noted that the vanes may be mounted on the shaft so as to be spaced apart from the impeller. Thus, rotation of the drive shaft by motor **78** serves to both introduce or replenish the melting media in bay **26** by action of the impeller and to create a vortex and therefore provide efficient ingestion and melting of metal scrap by action of the vanes.

It will also be seen that the melting media in bay **26** is shown generally as being pulled downward in the vicinity of the drive shaft by action of the vanes and as being pushed upward in the general vicinity of wall **48**. It should be noted that this may be a somewhat simplified representation of the flow pattern of the melting media in the vortex. That is, in addition to the up and down flow illustrated, there is substantial circular flow resulting from the action of the impeller and the vanes.

In the metal industries, and in particular the aluminum industry, one of the major problems with regard to reprocessing scrap aluminum is the processing oil that remains on the scrap metal after it has been worked. Large amounts of sheet and foil scrap are generated in the production of sheets and light gauge foil products. Generally these scraps are contaminated with oil and additives having a wide boiling point range of between about 150° and 1100°F.

There are usually two basic types of oily scrap generated by rolling processes: trim scrap and machine scrap. The trim scrap is produced when trimming the edges of the sheets during the rolling and coiling process or during the slitting process. Machine scrap is cut from the coils and consists of out-of-gauge materials, tails, rejected materials and returned goods. Scrap may also be generated due to unusable products such as wire, cable, automobile parts and machine shop turnings, etc., and can be of any shape or form. The scrap is usually briquetted for convenience of handling and storage with the trim scrap and machine scrap briquets having a density of approximately 30% and 40 to 60%, respectively, of virgin aluminum.

If the oil contaminated briquets are charged into a furnace, the oil catches fire and a black smoke is generated during the melting process. The fire generated by the burning oil increases the surface temperature of the briquets and creates additional oxides which result in dross and melt loss. The increased dross results in melt loss not only due to the formation of oxides but because pure metal is trapped in the dross. The amount of dross varies but, in an average melting operation, is approximately 20% recoverable aluminum by weight; therefore, the removal of the oil from scrap is most important because melt recovery is improved. An added advantage is that the air quality standards set by the Environmental Protection Agency (EPA) can be met more easily by reducing the amount of smoke generated during melting.

An improved technique developed by *J.M. Kartanson and R.M. Neel; U.S. Patent 4,141,373; February 27, 1979; assigned to RJR Archer, Inc.* provides a method of removing oil from metal scrap, comprising the steps of introducing oil-laden scrap into a hermetically sealed chamber, evacuating the chamber, heating the scrap to vaporize the oil, and removing the vaporized oil from the chamber.

It is well known that coatings, such as lacquers, and other surface contaminants on aluminum can scrap and the like should be removed prior to melting in order to avoid unacceptable melt loss. One preliminary cleaning process typically employs an organic solvent such as acetone which removes a substantial portion of the lacquer, oils and other surface contaminants.

The most common preliminary cleaning approach, however, is the delacquering furnace in which the material is continuously conveyed through controlled zones of increasing temperature, progressively heating the material to about 1000°F which volatilizes the impurities to effect their removal. After the scrap is held in the furnace for sufficient dwell time, the scrap eventually exits the furnace with very little remaining lacquer residue.

For aluminum can scrap, the current delacquering process significantly adds to the cost of a pound of recycled aluminum. Additionally, the delacquering furnaces are a continual maintenance problem resulting in excessive downtime and costly delays for necessary repair. For example, if the temperature in the delacquering furnace exceeds about 1050°F, the surface contaminants may ignite and cause a harmful fire to spread throughout the furnace. Thus, careful control is often needed to prevent these problems which, in turn, adds to the cost of recycling aluminum.

A technique which overcomes these problems has been developed by *R.E. Miller; U.S. Patent 4,147,531; April 3, 1979; assigned to Aluminum Company of America*. This improved process comprises the steps of (1) continuously depositing surface contaminated scrap onto the surface of a stream of molten metal; (2) directly propelling the floating scrap along the molten surface for a time sufficient for the heat from the molten metal to effectuate substantially complete removal of the contaminants from the scrap; and (3) forcibly submerging the remaining decontaminated scrap into the molten metal stream.

A system is also provided for removing impurities from the surface of scrap comprising a trough of molten metal, means for moving the molten metal through the trough, means for continuously depositing contaminated scrap onto the molten metal surface, means for propelling the floating scrap through the trough and means for forcibly submerging the remaining floating scrap into the molten metal after the contaminants are substantially completely removed.

A suitable form of apparatus for use in the conduct of such a process is shown in Figure 10. It shows a dual chamber recirculating melting system or furnace **2**. This arrangement includes a skim collection chamber **4** and a heating chamber **6** separated by a dividing wall **8**. The chambers are in fluid communication in order that molten metal may freely flow through the furnace from the skim collection chamber to the heating chamber.

The furnace shown also includes a side bay trough **10** through which molten metal may be circulated. It is a straight section **11** of this side bay trough or the like in which the process is preferably practiced.

A pump **12** is provided in the side bay trough to supply the driving force to continuously circulate molten metal through the furnace system and maintain a moving stream of molten metal through the trough.

Also illustrated in the melting system is an overhead hopper **14**, a vibrating conveyor **15** and a chute **16**. The hopper may be used to hold a large quantity of scrap which is to be delacquered and/or deoiled and then melted. The overhead hopper feeds the vibrating conveyor which, in turn, discharges the scrap onto the chute. The chute leading from the vibrating conveyor extends downward to a point at or near the surface **18** of the molten metal **20** in the trough. Preferably the chute is disposed at an angle or slope low enough to prevent excessive submergence of the scrap falling therefrom onto the molten metal surface.

Also, it is the vibration rate of the conveyor which controls the rate and therefore the amount of scrap continuously deposited onto the molten metal surface which determines the thickness of the body of scrap floating thereon. As explained in more detail below, the thickness of the body of scrap deposited on the surface of the molten metal should be such that the combustible contaminants on the scrap are substantially completely burned or vaporized prior to their reaching the location of the downstream pump **12**.

A variety of types of contaminated scraps may be used in the present process including skeletal scrap, lacquered can scrap, oily chips and turnings and borings. Since each of these scraps may have different density, size and weight as well as different amounts of combustible contaminants thereon, it may be helpful to provide a variable vibration control for the conveyor **15**, and perhaps a height and angle adjustable chute **16** in order to be able to adjust for and minimize surface turbulence and submergence as the different types of scrap are fed onto the molten metal surface **18**.

The process is practiced in a delacquering and/or deoiling zone **Z**, preferably in a straight section **11** of an open trough **10**.

Downstream of the end of the chute **16** is a rotatable cylinder **22**, also referred to herein as a feed gate igniter, which directs the scrap fed onto the surface of the molten metal **20** therethrough, holds the scrap close enough to the molten metal to ignite the combustibles on the scrap and then directly propels the scrap at a controlled rate along the molten metal stream independent of the flow rate of the molten stream. It is the feed gate igniter which controls the rate at which the floating contaminated scrap moves along the metal stream regardless of the rate at which the metal stream is circulating by means of direct propulsion applied to the scrap passing thereunder.

A preferred feed gate igniter is a rotatable refractory cylinder extending substantially completely across the width of the straight trough section and having its axis disposed perpendicularly to the direction of metal flow. The cylinder is located downstream of the discharge end of the chute and is preferably disposed adjacent the chute such that scrap exiting the chute is deposited on the molten metal surface under the outer peripheral surface of the cylinder. This arrangement lends itself to efficient control of the rate at which the scrap is propelled along the molten metal surface by the pinching action of the cylinder against the deposited scrap fed from the chute to force the scrap into intimate contact with the molten metal surface.

This cylindrical gate is disposed toward the molten metal surface such that the molten metal surface does not touch, but is substantially tangent to, the outer peripheral surface of the cylinder. The cylinder **22** may be able to be raised in instances where the scrap feed rate is increased. The cylinder is rotated about its axis, and rotation of the cylinder in the counterclockwise direction directs the floating scrap toward the downstream pump **12** in a controlled rate, independent of the flow rate of the molten metal stream.

A discharge or exit gate **24** may be located downstream from the feed gate igniter **22**. The exit gate may also be a rotatable refractory cylinder. The rotation rate of the feed gate igniter is not necessarily synchronized with the rotation rate of the discharge gate **24** because melting of the floating scrap occurring as the scrap floats from one gate to the other causes the volume of the scrap to continuously diminish. Usually, any cross section of the mass of floating scrap is retained on the molten metal surface in the delacquering zone for a mean residence time of from 30 sec to 2 min, during which time at least 30 wt % of the scrap may be melted.

The residence time for the floating scrap not only depends upon the type of scrap being charged and the amount and type of contaminants thereon, but the residence time is also proportional to the scrap charging rate. If more scrap is being charged, the mass of floating scrap is thicker. Thicker masses of scrap require more time to be decontaminated because the source of heat in the delacquering zone is from the circulating molten metal passing under the mass of scrap. The delacquering zone **Z** shown is the length of trough in which contaminants are being combusted, as defined by the flame therein having a decreasing height as the scrap proceeds in the downstream direction. The presence of the discharge gate **24** is primarily to provide a means for stopping any scrap in which the combustible contaminants have not yet fully volatilized from prematurely passing into the pump **12** where subsurface volatilization of uncombusted contaminants may adversely affect melt loss and skim formation.

The gates **22** and **24** may, at times, come into contact with the molten metal. When aluminum is being melted, the molten aluminum is preferably held at a temperature of 1300° to 1400°F. The gates, especially the feed gate igniter **22**, are subjected to very high temperatures upon combustion of the contaminants on the scrap passing thereunder. Therefore, it is preferable that the gates **22** and **24** are constructed of a durable, heat resistant material. To this end, stainless steel is an acceptable material but is often considered too costly. It has been found that a refractory material is acceptable for constructing the gates. Also, although the preferred configuration for the feed gate igniter **22** is cylindrical, a paddle wheel type arrangement may also be suitable.

The cylinders **22** and **24** could be rotated by a variety of known methods including motorized or hand-operated cranks, or by a series of gears. It should be noted that the particular mechanism used to turn such gates should also be heat resistant or otherwise protected from the area around the gate where adverse conditions, ambient heat and hydrocarbon emissions may hamper operations.

In the operation of the melting system described above, molten aluminum **20** is circulated from a skim collection chamber **4** to a heating chamber **6**, through a side bay trough **10** and back to the skim collection chamber **4**. The circulation of the molten metal **20** may be maintained and, simultaneously, a suitable vortex may be created by the operation of a centrifugal discharge pump. In a preferred arrangement, as illustrated, running a centrifugal discharge pump at approximately 100 rpm or operating an axial flow pump at about 200 rpm should be able to circulate molten metal at the rate of approximately 250,000 lb/hr. The velocity of the molten metal flowing through the straight trough section **11** should be fast enough to prevent the system from freezing in order that the molten metal maintains a temperature adequate to effect the removal of surface contaminants from the scrap metal. Preferably such velocity is between 2 and 2½ fps.

In this process, a quantity of contaminated metallic scrap is continuously deposited onto the moving stream of molten metal, as shown. Exemplary scraps include crushed beverage containers, skeletal scraps, turnings and borings, oily chips and compacted metallic briquettes. These scraps may contain lacquer coatings and rubber or plastic end seals or inserts in a container seam, along with certain grease and oil deposits and other surface contamination, all of which must be substantially removed prior to submerging the scrap metal in order to minimize melt loss and skim formation. The amount and type of surface contamination on the scrap may vary considerably depending upon the type of scrap being melted.

Figure 10: Aluminum Company of America Apparatus for Aluminum Scrap Reclamation in Surface Fired Molten Bath

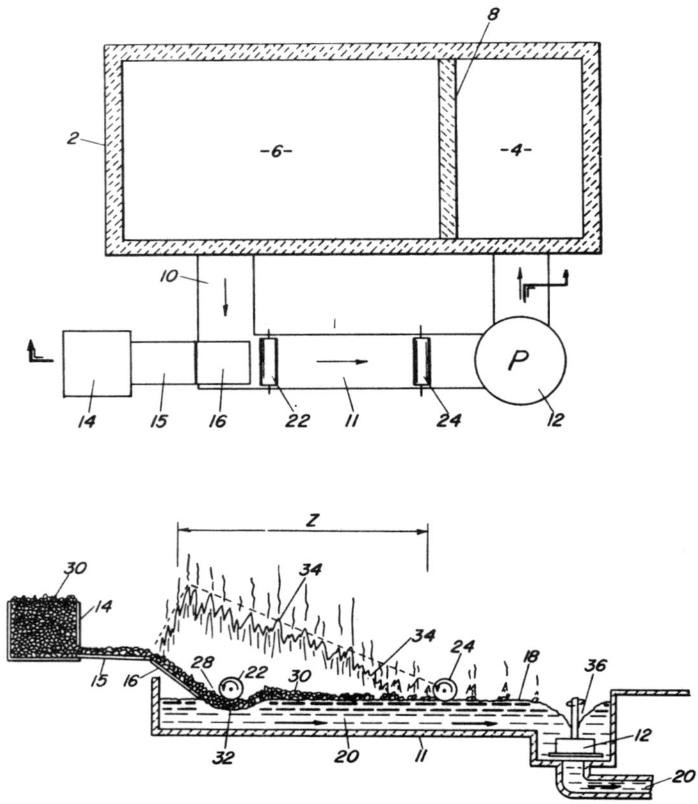

Source: U.S. Patent 4,147,531

Thus, the time required to volatilize and combust the contaminants can change with each type of scrap, and may vary considerably with each batch of the same type of scrap. It is preferred, therefore, to continuously charge the same type of scrap, at least for a period of time, in order to stabilize the melting operation as much as possible.

The overhead hopper **14** should be filled with sufficient scrap to maintain a constant scrap feed rate. Depending upon the size of the hopper, it may be necessary to fill the hopper during feeding of the scrap in order to sustain a certain charging and melting increment. The scrap may be fed from the hopper across a vibrating conveyor **15** and into a chute **16**, or the like, by any known method.

The scrap exits the chute and falls onto the surface **18** of the molten metal **20** just upstream of the feed gate igniter **22.** The sides of the metal trough and the gate serve as limits on the charging area **28** for the scrap. These dimensional limits, in conjunction with the scrap feed rate, determine the thickness of the cluster of scrap which is fed through and is directly propelled by the feed gate igniter **22**. The height of the feed gate igniter from the molten surface **18** may also be adjustable to control the thickness of the scrap cluster. The thickness of the cluster of scrap should be such that ample oxygen from the ambient air is supplied to the combustible contaminants located in the lower regions of the cluster scrap. If the raft of scrap is too thick, the combustibles in the lower regions of the scrap cluster are not able to be burned because of the lack of oxygen. It has been found that a cluster thickness of less than about 6" does not adversely affect burning of contaminants in bottom regions of the floating scrap.

In the preferred arrangement shown, the rotatable refractory cylinder **22**, disposed just above the surface of molten metal **20**, is utilized as the feed gate igniter **22**. Rotation of the cylinder causes the scrap **30** in the charging area **28** to be grasped and pushed downwardly under the cylinder but importantly not under the surface of molten metal **24**. The floating cluster of scrap which is forced under the rotating cylinder **22** for only a short period of time does not overcome the surface pressure of the molten metal. Rather, the unconfined surface of molten metal recedes under the cylinder and rises equivalently in other areas of the trough.

If the contaminated scrap is prematurely submerged into a molten aluminum bath, oxides and volatiles are released into the molten metal, trapping a certain amount of aluminum with the oxides as they rise to the surface. Such action adversely affects the melting operation by increasing the amount of metal loss and should, therefore, be avoided. If excessive premature scrap submergence is experienced in the operation of the process, the scrap feed rate should be reduced and/or the rotating cylinder **22** should be slightly raised above the molten metal surface **18** to insure efficient operation.

By the abovedescribed arrangement, the floating cluster of scrap, although not submerged into the moving stream of molten metal **20**, is disposed at a lower level than the upstream surface of molten metal at the location where the scrap passes under the feed gate igniter **22**. This slight depression creates enough of a pressure head behind the depressed scrap to propel all of the scrap clustered together downstream of the point of depression **32**.

The majority of the forward, or downstream, motion is maintained by the pinching and pushing action of the feed gate igniter **22** against the scrap. The surface of the many pieces of scrap which lie in or against the molten metal surface are continuously delacquered, deoiled and at least partially melted as the scrap moves downstream. The scrap melting, along with the heat released by the combustion of the surface contaminants, causes the scrap to agglomerate, forming a cluster of scrap which can be viewed as acting something like a raft.

Accretion of the continuously fed scrap **30** to the rearward, or upstream, end of the scrap raft ensures that the cluster of scrap is generally continuous in the lateral directions, i.e., without substantial voids, until the particular increment of scrap charging is complete. Minor pauses in charging of scrap are tolerable and in such instances, the discharge gate **24** may be utilized to stop the downstream cluster or clusters of scrap from floating until the voids or spaces between clusters, caused by the pause in charging, are filled by incoming scrap.

There are expected to be major pauses between charging of different types of scrap and in instances when the charging of cold scrap has reduced the temperature of the circulating molten metal below that desirable. In this latter instance, charging must be discontinued until the desired metal temperature has been recovered. In the apparatus shown in the drawing, metal temperature could be recovered in about 5 or 10 min as the molten metal passes through the fired heating chamber **6**.

Although the melting temperature for aluminum is about 1220°F, it will be understood that the temperature recovery necessary for optimum melting conditions for molten aluminum is between 1250° and 1700°F and preferably between 1300° and 1400°F. However, for delacquering purposes alone, much lower temperatures of about 1000°F would be adequate. To the extent that this process of removing contaminants from scrap is employed in conjunction with the melting operation, the melting temperature limitations must be followed. However, it is considered possible to delacquer alone by the method and in the system of this process by depositing contaminated scrap onto at least a 1000°F surface of molten salts.

The scrap aluminum **30** fed from the chute **16** hits the molten metal surface **18** having a temperature of approximately 1300° to 1400°F which is more than adequate to ignite the combustible contaminants. The majority of the contaminants are burned within the first few seconds, and minor amounts of remaining contaminants are burned as the scrap floats downstream toward the pump **12** or other submerging device. A graph of contaminants burned over the period of time that the floating scrap is propelled through a delacquering zone in a trough would parallel the flame height along the straight trough section, as shown by the dotted line, which can be observed upon charging of the scrap into the straight section of trough.

Even if the thickness of the cluster of charged scrap is held constant, the burning pattern for different types of scraps and for scraps having different quantities of combustibles thereon varies. The surface contaminants of some scraps ignite, and the burning flame height will diminish uniformly, as is the case with crushed lacquered can scrap. Other scraps ignite and the burning flame height remains high for a period of time before diminishing, as is the case with oily chips. Still other scraps ignite and the burning flame height diminishes relatively rapidly, as is the case with dried crushed skeletal scrap. The flame height, therefore, serves as a definite indicator of when the combustible contaminants have been substantially completely vaporized.

The preferred time for forcibly submerging the remaining decontaminated floating scrap **30** is within 20 sec after the flame is extinguished, i.e., after the surface contaminants are substantially completely volatilized and/or combusted. The time at which the flame is extinguished can easily be determined from a quick trial in the system described above. It will be found, however, that certain major fluctuations in amounts of contaminants may occur in each scrap charge increment. When such occurs, a higher than normal flame will be observed in the downstream areas of the trough **11**. The discharge gate **24** is utilized in such instances to stop the flow of scrap for a longer than normal time, until the flame has nearly extinguished, indicating that the higher quantity of contaminants has been substantially burned and volatilized.

The height of the flame **34** correlates with the amount of contaminants being burned at any location in the straight trough section **11**. Notice that the flame is nearly extinguished as the floating scrap reaches the discharge gate **24**. There may be minor, insignificant amounts of unburned contaminants which emit a flame even as the scrap **30** is drawn into the vortex **36** created by the pump **12** and is forcibly submerged as is typically the case even with scrap which has been pretreated in a delacquering furnace.

It should also be understood that the cluster of scrap floating on the surface **18** of molten metal **20** provides a covering over an otherwise exposed trough **11**. In fact, the ignition of the contaminants on the surfaces of the scrap helps minimize the heat loss from the circulating molten metal otherwise caused by the cooler ambient atmosphere.

In a preferred embodiment, a cover or hood may be provided above the trough to trap the emissions released as the combustible contaminants are burned from the scrap. Such a hood could prevent release of the hydrocarbons into the atmosphere. These gases could then be transported to a heating chamber, or the like, where they could be further combusted.

A method developed by *K. Amino; U.S. Patent 4,159,907; July 3, 1979; assigned to Amino Aluminium Industries, Co., Ltd., Japan* is one in which economically advantageous melting of aluminum scraps is obtained by first compressing the aluminum scraps under a magnitude of pressure exceeding the yield strength of the material making up the scraps for thereby converting them into a compressed mass having an apparent specific gravity greater than the specific gravity of normal aluminum melt and subsequently introducing the compressed mass into a molten aluminum bath. Figure 11 is a block flow diagram showing the application of such a method.

In this case, the scraps are assumed to be shredded twice. They are subjected to the first shredding and then, after separation of foreign particles such as debris of paper and paints thereform, are subjected to the second shredding, followed by magnetic separating which serves to remove iron pieces from the aluminum scraps.

The extent of these shredding and separating operations are variable with the kind of aluminum scraps under treatment and the kind of extraneous particles contained in the scraps. At any rate, these operations are desired to be carried out to the extent of converting the scraps, by the end of the second shredding, into particles most (more than 80%) of which have maximum diameters of between 2 and 20 mm.

The particles of aluminum scraps thus prepared are subsequently compressed with a magnitude of pressure equalling or preferably exceeding the yield strength of the material making up the aluminum scraps. A double-action type hydraulic molding press or a withdrawal molding press, for example, can be used for the purpose of compressing the particles of aluminum scraps.

ALUMINUM FROM SOLID PROPELLANT WASTE

See "Ammonium Perchlorate from Solid Propellant Waste."

Figure 11: Amino Aluminium Industries Process for Handling Scrap Aluminum

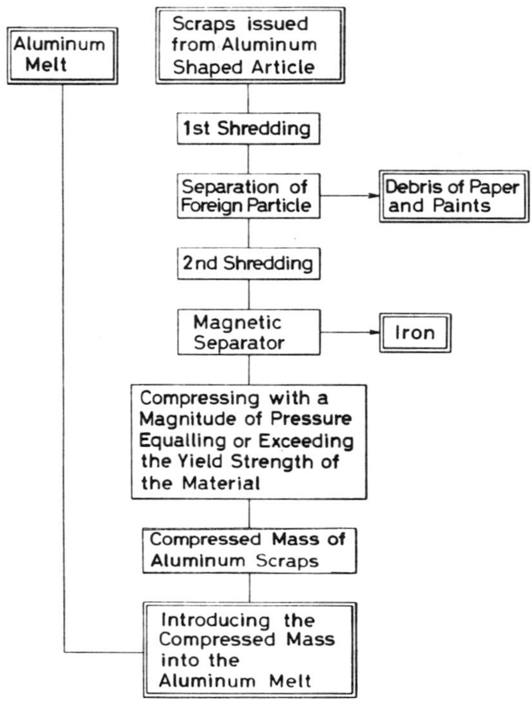

Source: U.S. Patent 4,159,907

ALUMINUM FLUORIDE FROM ALUMINUM REDUCTION PLANT WASTES

In a variant of processes for the recovery of HF (see "Hydrogen Fluoride from Furnace Pot Linings") a process has been developed for AlF_3 manufacture.

It has been found that the HF content of the pyrohydrolysis off-gas can be further increased by a considerable degree using as fluidizing medium and as a source for combustion of the carbon content of the waste material charge, a stream which contains at least about 90 vol % O_2. Using the essentially nitrogen-free stream for these purposes, in combination with the expanded reaction zone concept where a relatively finely divided source of Al_2O_3 is contacted with the vaporized Na-components of the off-gas, results in a more than 300% increase in HF content of the Na-free off-gas. This off-gas can be readily employed for the production of an AlF_3 product containing at least 85 wt % AlF_3 and/or a highly concentrated HF with a fraction of the energy required in comparison to prior art processes.

A process of *J.N. Andersen and N. Bell; U.S. Patent 4,160,809; July 10, 1979; assigned to Kaiser Aluminum & Chemical Corporation* is outlined in Figure 12.

Figure 12: Kaiser Aluminum & Chemical Corporation Modified Pyrohydrolysis Process for Aluminum Fluoride Production from Spent Aluminum Reduction Cell Linings

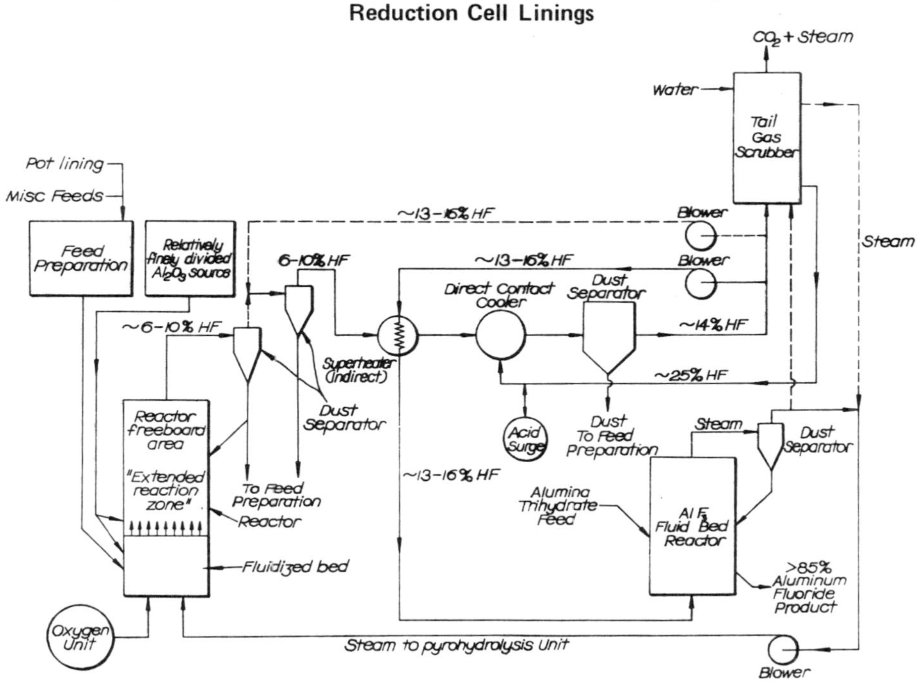

Source: U.S. Patent 4,160,809

As shown in Figure 12, the spent pot lining charge is introduced into the pyrohydrolysis reactor whereupon charging of oxygen (at least 90 vol % O_2 content) and steam, combustion of the carbonaceous constituents, and generation of hydrogen fluoride containing off-gas is achieved at temperatures in excess of 900°C, usually within the range of 900° to 1300°C. In order to convert the volatilized Na-containing components of the off-gas, a relatively finely divided source of Al_2O_3 is introduced either in the fluidized bed, or in the extended reaction zone or into both places.

The off-gases generated by the combustion and pyrohydrolysis of the spent pot lining will contain besides HF, also steam and CO_2, as well as entrained solids. Due to the use of O_2 for combustion, the diluting effect of nitrogen is minimized and the HF-content of the off-gas will be about 6 to 12 vol %. After removal of the entrained solids the off-gas is cooled. The hot gases exiting from the reactor can be cooled directly with a cold stream of HF-containing gas, such as shown by the dotted lines in Figure 12. Since this is a recycle stream and there is no heat removal from the system, the quantity of cooling gas used affects the intermediate concentration of HF, but has no effect on the HF concentration in the product gases. Subsequent cooling in a direct contact cooler, such as also shown in Figure 12, with a highly concentrated HF stream, for example, such as one recovered from the tail gas scrubber, can increase the concentration of the gas, for example, up to about 13 to 16 vol %. The cooled HF-containing stream is then introduced into a scrubbing unit where water is employed as a scrubbing medium. The scrubber off-gas will be free of HF and will contain mainly water vapor and CO_2. The aqueous HF stream recovered from the scrubber will have an HF content of about 25 wt %. A portion of the steam discharged from the scrubber unit may be recycled to the pyrohydrolysis unit as shown.

The aqueous HF recovered from the scrubber unit can be employed for making anhydrous HF as well as for the direct cooling of the HF-containing off-gas stream as shown in Figure 12. The higher than usual HF-content of the aqueous stream allows its ready conversion to anhydrous HF by conventional means. When the aqueous stream is used primarily for direct cooling of hot off-gas, a gaseous product which has an increased HF-content is obtained, for example, about 13 to 16 vol %.

This stream, containing about 13 to 16 vol % of HF, is utilized if desired for the production of AlF_3. While the prior art generally avoided the use of alumina trihydrate for the manufacture of AlF_3, the instant process prefers the use of this starting material. In the prior art processes the water content of the hydrate would have posed problems since the heat of reaction generated by the low HF content of the gas stream is insufficient to vaporize the water content of the trihydrate and to maintain the required temperature of the gaseous and solid reactants.

In this process the by-product steam of the AlF_3 production is also utilized. The steam discharged from the AlF_3 fluid bed reactor is used to provide a major portion of the steam requirement for the pyrohydrolysis reaction, thus providing a favorable heat and energy balance, surpassing the effectiveness of prior art systems. Use of the off-gas from the AlF_3 fluidized bed reactor as the source of steam for the pyrohydrolysis reactor recycles trace quantities of unreacted HF. This increases the overall HF recovery and eliminates a possible environmental problem. In addition, the aluminum hydrate starting material is less costly than calcined alumina; also the AlF_3 product, due to its high, more than 85 wt %, AlF_3 content, is an economically more valuable product than the prior art AlF_3 products of about 15% AlF_3 content.

Thus, it can be observed from the system shown that by using a fluidizing and combustion-promoting gas of at least about 90% O_2 content, the entire pot lining recovery process can be made more efficient both from a technical and an economical point of view.

ALUMINUM FOIL FROM PAPER/FOIL LAMINATES

Foil lined papers and paperboards find extensive use in moisture retardant packages, that is packages designed to either retain or exclude moisture from the product. The most commonly used metal for the foil is aluminum and this is backed by one or more layers of cellulosic fibers. Uses for foil lined materials include packages for a wide range of foodstuffs and also products such as cigarettes.

The cellulosic component of the foil lined papers and boards is frequently of a high quality and thus represents a valuable material for recycle and reuse in the further manufacture of paper products.

Conventional wastepaper reslushing units such as the Hydrapulper and similar machines rely largely upon high shear mechanical agitation for the redispersion of the recycled cellulosic fiber. When foil lined materials are treated in this manner, the foil is rapidly comminuted into small flakes by the action of the Hydrapulper impeller. Hence, the reslushed cellulosic fibers become contaminated with small flakes of metal which are difficult to separate from the cellulosic fibers in the subsequent stages of processing.

Consequently, the metal contaminated wastepaper pulp so produced is frequently relegated to a lower quality usage than would be the case if the flakes of metal foil were substantially absent.

A process developed by *H. Mamers and J.E. Rowney; U.S. Patent 4,168,199; September 18, 1979; assigned to Commonwealth Scientific and Industrial Research Organization, Australia* is one for the recovery of cellulosic fiber of papermaking quality from foil laminated papers and boards in which the laminate is heated and pressurized in the presence of an aqueous liquor, subjected to rapid discharge to a lower pressure environment under reduced pressure and optionally subjected to mechanical agitation. This treatment provides a product stream which can be separated into cellulosic fiber suitable for the further manufacture of fibrous cellulosic articles and a metallic foil suitable for subsequent manufacture of metallic articles.

AMMONIA FROM AMMONIA-SODA PLANT EFFLUENTS

It is known that in the method of manufacturing soda by the amino-soda process a sodium chloride brine saturated with ammonia is carbonated by means of carbon dioxide, this causing the precipitation of sodium bicarbonate, which is separated, for example, by filtration, and which is calcined to form sodium carbonate, while the mother liquor from the crystallization of the bicarbonate is treated to recover the ammonia for recycling. This mother liquor consists of an aqueous solution containing in solution principally ammonium chloride, but since the conversion of the sodium chloride is not complete the mother liquor also contains a relatively large proportion of sodium chloride. Free ammonia is also found in it, that is to say ammonia itself and volatile ammoniacal compounds, for example, ammonium bicarbonate and carbonate, as well as impurities from the brine fed into the process.

The operation of recovering ammonia, called distillation, is carried out in a plant comprising a column of which the upper portion functions as a heater and the lower portion as a still, a tank with agitator called a prelimer being interposed in the circulation of the liquor between these two portions to provide for the addition of a basic agent, for example lime or milk of lime. Steam is introduced at the bottom of the still to act as both heating agent and sweeping agent. In such a plant, the volatile ammoniacal compounds are decomposed in the heater and the decomposition products (ammonia and carbon dioxide) as well as the free ammonia are entrained by the rising current of steam.

In the prelimer, the ammonium chloride is dissociated by the basic agent liberating ammonia and chloride of the metal of the basic agent (calcium chloride when lime is used). The still serves to complete the dissociation of ammonium chloride and to bring about entrainment in the steam of the gaseous constitutent present in the dissolved state.

In order to ensure recovery of as high a proportion as possible of the ammonia, the basic agent has always been used in excess with respect to the amount stoichiometrically necessary for dissociation of the ammonium chloride.

The prior methods demand, therefore, always the use of an excess of lime to denude the mother liquor completely of ammonium chloride. If a lime of low reactivity is used, this excess can reach 10 wt % and is completely lost with the sludges removed at the foot of the still, sludges of which the volume is moreover the greater as the excess of lime used is larger. Furthermore, if the residual liquors are voided to the river, the residual excess of lime gives them a high pH which obviously contributes to pollution.

An improved process which overcomes this problem has been developed by *J. Verlaeten and P. Demilie; U.S. Patent 4,061,718; December 6, 1977; assigned to Solvay & Cie., Belgium*. In this method the liquor is heated to release the free ammonia, and the remaining liquor is then treated successively in a prelimer with lime in an amount insufficient to decompose the whole part of the ammonium chloride and in a still with steam and an alkali metal hydroxide in an amount sufficient to decompose the remaining part of ammonium chloride. Figure 13 compares the prior art apparatus with the improved apparatus.

Figure 13: Prior Art Apparatus (Left) and Solvay & Cie. Improved Apparatus (Right) for Ammonia Recovery from Ammonia-Soda Plants

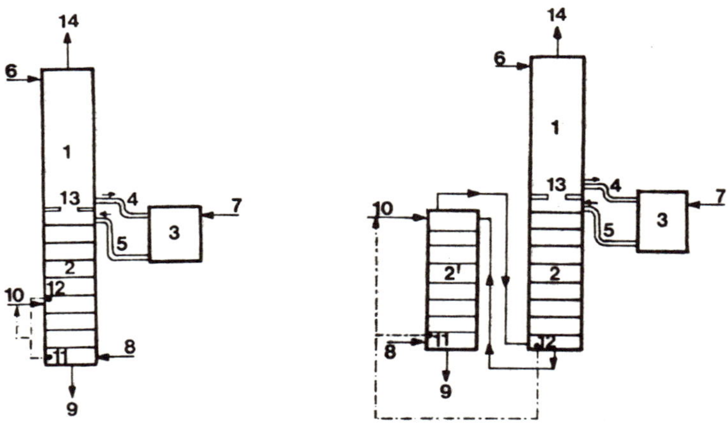

Source: U.S. Patent 4,061,718

In these views, the numeral **1** represents the heater, **2** and **2'** the still or stills, **3** the prelimer, **4** and **5** the pipes conveying liquor from the heater to the prelimer and conveying liquor from the prelimer to the still, **6** the inlet for liquor from the filters of the soda plant to the heating column, **13** the plate separating the heater and the still, **14** the outlet for gases and vapors from the heater, **7** the inlet for milk of lime to the prelimer, **8** the inlet for steam at the bottom of the still, **9** the outlet for liquor and sludges from the still, **10** the inlet for the solution of alkali metal hydroxide as balance of basic agent, and **11** and **12** the system of pH-meters and control equipment determining, as a function of the measured pH values, the amount of alkali metal hydroxide to be introduced as balance at **10**.

For operation of this installation, according to the improved process, the liquor from the filters of the soda plant, which contains, dissolved in water, in particular ammonium chloride, free ammonia and sodium chloride, passes at **6** into the heater **1** where the rising current of steam entrains the ammonia and carbon dioxide liberated towards the top **14** of the column. At the bottom of the heater, the liquor passes by way of pipe **4** into the prelimer **3** where it is submitted for some time, with agitation, to the action of a milk of lime introduced at **7**. The amount of lime added here is always less than the amount stoichiometrically necessary to liberate the ammonia fixed in the ammonium chloride, for example, 75 to 99% of the stoichiometric amount is added.

The control of this amount may be carried out through measurement of pH at **12** according to the method described in U.S. Patent 2,781,244. Next, the liquor leaves the prelimer **3** and passes by way of pipe **5** to the top of the distillation column **2** in which the dissociation of the fixed ammonia compounds (NH_4Cl) and the entrainment of the gases by the steam introduced at **8** have to be completed. The liquor, freed from ammonia, and the sludges leave the still at **9**. In order to allow completion of the dissociation of the ammonium chloride, there is introduced at **10**, a little below the center of the height of the still, for example, the necessary balance of basic agent, and this in the form of an aqueous solution of an alkali metal hydroxide, for example, of electrolytic caustic soda.

The calculation of the amount of the balance of basic agent to add into the still is carried out continuously and automatically by measurements of pH taken at two points in the distillation column, one at **11** being very close to the liquor outlet and the other at **12** preceding the point of introduction of the balance of alkali metal hydroxide into the column. The combined result of these two measurements is used as a signal transmitted to the valve operating the feed of the balance of basic agent at **10**.

According to the right hand view of Figure 13, the original still **2** being inadequate, it has been necessary to connect up a second column **2'** in series with the first one. The steam is introduced at the foot of column **2'**, at **8**, and the balance of basic agent is introduced at the top of this column, at **10**. The pH-meters **11** and **12** are placed respectively at the bottom of the two columns and the device controlling the introduction of the balance of basic agent at the top of the second column.

AMMONIA FROM AMMONOXIDATION PROCESS WASTES

See "Hydrogen Cyanide from Ammonoxidation Process Wastes" (1).

AMMONIA FROM TEXTILE TREATING EFFLUENTS

Fabrics constructed at least in part of cellulosic materials can be processed advantageously by exposure to liquid ammonia, to achieve improvement in shrinkage resistance and to provide greater affinity of the fabric to other process chemicals. In accordance with known liquid ammonia treating techniques, the fabric may be exposed briefly to liquid ammonia solution, as by immersion in a bath of the liquid. After a predetermined reaction time, advantageously less than about 9 sec, the fabric is heated, to vaporize and drive off the ammonia and terminate the reactions at a desired level.

In recent years, with such developments as permanent-press garments and demanding industrial requirements for textiles formed into specific shapes, there has been increased emphasis on such treatment of fabrics with liquid ammonia. Large quantities of ammonia find their way into such processes, so there is a consequent involvement in disposing of or reclaiming ammonia effluent from ammonia treatment chambers.

In a typical liquid ammonia process, only a small percentage, (for example, about 5%) of the ammonia is actually consumed in the process reactions, or otherwise lost. The balance is in the form of ammonia vapor. Because of the potentially hazardous and unpleasant nature of ammonia vapors, and also for obvious economical reasons, it is important in a practical liquid ammonia processing operation to recover, for reliquefaction and reuse, the spent ammonia vapors.

Broadly speaking, this can be accomplished by withdrawing the ammonia vapors from the fabric treatment chamber and compressing and condensing such vapors. The condensed vapors are returned to a liquid ammonia storage tank for eventual reuse in the system.

Unfortunately, recovery of ammonia from treatment chambers has been frustrated. Some difficulties encountered in ammonia recovery have been explosion hazards, unreliability, interference with (and worse, disruption of) the prime system of processing textiles.

Last, but certainly not least, expense has been prohibitive. Textile mills have been able to find more profitable ways to invest their capital than in ammonia recovery systems. A clear and present need has arisen for ammonia recovery from effluents of ammonia treatment chambers, within operational and economic parameters with which the textile industry can cope.

In many cases where large quantities of ammonia are used, methods of recovery and reuse are feasible economically with recovery efficiencies generally as high as 90%. Usually recovery of residual ammonia content beyond the 90% has been uneconomical and disposal thereof has been required. In other situations, however, where lesser quantities of ammonia are used, and in situations where the ammonia occurs with large quantities of air, water vapor, or steam, recovery of the spent ammonia is not economical. In such situations, disposal of the ammonia containing waste has been accomplished by discharge into natural streams or by incineration.

There is, therefore, a need for ammonia recovery methods and systems from ammonia treatment discharge, within the operational and economic parameters of the textile industry.

As pointed out by *J. Lawrence; U.S. Patent 4,058,375; November 15, 1977; assigned to Cluett, Peabody & Co., Inc.* there is a sizable industrial need for aqueous ammonia in concentrations of 20 to 30%. Ordinarily these ammonia solutions are made by diluting anhydrous ammonia with water. However, reclamation of ammonia from wastes otherwise having an ammonia content of marginal economic attractiveness may offer a new route to such solutions.

This process involves recovering ammonia from a waste gas mixture comprising ammonia with air, and water vapor or steam. The waste gas is bubbled upwardly through an aqueous bath in a condenser, and then through a scrubber in direct contact with a spray of fresh water. The scrubber and condenser are formed integrally so that fresh ammonia water from the scrubber joins the aqueous bath in the condenser. The temperature of the aqueous bath is controlled, preferably by sensing the temperature thereof to control flow of cooling fluid in noncontact heat exchange with the aqueous bath in the condenser. Aqueous ammonia from the bath is withdrawn as product, but it is recycled to the bath if its ammonia concentration is inadequate. Figure 14 shows a suitable form of apparatus for the conduct of such a process.

Figure 14: Cluett, Peabody & Co., Inc. Process for Aqueous Ammonia Recovery by Scrubbing and Condensing

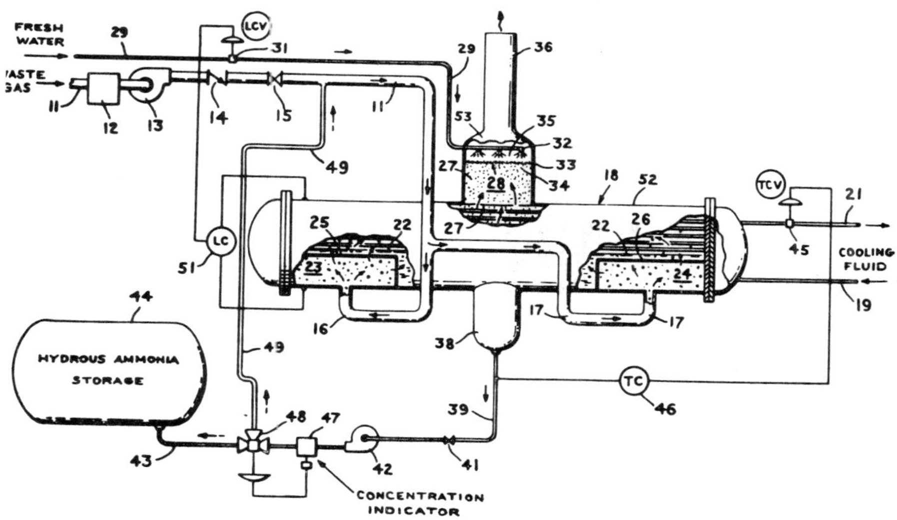

Source: U.S. Patent 4,058,375

As can be seen from the drawing, a waste gas feed (containing ammonia with air, water vapor and/or steam) enters the system via line **11**, a filter **12**, a blower **13**, a flow controller **14** and a damper **15**. The blower pressurizes the incoming waste gas to about 30" of water, so that the gas velocity is at a proper level to maintain the liquid level in the water and to minimize entrained liquid droplets in the discharged gas. The waste gas line **11** is divided into two lines **16** and **17** for entry into a condenser **18**. The condenser is divided into a tube side and a shell side. The tube side communicates with the lines **19** and **21** for circulation of a cooling fluid (usually water) therethrough. The shell side of the condenser contains a water bath **22**.

The waste gas from the lines **16, 17** is admitted into two chambers **23, 24** respectively (defined by two baffle plates **25, 26** respectively) for distribution along the length of the condenser **18** and for bubbling the waste gas upwardly through bath **22** so that a portion of the ammonia from the waste gas is absorbed by a bath **22**. A resulting waste gas **27** with a reduced ammonia content of about 10% passes upwardly to a scrubber **28** located above condenser **18**, and in flow communication therewith.

In scrubber **28** the waste gas is washed with fresh water. The fresh water is introduced via a line **29**, a control valve **31** and spray nozzles **32**. Atomizing means, shown as mesh **33**, is included to assure that the fresh water is broken up into many fine droplets **34**. The fresh water droplets pass downwardly through the scrubber **28** in direct physical contact with the waste gas mixture with reduced ammonia content so as to absorb additional ammonia therefrom resulting in a substantially ammonia free gas mixture **35** (containing about 0.50% ammonia) which passes upwardly via a vent stack **36**. The fresh water **34**, containing ammonia removed from the gas **27**, flows downwardly to join water bath **22** in the shell side of condenser **18**.

Hydrous ammonia product, with an ammonia content of 20 to 30 wt % or more, is withdrawn from the shell side of condenser **18** via a well **38**, a line **39** a valve **41**, a pump **42** and a line **43** to a storage tank **44**. Flow of cooling fluid via line **19**, the tube side of the condenser and line **21** is controlled by a control valve **45** which is regulated by a temperature controller **46**. The temperature range of bath **22** is maintained within the range of about 50° and 75°F.

A concentration indicator **47** is connected operatively with line **41** and measures a suitable electrical property of the product as a function of ammonia concentration. The concentration indicator **47** is connected to operate a three way valve **48** to channel the product either via the line **43** to the storage tank **44** or via a line **49** to the line **11** for recycle to the aqueous bath **22** until the product reaches the desired ammonia concentration. As noted above, the method and apparatus of the process are particularly appropriate for treating ammonia containing waste gas streams of relatively low ammonia concentration to, in turn, secure aqueous ammonia solutions with concentrations of between 20 and 30%, a typically useful concentration for many applications.

Such concentrations were secured, in the past by diluting fresh anhydrous ammonia. Thus, the process here serves two useful purposes, including securing a useful product from a waste gas considered economically unattractive in the past, and disposing of a potentially contaminating waste gas.

Input of fresh water via the line **29** and spray nozzles **32** is controlled by the level control valve **31** which is regulated by a level controller **51** connected operatively to condenser **18**.

Scrubber **28** and condenser **18** are fabricated preferably in a single housing **52**, for simplicity and economy. The condenser has a generally horizontal elongated cylindrical configuration and the scrubber is arranged midway relative to the condenser's length, and projects upwardly therefrom. The scrubber **28** has a generally vertical configuration terminating in an upper end **53** on which the vent stack **36** is mounted. The well **38** provides depth to the aqueous bath **22** so that enrichment of ammonia content can be achieved even as hydrous ammonia product is withdrawn via line **39** to storage tank **44**.

A recovery technique described by *J. Lawrence; U.S. Patent 4,074,969; February 21, 1978; assigned to Cluett, Peabody & Co., Inc.* provides a system for the recovery of spent ammonia, in connection with the processing of fabrics and the like with liquid ammonia, and concerns particularly the elimination from the recovered ammonia of undesired water.

Economic processing of fabrics by liquid ammonia requires recovery and reuse of substantial quantities of ammonia. In the course of processing, the ammonia unavoidably becomes contaminated with water. Separation of water from ammonia on a laboratory level or, in any kind of batch processing is a theoretically simple matter and can be coped with by conventional differential evaporation techniques, or otherwise. However, in a continuously operating processing line where large quantities of anhydrous liquid ammonia are being used as the treating medium, water accumulates rapidly, not only from the fabric being processed, but also from a certain inevitable amount of air leakage in the system.

Because so much of any given increment of the treating medium must be recycled, as compared to that actually used up in the treating process, water accumulates rapidly in the system and must be removed on a continuous basis. This process provides a feasible and highly efficient procedure for removal of water by effecting condensation of water and ammonia vapors, constituting the process effluent, by feeding the effluent to a desuperheating vessel, where it is brought into direct contact with a body of low temperature liquid ammonia. This is done in conjunction with a preliminary low temperature condensation of the effluent in a noncontact heat exchange stage.

The condensed body of liquid ammonia in the desuperheated vessel, including residual condenser water from the process effluent combined with reliquefied ammonia, forms the feed supply of liquid ammonia solution to the process. The condensed water, which in this process constitutes a portion of the feed supply, is applied to the fabric being treated, along with the liquid ammonia. Typically, some of the water is carried away with the processed fabric as a constituent of its moisture content. The remainder, which is driven off as steam in the process, is recycled.

A key factor in this process is that the reliquefied ammonia, instead of being sent directly back to the process, is directed into the desuperheated vessel, there being combined with the condensed process effluent. The combined solution, containing a minor fraction of condensed residual water is then fed back to the process. In this manner, the total water fraction in the process solution may be kept at a satisfactorily low level, typically on the order of 2 or 3% maximum, under extreme process conditions, and desirably much lower than that under more favorable process conditions.

The overall fabric treatment and ammonia recovery system are shown schematically in Figure 15.

Figure 15: Cluett, Peabody & Co., Inc. Process for Recovery and Reuse of Ammonia in a Liquid Ammonia Fabric Treating System

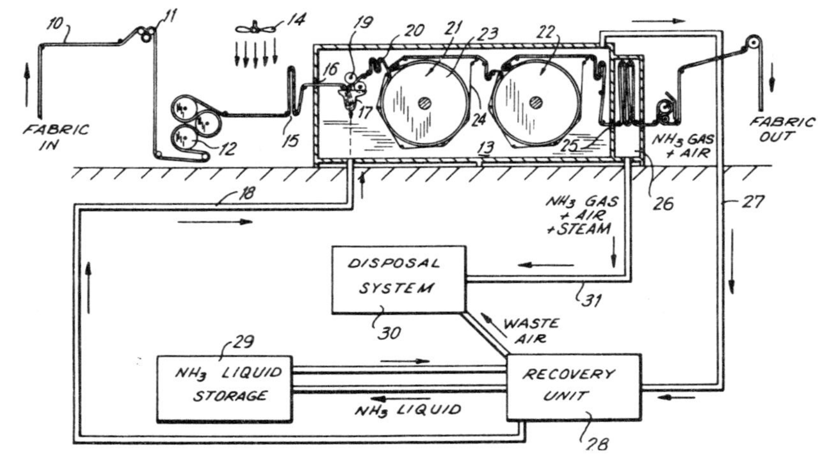

Source: U.S. Patent 4,074,969

As shown in Figure 15, a cellulosic fabric web **10**, from a suitable supply passes over tension control rollers **11** and is then directed about one or more heated rollers **12**, constituting a predrying section. Passing over the series of predrying rollers, the fabric is heated sufficiently to drive off excess moisture. In this respect, incoming fabric typically may contain as much as 7 to 10 wt % (of the fabric) of moisture. The amount of moisture in the fabric may unduly inhibit the desired reactions of the liquid ammonia process, which normally should be carried out in a liquid ammonia solution containing not more than about 10% water.

Although the weight of ammonia in relation to the weight of fabric during the reaction phase may vary widely, a relationship of one-to-one (e.g., one part by weight of the ammonia solution to one part by weight of fabric) is not untypical. In such cases, if the incoming fabric carries as much as 10% water, that amount of water will be present at the reaction site, and will constitute approximately 10% of the ammonia solution. This is an undesirably high level, particularly where the ammonia solution itself may contain some water, as contemplated in this process.

Accordingly, the predrying stage typically is controlled to drive off enough moisture from the fabric to leave a residual moisture content on the order of 3 to 5 wt % of the fabric. Of course, if the incoming fabric is sufficiently dry to begin with, the predrying stage may be omitted.

Fabric leaving the predrying stage will be at an undesirably elevated temperature and is thus cooled prior to entering the liquid ammonia treatment chamber **13**. Typically, suitable fan or blower means **14** is disposed downstream of the predrying section, to direct streams of cooling air on the fabric and return it to near ambient temperature levels.

The predried and cooled fabric, after passing over additional tension control rollers **15**, enters the treatment chamber **13** through a sealed opening **16**. Typically, the interior of the chamber is maintained at a slightly negative pressure, relative to ambient, and the entrance opening **16** is provided with a double seal. The intermediate chamber, between the double seals, is maintained at a slightly more negative pressure than the interior of the chamber, so that inevitable slight leakage of the seals will tend to be directed into the intermediate chamber. This minimizes leakage of ammonia vapors from the treatment chamber into the atmosphere. In a typical process, the main treatment chamber may be operated at a negative pressure of about 0.5" H_2O while the intermediate chamber may be kept at a negative pressure of about 0.75" H_2O.

In the simplified arrangement illustrated, a processing trough **17** is provided in the treatment chamber **13**. This trough, through appropriate controls, is supplied with liquid ammonia processing solution through an infeed line **18**. The controls of this may include a float valve, for maintaining the processing liquid at an appropriate level in the trough.

After entering the processing chamber, the fabric is guided into the trough **17** and thus immersed in the liquid ammonia solution which is at a temperature of about -28°F. It is then directed through padding rollers **19**, for extraction of excess processing solution, then about a series of adjustable timing rollers **20**. After a predetermined reaction time, the fabric is brought into contact with a source of heat, which flashes off the liquid ammonia.

In the illustrated system, a pair of Palmer-type dryer units **21**, **22** are provided. These include large heated drums **23** about which a confining blanket **24** is trained. For practical purposes, the ammonia reactions are substantially diminished soon after the initial contact between the fabric and the first dryer drum. To advantage, the time interval between initial immersion in the liquid ammonia and the initial contact with the first dryer drum is controlled to be within the range of 0.6 to 9 sec. This can be effectively controlled by regulation of the length of travel between the trough **17** and the first dryer unit **21**, as by adjustment of the rollers **20** to lengthen or shorten the path of the web, as may be appropriate.

However, except as relates to the control of water in the fabric and in the processing solution and as relates to the control of the operation of the dryer units, specific process conditions do not form a part of this process.

After leaving the second dryer stage **22**, the fabric leaves the main treatment chamber **13** through a discharge opening **25**. This opening, like the entrance opening **16**, advantageously is provided with a double seal with an intermediate chamber maintained at a slightly more negative pressure than the treatment chamber itself.

Fabric leaving the main treatment chamber, advantageously may be directed through a steam chamber **26**, after which the fabric may be conveyed away to a folder or batcher, for example.

In the processing of fabric in the main treatment chamber, only about 5% of the liquid ammonia supplied to the trough **17** is actually consumed. The remainder is flashed off as ammonia vapor. These vapors are not only potentially hazardous, but the reuse thereof is economically important in a continuous commercial process. Heretofore, recovery of the ammonia vapors has been carried out by withdrawing vapors from the treatment chamber, and compressing and condensing the vapors. Considerable amounts of air are normally contained in the withdrawn gases, but air is easily separated from the ammonia because of the relative noncondensability of air. The withdrawn gases also include quantities of water, which continually enter the process because of the basic moisture content of the fabric and also of the incoming air which, notwithstanding the efficiencies of the entrance and exit seals, is present in certain amounts in the interstices of the fabric and enters with fabric.

Such amounts of water have, in the past, proven difficult to remove as noted above, necessitating occasional discarding of quantities of the diluted liquid ammonia, or its use as a low grade material such as in fertilizer applications. This process is directed to the recovery of the spent ammonia in a manner that enables the water to be easily and effectively removed on a continuous basis, so that the process reactions are not inhibited by excessive water in solution with the otherwise relatively pure anhydrous liquid ammonia, and so that maximum utilization of the liquid ammonia in the processing operation may be realized.

In the simplified schematic representation shown, a suction line **27** leads from the main treatment chamber **13** for effecting continual withdrawal of gases from the interior of the chamber. These gases are directed initially to a recovery section, generally designated by **28**, in which the gases are processed to compress and condense the liquid ammonia and to separate air. A liquid ammonia storage vessel **29** is provided for temporary containment of the recovered liquid ammonia. The infeed line **18**, through which the processing trough **17** is supplied, is not directly connected with the storage vessel **29**, but rather leads from the recovery section **28**. The stored liquid ammonia is first directed from the vessel **29** back into the recovery system, where it is utilized and then is directed to the processing trough **17** along with an increment of water extracted from the recovered gases.

In accordance with conventional practice, air separated from the recovery system is directed to an incinerator or other disposal facility **30**. Likewise, the mixture of air and steam from the steam chamber **26**, containing some residual ammonia gas, is directed through a suction line **31** to the disposal facility. Because of the relatively small amounts of ammonia in these gases, it is considered uneconomical to attempt to recover it.

AMMONIA FROM UREA PLANT WASTEWATER

As is well known, urea is conventionally produced by reacting ammonia and carbon dioxide under high temperatures or pressures. The output of the reactor is directed to a vacuum crystallizer from which the desired products such as, for example, a 74% urea solution is withdrawn. The process also forms large amounts of water which are removed from the vacuum crystallizer as vapor along with relatively small but contaminative amounts of ammonia, carbon dioxide and urea. Heretofore this water and the contaminants have frequently been condensed and simply dumped. However, the dumping of untreated and contaminated urea plant condensate is a source of water pollution and efforts have been made to treat the condensate so as to remove the contaminants. Heretofore, such efforts have not been satisfactory since they have resulted in materially increased operating costs and/or significant additional costs for new equipment. therefore, the need for an economic recovery process is apparent.

A process developed by *W.H. Van Moorsel; U.S. Patent 3,922,222; November 25, 1975; assigned to CF Industries, Inc.* is one in which condensate from a urea crystallizer is directed to a low pressure boiler or heat exchanger and evaporated to accomplish thermal decomposition of urea and any biuret contained therein into ammonia and carbon dioxide which are subsequently removed and recycled to the urea plant, if desired, leaving substantially urea free water of sufficient quality to satisfy environmental regulations or guidelines or for use as boiler feed water. The flow diagram of such a process is shown in Figure 16.

Figure 16: CF Industries, Inc. Process for Ammonia Recovery from Urea Plant Crystallizer Condensate

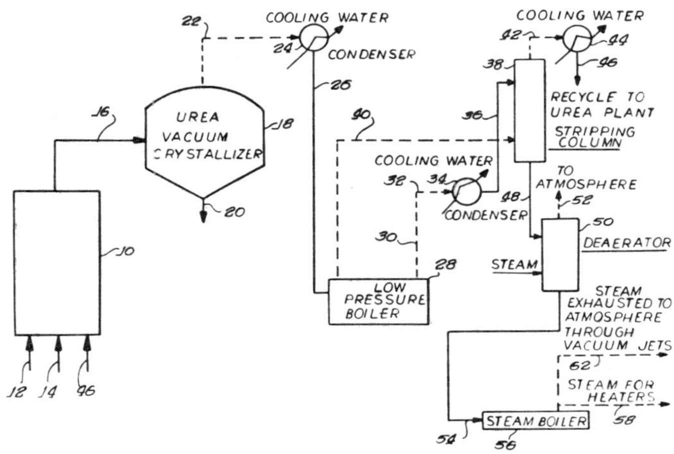

Source: U.S. Patent 3,922,222

Referring specifically to the drawing, the apparatus is shown in association with a urea plant which includes a conventional reactor **10** for receiving carbon dioxide and ammonia from suitable sources of supply **12** and **14.** The carbon dioxide and ammonia are reacted under high temperatures and pressure which may, for example, be on the order of 3,000 psi in the vessel **10** for producing urea and water.

The product of the reactor **10** is directed through conduit **16** to a vacuum crystallizer **18** of known construction. In a preferred embodiment, a 74% urea solution is withdrawn from the crystallizer through conduit **20.** Water vapor along with relatively small but contaminating amounts of ammonia, carbon dioxide and urea is discharged through conduit **22** and is directed to a cooler **24** of known construction which condenses the vapor.

The condensate from the cooler **24** is directed through conduit **26**. This condensate is principally water, but contains contaminating amounts of ammonia and urea. For example, when operating the urea plant at full rate under preferred conditions, a pressure of 1.5 psia and temperature of 125°F are maintained in the vacuum crystallizer **18** and the condensate in the line **26** from the crystallizer comprises 7,000 lb/hr water, 70 lb/hr ammonia, and 7 lb/hr urea. In accordance with this process, the condensate is processed as described below so as eventually to eliminate the urea and the ammonia from the water.

The condensate is directed by the conduit **26**, to a low pressure boiler or other heat exchanger **28** where it is heated and discharged as vapor to conduit **30**.

The low pressure boiler or heat exchanger may be of known construction having portions thereof which contact the condensate formed from stainless steel or some other suitable material which will not be adversely affected by the ammonia and urea present in the condensate. As indicated, the boiler is operated at relatively low pressures and temperatures. In a preferred embodiment, the boiler or heat exchanger is operated at a pressure of about 8.4 psia and at a temperature of 185°F. The condensate is evaporated and at the same time the urea is thoroughly decomposed into ammonia and carbon dioxide so that, in the example being described herein, the flow through the discharge conduits **30** and **40** from the boiler comprises 7,000 lb/hr of water, 74 lb/hr of ammonia and about 3 lb/hr of carbon dioxide.

A major portion of the vapor which in this example comprises about 6,500 lb/hr of water and a proportionate share of the ammonia and carbon dioxide is directed by a branch line **32** to a condenser **34** and from the condenser through line **36** to a stripping column **38** of known construction. The remainder of the vapor is directed to branch line **40** to the lower portion of the stripping column. The stripping column may comprise any known construction capable of functioning to remove most of the ammonia and carbon dioxide from the water.

For example, with the flow rates previously indicated, the stripping column functions so that approximately 64 lb/hr of ammonia and a proportionate amount of carbon dioxide along with about 500 lb/hr of water are discharged through line **42**. This discharge is in the form of vapor and is passed through a condenser **44** and is directed by line **46** for recycling back to the reactor **10** of the urea plant.

Water is discharged from the stripping column through line **48**. While this water has had most of the ammonia removed therefrom, a small amount remains. In the example given, condensed water is discharged through line **48** at the rate of about 6,500 lb/hr along with about 10 lb/hr of ammonia. Preferably this condensate is directed to a deaerator **50**. This unit is of known construction and is such that most of the ammonia and a portion of the liquid are vaporized and vented through the atmosphere through vent **52**, while the remainder of the water with a very small residual amount of ammonia is directed through line **54**.

This essentially uncontaminated water may be dumped, if desired, but it is preferably directed to a high pressure boiler **56** and provides at least a portion of the boiler feed water. In the example given herein, approximately 1,000 lb/hr of water vapor along with 8.5 lb/hr of ammonia are vented from the deaerator which is operated at 65 psia and 298°F, while 5,500 lb/hr or more and 1.5 lb/hr ammonia are directed to the high pressure boiler.

Steam from the high pressure boiler may be utilized for any desired purpose. For example, a portion of the steam may be directed from the boiler through line **58** for use in various heaters in the urea plant or elsewhere. Another portion of the high pressure steam may be directed to branch line **62** and exhausted through the atmosphere through vacuum jets which are used to maintain a vacuum in the various parts of the urea or other plant processes. It has been found that the ammonia concentration in the steam reaches an equilibrium of about 300 ppm and at the flow rate set forth herein about 250 lb/day of ammonia is exhausted to the atmosphere through the deaerator vent and the vacuum jet.

A process improvement developed by *I. Mavrovic; U.S. Patent 4,088,684; May 9, 1978;* consists of a urea synthesis process with liquid carbamate recycle, in which the reactor effluent is divided into major and minor streams. The major stream is reduced in pressure and subjected to heating and substantial carbamate decomposition to ammonia and carbon dioxide decomposer off-gas, either directly or after preliminary ammonia separation. The minor stream is reduced in pressure and is contacted with decomposer off-gas and fresh make-up carbon dioxide, either directly or after preliminary ammonia separation, and in indirect heat exchange with a relatively colder fluid, whereupon ammonium carbamate is formed from the mixture of the minor stream, fresh make-up CO_2 and decomposer off-gas in an aqueous solution of urea and the resulting solution is recycled to the urea synthesis reactor.

AMMONIA-BASE PULPING LIQUOR FROM SPENT PULPING LIQUOR

In the ammonia-base bisulfite process for pulping, wood chips are digested with ammonium bisulfite. During this digestion, a waste or spent liquor is produced which contains about 50 wt % of the original wood and substantial quantities of sulfur and ammonia. The spent liquor usually contains not less then 5%, and not more than 15% total solids. The sulfur and ammonia exist in the spent liquor largely as ammonium salts of lignosulfonic acid referred to generally as ammonium sulfolignate.

Disposal of ammonia-base bisulfite spent pulping liquors presents a serious problem because of air and water pollution which results when they are dumped in streams and lakes. Such disposal methods can no longer be tolerated so that a need exists for means to handle such waste liquors and dispose of them without pollution of the air or water. Furthermore, there is a need for nonpolluting disposal systems for ammonia-base bisulfite spent pulping liquors which permit recovery of at least some of the valuable chemicals therein so that they can be reused in making fresh ammonia-base bisulfite pulping liquor.

A process developed by *G.G. Copeland; U.S. Patent 3,864,192; February 4, 1975; assigned to Copeland Systems Incorporated* is one in which ammonia-base bisulfite spent pulping liquor is combined with magnesia thereby freeing ammonia and forming dilute magnesium-base bisulfite spent liquor. The mixture is concentrated by use of heat from hot exhaust gases from a fluidized bed reactor.

The concentrated magnesium-base bisulfite spent liquor is fed into a fluidized bed reactor, in which the bed contains magnesia, and burned therein. The hot exhaust gases from the fluidized bed reactor are used to concentrate additional magnesium-base bisulfite spent liquor formed from ammonia-base spent liquor.

The apparatus which may be used in the conduct of such a process is shown schematically in Figure 17. The ammonia-base bisulfite spent liquor is fed by conduit **10** to pump **11** and from it to conduit **12** which delivers the spent dilute liquor to the separator tower **14** of tower **20**. The tower also includes an intermediate cooling tower **15** and an upper or top absorption tower **16**.

Spent liquor is withdrawn from separator tower **14** by conduit **17** and pump **18** which conveys it to conduit **19** which feeds the spent liquor to contactor **30** which advisably is a wetted wall venturi scrubber. Hot exhaust gases containing sulfur dioxide and magnesia fines or particles are exhausted from the fluidized bed reactor in which magnesium-base bisulfite spent liquor containing magnesium sulfolignate is burned, the source of which will be described hereafter, by conduit **22** which feeds the hot gases and magnesia particles to contactor **30**.

The hot gases and magnesia particles are at about 1600°F or higher when they reach contactor **30** by conduit **22**. The spent liquor fed by conduit **19** to contactor **30** is at about 200°F so that upon contact with the hot gases a large amount of steam is formed. In addition, such contact results in a chemical reaction or neutralization that results in the magnesium ion replacing the ammonium ion in the sulfolignate products and compounds present in the waste liquor. In this way ammonia is freed and the spent liquor is converted in part or entirely from an ammonia-base bisulfite spent liquor to what can be appropriately called a magnesium-base bisulfite waste liquor containing magnesium sulfolignates and other magnesium compounds.

The spent liquor and exhaust gases, including ammonia, sulfur dioxide and steam are conveyed from contactor **30** by conduit **21** to separator tower **14** in which the liquid phase is separated from gases and vapors, including steam, ammonia and sulfur dioxide. Evaporation of water to steam leads to concentration of the spent liquor from its initial maximum of about 15 wt % solids to about 25 to 40 wt % solids. By circulating the liquor one or more times through the loop comprising conduit **17**, pump **18**, conduit **19**, contactor **30**, conduit **21** and separator tower **14** the liquor can be concentrated to a solids level determined by the available heat in the exhaust gases.

The spent liquor fed through conduit **19** can be substantially a magnesium-base bisulfite spent liquor or it may be mixed with ammonia-base bisulfite spent liquor. However, in practice it is expected that the spent liquor will comprise about 50/50 of each of these spent liquors. When the spent liquor in conduit **19** is concentrated to a level determined by the available heat in the exhaust gases, which concentration will usually be to at least 25 wt % solids, part of the stream of spent liquor is diverted to conduit **41** and sprayed into the freeboard space **42** of fluidized bed reactor **40**. The spent liquor falls to fluidized magnesia bed **43** where it burns, autogenously if sufficiently concentrated (35 wt % solids or higher) or by the aid of an auxiliary fuel.

Air is supplied by conduit **44** to pump **45** which pumps air to conduit **46** for delivery to wind box **47**. The air under pressure flows from the wind box **47** through nozzles **48** in constriction plate **49** and causes the bed of magnesia particles to be fluidized. The air oxidizes the spent liquor by combustion at about 1400° to 2000°F with the formation of hot combustion gases containing sulfur dioxide and entrained magnesia particles or fines which exhaust to conduit **22**. An auxiliary conventional heating system can be used to start up the reactor to bring it to combustion temperature.

Figure 17: Copeland Systems Inc. Apparatus for Producing Ammonia-Base Pulping Liquor from Spent Liquor

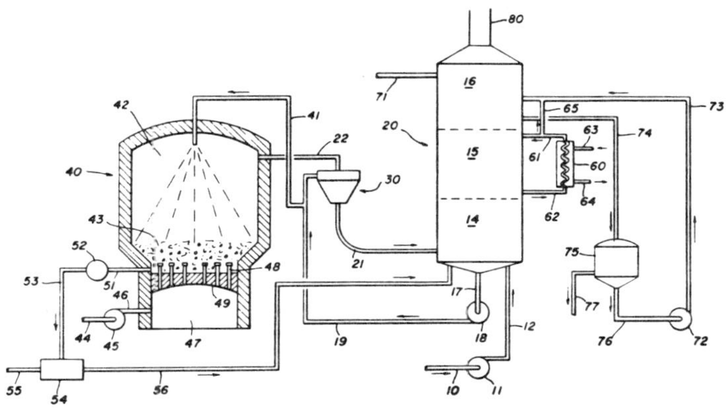

Source: U.S. Patent 3,864,192

Some of the magnesia from the spent liquor agglomerates in bed **43**. To maintain control of the volume of the bed and size of the bed particles, some of the bed is advisably withdrawn periodically by conduit **51** and sent to grinder **52** from which magnesia fines are delivered by conduit **53** to slurry tank **54** which is supplied with water by conduit **55**. Magnesium hydroxide aqueous slurry is conveyed from the tank by conduit **56** to the lower part of separator tower **14** to supply additional magnesia when and if needed to neutralize the ammonia-base bisulfite spent liquor.

It is expected to have at least 60%, and advisably 75%, by weight or more of the magnesia needed supplied by means of entrainment in the exhaust gases fed by conduit **22** to contactor **30**. The remainder of the magnesia requirements for neutralizing the ammonia-base spent liquor can be supplied by other means, one of which is magnesia withdrawn as described from the fluidized bed reactor and circulated to the separator tower.

The separated gases, including sulfur dioxide and ammonia, at about 180° to 200°F, pass upward from separator tower **14** into cooling tower **15** where they come into contact with cooling water at about 50° to 100°F supplied by conduit **61**. The cooling water condenses steam from the gases and dissolves some of the sulfur dioxide and ammonia. The resulting cooling water solution is withdrawn at about 90° to 150°F from the cooling tower by conduit **62** and passed through heat exchanger **60** to cool it and then to conduit **61**. Conduit **63** supplies cold water or some other cooling or refrigerating fluid to the heat exchanger and it is removed therefrom by means of conduit **64**.

Makeup aqueous ammonium hydroxide is supplied by conduit **71** to the top of the absorption tower **16**. Ammonium bisulfite solution is withdrawn from the lower part of the absorption tower by conduit **74** and conveyed to acid sump **75**. The acid is recirculated from the sump by conduit **76** to pump **72** and then back by conduit **73** to the middle part of the absorption tower. Some of the cooling water solution containing dissolved sulfur dioxide and ammonia is diverted from conduit **61** to conduit **65** and fed to conduit **73** for mixing with the liquid stream thereby supplied to the absorption tower.

The ammonium hydroxide solution together with the cooling water solution mixed therein flows downward countercurrent to flow of sulfur dioxide and ammonia which are absorbed to form additional amounts of the aqueous solution of ammonium bisulfite. By means of conduit **77** fresh ammonia-base bisulfite is removed from sump **75** and sent to a pulping mill as needed.

The nonabsorbed gases are vented from absorption tower **16** by vent **80** to the atmosphere. The vented gases are innocuous and comprise carbon dioxide, water vapor, nitrogen, and other inert gases from air and not more than minute amounts of sulfur dioxide.

The following is a specific example of the conduct of the present process. Ammonia-base bisulfite spent pulping liquor containing 12.25% by weight total solids is fed at 160°F by line **12** to separator **14** as shown. The liquor is fed at a rate of 273,566 lb/hr of which 33,528 lb is solids and 240,038 lb is water. Magnesia is fed by line **56** at 136 lb/hr in an aqueous slurry to the separator.

Air at 150°F is supplied to the air box **47** of reactor **40** at a pressure of about 5 psig and a rate of 248,270 lb/hr (55,550 cfm). Magnesium-base bisulfite spent liquor containing 31% by weight solids from separator **14** is delivered by conduits **19** and **41** at 121,000 lb/hr (37,462 lb of solids and 83,538 lb of water) to the freeboard **42** of the reactor. Autogenous combustion is effected in the reactor of the organic content of the spent liquor at 1600°F.

The exhaust gases containing magnesia particles are vented from the reactor at 351,328 lb/hr (188,400 lb N_2, 9,375 lb O_2, 48,350 lb CO_2, 4,945 lb SO_2, 544 lb MgO and 100,802 lb steam; vol %: 48.9 N_2, 21.1 O_2, 8 CO_2, 0.6 SO_2, 40.4 H_2O) to the venturi scrubber **30** where they are washed with a mixture of magnesium-base bisulfite and ammonia-base bisulfite spent liquors supplied by conduit **19**. The hot mixture of spent liquor and gases is conveyed by conduit **21** from the scrubber to the separator. Ammonia is freed in the separator and a dilute magnesium-base bisulfite spent liquor is formed.

A total of 508,310 lb/hr of gases at 189°F pass from the separator to the cooling tower **15**. The composition of the gases is 482 lb of NH_3, 188,400 lb N_2, 9,375 lb O_2, 48,350 lb CO_2, 4,945 lb SO_2 and 256,758 lb H_2O (vol %: 0.2 NH_3, 29.9 N_2, 1.3 O_2, 4.9 CO_2, 0.3 SO_2, 63.4 H_2O).

The gases are cooled in the cooling tower to 150°F and sent to the absorption tower **16**. Part of the cooling water solution containing dissolved sulfur dioxide and ammonia, is diverted from conduit **61** and fed to conduit **65** which delivers it to the ammonia-base bisulfite solution supplied by conduit **73** to absorption tower **16**. Makeup aqueous ammonium hydroxide containing 25% ammonia is fed by conduit **71** at a rate of 2,800 lb/hr to the upper part of absorption tower **16**. Product acid, i.e., ammonia-base bisulfite pulping liquor at 145°F, is removed from the absorption tower by conduit **74** at a rate of 26,283 lb/hr (4,450 lb SO_2, 1,182 lb NH_3, 20,651 lb H_2O). The gases vented to the atmosphere by conduit **80** from the absorption tower contain less than 0.1 vol % of sulfur dioxide.

AMMONIUM NITRATE FROM NUCLEAR FUEL PROCESSING WASTE

In a variety of processes for producing nuclear fuels, ammonium nitrate is separated as a by-product. Thus, for example, according to the reactions set forth below in which solutions of uranyl nitrate or of thorium nitrate are treated with aqueous ammonia, the so-called ammonium diuranate or thorium hydroxide is precipitated and ammonium nitrate is formed.

$$2UO_2(NO_3)_2 + 6NH_4OH \rightarrow (NH_4)_2U_2O_7\downarrow + 4NH_4NO_3 + 3H_2O$$

$$Th(NO_3)_4 + 4NH_4OH \rightarrow Th(OH)_4\downarrow + 4NH_4NO_3$$

The precipitated uranium or thorium compounds are further treated to produce oxide fuel or breeder materials, such as are required for placement in nuclear power reactors.

The ammonium nitrate formed as a by-product in the reactions set forth above is produced in radioactively contaminated form, so that any of the usual ways of utilizing the ammonium nitrate in industry, for example as a fertilizer, is out of the question. Even disposal of solutions containing this ammonium nitrate in wastewater drains and conduits is not permissible because of the risk to the environment resulting from the radioactivity. In the usual decontamination of radioactively poisoned water by evaporation and condensation, the ammonium nitrate is a troublesome factor for the further treatment and solidification of the residue after distillation, because under certain circumstances it is inclined to decompose explosively and, at any rate, has an oxidizing effect.

A process has, however, been developed by *P. Morschl and E. Zimmer; U.S. Patent 4,187,280; February 5, 1980; assigned to Kernforschungsanlage Jülich GmbH, Germany* in which radiation contaminated ammonium nitrate is heated in solution to about 100°C in the presence of finely powdered calcium oxide or lithium hydroxide. Ammonia and water vapor are given off leaving an alkaline or alkaline earth nitrate which can then be safely decomposed by calcination into a metal oxide and oxides of nitrogen.

The metal oxide can be recycled in a continuation of the process. The oxides of nitrogen can be passed through water to produce nitric acid useable in dissolving oxides of fissionable materials and the ammonia may be used in aqueous solution to react with nitrates of nuclear fuel or breeder metals in the very process that produces the by-product ammonium nitrate. Thus, all by-products and reagents can be reconverted and recycled.

AMMONIUM PERCHLORATE FROM SOLID PROPELLANT WASTE

Pyrotechnic compositions, particularly compositions which comprise a polymerically bound binder containing an inorganic oxidizer, finely divided metal fuel and optionally a burning rate modifier and other specialized additives, are usable principally in the field of rocket and missile propellants, smoke and gas generators, and illuminants.

The most common past method of disposal of such compositions which were overage, surplus, or otherwise unusable, was by open pit burning. This method while reasonably safe, suffered from the drawbacks that it contributed a great deal of atmospheric pollution, resulted in total destruction of the propellant precluding recovery of any economically valuable or scarce components thereof, and prevented reuse of hardware components associated with the composition.

However, a technique developed by *G.C. Shaw, E.A. Lawton, and L.L. Jones; U.S. Patent 4,057,442; November 8, 1977; assigned to Thiokol Corporation* eliminates the pollution and a large portion of the economic waste associated with open pit burning, allows recovery of the oxidizer, metallic fuel and any associated hardware, does not require prior removal of the pyrotechnic compositions from any hardware to which they may be attached, and does not require the energy expenditure of milling.

In general, the process comprises the following steps:

(a) Disrupting the matrix structure of the polymeric binder by treating the pyrotechnic composition with swelling agents or depolymerizing agents, substantially chemically unreactive with the inorganic oxidizer and the metallic fuel;

(b) Separating the inorganic oxidizer and metallic fuel from the disrupted polymeric matrix of step (a);

(c) Contacting the inorganic oxidizer and metallic fuel of step (b) with a fluid selected from water, or a fluid having a density intermediate between that of the inorganic oxidizer and metallic fuel, and being substantially chemically unreactive therewith; and

(d) Separating the mixture of step (c).

The following is one specific example of the application of the process. An 88% solids high energy propellant containing a binder consisting of an isophoronediisocyanate-cured hydroxyl terminated polybutadiene is treated with about 1 to 15 wt % of aluminum chloride and tetrahydrofuran. The solution is either refluxed for about 4 hr or allowed to stand for an extended period, conveniently about 18 to 40 hr. Once the binder has dissolved, the resulting mixture is filtered, washed with fresh tetrahydrofuran and dried.

The recovery of mixed aluminum fuel and ammonium perchlorate is quantitative. Treatment with water gives a solution of ammonium perchlorate which after separation from the undissolved aluminum is suitable for recrystallization. The aluminum powder is in a form suitable for recycling.

AMMONIUM SULFATE FROM ACRYLONITRILE PROCESS EFFLUENTS

A process developed by *A.T. Guttmann, M.G. Konicek and A.J. Grigsby; U.S. Patent 3,576,607; April 27, 1971; assigned to The Standard Oil Company* involves the recovery of fertilizer-grade ammonium salts in high yields from acrylonitrile wastewater process streams containing water-soluble organic tars. The removal of water and volatile organic matter from the wastewater process streams is carried out in the presence of a water-miscible solvent, recovering the solid ammonium salt from the residual ammonium salt-solvent-tar mixture and recovering the solvent for reuse.

A process developed by *J. Bonnema and H.J.H. Simon; U.S. Patent 4,015,946; April 5, 1977; assigned to Stamicarbon NV, Netherlands* is one for recovering ammonium sulfate from an aqueous ammonium sulfate solution containing organic contaminants. The aqueous solution is contacted with an organic substance which is water-soluble and which is capable, in combination with a saturated ammonium sulfate solution, of forming a two-layer system. The organic substance can be lactams of 4 to 20 carbons, and mixtures thereof. The resultant two layers may be separated into an aqueous phase containing the ammonium sulfate, and an organic phase containing an organic substance and the organic contaminants.

Typical organic contaminants include acrylonitrile prepared from propylene, methyl methacrylate prepared from acetone cyanohydrin, and methyl methacrylate prepared from acrylonitrile. The ammonium sulfate is readily crystallized from the aqueous phase. Alternatively, the ammonium sulfate may be directly crystallized from the aqueous solution while the solution is in contact with the organic substance.

See also "Acrylonitrile from Acrylonitrile Process Effluents" in Reference (1).

AMMONIUM SULFATE FROM CAPROLACTAM MANUFACTURE

In the well-known Beckmann rearrangement, cyclohexanone oxime is rearranged to obtain caprolactam which is the monomer for making nylon-6. In commercial practice, following completion of the rearrangement, the reaction mixture is neutralized and the lactam separated from the mixture by solvent extraction or by other appropriate means. The most common neutralizing agent used is aqueous ammonium hydroxide. Thus, when sulfuric acid is used as the rearrangement catalyst, the by-product of the neutralization is ammonium sulfate which cannot be reused in the process. The ammonium sulfate therefore must be sold as for fertilizer use; however, this material is often in oversupply on the market and thus may bring only low prices.

Moreover, about three pounds of ammonium sulfate is produced per pound of caprolactam, so that disposal of the by-product becomes a severe problem as the quantity of caprolactam produced is increased. Besides leading to a low priced by-product, this neutralization process also involves large volumes of water which are required since the neutralization is exothermic and the heat released must be dissipated to control the process. The large amount of neutralization medium involved keeps high the cost of separating the lactam and recovering the by-product, ammonium sulfate.

Neutralization of the sulfuric acid with other bases leads to materials which are of less value or used in lesser quantity. Calcium hydroxide, for example, which is a low cost base would lead to calcium sulfate which, in addition to having a low market value, is insoluble and would clog up the reaction system. Thus, the desirable alternative is not to recover ammonium sulfate (or other sulfates) but is to develop a process which avoids this problem.

A discussion of the manufacture of caprolactam without concurrent ammonium sulfate production has been presented by R. Mattone, G. Sioli and L. Giuffre of Snia Viscosa in *Hydrocarbon Processing,* January 1975, pp 83-87.

The reader is also referred to the preceding discussion of U.S. Patent 4,015,946 under "Ammonium Sulfate from Acrylonitrile Process Effluents" which also deals with caprolactam process wastes.

AMMONIUM SULFATE FROM STORAGE BATTERY PLATE WASH WATER

A process by *Z.L. Burke; U.S. Patent 3,882,019; May 6, 1974; assigned to Hill Brothers Chemical Co., Inc.* involves converting effluent water which has been contaminated with lead and sulfuric acid as a result of a manufacturing process into water suitable for reuse in such manufacturing process. The process involves the steps of adding lead carbonate to the effluent water in a first agitated enclosure thereby forming water, carbon dioxide, and insoluble lead sulfate, filtering the water from the lead sulfate, and conducting the filtered water back to the manufacturing process for reuse. The filtered lead sulfate is added to a solution of ammonia, ammonium sulfate, carbon dioxide, and water in a second agitated enclosure to form ammonium sulfate and lead carbonate.

The ammonium sulfate liquid is filtered from the lead carbonate and conducted to a holding tank for reuse in the second agitated enclosure. The filtered lead carbonate is dried to remove residual ammonia therefrom whereupon the dried lead carbonate powder may be conducted back to the first agitated enclosure for reuse. The carbon dioxide from the first agitated enclosure and the liberated ammonia from the dryer may be conducted back to the ammonium sulfate holding tank. Finally, excess ammonium sulfate from the holding tank may be conducted to a crystallizer for generation of ammonium sulfate crystals. Such a process is shown in Figure 18.

Referring to the drawing, the manufacture of wet cell batteries requires large amounts of process water that becomes contaminated with lead, Pb, and sulfuric acid, H_2SO_4, in varying quantities. The contaminated process water from the manufacturing process, appearing in a conduit **10** and containing Pb in various forms and H_2SO_4, is conducted to a rough filter **11**. The filter is a conventional filter for removing the larger undissolved solids in the effluent water, such as a good deal of the free lead and miscellaneous other contaminants. The output of the rough filter, in a conduit **12**, therefore contains sulfuric acid and lead.

The effluent water in conduit **12** is conducted to a holding tank **13** for accumulation. Since the process is a batch process, the tank permits the effluent water to accumulate while the previous batch of water is being processed. At an appropriate time, the output of the holding tank is conducted to a first reaction vessel **14**. Vessel **14** may simply be a large tank, the contents of which are adapted to be agitated by a blade **15** connected by a shaft **16** to a drive motor **17**. After tank **14** is filled with the effluent water from holding tank **13**, fine, powdered lead carbonate is added from a bin **18** and the mixture is agitated.

Figure 18: Hill Brothers Chemical Co. Process for Effluent Water Recovery and Ammonium Sulfate Recovery from Battery Manufacture

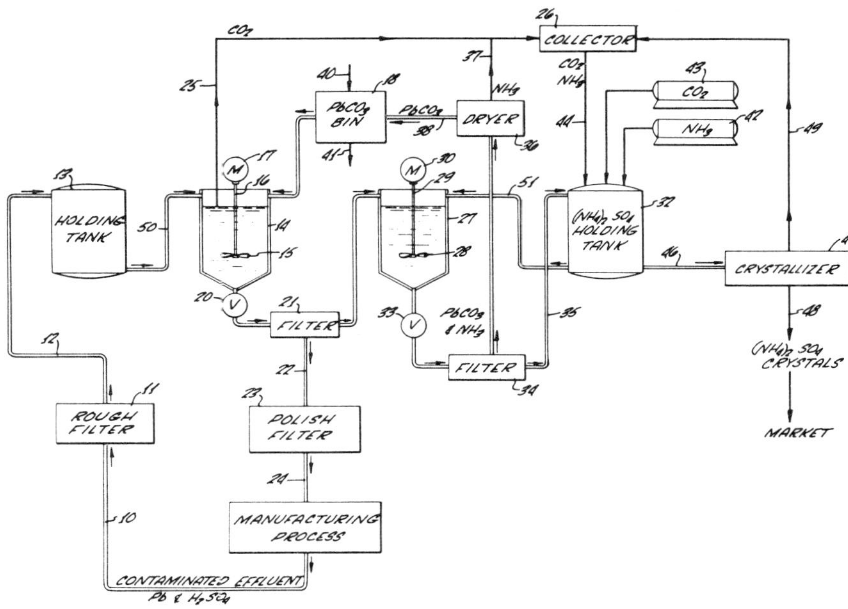

Source: U.S. Patent 3,882,019

Lead carbonate is soluble in sulfuric acid and is converted to lead sulfate which precipitates out as an insoluble sulfate, thus stripping the sulfate molecules from the water molecules. Another product of the reaction is carbon dioxide. The complete reaction is represented by the formula

$$PbCO_3 + H_2SO_4 \longrightarrow PbSO_4 + CO_2 + H_2O$$

The amount of $PbCO_3$ needed in tank **14** is equal to the amount of total dissolved solids (TDS) in the effluent water. For example, if the rough filtered rinse water from tank **13** contains 15,000 ppm TDS, then 15,000 ppm or 15 g/ℓ of $PbCO_3$ is needed to effect the reaction. Although a 1:1 ratio is sufficient, any increase in the amount of $PbCO_3$ over this ratio proportionately increases the rate of the reaction.

In a typical case, wherein the rough filtered rinse water contains 15,000 ppm TDS, 25 ppm lead, and a pH of approximately 2.3, after only 10 minutes, the pH has risen above 6, the total dissolved solids is approximately 1,200 ppm, and the quantity of lead has been reduced to 3 ppm. While an additional 30 minutes is required to reach a pH of 7 and a concentration of total dissolved solids of 1,000 ppm or less, the water is of sufficient quality after 10 minutes that it may be reused in the manufacturing process. Thus, after approximately 10 minutes, when the pH has risen above 6 as indicated by a pH probe, not shown, positioned within tank **14**, a valve **20** at the outlet of tank **14** is opened and the solution is conducted to a filter **21**.

Filter **21** can be a rotary drum filter, a centrifuge, or any other conventional type of filter for recovering the solids from a solution, thereby separating the water from the lead sulfate. The filtered water containing 1,000 ppm TDS or less and 3 ppm lead or less is conducted via a conduit **22** to a polish or precoat filter **23** for further removal of fine particles. The output of filter **23**, in a conduit **24**, is then conducted back to the manufacturing process for reuse.

Although it requires 40 minutes to obtain a pH 7 in tank **14**, the reaction shown in the above equation continues after the agitation is stopped and all throughout the filtration cycle in filter **21** or as long as there is any $PbCO_3$ and/or H_2SO_4 available to effect a reaction. Therefore, the solution need remain in tank **14** for only approximately 10 minutes before beginning filtration to obtain a satisfactory water for recirculation as process water.

The carbon dioxide, in the form of a gas, from tank **14**, is conducted via a conduit **25** to a holding tank or collector **26**. The lead sulfate from filter **21**, in the form of a wet cake, is next added to a solution of ammonia, ammonium sulfate, and carbon dioxide, in a second reaction vessel **27**. Vessel **27** is identical to tank **14** and includes an agitating blade **28** connected by a shaft **29** to a drive motor **30**. Tank **27** is first filled with the $(NH_4)_2SO_4$, CO_2 and NH_3 solution from a holding tank **32**. Lead sulfate is readily soluble in ammonium sulfate, but in the presence of CO_2, the lead immediately reacts, forming the insoluble carbonate $PbCO_3$. The reaction is represented by the formula

$$PbSO_4 + (NH_4)_2CO_3 \longrightarrow PbCO_3 + (NH_4)_2SO_4$$

The reaction proceeds more readily in an excess of ammonium sulfate.

A slurry containing approximately 15 g $PbSO_4$ per liter of $(NH_4)_2SO_4$, NH_3 and CO_2 solution will result in a reaction time of approximately 10 to 15 minutes. The $(NH_4)_2SO_4$, NH_3 and CO_2 solution may contain 15 to 100 grams $(NH_4)_2SO_4$, a minimum of 2 g CO_2 and sufficient NH_3 to obtain an alkaline pH. Thus, tank **27** may be processing the output of filter **21** while a second batch of effluent water is being processed in tank **14**. After approximately 10 minutes, a valve **33** at the outlet of tank **27** is opened and the solution is conducted to a filter **34**, which may be identical to filter **21**, for recovering the solids from a solution. Thus, filter **34** separates the liquid ammonium sulfate solution from the lead carbonate precipitate, conducting the ammonium sulfate solution via a conduit **35** to holding tank **32**.

The lead carbonate precipitate in filter **34**, in the form of a wet cake, is conducted to a dryer **36**. The reason for dryer **36** is that the filter cake traps ammonia, thereby contaminating the lead carbonate. In dryer **36**, the cake is heated to approximately 212°F, thereby driving off the ammonia which is conducted via a conduit **37** to collector **26**. The dried, powdered lead carbonate from dryer **36** is conducted via a conduit **38** back to bin **18** for reuse in tank **14**.

If the amount of lead carbonate recovered in dryer **36** equals the amount of lead carbonate needed in tank **14**, the lead carbonate is simply recirculated, forming a minor closed loop system within the total system of the process for effluent water recovery. As a practical matter, there may be an excess or an insufficient amount of lead carbonate recovered in dryer **36**. If there is an insufficient quantity of $PbCO_3$ recovered, an additional quantity is conducted into bin **18** from an outside source via conduit **40**. Any excess $PbCO_3$ may be removed via a conduit **41** for marketing.

It can be seen that the sulfate molecules which are stripped from the sulfuric acid in tank **14** and combined with lead to form lead sulfate and then again stripped in tank **27**, accumulate in holding tank **32** in the form of ammonium sulfate. The alkaline pH of the solution in tank **32** is maintained with additional quantities of ammonia and carbon dioxide derived from holding tanks **42** and **43**, respectively. Furthermore, the ammonia on line **37** from dryer **36** and the carbon dioxide on line **25** from tank **14** may also be conducted via collector **26** and a conduit **44** back to tank **32** to control the pH of the solution contained therein.

If left undisturbed, the solution in tank **32** will become so saturated with ammonium sulfate that crystals will start to precipitate out, such crystals clogging the lines, etc. Thus, excess $(NH_4)_2SO_4$ accumulating in tank **32** is continuously bled via a conduit **46** to a crystallizer **47** where $(NH_4)_2SO_4$ crystals are constantly generated. Such crystals may be removed, at **48**, and marketed since being in a commercially usable form. The remaining weak solution may be conducted via a conduit **49** to collector or holding tank **26** for return via conduit **44** to holding tank **32**.

It can therefore be seen that there is provided a process for effluent water recovery which completely eliminates the necessity for disposing of the contaminated effluent water appearing on conduit **10**. Rather, the large quantities of water used for battery manufacturing, contaminated with lead and sulfuric acid, are processed for recirculation, thus converting the manufacturing process to what virtually amounts to a closed loop system. The by-products of the process are ammonium sulfate crystals and lead carbonate, both of which have certain economic value. By the application of this process, literally thousands of tons of polluted water per day are kept out of the environment and all its contaminants converted to reusable products.

The drying of the filter cake from filter **34** in dryer **36** to liberate the ammonia trapped therein is critically important to the process. If this were not done, ammonia would be added with the lead carbonate to tank **14** thereby forming ammonium sulfate in conduit **22**. Such ammonium sulfate would render the water unsuitable for reuse in the manufacturing process.

The reaction times in tanks **14** and **27** are sufficiently compatible that both reactions may take place simultaneously. Thus, the tanks may be drained simultaneously, the liquid from filter **34** being conducted to holding tank **32** and the liquid from filter **21** being conducted via polish filter **23** back to the manufacturing process. As soon as tanks **14** and **27** have been completely drained, valves

20 and 33 are closed and tanks **14** and **27** are refilled, tank **14** being filled with effluent water over conduit **50** from tank **13** and tank **27** being filled with ammonium sulfate, ammonia and carbon dioxide solution over conduit **51** from tank **32**. As soon as tanks **14** and **27** are filled, a measured quantity of $PbCO_3$ from bin **18** may be dumped into tank **14** and the lead sulfate from filter **21** may be dumped into tank **27**. At this time, the filter cake from filter **34** may be dumped into dryer **36** permitting tanks **14** and **27** and dryer **36** to operate simultaneously.

ANTIMONY FROM POLYESTER STILL BOTTOMS

While the production of polyethylene terephthalate resins is carried out by a number of proprietary processes, there are areas in which the methods are essentially the same. Normally dimethyl terephthalate (DMT) or terephthalic acid (TPA) is condensed with ethylene glycol to form the diglycol ester. Thereafter, in the presence of a catalyst and an increased temperature and pressure the polyethylene terephthalate is produced by splitting of methanol or water. One of the most widely used catalysts employed is antimony trioxide.

Inherent in most of the processes is the recovery of the excess or spent glycol, normally by distillation, which, by this time is contaminated with, among other things, low molecular weight polyester chains, antimony oxide, DMT and terephthalic acid.

As the ethylene glycol (in the mono-, di- and tri- forms) is fractionally distilled, most of the residual impurities remain in the bottom of the still taking the form of a waxy granular mass which is then removed as a waste material for disposal.

A typical analysis of such a still bottom, including added water and a small amount of sodium hydroxide to make the material handleable is as follows:

Water, %	26
Solids, %	32
Ethylene glycol, %	36
Diethylene glycol, %	4.25
Triethylene glycol, %	0.75
Antimony, ppm	7,200
Sodium, ppm	4,300

While the still bottoms described have, in the past, been disposed of by burying or dumping into a waterway, the Environmental Protection Agency has taken the position that the presence of antimony constitutes an environmental and subsequent health hazard. A maximum acceptable limit of 5 ppm of antimony has been established by the Environmental Protection Agency for effluent discharged into a waterway.

Burying can no longer be considered an acceptable disposal because the hazardous metals can leach or drain into a water supply. Pretreating for burying to insure against this or, alternatively, encasing the material becomes enormously expensive and unacceptable as a solution.

Incineration without costly accumulators and stack scrubbers simply puts the hazardous metal into the atmosphere. Ocean dumping is simply another form of waterway contamination and has a deleterious effect on marine life as well as human environment.

A process developed by *B.N. Hoppert and A. Berghausen, III; U.S. Patent 4,013,519; March 22, 1977; assigned to Safetech, Inc.* provides a recovery method by means of which an antimony compound is separated from polyester still bottoms. If desired, the method can be employed so as to recover in usable form, the antimony compound, and the monomers polymerized to produce the polyester.

The process may be described in somewhat more detail as follows:

(1) A predetermined amount of still bottom waste material taken with a predetermined amount of water, if required, is subjected to an alkaline hydrolysis step at a temperature of about 185° to 212°F to split the low molecular polyester material present into the salts of terephthalic acid and ethylene glycol (as used herein to mean mono-, di- or triethylene glycol).

(2) The material resulting from step (1) is acidified to a pH of about 6.2 to 6.5 forming a precipitate of antimony and/or antimony compounds which are removed and recycled to an antimony production facility.

(3) The filtrate from step (2) is subjected to a color removal step using, but not limited to, a material such as carbon.

(4) The filtrate from step (3) heated to about 170°F is acidified to a pH of about 1 to 3 so as to precipitate terephthalic acid out of the filtrate and the precipitated terephthalic acid is filtered, washed and dried. The terephthalic acid is substantially pure and may be sold in commerce, recycled for use in the polyester manufacturing process, etc.

(5) To the filtrate from step (4) heated to about 170° to 212°F is added a material such as, but not limited to, sodium thiosulfate to convert the antimony and/or antimony compound in the filtrate to antimony sulfide. As used hereinafter antimony compounds is meant to include antimony and/or one or more compounds thereof.

(6) The material is then filtered to remove the antimony sulfide precipitate leaving a filtrate consisting essentially of ethylene glycol, water, sodium sulfate, and perhaps, sodium bisulfate.

(7) Some of the water from the material under step (5) is then removed.

(8) To the filtrate from step (7) is added a base to provide a pH of about 6.5 to 8. At this point, if desired, step (3) may be repeated.

(9) One of the many ways to those skilled in the art of separating and purifying ethylene glycol is to add a solvent to the material from step (8) to precipitate sodium sulfate.

(10) The sodium sulfate precipitate is removed by filtering, for example, and is washed with a solvent and recovered.

(11) The filtrate from step (10), consisting essentially of a solvent, ethylene glycol and water, is heated to evaporate a part of the solvent.

(12) The solution remaining from step (11), consisting essentially of ethylene glycol and water is then distilled to separate the ethylene glycol from the water. The ethylene glycol so provided may be used for its customary uses. If it is not desired to recover the ethylene glycol, the process can be stopped after step (7) above.

A process developed by *S.J. Dougherty and K.J. Garska; U.S. Patent 4,100,253; July 11, 1978; assigned to Union Carbide Corporation* can be used for recovering antimony and sodium compounds from antimony-containing spent ethylene glycol residues resulting from the manufacture of polyethylene-terephthalate polyester. The process comprises combusting the spent glycol residues to produce an ash, and then contacting the ash with water to form an ash-water mixture having a liquid phase and a solid phase, the water-soluble sodium compounds being essentially contained in the liquid phase and the water-insoluble antimony compounds being essentially contained in the solid phase. The liquid phase and solid phase are then separated in order to recover therefrom sodium and antimony compounds, respectively.

See also "Ethylene Glycol from Polyester Process Wastes" in Reference (1).

ANTIMONY FROM SCRAP ANTIMONY

Approximately 50% of the domestic supply of antimony is normally derived from scrap. Secondary antimony recovered in 1977 amounted to 30,601 tons. Recovery of antimony from battery-plate scrap comprised about 84% of the total secondary output (16). Type-metal scrap amounted to about 8% of the output, and bearing-metal scrap and miscellaneous scrap contributed the remainder.

The United States has a large quantity of antimonial alloys now in use that are in a continuous recycling process. Consideration of the number of automobiles in use leads to an estimate of 91,000 tons of recoverable antimony of this type.

ANTIMONY PENTACHLORIDE FROM SPENT CATALYSTS

A process developed by *E. Joerchel and R. Kohlhaas; U.S. Patent 3,784,671; January 8, 1974; assigned to Farbwerke Hoechst AG, Germany* involves recovering antimony pentachloride from catalyst solutions used for the fluorination of chlorine containing hydrocarbons. The process is based upon a thermal decomposition of antimony pentachloride into antimony trichloride and chlorine at a temperature of 60° to 300°C.

The organic impurities are extracted from the precipitated antimony trichloride by means of halogenated hydrocarbons. The remaining antimony trichloride is reoxidized with chlorine, which may be recycled from the step of the thermal decomposition, at a temperature of from 60° to 150°C and a pressure of 0 to 5 atm to yield antimony pentachloride which is refined by distillation.

A process developed by *Y. Osaka and T. Tohzuka; U.S. Patent 4,070,439; January 24, 1978; assigned to Daikin Kogyo Co., Ltd., Japan* is one in which antimony pentachloride is recovered from a spent antimony catalyst mixture in a simple manner by bubbling through the spent catalyst mixture an inactive gas such as trichlorotrifluoroethane at an elevated temperature to convert and to remove high melting point or high sublimating point substances as lower boiling point substances.

The process then involves converting fluorinated antimony into chlorinated antimony by means of a chlorinating agent such as carbon tetrachloride; converting antimony(III) into antimony(V) by oxidation with chlorine; and recovering antimony pentachloride by distillation. The antimony pentachloride thus recovered may readily be used as a catalyst for fluorination of a chlorinated hydrocarbon by reaction with hydrogen fluoride.

ARSENIC FROM COPPER SMELTER FLUE DUSTS

As described by A. Patrick, Jr., H.J. Bennett, K.E. Starch and R.C. Weisner in *The Economics of By-Product Metals* (17), arsenic is produced as a by-product from flue dusts from copper smelters.

The arsenic refining process begins with roasting of the flue dusts and their collection in the form of crude arsenic trioxide in an operation consisting of roasting furnaces and collecting chambers connected in series. Furnaces are operated at 650°F. Higher temperatures would cause fusion in the condensers. The roasting furnaces have a capacity of 50 tons per hour of charge.

Refining furnaces of the reverberatory type are of refractory brick construction with cast iron hearths, each handling 30 tons per day. High-grade material is dropped into the furnace on each of three shifts, and residues are removed once every 2 to 3 days. The condensing chambers for the arsenic trioxide consist of a series of kitchens or subchambers.

The first of these operate at a temperature above the condensing temperature of white arsenic (218°C), and thus it collects the nonarsenical dust. The other kitchens near the furnace precipitate black amorphous impure arsenic (95% As_2O_3) which is re-refined. The succeeding kitchens precipitate white crystalline arsenic trioxide of 99.9% purity which is packaged and marketed.

A flow sheet for the production of arsenic from copper smelter flue dusts is shown diagrammatically in Figure 19 on the following page.

Figure 19: Arsenic Production from Flue Dusts

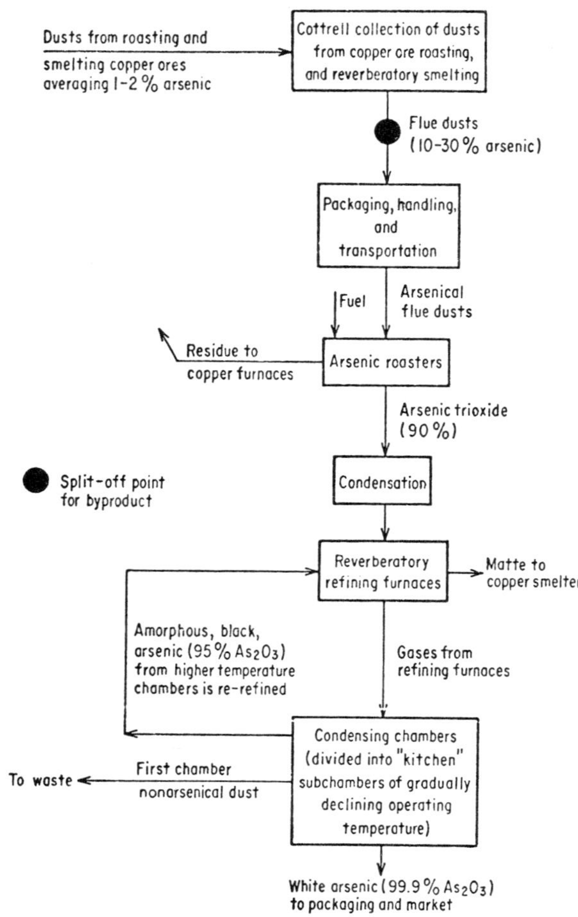

Source: Reference (17)

ASBESTOS FROM WASTE ASBESTOS SLURRIES

It is well known that macroscopic aggregates or bundles of asbestos, especially of the chrysotile variety, can be converted into a stable dispersion or colloidal form by treatment with an aqueous solution of organic detergent surface active agent.

It is also known that diluted asbestos dispersions may be converted into solid form by bringing a sheet, strand or other body of the dispersion into contact with an agent that renders the soap insoluble. In most commercial processes, the dispersion is passed through a nozzle or other forming channel and is brought into contact with an aqueous bath containing divalent or trivalent metal ions, such as calcium, magnesium, aluminum or the like. The metal ions react with the surfactant and cause the dispersion to be converted into a solid strand. The strand may be produced continuously and may be collected in a package such as by pot spinning. The package is then dried, cleaned to remove the insoluble soap residue, and then may be woven into a fabric or utilized in other applications.

After the sheet or strand has been subjected to the precipitating agent, and up to the time of cleaning, several intervening operations take place that may result in the generation of waste. For example, the strand may break during the collection, winding, twisting, weaving or beaming operations, resulting in losses of asbestos mineral fibers which heretofore had to be disposed of in accordance with hazardous materials disposal regulations.

Such waste asbestos containing the insoluble soap, resulted in a direct loss to production efficiency and a gain in product cost. Also, since asbestos dust is considered dangerous to human health, proper disposal of the waste over a period of time can present serious difficulties to the producer.

These problems can be overcome, it is claimed by *E.K. Spitz and L.E. Cox; U.S. Patent 4,131,661; December 26, 1978; assigned to Raybestos-Manhattan, Inc.* In this process waste asbestos generated from a production process utilizing a surfactant as the dispersant and a coagulant as a solidifying agent is recovered for reprocessing by treating the waste with an inorganic acid to react with the precipitated surfactant and solubilize the precipitating ions. The precipitating ions are removed by washing. The insoluble residue, containing asbestos and the surfactant, is mixed and reacted with an alkali to form a dispersion of asbestos that may be reprocessed by conventional methods.

It should be noted that asbestos fibers cannot be recycled in their common role of reinforcing a host matrix because removal from the matrix would destroy the fibers.

There are few asbestos uses (mainly textile) that have uncombined fibers in the end product. When these products are no longer useful, neither is the asbestos because of the physical and/or chemical changes that have made it a different mineral with less strength or shorter fibers of less value than reclamation costs. Thus, asbestos is a nonrenewable natural resource.

B

BARITE FROM DRILLING MUDS

Barite is consumed or incorporated into the product in most of its uses and therefore is not recoverable; however, in its use as a drilling mud most of the barite is returned to mud ponds on the surface where it is dewatered and then recycled in the well. Upon completion of the well, the barite in the pond is usually abandoned. The low cost of barite makes recovery for reuse at another location economically unattractive (16).

BERYLLIUM FROM SCRAP BERYLLIUM

Scrap beryllium is treated by pyrometallurgical methods to produce pure metal. Typically, vacuum casting followed by hot pressing and grinding are used. Excellent hooding and bag filters where necessary are used to prevent escape of beryllium. Beryllium inhalation produces effects akin to silicosis and, therefore, extreme care is taken to minimize the indoor and ambient concentration of this metal.

BERYLLIUM FROM STAINLESS STEEL PICKLE LIQUORS

See "Copper from Nitric Acid Pickle Liquors."

BISMUTH FROM DEBISMUTHIZING DROSS

It is known in the prior art to remove bismuth from lead by treating the molten bath of bismuth-containing lead with appropriate alkaline earth or alkali metal as debismuthizing reagent. The bismuth forms an alkaline bismuthide with the debismuthizing reagent and the alkaline bismuthide forms a dross on the molten bath surface, the dross being referred to in the art as alkaline bismuthide dross or as "debismuthizing" dross. The dross contains an appreciable amount of lead in addition to the alkaline bismuthide. Calcium and magnesium are utilized as debismuthizing reagents for removing the bismuth from the lead. Sodium and potassium have also been utilized heretofore as the debismuthizing reagent. The dross is then separated from the lead, for instance by skimming.

In the recovery of the bismuth from the debismuthizing dross, it is important that the alkaline bismuthide be concentrated in the dross in preparation for further refining to obtain pure bismuth. In this regard, it would be highly desirable to at least double the bismuth content of the debismuthizing dross in preparation for further refining. For example, if the debismuthizing dross contained 6% of bismuth as alkaline bismuthide, it is highly desirable to concentrate the bismuthide in the dross so as to increase the bismuth content to 12% or higher.

Heretofore, the bismuthide has been concentrated in the debismuthizing dross by either mechanically pressing the dross, or by liquating the dross under a flux. In the pressing of the debismuthizing dross, the dross was pressed to expel molten lead from the bismuthide particles by steel plates by means of hydraulic pressure applied to the plates. Some molten lead is expelled but the pores of the alkaline bismuthide solid mass or cake are soon blocked due to significant physical compression of the bismuthide cake by reason of the mechanical pressing. Consequently, additional lead cannot be pressed out of the cake to any considerable extent, and the cake retains considerable molten lead occluded within the bismuthide cake, with molten lead surrounding partially or completely the bismuthide particles in the interior of the cake.

Further, the interior of the compressed cake is substantially impermeable or impervious to air due to the mechanical pressing, and it is necessary that the alkaline bismuthide cake be air permeable or pervious for a subsequent bismuth recovery operation in accordance with the Asarco process. In the liquating method, the debismuthizing dross is liquated under a flux, usually a chloride flux. Liquating the dross under a flux is costly for the reasons considerable bismuth is diverted back to the lead and the chloride flux is expensive. Further, liquating under a flux presents a problem of disposal of slag, for the reason a salt slag is produced that is usually difficult to dispose of. Moreover, the liquation of the dross under a flux is a high temperature operation, involving temperatures of about 700°C.

In a process developed by *C.R. DiMartini and W.L. Scott; U.S. Patent 4,039,323; August 2, 1977; assigned to Asarco Incorporated,* bismuth is recovered as elemental bismuth or as a Pb-Bi alloy from an air-pervious residue cake of debismuthizing dross of concentrated alkaline bismuthide content, by heating the cake in air to the ignition temperature of the cake to ignite the cake, and burning autogenously in a flameless or substantially flameless combustion the thus ignited cake to oxidize the alkali and a portion of the lead to the exclusion of the bismuth. A powder-like residue, formed by the burning, comprises bismuth, lead, lead oxide and alkali oxide.

The lead oxide and alkali oxide of the residue are then selectively dissolved in an oxide-dissolving flux, e.g., molten lead chloride, and the flux containing the lead oxide and alkali oxide is skimmed from the resulting lead bismuthide alloy. The lead of the alloy can, if desired, be separated to recover elemental bismuth, for example, by reacting the Pb-Bi alloy in molten droplet or molten particulate form with Cl_2 to form $PbCl_2$, by electrolysis, or by blowing the molten Pb-Bi alloy with air in a cupel to form litharge which is skimmed off.

The process is shown in block flow diagram form in Figure 20. As shown there, an air-pervious or permeable cake containing alkaline bismuthide, e.g., calcium magnesium bismuthide ($CaMg_2Bi_2$), and lead is ignited by heating the cake in an open vessel in the presence of air to the ignition temperature of the alkaline bismuthide or higher, and typically in the range of about 850°F to about 900°F for the calcium magnesium bismuthide. Consequently the cake is ignited and burns autogenously to a powdery residue of metallic bismuth, metallic lead, lead oxide, e.g., PbO, and alkali oxides, for example alkaline earth metal oxides, e.g., CaO and MgO.

Figure 20: Asarco Inc. Process for the Recovery of Bismuth from Debismuthizing Dross

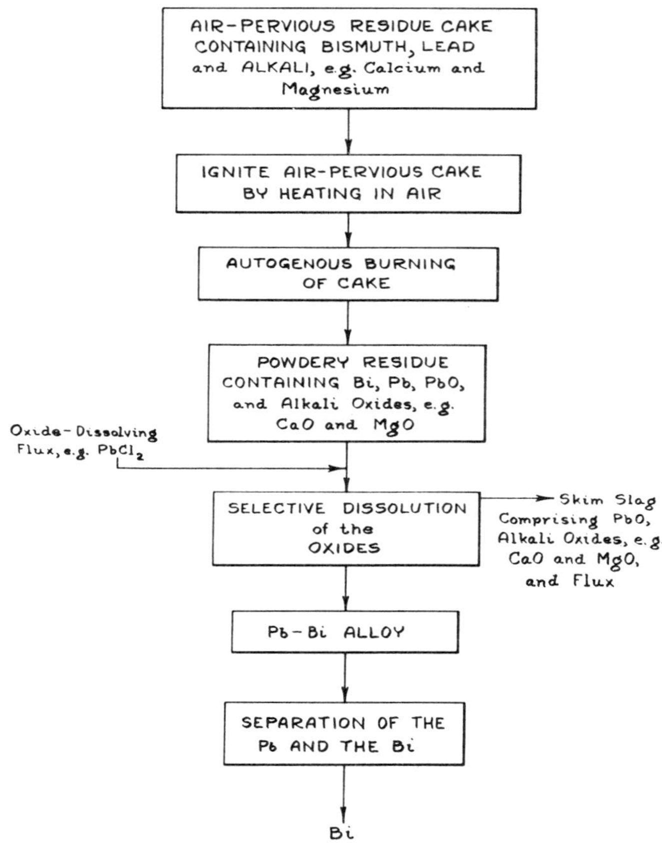

Source: U.S. Patent 4,039,323

Due to the porous, air permeable nature of the cake, air is able to pass into the interior of the cake to support the combustion therewithin so that all or substantially all of the cake is combusted to the powdery residue.

An oxide-dissolving flux, e.g., lead chloride, $PbCl_2$, is then admixed with the powdery residue from the combustion in a suitable vessel, and the lead oxide and alkali oxides are dissolved in the flux to the exclusion or substantial exclusion of the metallic bismuth and metallic lead. The oxide-dissolving flux can be admixed in solid or molten liquid state with the powdery residue. Such flux, if admixed in solid state with the powdery residue, is melted in the presence of the powdery residue to liquefy the flux by heating to a temperature above the melting point of the flux, which is a temperature above 501°C when lead chloride is the flux. The molten flux containing the dissolved lead oxide and alkali oxides forms a separate layer of oxide-containing flux phase or slag phase on the surface of a liquid pool of the molten metallic phase containing the lead and bismuth as a Pb-Bi alloy.

The oxide-containing flux or slag phase is then separated from the Pb-Bi alloy, for instance by skimming the oxide-containing flux phase layer from the surface of the molten pool of Pb-Bi alloy.

When elemental Bi is the desired product, the Bi is separated from the Pb of the molten Pb-Bi alloy by treating the Pb-Bi alloy with chlorine gas. In carrying out this treatment, Cl_2 gas is supplied into a lower portion of a closed chamber, and the molten Pb-Bi alloy is showered or "rained" through the gas in conventional manner. The Cl_2 reacts selectively with the Pb to form $PbCl_2$ which separates as a separate phase layer on the surface of a pool of molten Bi. The $PbCl_2$ is skimmed off the surface of the pool of molten Bi. Alternatively, the Pb can be separated from the Bi by electrolysis, or by blowing the molten pool with free oxygen-containing gas, usually air, to selectively oxidize the Pb to form litharge. The litharge forms on the surface of the molten Bi, and is skimmed off the surface of the pool of molten Bi.

BLEACH-FIX SOLUTIONS FROM SPENT PHOTOGRAPHIC SOLUTIONS

The processing of a silver halide color photographic material includes the two steps of color development and silver removal. When a color photographic material is subjected to color development, the exposed silver halide is reduced to silver, and simultaneously the oxidized color developing agent (oxidized aromatic primary amine) reacts with a coupler to form a dye image. The silver formed by the development is oxidized with a bleaching agent (oxidizing agent) in the silver removal step to convert the silver to silver halide which is then dissolved with a silver halide solvent and thus removed.

As described, the silver removal step includes the two steps of bleaching and fixing. In photographic processing methods that are being used, the bleaching step and the fixing step are either separately carried out, or performed simultaneously.

An iron(III) chelate compound present in the bleaching-fixing solution oxidizes the developed silver during processing, and is itself reduced to a divalent iron salt. On the other hand, the silver ion formed by oxidation is dissolved by the fixing agent. Hence, as the processing is continued, the iron(II) salt and a silver complex salt gradually accumulate in the bleaching-fixing solution. Thus, the activity of the bleaching-fixing solution is decreased, and fatigue is exhibited.

In order to prevent this, the bleaching-fixing solution is replenished with a suitable amount of a fresh supply of the bleaching-fixing solution at the time of development, and the used solution is discharged by an overflow from the tank. To discard the used solution directly is undesirable not only from the viewpoint of environmental pollution but also from an economical viewpoint. It would be very advantageous if the activity of the solution could be revived and the solution reused.

A process developed by *H.C. Baden, C.C. Bard and D.J. Brugger; U.S. Patent 4,036,715; July 19, 1977; assigned to Eastman Kodak Company* offers a practical and highly efficient electrolytic method for concurrently recovering silver from photographic bleach-fix solutions which contain iron, and regenerating the bleach-fix for reuse. A substantial improvement in current efficiency at low current densities is realized by a combination of (a) adding a reductant such as sodium dithionite before electrolysis, as a reducing agent for ferric ions in the bleach-fix solution, (b) maintaining a nonoxidizing atmosphere in the cell, and (c) providing either a cation permeable membrane or a porous alundum membrane between the electrolytic cell anode and cathode, to separate a first pool of bleach-fix catholyte from a second pool of a different salt solution such as sodium carbonate as anolyte.

Figure 21 is a schematic flow diagram showing the passage of spent bleach-fix solution from a film processing tank to the electrolytic cell for silver recovery, followed by bleach-fix regeneration; and then passage of the regenerated bleach-fix solution back to the processing tank.

Figure 21: Eastman Kodak Co. Method of Recovering Silver from Photographic Bleach-Fix and Concurrently Regenerating the Bleach-Fix

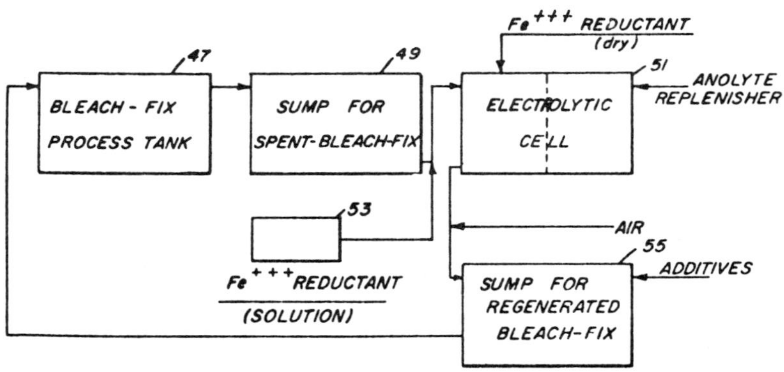

Source: U.S. Patent 4,036,715

The bleach-fix can be regenerated by continuously flowing it from a photographic processing tank **47** into a sump **49**, and then into the top of the cell **51.** Before the bleach-fix enters the cell a stream of ferric ion reductant, such as an aqueous solution, is bled in from a vessel **53.** Alternatively, the reductant can be added as a dry powder directly to the cell, this being particularly advantageous for sodium dithionite which tends to decompose in water.

Electrolysis conditions are maintained continuously in the cell, and the bleach-fix, minus silver, is continuously withdrawn to a sump **55** from which it can be bled back into the processing tank as needed.

Regeneration in both batch and continuous operations is accomplished by introducing oxidizing air into the bleach-fix to reoxidize iron(II) to iron(III) after electrolysis, as described in U.S. Patent 3,634,088. For example, air is introduced into the conduit between cell **51** and tank **47,** or into the sump **55.** Chemicals may be added to replenish any of the bleach-fix ingredients that have been depleted below optimum levels.

A process developed by *Y. Idota; U.S. Patent 4,128,464; December 5, 1978; assigned to Fuji Photo Film Co., Ltd., Japan* is one for electrolytically regenerating a used bleaching-fixing solution resulting from the processing of a color developed color photographic material with a bleaching-fixing solution containing an iron(III) chelate compound and a silver halide solvent, which comprises bubbling an oxygen-containing gas into an anode compartment containing the used bleaching-fixing solution while electrolyzing the used bleaching-fixing solution in an electrolytic cell.

Figure 22 shows a suitable form of electrolytic cell for the conduct of such a process.

Figure 22: Fuji Photo Film Co., Ltd. Process for Regenerating Bleaching-Fixing Solution

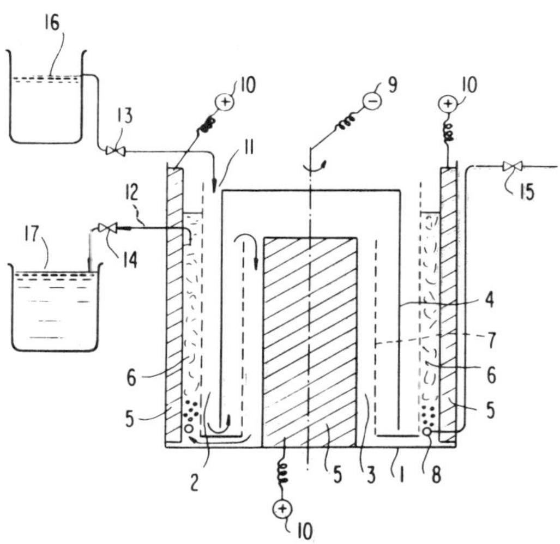

Source: U.S. Patent 4,128,464

As shown there, an electrolytic cell **1** is made up of a cathode compartment **2,** an anode compartment **3,** a cathode **4,** anodes **5** and **6,** a diaphragm **7** separating the cathode compartment from the anode compartment, and a tube **8** for supplying an oxygen-containing gas. The cathode and the anodes are connected to a direct current source by two lead wires **9** and **10.** The anode **6** is made of carbon fibers (felt-like fibers). An inlet **11** and an outlet **12** of the electrolytic cell are capable of being opened and closed by valves **13** and **14.** The arrows in the drawings show the direction of cathode rotation or the direction of the flow of the solution.

The used bleaching-fixing solution is introduced from a tank **16** into the cathode compartment **2** via the inlet **11,** and the silver complex in the used bleaching-fixing solution deposits as silver onto the cathode and is removed from the solution. Then the solution is introduced into the anode compartment **3** past the diaphragm **7.** The solution contacts an oxygen-containing gas in the anode compartment and is oxidized and regenerated. The regenerated solution is conveyed through the outlet **12,** and collected in a tank **17,** thus completing the entire regeneration process.

BLEACH LIQUOR FROM SPENT BLEACHERY EFFLUENT

An integrated pulp and fine paper mill combines, broadly, wood pulping, pulp washing, pulp bleaching, and paper-making operations. In each of these operations, large quantities of water are necessary and, in each, the resulting water effluent contains pollutants necessitating treatment of the effluent prior to recycling for use or prior to returning the water into a watercourse.

Of all these operations, the major source of pollution is the outfall effluent waste stream from the bleachery. The bleachery outfall, in fact, of an integrated pulp and fine paper mill, as described above, constitutes about 40% of the total mill effluent. In more specific terms, this bleachery effluent, in the case of a chlorine-based bleachery, contains most of the chloride content rejected by the mill, the major portion of the total mill color outfall (greater than 60%), and a substantial amount of total mill Biological Oxidation Demand (B.O.D.). From this effluent comes a major portion of the primary sludge separated from the mill effluent.

Procedures for treating such bleachery effluents to purify the same, act to reduce some of the pollutants, such as those primarily contributing to the high B.O.D., but they are not capable of reducing the chloride content or the color. Both levels remain unsatisfactorily high. There are governmental environmental requirements calling for lowering the chloride content of an effluent stream prior to discharge to about 250 parts per million. In addition to meeting the environmental requirements of federal, state, and local governments, it has long been desired to recover the outfall constituents in the effluent as well as to be able to recycle the water for reuse in the paper-making operations in order to lower the large amount of water that is necessary for paper-making and to reduce the energy required.

A process developed by *H.A. Fremont; U.S. Patent 3,986,951; October 19, 1976; assigned to Champion International Corporation* is one for treating chlorine-based bleachery effluents so as to remove the undesired constituents and to enable reuse of the water, recovery of the useful constituents, and reduction in the amount of primary sludge.

The process uses only "a chlorine" (Cl_2 gas or ClO_2) and sodium salts in the bleach sequence followed by treating such bleachery effluent with aluminum ion to precipitate the organics and suspended solids, separating the solids from the liquid, electrodialyzing the treated effluent to provide one effluent stream having 200 to 600 ppm sodium chloride, which can then be returned to the bleachery and a second effluent stream containing 5 to 10% NaCl, evaporating the concentrated second stream from the electrodialysis unit to form a 25% by weight sodium chloride solution, and electrolyzing the solution to produce chlorine gas, sodium hypochlorite and sodium hydroxide to be returned for use in the bleachery.

Figure 23 shows this process in diagrammatic form.

Figure 23: Champion Int'l. Corp. Chlorine-Based Bleachery Effluent Treatment for Bleach Recovery

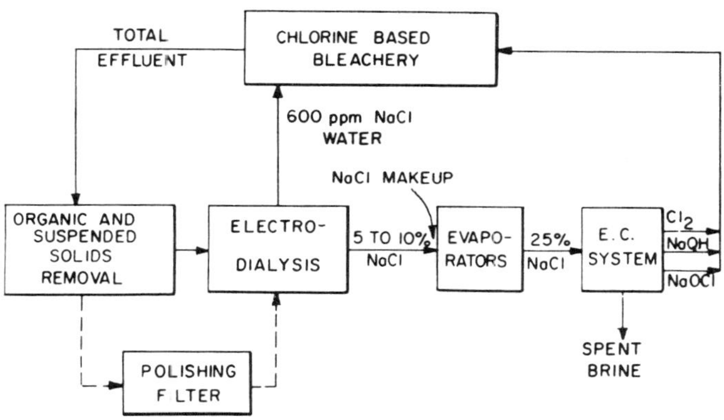

Source: U.S. Patent 3,986,951

BORIC ACID FROM CYCLOHEXANE OXIDATION WASTES

Boric acid and combustible organic materials constitute a substantial portion of the waste stream of a commercial process for the preparation of cyclohexanol and cyclohexanone by boric acid catalyzed oxidation of cyclohexane. The loss of the boric acid is not only wasteful but it is environmentally unacceptable, causing extremely offensive air and surface water pollution. Prior art methods of recovery involve the so-called "wet air oxidation", a liquid phase oxidation process conducted under high pressure and consequently at considerable expense.

An atmospheric pressure disposal system with boric acid recovery as an added feature is provided in a process developed by *O.A. Meyer; U.S. Patent 4,183,904; January 15, 1980; assigned to Monsanto Company*.

The process involves incineration of the combustible organic materials and dehydration of the boric acid to boric oxide. Quenching of the boric oxide, depending upon temperature and water content, yields orthoboric and/or metaboric acids in vapor phase and polyboric acid in liquid phase which is cooled along with the products of the combustion of the organic materials and, in the presence of water, converted to a mixture of boric acids in solid and liquid phases leaving the products of combustion in vapor phase. The mixture of boric acids is then separated from the products of combustion.

A suitable arrangement of apparatus for the conduct of this process is shown in Figure 24.

Figure 24: Monsanto Process for Recovery of Boron from Waste Streams from Cyclohexane Oxidation

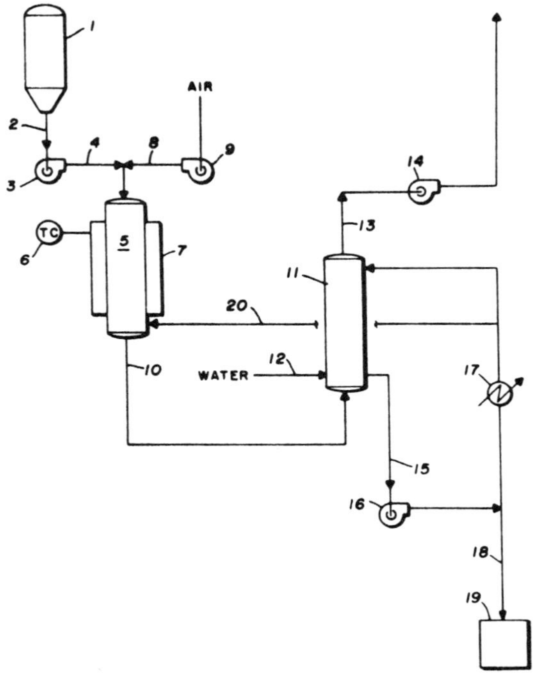

Source: U.S. Patent 4,183,904

Referring now in detail to the drawing, a waste stream containing boric acid and combustible organic materials is fed from feed vessel **1** via line **2**, pump **3** and line **4** to tubular reactor **5**. The tubular reactor passes through tube furnace **7**, the temperature of which is controlled by temperature controller **6**. Air passes into the system from line **8** employing blower **9** for pressure. From the tubular reactor, the combustion products and the converted boric compounds pass through line **10** to scrubber **11** to which a cold water supply is fed through line **12**. Vapors containing most of the products of combustion pass from the scrubber **11** to the atmosphere via line **13** and induced draft fan **14.** The liquid emerging from the scrubber passes through line **15,** moved by recirculating pump **16,** and is cooled in heat exchanger **17.** A portion of the liquid from the recirculating system is drawn from line **15** through line **18** from the base of tubular reactor **5** for recovered products collecting in vessel **19.**

Cooled quench solution is also provided in line **20** to keep the bottom of the reactor and line **10** clean of boric compounds and products of combustion. Make-up water is furnished via line **12.** Heat exchanger **17,** although not essential, is desirable in that it provides for heat recovery of a portion of the high heat of combustion evolved in the process and also serves as temperature control device to prevent pollution of the atmosphere with the volatile boric acid. Heat exchangers may also be placed in line **10** for more effective heat recovery.

In operation, the combustible organic materials are incinerated in the tubular reactor, and orthoboric acid is converted to boric oxide in equilibrium with the water in the vapor phase from the feed and product of combustion. At the base of the tubular reactor, water is injected to carry the waste stream through line **10** into the scrubber. At the base of the reactor, quench streams of water impinge upon the boric oxide and polyboric acids so as to hydrate to equilibrium the boric oxide and water system. In the scrubber, the metaboric and polyboric acids are, for the most part, reconverted to orthoboric acid. Gaseous products of combustion are eliminated through line **13.**

BUILDING BRICKS FROM RED MUD FROM BAYER PROCESS

In producing alumina from bauxite according to the Bayer process, up to two tons of filter-wet red mud, having a water content of from 40 to 50 weight percent, may be obtained for every ton of alumina produced in the process. The amount of red mud actually obtained depends on the particular composition of the bauxite, i.e., on the geographical origin of the bauxite.

With such significant quantities of red mud occurring as a by-product, many have worked to find ways to use the red mud. Among other things, people have attempted to use the red mud for making bricks. However, these attempts at brick making have not been successful, because the high alkali content of up to 10 weight percent based on the dried red mud has caused the resulting bricks to show signs of efflorescence. Moreover, considerable difficulties have been experienced even at the stage of shaping the raw (yet unfired) bricks. For example, the raw bricks have shown an unsatisfactory ability to maintain their shapes. Additionally, there has been crack formation due to high shrinkage during drying and burning.

It has been claimed that a small amount of red mud may be added to clay to provide a desired color in clay bricks. This has, however, given no solution of the problem of what to do with the enormous quantities of red mud produced by the Bayer process.

A process has been proposed by *G. Bayer, E. Cherdron, M. Haerter and E. Hecht; U.S. Patent 3,886,245; May 27, 1975; assigned to Gebr. Giulini GmbH, Germany* for producing bricks from red mud, including mixing red mud having a water content of 15 to 35 weight percent with clay to provide a resulting mixture containing 50 to 92 weight percent red mud based on the dry weight of the resulting mixture, forming the resulting mixture at a water content of 18 to 25 weight percent into raw bricks, drying the raw bricks with heated gases at a temperature below 70°C, and burning the raw bricks at a temperature from 900° to 1000°C.

In a similar process by *G. Bayer, E. Cherdron, M. Haerter and E. Hecht; U.S. Patent 3,886,244; May 27, 1975; assigned to Gebr. Giulini GmbH, Germany,* the red mud is first mixed with water-fixing inorganic and/or organic substances before admixture with the clay. Inorganic water-fixing substances may be bentonite, silica gel, alumina gel, and the like. The preferred organic water-fixing substances are cork, peat, sawdust, coal dust, paper and textile waste, or mixtures of these substances. It is also possible to use dried red mud as the water-fixing material.

A process developed by *G.O. Iwu; U.S. Patent 3,985,567; October 12, 1976; assigned to University of Guyana, Guyana* is one in which bauxite waste red mud is treated with hydrochloric acid to dissolve only the sodium content. The treated material is mixed with kaolinitic clay and formed into construction bricks.

C

CADMIUM FROM BATTERY SCRAP

A process developed by D.A. Wilson and B.J. Wiegard, Jr., Report R.I. 7566, Washington, D.C., U.S. Bureau of Mines (1971) involves the recovery of cadmium and nickel from scrap batteries. Such cadmium-bearing waste occurs in the electrical battery industry as waste from the manufacture of alkaline nickel-cadmium batteries and from such scrap batteries. The waste contains mainly the metals cadmium, nickel and iron, but small quantities of other metals may also occur. Figure 25 is a flow diagram of the overall recovery process.

Figure 25: Process Flowsheet for Cadmium and Nickel Recovery from Scrap Batteries

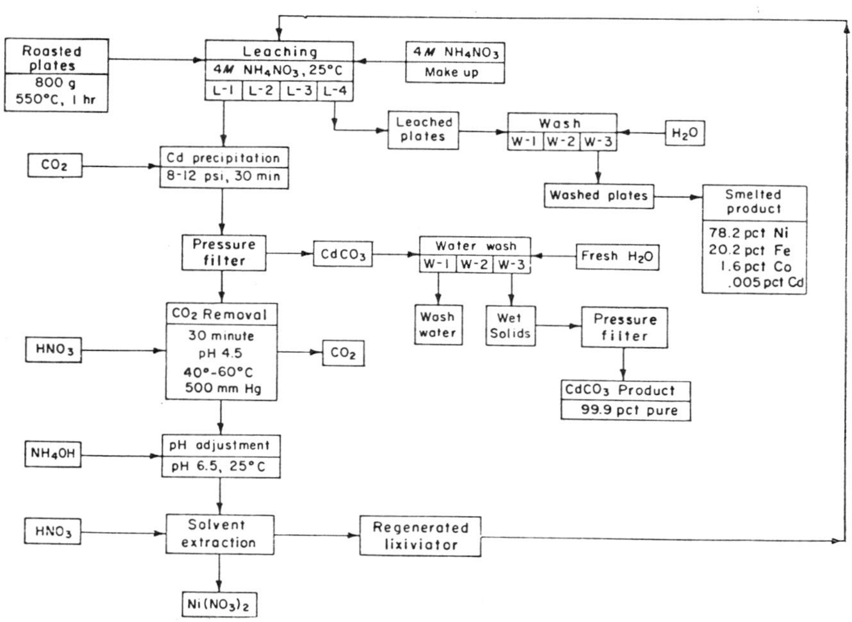

Source: Bureau of Mines R.I. 7566

A cyclic leaching process was developed for selectively leaching and recovering cadmium from nickel-cadmium scrap battery waste. The six major steps in the process are the following: washing the plates to remove KOH electrolyte, roasting at 550° to 600°C to oxidize metallic cadmium and decompose cadmium and nickel salts, leaching with an ammonium nitrate solution, precipitation of the leached cadmium from solution as cadmium carbonate using CO_2 gas, removal of CO_2 from the lixiviator by adjusting to pH 4.5, heating, and evacuating the system, and removal of small amounts of leached nickel from the lixiviator using a solvent extraction technique. The regenerated lixiviator is then recycled to leach additional plates. Cadmium extraction was 94% and the cadmium carbonate product contained 0.14% nickel and 0.12% cobalt impurity. The leached plates were smelted to a high nickel-low iron alloy with no detectable amounts of cadmium.

The disadvantage of the method just described, apart from the costly furnace apparatus, is that only about 94% of the cadmium is leached out of the waste and that the iron and the main quantity of the valuable nickel are not separated from one another. The CO_2 gas expelled is also lost, which adds to the expense of the process.

Thus, an improved process has been suggested by *H. Reinhardt, H.D. Ottertun and J.H.A. Rydberg; U.S. Patent 4,053,553; October 11, 1977; assigned to Nife Jungner, AB, Sweden.*

This process consists of the following steps:

(a) Leaching the waste with an ammoniacal carbonate solution to form an aqueous ammoniacal carbonate solution containing cadmium, nickel and cobalt(II) ammine complexes and a leaching residue. Any iron in the waste is in the leaching residue.

(b) Adding air to the aqueous ammoniacal carbonate solution to oxidize the cobalt(II) ammine complex to cobalt(III) ammine complex.

(c) Contacting the resulting aqueous ammoniacal carbonate solution with a substantially water-insoluble organic solution which contains a hydroxyoxime which forms a nickel compound readily soluble in the organic solution with any nickel present but which does not affect cadmium and cobalt(III). Nickel is thereby removed.

(d) Precipitating the cadmium as a carbonate by removing ammonia from the aqueous ammoniacal carbonate solution and then removing the cadmium in the form of a carbonate.

(e) Withdrawing a portion of the aqueous ammoniacal carbonate solution from the preceding step after precipitation of cadmium and precipitating the cobalt by removal of ammonia.

(f) Returning the ammonia from the preceding step to the cadmium free solution in step (d) and replacing lost carbonate ions by adding carbon dioxide in gaseous form. The resulting solution can then be reused in step (a).

(g) Stripping the nickel compound in step (c) by washing with an aqueous solution of sulfuric acid to strip the nickel as nickel sulfate. The organic solution stripped of nickel can thereafter be reused in step (c).

The flow diagram of the process is presented in Figure 26.

Figure 26: Nife Jungner AB Process for Separating Iron and Nickel and Recovering Cadmium from Nickel-Cadmium Battery Wastes

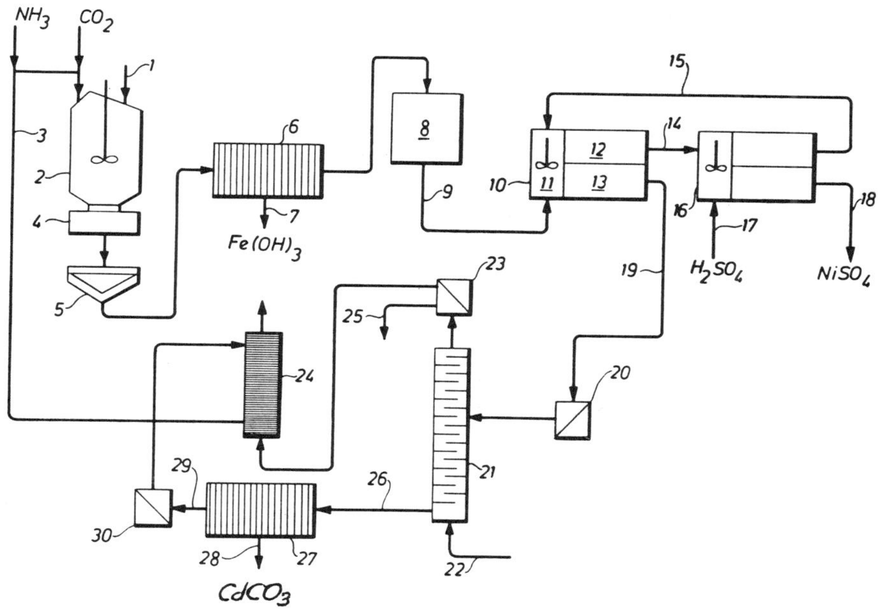

Source: U.S. Patent 4,053,553.

The cadmium-bearing waste **1** is leached in a leaching vessel **2** containing an ammoniacal carbonate solution **3**, the composition of which is regulated by addition of NH_3 and CO_2. Fragmentation of the material is done preferentially below the surface of the liquid in an appropriate crushing or grinding device **4**, so avoiding exposure to the air. Prewashing with water to eliminate accumulation of potassium hydrate may be adopted if desired.

The need for such a processing stage depends on the quantity of potassium hydrate in the leaching material in relation to the content of nickel and cadmium.

In the leaching operation nickel and cadmium are obtained in the solution, while iron remains in the leaching residue. Under optimal conditions yields of >95% for nickel and >99% for cadmium are obtained.

The leaching vessel **2** is connected to a screen **5** in which the larger particles in the leaching residue are separated, and thereafter to a filtering press **6**. The filtered residue **7**, consisting chiefly of iron hydroxide, is washed with fresh leaching solution. The leaching solution passes via a buffer tank **8** and a pipe **9** to the liquid extraction operation of the process where, in an extraction unit **10**, nickel is extracted selectively to an organic solution containing the reagent LIX 64N or Kelex 120 dissolved in a diluent. LIX 64N is a chelate former, as is Kelex 120.

Each stage in the extraction unit is of mixer-settler type and comprises a mixing chamber **11**, in which the leaching solution is mixed with the organic solution, and a separation chamber **12, 13** in which the two liquid phases, which must be essentially insoluble in one another, separate owing to their different densities. The lighter phase, generally the organic solution, collects in the upper portion **12**, and the heavier phase in the lower portion **13**. An extraction unit consists of one or more mixer-settler stages. If multistage extraction is used, it is done in counterflow.

The extraction unit **10** is connected by pipes **14, 15** with a stripping unit **16**, each stage of which is a mixer-settler of the same type as in the extraction unit. The stripping unit as well may consist of one or more stages. In the stripping unit **16** the organic solution **14** from the extraction unit **10** is mixed with a sulfuric acid solution **17** containing 10 to 800 g/ℓ free sulfuric acid. Under the influence of the sulfuric acid in the metal organic compound, NiR_2, is split in the organic solution, and, therefore, Ni is transferred to the sulfuric acid solution.

The flow of the sulfuric acid solution should preferably be adjusted so that the solution leaving the separation chamber through pipe **18** is a strong nickel sulfate solution. The organic solution freed from nickel circulates through pipe **15** to the extraction unit **10** for renewed mixing with the leaching solution and thus for extraction of nickel. Centrifuges can be used in place of all settlers in the plant.

The leaching solution from the extraction unit **10**, which now contains cadmium as the only metal, passes through a pipe **19**, on to a processing stage for precipitation of $CdCO_3$. The leaching solution should preferably be heated in a heat exchanger **20**. The precipitation is accomplished in an ammonia separator **21** by thermal expulsion of NH_3 with water vapor **22**. The expelled ammonia is taken to a condenser **23**. The condensate contains predominantly NH_3 but also some CO_2 and is recirculated via an absorption tower **24**. The treatment of the condensate depends on the total water balance in the process, so that it may be necessary to tap off part of the condensate through a pipe **25**.

The cadmium-carbonate-bearing solution from the ammonia separator **21** is led through pipe **26** to a filter **27**, in which $CdCO_3$ is filtered out. After filtration for separation of the precipitation product **28** the filtrate **29** is cooled in a heat exchanger **30** and is recirculated through the absorption tower **24** for NH_3 to the leaching operation.

The processing arrangement described above yields the best degree of purity of recovered cadmium. The precipitation of cadmium from the leaching solution can, however, also be done before the liquid extraction, yielding a higher degree of purity of recovered nickel.

CADMIUM FROM ZINC REFINERY FLUE DUSTS

Cadmium-bearing materials obtained as by-products by zinc, lead, and copper processing include the dusts from roasters, sintering machines, furnaces, fuming, and Waelz kilns. They also include the blue powder from the distillation of zinc, the cadmium fraction in the rectification of zinc, and the copper-cadmium filter cakes derived from purification of sulfate by cementation. The hydrometallurgical process shown in Figure 27 may be used on any of these materials.

Figure 27: By-Product Cadmium Subsystem

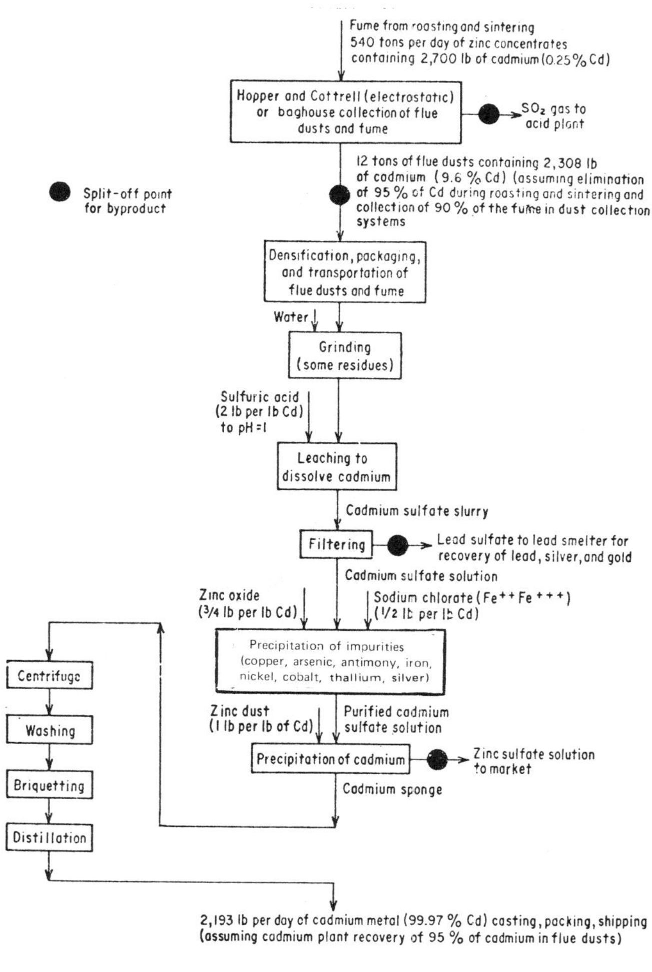

Source: Bureau of Mines IC 8570

The hydrometallurgical process for cadmium involves leaching the cadmium-bearing materials with sulfuric acid to transfer the cadmium to solution. Subsequently the cadmium sulfate solution is treated to remove impurities such as arsenic, antimony, and iron. The cadmium sponge is then precipitated, usually with zinc, or it is electrodeposited from the solution. The sponge is then further treated by electrolytic or distillation methods, yielding a final product of high purity cadmium. Special high purity cadmium may be obtained by vacuum distillation, zone recrystallization, ion exchange, and other techniques.

It is perhaps worth noting that the metallurgy of cadmium pinpoints the split-off point for processing cadmium residues at the dust collectors. This is the point of economic decision where additional costs must be balanced by expected revenues from the cadmium metal. However, this point is not entirely independent of what has gone before. The amount of cadmium in the zinc concentrate is determined not only by the cadmium grade in ores but also by prior metallurgical processes.

Selective flotation practices will determine how much cadmium passes to the zinc concentrate during the milling of lead-zinc ores. For maximum transfer of cadmium into the zinc concentrate it is necessary that the depression of the zinc-cadmium minerals in the lead flotation cycle should be as effective as possible.

It is then equally necessary that vigorous prolonged activation take place before zinc flotation. Likewise, the degree of saturation of individual particles in the flotation process affects the concentration of cadmium in the separate products of flotation. The metallurgy and the economics of cadmium, therefore, begin at the mill though the actual economic decision to produce cadmium is made at the smelter or refinery.

The first step in the processing of cadmium-bearing materials shown in Figure 27 is the densification, packaging, and transportation of the cadmium fume. Inclusion of this step may be questioned because many cadmium plants are in close proximity to the source of raw material. Nevertheless, there is some transportation of the flue dusts and concentration of cadmium production. Cadmium dusts produced at the New Jersey Zinc Smelter at Depue, Illinois, for example, were shipped to Palmerton, Pennsylvania, for recovery of cadmium. Also, a considerable portion of the United States cadmium supply originates in flue dusts imported from smelters in Mexico.

The cadmium-bearing flue dusts are treated with sulfuric acid to dissolve the cadmium. The method shown is that of the acid process. Sulfuric acid is added in amounts to insure extraction of cadmium and other acid soluble components into solution. Lead remains as a solid residue as sulfate, which is subsequently filtered and shipped to a lead smelter for recovery of lead, gold, silver, and selenium, tellurium, and indium, if present.

The cadmium sulfate solution is purified by neutralization. Neutralization of the solution causes iron, arsenic, and various other elements to precipitate. The neutralizing agent in the hypothetical flowsheet shown is zinc oxide, which will precipitate iron hydroxide. The ferrous compounds are converted to ferric compounds by oxidation with sodium chlorate, thus promoting complete precipitation of the iron. The zinc oxide also precipitates arsenic, antimony, indium, gallium, germanium, and thallium, if they are present. Other impurities may require additional processing. Copper, for example, may be precipitated with zinc dust.

After the cadmium sulfate solution is purified, cadmium is precipitated by the addition of metallic zinc dust. The cadmium deposit is a spongy mass that can be separated from the zinc sulfate solution by centrifuging or by filtering. In some cases, the cadmium sponge may be redissolved in sulfuric acid and reprecipitated to obtain a cleaner product for distillation. The zinc remains in solution as zinc sulfate. In the hypothetical flowsheet shown in Figure 27 this step of the process is carried out by the addition of zinc dust. In some operations zinc plates are used, but this procedure is much less common.

In the cadmium subsystem, the cadmium sponge is purified by distillation rather than by electrolytic methods. The cadmium sponge is washed, compressed into briquets, and placed in oil-fired retorts. At the exit of the retort is a cast-iron cylinder in which the cadmium vapors condense. The distillation is a batch process carried out in a 20 hr cycle. The last 10% of the charge is redistilled.

The subsystem shown is assumed to handle 12 tons of flue dust per day containing 2,308 pounds of cadmium. These flue dusts are the result of roasting and sintering 540 tons per day of zinc concentrates containing 2,700 lb of cadmium. The greatest losses in the system shown are in the elimination of cadmium from the concentrates (95% recovery of cadmium) and the capture of cadmium-bearing material in the dust collection system (90% recovery of cadmium). Thus overall recovery of cadmium at the first stage is 85%. A major factor in this recovery is the efficiency of the dust collection system.

The recovery in the cadmium-processing system itself is assumed to be 95%, which is believed representative in the industry (17).

CALCIUM FLUORIDE FROM PHOSPHATE POND WATERS

In a wet process phosphoric acid plant using the dihydrate technique, about 50 gpm per ton of product P_2O_5 per day of waste water is directed to a waste pond. The pond water contains from about 0.1 to 5% fluorine, from about 0.1 to 5% P_2O_5, from about 0.1 to 2.5% SiO_2, from about 0.1 to 0.5% dissolved calcium and from about 0.1 to 0.5% soluble sulfate salts. The fluorine in pond waters is a major concern because it can present an ecological hazard. The loss of P_2O_5 values in pond waters is considered a serious problem not only because it is a valuable product but also because it can present an ecological hazard.

Traditionally pond waters have been passed through settling basins prior to their discharge into rivers and streams. At times, pond waters are treated with limestone and lime to precipitate out fluorine and other values before discharge to meet pollution control laws and regulations. In the settling basins, the various chemical values, such as fluorine, P_2O_5, calcium, etc., decrease so that the pond water when discharged into the streams contains lesser, but appreciable, amounts of these materials. Not only do discharged pond waters add chemical values to streams, but they also cause a decrease in the pH of the streams. Pond water typically is acidic and has a pH from about 1 to 3.

Workers in the art have recognized the economic loss and ecological problem of pond waters and have developed methods of treating pond waters. However, it appears that none of these methods have been economically attractive or feasible since none of the methods are known to be in commercial use in the United States.

However, a process developed by *R.J. Hirko and H.E. Mills; U.S. Patent 4,171,342; October 16, 1979; assigned to Occidental Chemical Company* is one for recovering calcium fluoride and calcium orthophosphate from phosphate waste waters and provides a method of neutralizing the waste waters so they can be discharged into rivers and streams without fear of pollution or recycled as process waters in the production of phosphoric acid.

Such a process is shown schematically in Figure 28.

Figure 28: Occidental Chemical Co. Process for Recovery of Calcium Fluoride from Phosphate Operation Wastewater

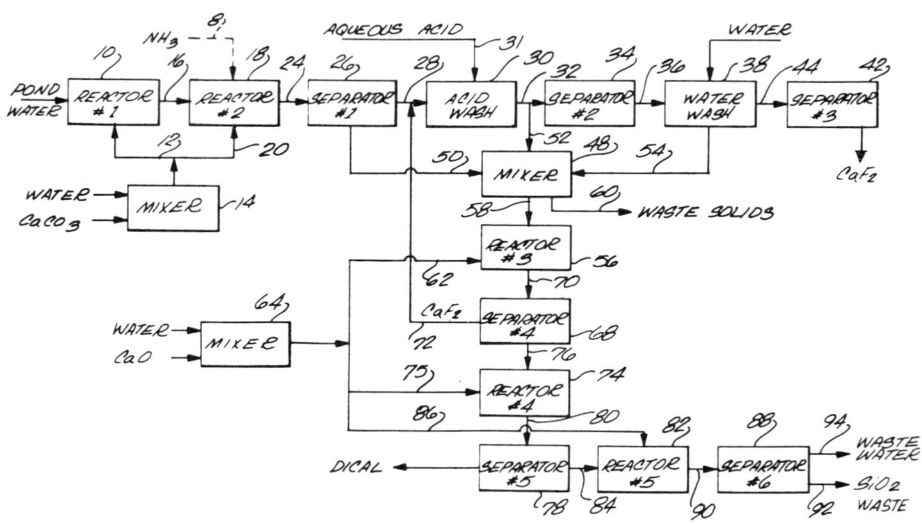

Source: U.S. Patent 4,171,342.

Referring to the drawing, pond water is passed to the first reactor stage **10.** Simultaneously, calcium carbonate such as a stream of slurried calcium carbonate **12** is also passed to reactor **10.** The stream **12** originates in a mixer **14** wherein a source of ground calcium carbonate, such as limestone, is mixed with fresh water, not pond water or process water containing F or P_2O_5 values, to form the calcium carbonate slurry. The calcium carbonate source is ground to at least a -10 mesh, preferably to a 90% -200 mesh. The calcium carbonate slurry is formulated with a solids content of between about 5 and 50%, preferably about 10 to 35%. The calcium carbonate slurry and pond water are agitated in reactor **10** for about 5 minutes, although shorter reaction times or long reaction times can also be employed where necessary. The amount of calcium carbonate added to reactor **10** is designated as Z_1. Z_1 is equal to the equivalents of calcium added to the pond water per equivalent of fluorine in the pond water. Z_1 typically has a value of from about 0.3 to 0.8, preferably a value of about 0.4. The reaction that occurs in reactor **10** normally produces soluble calcium values such as $CaSiF_6$, although in some instances insoluble values will be formed. Preferably no solids will be formed in reactor **10**.

The reaction mixture from reactor **10** passes as a stream **16** to a second reactor stage **18** wherein the stream **16** is mixed with additional calcium carbonate such as slurried calcium carbonate via stream **20**. The stream **20** also originates from the mixer **14**.

The reaction mixture in reactor **18** is agitated for a period of about 30 minutes, although shorter or longer residence times as described above can be employed. The amount of calcium carbonate added to reactor **18** is designated as Z_2. Z_2 is equal to at least Z_R minus Z_1. Z_R designates the minimum equivalence of calcium that should be added to reactors **10** and **18** to remove from about 85 to 100% of the fluorine values from the feed pond water. Z_R, on a mineral acid free basis, has a value equal to $1 + 0.1927\ R_W$, wherein R_W is equal to the weight ratio of P_2O_5 to fluorine in the pond water. Z_R usually has a value of from about 1 to 2. Z_2 has a value of from about 0.1 to 2, preferably from about 0.4 to 1.4; and an especially preferred range of from 1 to 1.4.

In theory the Z_R value should provide sufficient calcium to remove the bulk of the fluorine from the pond water. However, as explained above, the pond water contains other values besides fluorine, such as P_2O_5 and sulfate, which can complex with the calcium and limit the amount of calcium available for reaction with the fluorine values. The Z_R value takes account of those other values. However preferably more calcium is added, such as 5% excess, than required according to the Z_R value. However, excessive amounts of calcium are not necessary and interfere with the acid wash stage described below.

Dry or slurried calcium oxide can be used to replace all or part of the calcium carbonate added to the first reactor stage **10** and/or the second reactor stage **18**. However, it is preferred that calcium carbonate be used because calcium carbonate is a less expensive source of water-soluble calcium than is calcium oxide, and it is easier to control the pH in the reactor stages with calcium carbonate than with calcium oxide. This is because of the buffering effect of the carbonate ion.

The reaction mixture from reactor **18** is passed as a stream **24** to a first separator stage **26**. The major part of the fluorine values in stream **24** are solid calcium fluoride, more preferably at least 95% of the fluorine values are solid calcium fluoride. In addition, substantially all the iron and aluminum values in stream **24** are solid values, and about 50% of the sulfate values and about 25 to 50% of the P_2O_5 values are solid values.

The separator **26** can be a conventional liquid-solid separator, such as a filter, centrifuge or decanting tank. Preferably the separator **26** is a mechanical continuous thickener with a sludge-raking arm. A flocculent such as Dow AP 30 or AP 273 can be introduced to the thickener to aid in separation of the calcium fluoride containing solids from the liquid.

The calcium fluoride solids are separated from the liquid phase of stream **24** in separator **26** and passed as a stream **28** to an acid washer **30**. The solids, which can contain about 15 to 35% fluorine, are washed with a stream **31** of a dilute aqueous solution of a strong mineral acid, such as sulfuric acid, hydrochloric acid, nitric acid or the like in the washer **30**. In washer **30**, the solid is washed with from about 1 to 50% (w/w) mineral acid solution, preferably from about 5 to 25% (w/w) acid solution. The calcium fluoride solids are preferably washed with an equivalent of mineral acid about equal to the equivalents of P_2O_5 and CO_2 values contained in the solids, plus or minus 10%.

The acid concentration and wash temperature are controlled so that reaction between the fluorine values and the acid is minimized to prevent volatilization of hydrogen fluoride. The maximum wash temperature is about 180°F. Although any strong mineral acid, such as hydrochloric, perchloric , nitric, sulfuric, hydrofluoric or other strong mineral acids, can be used in the wash step, sulfuric acid is preferred. In the acid wash step, a substantial portion of the P_2O_5 values are removed from the solid product. In addition, other solid values, such as $CaCO_3$, MgO, Fe_2O_3, Al_2O_3 are also removed by the acid wash in varying amounts.

The wash slurry stream **32** is passed to separator **34** wherein the solids are separated from the aqueous phase. Separator **34** preferably is a thickener of the same type as separator **26**. The solids from separator **34** are passed as stream **36** to a water washer **38** wherein the calcium fluoride stream **36** is washed with fresh water, not process waters or pond waters. The solid is washed with at least an equal weight amount of water, preferably about 2 weights of water, although larger amounts of water can be used in the wash. The water wash removes many of the soluble P_2O_5 values from the solids.

The acid wash and water wash up-grade the calcium fluoride solids with respect to the percentage of contained fluorine and the F/P_2O_5 weight ratio. The upgrade calcium fluoride solids can be converted to HF by conventional processes with conventional apparatuses.

The washed solid, as a slurry, is passed to a third separator stage **42** as stream **44**. Separator **42** is a conventional liquid-solid separator such as separator **26** described above. Solids are separated from stream **44** in separator **42** to yield a cake containing up to about 45% fluorine as calcium fluoride. The cake can be processed by conventional means (not shown) into hydrogen fluoride or it can be dried by conventional means (not shown) at a temperature between about 90° and 150°C and stored or transported as a dry solid.

The calcium fluoride cake is substantially made up of calcium fluoride; however, the cake could also contain a small amount of P_2O_5 values, less than about 5%, silicon dioxide values less than about 3%, and sulfate values.

The calcium fluoride solids prepared by this process can be used as a precipitant to precipitate impurities from phosphoric acid. These calcium fluoride solids are of sufficient purity to prepare hydrogen fluoride directly therefrom by heating and treating the solids in the presence of sulfuric acid by conventional methods. If the P_2O_5 concentration is greater than about 5 to 10% or if the silicon dioxide concentration is greater than 5% in the calcium fluoride solids, it is difficult to prepare good quality hydrogen fluoride from the calcium fluoride solids by treatment with sulfuric acid.

At high temperatures in an acid environment phosphorus and silicon will react with fluorine to produce gaseous fluorine values such as PF_3O, and SiF_4 which contaminate the hydrogen fluoride produced and are difficult to remove. Accordingly, it is one of the achievements of the process that a calcium fluoride solid can be produced which can be directly converted into hydrogen fluoride by conventional means.

The aqueous phases from the separators **26, 34** and **42** are passed to a mixer **48** as streams **50, 52** and **54** respectively. In the mixer **48**, the various streams are thoroughly mixed together and passed to a third reactor stage **56** as stream **58.** In the mixer **48** some solids, mainly gypsum, may settle out and are removed as stream **60** as waste solids.

In reactor **56**, stream **58** is reacted with a slurry of calcium oxide from stream **62**. The origin of stream **62** is a mixer **64** wherein, a ground calcium oxide feed source, such as lime, and water are mixed to form a calcium oxide slurry. The calcium oxide feed source is ground to at least -10 mesh, preferably at least 90% -200 mesh. The calcium oxide slurry prepared in mixer **64** contains from about 5 to 50% solids, preferably from about 10 to 35% solids. The calcium oxide slurry is prepared from fresh water and not pond waters or process waters.

The stream **58** has a pH between about 1.8 and 2.6. Sufficient slurried calcium oxide is added to reactor **56** to raise the pH of the resulting reaction mixture from about 3 to 4, preferably from about 3.6 to 3.8. The retention time of stream **58** in reactor **56** is from about ½ to 60 minutes, preferably about 15 minutes.

The reaction temperature within reactor **56** is from about the freezing point to the boiling point of the reaction mixture, preferably about ambient temperature. The reaction mixture contains about 0.01 to about 0.2% fluorine, a substantial portion of which is precipitated out as calcium fluoride upon the addition of the calcium oxide.

The reaction mixture from the reactor **56** is passed to a fourth separator stage **68** as stream **70**. The separator stage **68** is a conventional liquid-solid separator such as the separator **26** described above. The solids separated out in the separator **68** consist of calcium fluoride and phosphate values and are recycled as stream **72** to stream **28** wherein the solids undergo the acid wash and the water wash as described as part of stream **28**.

The aqueous phase separated in separator **68** is passed to a fourth reactor stage **74** as stream **76**. In reactor **74**, the aqueous phase is treated with slurried calcium oxide from stream **75** which has its origin from mixer **64**. The reaction temperature and reaction residence time in the reactor **74** are similar to those in reactor **56**. The pH of the stream **76** is from about 3 to 4. Sufficient calcium oxide is added to reactor **74** to raise the pH of the reaction mixture to between about 5 and 7 to cause the precipitation of dical (calcium orthophosphate or dicalcium phosphate).

The reaction slurry from reactor **74** is passed to a fifth separator stage **78** as stream **80**. The separator **78** is a conventional liquid-solid separator such as the separator **26** described above. The solids separated in separator **78** consist principally of dical. The dical produced by this method is of sufficient purity to employ in animal feed. That is, the P/F weight ratio in the dical is greater than 100.

The aqueous phase from the separator **78** is passed to a fifth reactor stage **82** as stream **84**. Stream **84** is combined with a calcium oxide slurry from stream **86** in the reactor **82**. Stream **86** has its origin from mixer **64**. Sufficient calcium oxide is added to the reaction mixture in the reactor **82** to raise the pH to a value of between about 8 and 11 to cause precipitation of a substantial portion of the remaining soluble minerals as insoluble values, such as silicon dioxide solids and other solids from the reaction mixture.

The reaction mixture in the reactor **82** is passed to a sixth separator stage **88** as stream **90**. The separator **88** is a conventional liquid-solid separator such as the separator **26** described above. Separated solids from the separator **88**, comprised chiefly of silicon dioxide solids, are sent to waste as stream **92**. The aqueous phase from separator **88** consists of a nonacidic water which is substantially free of calcium, fluorine, P_2O_5 and sulfate values. This water is discharged as stream **94** and can be used as a fresh water for industry or it can be discharged into streams or lakes.

In another embodiment of the process, ammonia is added to the reaction stream before stream **80**, preferably in stream **8** or reactor **10**. The ammonia is maintained at a concentration between about 10 and 5,000 ppm, preferably a concentration of about 600 to 800 ppm ammonia. Presence of ammonia in the process stream optimizes the yield of calcium fluoride and the P/F ratio of the dical produced in reactor **74**.

The reason for this is not thoroughly understood. However, it is believed that the ammonia complexes with the fluorine values furnishing soluble fluorine values which can readily react with soluble calcium values to form the insoluble calcium fluoride. Thus the presence of ammonia not only enhances the quality of dical but also increases the removal of fluorine from the pond waters and the overall yield of calcium fluoride.

CALCIUM FLUORIDE FROM UO_2 PROCESS WASTEWATERS

See "Uranium from UO_2 Process Wastewaters".

CAST IRON FROM SCRAP AUTOMOBILES

See "Steel from Scrap Automobiles".

CAUSTIC FROM CAUSTIC SLUDGES

See "Mercury from Caustic Sludges".

CEMENT FROM CEMENT KILN DUST

As pointed out by M. Sittig in *Pollution Control in the Asbestos, Cement and Glass Industries,* Noyes Data Corporation, Park Ridge, N.J. (1975), dust is a major problem in the cement industry and dust reclamation is an economic necessity. The reader is referred to that volume for details of air pollution control equipment as applied to cement plants.

T.E. Kreichelt, D.A. Kemnitz and S.T. Cuffe in Publication 999-AP-17, Cincinnati, Ohio, U.S. Public Health Service (1967) state that the economic benefits resulting from the reuse of collected dust do not completely offset the cost of the dust collection equipment. However, the esthetic value to the community plus compliance with air pollution control ordinances make dust recovery worthwhile. There are also process problems related to reuse of the dust, however.

Efforts to reuse the dust by returning it to the kiln in the form in which it is collected have not been entirely satisfactory for two basic reasons.

Firstly, it is in such fine form that a high proportion of it is again carried up the exhaust gases before it enters the cement-forming reaction and thus lost to the operation, as well as placing an excessive burden on the stack dust-collecting system.

But more importantly, the waste dust contains excessive amounts of alkali metal sulfates and chlorides which cannot be tolerated in the making of cement of acceptable quality. As pollution standards are raised and/or stack collecting equipment is increased in effectiveness so that dust of greater fineness particle size is collected, this latter problem is intensified, as it is known from analyses that generally the alkali content increases with any increase in fineness of the dust.

Procedures heretofore proposed for changing the physical form and/or lowering the alkali content of cement kiln dust have been unattractive or impractical for economic or other reasons.

For example, over 50 years ago U.S. Patents 1,354,642 and 1,402,173 proposed to treat cement kiln dust with a calcium chloride solution to convert the alkali metal sulfates to alkali metal chlorides, separating the solution from the solids, and recovering the potassium chloride from the solution. However, the proposed process has never been well received, presumably because it is uneconomical, presenting the difficulty of separating the solids from the waste liquor, as well as the environmental problem of disposing of the large volume of waste liquor.

An improved process has been proposed by *A.T. McCord, U.S. Patent 4,031,184; June 21, 1977* which claims that it can treat economically waste cement kiln dust, regardless of its degree of fineness and its alkali metal content, to convert it to both a physical form and chemical content suitable for recycling in the cement-making process, with the facile recovery of the alkali metal content as potassium chloride in relatively pure, crystal form.

The improved process comprises (1) leaching the cement kiln dust at elevated temperatures with an aqueous solution of potassium chloride, (2) treating the leached slurry of cement kiln dust with a relatively small amount of oil and a fatty acid to flocculate, and preferably pelletize, the solids, in the aqueous phase, (3) extracting the flocculated or pelletized dust from the aqueous phase, (4) desirably lightly washing or rinsing the flocculated or pelletized material to still further reduce the alkali content, (5) cooling the leaching solution to throw down or precipitate potassium chloride crystals by crystallization, and (6) removing the crystallized material therefrom. The residual leaching solution, after removal of substantial amounts of potassium chloride crystals, is still saturated with potassium chloride in solution, and after reheating is available for recycling with additional cement kiln dust.

CEMENT FROM READY MIX CONCRETE RESIDUES

Formerly, the construction of concrete structures was accomplished by mixing the individual ingredients: aggregate, sand, cement, and water, at the site, in a stationary mixer, then transferring it to a container and pouring it from the container into the forms. At the end of each day, or periodically during the day, the mixers and containers were flushed with water to prevent hardening of remnant concrete therein, the wash water and contained solids going down the nearest drain.

Recently, the stationary mixer has been replaced by the so called "transit mix" truck, which is charged with the raw materials at a central plant, which mixes them into concrete while in transit, and discharges the product upon arrival at the site. Upon return from each delivery, each truck is flushed with water for the same reasons as stated.

Still more recently, very large industrial plants have been constructed for the sole purpose of producing precast concrete products, ranging from modest residential type concrete blocks to elaborate building facing panels and structural elements. Again, every piece of equipment contacted by the wet concrete must be regularly flushed with water to prevent hardening thereof and consequent equipment breakdown.

Going back to the transit mix plant problem and considering the operation of a normal yard having approximately 150 trucks, it will be appreciated that each truck at the end of the day having some overage that may vary from a half a yard to a yard or even two yards represents a problem of disposal since the material must be dumped and disposed of and the truck cleaned out for the next day's operation.

By assuming at least 75 trucks having an overage of at least one yard, it can be appreciated that 75 yards of concrete must be disposed of and that 150 trucks must be cleaned for the next day's operation.

Formerly, some operators would simply charge their trucks with excess water and dump the remaining material in the sewer. This practice has been frowned upon by many local authorities and is now against the law in many communities.

In some locales where the terrain permits, a cement truck at the end of the day would travel to a suitably located quarry or other area where the load could be dumped without objection and then returned to the yard for the ultimate washout procedures mentioned above. This procedure, unfortunately, is expensive since the operator must pay for the cost of the truck being moved to the remote location and the return to the yard. In addition, the costs include the driver time and his charges covering the trip to the grounds and the return to the yard for cleanup procedures. It is quite obvious that the cost of operating the truck to the dump and return and the cost of the driver for this period of time is unproductive and in reality is dead time that must be borne by the operator.

As simple as the procedure sounds, it has problems since available areas for dumping the unused charge at the end of the day are not readily available to the average operator. In addition, areas that are available are usually under private control requiring a charge of admission for the privilege of dumping the waste material. The added charge of admission to the dump area and the delay associated with the dumping procedure must all be added to the costs and the overhead of operation.

In considering the overall operation, it becomes quite apparent that the rock, sand, cement and water comprising the residual charges are all wasted and hence not available to the operator by way of reclamation or saving. The loss and resources of the material can only increase the cost of operation and the loss to the environment in accepting waste concrete is enormous each year, as dumping areas are gradually made unavailable to the concrete industry. The waste of natural resources can only be considered a disaster to the whole country and to the world.

A technique has been developed by *I.P. Macauley, Jr.; U.S. Patent 3,997,434; December 14, 1976; assigned to Jetomatic Systems, Inc.* that is applicable to both the transit mix and precast type of concrete plants, necessary differences relating only to nonessential details.

It is based, in one part, on the realization that a thickened (e.g., settled) cement slurry is essentially useless; if it is to be at all useful, it must be kept agitated and not allowed to settle. It is further based on the realization that equipment handling such a slurry must be kept in substantially continuous operation lest it become accreted with cement and plugged up. Lastly, it is based on the provision of means whereby the continuous operation of a slurry recycling system can be adapted to interface with a concrete producing system, which is inherently a batch operation.

Figure 29 shows the application of such a system to a transit mix plant.

A truck cleaning area **10** is provided with one or more bays for trucks **12** as needed. Each bay has washout jet sprays **14** which clean the truck, effluent passing out the truck's chute **16** into a hopper **18**. The latter passes the effluent onto a plurality of screens or other separatory devices, indicated generally at **20**, such as are disclosed in the prior art. The largest fraction, stone, is picked up in a conveyor **22** and taken to a first reclamation area, and the sand fraction is similarly conveyed as at **24** to a separate reclamation area. The last fraction, the water-cement slurry, flows by gravity through a chute **26** into a basin or tank **28**. A second basin or tank **30**, connected to tank **28** via a weir **32**, may be provided.

Figure 29: Jetomatic Systems, Inc. Concrete Reclamation System

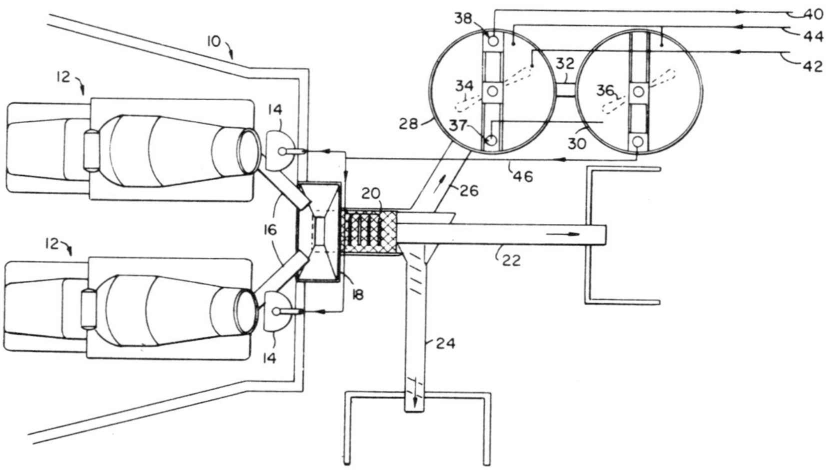

Source: U.S. Patent 3,997,434.

The system then provides agitators **34**, **36** in tanks **28**, **30**, a slurry pump **38** and line **40** going to the batching facility (not shown in the figure), return slurry line **42**, and make-up water line **44** having outlets in each tank.

The provision of second tank **30** is preferred, as it provides a local, closed circuit source of wash water for trucks **12** via line **46**. As an alternative to agitator **36**, it may be preferred to omit it and provide instead a feed back loop in the form of limited flow transfer pump **37** sucking relatively thick slurry off the bottom of quiescent tank **30** back into agitated tank **28**.

The same system as described above is applicable to precast plants, except that simpler means, e.g., sump pumps, may be used to separate sand and gravel portions.

An apparatus developed by *W.F. Miller; U.S. Patent 4,127,478; November 28, 1978; assigned to Hy-Way Heat Systems, Inc.* is one for receiving mixed concrete from a ready mix truck which provides a receiving hopper and an elevator for lifting the material therefrom arranged to discharge it into a classifying rotating drum incorporating a spiral conveyor into which water is sprayed during use. A sump and a secondary conveyor are positioned beneath the classifying drum so that the separated sands, aggregate and other material, washed and separated from the concrete mix, introduced with the cement and waste water are confined to the sump and a separate settling tank from which the fine sands and similar materials can be salvaged.

A system developed by *A.R. Borges; U.S. Patent 4,154,671; May 15, 1979* involves recycling undelivered cement left as residue in cement trucks. Undelivered cement at the end of the working day is deposited on a shaker screen under action of high pressure water jets which trap the large aggregate and force the large aggregate onto a first belt for collection. Sand, cement and water fall through the first screen onto a second screen which traps sand under action of high pressure jets that force the sand onto a second belt for delivery to a collection area. The remaining cement, water and slurry is collected in a trough and pumped to a settling tank having an auger that periodically mixes the water and cement to prevent hardening. A skimmer located at the uppermost portion of the settling tank collects clear water used to clean the aggregate and cement previously mentioned. The cement, water and slurry collected in the settling tank is used in combination with fresh sand and cement to provide a fresh charge to waiting cement trucks. Sand, cement, water and aggregate previously lost is now recycled and reused rather than being wasted.

CHLORIDED ALUMINA CATALYST FROM SPENT CATALYST

Chlorided alumina catalysts are useful in the alkylation of paraffins with olefins, as for example the alkylation of isobutane with butene-2 or ethylene, to produce high octane gasoline blending components. Descriptions of chlorided alumina catalysts useful in alkylation are set forth in U.S. Patents 3,240,840 and 3,523,142 where the freshly prepared catalyst possesses a high degree of initial activity. Unfortunately, the activity of the catalyst declines after a relatively short period of time on stream and detracts from the commercial attractiveness of an alkylation process employing the same. This decrease in activity or deactivation is related to the concomitant formation of coke which deposits upon the catalysts's surface along with some loss of chlorine associated with the catalysts's active surface.

The substantial deposition of coke on the catalyst's surface interferes with the catalytic sites thereby reducing the catalyst's alkylation activity and selectivity leading to a reduction in desired product formation. To partially overcome the deactivating effect of coke upon the catalyst and to maintain the rate of alkylation, more severe alkylation reaction conditions are employed including the use of higher reaction temperatures and the introduction of chloriding agents in the feedstock.

Such procedures, while costly, to some extent enable the process to operate on stream for a longer period of time. However, alkylation is preferably carried out at low temperatures to favor conversion and product selectivity and raising reaction temperatures to overcome catalyst deactivation is counter-productive. Ultimately the catalyst will become substantially deactivated by the rising level of coke deposited thereon and continued alkylation only results in diminished conversion and reduced selectivity to desired products.

A process developed by *S. Herbstman and J.H. Estes; U.S. Patent 4,083,800; April 11, 1978; assigned to Texaco, Inc.* provides a regenerated catalyst having improved activity. The regenerated catalyst is prepared by contacting a deactivated chlorided alumina catalyst in a nonoxidizing atmosphere at a temperature between 400 and 1200°F, thereafter contacting the catalyst in an oxidizing atmosphere at a temperature between 700 and 1200°F and subsequently rechloriding the catalyst at a temperature between 200 and 800°F.

CHLORINE FROM ALUMINUM CHLORIDE ELECTROLYSIS

While the production of aluminum by electrolysis of aluminum chloride in a halide melt or electrolyte has been a long desired and theoretically feasible objective of the art, its economic attainment has never become a commercial reality. Among the many reasons thereof are numerous unsolved problems occasioned, for example, by the highly corrosive nature of an alkali and/or alkaline earth metal halide electrolyte or melt, the necessity of maintaining the purity and concentration of aluminum chloride in the cell employed within critical values, and the like.

All of these pose both technical and economic barriers to the necessary continuity of low-cost operations requisite for commercial production of aluminum by such methods. One of the many problems presented involves the handling and disposition of effluent gas from the electrolytic reduction cell.

Such problems are particularly acute when alkaline earth metal halides or alkali metal halides or mixtures thereof containing the aluminum chloride are used as the electrolyte, since the effluent gas will be made up principally of chlorine. This chlorine is highly reactive, poisonous and corrosive, with minor amounts of such components as nitrogen and traces of CO_2 together with melt values such as sodium aluminum chloride, aluminum chloride, and combinations of aluminum chloride with the alkali metal and/or alkaline earth metal halides in the form of condensable gas.

The presence of such melt constituents in the effluent gas, apart from constituting undesired and uneconomic losses of melt values and being of highly corrosive character, has effectively precluded the direct reuse of the chlorine values because of the detrimental presence of such impurities and because of attendant handling problems created thereby, such as clogging of the conduits through which the effluent gas is conducted, for example, by condensation therein of alkali and/or alkaline earth metal halides or some of the other aforementioned compounds.

Still other problems encountered with the use of such effluent gases include the difficulty of condensing the gases for use in liquid form in storage or chemical process use.

An improved recycle and recovery process has been developed by *S.C Jacobs and R.C. Schoener; U.S. Patent 3,904,494; September 9, 1975; assigned to Aluminum Company of America* and is shown schematically in Figure 30.

Aluminum is produced in cell **14** by the electrolysis of aluminum chloride dissolved in molten alkali or alkaline earth metal chlorides or mixtures thereof, the temperature within the cell **14** normally being about 700°C. In operation of such a cell, aluminum is removed in molten form as indicated by the arrow **10**, and a gaseous effluent, comprising principally chlorine, together with a small amount of nitrogen and containing alkali metal chloride and/or alkaline earth metal chloride from the electrolyte and aluminum chloride, is evolved therefrom as indicated by the arrow **12**.

Such a gaseous effluent may be constituted of about 91.5% chlorine, 1.8% nitrogen, 4.8% alkali and/or alkaline earth metal chlorides, 1.9% aluminum chloride and traces of oxygen, and with the alkali and/or alkaline earth metal chlorides being in the form of a gas or in such finely divided form as to constitute a fume.

Figure 30: Aluminum Co. of America Process for Effluent Gas Recycling and Recovery in Electrolytic Cells for Production of Aluminum from Aluminum Chloride

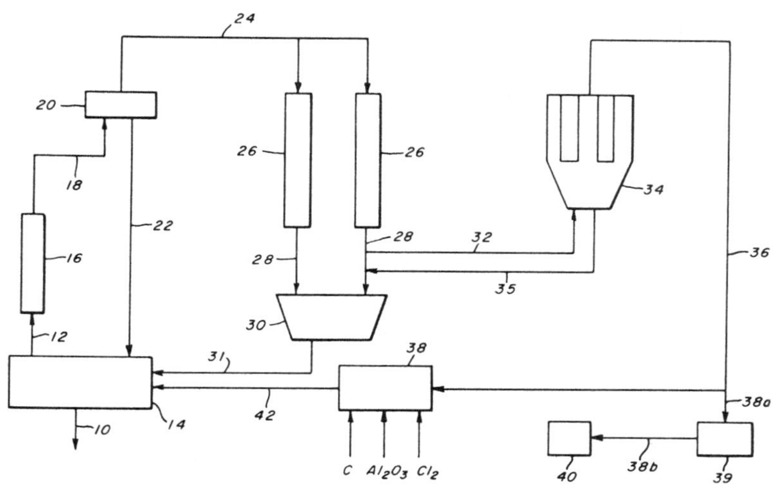

Source: U.S. Patent 3,904,494.

After evolvement and removal from the cell **14**, the effluent gas as above constituted, and at a temperature of approximately 700°C, is cooled, preferably by passage through a heat exchanger **16**, to reduce its temperature to a first predetermined value range that is sufficiently low as to effect selective condensation of all of the composite alkali and/or alkaline earth metal aluminum chloride values without appreciable condensation or desublimation of the aluminum chloride values contained therein.

The effluent gas may be reduced by the heat exchanger to a temperature range of from about 150 to 200°C, thereby effectively condensing such values into small liquid droplets or a mist. Such cooled gaseous effluent is then conducted via line **18** to a coalescing zone such as the one illustrated by the demister **20**, wherein the now condensed liquid droplets of alkali and/or alkaline earth metal chloride-aluminum chloride combinations are coalesced and separated from the effluent gas.

The coalesced liquids, including any aluminum chloride values dissolved therein, and for example, amounting in total to about 0.01 to 0.2 lb/scf of effluent gas under standard conditions of 1 atmosphere and 25°C, are then directly returned through the line **22** to the electrolytic cell **14**, to thus continually replace the melt constituents being lost therefrom.

The now depleted effluent gas, which still contains about 0.004 lb/scf effluent gas of aluminum chloride values in gaseous form, is then passed via line **24** to a condensation zone or condensing device **26**, which may be, for example, a shell-and-tube heat exchanger or a fluidized bed of aluminum chloride operated at a second and lower predetermined temperature range that is often less than about 100°C, to desublime the remaining aluminum chloride values thereof and effect a preliminary separation and transfer thereof via lines **28** to an appropriate collecting device, such as **30**, in crystalline form.

Such collected aluminum values may then be reintroduced via line **31** into the same or another electrolytic cell for electrolytic decomposition thereof. The now further depleted and relatively pure gaseous effluent is then conducted via line **32** to one or more bag filter units or assemblies **34**, by which any remaining or entrained solid impurities, and particularly those of small size, are removed.

The solids collected by bag filter unit **34** will essentially comprise aluminum chloride and, if of sufficient purity, may be returned via line **35** to the collector **30** for eventual reintroduction, if desired, into an electrolytic cell **14**.

The residual gaseous effluent from the filter **34**, now in the form of relatively pure chlorine gas and some nitrogen, may then be conducted via line **36** to a locus of utilization thereof. Such chlorine gas, possibly along with other chlorine values, may be introduced into an aluminum chloride preparation zone **38** in which it may be reacted with alumina-bearing material in the presence of a reducing agent, e.g., carbon. At least some of the chlorine values and carbon in such reaction may be employed in combined form, e.g., as carbon tetrachloride or carbonyl chloride. Alternatively, such chlorine may be conducted via line **38a** to a condenser **39** and the resultant liquid chlorine transferred via line **38b** to storage as at **40**. As indicated at **42**, the aluminum chloride produced in zone **38** may be utilized in the same, or in another electrolytic cell **14**.

CHLORINE FROM ILMENITE CHLORINATOR DUST

See "Iron Oxide from Ilmenite Chlorinator Dust."

CHLORINE FROM WASTEWATERS

A process developed by *R.W. Lynch; U.S. Patent 4,196,140; April 1, 1980; assigned to Olin Corporation* is one for recovering recoverable chlorine from chemical plant waste streams containing alkaline earth metal compounds.

An aqueous stream containing recoverable chlorine is reacted with an alkali metal hydroxide, such as sodium hydroxide, to form a slurry of solid particles of alkaline earth metal hydroxide, such as calcium hydroxide, suspended in a liquid. The calcium hydroxide is filtered or otherwise separated from the liquid.

The liquid is admixed with an organic alcohol to form an organic-aqueous solution. A halogenating agent, such as chlorine, is reacted with the organic-aqueous solution to form a solution of organic hypochlorite in an organic phase and an aqueous phase.

The solution of organic hypochlorite is phase separated to form aqueous and organic phases. The aqueous phase containing sodium chloride may be recycled for use as a reactant in a chlor-alkali electrolytic cell.

The organic phase containing organic hypochlorite may be used as a chlorinating agent or may be treated with an acid, such as hydrochloric acid, to reclaim free chlorine.

CHROMIC ACID FROM SPENT ETCHING SOLUTIONS

A process developed by *R.R. Dougherty; U.S. Patent 3,531,262; Sept. 29, 1970; assigned to Control Data Corp.* is one in which spent or used aqueous chromic acid etchants are treated to separate chromic acid from by-products of the etching process permitting reuse of the chromic acid and concentration or isolation of the by-products for waste disposal and/or recovery of the metal values contained therein.

In a typical process, chromic acid and water are separated from the by-products by acetone extraction. The extract is then stripped of the acetone to produce an aqueous solution of chromic acid which can be reused in the etching process. The raffinate is rich in the by-products of the etching process.

An aqueous solution of sulfuric acid and a hexavalent chromium ion-containing compound is often used as an etchant for removal of metal. This etchant is especially useful in dissolving copper from copper clad laminates in forming electronic printed circuit boards. During the etching process, the hexavalent chromium is converted to trivalent chromium and the dissolved copper forms as a copper sulfate in the etchant. With continued use, the rate of copper removal is slowed due to the increased concentration of the copper in the solution and the reduction of the hexavalent chromium until eventually the etchant must be replaced. The spent etchant presents a disposal problem due to the toxicity of the copper and the chromium.

A technique developed by *H.C. Hicks, Jr.; U.S. Patent 4,073,708; February 14, 1978; assigned to The Boeing Co.* is one in which a chromic-sulfuric etching solution for copper is regenerated in an etching container with a permeable diaphragm dividing the container into an etching chamber and a sulfuric acid-containing chamber. Electric current passes between an anode electrode located in the etching chamber and a cathode electrode in the sulfuric acid chamber to regenerate the etchant in the anode-containing chamber and to deposit copper on the cathode.

A suitable apparatus for the conduct of the process is shown in Figure 31.

The apparatus **10** has a tank **12** which is divided into compartments **14** and **16.** Compartment **14** contains an aqueous etching solution **18** of a hexavalent chromic oxide-sulfuric acid. The hexavalent chromic acid is obtained from an alkali metal dichromate or from chromic acid, but preferably from sodium dichromate. Compartment **16** contains an aqueous solution **20** of sulfuric acid or of an alkali metal sulfate.

The tank is divided into compartments by a barrier **22** which holds a permeable membrane or diaphragm **24**, communicating between the 2 compartments.

Figure 31: Boeing Co. Apparatus for Regeneration of Chromo-sulfuric Acid Etchants

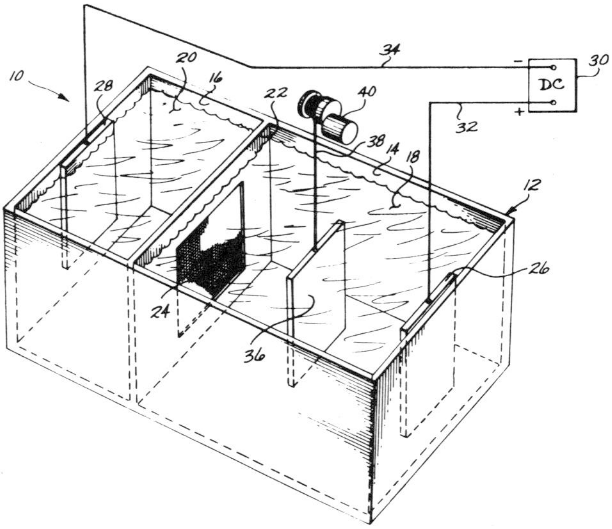

Source: U.S. Patent 4,073,708.

This ion selective member preferably is of polyester or perfluorosulfonic acid; however, it is not desired to limit it to those materials as any material resistant to the chemicals involved may be used.

The diaphragm preferably is of a nonwoven spun-bonded fabric having interstices to impart permeability to chromium ions of from 0.02 to about 17.4 g/sq ft/hr when a voltage is impressed across the diaphragm. An anodic electrode **26** is placed in compartment or chamber **14** which contains the etchant, and a cathodic electrode **28** is located in compartment **16** which contains the aqueous solution of sulfuric acid.

A direct current power supply **30** impresses a voltage to impart a dc current between the electrodes through conductors **32** and **34**. A copper or copper-containing article such as a circuit board **36** is inserted by use of cable **38**, which is controlled by a hoist **40** which is supported overhead by a means not shown.

In operation, the circuit board is lowered into the etchant **18** for a period sufficient to remove the desired amount of copper. As the copper is dissolved and goes into the solution, the following chemical reaction is taking place:

$$3Cu + 6H_2SO_4 + 2CrO_3 \longrightarrow 3CuSO_4 + Cr_2(SO_4)_3 + 6H_2O$$

The etchant may be regenerated after the circuit board has been removed or regeneration may be performed at the same time etching is taking place. It is preferred, however, to etch and regenerate at one and the same time as this makes it possible to maintain a nearly constant etching rate with a resultant better control of the etching. In either condition, the etching and the regeneration preferably take place in the same container. To regenerate, a direct current is passed through the electrodes and the following reactions take place:

$$Cr_2(SO_4)_3 + 6H_2O \rightarrow 2CrO_3 + 3H_2SO_4 + 3H_2\uparrow$$

$$2CuSO_4 + 2H_2O \rightarrow 2Cu + 2H_2SO_4 + O_2\uparrow$$

During regeneration, the trivalent chromic ion is regenerated back to the hexavalent state, the sulfuric acid is reformed and copper dissolved in the anodic chamber **14** passes through the membrane and is deposited on the cathodic electrode **28.** It is preferred to use a current density across the diaphragm of from about 110 to 150 A/sq ft of diaphragm area.

CHROMIC ACID FROM WASTE CHROMIC ACID SOLUTIONS

A process developed by *M. Watanabe and S. Nishimura; U.S. Patent 4,144,145; March 13, 1979; assigned to Solex Research Corp., Japan* is shown in Figure 32.

Figure 32: Solex Research Corporation Process for Recovery of Reusable Chromic Acid from Waste Chromic Acid Solution Containing Impurities

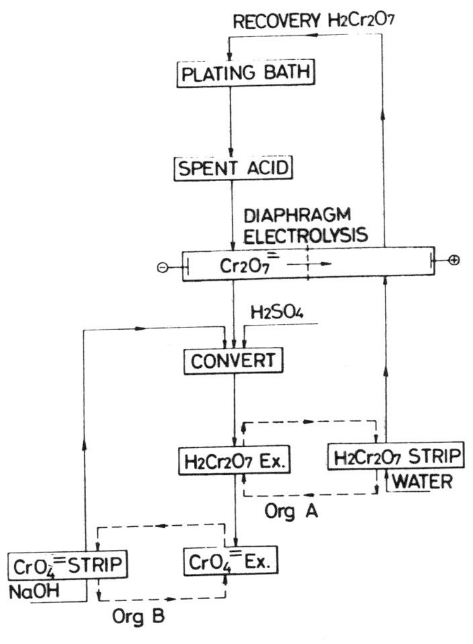

Source: U.S. Patent 4,144,145.

In the process hexavalent chromium values are recovered from an aqueous solution containing hexavalent chromium values and heavy metallic ions comprising:

(a) subjecting the aqueous solution to an electro-dialysis process wherein the aqueous solution containing the hexavalent chromium values is introduced into the cathode compartment of the process and free hexavalent chromium ions are transferred into the anode compartment of the process;

(b) removing the liquid in the cathode compartment from the electro-dialysis process and adding sulfuric acid thereto in an amount at least stoichiometrically equivalent to the amount of heavy metal ions therein.

(c) extracting the hexavalent chromium values in the acidified solution by liquid-liquid extraction using an organic solvent (A) containing one or more compounds selected from the group consisting of phosphoric acid esters together with a petroleum hydrocarbon as a diluent to produce an organic phase containing hexavalent chromium values and an aqueous raffinate;

(d) extracting any remaining hexavalent chromium in the aqueous raffinate by liquid-liquid extraction with an organic extractant (B) which contains one or more compounds selected from the group consisting of primary, secondary, tertiary and quaternary amines together with a petroleum hydrocarbon as a diluent; and

(e) regenerating extractant (A) and (B) by stripping the organic phase with a stripping reagent.

CHROMIUM FROM CHROMIC ANHYDRIDE PRODUCTION

As well known, in the process of producing chromic anhydride (CrO_3) from sodium bichromate and sulfuric acid, a waste residue is obtained which is highly polluting and can be disposed of only with great difficulty, essentially comprising sodium bisulfate, sulfuric acid and more or less significant values of trivalent and hexavalent chromium compounds.

When treated with water, this waste mass will form a slurry comprising a solution of sodium bisulfate, sulfuric acid, chromic acid and a substantially insoluble compound of trivalent chromium of a varying composition, which in the following will be referred to as chromium bisulfate.

Because of its high degree of acidity, due to the presence of sulfuric acid and sodium acid sulfate (or bisulfate), the slurry thus obtained could be used, through recycling to the processing of sodium bichromate, as acidificant in conversion of sodium chromate from standard chromite processing, into sodium bichromate.

However, this could not hitherto be carried out due to the presence of water-soluble trivalent chromium. Thus, even assuming a possible economical separation by filtration of the insoluble fraction (chromium bisulfate), the remaining solution containing soluble trivalent chromium would anyhow contaminate the sodium bichromate line of production, giving rise to serious disadvantages, such as alterations in the usual characteristics of the final product, and difficulties in filtration.

These problems are avoided in a process developed by *D. Perrone; U.S. Patent 4,045,340; August 30, 1977; assigned to Luigi Stoppani di P. Stoppani & Co.* This process provides complete separation of trivalent chromium in the waste, through precipitation in the form of a compound having the formula

$$xCrO_3 \cdot yCr_2O_3 \cdot nH_2O$$

which will be referred to as chromium chromate, which compound can be readily and completely filtered and washed with water and which is also soluble, particularly in acids.

The slurry comprising a water suspension of solid waste residue from chromic anhydride processing is subjected to a first reaction stage with sodium chromate solution at a constant flow rate and such a density as to avoid precipitation of sodium sulfate, setting pH to 2 to 8 by adjusting the slurry flow rate, and finally separating by filtering insoluble chromium chromate formed from the sodium bichromate and sodium sulfate solution.

CHROMIUM FROM COOLING TOWER BLOWDOWN WATER

In many industrial operations it is necessary to remove large amounts of heat and circulating water is commonly used as a heat transfer medium. Various chemical treating agents are added to the water used for this purpose including chromium-containing corrosion inhibitors in the form of chromate salts. In operation, the chemically treated process water is circulated through heat exchangers at the heat source and then passed to a cooling tower where it is cooled and recycled.

Since evaporation occurs in the cooling tower, there is a resulting increase in concentration of harmful scaling compounds and other dissolved solids present in the process water due to the losses in water volume. In order to maintain the solids concentration at acceptable levels, a portion of the process water is continuously withdrawn from the cooling tower and discharged as waste.

This flow of waste water is generally referred to as the blowdown stream. It is, of course, necessary that fresh make-up water be added to the system to compensate for the volume lost by evaporation and blowdown. Also, chromate salts and other treating agents must be added to the make-up to compensate for the losses incurred in the blowdown flow.

The need to replace the relatively expensive corrosion inhibitors lost in the blowdown obviously adds to the cost of the cooling operation. In addition, the relatively high concentration of these materials in the blowdown stream causes difficult disposal problems. The reason for this is that hexavalent chromium is known to be toxic and its discharge into sewers and waterways has come under strict environmental control.

A process by *A. Geinopolos, M.K. Gupta and W.J. Katz; U.S. Patent 3,625,885; Dec. 7, 1971; assigned to Rex Chainbelt Inc.* is one in which process water containing hexavalent chromium is pretreated to bring it to a state of dichromate dominance and the water is then subjected to reverse osmosis using a loose membrane.

The membrane allows passage of most of the dichromate ions with the product water while concurrently effecting a significant reduction in the hardness and dissolved solids levels. The softened product water containing dichromate ions is recyclable with the recirculated process water and the waste fraction contains substantially lower amounts of potentially harmful chromium.

In a process developed by *T.A. Orofino; U.S. Patent 3,986,950; October 19, 1976; assigned to Monsanto Co.,* a dialysis method for selectively removing chromates from cooling tower blowdown streams while retaining calcium and magnesium scaling compounds is provided. Preferential transport of chromates through the walls of hollow fiber assemblies is achieved by control of the chromic acid dissociation equilibrium, through feed and product pH adjustment. In a preferred method, blowdown waste water is brought into contact with the exterior surface of the fibers while chromate-free make-up water is simultaneously circulated through the fiber bores. The permeating species is collected by the makeup water within the fiber bores for re-use in corrosion inhibition.

CHROMIUM FROM ELECTROPLATING RINSE WATERS

Waste waters containing chromium, e.g., in the form of chromic acid, metallic chromate salts, etc., can no longer be discharged directly into sewers, rivers or streams due to various legal requirements imposed by the environmental authorities. Also, chromium is an expensive metal and the treating solutions and salts are desirably recovered from the economic point of view. An economical process for not only removing chromium from waste waters but also recovering it efficiently has long been desired.

A process developed by *E.J. Feltz and R. Cunningham; U.S. Patent 3,969,246; July 13, 1976;* is one for removing and recovering chromium from waste water in the form of chromic acid and/or metallic chromate salts by the direct precipitation of chromium using barium carbonate. The process is carried out in aqueous solutions acidified with glacial acetic acid at an acidic pH preferably ranging from 4.5 to 4.7 followed by filtering the resultant chromium material with an acid resistant filter media having openings preferably in the 2 to 4 μ size range and acid resistant within the pH range of 2 to about 5. The weight ratio of barium carbonate to chromium material contained in the waste liquid is preferably 2:1 and the preferable weight ratio of the barium carbonate to the acetic acid is 3:1. The barium carbonate and acetic acid are freshly mixed in aqueous media prior to either addition to the chromium waste liquid or addition of the chromium waste liquid to the freshly prepared treating material in aqueous media.

Such a process is shown in block diagram found in Figure 33.

As a result of a chromium plating process, the plated parts are rinsed in one or more rinse tanks, the last of which is usually a hot rinse tank.

As parts are rinsed in the rinse tank(s) **1**, the solution therein builds up with contaminating chromium ions. In this process, the contaminated rinse water from one or more rinse tanks is removed for treatment through line **2** or by free flow from the rinse tank(s) to treatment tank. The rinse water in the treatment tank is then treated with the barium carbonate and acetic acid constituents in the proper concentration relative to the amount of chrome in the contaminated water as indicated.

Figure 33: Cunningham Process for Chromium Removal and Recovery from Plating Wastewaters by Barium Precipitation

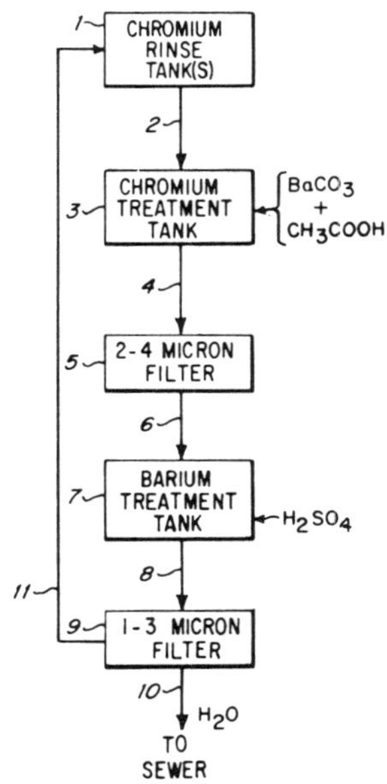

Source: U.S. Patent 3,969,246

The treatment tank **3** is equipped with conventional agitation, preferably air, and the agitated solution in the treatment tank is then fed through line **4** to a 2 to 4 μ filter **5**. If further processing is required, the filtered effluent is then fed through line **6** to an additional holding tank **7** or self-contained filtering system where sulfuric acid or similar material is added to precipitate the slight remaining barium, e.g., as barium sulfate, as indicated by the arrow.

The resulting effluent is fed through line **8** to a 1 to 3 μ filter **9** to capture the barium sulfate. The resulting effluent can be removed directly to the sewer, river, or stream through line **10** or, in a continuous operation, back to the original rinse tank(s) through line **11** for reuse.

A recovery process developed by *Y. Senoo, U.S. Patent 3,961,029; June 1, 1976; assigned to Tokico, Ltd., Japan* is one in which the chromate ion can be advantageously recovered in the form of chromic acid by the process comprising the steps of adding a reducing agent to the chromate ion-containing aqueous waste liquor to precipitate the chromate ion in the form of chromium (III) hydroxide, adding an alkali and aqueous hydrogen peroxide to the precipitated chromium (III) hydroxide to produce a chromate ion-containing aqueous solution; and passing the resultant chromate ion-containing aqueous solution through an H-form acidic cation-exchange resin.

A process developed by *S. Takada and K. Sato; U.S. Patent 4,049,772; Sept. 20, 1977; assigned to Tokico, Ltd., Japan* is one in which a chromic acid solution can advantageously be recovered from the waste water by a process which comprises passing the waste water successively through an anion-exchange column system consisting of two or more serially interconnected anion-exchange resin columns each packed with an anion-exchange resin thereby deionizing the waste water.

In the course of this deionization treatment, the first anion-exchange resin column is backwashed when the anion-exchange capacity thereof has fallen to the point where the influent to and the effluent from the first anion-exchange resin column have become substantially equal in terms of chromate ion content.

In the course of the regeneration treatment, the waste water is diverted to enter the anion-exchange column system through the second anion-exchange resin column. After the regeneration treatment has been completed, the first anion-exchange resin column is put back into service in the anion-exchange system so as to serve as the last stage.

Thereafter the procedure is repeated with respect to the successive anion-exchange resin columns composing the system thereby permitting passage of the waste water to be carried out continuously. The eluate which results from the regeneration treatment is recovered as a reusable chromate solution.

A process developed by *T. Hayashi; U.S. Patent 4,012,318; March 15, 1977; assigned to Kayabakoyo-Kabushiki-Kaisha, Japan* is one for the recycle treatment of chromium plating waste water (obtained by washing the chromium plated products) with ion exchange resins. For the purposes of improving the rate of recovery of chromic acid, of extending the life of the ion exchange resins, and of preventing the formation of sludge, the chromium-plating waste water is adjusted almost to neutrality by the addition of an alkaline substance so as to precipitate the metal components contained therein in the form of hydroxides, which are then removed by filtration.

The filtrate so obtained is kept neutral during the subsequent treatment for the removal of alkaline components by adsorption in a pretreatment tower packed with cation exchange resins. The dilute dichromic acid solution obtained is also treated for concentration in an adsorption tower packed with anion exchange resins and a purification tower packed with cation exchange resins so as to recover pure dichromic acid solution. The pretreatment tower and purification tower are regenerated by treatment with sulfuric acid.

The flow diagram of the process is shown in Figure 34.

The chromium plated products in the plating tank **10** are first washed with water in the primary washing tank **12** and again washed with water in the secondary washing tank **14**. The water used for washing the products is first sent to the secondary washing tank **14**, the washing water which has become rich in chromic acid in the secondary washing tank **14** is then transferred into the primary washing tank **12**, and the washing water which has become even richer in chromic acid in the primary washing tank **12** is then sent to the storage tank **16** for storage. In general, the chromium plating waste water in the storage tank **16** is acidic, with a pH value of 2.3 to 3.

Figure 34: Kayabakoyo-Kabushiki-Kaisha Process for the Recycle Treatment of Waste Water from Chromium Plating by Ion Exchange

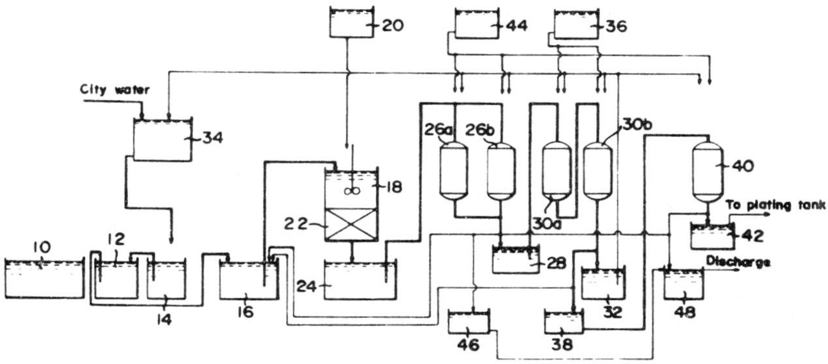

Source: U.S. Patent 4,012,318.

The chromium plating waste water stored in the above-mentioned storage tank **16** is transferred into the neutralization tank **18** where an aqueous sodium hydroxide solution is added for neutralization from the sodium hydroxide tank **20** to adjust the pH value to almost neutral, or pH 7 to 9, in order to precipitate out the iron, copper and other metallic components in the form of their hydroxides. The mixture is passed through the filter **22** positioned below the tank **18**, and the filtrate is transferred into the filtrate tank **24**. Through such treatments the metal ions in the chromium plating waste water are replaced by sodium ions, which hardly contaminate the cation and anion resins.

The filtrate in the filtrate tank **24** is, while being kept almost at neutrality, or pH 7 to 8, then allowed to pass through the pretreatment towers **26a** and **26b** packed with cation exchange resins to remove sodium ions by adsorption so as to obtain a dilute dichromic acid solution, which is temporarily stored in the buffer tank **28**.

The purpose of lowering the pH value in the dilute dichromic acid solution and removing the sodium ions from the filtrate in the filtrate tank **24** by adsorption over cation resins is to improve the adsorption efficiency of the subsequently used anion resins for chromic acid ions; in particular, when the pH value is below 3.5, the chromic acid ion is changed from $CrO_4^=$ to $Cr_2O_7^=$, and thus the adsorption efficiency of the anion resins is improved by about 30%.

While the pH is kept below 3.5, the dilute dichromic acid solution in tank **28** is allowed to pass through the adsorption towers **30a** and **30b** packed with anion resins to remove dichromic acid ions; the effluent from the adsorption towers is stored in the primary water recovery tank **32**. The treated water has thus become an innocuous liquid and can be reused via the secondary water recovery tank **34**.

When the adsorption capacity of the adsorption towers **30a** and **30b** has been reached, the following operations are performed:

(1) Purging – The water remaining on the anion resin particles is first blown out by passing air over them.

(2) Prewashing – The resins are then washed by passing about 20 volumes of water per volume of resins downwardly and upwardly through the latter. As washing water, the recycled water in the primary water recovery tank **32** is used.

(3) Purging – Air is again passed through the resins so as to purge the water remaining among the resin particles, and all such water as well as the washing water is returned to the storage tank **16**.

(4) Reagent injection – An amount of aqueous sodium hydroxide solution equivalent to the amount of resin is injected from the caustic soda tank **36** per volume of the resins so as to desorb the dichromic acid ions adsorbed on the anion resins in the form of a sodium chromate solution.

(5) Purging – Air is then blown in so as to purge the sodium chromate solution remaining on the resin particles and all of the solution is transferred into and stored in the primary solution recovery tank **38**.

(6) Afterwash – In a manner similar to that described above for the prewashing, the resins are washed by passing about 20 volumes of the treated liquid stored in the primary water recovery tank **32**, per volume of resins upwardly and downwardly through the latter.

(7) Purging – Air is then blown in so as to purge the water remaining on the resin particles, and the washing water as well as this water is returned to the storage tank **16**.

(8) Water filling – Water is then added up to a level just sufficient to cover the resin beds in the adsorption towers **30a** and **30b**.

Through the treatments described, the dichromic acid ions adsorbed on the anion resins are removed as sodium chromate solution and admitted into the primary solution recovery tank **38**; concurrently, the anion resins are washed to regenerate the towers. Then the sodium chromate solution in tank **38** is fed into the purification tower **40** packed with cation resins, in a manner similar to that described for towers **26a** and **26b**, to obtain pure dichromic acid solution by adsorption of the sodium ions, and this dichromic acid solution is then transferred into the secondary solution recovery tank **42** and used as a plating solution in the plating tank **10**.

Further when the adsorption capacities of the pretreatment towers **26a** and **26b** and the purification tower **40** are reached, the following operations are performed, as in the case of the adsorption towers **30a** and **30b** as described.

(1) Purging – Air is blown in so as to purge the dichromic acid solution remaining on the cation resin particles.

(2) Prewash – About 20 volumes of water per volume of the cation resins are then passed upwardly and downwardly through the latter for washing. As the washing water, the treated solution in the previously described primary solution recovery tank **38** is utilized.

(3) Purging – Again, air is blown in so as to purge the washing water remaining on the resin particles, and all of the above dichromic acid solution and the washing water is returned to the tank **16**.

(4) Reagent injection – A volume of sulfuric acid equivalent to the amount of resin is supplied from the sulfuric acid tank **44** so as to desorb the sodium ions adsorbed on the cation resins and is removed in the form of innocuous sodium sulfate solution.

(5) Purging – Air is then blown in so as to purge the sodium sulfate solution remaining on the resin particles.

(6) Afterwash – As in the prewash, about 20 volumes of the treated solution in the primary water recovery tank **32** is used for the upward and downward washing of the resin.

(7) Purging – Air is then blown in so as to purge the washing water remaining on the resin particles, and the sodium sulfate solution and the washing water are stored in the waste sulfuric acid tank **46**; this is discarded after adjusting the pH value in the pH control tank **48**.

(8) Water-filling – Water is then added to the pretreatment towers **26a** and **26b** and the purification tower **40** to a level just sufficient to cover the resins.

Through the treatment described above, sodium ions adsorbed on the cation resins are recovered in the form of sodium sulfate solution, and the pretreatment towers **26a** and **26b** and the purification tower **40** are regenerated through washing of the cation resins.

In the example shown, two of the pretreatment towers **26a** and **26b** are installed in parallel in order to make continuous operation possible by using one for adsorption while the other is being regenerated. The reactions in these pretreatment towers **26a** and **26b**, adsorption towers **30a** and **30b**, and purification tower **40** are shown below, the resin moiety being expressed as R.

In the following formulas, (a) is pretreatment tower; (b) is adsorption tower; (c) is adsorption tower (regeneration); (d) is purification tower; and (e) is regeneration of pretreatment and purification towers.

$$\text{(a)}\quad 2R{-}H + Na_2Cr_2O_7 \rightarrow 2R{-}Na + H_2Cr_2O_7$$

$$\text{(b)}\quad 2R{-}OH + H_2Cr_2O_7 \rightarrow R_2{-}Cr_2O_7 + 2H_2O$$

$$\text{(c)}\quad R_2{-}Cr_2O_7 + 2NaOH \rightarrow 2R{-}OH + Na_2Cr_2O_7$$

$$\text{(d)}\quad 2R{-}H + Na_2Cr_2O_7 \rightarrow 2R{-}Na + H_2Cr_2O_7$$

$$\text{(e)}\quad 2R{-}Na + H_2SO_4 \rightarrow 2R{-}H + Na_2SO_4$$

A process developed by *F. Tuznik and A. Lis; U.S. Patent 4,171,255; October 16, 1979; assigned to Instytut Mechaniki Precyzyjnes, Poland* is one for thorough neutralization of all plating bath solutions and effluents and for full recovery of used plating metals. Hexavalent chromium is reduced to trivalent chromium with production of chromium hydroxide. Metals such as zinc and cadmium are recovered as their respective hydroxides; silver and copper are recovered as free metals. Additionally, this process provides for neutralization of cyanide and chromic fumes exhausted from the reclaiming tanks.

The process involves immersing plated articles in a reclaiming bath containing an aldehyde or an aldehyde precursor, and optionally an oxidizing agent, to precipitate the excess metal in the residual plating bath in the form of free metal or metal hydroxide and to decompose the cyanide and to reduce the hexavalent chromium to trivalent chromium. The plating bath and reclaiming bath are regenerated.

CHROMIUM FROM NONFERROUS SLUDGES

See also "Metals from Mixed Metal Slimes".

A process developed by *C.O. Weiss; U.S. Patent 3,552,917; January 5, 1971; assigned to M & T Chemicals Inc.* involves recovering the chromium content from waste sludge produced from the separation of precipitated solids from the waste effluent streams of metal finishing processes, the chromium content of the sludge being in the form of barium chromate. The above is achieved by forming an aqueous slurry of the sludge, adding sulfuric acid thereto, agitating the resulting admixture for a period of time at ambient temperature, separating the insolubles therefrom, adding to the filtrate from the separation an amount of a carbonate selected from the group consisting of Ca^{++}, Sr^{++}, Ba^{++}, and Pb^{++} to precipitate out a substantial portion of the sulfate ion from the solution, again separating the insolubles from the reacting environment, adding to the filtrate of the second separation step strontium carbonate to precipitate out the remaining sulfate ion, and separating the insolubles from the reacting environment to obtain a solution containing all the CrO_3 content from the initial extraction.

Example: In this example, a representative sludge was prepared of a 55 kg sample of a mixture of $BaCrO_4$ (40%) and $BaCO_3$ (60%). This mixture was slurried with water to give a readily stirrable mixture. To the slurry was added 25.1 kg of H_2SO_4 (13.9 ℓ), and the admixture volume was adjusted to about 85 ℓ. The mixture was agitated with no external heating for 2½ hours.

The insolubles were then filtered off and the cake washed with several portions of water until the solution analyzed 5.5 g/100 ml of CrO_3 and 7.8 g/100 ml of H_2SO_4. The solution was then charged with 8.0 kg of $CaCO_3$ and agitated at room temperature for about 10 minutes. The insolubles were filtered off and the cake washed as described above to give a solution volume of about 115 ℓ. This solution still contained 2.0 kg of H_2SO_4, and this solution in turn, was charged with 4.5 kg of $SrCO_3$ and agitated at room temperature for about 15 minutes.

The insolubles were filtered off and the cake washed to produce a final volume of 130 ℓ. The resulting solution contained no H_2SO_4 and virtually all the CrO_3 obtained in the initial extraction.

Thus the process is effective for obtaining all of the chromium content from a slurried mixture containing barium chromate solids and with the solution obtained having substantially no sulfuric acid contained therein, thus making the final solution from the process particularly appropriate as a raw material for recirculation into a chromium electroplating process.

The recovery is performed in the absence of any high temperature operating conditions which reduces the need for expensive evaporative equipment, and in the presence of relatively dilute solutions for carrying out the various extraction steps so as to eliminate the requirement of expensive corrosion-resistant equipment.

A process developed by *W. Müller and L. Witzke; U.S. Patent 4,151,257; April 24, 1979; assigned to Th. Goldschmidt AG, Germany* is one for separating all the various metals in mixed hydroxide sludge wastes containing these by first chlorinating the aqueous waste to oxidize the chromium therein, separating the insoluble components therefrom and treating the resulting Cr(VI) solution in a fixed bed anion exchanger to separate the Cr(VI) from the solution. The copper is then separated from the aqueous solution by liquid-liquid extraction, the zinc from the copper-free solution by liquid-liquid extraction. The aluminum is precipitated and separated in the form of hydroxide from the zinc-free solution and the nickel, then separated from the aluminum-free solution by liquid-liquid extraction. The nonferrous metals are recovered from the respective solutions and precipitates by conventional procedures.

Another process for treating nonferrous metal hydroxide sludge waste to separate and recover the nonferrous metals therein has been described by *L. Witzke and W. Müller; U.S. Patent 4,162,294; July 24, 1979; assigned to Th. Goldschmidt AG, Germany.* The waste is first mixed with an alkali hydroxide or carbonate, dried, and then roasted and the water-insoluble residue in the roasted material is treated with sulfuric acid to produce an acid-insoluble residue and a sulfuric acid solution containing copper, aluminum and chromium and thereafter, separating the copper from this solution by liquid-liquid extraction and then, separating the aluminum, chromium, zinc and nickel from the remaining solution by separate liquid-liquid extraction steps.

CHROMIUM FROM SCRAP CHROMIUM

Chromium-containing alloys and industrial wastes return small quantities of secondary chromium to the market. However, only stainless steel scrap is a sizable source of supply. In 1974, stainless steel producers shipped about 63% of production. Neglecting inventory changes, most of the balance was returned to the melting cycle as home scrap, which is a permanent recirculating load. Purchased scrap, which is considered new supply, represented 10% of domestic chromium demand in 1974. Most of this material was secured from fabrication plants of manufacturers of end products or from dealers in obsolete equipment.

CHROMIUM FROM TANNERY WASTES

Hides are subjected to treatment with various tanning compounds to preserve the structure of the hide and prevent putrefaction. Trivalent chromium, typically hydrated basic chromium sulfate, is widely used. The trivalent chromium is often prepared at the tannery by reducing hexavalent dichromate with sulfuric acid and sugar.

The loss of soluble compounds in tannery effluents has gradually increased over a period of years because it was cheaper to buy fresh supplies of chrome salts than to incur the expense of chrome recovery.

Another factor contributing to this discharge of chromium salts was the desire to speed up the tanning process by forcing penetration of chromium salts by using higher concentrations of chromium salts. This results in an increased tanning rate with commensurate economics in equipment cycling and labor per unit of production. At the same time, however, it increases the amount of chromium discharged in the tannery effluent as well as the percentage of chromium lost from the leather making process.

From a number of surveys made for various tanneries, it has been found that the most efficient tanneries use about two-thirds of the total chromium purchased and discard one-third in their plant effluent. Less efficiently managed tanneries may discard up to one-half of the chromium purchased.

Large quantities of chromium are also carried away with leather waste produced as scrap from trimming, shaving, buffing and the like. Such losses can amount to as much as two-thirds of the chromium employed in the process. Such loss is also an environmental hazard since all the ions of chromium are toxic, particularly the hexavalent ions. A typical tannery processing from 2,000 to 3,000 hides per day will lose about 2,500 lbs of chromium (as Cr_2O_3) per day, about 50 to 60% in the waste water and about 40 to 50% in the leather scrap.

A continuous chromium recovery process for application to tannery wastewaters has been developed by *H.H. Young; U.S. Patent 3,950,131; April 13, 1976; assigned to Hoffman-Stafford Tanning Co.* The method includes continuous agitation of chrome sludge along with continuous filtration and movement through successive filtering zones. This method is both economically attractive and vital from an ecological standpoint. The reclaimed chromium values are in condition for reconstitution into fresh chrome-tanning liquors.

Such a process is shown in Figure 35.

In the drawing, **10** is the sump or well into which all spent chrome liquor is conveyed as collected in various parts of the plant. It is pumped by pump **12** over the Hydrasieve **14** (a cascade sieve screen), screenings (solids) being caught on an endless belt conveyor **16** for discharge into a receptacle (not shown). The screened liquor flows into a separator **18** where fat and grease accumulate at the surface, eventually overflowing through port **20**. Any sediment sinks to the cone bottom **22** and is drawn off through line **24** by pump **26** and returned to sump **10** for rescreening.

The intermediate layer of clear spent chrome liquor in tank **18** exits under the baffle **28** into a mixing tank **30** where liquid caustic soda or other suitable alkali solution is introduced from feeder **32** to adjust the pH in tank **30** to the range of 8.0 to 9.5. If the concentration of precipitated chromium hydroxide is high enough to preclude settling in the clarifier **34**, it by-passes directly to the continuous pressure filter **36**. But if the suspension of chromium hydroxide is sufficiently dilute to permit settling, it is detained in the clarifier **34** so as to permit the precipitate to collect in the cone bottom.

Figure 35: Hoffman-Stafford Tanning Co. Continuous Method for Reclaiming Chromium Hydroxide from Spent Chrome Tanning Liquors and Reuse Thereof in Subsequent Tanning

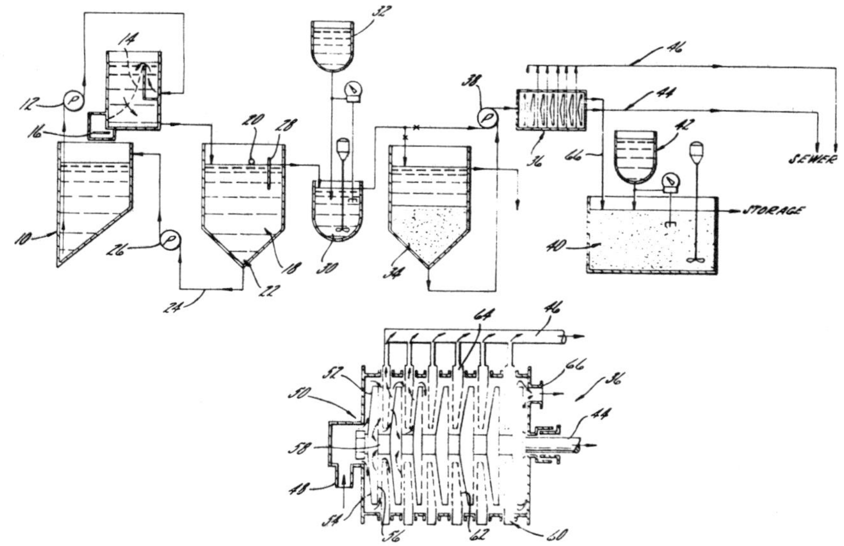

Source: U.S. Patent 3,950,131.

Simultaneously, clear supernatant salt solution is decanted from the clarifier at predetermined levels by a simple decanting line (not shown), thereby reducing to a minimum the volume of chromium hydroxide suspension which must be filtered. In either case, the suspension is pumped under pressure by pump **38** into the continuous filter **36** where the filter cake is maintained at a constant thickness and the chromium hydroxide is concentrated to a predetermined concentration.

At this desired concentration (14 to 15%), the concentrated paste of chromium hydroxide is ejected into holding tank **40** where a proportionate quantity of sulfuric acid from feeder **42** is mixed in until the desired basicity is achieved and the precipitate has been redissolved. The clear filtrate is conveyed through lines **44** and **46** to the plant sewers, having less than two ppm of total chromium equivalent.

The lower detail view in the figure is an enlarged view of the filter **36** as depicted in the flow diagram. As shown this embodiment of a multisegment continuous pressure filter has an inlet port **48** feeding into a housing **50**. Within the housing is a series of rotary filter plates **52** comprised of a metal disk **54** adapted to have a filter surface **56**. Liquid passes through the filter **56** and out through a collector pipe **58**. Stationary plates **60** are held within the filter housing and consist of circular hollow frames having filter surfaces **62** through which filtrate flows and is withdrawn through collector lines **64**. Filtrate passes through discharge lines **44** and **46** to the sewer. Filter cake ejects from the line **66** which may have a pressure actuated valve to eject when pressure builds up to a desired level.

A process for handling both liquid and solid tannery wastes to maximize chrome recovery has been developed by *B.H. Jones; U.S. Patent 4,086,319; April 25, 1978.* It minimizes environmental hazards, is relatively economical to operate, utilizes the fuel value of the organic material in such wastes, and recovers all or a substantial portion of the waste chromium for reuse, or renders it to a stable form with a 90 to 95% reduction in volume.

Such a process is shown in Figure 36.

Figure 36: Jones Process for Recovery of Chromium from Mixed Tannery Wastes Including Leather Scrap

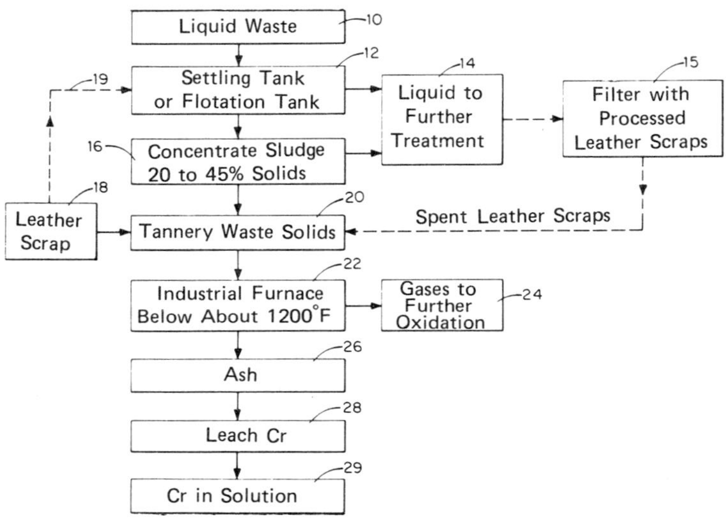

Source: U.S. Patent 4,086,319.

Referring to the flow chart illustrated liquid waste indicated at **10** is passed to either a settling or flotation tank indicated at **12**. The liquid waste **10** can comprise the liquid from the chrome tanning operations alone, but preferably is a mixture of such tanning wastewater with the wastewater from the Beamhouse (hide dehairing) operations which produce a $Cr(OH)_3$ precipitate. While only the tanning water contains chromium, both liquid effluents must be treated and are conveniently treated in a single system in which the acidic tanning wastes and the alkaline Beamhouse wastes are at least partially self-neutralizing.

The settling of suspended material and the precipitation of some dissolved material is enhanced in the separation tank by addition of settling and precipitating agents, such as the ground leather scraps indicated by the broken line **19** or by polymers which are well known for this purpose. Separation using flotation principles is also known.

Following separation, the separated sludge is removed. As illustrated at **14**, the supernatant liquid is withdrawn and preferably given additional treatments to purify it before discharge. Secondary and tertiary treatments are well known. Alternatively, the need for such treatments can be reduced or eliminated by passing the liquid through a filter body of processed leather scrap, either ground leather scrap or carbonized leather scrap, as indicated at **15**.

The sludge initially separated from the settling tank **12** normally contains from 85% to 95% water and is too wet to support combustion. While it can be dried in the furnace, this requires addition of fuel and it is preferred to concentrate it prior to introduction to the furnace. Concentration can be performed by evaporation, but the wet sludge is preferably mechanically dewatered by compaction, centrifuging, or the like.

Compacting is preferred by use of a filter press or belt press. Such concentration of sludge or mixture of sludge and ground leather scraps can produce material which will support combustion, preferably 20 to 45% solids by weight, and more preferably 30 to 45% solids.

The concentrated sludge, indicated at **16**, can be separately incinerated or the leather scrap **18** separately incinerated, the leather scrap having typically from 40 to 50% of the available chromium for recovery and high fuel value. However the acidic leather scrap is preferably mixed with the alkaline sludge in one or more of the procedures previously described.

The tannery waste solids **20** are fed into an industrial furnace indicated at **22**. The furnace is operated at relatively low temperature, below about 1200°F, preferably between about 800° and 1200°F, and most preferably about 1000°F. The incombustible material including the trivalent chromium is reduced to an ash in the furnace while the organic material in the waste is converted to gases or a mixture of gases and carbon as previously described.

At the low temperatures employed, these gases are not completely oxidized and are removed from the combustion zone and subjected to further treatment to complete oxidation. They may also be subjected to additional treatments to remove other harmful ingredients. By thus employing low temperature combustion to produce an ash, and by subjecting the organic material and resulting gases to at least two oxidation steps, the evolution of chromium in the gases is minimized and the organic material can be burned to products, principally CO_2 and H_2O, which can be exhausted to the atmosphere. The secondary oxidation of the furnace gases is indicated at **24**.

The ash **26** containing the chromium in trivalent or hexavalent form as previously described is removed from the furnace. The hexavalent chromium can be leached therefrom at **28** to produce a chromium solution **29** by a leaching solvent. Water or aqueous acids such as nitric acid (HNO_3) or sulfuric acid (H_2SO_4) or mixtues thereof can be employed and the chromium recovered either in dilute or concentrated form as preferred for the operations of the particular tannery employing the process.

If the tannery operations permit, the chromium in solution may be used in the form obtained as a source of bichromate. However, if required, the chromium may further be treated in any suitable manner, for example, by concentration with heat or purification by filtration, precipitation, washing, redissolving or the like. Filtering is normally desirable. Heat from the furnace may also be recovered.

A technique whereby chromium can be recovered from various sources including tannery wastes involves magnesia precipitation as described by *D.C. Hemming, R.E. Hahn and J.W. Robinson, Jr.; U.S. Patent 4,108,596; August 22, 1978; assigned to Revere Copper and Brass, Inc.*

The process comprises contacting an acidic solution containing sulfate ion and trivalent chromium ion with at least about 4 molar equivalents of MgO or $Mg(OH)_2$ per 3 molar equivalents of trivalent chromium in addition to the amount required to neutralize the free acid to a pH of about 4 to form an amorphous, dense solid, grainy, easily settleable, trivalent chromium-containing precipitate in an alkaline solution according to the equation:

$$3Cr_2(SO_4)_3 + 8MgO \longrightarrow Cr_2(OH)_4SO_4 \cdot 4Cr(OH)_3 \cdot 4H_2O + 8MgSO_4$$

The trivalent chromium-containing product can be readily separated from the water and is a useful source of chromium for subsequent processing. The water remaining after separation of the precipitate contains less than 0.5 mg/ℓ chromium.

The following is an example of the operation of such a process. A 4 liter sample of exhausted tanning liquor containing 4.66 g/ℓ Cr at a pH of 3.6 was filtered to remove skin fragments and other extraneous matter accumulated during processing. A quantity (21.6 g) of MgO (10% by wt excess) was added with constant agitation. The resultant pH was 9.5 and a grainy precipitate rapidly formed and settled. The precipitate was easily vacuum filtered (3" Hg) and the filtrate contained 0.062 mg/ℓ of Cr. The filter cake was transferred to a 1 ℓ container and was dissolved in H_2SO_4. The concentrated acidic chromium solution had a pH of 2.5 and a volume of 290 ml. The solution (evaporated to final volume of 233 ml) analyzed as 82.1 g/ℓ of Cr and 1.2 g/ℓ Mg.

CHROMOXYFLUORIDE CATALYSTS FROM SPENT CATALYSTS

Fluorination catalysts belonging to the group of chromoxyfluorides obtainable, e.g., by fluorination of hydrated chromic oxides with hydrogen fluoride or by heating water-containing chromic trifluoride in the presence of oxygen are excellently suitable for the fluorination of chlorinated and/or brominated aliphatic hydrocarbons in the gaseous phase.

These catalysts have the disadvantage that their exchange activity gradually decreases and that the amount of hydrogen fluoride required to reactivate them increases, when they are contacted for a prolonged period of time with halogen-containing carbon compounds, especially those which contain in addition hydrogen and/or oxygen.

Regeneration with oxygen is not long-lasting. Regeneration with chlorine requires excessively high temperatures which damage the catalyst. Regeneration with HF may require high temperatures and may lead to gradual deactivation.

A process developed by *S.P. von Halasz and J. Korinth; U.S. Patent 4,145,311; March 20, 1979; assigned to Hoechst AG, Germany* uses fluorine as a reactivating agent. It can be carried out at moderate temperatures.

It enables even highly deactivated catalysts, which could be activated only with difficulty hitherto, to be reactivated economically and be used again in the industry, with a low expenditure of material and energy.

COBALT FROM SCRAP COBALT

Foreign processors recover significant quantities of cobalt contained in mixed scrap. However, with rising price of cobalt, domestic interest has heightened. There are only a few significant sources of cobalt-bearing scrap. One of the most important is waste material discarded in forming and fabricating cobalt-bearing superalloys. Moreover, high temperature alloy components of jet engines are replaced periodically, providing a major source of recycled material.

A significant new source of secondary meterial could be spent petroleum catalysts, those used in desulfurization of crude oil. In 1978, one U.S. corporation attempted to recover cobalt from these catalysts but was unsuccessful, while another began producing significant quantities of cobalt oxide from other chemical wastes.

Another significant development involving recycling was a 1977 announcement by GTE Sylvania of plans to produce ultrafine cobalt powder, by a proprietary method, for use in cemented carbide cutting tools, using secondary material as feedstock. The new operation is housed in a 25,000 sq ft addition to the company's existing plant in Towanda, Pa. Production began in 1978.

Prior to this development, this special powder was mainly produced by Metallurgie Hoboken Overpelt in Belgium. Relatively small quantities were also produced in France and West Germany. Several companies are using a process developed and patented by the Bureau of Mines to recover the cobalt binder from cemented carbide scrap. In the process, molten zinc is used to dissolve the cobalt. The molten zinc is then recovered by distillation, and refractory cobalt and the carbides remain in a mass that can be recycled.

One company has used this process to treat 750 lb per day of scrap in a specially designed furnace(16).

COBALT FROM SINTERED METAL CARBIDES

See "Tungsten from Sintered Metal Carbides."

COBALT FROM SPENT CATALYSTS

A process developed by *Y. Ichikawa; U.S. Patent 3,525,762; August 25, 1970; assigned to Teijin Limited, Japan* involves recovering cobalt salt catalysts of lower aliphatic monocarboxylic acids of 2 to 4 carbon atoms, and comprises heating such cobalt salt catalysts at a temperature from 50 to 300°C in the presence of an aqueous solution of a lower aliphatic monocarboxylic acid of 2 to 4 carbon atoms in which (a) not less than 2 mols of the aliphatic monocarboxylic acid, and not less than 4 mols of water, per gram-atom of the cobalt are present, and (b) the concentration of the aliphatic monocarboxylic acid in the total aqueous solution present in the system is not less than 0.2 mol % to not more than 80 mol %.

Such cobalt salt catalyst is one used for the oxidation of hydrocarbons or their oxidation derivatives with molecular oxygen, the used catalyst being in the form of a trivalent form of the cobalt salt.

A process developed by *P.J.V. Jones; U.S. Patent 4,162,991; July 31, 1979; assigned to Imperial Chemical Industries Ltd., England* is one in which cobalt catalyst values are removed from mother liquors remaining from oxidations of aromatic hydrocarbons to mono- or polycarboxylic acids (such as xylene to terephthalic acid) in lower aliphatic monocarboxylic acids in the presence of catalysts comprising cobalt and bromide ions.

The process is carried out by:

(1) if necessary, adjusting the cobalt and bromide concentrations so that they are within specified limits;

(2) adsorbing cobalt and bromide ions (differentially with respect to extraneous ions) on a strongly basic anion exchange resin; and

(3) desorbing cobalt and bromide ions with lower aliphatic monocarboxylic acid containing more than 10% by wt of water.

A process by *S. Toida, A. Ohno and K. Higuchi; U.S. Patent 4,145,397; Mar. 20, 1979; assigned to Marubeni Corp. and Fuji Fine Chemical Co., Ltd., Japan* is one in which valuable molybdenum, vanadium, cobalt and nickel are recovered easily from roasted product of spent hydrodesulfurization catalysts and a high percent recovery achieved by means of a combination of simple chemical procedures and also inexpensive chemicals. The recovered metal components can be reused for preparing new catalysts and the like.

Figure 37 is a schematic flow diagram illustrating the processing steps for effecting hot water alkaline leaching of roasted used catalysts from hydrotreatment desulfurization of petroleum, to separate and collect mother liquor containing most molybdenum component and some vanadium and aluminum components and solids including most of cobalt and nickel components and some aluminum component, etc., respectively, from the leaching liquid, as well as the processing steps for separating and collecting molybdenum and vanadium components respectively, from the mother liquor.

Then, Figure 38 is a schematic flow diagram illustrating the processing steps for effecting hot water acid leaching of solids including most of cobalt and nickel components and some aluminum component, etc., to separate and collect cobalt component and nickel component respectively, from the leaching liquid.

Figure 37: Marubeni Corp. and Fuji Fine Chemical Co. Process for Spent Hydrodesulfurization Catalyst Recovery

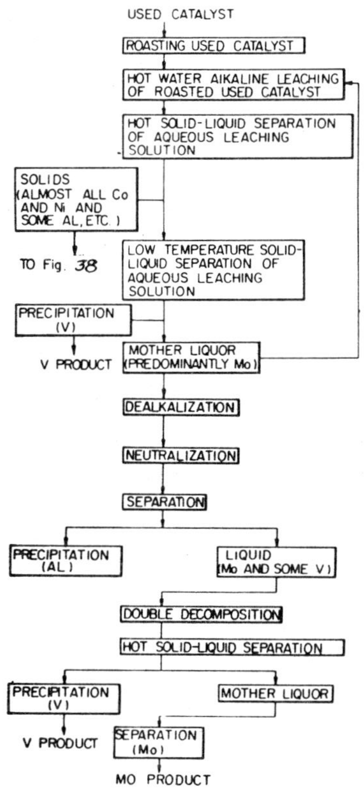

Source: U.S. Patent 4,145,397.

Figure 38: Portion of Marubeni Corp. and Fuji Fine Chemical Co. Process for Treating Solid Residue from Hydrodesulfurization Catalyst Recovery Process of Figure 37

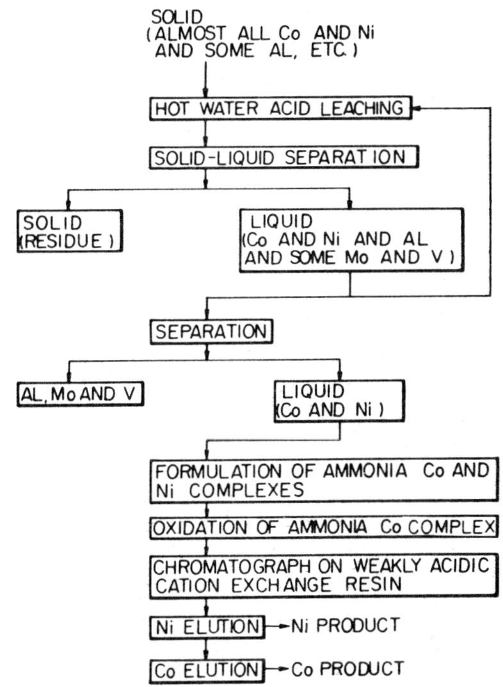

COLORED GLASS FROM MIXED GLASS WASTE

See also "Glass from Municipal Waste."

Although the particulate glass recovered by froth flotation from shredded municipal trash can be made clean enough for recycling, wide adoption of such a process is limited because the particulate glass is generally of mixed colors. For example, particulate glass thus obtained and contaminated with relatively small amounts of green glass, e.g., of 5 to 8%, can only be melted to a green or amber glass.

Although it was found that particulate glass can be color-sorted by magnetic separation, an economically feasible method employing this principle has not as yet been fully developed. Other magnetic sorting approaches using paramagnetic fluids to establish magnetically sensitive density gradients have also been tried without success.

A process developed by *T. Kumar and B.W. Morey; U.S. Patent 4,077,871; Mar. 7, 1978; assigned to Occidental Petroleum Corp.* is one for the color sorting of multicolored glass particles in particulate glass, especially particulate glass recovered by froth flotation from shredded municipal trash or waste, by a procedure which involves differential heating of particles of glass of a preselected color or colors, by light absorption of such glass particles, and contacting such glass particles with an organic thermoplastic material in the form of a coating on the glass particles or on a separate surface, causing melting of such organic thermoplastic material in contact with the thus differentially heated colored glass particles and permitting separation thereof from the remaining glass particles as by flotation or adhesion.

In one embodiment of this process, a mixture of particulate colored glass such as amber and flint glasses are coated with 1-tetradecanol, and the amber glass is differentially heated by irradiation to a temperature in excess of 38°C, causing melting of the tetradecanol, the temperature of the flint glass being insufficient to melt the coating. The mixture is placed in a froth flotation apparatus in which the amber glass containing the previously melted alcohol coating floats, while the flint glass on which the alcohol coating was not melted, does not, permitting removal of the amber glass, the flint glass remaining in the pulp in the machine.

Alternately, a surface in the form of a glass sheet is coated with 1-tetradecanol, and a mixture of amber and flint glasses are irradiated causing heating of the amber glass absorbing the irradiation to above 38°C while the flint glass which does not absorb such irradiation is not heated to above 38°C.

The resulting glass particles are poured upon the tetradecanol coated glass sheet, and following cooling to room temperature the differentially heated amber glass adheres to the alcohol coating while the flint glass does not so adhere and falls off the sheet.

In the flotation aspect, the organic thermoplastic compound, e.g., tetradecanol, which is water insoluble, displays the property of having a sharp melting range and of wetting glass when it is melted onto a surface of the amber glass particles by differentially heating the glass by irradiation. Flotation is selective and temperature dependent, i.e., the use of a temperature which causes melting of the alcohol as by differential heating or irradiation of the amber glass and absorption of light thereby, and adhesion of such melted alcohol to the amber glass.

COLUMBIUM FROM SCRAP COLUMBIUM

Columbium scrap is recovered by embrittling the metal through hydrogen absorption after which it is crushed and reprocessed by sintering or melting in a vacuum. Columbium-bearing alloy scrap is chemically treated to separate the alloy components, then the columbium is recovered separately. This process varies according to the composition of the particular alloy. At the present time, the recycling of old scrap is not significant; only about 100,000 lb of home or runaround scrap are recycled annually (16).

COLUMBIUM FROM SPENT CATALYSTS

See "Tantalum from Spent Catalysts".

COPPER FROM ACETYLENE SCRUBBING SOLUTIONS

See "Aluminum from Acetylene Scrubbing Solutions" and specifically *U.S. Patent 4,153,452* cited there.

COPPER FROM BRASS MILL DUSTS

A process developed by *R.N. Kust; U.S. Patent 4,149,945; April 17, 1979; assigned to Kennecott Copper Corp.* is one for winning copper and zinc from brass mill waste products containing oxidic copper and zinc, iron, nickel, or cobalt. The feed material is leached in sulfuric acid, neutralized to a pH between about 2.5 and 3.0, and treated with zinc metal to precipitate copper. After separation of the copper product, residual copper as well as iron and cobalt are removed in a second, oxygen-free, cementation on zinc metal. The liquor is then raised to about pH 4.0 or greater and flushed with oxygen to precipitate iron. Lastly, the purified electrolyte is subjected to zinc electrowinning wherein acceptably pure zinc cathode is produced and sulfuric acid is regenerated.

Such a process is shown schematically in Figure 39.

Acid leach tank **10** is serviced by aqueous sulfuric acid solution line **12** and air line **14**, and is provided with feed material from lines **16** and **18**. The pH of the solution from line **12** is below 1.0. Zinc oxide, cupric and cuprous oxides, and small quantities of oxidic iron, nickel, and cobalt material from line **16** react in tank **10**, for example, at 80°C, resulting in a Zn^{++}, Cu^{++}, Fe^{+++}, Ni^{++}, and Co^{++} sulfate solution. The solids from line **18** are similar to those in line **16**, except that the zinc oxide content is reduced as a consequence of the neutralization reaction taking place in leach tank **20**.

Figure 39: Kennecott Copper Co. Process for Hydrometallurgical Brass Dust Reclamation

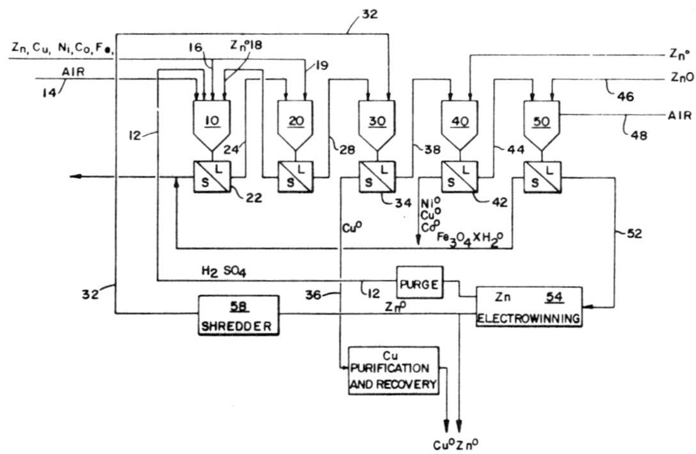

Source: U.S. Patent 4,149,945

After the unleachable solids are removed in liquid-solid separator **22**, the leach liquor is delivered to neutralization leach tank **20** via line **24**. In tank **20**, the pH of the liquor is raised to between 2.5 and 3.0 by acid-consuming reaction with zinc oxide, and to a lesser extent, copper oxide, both introduced via line **19**.

Since significant quantities of the metal values of interest remain as solids in this pH range, after attaining the desired pH, the solid residue is delivered to tank **10** via line **18** for further leaching. The pregnant, pH controlled leach liquor is then delivered to copper cementation reactors **30** via line **28**.

The cementation reactor **30** is serviced with a zinc metal feed line **32** which transports shredded zinc cathode for reaction with the copper values in the liquor. When a 10% molar excess (relative to copper content) of zinc is used, a pH within the range of 2.5 to 3.0 is present, and the zinc employed is a relatively low surface area to mass ratio material (i.e., as opposed to a fine powder), hydrogen evolution is kept to a minimum, essentially no nickel, cobalt, or iron precipitates out, the precipitation of copper hydroxide is avoided, and greater than 90% of the copper values are recovered as an acceptably pure cemented product. Oxygen is excluded from reactor **30**.

The cemented copper is separated from the other solubilized metal values in liquid solid separator **34**, and delivered through line **36** for repulping, redissolving, electrowinning, or other conventional treatment designed to purify and recover copper. The substantially copper-barren leach liquor is delivered through line **38** to cementation reactor **40**.

In reactor **40**, zinc dust is added, preferably in substantial molar excess of that required to cement all nickel, cobalt, iron, residual copper and other metals more noble than zinc which remain in the leach liquor. In the presence of ferrous iron, nickel and cobalt levels are decreased to about 1.0 mg/ℓ or less, and the concentration of other cementable metals is suitably reduced.

After separating the solids in liquid-solid separator **42**, the iron and zinc containing liquor is delivered through line **44** to iron precipitator **50**. Zinc oxide (free of impurities which would contaminate the leach liquor) is introduced through line **46** to the leach liquor to raise the pH to levels at which iron values precipitate as oxides, hydroxides, or jarosites. Air is introduced via line **48** to facilitate oxidation of iron to the ferric state. A typical pH range for this precipitation is between 4.0 and 5.0.

The now purified leach liquor is then delivered via line **52** to zinc electrowinning cells **54** where zinc is plated out and sulfuric acid manufactured in a manner known per se. After purging the zinc-barren electrolyte (if necessary), the acid byproduct is delivered via line **12** to reactor **10**. A portion of the zinc cathode is shredded at **58**, and used to precipitate copper in reactor **30.**

COPPER FROM COPPER REFINERY ANODE SLIMES

Electrolytic refining of copper results in the formation of an anode mud or slime which will range anywhere from about 0.5 to 3% or more of the original anode weight. The composition of the slime will vary dependent upon the components of the anodes which are not soluble in electrolyte. The slimes themselves are of fine particle size, about -200 mesh, and ordinarily grayish black in color. Slimes usually contain varying quantities of copper, silver, gold, sulfur, selenium, tellurium, antimony, arsenic, nickel, iron, silica, bismuth, and the like.

Ordinarily, the largest component of the slimes is copper, which ranges from about 15 to 30% by weight, or more, of the slime. In addition to the valuable copper present in the slimes, there is a significant amount of precious metals, such as silver and gold, as well as other metals which are of economic value, such as selenium and tellurium.

Because of the large amounts of copper in the slime as well as of the precious metals and other valuable metals, a variety of techniques have been utilized to recover such metals from the slime. However, none have been commercially successful because of the large cost involved and the fact that it is difficult to recover the precious metals from the slime.

One type of technique involves direct smelting of slimes. This is disadvantageous since excessive formation of matte and slag cause heavy recirculation of precious metals, expecially of silver. Such techniques involve roasting to convert copper to copper oxide followed by leaching of the roasted slimes with sulfuric acid. The leached slimes are then melted down in a small reverberatory (dore) furnace in which the impurities are oxidized and to leave a dore bullion of the precious metals silver and gold.

The dore is then parted to recover the fine gold, fine silver, and any other precious metal present in the original slime. This particular technique is, as noted, unsuitable because of the excessive formation of matte and slag.

A variety of other pyrometallurgical techniques combined with acid leaches and acid leach procedures have also been attempted, but they are equally unsuitable mainly because it has not been possible to initially remove substantially all of the copper from the slime. It has been found that copper levels at or above 5% by weight of the material to be treated greatly inhibit the known procedures for recovery of selenium and precious metals, particularly silver, from slime.

In addition, in leaching slimes with an acid, even strong sulfuric acid leach solutions (400 g/ℓ), a leach time under agitation of up to 72 hours is required. Also during such leach, it is necessary to periodically stop the agitation, heat, and aeration to permit the solids to settle and then to decant the clear solution. After decantation, additional leach solution (water and acid) is added and heat, air, and agitation started to continue the leach. As a consequence it has heretofore been costly in terms of material, labor, maintenance, and time to recover the copper, precious metals, and other metal values from copper anode slimes.

An improved process for copper recovery from copper anode slimes has been developed by *E.A. Bilson; U.S. Patent 4,076,605; February 28, 1978; assigned to Inspiration Consolidated Copper Co.* It involves leaching the slimes with a solution containing from 20 to 200 g/ℓ H_2SO_4 and from 2 to 25 g/ℓ hexavalent chromium at a temperature not exceeding 80°C, whereby copper dissolution is maximized and selenium and silver dissolution are minimized, separating the leach solution from the undissolved slimes residue, precipitating selenium from the leach solution by treatment with metallic copper, and electrowinning copper from the purified solution. Such a process is shown schematically in Figure 40.

Figure 40: Inspiration Consolidated Copper Co. Process for Dichromate Leach of Copper Anode Slimes for Copper Recovery

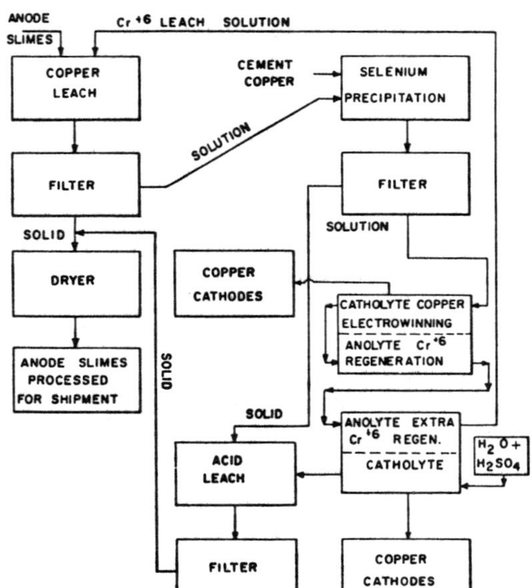

Source: U.S. Patent 4,076,605

When leaching the slimes in a continuous circuit with provisions for electrolytic recovery of copper and regeneration of Cr^{+6}, a water balance must be taken into consideration. Thus the slimes fed to the leach should, at least, be filtered, and better yet, be dried prior to addition to the leach.

The slimes will be added to the leach at a constant rate which will allow a total leach time of between 1 and 3 hours.

In order to minimize the dissolution of selenium, the Cr^{+6} solution used to leach the slimes is added to all but the last leach tank. Thus, in a four-tank, continuous circuit, one half of the leach solution could be added to the first tank, one third to the second tank, and one sixth to the third tank. The fourth tank will allow the dissolution of copper to go to completion.

It is desirable to maintain a temperature of 75°C throughout the leach. Thus, a small amount of heat may be needed in the latter stages of the leach, i.e., tanks 3 and 4.

The solution fed to the leach could have the following analysis:

Component	Concentration, g/ℓ
Cu	25
Cr	40
Cr^{+6}	20
H_2SO_4	150
Se	0.05
Te	0.2
As	0.02
Sb	0.02
Bi	0.02
Fe	0.1
Ag	trace

The slurry from the leach can be thickened prior to filtration or filtered directly from the leach. In either case, the solids are washed, dried, and prepared for sale. The solids will have substantially the following analysis:

Component	Concentration
Cu	4.0%
Se	35.0%
Ag	6,000 oz/ton
Au	65 oz/ton
SiO_2	30%
Al_2O_3	2%
As	0.1%
CaO	2%
Pb	1.5%
S	3.0%
Co	trace
Te	0.6%
Sb	0.1%
Fe	0.1%
Zn	0.1%

The filtered solution will have the following analysis:

Component	Concentration, g/ℓ
Cu	55
Cr	40
Cr^{+6}	3
H_2SO_4	20
Se	3
Te	0.4
As	0.02
Ag	trace
Sb	0.02
Bi	0.02
Fe	0.2

The clarified, or filtered, solution is purified by the addition of relatively pure cement copper. The purification can be made in a batch step or in a continuous step similar to the leach. The reaction is allowed to continue for 1 hour at the boiling point of the solution.

The slurry from the purification step is filtered. The solids which contain the previously dissolved selenium and tellurium can be processed by the conventional sulfuric acid, heat, air leaching to dissolve the copper and tellurium. Thus far, attempts to add the selenium precipitate to a dichromate leach have resulted in total dissolution of selenium and tellurium.

The sulfuric acid leach will dissolve the copper and tellurium, leaving a selenium product that can be sold separately or added to the processed slimes.

Actually the selenium dissolved by the chromic acid leach is less than 10% of the selenium fed to the leach. The copper selenium product produced during the purification of the solution could easily be sent to a smelter to be reprocessed into anodes.

The purified solution is then subjected to copper electrowinning in a diaphragm cell. The solution will have the following analysis:

Component	Concentration, g/ℓ
Cu	60
Cr	40
Cr^{+6}	0
H_2SO_4	5
Te	0.2
As	0.02
Ag	trace
Sb	0.02
Bi	0.02
Fe	0.2

The copper produced during the electrolysis of the purified solution is not of refined cathode quality. A typical analysis indicates higher than desirable quantities of tellurium, antimony and arsenic. A typical analysis is given below in ppm.

Component	Concentration, ppm
Ag	<1.0
Fe	<1.0
Se	<2.6
Pb	1.0
Zn	3.0
Sn	1.2
Ni	<1.0
Bi	0.9
Sb	15.2
As	4.8
Te	>4.6

While further purification of the solution may allow the production of refined quality cathode copper, this process is not in any way dependent on the quality of the cathode copper produced during copper electrowinning.

Removal of the desired amount of copper will have a solution with the following analysis:

Component	Concentration, g/ℓ
Cu	25
Cr	40
Cr^{+6}	0
H_2SO_4	50
Se	0.01
Te	0.1

This solution is ready for Cr^{+6} regeneration. This is accomplished by passing the solution through the anode compartments of the same diaphragm cells that the copper electrowinning took place in. Generally, sufficient Cr^{+6} will be regenerated in this operation. However, in cases where the chromium consumption greatly exceeds the theoretical chromium(VI) requirements, extra Cr^{+6} regeneration cells will be required. This will generally be accomplished while producing hydrogen at the cathode, but any solution, including the copper solution obtained by leaching the copper selenide produced when purifying the solution before copper electrowinning, can be used.

The regenerated solution will have the following analysis and is ready to leach additional slimes:

Component	Concentration, g/ℓ
Cu	25
Cr	40
Cr^{+6}	20
H_2SO_4	150

The operation of this process is controlled in such a way that the copper is reduced to a no penalty level while minimizing the dissolution of selenium. Silver is not dissolved when leaching at 80°C and a sulfuric acid concentration of 200 g/ℓ.

A variant of the process which gives both selenium and copper recovery is shown in Figure 41.

Figure 41: Alternative Inspiration Consolidated Copper Co. Process for Treatment of Anode Slimes to Recover Both Copper and Selenium

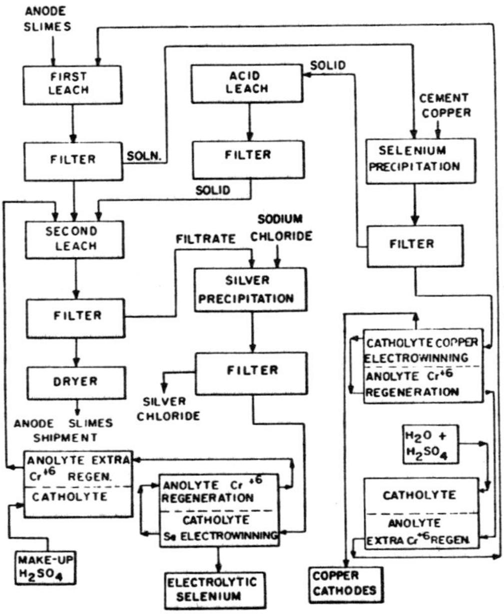

Source: U.S. Patent 4,076,605.

This process involves two leaches with separate solutions. The process proceeds utilizing a copper leach with chromic acid solution.

The first leach is similar to the copper leach described above in which the leach was continuous and the copper was recovered by electrolysis and the Cr^{+6} regenerated for reuse. This process produced a leach tail with between 2 and 5% copper. The process described now requires that the copper be reduced to as low as practical, i.e., 0.2 to 0.5%. In order to do this, the leach must be performed at the boiling point of the solution.

Aside from the increased temperature, the copper leach proceeds as previously described. Due to the increased temperature, more selenium is dissolved, up to 20% of the selenium in the raw slimes. The leach will continue for up to 2 hours. The slurry can go to a thickener or be filtered directly from the leach. The filtered and/or the clarified solution is purified with cement copper and the purified solution subjected to copper electrowinning. The purification step, when performed at the boiling point of the solution, will remove the selenium to a low level.

The solid produced during the purification step contains copper selenides as well as unreacted metallic copper. When this product is leached in a sulfuric acid solution containing between 200 and 300 g/ℓ acid, with air, at a temperature of up to 100°C, the copper is dissolved to a low level, leaving a fairly pure elemental selenium. The tellurium precipitated during the purification step is also dissolved. Little selenium is dissolved.

The tail from the leach of the copper selenide purification product is added to the tail from the first chromic acid leach. No other method of handling the copper selenide product has yet been developed. For some reason, the product dissolves completely in chromium acid solutions, even at room temperature.

The combined tails are then subjected to a second chromic acid leach. The solid fed to this leach will have the following analysis:

Component	Concentration
Cu	0.2 g/ℓ
Se	32.1 g/ℓ
Te	0.14 g/ℓ
Ag	6132 oz/ton
Au	68.2 oz/ton
SiO_2	30%
CaO	0.8 g/ℓ
Al_2O_3	2.4 g/ℓ
Pb	1.7 g/ℓ
S	3.8 g/ℓ

The solution for this leach will have the following analysis:

Component	Concentration, g/ℓ
Cu	0.2
Se	14.0
Cr	40
Cr^{+6}	25
H_2SO_4	350
Te	0.1

The major difference between this solution and the one to dissolve the copper is the sulfuric acid concentration. While 150 to 200 g/ℓ H_2SO_4 is satisfactory to dissolve the copper, 350 g/ℓ H_2SO_4 is needed to dissolve the silver and selenium. Also, due to the method of recovering the selenium, the copper in solution and thus, the copper in a solid feed to the leach, must be kept at a minimum.

The leach, which lasts 4 hours, takes place at the boiling point of the solution. The leach can either be a batch process in which the solids (filtered from the copper leach) are added to all of the solution needed for the leach, or a continuous cocurrent leach in which all of the solids and all of the solution are added to the first of three or four agitated leach vessels. During the leach, 99% of the selenium and 92% of the silver are dissolved.

The slurry from the leach should be filtered hot, to minimize the formation of a complex silver chromate selenite product. The hot, but not boiling, filtered solution is treated with sodium chloride to precipitate the silver as silver chloride. Excess chloride is undesirable because of its adverse effect on Cr^{+6} regeneration.

The silver chloride slurry is cooled, and the silver chloride filtered from the solution. This silver chloride product is quite pure and is ready for processing to produce elemental silver by known processes. The silver chloride product contains about 92% of the silver in the slimes.

The solution fed to the silver removal step would have the following analysis:

Component	Concentration, g/ℓ
Cu	0.4
Se	30.0
Cr	40
Cr^{+6}	3
H_2SO_4	250
Ag	5.5
Te	0.2

After removal of the silver, the solution, which has the following analysis, is subjected to cathode compartment electrowinning to recover elemental selenium.

Component	Concentration, g/ℓ
Cu	0.4
Se	30
Cr	40
H_2SO_4	250
Ag	trace
Cl^-	trace
Cr^{+6}	3
Te	0.2

It is recommended that a titanium cathode be used because it produces a higher grade selenium product. The elemental selenium produced has the following analysis:

Component	Concentration
Se	97.25%
Cu	0.8%
Cr	0.1%
Fe	0.1%
Cl^-	0.3%
Pb	trace
$SO_4^=$	0.6%
Te	0.04%
Ag	45.80 oz/ton

The selenium product produced during electrowinning is a nonadherent deposit which tends to float to the surface of the cell. Thus, the titanium cathodes need not be pulled to recover the selenium. If the cell flow rate is maintained at a high enough level, the elemental selenium is flushed from the cell to a thickener or directly to filtration.

After sufficient selenium has been removed from the solution, and the particulate selenium is removed by filtration, the solution is subjected to Cr^{+6} regeneration in the anode compartments of the same cells. Extra regeneration is needed and accomplished in additional diaphragm cells which produced hydrogen at the cathodes. The regenerated solution will have the following analysis, and is ready to leach additional slimes.

Component	Concentration, g/ℓ
Cu	0.2
Se	15.0
Cr	40.0
Cr^{+6}	25
H_2SO_4	350
Te	0.1

During the selenium electrolysis, other elements and compounds are produced which must be removed from the selenium to make a high purity product. This can be accomplished using known methods, such as an inert gas distillation. As soon as sufficient electrolytic selenium is produced, a solvent extraction technique will be tried which is to dissolve the selenium and leave the impurities behind. The solvent would be removed using distillation technique to produce high purity selenium.

The tail from the selenium silver leach would have the following analysis:

Component	Concentration
Ag	1100 oz/ton
Au	135 oz/ton
Cu	0.13%
Se	0.1%
SiO_2	57%
Te	trace
Fe	0.5%
S	2.3%
Sb	0.05%
As	trace
Bi	trace
Pb	5%

This product is ready for precious metals recovery by known methods. The slimes processed in either of the above examples had the following analysis:

Component	Concentration
Cu	34.98%
Se	23.04%
Ag	3412.2 oz/ton
Au	37.0 oz/ton
SiO_2	16.2%
Al_2O_3	1.1%
CaO	1.0%
Pb	0.7%
S	5.5%
Co	0.1%
Te	0.5%
As	0.075%
Sb	0.13%
Fe	0.08%
Zn	0.02%
Ni	trace
Cr	trace
Bi	trace

COPPER FROM ELECTRICAL SCRAP

See "Aluminum from Electrical Conductor Scrap" and specifically U.S. Patent 3,582,004.

While in most cases copper is recovered from electrical scrap by chemical treatment or by destroying the insulation (e.g., by burning it off), it may be possible to recover both the base copper and the insulation in a mechanical separation process. See "Polyvinyl Chloride from Electrical Scrap". Recovery of both copper and insulation may also be carried out by cryogenic processing.

A process developed by *B.J. Weston; U.S. Patent 3,990,641; November 9, 1976; assigned to Jeno, Inc.* is one utilizing cryogenic means for separating metallic conductor wire from its insulation for reclamation of the metal and secondarily the insulation material. Random size, nonuniform masses and pieces of the insulated wire are conveyed to an entrance end of a shaftless rotatable drum having helically deployed baffles along the inner surface thereof. The masses and pieces introduced to the drum initially and continuously are exposed therein to cryogenic coolant to freeze them and render them brittle.

The baffles continuously turn the frozen masses and pieces to ensure that all surfaces and areas thereof are exposed to the coolant for thorough freezing. Simultaneously the actions of the baffles and drum move the masses and pieces from the entrance end to an exit end of the drum. The supercooled brittle masses and pieces are delivered from the exit end to a grinding apparatus where they are reduced to particles of metal and insulation material. Thereafter, the particles are moved to a separating apparatus for separate accumulation of ground metal and insulation.

The drum is mounted in a cage formed by longitudinal rods extending the length thereof and transverse posts between the rods and the exterior surface of the drum so that physical connection between the drum and cage is minimal and present only at the post ends terminating on the drum surface.

The minimal drum and cage connection permits drastic expansion and contraction of the drum during introduction of cryogenic material therein without hindering mechanical operation of the drum. Figure 42 shows the essential steps in such a process.

Figure 42: Jeno, Inc. Process for Cryogenic Treatment of Electrical Scrap to Permit Recovery of Both Metal and Insulation.

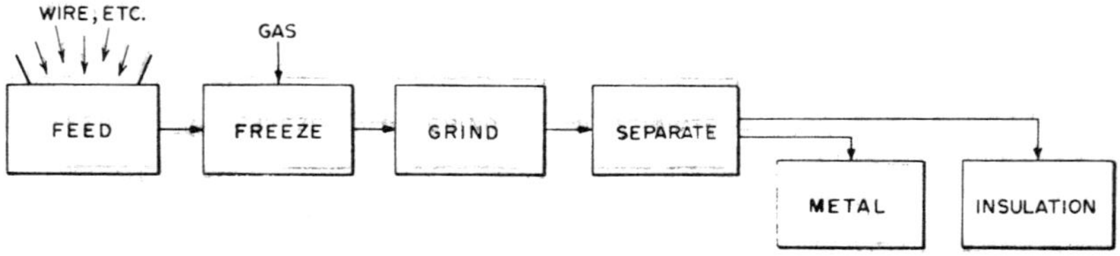

Source: U.S. Patent 3,990,641.

A process developed by *J.H. Weet; U.S. Patent 4,022,638; May 10, 1977; assigned to Dart Industries, Inc.* is one providing on a continuous, commercially practical basis for recovery of a nonferrous base metal, such as copper or aluminum in the form of clean, bright metal pieces or bits and for enabling recovery of different nonferrous coating metals, such as of lead, tin or alloys thereof, and under conditions in which a resin insulating coating may be present.

Mechanical shredding means is used for reducing insulated wire to an aggregate containing bits, particles or pieces within a requisite size range, extraneous magnetics are removed, and the content of resin coating material is reduced under dry conditions to within a maximum of 2%.

A preliminary step for attaining a less than 2% resin content importantly includes moving the aggregate as a vibration-activated, dry, fluid mass of substantially uniform thickness upwardly along a spiral path while separating the material into heavier and lighter portions, with the lighter innermost portion being recycled to further concentrate the metal bit content thereof and to further minimize the resin content thereof.

After the resin content has been thus reduced, and the metal bit content maximized, the aggregate is again moved along a fluidized helical path in a vibration activated stream of substantially uniform thickness upwardly, while applying a reactive chemical solution thereto in a counterflow path downwardly, with a substantially full area of exposure of the metal bits to the solution during its upward movement.

A process described by *F.W. Eggers III, R.O. Wigger and R.L. Coan; U.S. Patent 4,040,865; August 9, 1977; assigned to Cerro Corporation* provides a continuous method of removing polyvinyl chloride insulation or other volatilizable material from items such as copper wire without producing undesirable by-products.

The method involves pyrolysis of the insulation in a pyrolysis chamber, passing the products of pyrolysis through an afterburner and thence to a fluidized bed reactor which uses fluidized particles of $CaCO_3$ to remove any chlorine from the products of combustion of the afterburner.

The calcium carbonate reacts with the gaseous chlorine from the insulation to produce calcium chloride which then is collected and can be used for snow removal, road construction, and the like.

Such a process is shown schematically in Figure 43.

This system comprises a decomposition chamber **100** including an afterburner **111**, a fluidized bed reactor **112**, a dust collector **113**, and an induced draft fan **114**. The chopped input material (usually insulated copper wire containing polyvinyl chloride insulation) is continuously fed at a rate controlled by a conveyor **115** through a water trap **116** to the chamber **100**.

The cut feed material is moved on the bed **119** by a conveying means **119a** through the decomposition chamber **100** where it is ignited and decomposed in a controlled atmosphere void of excess oxygen and controlled with respect to analysis and temperature by the burners **117**.

Figure 43: Cerro Corp. Process for Pyrolyzing Vinyl Chloride Insulation from Copper Wire and Recovery of Calcium Chloride by Calcium Carbonate Scrubbing of Effluent Chlorine

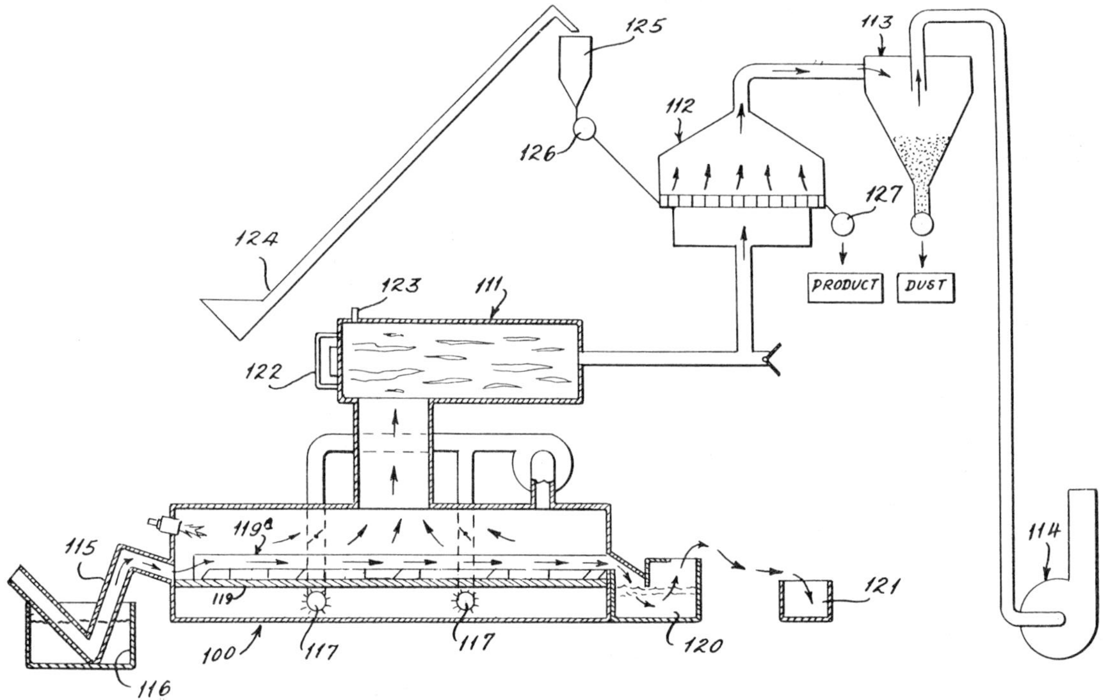

Source: U.S. Patent 4,040,865.

The feed material generally is heated to a temperature from about 600° to 1200°F for best decomposition.

The insulation on the feed material decomposes and burns in the decomposition chamber **100**. The clean copper wire leaves the decomposition chamber **100** through an exit water seal **120** which also acts as a quench for the wire scrap, thereby cooling and further cleaning the product. The product from the water seal **120** is collected in a receptacle **121** or a bailer for further handling.

The gases and smoke generated during decomposition, which contain chlorine, pass into an afterburner **111** where all of the remaining combustible products are consumed.

The afterburner includes a burner **122** and an excess combustion air inlet **123**. The afterburner operates at a temperature of about 1400° to 2000°F. The burner gases from the afterburner **111** still contain chlorine as well as the products of combustion.

The products of combustion, including the chlorine from the afterburner **111**, are passed through a fluidized bed reactor **112** where the fluidized bed material is a calcium-containing substance in a granular form of about -10 to -200 mesh. The substance preferably is calcium carbonate ($CaCO_3$).

The gases from the afterburner **111** act as the fluidizing medium. The intimate contact of the exhaust gases with the calcium carbonate causes a chemical reaction to take place in which the chlorine is reacted with the calcium carbonate to produce calcium chloride.

The calcium carbonate is deposited in a conveyor **124** where it is carried to a hopper **125** and then passed through a rotary air lock feeder **126** to the fluidized bed reactor **112**.

The calcium carbonate passes across the bed in fluidized condition to the exit rotary air lock **127**. The air lock feeder **126** controls feed of make-up calcium carbonate to the reactor **112** to replace that consumed in the reaction.

From the fluidized bed reactor **112** the gases, less the chlorine, are passed through the mechanical dust collector **113** and then exit through the induced draft fan **114** which provides the pressure differential to flow the gases through the entire system.

One type of electrical scrap which may be processed for reclamation is petroleum jelly filled communication cable of the type employed for overhead and underground telephone cables. These cables comprise a large number of individual copper wires insulated with polyurethane.

The insulated individual wires are grouped together and wrapped with a copper shield. This assembly is coated with an outer sheath of polyurethane insulation and the spaces inside the composite cable are filled with petroleum jelly.

Such scrap may be processed by a process developed by *A.B. Baker; U.S. Patent 4,091,825; May 30, 1078; assigned to International Recycling Enterprises, Ltd.*

The scrap cable may be processed to recover a clean, unoxidized metallic substrate and various organic by-products by thermally decomposing the organic materials under pressure in a closed, externally heated vessel. Volatile decomposition products are periodically vented to a vapor recovery system and nonvolatile decomposition products, primarily carbonaceous char adhering to the clean scrap, are separated by mechanical action such as vibration or tumbling.

COPPER FROM MIXED SCRAP

In the recycling of scrap for metal recovery, much of the scrap available contains various metals and nonmetals which are in such intimate mechanical association as to be not easily separable. The cryogenic process is very effective for rendering certain types of scrap brittle, and thus enabling it to be fragmented. However, problems are encountered where the scrap contains significant amounts of metals which remain ductile at cryogenic temperatures. As one example, the copper in a scrap material containing copper and ferrous metal does not experience brittle fracture following cryogenic cooling. Instead, it tends to become flattened, sheared, stretched or otherwise plastically deformed during the self-same crushing operation which shatters the ferrous material. As a consequence, substantial portions of ferrous material are not liberated and amenable to economic recovery. They are, instead, mechanically entrapped by nonferrous metal aggregates which encapsulate, entangle and make conventional sorting impossible.

A process developed by *D.J. Drage; U.S. Patent 3,905,556; September 16, 1975; assigned to Air Products & Chemicals, Inc.* is applicable to metallic scrap material of widely varying composition. It will be described with specific reference to a selected scrap having a relatively high copper content. Other nonferrous metals such as aluminum, zinc or precious metals may also be present. Typical of the scrap compositions that may be treated is the composite obtained by initial fragmentation of a mixture of discarded automobile generators, voltage regulators, electric motors, armatures, stators, electrical wire, electronic devices, relays and the like having a relatively high copper content, an appreciable quantity of aluminum, organic insulating material, together with ferrous materials and usually small quantities of other metals.

The original feed stock of scrap may be initially subjected to cryogenic cooling to effect embrittlement of ferrous and organic components following which it is broken up by impact and crushing into pieces, which may be in the order of about up to 3" in smallest dimension.

The scrap, after such first size reduction, is separated into light and heavy fractions and the heavy fraction is separated into magnetic and nonmagnetic portions, the latter portion being combined with the light fraction from the first separation and the combined stream subjected to a second crushing operation by impact and crushing, whereby the scrap is now reduced to about 2" size in smallest dimension.

After the second size reduction step, the thus comminuted scrap is subjected to a second air separation with the stream of heavier particles being subjected to magnetic separation and the resulting nonmagnetic fraction being recycled to the second crushing mill. The light fraction from the air separation step is further reduced in size to particles of about 1" in smallest dimension and the entire product of such milling subjected to one or more magnetic separation steps to remove magnetic components therefrom. The residual nonmagnetic portion may at this point be sent to further processing for recovery of the individual metal components.

A flow diagram showing the metal recovery steps is presented in Figure 44.

Figure 44: Air Products and Chemicals Inc. Process for Recovery of Copper and Other Metals from Mixed Scrap

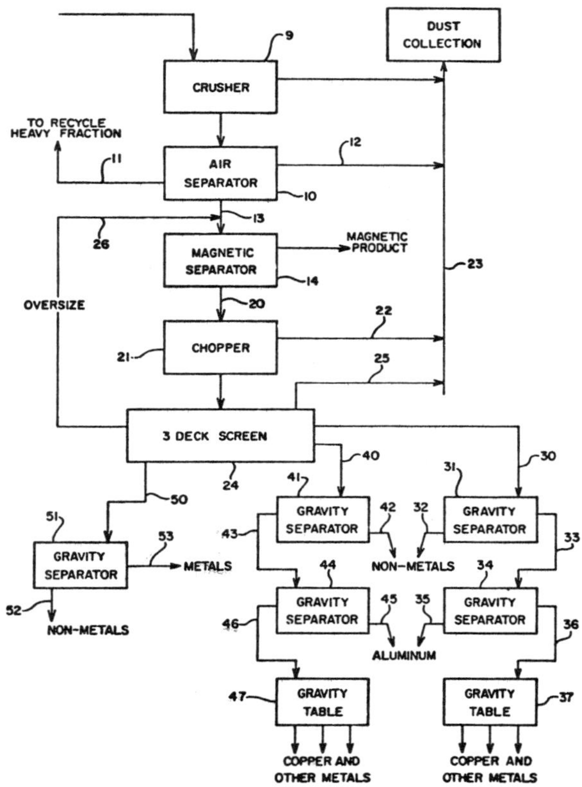

Source: U.S. Patent 3,905,556.

The material in line **20** being fed to the granulator or wire chopper **21**, is composed of small pieces of scrap resulting from previous successive comminution in hammer mills or the like, and comprising a mixture of metals rich in copper. While the particular size of the pieces is largely optional, convenient size pieces for feeding to commercially available choppers may be of about ½" in cross section and up to about 1" lengths, the size being determined by the grate openings in the last preceding hammer mill.

This fraction being composed chiefly of ductile metals mixed with some nonmetallic substances is not readily susceptible to comminution to controlled uniform size range by milling operations such as crushing or impacting as in the case of brittle or embrittled materials. Accordingly, to obtain further reduction to uniform size range in order to facilitate separation and recovery of the individual metal components therein, the composite stream of metal pieces in line **20** is fed to the wire chopper or granulation mill **21** of the type employing chopping or shearing action.

In this granulating mill **21**, the pieces are reduced to predominantly about ¼" size and smaller or about 5⁄16" size and smaller depending upon subsequent separation operations to be employed. Dust is removed overhead from the mill and transported, as indicated at **22**, to a dust collection duct **23** leading to a collection system. The dust which is collected in cyclones, bag filters or the like, may be subjected to recovery of metallic and nonmetallic components contained therein.

The comminuted product from granulator **21** is discharged to a screening operation for classification into 3 or more cuts of different particle size ranges, the particle size in each such fraction essentially lying within a narrow range. In the embodiment illustrated, a conventional three deck screen is employed at **24**, in which instance four distinct size graded fractions will be obtained. While not limited to the precise ranges, by way of example a suitable size distribution for the four fractions could be: (1) largest size, above ¼"; (2) first middle fraction, ¼ to ⅛"; (3) second middle fraction, ⅛ to 1⁄16"; and (4) fine fractions, below 1⁄16".

Dust particles released during the screening operation will be picked up by the dust collecting system through line **25**. The oversize fraction (1) is returned by line **26** for further granulation in chopper **21**. If significant amounts of magnetic materials are entangled or encapsulated in the product in line **26**, it is advisable to subject it to magnetic separation to remove such ferrous materials prior to feeding the nonmagnetic residue to the chopper **21**. Each of the remaining three fractions (2), (3), and (4) from screening at **24** is separately subjected to density separation.

Multideck screens are provided for protecting the cutters in granulator **21** against damage by large particles in the form of nuggets or the like. The original scrap material having been reduced in size by previous crushing, the last stage of which is indicated at **9**, is subjected to separation in an air separator system **10**. In the separator, the scrap falls through a generally transverse air stream and is thereby sorted into light and heavy fractions.

As illustrated in the accompanying drawing, the heavy fraction from air separator **10** is removed by line **11** and may be recycled to an earlier crushing or grinding stage with or without further intermediate separation or sorting. The dust is picked up by the dust collection system, being fed to duct **23** by connecting line **12**, while the light material is conveyed by line **13** en route to granulation at **21**. Although the scrap entering the last stage of crushing at **9** is composed essentially of pieces which are not ordinarily subject to magnetic attraction, as a result of such last crushing stage at **9** and subsequent air separation at **10**, magnetically-responsive materials earlier shielded by entanglement and/or encapsulation with nonmagnetic covering materials, may be now freed from such association. These released magnetically responsive particles can now be removed from the product stream in line **13** by magnetic separation as indicated at **14**, obtaining the nonmagnetic product fed by line **20** to granulator **21**.

Treatment of the three cuts resulting from size classification at **24** will now be described. The coarsest fractions are each separately treated in similar manner to separate the components thereof according to particle density. Thus the fraction containing the largest size particles discharged from screening to line **30** is initially charged to a dry specific gravity separator **31** in which nonmetallic particles and materials lighter (lower in density) than aluminum are removed.

This may be accomplished, for example, in a conventional apparatus known in the art as a destoner. In this type of device, as designed, the charge of dry granular material flows over an inclined vibrating screen-covered deck. Air is blown or sucked through the deck and through the layer of material. The steady flow of air holds the material in the bed in stratified flotation, with the lighter material being in the upper strata, as it flows or bounces down the inclined vibrating deck. The heavier material travels uphill on the screen to discharge.

In operations for which this type of gravity separator was particularly designed, the heavy fraction usually contained stone, glass and the like, hence the designation destoner. A destoner will effectively separate any two components differing in density by at least 100%. Airborne particles are lifted from the upper strata by the air flow therethrough and are discharged to a dust collection system. At the discharge end for the lighter fraction which is at the low point of the inclined deck, a skimmer knife may be provided, to skim off and divert the top layer (containing relatively lighter materials) from the lower layer (containing relatively heavier materials).

For the required density separation of materials in line **30**, the operation is conducted in separator **31** to remove fluff and materials lighter than aluminum, discharged through line **32**. The heavier materials are sent by line **33** to a second density separator **34**. For this initial separation in **31** there may be employed a Forsberg model H-2 vacuum destoner, or any other similar apparatus designed to separate by density dry granular materials of substantially uniform particle size, into portions of widely differing density.

Separator **34** may be of the same type as that employed at **31**, but operated under adjusted conditions, as to inclination of the deck, speed of vibration, throw and air flow intensity, to separate aluminum from the remaining heavier metals. Since the density of aluminum is considerably less than that of the other common metals likely to be encountered in the scrap (differing in density by at least 100%) it is readily recovered from separator **34** in highly pure form, and discharged through line **35**. The residual metals from separator **34**, some of which are close to one another in density are not readily subject to sharp separation by devices of the destoner type.

To effect sharp separation of the remaining metal components from separator **34**, these are sent by line **36** to a gravity table type of separator. This type of separator is effective in separating uniformly sized particles having a density difference of 25% or more. The common form of gravity table comprises a meshed deck screen inclined in two directions. A stream of air is passed through the pervious deck stratifying the material thereon by fluidization so that the fluidized lighter materials float in the upper strata, while the heavier materials below are moved further upwardly by the riffled surface of the deck. The charged product is thus divided along the deck into several fractions progressively differing in density. By suitable placement of adjustable dividers, sharp cuts can be effected to provide individual fractions each having its own substantially uniform density, thus obtaining and recovering from the mixed feed three or more products, each of substantially uniform composition.

Products of intermediate density between the streams isolated by the dividers, can be separately discharged and recycled to the deck for further separation thereon. Flow of air through the deck may be achieved by blowers below, as in the so-called pressure type gravity tables or induced by suction applied above the deck as in the case of vacuum gravity tables.

The product in line **36** will contain copper and may also contain one or more of the common nonferrous metals or alloys such as brass, bronze, lead and/or zinc or the like, and perhaps a small quantity of silver. The various nonferrous metals named thus far rank in ascending order of density:

$$\mathrm{Al} < \mathrm{Zn} < \mathrm{Cu} < \mathrm{Ag} < \mathrm{Pb}$$

In the gravity table **37** fed by line **36**, the zinc component can be readily separated and recovered from the copper and heavier metals. Depending upon the desired purity and the proportional content of other metals, after recovery of the zinc and at least a major portion of the copper at **37**, the remaining metals may be recovered as separate divided fractions at **37**, or these may be sent to another similar gravity table for separation into the individual metal components. If desired, the individual fractions can be further purified chemically or in other known manner, forming no part of the process.

The middle cut from screen deck **24** is discharged by line **40** to the separator **41**, which may be similar in design and operation to separator **31**. Materials lighter than aluminum are separated out and discharged through line **42**, the remaining metal components being sent by line **43** to separator **44**, similar in design and operation to separator **34**.

From separator **44** the aluminum component is recovered in line **45**, while the remaining metal components are forwarded by line **46** to one or more gravity tables **47**, and subjected to treatment similar to that applied to the mixed metals in separator **37**. Some aluminum, not recovered in line **45** may be present in the mixed stream charged to separator **47**, in which event it can be readily recovered as a light cut on the gravity table **47**.

The fines passing through the lowermost screen at **24** are conveyed by line **50** to a specific gravity separator, which may be similar to the separators **31** and **41** previously described, or a gravity table similar to separators **37** and **47**, previously described.

As before, the nonmetallic materials are removed from the remaining metals and discharged by line **52.** The metal product ordinarily remaining will be chiefly copper and may be recovered as such without further separation. If economically warranted, however, this metal residue in line **53** may be subjected to further separation into its individual metal components by means similar to that employed in connection with the mixed metal streams in lines **33** and **43**, or in other desired manner.

Each of the separators **31, 34, 37, 41, 44, 47** and **51** may be equipped with means for collecting dust and discharging it to a common dust collection system, or they may be equipped with individual provisions for dust collection. The nonmetallic portions discharged from the metals recovery system, in lines **32, 42,** and **52** may be combined and salvaged.

COPPER FROM NONFERROUS SLUDGES

See "Chromium from Nonferrous Sludges" and also "Metals from Mixed Metal Slimes".

M. Kawasaki, S. Mizumoto and H. Nawafune; U.S. Patent 4,064,022; December 20, 1977; assigned to Ebara Udylite Co., Ltd., Japan describe a process whereby copper may be recovered from nonferrous sludges using an apparatus described under "Nickel from Nonferrous Sludges".

The following is an example of such an operation: Using a cupric hydroxide sludge (containing partially cupric oxide) obtained by neutralizing a solution containing cupric ion, an electrolytic treatment was carried out to recover copper by way of electrolysis.

One liter of an aqueous solution of cupric sulfate with a concentration of one mol/ℓ having dissolved therein 10 g of sulfuric acid was used as an electrolytic solution and a cupric hydroxide sludge (containing partially cupric oxide) having an average moisture content of 90% was supplied to the anode compartment at a rate of about 120 to 150 g/hr. The electrolytic treatment of the sludge was conducted at an electric current of 5.4 A whereby fine metallic copper with good adhesiveness was built up on the surface of the cathode at a rate of 6.3 g/hr. In this case, smooth dissolution of cupric hydroxide in the sludge added to the anode compartment is necessary for smooth supply of cupric ion to the cathode compartment. Since dissolution of $Cu(OH)_2$ was promoted by oxygen generated at the anode and this purpose was also attained by the reaction between sulfuric acid and cupric hydroxide, dissolution of the copper components contained in the sludge was attained in the anode compartment with a quantity of electricity considerably smaller than the theoretically needed quantity. According to this method, a quantity of electricity as small as about 3 kW was sufficient enough to attain recovery of 1 kg of copper.

COPPER FROM SCRAP AUTOMOBILES

See "Steel from Scrap Automobiles".

COPPER FROM SMELTER FLUE DUST

In a process developed by *J.D. Prater and B.A. Wells; U.S. Patent 4,149,880; April 17, 1979; assigned to Kennecott Copper Corp.* highly toxic metallurgical waste materials from copper smelting and/or refining operations are processed on a cyclic basis for recovering copper values therefrom while protecting the environment from contamination.

The waste materials, usually smelter flue dust and/or refinery sludge containing copper values and principally arsenic, bismuth, lead, antimony, and cadmium, as toxic values, with or without added sulfuric acid, are reacted in an autoclave at an overpressure atmosphere of O_2. The resulting solution, pregnant with copper and still containing a significant amount of arsenic, e.g., from about 0.5 to 2.0 g/ℓ, is subjected to copper cementation on metallic iron to recover copper and to provide iron values in the solution while substantially eliminating residual toxic constituents, surprisingly without the evolution of arsine gas.

The copper-barren but iron-containing solution is then recycled to the leaching step. The cement copper precipitate is smelted along with copper ore concentrates, while the solid residue from the leaching step is discharged to a tailings pond or the like as an environmentally safe, insoluble residue. Such a process is described schematically in Figure 45.

Figure 45: Kennecott Copper Corp. Process for Recovery of Copper from Arsenic-Containing Metallurgical Waste Materials

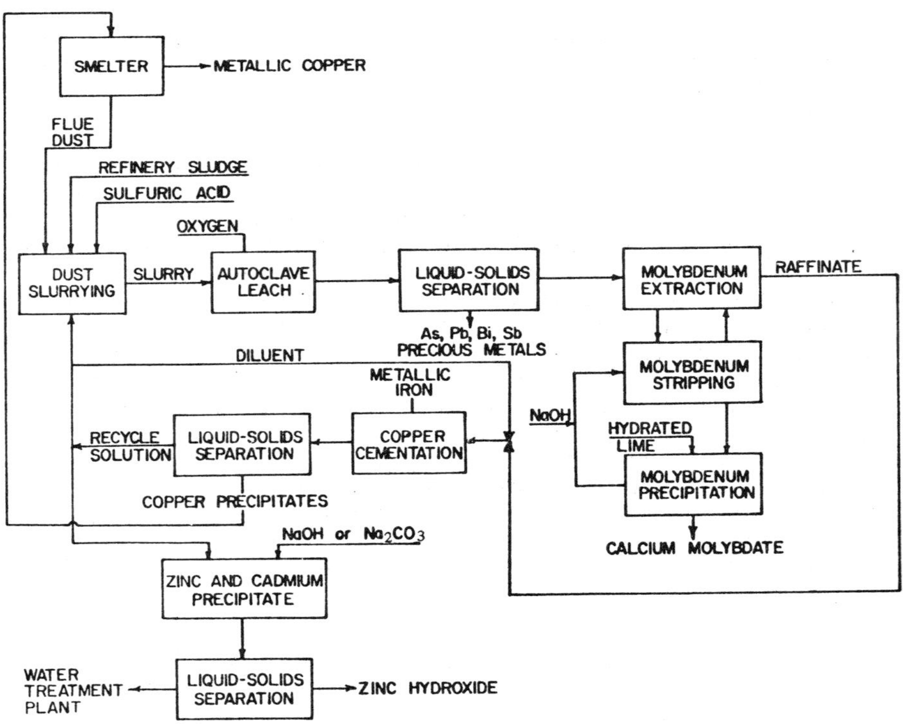

Source: U.S. Patent 4,149,880.

A typical flue dust will analyze, by weight: 13.6% copper; 0.9% molybdenum; 2.1% zinc; 13.5% arsenic; 1.71% bismuth; 14.7% lead; 0.2% antimony; 0.2% cadmium; and 3.8% iron.

The flue dust and refinery sludge if desired are formed into a slurry by the addition thereto, in appropriate mixing equipment, of the aqueous recycled solution from the copper cementation stage of the process. The slurry will usually have an iron content of from 15 to 35 g/ℓ due to iron in the recycled solution. It will also have a pH that is usually below 2 due to the acid formed when the aqueous solution is added to the flue dust and because of the acid in the recycled solution. Additional sulfuric acid may be added to the slurry, if necessary, to bring it to the desired level for satisfactory copper recovery.

The aqueous, acidic slurry is introduced into an autoclave. Leaching is carried out in the autoclave within the lower temperature end of the range previously mentioned for the prior art proposal, typically at about 140°C (284°F) and at a considerably lower oxygen overpressure than is within the range previously mentioned, typically at about 50 lb per square inch. The resulting slurry is subjected to liquid-solids separation, usually by filtration, to separate the copper-bearing solution from residual solids, which include precious metal values along with arsenic, bismuth, lead, and antimony values in insoluble form. These are washed and passed to waste, e.g., to the usual tailings pond following recovery of the precious metals by conventional methods. Because of their aqueous insolubility, there is no danger of harming the environment.

It is usually desirable, though not necessary, to pass the pregnant leach solution through a molybdenum recovery circuit (see U.S. Patent 4,026,988), as indicated, for salvaging molybdenum values. In any event, it is passed to a conventional copper cementation stage, where the copper is precipitated on metallic iron. Surprisingly, despite the fact that the flue dust leach stage described above leaves a significant amount of the arsenic values, e.g., from about 0.5 to 2.0 g/ℓ and typically 1.0 g/ℓ, in the leach solution subjected to cementation of copper on metallic iron for recovery of the copper values, there is substantially no evolution of highly toxic arsine gas during the cementation stage.

Some of the arsenic remaining in solution is precipitated with the copper in the cementation stage. Copper is generally precipitated first, however, so that the amount of arsenic precipitated with the copper can be held at a very low level by holding copper recovery in the cementation stage to about 95%. Overall copper recovery is not affected by this, since the solution is recycled.

Following separation of the liquid and solid phases, as by a conventional filtration step, the precipitate copper solids are preferably recycled to the smelter, while barren solution is recycled for use in slurrying the smelter flue dust and as a diluent for the copper-bearing solution sent to the cementation stage. It should be noted that this supplies iron values additional to those contained in the autoclave feed material, which is desirable.

It is usually advantageous to control impurity buildup in the solution and to recover zinc and cadmium values by establishing a quantitatively minor bleed stream of the barren solution and neutralizing the bleed stream by the addition of sodium carbonate or hydroxide thereto to precipitate zinc values as zinc hydroxide. The cadmium values come out with the zinc values. Following filtration, the liquid phase may be sent to waste or to a water treatment plant.

COPPER FROM SMELTER SLAGS

Slag material resulting from the smelting of ores, such as sulfide copper ores, is generally considered a waste product. The slag material normally contains appreciable amounts of various metal values, including (depending on the nature of the ore) copper, iron, zinc, lead, and other materials all in combination with alkaline earth metals in a matrix of metallic and nonmetallic silicates. The recovery of these metal values has not been found attractive up to the present due to the complex nature of the slag material and the tendency of it to form gelatinous products upon processing.

An improved process for copper recovery from smelter slags has been developed by *K.O. Lindblad and R.E. Dufresne; U.S. Patent 3,868,440; February 25, 1975; assigned to The Anaconda Company.*

It is a hydrometallurgical process for treating slag material, especially copper smelter slag material which comprises vigorously mixing the slag material with about 1 pbw of H_2SO_4 and about 1 pbw of water for each part by weight of the slag, and allowing the mixture to react to produce a dry, solid material, containing the metal values of the slag in water-soluble form. These solubilized metal values may be separated by leaching from the insoluble silicious residue of the dry solid material. Such a process is shown in block diagram form in Figure 46.

A process developed by *J.W. Donaldson, S.N. Sharma and N.J. Themelis; U.S. Patent 4,032,327; June 28, 1977; assigned to Kennecott Copper Corporation* is a cyclic, two stage process for pyrometallurgical recovery of copper as a high-grade matte from molten slag material containing both dissolved and/or entrained copper values. A charge of the copper-bearing slag material is treated in the first stage of the process with a hydrocarbonaceous or hydrogenous reductant in a rotary furnace, such as a top blown rotary reactor, rotating at about 10 rpm, to form a high-grade copper matte phase and a slag phase of reduced copper content.

The rotation of the reactor is discontinued and the high-grade copper matte phase separates from the slag phase and is recovered from the reactor. The reactor containing the remaining slag of reduced copper content is again rotated at about 10 rpm, and iron sulfide, e.g., pyrite, is added in the second stage of the process to the slag in the rotating reactor to provide a low-grade matte phase which literally washes out most of the remaining copper contained in the slag.

The rotation of the reactor is again discontinued and the low-grade matte phase separates from the remaining slag phase. The remaining slag phase is removed from the reactor, and the low-grade matte produced in the second stage is left in the reactor to mix with an additional charge of slag material in a subsequent cycle of the process. Figure 47 shows a suitable form of apparatus for the conduct of such a process.

The reactor used in the process is preferably a top-blown, rotary reactor as developed and used in the Kaldo process for producing steel.

Figure 46: Anaconda Co. Hydrometallurgical Process for Recovery of Metal Values from Copper Slag

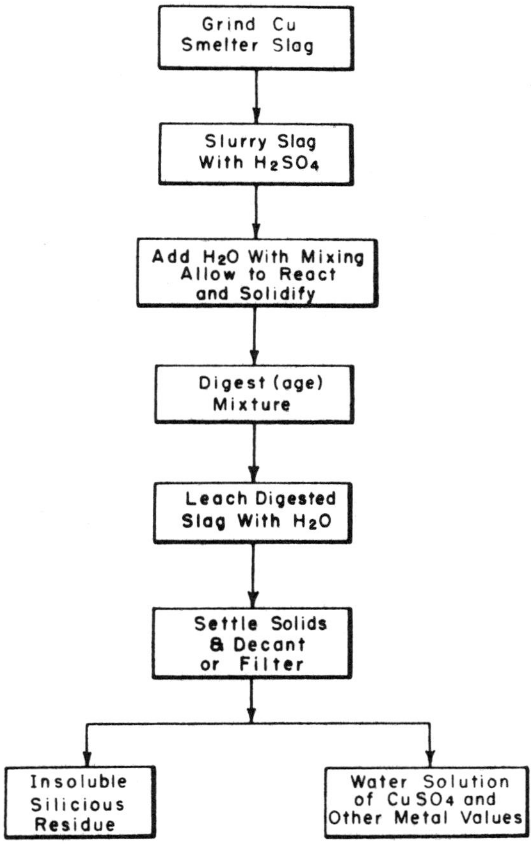

Source: U.S. Patent 3,868,440.

A side view of such a reactor is shown in the drawing. The reactor comprises a vessel **10** having a cylindrical wall **11**, a generally circular end wall **12**, and an opposite, open, frustoconical end **13**. The vessel is preferably fabricated from steel plate and is lined with refractory brick, as shown. It has steel track rings **14** positioned around cylindrical wall **11** for riding on motor driven, support rollers **15**. These support rollers **15** are mounted on annular frame **16**, which circumscribes cylindrical wall **11** of the vessel. Thrust rollers **17** are mounted on frame **16** so as to engage the side of the track ring **14** nearest end wall **12** of the vessel. Vessel **10** is rotatable about its cylindrical axis, with the motor-driven support rollers **15** providing the means for rotating it.

Figure 47: Kennecott Copper Corp. Apparatus for Pyrometallurgical Recovery of Copper from Slag Material

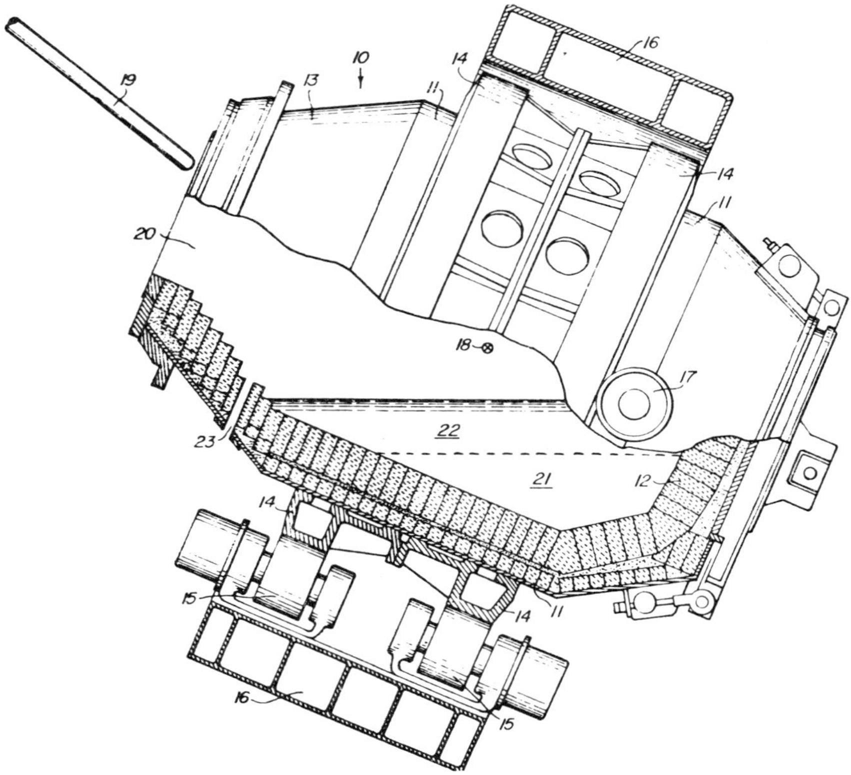

Source: U.S. Patent 4,032,327.

Frame **16** can be tilted about a horizontal pivot axis **18**, so vessel **10** can be tilted to a small angle below and above the horizontal. Means are provided for heating the reactor and contents thereof. As shown, a fuel burner **19** is positioned near the open mouth **20** of vessel **10**, so as to provide for combustion of liquid or finely divided solid fuel in the vessel **10** above the level of material therein. During operation, the rotary reactor is inclined at about 15 to 20° to the horizontal, with its open end uppermost. Speed of rotation is as found convenient, from about 10 to 40 rpm.

At the conclusion of the first stage of the process, rotation of the reactor is stopped and a high-grade matte phase **21** rapidly separates from a supernatant slag phase **22**. The high-grade matte **21** is recovered, leaving the slag phase **22** in the vessel for further treatment in the second stage of the process.

The high-grade matte **21** can be recovered from the reactor vessel **10** in several ways. Thus, the vessel **10** can be tilted so that its open end **20** faces downward, allowing the slag **22** to be poured into a holding ladle (not shown). The matte **21** is then poured into a recovery vessel, and the slag in the holding ladle is returned to the reactor vessel **10**.

Transferring the slag to a holding ladle can be avoided by providing the reactor vessel **10** with means for undertapping the matte layer **20**. As shown, a normally plugged taphole **23** is located in the frustoconical section **13** of the vessel **10** so as not to be submerged by the molten bath during normal rotation of the vessel. When the matte **21** is to be undertapped, the open end of the vessel **10** is tilted downwardly until the taphole **23** is covered by matte.

If found necessary, a refractory dam can be built up in the vessel about the open mouth **20** to prevent molten slag from pouring out. The refractory material plugging the taphole is then removed and the matte **21** is drained from under the slag. Following removal of the matte **21**, taphole **23** is again plugged, and the vessel **10**, containing only the slag **22**, is returned to its normal operating position.

Although not shown in the figure, the reactor vessel **10** could also be provided with a sliding gate or other mechanical valve to eliminate the necessity of removing the slag phase from the vessel at the end of the first stage of the process. This sliding gate or other mechanical valve would preferably be positioned in the cylindrical wall **11** of the vessel **10** near end wall **12**. Sliding gate valves have been used extensively in the steel industry to empty ladles and other similar vessels. In use, the vessel **10** would be rotated until the sliding gate or other mechanical valve is at its lowest position. The valve would then be opened to undertap the matte phase **21** from the vessel **10** and closed thereafter to leave the slag phase **22** in the vessel **10** for further treatment.

COPPER FROM SPENT CATALYSTS

The oxidation of ethylene in the homogeneous phase in a solution containing hydrochloric acid, copper and palladium chloride, represents an important process for producing acetaldehyde. It is characterized by a high level of economy and technical efficiency but a solid precipitate which forms in the contact solution during catalysis constitutes a serious disadvantage.

This leads to deactivation of the content fluid so that the contact fluid cannot be used for the desired length of time but has to be periodically clarified by filtration or decanted from the precipitated metal containing substances. Also it must be enriched with new metal salts to prevent any blockages in the supply lines, in addition, to the output drop of the reactor. A particular disadvantage is the wastage or loss of costly noble metal which occurs.

A process developed by *R. Thiel, C. Zinsstag and G. Faschinger; U.S. Patent 3,947,543; March 30, 1976; assigned to Lonza, Ltd., Switzerland* is one for recovering copper and noble metals from such organic precipitates, residues and polymerizates which involves subjecting the organic components to a slow, low-temperature carbonization process without the use of any additional heat supplying elements. The metals remain in the ash in the form of salts. Normally the metals in the resulting solution are not separated out but can be recycled directly to the process as active components.

No additional components need be added. Typically, the burning process takes place with a conversion rate of approximately 1%/hr. Preferably the burning process takes place at a temperature between 300° and 500°C. The ash is obtained in a form in which it almost completely dissolves in suitable acids, preferably hydrochloric acid. Preferably the process is used to burn the copper and palladium chloride containing residues obtained from the process of the oxidation of ethylene to form acetaldehyde.

A process developed by *R. Baatz, G. Beilstein, D. Grenner, W. Keller and D. Steinbach; U.S. Patent 4,070,405; January 24, 1978; assigned to Bayer Aktiengesellschaft, Germany* is one for working up a liquid copper complex-containing catalyst waste stream from the isomerization reaction of dichlorobutene wherein the copper complex dissolved in dichlorobutene is precipitated at a temperature of from 0° to 250°C by the addition of an apolar solvent to the waste stream and the precipitated copper complex is recovered by filtration.

Such a process is shown both as batch and continuous versions in Figure 48.

Figure 48: Bayer AG Process for Working Up Liquid Copper-Containing Catalyst Waste from Dichlorobutene Isomerization by Batch Process (Above) and Continuous Process (Below)

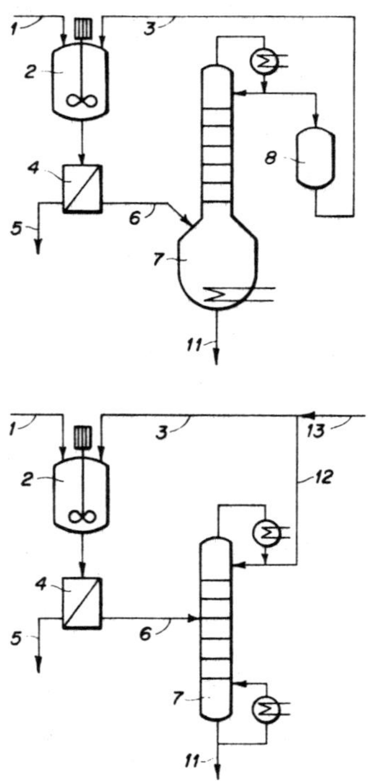

Source: U.S. Patent 4,070,405.

In the batch version the waste stream **1** containing the copper complex is mixed with from 2 to 20 times the quantity of an apolar solvent **3** in a mixing vessel **2**, which may be provided with a stirrer. A deposit is precipitated, containing substantially the total quantity of copper complex and by adsorption significant proportions of the high-boiling fractions present. The deposit is filtered off in a filter **4** and as a solid filter cake **5** is either dumped or treated to recover the copper present in it.

The filtrate **6** is substantially free from copper. It is transferred into the still of a distillation column **7** and rectified. Virtually all the petroleum ether distils off overhead and is collected in the receiver **8** and introduced back into the precipitation vessel **2**. Any fractions of dichlorobutene which may be present are harmless. In addition to dichlorobutenes, the distillation residue **11** contains higher-boiling compounds which, after recycling into the chloroprene process, are separated off from the dichlorobutenes and delivered to a suitable destruction plant. They only make up about 1 to 10% of the original quantity of waste and moreover are free from copper.

In the continuous version the copper complex is precipitated from the waste stream **1** in the container **2** by means of an apolar solvent **3** in the same way as described above. The deposit **5** is discharged from a continuously operating filter **4** or from a filter battery designed to be operated in alternation. The filtrate **6** is delivered to the middle of a distillation column **7**. The petroleum ether **3** which may be reused for precipitation is run off overhead from the distillation column **7**. The high-boiling residue accumulates in the sump in admixture with dichlorobutenes **11** which are recovered after separation by distillation in the chloroprene process while the copper-free, high-boiling residue is removed and destroyed.

By means of the process, it is possible to reduce the accumulation of residues to be destroyed from the copper-catalyzed isomerization of dichlorobutenes to between 1 and 10% of the original quantity and at the same time to free them from copper (copper content less than 5 ppm). This represents a significant contribution towards keeping air and water clean. The copper catalyst accumulates in a form in which it can be safely transported and subsequently used, e.g., for the production of copper. A significant economic advantage of the process is the fact that most of the dichlorobutene present as solvent for the copper complex is recycled into the isomerization process.

COPPER FROM SPENT ETCHING SOLUTIONS

Solutions containing copper values are produced when metallic copper is subjected to the etching action of alkaline etching solutions. Efforts have been made in the past to remove the dissolved copper values from spent etching solutions, either to regenerate such solutions for reuse or to recover the metal values or both.

Considerable quantities of copper-containing, ammoniacal etching solutions are sent to waste by manufacturers of printed electronic circuit boards despite recognition by the art that it is advantageous to regenerate etching solutions and recover the etched copper values therefrom. This has created pollution problems and considerable expense for etching solution. Although in some instances copper values have been precipitated from such solutions with lime, as copper oxides, and have been used in food processing and wood treating industries, this has not proven to be an entirely satisfactory solution to the problems presented.

A process developed by *W.D. Hamby and M.D. Slade; U.S. Patent 4,083,758; April 11, 1978; assigned to Criterion* is one in which copper values are recovered in metallic form from a solution containing such values along with chloride ions by, first, selectively removing the chloride ions from the solution, and then electrowinning the copper from the chloride-free solution. The process is particularly useful in recovering copper values from etching solutions in a metallic form, such as wire grade metallic copper, that is available for fabrication into commercial copper products.

The process is preferably applied to ammoniacal etching solutions derived from the manufacture of printed electronic circuit boards. As so applied, the chlorides are selectively removed by liquid-liquid extraction procedures, resulting in a substantially chloride-free, acidic stripping solution containing the copper values, which solution is subjected to electrolysis for winning copper metal. The aqueous raffinate from the extraction stage is treated for the removal of residual organic and is available for reuse as an etching solution when brought up to the necessary strength by addition of ammonia. Figure 49 shows such a process.

Figure 49: Criterion Process for Regenerating and for Recovering Metallic Copper from Chloride-Containing Etching Solutions

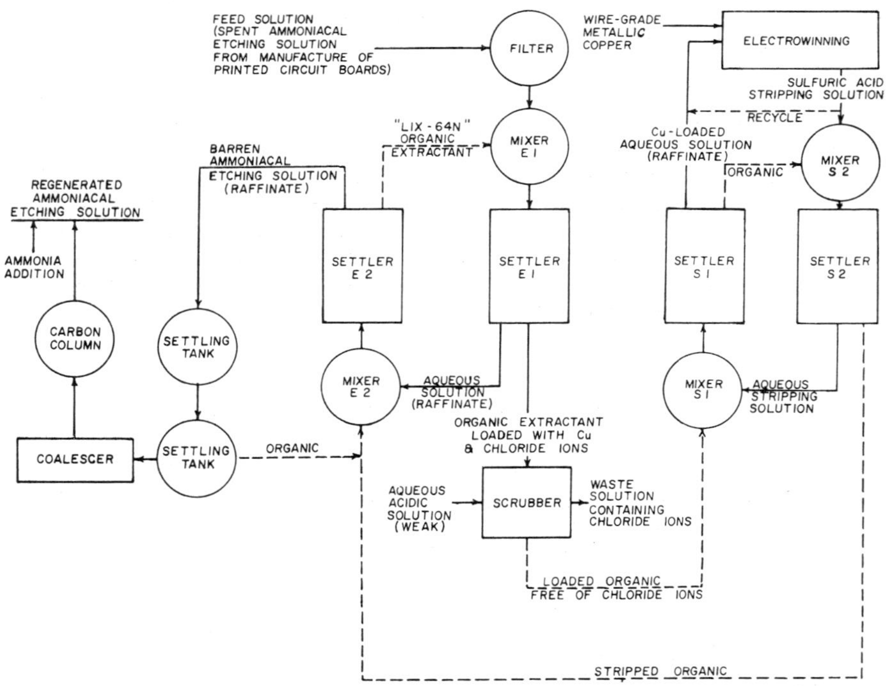

Source: U.S. Patent 4,083,758.

The specified feed solution is made up primarily of ammonium hydroxide and ammonium chloride and typically contains from 16 to 24 oz/gal of copper values, with about 10 to 50 ppm of lead and traces of tin.

As illustrated, the feed, constituting an ammoniacal etching solution containing copper values intermixed with chloride ions, is run through a filter to the usual mixer and settler units of liquid-liquid separation equipment. Such equipment may be utilized in as many extraction and stripping stages as deemed proper in any given instance, there being here shown two extraction stages designated **E1** and **E2**, respectively, and two stripping stages designated **S1** and **S2**, respectively.

The extractant employed as the organic medium is an alpha-hydroxy oxime, LIX 64N (General Mills Chemicals, Inc.), in a liquid hydrocarbon vehicle pursuant to the process disclosed by U.S. Patent 3,224,873. The aqueous feed solution is contacted by this organic extractant in units **E1** and **E2** to form a dispersion.

The organic phase extracts both the copper and the chloride ions from the spent ammoniacal etching solution and the chloride ions can be selectively removed therefrom by scrubbing the loaded organic extractant with a weak sulfuric acid solution. Accordingly as shown, the loaded organic phase is passed to a conventional scrubbing tank and scrubbed by a weak sulfuric acid solution for the selective removal of chloride ions in an aqueous solution which is normally sent to waste.

The Cu-loaded, organic extract, substantially free of chloride ions, is passed to the first stripping stage **S1**, from which the copper-ion-loaded, aqueous raffinate is passed to an electrowinning cell or cells as the electrolyte therefor.

The stripping medium is a relatively strong sulfuric acid solution taken from the electrowinning cells and successively mixed with the scrubbed, chloride-free, organic extract in the stripping units **S2** and **S1**. Some of such sulfuric acid solution (spent electrolyte) may be recycled directly to the electrowinning cells, as indicated, for building up sulfuric acid strength. The end product of the electrowinning stage is a commercial wire grade metallic copper as widely used in copper fabrication.

The aqueous effluent from the extraction stages **E1** and **E2**, i.e., the raffinate, is barren ammoniacal etching solution which may carry residual amounts of the organic extractant. To complete the regeneration aspect of the process, such solution is treated for the removal of any organic residual. Such treatment may take a variety of forms.

As here shown, the raffinate solution from the final extraction stage **E2** is passed into a series of settling tanks, indicated as such on the figure, from where any organic that is separated from the aqueous phase is recycled to the extraction stages. It is preferred to then pass the aqueous phase into a coalescer of standard type and to pass the effluent from that through one or more carbon columns to rid the aqueous solution of the last traces of organic. The resulting effluent constitutes the completely regenerated ammoniacal etching solution.

Since the regenerated ammoniacal etching solution is ordinarily somewhat depleted in its ammonia content it may be necessary to add ammonia, as indicated, to bring the solution up to appropriate etching strength. The end product is a regenerated etching solution that may be reused for etching purposes.

COPPER CATALYST FROM ADIPIC ACID MANUFACTURE WASTES

See "Vanadium Catalyst from Adipic Acid Manufacture Wastes".

CRYOLITE FROM POT LINE RECOVERED PARTICULATES

Aluminum is produced by the electrolytic reduction of pure alumina in a bath of fused cryolite ($3NaF{\cdot}AlF_3$). The reaction is carried out in an electrolytic cell called an aluminum reduction pot in which the alumina is dissolved in a molten cryolite bath. The solution of alumina in molten cryolite is electrolyzed to form metallic aluminum. The aluminum produced in the reaction is heavier than the electrolyte and so forms a molten layer at the bottom of the reduction pot which serves as the cathode of the cell. Carbon anodes extend into the bath and oxygen liberated at the anode oxidizes carbon from the electrode.

Control of the electrolyte composition is an important consideration in the aluminum production process. The electrolyte contains cryolite and fluorspar (calcium fluoride, CaF_2). Some excess aluminum fluoride is also present in the electrolyte which, with the dissolved alumina, reduces the liquidus temperature to the point where the cells can be operated in the 940° to 980°C range.

A large volume of gas, with gas-entrained dust, is emitted by the batteries of reduction pots used in the production of aluminum. The gases emitted include carbon dioxide, volatilized fluorides and gaseous HF. Particulate solids, including fine alumina, cryolite, carbon and organics, are elutriated from the electrolytic bath with the gases. Small amounts of lithium, calcium, silica, iron, sodium sulfate and magnesium are solids also present in the gas stream.

The volatilized fluorides and the gaseous hydrogen fluoride are collected, together with other gases and entrained solids evolved from the cells, by gas-collecting hoods or manifolds and are passed through large ducts to central gas treatment and recovery facilities. The loss of these materials from the electrolyte necessitates periodic additions of aluminum fluoride to maintain the desired bath composition.

In some cases, the gaseous products with entrained particulates have simply been discharged into the atmosphere, the recovery of the particulates being judged too difficult because of their fine particle size (less than 2 microns) and contamination. However, at best, simple discharge of the waste fumes from aluminum reduction pot lines into the atmosphere is no longer feasible in view of environmental considerations.

Accordingly, rather elaborate and expensive gas treatment equipment including a wet scrubbing system has been provided or is contemplated at many facilities. The wet scrubbing is quite efficient in removing solid particulates from the gas stream, and accordingly, a substantial quantity of very fine, potentially valuable but contaminated solids are available and must be disposed of or utilized in some fashion.

A process which has been developed by *E.J. Roberts, S. Bunk and P.A. Angevine; U.S. Patent 4,053,375; October 11, 1977; assigned to Dorr-Oliver, Inc.* is one in which a slurry of alumina-cryolite waste from aluminum pot lines is subjected to oxidation in a fluidized bed reactor to remove carbon and organic contamination. Agglomeration of the alumina-cryolite product occurs in the fluidized bed and a pellet product suitable for return to the aluminum pot lines is produced. Such a reactor is shown in Figure 50.

Figure 50: Dorr-Oliver Inc. Process for Recovery of Alumina-Cyrolite Waste in Aluminum Production

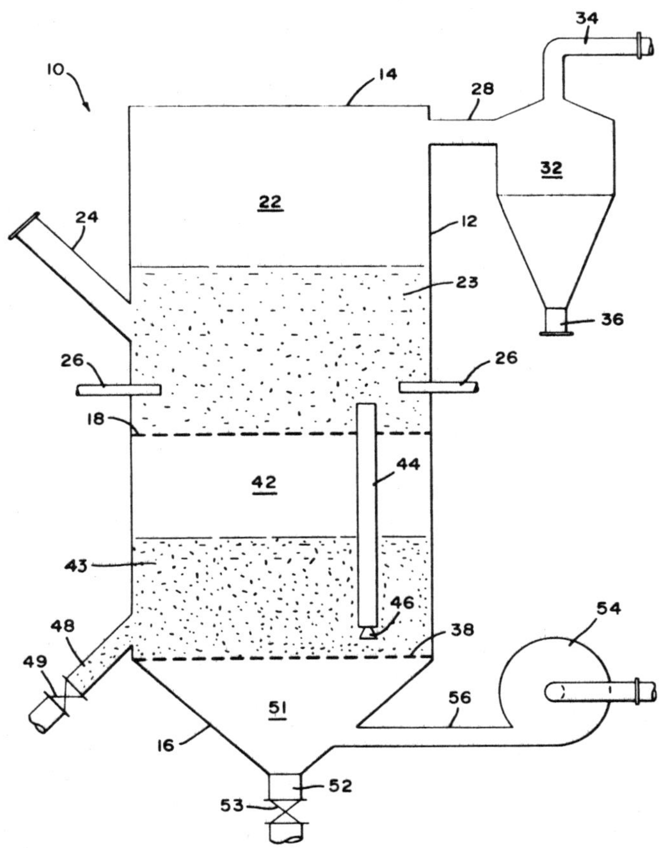

Source: U.S. Patent 4,053,375.

Referring to the figure, the reactor **10** comprises an outer shell **12** which is capped by a cover **14**. The conical portion **16** forms the bottom of the reactor **10** and is provided with a cleanout port **52**. Within shell **12** the reactor **10** is provided with an upper constriction plate **18** and a lower constriction plate **38**. Constriction plate **18** divides the interior of reactor **10** into a treatment chamber **22** located in the upper portion thereof and a cooling chamber **42** located in the lower portion of the reactor. Constriction plate **38** separates the cooling chamber **42** from the windbox **51** which is located between the constriction plate **38** and the conical portion **16**. The constriction plate **18** supports a fluidized bed **23** composed of particulate solids. A feed pipe **24** provides access to the treatment chamber **22** and a plurality of fuel guns **26** penetrate the shell **12** above the constriction plate **18**. An exhaust gas conduit **28** is provided near the top of treatment chamber **22** and is connected to a cyclone **32** having an off-gas conduit **34** and a solids underflow conduit **36**.

Constriction plate **38** supports a fluidized bed **43** in the cooling chamber **42**. A transfer conduit **44** having a control valve **46** therein communicates between fluidized bed **23** and treatment chamber **22** and fluidized bed **43** and cooling chamber **42**. A product conduit **48** communicates with fluidized bed **43** and has a control valve **49** therein. A blower **54** communicates with windbox **51** through air conduit **56**.

In operation, fluidized beds are established in both the treatment chamber **22** and the cooling chamber **42**, both sustained in a fluidized condition by air supplied by blower **54** through conduit **56**. A feed of waste particulate solids separated from the off-gases of an aluminum pot line is supplied in a moist or wet state through the feed pipe **24** into treatment chamber **22**. In this chamber the moisture in the charge is rapidly evaporated and the solids are heated to the treatment temperature in the violently agitated environment provided by the fluidized bed **23**. The fluidized bed **23** is maintained at a temperature in the range from 770° to 800°C or preferably, at a temperature of 790±5°C.

While this temperature can often be achieved simply by combustion of the carbon and organics present as impurities in the charge, it may sometimes be necessary to introduce auxiliary fuel into the fluidized bed **23** through the fuel guns **26**. The oxygen for this combustion is supplied by the fluidizing air. The off-gases of the reactor comprise HF, CO_2 and water vapor and may be fed into the pot line gas collection system.

At the specified temperature it is found that at least a small amount of a liquid phase is present in the bed constituents and there is a strong tendency for the extremely fine solid particles to agglomerate or pelletize. The pellets from the bed **23** move down the transfer conduit **44**, into the cooling bed **43**.

In the cooling chamber **42** the pellets which form the fluidized bed **43** are cooled to the temperature range from 200°F (93°C) to 250°F (121°C) and may be continuously removed from the reactor **10** through the product outlet **48** which is controlled by valve **49**. It should be understood that the fluidizing air provided by blower **54** not only cools the product pellets in traversing the cooling chamber **42**, but is itself preheated to a temperature in the range from 93°C to 121°C so that it is a preheated fluidizing gas which passes through constriction plate **18** to fluidize the particulate bed in treatment chamber **22**. The fluidized cooling bed is convenient but not essential to the process. Other types of cooling procedures can be utilized satisfactorily in the process.

For the purposes of illustrating the advantages of the process, the following example is given. A feed of solids elutriated from an aluminum pot line has the following composition (dry weight):

Constituent	Wt %
Cryolite	49.2
Carbon, organics	26.4
Al_2O_3	12.0
Na_2SO_4	5.5
CaF_2	1.8
AlF_3	1.5
Na_2CO_3	1.0
Fe_2O_3	0.7
LiF	0.4
SiO_2	0.3
P_2O_5	0.04
NiO	0.2
MgO	0.2
Miscellaneous	Balance less than 1%

This feed is introduced into a fluidized bed reactor operating at a temperature of 790 ±5°C. The starting bed is composed of alumina particles. The product withdrawn from the bed has an average particle size such that at least 90% of the particles produced are retained on a 65 mesh (Tyler mesh) screen; i.e., a particle size of 0.016 inches or greater. The product composition (dry weight) is as follows:

Constituent	Wt %
Cryolite	50.4
Al_2O_3	30.4
NaF	7.5
Na_2SO_4	6.6
CaF	1.9
Fe_2O_3	1.0
LiF	0.5
SiO_2	0.4
Na_2CO_3	0.2
MgO	0.2
Miscellaneous	Balance less than 1%

The product is in the form of very regular spherical pellets readily handled for feeding to the aluminum pot lines. The product composition is such that it is highly suitable for pot line feed.

D

DICALCIUM PHOSPHATE FROM PHOSPHATE POND WATERS

See "Calcium Fluoride from Phosphate Pond Waters."

DINITROGEN MONOXIDE FROM ADIPIC ACID PROCESS OFF-GASES

In addition to dinitrogen monoxide, the oxidation off-gases formed during the preparation of dicarboxylic acids, for example, adipic acid by oxidation of the corresponding cycloalkanols and/or cycloalkanones also contain other gaseous constituents, such as nitrogen monoxide, nitrogen dioxide, nitrogen, carbon dioxide and steam.

In order to isolate dinitrogen monoxide, which can be used, for example, in large amounts as an anaesthetic, from the off-gases of the cycloalkanol/cycloalkanone oxidation, after most of the nitrogen monoxide and nitrogen dioxide have been removed, it is possible, for example, to subject the oxidation off-gases to a pressure wash.

However, this process is not very suitable for isolating dinitrogen monoxide since dinitrogen monoxide is only slightly soluble in water. Even under high pressure, only a few percent by weight of dinitrogen monoxide dissolve in water, so that large amounts of water are required for absorbing the dinitrogen monoxide. Such a separation process would be uneconomical because of the large amounts of water which must be circulated, for example, in a recycle procedure.

An improved process for N_2O recovery has been developed by *H-H. Schwarz; U.S. Patent 4,177,645; December 11, 1979; assigned to Bayer AG, Germany.* This process comprises:

(a) Drying the oxidation off-gas and, if necessary removing nitrogen monoxide, nitrogen dioxide and carbon dioxide.

(b) Compressing the resultant off-gas from which water has been removed to a pressure in the range of 15 to 300 bars and cooling the same to a temperature in the range of 0° to -88°C.

(c) Removing the liquid phase obtained as a result of the compression and cooling; and

(d) Cooling the resultant gas phase by expansion and passing the same in indirect heat exchange with off-gas which has been prepurified by removal of water content thereof.

F

FERRIC CHLORIDE FROM SPENT PICKLE LIQUOR

Ferric chloride is a product used in large quantities, particularly for the purification of drinking, industrial and waste water; for this kind of application it is necessary for the ferric chloride to be of great purity, failing which it will in turn become a contamination agent for water.

It is known to produce anhydrous ferric chloride by the action of gaseous chlorine on granulated iron at bright red heat, or on iron oxide and iron phosphate at a temperature of about 1000°C, or else from iron oxides and hydrochloric acid.

Similarly, it is known to operate in aqueous solution by dissolving granulated iron in hydrochloric acid and oxidizing the resulting ferrous chloride with gaseous chlorine or nitric acid in the presence of hydrochloric acid.

Processes also exist in which the raw material used is ferrous chloride, which is oxidized with gaseous chlorine, either at high temperature or in aqueous solution.

An attractive alternative route which would seem economically attractive would be to produce ferric chloride from spent hydrochloric acid pickling liquors which contain large amounts of ferrous chloride, generally between about 20 and 25% by weight, when they are exhausted. However, one disadvantage of this raw material is that it is heavily contaminated, particularly by free hydrochloric acid and by the auxiliary substances used in processing iron, particularly passivating agents, lubricating oils or greases, tars, etc. Consequently, a first condition is the development of a process making it possible to obtain ferric chloride free from these contaminants, so that the resulting aqueous solution of ferric chloride can be used directly for the purpose of purifying water.

Furthermore, if the process is to be economically viable it is also important to be able to carry out this process in a continuous manner so as to obtain an aqueous solution of ferric chloride of constant quality.

A process which is designed to meet these requirements has been developed by *J.-M. Lietard and G. Matthijs; U.S. Patent 4,066,748; January 3, 1978; assigned to UCB, SA, Belgium*. In this process an aqueous solution of ferric chloride directly usable for purifying water is produced from a pickling liquor containing ferrous chloride by a process comprising the steps of:

(1) continuously concentrating the pickling liquor by evaporation to at least about 34.25% by weight ferrous chloride concentration;

(2) continuously neutralizing the resulting concentrated liquor, which still contains free hydrochloric acid, by means of a fluidized bed of iron oxide particles (e.g., mill scales); and

(3) continuously countercurrently chlorinating the resulting neutralized concentrated liquor in two successive reaction zones, to the extent of from 73 to 86% at a temperature of from 35° to 75°C in the first reaction zone and the balance at a temperature of from 40° to 100°C in the second reaction zone.

The product of the process is an aqueous solution containing per liter at least 40% by weight of ferric chloride, less than 0.1% by weight of ferrous chloride and less than 0.1% by weight of free hydrochloric acid. Chlorination yield is practically quantitative.

Such a process is shown in Figure 51 and the process will be described in further detail with reference to that figure.

Figure 51: UCB Continuous Process for Producing an Aqueous Solution of Ferric Chloride from Spent Pickling Liquor

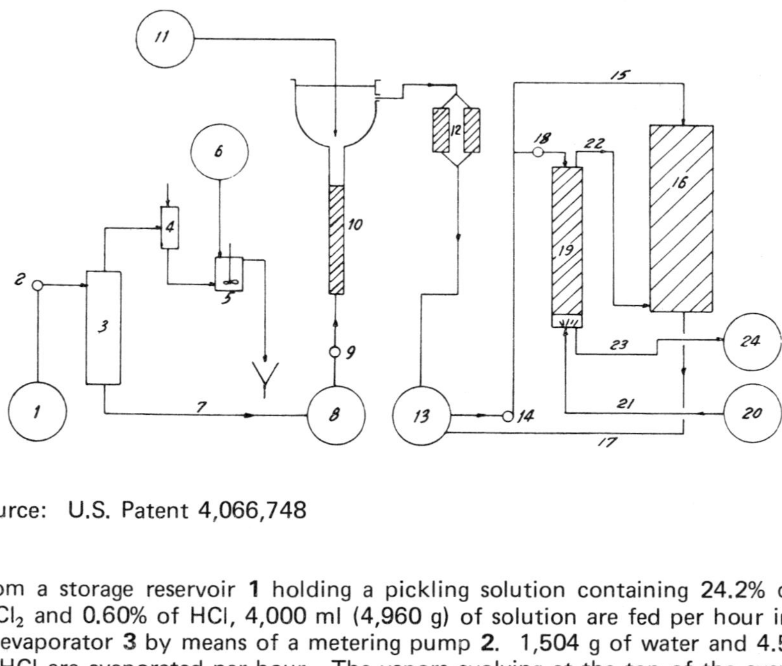

Source: U.S. Patent 4,066,748

From a storage reservoir **1** holding a pickling solution containing 24.2% of $FeCl_2$ and 0.60% of HCl, 4,000 ml (4,960 g) of solution are fed per hour into an evaporator **3** by means of a metering pump **2**. 1,504 g of water and 4.5 g of HCl are evaporated per hour. The vapors evolving at the top of the evaporator are condensed in a washer **4** by a current of cold water. The water then enters a mixer **5** in which it is neutralized by a 30% NaOH solution coming from the reservoir **6**; it then passes out by overflowing and is discharged to the drain.

At the bottom of the evaporator **3** there are obtained per hour 2,550 ml (3,450 g) of a solution containing 34.75% of $FeCl_2$ and 0.74% of HCl; this solution is passed through a thermally insulated pipe **7** to a thermally insulated tank **8**. The still hot solution (80°C) is delivered by the pump **9** into the thermally insulated neutralization reactor **10** in which it passes from bottom to top through a bed of iron oxide mill scales (height 40 cm–640 g of mill scales) at a space velocity of about 42.5 cm^3/min, so that at least part of the mill scales is fluidized. The widened top portion of the reactor reduces the ascending speed by the factor 15, thus substantially reducing the entrainment of small particles by the neutralized solution passing out through a side branch situated below the level of the liquid.

A make up of 27.5 g of mill scales coming from the reservoir **11** is added per hour. Grease and part of the foreign bodies accumulate on the surface of the liquid, from which they are skimmed from time to time. The neutralized solution still contains very fine solid particles, which are eliminated by two glass wool filters **12** used alternately; it is then passed to the thermally insulated storage reservoir **13**. The temperature is about 60°C. Mean analysis gives: $FeCl_2$, 35.25%; $FeCl_3$, 0.47%; and HCl, zero.

The yield of the neutralization is 87.5% based on the weight of the mill scales, and 94.3% based on their iron oxide content. The remainder of the iron, which is found in the skimmed material and on the filters, is not reusable.

From the reservoir **13** the neutralized ferrous chloride solution is passed at a flow of 100 ℓ/hr by means of the circulation pump **14** through the pipe **15** to an absorption tower **16** filled with Raschig rings, and returns to the storage reservoir **13** through the pipe **17**. From this circuit 2,675 ml of $FeCl_2$-$FeCl_3$ solution are taken per hour by means of a metering pump **18** and delivered to the top of the reactor-finisher **19** filled with Raschig rings, in which it is chlorinated in countercurrent.

The gaseous chlorine coming from the cylinder **20** through the pipe **21** is injected at the base of the reactor-finisher. The excess of chlorine leaves the reactor-finisher **19** through the pipe **22**, and passes to the bottom of the absorption tower **16**, in which it is completely absorbed.

On starting up the installation, no $FeCl_2$-$FeCl_3$ solution is fed to the reactor-finisher and all the chlorine is supplied to the absorption tower circuit. As soon as 50% of the $FeCl_2$ has been converted into $FeCl_3$, the feeding of the reactor-finisher is started. The flow of chlorine is adjusted so as to obtain a conversion of $FeCl_2$ to $FeCl_3$ of about 75 to 80% in the absorption tower circuit. When stationary operating conditions are reached, the flow of chlorine is 107 ℓ/hr (342.5 g), and that of the $FeCl_2$-$FeCl_3$ mixture to the finisher amounts to 2,675 ml/hr (3,746 g).

The chlorinated solution leaving the bottom of the finisher **19** (3,814.5 g or 2,680 ml/hr) is passed to the storage vessel **24** through pipe **23**. Mean analysis: $FeCl_3$, 41.45%; $FeCl_2$, 0.02%; HCl, 0.05%; and Cl_2 nil. The chlorination yield is practically quantitative. The assembly comprising the absorption tower circuit and reactor-finisher and all the pipes are thermally insulated in order to maintain the solution at a fairly high temperature and to obtain a good speed of chlorination (50° to 70°C).

FERROALLOYS FROM PLATING SLUDGES

When waste liquor of metal plating is neutralized and then subjected to dehydration and drying treatments, a plating sludge can be separated. However, there have not hitherto been known effective methods of treating the resulting plating sludge, and almost all of the sludge is used in land-fill as an industrial waste. Such industrial waste causes public nuisance and various methods are proposed to solve this problem. However, all of the hitherto proposed methods have drawbacks in the treatment of the plating sludge on a commercial scale, and therefore almost all of the sludge is stacked and left as it is or is solidified by concrete and used for landfill.

While, the plating sludge contains a large amount of valuable metals, such as Ni, Cr, Fe and the like, and it is not preferable from the viewpoint of resource conservation to discard these metals after these metals are converted into harmless state. It is, therefore very important to establish an effective method of treating plating sludge, which can recover the valuable metals.

A process developed by *H. Fukuoka, T. Kameyama, Y. Tanaka, H. Suzuki and K. Imai; U.S. Patent 4,123,257; October 31, 1978; assigned to Nisshin Steel Company Ltd. and Japan Metals & Chemicals, Co., Ltd., Japan* is one in which a plating sludge, whose water content is decreased to not higher than 10% by weight, is mixed with scale, dust, both of which are formed in the alloy steel-making step, a binder and a carbonaceous material, and water is added to the mixture to adjust the water content of the mixture to 7 to 30% by weight, the mixture is kneaded and then molded into a briquet, and the briquet is subjected to a strength-improving treatment.

The resulting briquet can be used as a raw material for ferroalloy. When the briquet is subjected to a reduction smelting to recover a metal, the metal can be used as a raw material for alloy steel.

Such a process is shown in Figure 52 and the process will now be described in more detail with reference to that figure.

Figure 52: Nisshin Steel Co. Ltd. and Japan Metals & Chemicals Co., Ltd. Process for Converting Waste Plating Sludges to Ferroalloys

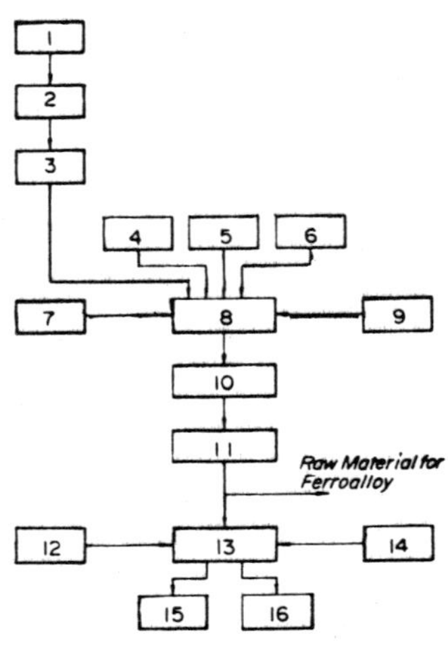

Source: U.S. Patent 4,123,257

When a hydrated plating sludge **1** formed by neutralizing a waste liquor of metal plating and agglomerating the resulting precipitate is intended to dry as such, a large amount of fuel must be used in a dryer **3**, and moreover the feeding of the hydrated plating sludge **1** into the dryer **3** is difficult. Therefore, after the hydrated plating sludge **1** is dehydrated by a dehydrator **2** to decrease the water content to about 50% by weight, the sludge **1** is fed into the dryer **3** and dried therein to produce a powder or granular plating sludge having a water content of not higher than 10% by weight.

To 100 parts by weight of the dried plating sludge having a water content of not higher than 10% by weight is added 10 to 200 parts by weight of dust **4** and 10 to 100 parts by weight of scale **5**, both of which are formed in the alloy steel-making step. There are the following objects in the addition of the dust **4** and scale **5** to the dried plating sludge having a water content of not higher than 10% by weight. The first object of this addition is as follows. The plating sludge contains not only useful metals as a raw material for ferroalloy, but also a relatively large amount of Zn, S, Cu, P and the like, which are harmful elements as a raw material for ferroalloy, as compared with natural ore for ferroalloy, and the plating sludge cannot be used alone as a raw material for ferroalloy. Therefore, the harmful elements are diluted into harmless state.

The second object of the addition of the dust **4** and scale **5** to the dried plating sludge is as follows. The dried plating sludge is present in the form of fine particles consisting mainly of metal hydroxide formed by the neutralization of waste liquor of metal plating, and the sludge itself is poor in the moldability and cannot be made into large briquets having a high strength. Therefore, the dried plating sludge is mixed with scale **5** as an aggregate and further with dust **4** in order to adjust and improve the grain size distribution, whereby the strength of molded briquet is increased.

In addition to the abovedescribed two objects, the dust **4** and scale **5** are added to the dried plating sludge in order to adjust the ingredients of the plating sludge and to treat the sludge on a larger scale, so that the treatment can be carried out on a commercial scale.

In the dust **4** and **5** to be mixed with the dried plating sludge, the dust **4** had been formed in a dry state having a water content of not higher than 10% by weight, and it is not necessary to dry the dust **4**. Further, the water content of the scale **5** can be easily decreased to not higher than 10% by weight through a simple filtering, and it is not necessary to dry the scale **5**. The reason why the water content of the dried plating sludge is limited to not higher than 10% by weight is that, when the water content exceeds 10% by weight, it is difficult to adjust the amount of water in the following kneading step, and the molding operation becomes difficult.

Into a kneader **8** are charged the dried plating sludge, whose water content has been decreased to not higher than 10% by weight as described above, the dust **4** formed in dry state, the scale **5**, whose water content has been decreased to not higher than 10% by weight by a simple filtering, and further a carbonaceous material **6** and a binder **7**. Then, water **9** is added to the resulting mixture to adjust the water content of the mixture to 7 to 30% by weight, and the resulting mixture is fully kneaded in the kneader **8**. In this case, as the dried plating sludge, a mixture of a plating sludge and a pickling sludge produced in the following manner can be also effectively used. When a dried plating sludge obtained by merely dehydrating and drying a hydrated plating sludge alone has a high phosphorus content, after a hydrated pickling sludge having a low phosphorus content, which has been formed in the alloy steel-making step, is dehydrated by a dehydrator and dried by a dryer to decrease the water content to not higher that 10% by weight in the same manner as described in the case of drying the plating sludge **1**, the dried pickling sludge may be mixed with the dried plating sludge **1** having a water content of not higher than 10% by weight.

Alternatively, after a hydrated plating sludge **1** is mixed with a hydrated pickling sludge, then the resulting mixture may be dehydrated by a dehydrator **2** and dried by a dryer **3** to decrease the water content to not higher than 10% by weight. When a plating sludge is produced from a hydrated plating sludge and a hydrated pickling sludge, the amount of the hydrated pickling sludge to be mixed with the hydrated plating sludge should be determined depending upon the phosphorus content of the hydrated plating sludge.

In the abovedescribed kneading, the scale **5** is used as an aggregate of the resulting homogeneously kneaded mixture. The addition amount of the scale **5** is 10 to 100 parts by weight based on 100 parts by weight of the plating sludge charged into the kneader **8**. The reason why the addition amount of the scale **5** is limited to 10 to 100 parts by weight is as follows. When the amount of the scale **5** is less than 10 parts by weight, the briquet obtained by molding the homogeneously kneaded mixture is poor in the strength, and the scale **5** is not effective as an aggregate. However, even when the amount of the scale **5** exceeds 100 parts by weight, the effect of the scale **5** as an aggregate does not increase.

The dust **4** is used in order to give a sufficiently high denseness and a high strength to the molded briquet. The addition amount of the dust **4** is 10 to 200 parts by weight based on 100 parts by weight of the plating sludge charged into the kneader **8**. The reason why the addition amount of the dust **4** is limited to 10 to 200 parts by weight is as follows. When the amount of the dust **4** is less than 10 parts by weight, a briquet having a sufficiently high denseness cannot be obtained, and the briquet is poor in the strength. However, when the amount of the dust **4** exceeds 200 parts by weight, the homogeneously kneaded mixture is somewhat poor in the moldability and the strength of the molded briquet lowers.

In general, the carbonaceous material **6** consisting of coal powder and/or coke breeze is used in order to improve the strength of the resulting briquet and to improve the reactivity of the briquet at its reduction smelting by incorporating the carbonaceous material **6** into the briquet. The addition amount of the carbonaceous material **6** is not more than 40 parts by weight based on 100 parts by weight of the plating sludge charged into the kneader **8**. The reason why the addition amount of the carbonaceous material **6** is limited to not more than 40 parts by weight is that, when the amount of the carbonaceous material **6** exceeds 40 parts by weight, the strength of the resulting briquet is considerably low.

The binder **7** is used in order to improve the bonding ability of the resulting briquet. The addition amount of the binder **7** is 10 to 100 parts by weight based on 100 parts by weight of the plating sludge charged into the kneader **8**. The reason why the addition amount of the binder **7** is limited to 10 to 100 parts by weight is as follows. When the amount of the binder **7** is less than 10 parts by weight, the binder does not fully exhibit its binding effect, and the bonding of the resulting briquet is insufficient, and further the moldability of the kneaded mixture is poor, and the resulting briquet is poor in strength. While, when the amount of the binder **7** exceeds 100 parts by weight, the molding operation of the kneaded mixture is difficult and the effect as a binder does not improve effectively, and further the treating cost becomes high, and the use of such large amount of binder is not economic.

Water **9** is used in order to improve the kneading ability of the kneaded mixture and to improve the moldability of the mixture into briquet. The water **9** is used in an amount of 10 to 200 parts by weight based on 100 parts by weight of the plating sludge so that a mixture to be kneaded contains 7 to 30% by weight of water. When the water content of a mixture to be kneaded is adjusted to 7 to 30% by weight, the mixture can be fully kneaded to produce a homogeneously kneaded mixture having proper adhesion and hardness suitable for the production of briquet.

The reason why the water content of a mixture to be kneaded is limited to 7 to 30% by weight is as follows. When the water content is less than 7% by weight the mixture to be kneaded cannot be fully kneaded and the kneaded mixture is molded difficultly into a briquet, and further dust is scattered to make the working environment worse. While, when the water content of the mixture to be kneaded exceeds 30% by weight, the kneaded mixture adheres on the surface of a roll of a briquet-molding machine **10** to make the molding operation difficult, and the resulting briquet is poor in strength.

The homogeneous mixture obtained by kneading under the abovedescribed condition is press molded into a briquet having a certain predetermined shape by a briquet-molding machine **10**. The briquet is somewhat dried during the molding, but the resulting briquet still contains about 7 to 30% by weight of water, and is poor in strength at low temperature. Therefore, the strength of the briquet is improved by a strength-improving machine **11** in order to use the briquet as a raw material for ferroalloy, or in order to prevent the difficulty in the smelting operation in a reduction smelting furnace **13** due to the breakage of the briquet caused by the sudden evaporation of water at the recovery of metal **15** in the furnace **13**.

The strength-improving treatment of the briquet by the strength-improving machine **11** can be carried out as follows. When at least one of pulp waste liquor, bentonite, molasses and CMC is used as a binder, the strength of the briquet may be improved by a drying treatment only. When cement is used as a binder, the briquet is required to be cured by natural curing at low temperature or by steam curing at low temperature in a tunnel furnace.

When a briquet is subjected to a drying treatment as a strength-improving treatment by a strength-improving machine **11** so that the water content of the briquet is not higher than 2% by weight, the crushing strength of the briquet is improved to not lower than 60 kg per one briquet, and the briquet can be easily handled in the following steps.

The strength-improved briquet as described above is used as a raw material for ferroalloy. Alternatively, the briquet is further charged into a reduction-smelting furnace **13** together with a carbonaceous reducing agent **12** and subjected to a dry system reduction smelting in the furnace **13** to recover a metal, and the metal is separated into metal **15** and slag **16**, and the metal **15** is used as a valuable metal. In the reduction of a briquet by the use of the carbonaceous reducing agent **12**, it is preferable to use flux **14**, such as silica, quick lime and the like. As the carbonaceous reducing agent **12**, charcoal, coke and the like can be used.

FERRONICKEL FROM SPENT CATALYSTS

See "Nickel from Spent Catalysts" and specifically U.S. Patent 4,120,698.

FERROUS METALS FROM MUNICIPAL WASTE

See also process described under "Paper from Municipal Waste."

Figure 53 shows a process developed by *P.G. Marsh and D.H. Kohlhepp; U.S. Patent 4,113,185; September 12, 1978; assigned to Black Clawson Fibreclaim, Inc.* for the production of boiler fuel along with ferrous metals recovery from mixed municipal wastes.

The collection trucks dump directly into a pit **100** from which the garbage is fed into a pulper **101** equipped with a junk remover **102** dumping into a separator **103** which removes the ferrous metals at **104** and returns the lights by line **105** to the pulper **101**.

The slurry extracted from the pulper **101** by line **109** is supplied by a pump **110** to a cyclone cleaner **111** from which the heavies are delivered to a minerals recovery system **112**. The lights pass from cleaner **111** to a surge chest **113**, thickeners **114** and fuel storage bin **115**.

The trash fraction of the solid waste which is collected separately from the garbage is preferably first delivered to a station **120**, such as a conveyor, where it can be sorted into a metals rich fraction **121** and an organics rich fraction **122**, for example, by means of a manually controlled cherry picker or the like. The metals rich fraction is delivered to a shredder **123** from which the output goes to a first magnetic separator **125**. The heavies sorted out by the magnetic separator **125** are delivered by line **126** to a second magnetic separator **127** which in turn delivers the separated heavies to a ferrous metals recovery station **130**.

The nonmagnetic materials separated by each of magnetic separators **125** and **127** are delivered to a line **131** by which they are transported to the same pit **100** to which the garbage is initially delivered.

The organics rich fraction from the station **120** is also delivered to a shredder **135** the output of which is delivered to an air classifier **136**. The heavies from the air classifier **136** are delivered by a line **137** to the line **126** connecting the magnetic separators **125** and **127**, so that any nonmagnetic materials entrained therewith will be separated out at the magnetic separator **127** and delivered to the line leading back to the pit **100**.

Figure 53: Black Clawson Fibreclaim, Inc. Process for Recovery of Ferrous Metals and Producing Fuel from Mixed Municipal Wastes

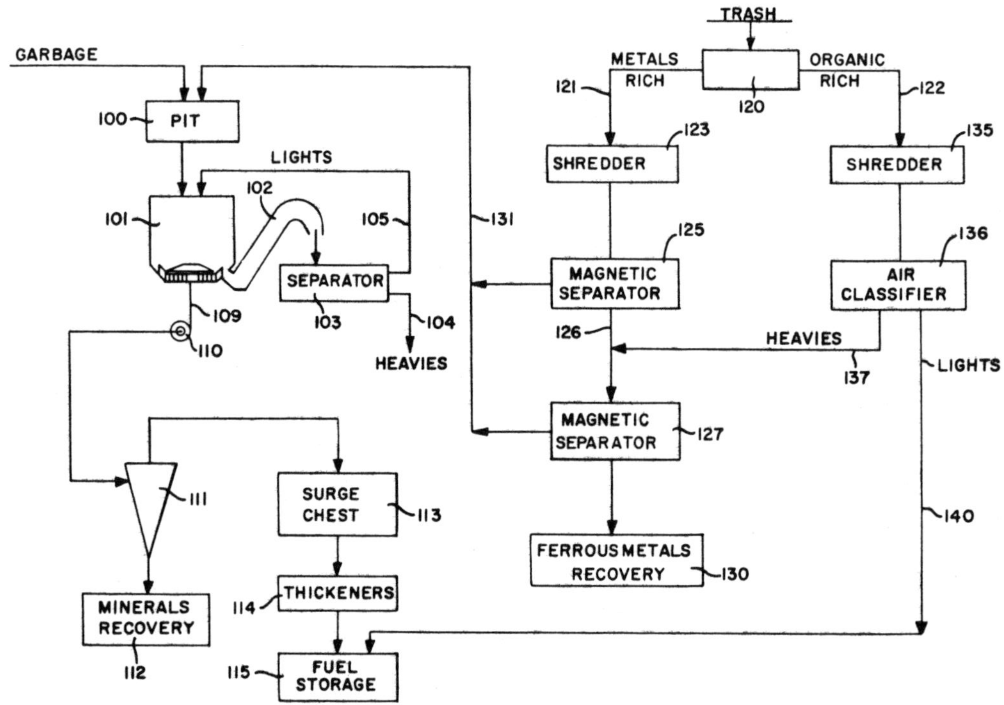

Source: U.S. Patent 4,113,185

The lights from the air classifier **136** are taken directly by a line **140** to the fuel storage bin **115** for mixing with the fuel delivered thereto from the thickeners **114**.

A process developed by *D.E. Garrett; U.S. Patent 4,129,440; December 12, 1978; assigned to Occidental Petroleum Corporation* is one in which solid trash is comminuted and classified into an organic fraction and a metals-inorganic fraction. The organic fraction is processed for recovery of paper pulp and the residue pyrolyzed. The metals-inorganic fraction is processed for recovery of each major metal and inorganic constituent. Such a process is shown in Figure 54.

With reference thereto, the solid waste is fed to a first comminuter **10**, illustrated as a chopping mill, which provides an initial trash expansion, fluffing, or mild size reduction. The effluent from comminuter **10** is then brought into contact with a gas stream, preferably a hot flue gas supplied from another part of the process, which dries the comminuted waste and allows a majority of the paper stock present to pass by gas elutriation to collection chamber **12**. When this procedure is followed, the degree of comminution carried out in chopping mill **10** is generally kept to a minimum in order that fairly large paper stock will be collected in chamber **12**. The paper collected in this manner will generally be comparatively free of contaminants and may readily be disintegrated in pulping tank **14**.

The material which is not disintegrated to fiber size in the pulping tank **14** is cycloned, screened or otherwise removed from the pulp in **16** and passed to a second pulping tank **18** or sent to other recovery equipment such as a system for removing plastics by density differential. The disintegrated pulp is then passed to a second classifier or purification stage **20** where it is further washed and purified and passed to dewatering equipment **22** where it leaves as paper pulp for sale or for deinking and bleaching treatments in a manner conventional to waste paper operations. Impurities from the purifier **20** are then passed to the second pulping tank **18**.

The solids which are not carried by the hot gas stream to paper collection chamber **12** are passed to a second comminuter **26**, again shown as a chopping mill, for a further size reduction and passed to gravity separator **26** where segregation occurs by density differentials.

From there, particles then pass to elutriator-classifier **28** or a similar separatory vessel where there is separated the organic fraction of materials having an effective density less than or equal to that of the upflowing water or which is light enough to be carried upwards by current flow. The organic fraction containing the balance of the pulp and pulp-like materials is passed to pulping tank **18** for disintegration with the impure-pulp fractions from strainer **16** and purifier **20**. After pulping the mass is cycloned and passed to strainer **30** for separation of coarse matter, and a secondary pulp purifier **32** where the pulp is sent to sale or further processing, or returned to pulping tank **14** for reseparation and processing with the dry-separated paper pulp.

The organic residue which remains, including any unwanted pulp, is dewatered in unit **34** and passed to a pyrolysis furnace **36**, either after or without drying. A multistage pyrolysis furnace is shown for conversion to a gas stream containing recoverable chemicals and solid char, as hereinafter explained.

Figure 54: Occidental Petroleum Corp. Process for Recovery of Ferrous Metals and Other Valuable Materials from Municipal Waste

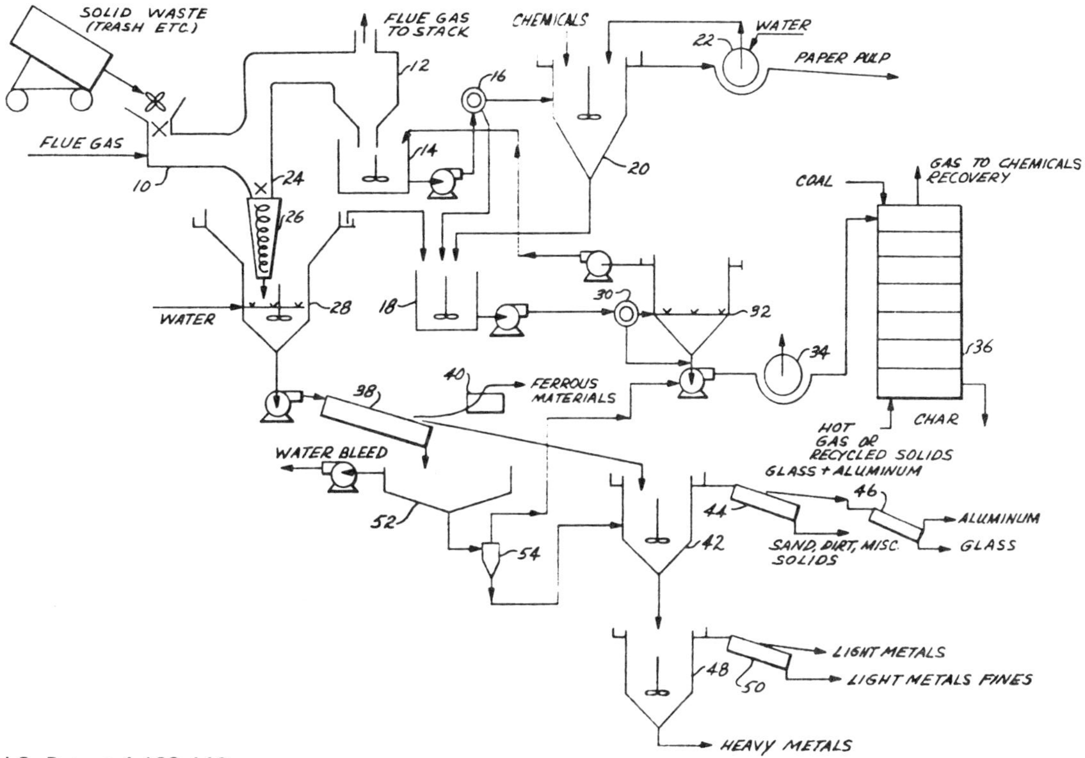

Source: U.S. Patent 4,129,440

A flash pyrolysis unit may also, and preferably be used, as well as a rotary kiln or other carbonizing equipment. In the alternative, and where hot gas elutriation removes most of the paper pulp values from the trash, the organic fraction may be passed from classifier **28** to filter or screen **34** and then to pyrolysis furnace **36**, or the dry solid waste may be sent directly to the pyrolysis unit.

The underflow from classifier **28** is first passed over screen **38** to permit the fine and silt-like materials in the waste to separate from the metals-inorganic fraction. The resultant fraction is then passed through magnetic separator **40** where the magnetic ferrous materials are separated from the metals-inorganic fraction for processing as iron scrap.

The residual metals and inorganics are then passed through to heavy media separator **42** where the inorganics and metals of common density or a metal having density lower than the inorganics such as magnesium, are collected at the top and the heavier metals settle to the bottom as underflow. In heavy media separation, effective media specific gravity may be precisely adjusted by the addition of iron oxide and the like to water to form a slurry. At a selected effective specific gravity, substances having a lower specific gravity will float whereas substances having a higher specific gravity will sink.

The grouping which collects at the top of heavy media separator **42** may then be passed through screen separator **44** where crushed glass and aluminum are separated from sand, dirt, etc., by particle size differential.

The glass may then be separated from the aluminum by particle size differential in a separate crushing and screening operation in **46**. If the particles are close in size, the particle size differential required for screening can be established by passing the group to a ball mill or the like where aluminum and magnesium, if present, are flattened and the glass further crushed. The recovered glass may then be sold for reformation into bottles and the like.

If magnesium is present it may in the alternative be separated from the inorganics and/or aluminum by heavy media separation.

The heavy metals fraction which is the underflow from heavy media separator **42** may then be segregated in one or more heavy media separators **48** where by sink-float separation, the heavier metals are separated from the lighter metals. For the separation illustrated, the lighter metals after separation from the heavier metals are screened in screen separator **50** to remove light metals fines.

Alternatively, steps of melting, leaching and flotation can be used to fine separate the several metals in the heavy metals group into groupings of relatively pure metals for further processing and purification.

The slimes contained in the various streams are then sent to thickener **52** where process water is recovered and where the separated solids are sent to cyclone **54** to separate inorganic materials for return to the heavy media separation system.

As the process water will, over time, show an attendant buildup of soluble impurities it would have to be discarded or processed by evaporation for reclamation

FERROUS SULFATE FROM SPENT PICKLE LIQUOR

See "Sulfuric Acid from Spent Pickle Liquor," especially U.S. Patent 3,900,955 and 4,153,628.

FERROVANADIUM FROM PIG IRON

Vanadium in the form of its low-melting ferrovanadium alloys, is widely used as an alloying agent to impart toughness and other desirable properties to iron and steel. Among the most useful vanadium alloying agents are the alloys produced according to the process of Rathmann and Rasmussen, U.S. Patent 3,420,659; January 7, 1969. This process is carried out in two steps. In the first step a vanadium-containing material is smelted with silica, a flux, and a carbonaceous reducing agent to produce a primary vanadium silicide having a silicon content of 25 to 60%.

Inasmuch as such a silicon content is too high for an alloying agent for iron or steel, the primary silicide is refined in the second step of the process by melting it with lime and a vanadium-containing refining agent to increase the vanadium content of the alloying agent and reduce its silicon content to less than 20%. While the richest source of vanadium for the refining step is vanadium pentoxide, this material is very expensive due to the chemical processing required in its manufacture. For that reason it is preferred to look elsewhere for the necessary vanadium. A number of relatively inexpensive sources of vanadium exist such as spent catalysts, petroleum residues and vanadium-bearing slags from the refining of pig iron.

To be useful in the refining step of the Rathmann et al process, a refining agent should contain at least about 8% of vanadium. Materials containing up to about 34% vanadium may be used with advantage. Materials having a higher content can be economically processed by aluminum reduction and so are not used in this process normally. Raw materials containing less than 8% or more than 34% vanadium may be blended to produce a refining agent having a suitable concentration.

The preferred refining agents for use in the Rathmann et al process are vanadium-containing slags derived from the refining of vanadium-containing pig iron. Such slags are available in huge quantities and generally contain more than the required 8% of vanadium. However, it is also necessary to the process that the slag have a vanadium to iron weight ratio of at least about 0.5, and preferably from about 0.75 to 1.0, to be useful directly as a refining agent without prior deironing. Available slags are generally deficient in this respect.

A process developed by *J.O. Staggers; U.S. Patent 4,071,355; January 31, 1978; assigned to Foote Mineral Company* is one for the refining of pig iron containing at least about 0.5% vanadium to obtain a slag containing at least about 8% vanadium and having a vanadium to iron ratio of at least about 0.5.

This process consists of simultaneously introducing a highly oxidizing gas and a protective fluid below the upper surface of a bath of the molten metal maintained at a temperature of 2600° to 2900°F. Preferably, the process is continued until the carbon content of the metal is within the range of from about 2 to 0.02%. It has been unexpectedly discovered that this results in an increased recovery of vanadium and a marked increase in the vanadium to iron ratio of the slag, thus producing directly a slag suitable for use in the refining step of the Rathmann et al process without the need for beneficiation or deironing.

FERTILIZER FROM FLUE GAS SCRUBBER SLUDGE

When pollutants are removed from stack gases with some processes, lime and limestone wet scrubbing, for example, disposal of sludge discharged from the scrubber system is a significant and often costly problem. When lime or limestone is used for reaction with the sulfur dioxide in the flue gases, the resulting sludge is an aqueous mixture of calcium-sulfur reaction products, unreacted calcium compounds and fly ash.

Thickening and/or vacuum filtration techniques are effective in dewatering the sludge to about 40 to 60% solids by weight. The dewatered sludge is thixotropic and it cannot be handled with conventional dry material handling equipment. However, it can be easily fluidized by agitation and then pumped.

The dewatered sludge does not dry or harden during storage and it has essentially no mechanical or bearing strength. Because of these properties, and the large amount of sludge produced, pond-like enclosures are normally used for disposal. About 7,000 tons/day of dewatered sludge (containing 50% solids by weight) would be produced by a limestone wet scrubber serving a 1,000 MW electric generating unit that burned coal containing 4 to 5% sulfur and about 20% ash.

Methods for stabilizing (hardening) the sludge with additives such as lime, plaster of Paris, Portland cement, fly ash, or various other ingredients have been developed (U.S. Patent 3,785,840; Minnick et al) and are reported to be commercially available. The stabilized sludge has properties that make it suitable for landfill or other disposal techniques that do not require pond storage. Stabilization processes are costly due to large equipment and space requirements, and large quantities of relatively expensive additives. The stabilized sludge has little or no commercial value.

While sludge stabilization processes have useful applications in some situations, they are usually not a desirable treatment technique for sludge produced by an electric-generating scrubber system because of (1) the large amount of land space required for the treatment facilities and ultimate disposal of the treated sludge, (2) the large quantities of relatively expensive additives required to obtain adequate properties for stabilization, and (3) the cost of these additives and the cost of processing cannot be recovered because the large quantity of stabilized material has little or no commercial value.

A process developed by *J.J. Schultz and V.J. Van Pelt; U.S. Patent 4,028,087; June 7, 1977; assigned to Tennessee Valley Authority* overcomes most of these disadvantages. It is a process for the production of a granular fertilizer material from sludge materials discharged from scrubbing processes that remove pollutants from boiler flue gases. Waste sludge discharged from flue gas scrubber systems often contains large amounts of calcium and sulfur as well as lesser amounts of other compounds known to have plant nutrient value. The sludge is treated with an acid such as phosphoric, sulfuric or nitric, or a mixture of these, and ammonia. The resulting sludge-ammonia-acid salt mixture is combined with partially dried, previously processed solid material (recycle) in a horizontal rotating vessel. Additional ammonia and/or acid is added to the rotating vessel to cause the sludge-salt mixture to solidify and granulate. Other plant nutrients may be added to the rotating vessel during granulation. The granulated material discharged from the rotary vessel is further processed (dried, cooled, and screened) to obtain desirable moisture content and particle size. Such a process is shown in Figure 55.

Figure 55: TVA Process for the Production of Solid Granular Fertilizer Material from Scrubber Sludge

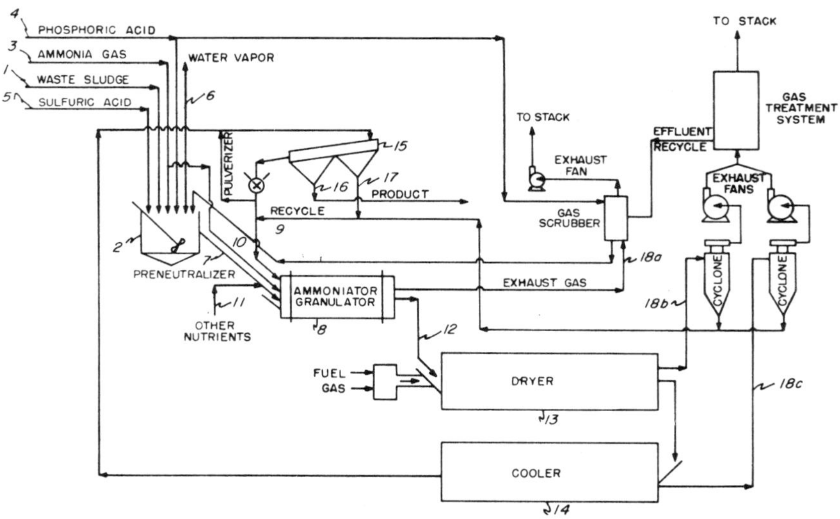

Source: U.S. Patent 4,028,087

As shown there, waste sludge from the boiler flue gas treatment system (not shown), either reclaimed from a storage pond or obtained direct from the flue gas treatment system, is fed via line **1** to reaction vessel (preneutralizer) **2**, along with a quantity of ammonia gas via line **3** and phosphoric acid via line **4** and/or sulfuric acid via line **5**. The exothermic chemical reaction between the waste sludge, acid, and ammonia causes a large amount of water contained in the sludge to evaporate and off-gas via line **6**.

Decomposition of the calcium-sulfur compounds in the waste sludge, especially calcium sulfite, causing the liberation of sulfur dioxide, is prevented by the method of introduction of acid, ammonia, and waste sludge to preneutralizer vessel **2**.

The mol ratio of the ammonia to phosphoric acid (N/P) is maintained between 0.5 and 0.7 or 1.4 and 1.5 to obtain a fluid slurry in preneutralizer vessel **2** that can be easily conveyed via line **7** to rotary ammoniator-granulator vessel **8**. Partially dried, previously processed solid material (recycle) along with fluid slurry from the preneutralizer vessel **2** is fed via line **9** to ammoniator-granular vessel **8**.

Additional ammonia is added via line **10** to the material in ammoniator-granulator vessel **8** to increase the N/P mol ratio to about 1.0 or 2.0 depending on the product desired. This addition of ammonia to the tumbling and rolling mass of slurry and recycle material causes the slurry to solidify and form solid granules. Other nutrients, (for example, potassium salts) can also be added via line **11** during the ammoniation-granulation step in vessel **8**.

The granulated product discharged via line **12** from the ammoniator-granulator is subsequently dried in dryer **13**, cooled in cooler **14**, and screened in screens **15** to specified particle size. A portion of the product and/or off size material is fed as recycle via lines **16** and **17**, respectively, from the screening, drying, and cooling operations to ammoniator-granulator **8** to aid in granulation of fresh slurry from preneutralizer **2**.

Exhaust gases from ammoniator **8**, dryer **13**, and cooler **14** are collected via lines **18** (**a** to **c**) and treated to remove pollutants before being discharged to the atmosphere. As shown, effluents from the treatment process are recycled to the granulation process or further treated to obtain acceptable discharge standards.

The process as described is operated on a continuous basis. However, it can also be performed on an intermittent or batch basis.

The amount of recycled in-process material required per unit of fresh slurry from the preneutralizer vessel is primarily dependent upon the initial N/P mol ratio in the preneutralized slurry and the desired N/P mol ratio in the finished fertilizer product. Other factors that affect the recycle/fresh slurry ratio and therefore the ultimate capacity of a given production unit are the ratio of waste sludge to ammonia and acid and the amount of free water contained in the waste sludge.

FERTILIZER FROM MUNICIPAL WASTES

A process developed by *R.A. Penque; U.S. Patents 3,933,577; January 20, 1976; and 4,006,006; February 1, 1977; both assigned to Biocel Corporation* is one in which municipal solid waste is treated by a method in which pollution of air, water and soil is minimized. Light, long fiber paper pulp which is in great demand and fertilizer which controllably releases its nutrients are recovered which render the treatment efficient and economical.

The solid waste after removal of ferrous and nonferrous metals, glass and other aggregates, is sonically pulped to prevent shearing of fibers using such chemicals that when subsequently phosphated, and xanthated if desired, will cause salts to precipitate with inks, coatings and waxes removed from the fibers, to provide fertilizers with controlled nutrient release on the one hand and high quality paper pulp on the other.

As shown in Figure 56, a light stream **11c** from an elutriation step (not shown) which constitutes in some excess of 40% by weight of the original waste input, about 88% by weight paper products and about 2% by weight other combustibles, such as plastic film, is acted upon in a series of sequential and branch steps to produce as end products fertilizers of desired composition and controlled release and high quality, long, light fiber paper pulp.

Figure 56: Biocel Corp. Process of Sonically Treating Municipal Waste to Produce High Quality Paper Pulp and Fertilizer

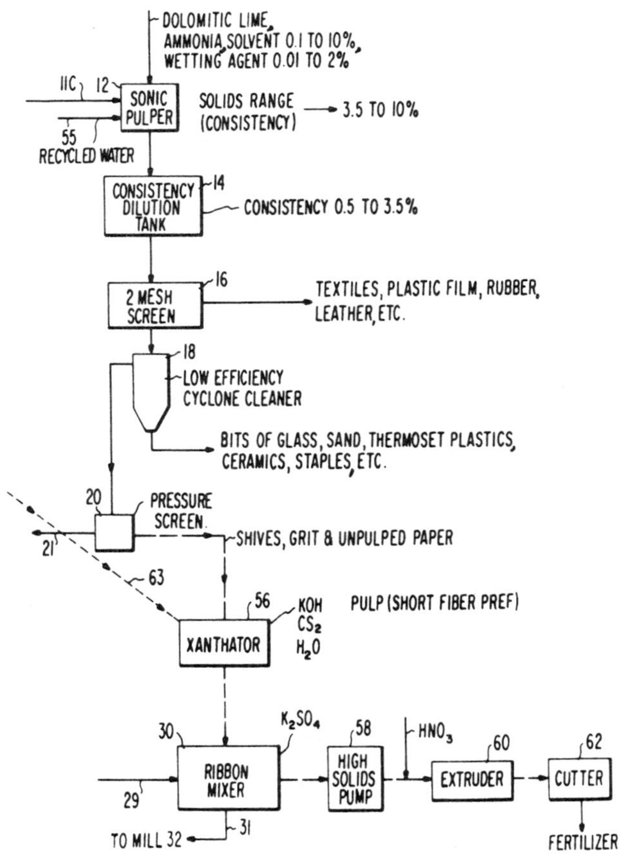

Source: U.S. Patent 3,933,577

The float stream from a vertical elutriator merged with the light stream from a zig-zag elutriator **11c** is fed into zones in which the pulp fiber is not broken and ink, coatings, and waxes are lifted and separated from the fibers and precipitated with chemicals which act to provide fertilizer nutrients on the one hand and light long fiber pulp on the other.

The merged float stream is fed into a sonic pulper **12**, a device which includes a large vertical tank open at its top to receive the stream, chemicals and water and a large rotating disk adjacent its bottom having vanes on its circumference designed to cause cavitation in the liquid medium when the disk rotates at high speed, i.e., peripheral speed of about 145 ft/sec minimum.

Tearing and shearing action are minimized because the vanes are relatively small and, when operated at the aforementioned high peripheral speed, produce sonic shock waves which defiber the paper products without shearing the fibers. Such apparatus is described in U.S. Patent 3,420,454. Although the paper and other fibrous material are completely pulped by the sonic waves, very little if any breakdown of the nonfibrous materials present, such as plastic pieces and films, occurs. This simplified the separation of the defibered pulp from the plastics.

The chemicals added to the sonic pulper perform the dual function of deinking, decoating and dewaxing the fibers and removing all other contaminants and of eventually forming soil nutrients or fertilizers upon subsequent reaction with phosphate ions by causing a precipitate to form which contains the soil nutrients, the removed inks, coatings, waxes and other contaminants including putrescibles which remained on the paper products when the light stream entered the sonic pulper.

At this point it would be well to set forth a specific nonlimitative example of a pilot run of the process starting at the sonic pulper **12**. Into a 250 hp sonic pulper was charged 3,000 lb of the light stream from the vertical elutriator representing about 61.3% by weight of the total input weight of the waste or garbage containing in excess of 95% by weight of paper and paper products.

Also charged into the sonic pulper were the following: 921 lb, dolomitic lime (325 mesh); 668 lb, ammonium hydroxide (28%); 200 lb, calcium hypochlorite; 500 lb methylpyrrolidone; 25 lb, Tergitol 15S9 (Union Carbide) (a linear ethoxylate having 9 mols of ethylene oxide) and 60,000 lb water (recycled from pulping).

The pH of this composition, for the particular dolomitic lime (Charles Pfizer, Ohio deposit), was 12. The pH may be further reduced to a lower alkalinity, (e.g., 9.5 to 10) by a suitable acid such as phosphoric, sulfuric or nitric.

The water is added first to the sonic pulper, then the light or float stream from the vertical elutriator enters, agitation is initiated, and finally the rest of the aforementioned chemicals are added. The pulper was run for about 12 min. During the last 2 min, 400,000 lb additional water was added to the sonic pulper **12** while the bottom drain was open so that the entire charge was discharged into a consistency dilution holding tank **14**. The initial consistency in the sonic pulper **12** was about 5% and was reduced to a consistency of about 3% in the holding tank **14**.

Additional water was added to the consistency dilution tank to reduce the consistency to about 0.5% and then the pulp slurry was passed through a 2 mesh screen **16** which removed the textiles, plastic film, rubber and leather amounting to about 68 lb. The material passing through the screen was then sent into a conventional low efficiency (low pressure drop–5 psi) cyclone precleaner **18** where bits of glass, sand, thermosetting plastics, ceramics, staples, etc. fell to the bottom and were removed.

The stream was then passed through a pressure screen **20** to remove shives, grit and a small amount of unpulped paper.

The stream **21** was then passed through a screw press where the consistency was increased to 40% by weight of pulp and the liquid effluent was discharged into a precipitator tank which was maintained at about 60°C.

Into the precipitator tank was charged about 1,884 lb of phosphoric acid (52% solution) which gave a pH of about 7.5 and caused extensive precipitation. The contents of the precipitator tank were pumped into a settling tank equipped with a screw discharge to remove the settled solids which (together with the further settled solids removed by centrifuge) then went as at **29** to a ribbon mixer **30** to which was added about 270 lb of potassium sulfate, which ribbon mixer was steam jacketed to dry the product.

The contents of the ribbon mixer were milled as at **32** to about 20 mesh as a fertilizer yielding about 2,050 lb, the approximate analysis of which was about 7% N, 40% P_2O_5, 6% K, 15% MgO, 1.67% S, all by weight. The fertilizer also contained about 6.5% by weight inks, coatings, waxes, adhesives and putrescibles which came down with the precipitate in the precipitator, none of which are soil pollutants. Alternatively, the settled solids from the screw discharge can go to a dryer other than a steam jacketed ribbon mixer and then via line **31** to a mill **32** to provide the fertilizer.

Instead of the dolomitic limestone ($CaCO_3$ and $MgCO_3$ mol for mol) CaO, MgO, $CaCO_3$ or mixtures thereof may be substituted in stoichiometric amounts necessary to form calcium or magnesium ammonium phosphate in the neutralizer.

Also, in the sonic pulper, stoichiometric amounts of KOH, K_2CO_3, $KHCO_3$ or any other soluble potassium salt can be added in which case the subsequent addition of K_2SO_4 or any other potassium salt to the ribbon mixer becomes unnecessary. The potassium salt may be used with calcium and/or magnesium salts, in which case the double salt of $CaKPO_4$ or $MgKPO_4$ will precipitate out. Technically, alkaline potassium salts can replace the ammonia all or in part, but the N required for the fertilizer would have to be added at a later stage. It is preferred that ammonia be added in the sonic pulper and potassium ion elsewhere.

As a source of nitrogen to be added to the sonic pulper any water-soluble ammonium salt may be used such as $(NH_4)_2CO_3$, NH_4HCO_3, NH_4Cl, NH_4NO_3, $(NH_4)_2SO_4$, NH_4COOCH_3, except that where the ammonium salt has an acid pH, the pH of the charge should be rendered alkaline up to a pH of 10.5 with other soluble compounds, preferably KOH to add the K ion. The quality of ammonium salt required is that stoichiometric amount needed to form the metal ammonium phosphate.

The type of solvent required is one which is water soluble and will dissolve polymeric, asphaltic, lipophilic and normally water-insoluble contaminants of the fibers. Highly aprotic solvents, such as dimethylformamide, dimethyl sulfoxide, methylpyrrolidone, etc., have the broadest solvating action and should be used from about 0.1 to 10% by weight of the water initially charged into the sonic pulper. Other solvents found useful are methyl, ethyl and isopropyl alcohols, water soluble glycol ethers, and ethoxy alcohols such as Carbitol and Cellosolve.

The action of the solvent is to penetrate the fibers and lift off the inks, coatings, waxes, adhesives and other contaminants including putrescibles rendering them emulsifiable by solvation and coupling with the wetting agent.

Any wetting agent may be used provided it is salt stable and does not precipitate insoluble metal soaps, i.e., calcium, magnesium, etc., and will tolerate high pH such as, for example, the nonionic ethoxylated nonyl and octyl phenols, such as Triton X-100 (Rohm and Haas), Tergitol NPX (Union Carbide) and the linear ethoxylated nonionic wetting agents, such as Tergitol 15S9 (Union Carbide), selected alkylaryl sulfonates having 4 to 30 C atoms in the alkyl substituent such as sodium dodecylbenzylsulfonate, sorbitol base surfactants, etc.

The concentration of the wetting agent should be about 0.01 to 2% by weight of the water added to the sonic pulper. Thus, the wetting agent and the coupling solvent form a stable emulsion of the water insoluble fiber contaminants such as asphaltic materials, polymeric coatings, waxes, hot melt adhesives, oil, etc. The wetting agent also serves to disperse the particulate matter such as clays from coatings, carbon black and pigments from the inks, etc.

The oxidizing agent serves to decompose organic matter and thereby render it soluble or emulsifiable, such as proteinaceous glues and coatings, starches, gums, natural resins, etc., which are normally neither solvent or water soluble. It also serves to kill microorganism contamination and to decolorize contaminants and dyes and thus lighten the pulp.

The oxidizing agents may be selected from hypohalites, such as calcium hypochlorite, sodium hypobromite, etc., peroxides such as H_2O_2, Na_2O_2, K_2O_2, urea peroxide, persulfates, perborates, and so forth, which are water soluble and oxidize in alkaline pH.

Calcium hypochlorite is preferred for the sake of economy. However, when processing pulp with high ground wood content in the past, the use of hypochlorite was prohibited because of chromophore development in the lignin badly discoloring the pulp. The employment of the mildly alkaline MgO, CaO, NH_4OH in the sonic pulper and the ease of deinking could eliminate this color development problem.

An additional cost saving is realized by on site hypochlorite production with a portion of the dolomitic lime.

The optimum range of hypochlorite or other oxidizing agent utilizable is from 0.2 to 2.0% by weight based on the weight of the fiber.

In addition to the oxidizing agent being added in the sonic pulper, the oxidizing agent may be added in the weir chest (not shown on diagram) or in either place, depending on the degree of whiteness required in the end product pulp. Because all input chemicals other than the oxidizing agent are either recovered for sale at a profit, or recycled in the system, additional oxidizing and bleaching steps heretofore uneconomical are possible because the oxidizing agent is the only significant expended chemical used in the process.

The use of urea peroxide was found to be especially advantageous because it dissociates to H_2O_2 and urea, the H_2O_2 becoming water and the urea contributing to the total N content of the fertilizer.

Xanthation can be used alternatively with regard to fertilizer formation between the pressure screen **20** and the ribbon mixer **30**. The xanthation is preferably carried out as at **56** with KOH to add K to the fertilizer.

Preferably, the material is pumped by a high pressure pump **58** to an extruder and, prior to entrance in the extruder, acid is added in line to change the pH of the material from alkaline to acid which also causes the formation of alpha cellulose which acts as a binder and coater for the fertilizer compound particles. The mass is such that it can be extruded as at **60** and cut as at **62** into short rods or other forms having predetermined concentrations and amounts of the nutrient atoms N, P, K, Mg and S.

While the output from the ribbon mixer **30** may be so controlled that one portion goes directly to the mill **32** and the other through the extruder, it is preferred from a practical standpoint that the two processes be carried out alternatively (as indicated by broken lines in the figure) that is, when the xanthation takes place the output from the ribbon mixer **30** goes to the extruder and not directly to the mill **32**.

Xanthation also enables one to use all of the rejected grits and aggregates **63** as a diluent bound with the fertilizer chemicals. By varying the amount of xanthate and the density of the extruded cellulose-bound fertilizer (i.e., varying extrusion pressure primarily) additional control of the rate of nutrient release, other than that inherently present because of the limited solubility of the phosphates, can be effected.

FERTILIZER FROM SPENT SULFURIC ACID

A process developed by *S.J. Shirley; U.S. Patent 3,888,653; June 10, 1975; assigned to Shamrock Chemicals Limited, Canada* involves producing a potassium sulfate fertilizer from spent sulfuric acid having impurities therein comprising the steps of: (a) selecting as a first feed material spent sulfuric acid; (b) selecting, as a second feed material, one consisting essentially of potassium chloride; (c) mixing the first and second feed materials to achieve a feed mix; (d) heating the feed mix at an elevated temperature in excess of 300°F whereby the potassium sulfate fertilizer is created in fine crystalline powder; (e) mixing the recovered crystalline potassium sulfate with granulating agents and small amounts of water until the potassium sulfate congeals into small pellets; and (f) grading the pellets into a predetermined range of sizes.

Such a process is shown in Figure 57. The apparatus used includes large rotary kiln **10**, a granulator **11**, and a rotary cooler **12**. Feed materials include water (H_2O), **20**, spent sulfuric acid (H_2SO_4), **21**, potash (KCl), **22**. The kiln **10** is approximately 35 ft long, 6 ft in diameter, and is fired by natural gas burners generating approximately 12 to 16 MM Btu/hr. The kiln **10** is rotated by a gear means and a chemical reaction takes place such that the resultant product, potassium sulfate (K_2SO_4) fertilizer is created and discharged at **26** into the granulator **11**, while hydrogen chloride (gas) (HCl) is given off at **34**.

Figure 57: Shamrock Chemicals Ltd. Process for Producing Fertilizer from Spent Sulfuric Acid

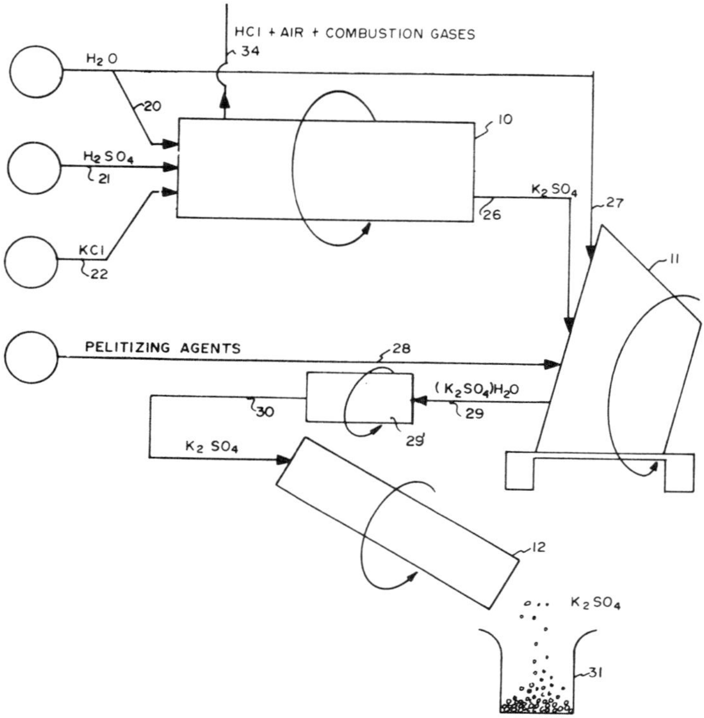

Source: U.S. Patent 3,888,653

The reaction may be represented by the following equation:

$$H_2SO_4 + 2KCl \rightarrow K_2SO_4 + 2HCl\uparrow$$

or more accurately,

$$H_2SO_4 + 2KCl + Heat \rightarrow K_2SO_4 + 2HCl\uparrow$$

The resultant fertilizer, K_2SO_4, at **26** is normally dry and a fine crystalline powder. In order to make it commercially acceptable and to aid in delivery and packaging, it is pelletized by the use of the granulator **11**; thus, the output **26** is mixed in the granulator **11** with water **27**, and granulating or pelletizing agent **28**, as those skilled in the art will know.

The resultant range size of pellets, leaving the granulator **11**, are preferably in the range of 6 to 10 mesh and are moist at the output **29**. They are then sent through a rotary dryer **29'** and are discharged hot at **30** and thereafter are cooled in the rotary cooler **12** and placed into bulk storage at **31** for shipment as required.

FLOCCULATING AGENT FROM RED MUD FROM BAYER PROCESS

In the production of alumina by the Bayer process, up to two tons of a filter-wet residue having a water content of from 40 to 50% by weight may be obtained for every ton of alumina produced, and it has been estimated that more than 5 million tons of such residue accumulate annually.

This residue is a deeply-colored clay-like substance which is highly hydrated (about 35 to 50% water by weight), the mineral content of which is principally compounds of iron and aluminum with lesser proportions of alkalis and compounds of other metals. Because of its color and consistency, this material is commonly referred to as red mud.

The substantial amounts of such residue from the production of alumina have posed a substantial ecological as well as an economic problem for many years and various proposals have been made for the industrial utilization of this otherwise waste product, none of which have proven satisfactory.

A process has been proposed by *G. Bayer and E. Cherdron; U.S. Patent 3,986,975; October 19, 1976; assigned to Gebr. Giulini GmbH, Germany* in which red mud from the production of alumina from bauxite by the Bayer process is converted to a flocculating agent by reaction with sulfuric acid and drying the reaction product by direct heat transfer thereto by contact with hot solid bodies and contact with hot gases. The resulting solid product is a mixture of water-soluble sulfates of iron and aluminum with minor proportions of other water-soluble metal sulfates in the form of free-flowing discrete particles.

The apparatus which may be used in the conduct of this process is shown in Figure 58.

Figure 58: Gebr. Giulini GmbH Process for Continuous Production of Flocculating Agent from Red Mud

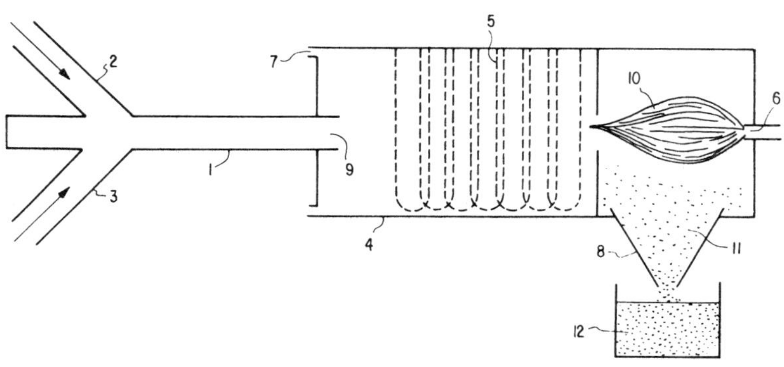

Source: U.S. Patent 3,986,975

The red mud and sulfuric acid are supplied to the mixing chamber **1** through reactant inlets **2** and **3** respectively, the resulting slurry passes into drying kiln **4** and the dried reaction product is discharged through outlet **8** and so collected in container **12**.

Referring to the figure, the flow pipe **1** is of acid-resistant steel or other suitable material, and the red mud slurry is fed thereto through line **2**, at a predetermined rate by pumping and metering means not shown, with the sulfuric acid being fed through line **3**, likewise by pumping and metering means not shown.

The heat of mixing and reaction generates considerable water vapor which serves to eject the reaction mixture into the drying means **4**, through inlet **9**. The dryer is preferably a rotary kiln provided with hanging chains **5** and a combustion device or other means **6** for supplying a stream of hot gases **10** which provide the heat for evaporating the water in the reaction mass. These gases, along with water vapor from drying the reaction mass, are discharged through outlet **7** at the opposite end of the kiln. The temperature in the kiln will be in the range of 120° to 180°C, and preferably 130° to 150°C.

As the reaction mixtures passes through kiln **4**, the water is evaporated therefrom and the solid reaction products are dried and comminuted by the resulting agitation and contact with the hot chains and hot gases. This material **11** is discharged from the kiln through solids outlet **8** and is collected in container **12**.

The chains **5** thus serve to agitate the reaction mass to prevent caking thereof on the kiln wall, as well as to assist in the heat transfer. During movement through the hot rotating drum, the solid reaction products are reduced to discrete dry particles which are especially suitable for treating of raw and waste water. The size of this particular material can be controlled from granular to pulverulent by appropriately regulating the speed of rotation of the drum and associated heating elements. The process and equipment are particularly suited for continuous operation, and the discharge of the dried and comminuted products also takes place continuously.

The reaction between the red mud and the sulfuric acid starts immediately upon the mixing and, depending upon the feed rate and length of chamber **1**, the reaction may be entirely or partially completed by the time the reaction mixture passes into the dryer **4**. If any unreacted mud passes into dryer **4**, conversion will be completed within the dryer.

Instead of countercurrent heating as indicated, the reaction mass may also be heated with the gases flowing concurrently therewith during its passage through the dryer.

FLUOROSULFURIC ACID FROM SPENT CATALYSTS

Alkylation of paraffins with olefins to produce gasoline boiling range alkylates may be carried out with a variety of catalysts including sulfuric acid, aluminum chlorides and fluorosulfuric acid.

A process developed by *I. Mayer; U.S. Patent 4,096,197; June 20, 1978; assigned to Exxon Research & Engineering Co.* is one for regenerating a catalyst comprising fluorosulfuric acid, at least a portion of the catalyst having become deactivated due to the formation of stable catalytically inert species during contact with a hydrocarbon feedstock in an alkylation process.

It is well known in the prior art that as the alkylation reaction proceeds, an organic material will form and will accumulate in the fluorosulfuric acid catalyst phase. The material has been given a variety of names including red oil, sludge, organic sludge, acid oil and the like. The process consists of:

(1) removing a portion of the fluorosulfuric acid from the catalyst by contacting same with a paraffin to form a liquid acid phase containing fluorosulfuric acid and an organic sludge formed during the alkylation and a gas phase containing the paraffin and fluorosulfuric acid;

(2) contacting the liquid acid phase formed in step (1) with water to form an acid-water mixture, thereby converting at least a portion of the fluorosulfuric acid contained in the liquid acid phase to hydrogen fluoride and sulfuric acid;

(3) removing at least a portion of the hydrogen fluoride from the acid-water mixture of step (2) by contacting same with a paraffin so as to form a gas phase containing paraffin and hydrogen fluoride; and

(4) treating the gas phases from steps (1) and (3) with sulfur trioxide so as to regenerate the fluorosulfuric acid.

Use of the process permits recovery of a portion of the fluorosulfuric acid present in the deactivated or partially deactivated catalyst prior to undergoing the reaction with water in step (2) above. This results in a reduced consumption of sulfur trioxide since one mol of sulfur trioxide is conserved for each mol of fluorosulfuric acid that is not reacted in step (2). This results in a significant reduction in regeneration costs because sulfur trioxide is fairly expensive. In a preferred embodiment, at least a portion of the regenerated fluorosulfuric acid is recycled to the alkylation process. Such a process is shown in Figure 59.

There is shown an olefin stream in line **2** which is, preferably, admixed with a paraffin stream in line **4** before introducing the combined stream into alkylation zone **6**. If desired, however, the olefin and paraffin streams can be fed directly into alkylation zone **6**. The olefin concentration in the feed ranges from 0.5 to 25 volume percent based on total feed and preferably below 10 volume percent.

A catalyst comprising fluorosulfuric acid and one or more moderators is shown being introduced into alkylation zone **6** via line **8**. Generally, the moderator contains at least one oxygen atom per molecule and includes water, aliphatic and cycloaliphatic alcohols and ethers, aliphatic, cycloaliphatic and aromatic sulfonic and carboxylic acids and their derivatives, inorganic acids and other oxygen-containing organic compounds. By the term moderator is meant a compound which, in combination with fluorosulfuric acid, produces a catalyst system of reduced acidity vis-a-vis the fluorosulfuric acid, and thereby decreases the probability of undesirable competing side reactions which have a detrimental effect on product quality, while increasing catalyst selectivity to desirable highly branched paraffinic products, thus resulting in high quality alkylate product.

Figure 59: Exxon Research & Eng. Co. Process for Regenerating Fluorosulfuric Acid Catalyst

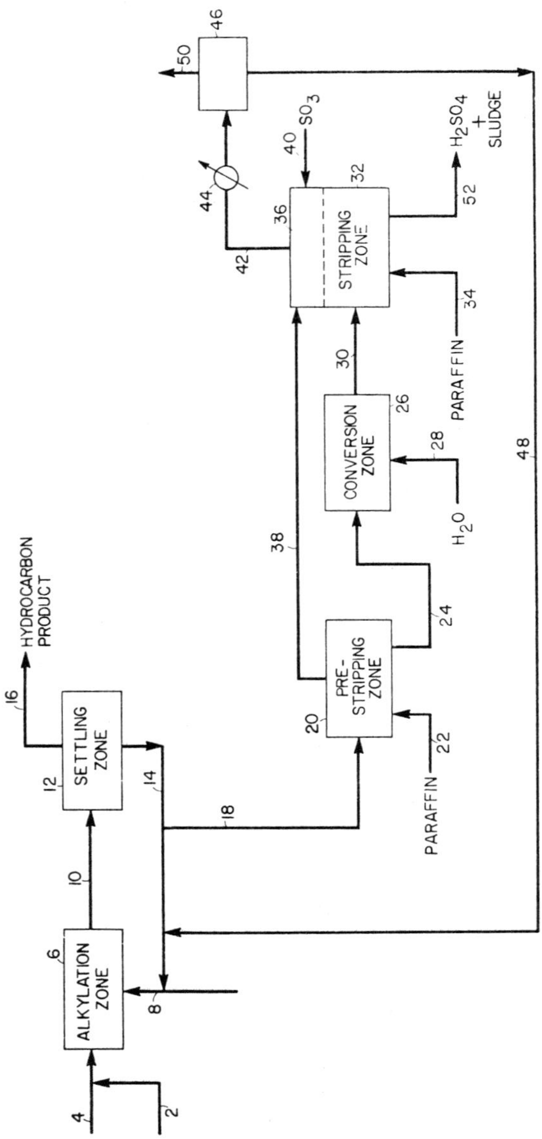

Source: U.S. Patent 4,096,197

After allowing sufficient residence time for the reaction to progress, typically on the order from about 1 min to 1 hr or more, the reaction mixture which contains hydrocarbon and deactivated or partially deactivated catalyst (often referred to as the emulsion mixture) is withdrawn from the alkylation zone **6** via line **10** and passed into a settling zone **12**.

The reaction mixture will separate in zone **12** into a heavy acid phase containing the fluorosulfuric acid, sulfuric acid, hydrogen fluoride and moderator (assumed to be water for the purpose of illustration) as well as organic sludge formed during the alkylation process, and a hydrocarbon phase containing the alkylate product along with smaller amounts of fluorosulfuric acid, hydrogen fluoride and water which are dispersed and/or dissolved in the alkylate product.

The acid phase is withdrawn from settling zone **12** via line **14** and at least a portion thereof can be recycled to alkylation zone **6** via line **8** or charged to another alkylation zone, if desired. The hydrocarbon phase is withdrawn from settling zone **12** via line **16**.

As shown a purge stream **18** of the heavy acid phase is withdrawn from line **14** and passed into the prestripping zone **20** and intimately contacted with a vaporized paraffin introduced via line **22**. Preferred paraffins are C_{3-6} paraffins, more preferably C_4 paraffins. Normal butane is the preferred paraffin. As a result of the contacting, a portion, preferably a major portion, more preferably from about 60 to 90%, of both the hydrofluoric acid and the fluorosulfuric acid is stripped from the purge stream, thereby forming a gas phase containing paraffin, fluorosulfuric acid and hydrogen fluoride and a liquid phase containing fluorosulfuric acid, organic sludge and sulfuric acid and traces of hydrogen fluoride.

The amount of stripping agent employed is that required to remove the desired amount of fluorosulfuric acid. It should be pointed out that hydrofluoric acid and sulfuric acid are present in streams **14** and **18** because the fluorosulfuric acid is partially dissociated when contacted with the moderator, e.g., water. If no moderator is employed, small amounts of water are normally introduced into the alkylation zone (e.g., with the feed) such that the partial dissociation will occur. Be that as it may, however, the present regeneration process is also applicable to a fluorosulfuric acid catalyst that has not been hydrolyzed.

The liquid phase then passes from prestripping zone **20** via line **24** to conversion zone **26** wherein it is contacted with water injected via line **28** in an amount sufficient to convert the fluorosulfuric acid to free hydrogen fluoride and sulfuric acid according to the reaction:

$$H_2O + HSO_3F \rightleftharpoons H_2SO_4 + HF\uparrow + \text{Heat}$$

In one embodiment of the process, it may be desirable to add up to a mol of water in excess of the stoichiometric amount required. Preferably, less than about 0.5 mol excess water is used. The resulting stream of water, hydrogen fluoride, sulfuric acid and organic sludge is then passed from conversion zone **26** via line **30** into stripping zone **32** and intimately contacted therein with a paraffin introduced via line **34**, thereby stripping at least a portion, preferably a major portion, more preferably substantially all (i.e., about 95% or more) of the hydrogen fluoride from the stream. The gas phase from prestripping zone **20** is introduced into the upper section **36** of stripping zone **32** via line **38** wherein the hydrogen fluoride present therein, as well as that removed in the lower section of stripping zone **32**, is reacted with at least a stoichiometric amount of sulfur trioxide, based on HF, so that at least a portion, preferably substantially all, of the hydrogen fluoride present is converted to fluorosulfuric acid according to the reaction:

$$HF + SO_3 \rightarrow HSO_3F + Heat$$

The sulfur trioxide, which is introduced via line **40**, thus regenerates the fluorosulfuric acid catalyst which, together with the paraffin and perhaps a trace of water is taken overhead via line **42**, condensed in condensation zone **44**, and passed to separation zone **46** wherein the fluorosulfuric acid is separated from the paraffin present in the stream.

The regenerated fluorosulfuric acid stream, which may contain negligible amounts of water (typically less than 100 ppmw since most all of the water present will react with sulfur trioxide to form sulfuric acid), is withdrawn from the separation zone via line **48** and at least a portion thereof may be combined with the recycle stream **14** for return to alkylation zone **6** via line **8**.

The paraffin stream is removed from separation zone **46** via line **50**. If desired, at least a portion of the paraffin stream may be recycled to conversion zones **26** and **36** for temperature control purposes or may be used as part of the stripping agent in zone **20**. Additional hydrocarbon stripping agent can be introduced into the stripping zones if desired. Sulfuric acid and the sludge formed during the alkylation process can be removed from the bottom of stripping zone **32** via line **52** and sent to sulfuric acid regeneration (not shown) for sludge removal and reconcentration, or it can be discharged. Alternatively, the sulfuric acid-sludge stream can be employed for removing dissolved and/or dispersed fluorosulfuric acid from a hydrocarbon phase **16**.

FLUOSILICIC ACID FROM PHOSPHORIC ACID MANUFACTURE

In the manufacture of concentrated phosphate fertilizer, the phosphate ore is initially subject to extraction with sulfuric acid by the wet process. The resulting dilute phosphoric acid is concentrated by boiling and then used for the production of concentrated liquid and granular phosphate fertilizers. During the extraction of phosphoric acid from the phosphate ore by sulfuric acid, the reaction produces hydrogen fluoride. In the presence of silica, the hydrogen fluoride is converted to fluosilicic acid.

A large quantity of the fluorine remains in the dilute phosphoric acid as hydrogen fluoride and fluosilicic acid. When the dilute phosphoric acid is concentrated by boiling, much of the fluorine is evolved with the steam as hydrogen fluoride and silicon tetrafluoride vapors in variable amounts depending upon the variety of the crude phosphate ore. The toxicity of both compounds in itself demands careful purification of the waste gases. Furthermore, the waste gases form a valuable raw material source for obtaining fluorine values in the form of inorganic fluorine compounds. From some phosphate plants, as much as 10,000 to 30,000 tons/yr of fluorine compounds may be liberated in gaseous plant effluents, which should be captured to prevent air and water contamination.

A process developed by *A.W. Petersen and J.M Stewart; U.S. Patents 4,152,405; May 1, 1979; and 3,893,830; July 8, 1975; both assigned to Stauffer Chemical Company* describes a cocurrent absorber for recovering fluorine values as low phosphorus fluosilicic acid from phosphate plant effluents. The device is an improved cocurrent absorber utilizing multiple cyclonic entrainment separators, cocurrent flow and renewed droplet surfaces. The absorber and its place in the process are shown in Figure 60.

Figure 60: Stauffer Chemical Co. Process for Recovering Low-Phosphorus Fluosilicic Acid from Phosphate Plant Effluents

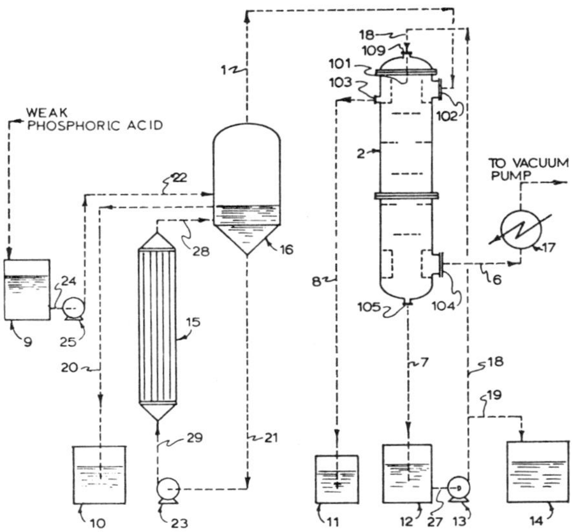

Source: U.S. Patent 4,152,405

Referring to the flow diagram at the start-up of the system, the heater **15**, the evaporator **16** and circulation pipe **21** are filled with phosphoric acid to a designated level. The level of the acid in the evaporator **16** is kept constant and is regulated by overflow from a barometric leg, pipe **20**. Then, water is introduced into the condenser **17**, and with a vacuum pump (not shown), a vacuum is created in the system.

The phosphoric acid circulation pump **23** is started and steam is then introduced into the heater **15**. With this, the acid in the tubes of the heater overheats and changes into a vapor-acid mixture, which flows in a continuous stream from the heater **15** into the evaporator **16**. In the evaporator **16** the acid loses vapors via line **1** and drains into the circulation pipe **21**. In this way, the system has an uninterrupted circulation of acid from the circulation pipe **21** into the heater **15** through pump **23** and line **29**, then into the evaporator **16** via line **28** and from the evaporator **16** again into the circulation pipe **21** to have the process repeated.

When the acid in the circulation pipe **21** reaches a given concentration, dilute acid is introduced continuously into the system from the feed vessel **9** for the weak phosphoric acid source (not shown) through pipe **24**, pump **25** and pipe **22**.

The concentrated acid, having reached the overlow level, continuously drains through pipe **20** from the evaporator **16** into the collecting vessel **10** for concentrated acid. In the process prior to this development the vapor separating in the evaporator **16** continuously flowed into the condenser **17** with the aid of the vacuum pump (not shown). In this manner, the system worked continuously, but resulted in fluosilicic acid recovered which contained varying amounts of entrained phosphoric acid.

To recover fluorine from vapor from plant effluent such as a phosphoric acid plant, a vessel **2**, hereinafter called the cocurrent absorber, is added to the concentration flow diagram and apparatus described above. It is preferably placed between the evaporator **16** and the condenser **17**.

The vapor separating in the evaporator **16** goes from the evaporator **16** through pipe **1** to the top of the cocurrent absorber **2**, through entrance port **102**, and is contacted with a recirculating liquor. The vapor is contacted with a recirculating liquor in the cocurrent absorber **2** and exits through pipe **6** from the cocurrent absorber **2**. The vapor enters the condenser **17** where it is condensed. The condenser **17** is maintained at a temperature sufficiently low enough to condense all the vapors. Thus, the possibility of contamination of air has been effectively removed.

At the start-up of the recovery system, the circulating liquor vessel **12** is filled with water and the centrifugal pump **13** is activated. Then the cocurrent absorber **2** is irrigated with liquid brought into the cocurrent absorber **2** by the centrifugal pump **13** through pipe **18** from the circulating liquor vessel **12**. The vapor entering the upper section of the cocurrent absorber **2** from the evaporator **16** is drawn through the cocurrent absorber **2** and irrigated with liquor from pipe **18** by outlet **101** with the aid of the vacuum pump. The water soluble components of the vapor are absorbed by the recirculating liquor.

In the case of plant effluents from a wet process phosphoric acid plant, the vapors are absorbed to make fluosilicic acid according to the following reactions. Silicon tetrafluoride reacts with water as shown by the equation:

$$3SiF_4 + 2H_2O \rightleftharpoons 2H_2SiF_6 + SiO_2$$

Hydrogen fluoride reacts with silica to form fluosilicic acid:

$$6HF + SiO_2 \rightarrow H_2SiF_6 + 2H_2O$$

The liquid in the cocurrent absorber **2** is transferred by pipe **7** to the circulating liquor vessel **12**, from where it is fed into the cocurrent absorber for irrigation by the centrifugal pump **13**. As it attains a concentration of up to about 30% H_2SiF_6, the liquid is transferred from the circulating liquor vessel **12** via **19** into the fluosilicic acid product vessel **14**.

The temperature of the unit for recovering fluorine is the equilibrium temperature of the vapor at a given pressure. The cocurrent absorber **2** operates at a vacuum of about 100 to about 740 mm Hg, and the temperature of the vapor under these conditions is about 20° to 95°C.

Also shown in the figure is an entry port **102** for the introduction of vapors through pipe **1** from evaporator **16.** In the upper portion of the absorber is an exit port **103** for the removal through pipe **8** into vessel **11** of any collected liquid in the upper portion of the cocurrent absorber **2** prior to passage into and through the cocurrent absorber. In the lower portion of the housing there is an exit port **104** for removal through pipe **6** of vapors remaining after passage through the cocurrent absorber **2.** At the bottom of the lower portion of the housing, there is provided a liquid outlet **105** attached to pipe **7** leading to the circulating liquor vessel **12.**

FOUNDRY SAND FROM USED FOUNDRY SAND

In the foundry industry, it has long been known to use sand molds for casting. Typically, the sand molds are a mixture of sand and resin (or sand otherwise chemically bonded) which is tailored to form a mold for a particular item to be cast. Once an item has been cast in a mold, the mold must be broken to remove it from around the item.

Rather than discard the broken molds, it has been known in the foundry industry to recycle these molds to reclaim the sand. Once that is accomplished, the reclaimed sand may be used in further molds.

Binders, such as sodium silicate (water glass), may be used in foundry molds and cores to impart desirable strength and surface characteristics in sand casting processes. Use of sodium silicate binders allows a foundry to obtain closer tolerances in sand castings. The relatively high strength of the molds allows them to be used without back-up aids, such as jackets, flasks, and bottoms. Sodium silicate-sand mixtures strengthened by carbon dioxide are used widely for cores, because the cores can be quickly made and are ready to use without baking.

Expendable cores used in permanent mold casting are commonly made by the sodium silicate-carbon dioxide process because of the dimensional accuracy obtainable. The process is suitable for all common castings alloys, including aluminum and magnesium. It is most widely used, however, for casting of steel, cast iron and copper alloys.

The cost of operating a foundry can be substantially reduced if the sand used for molding is reclaimed and used again. It has been recognized that it is important to thoroughly remove the resin coating on the recycled sand when organic binders are used in order to achieve proper molding characteristics for the formation of cores.

One method of achieving this is to burn off the resin by subjecting the sand to heat in rotary kilns and various type furnaces. These methods suffer from various objections including exposure of the sand to the contaminants in flue gases, the use of moving parts in a highly contaminated sand-filled atmosphere, and the requirement of considerable heat input and high sand outlet temperatures requiring cooling of the sand for reuse.

In any event, much effort has been put into the development of processes for reclaim and recycle of foundry sand.

A process developed by *V. Devé; U.S. Patent 3,685,165; August 22, 1972; assigned to Combustion Engineering, Inc.* utilized an apparatus for thermally reclaiming resin coated sand comprising a plurality of vertical chambers in which the sand is fluidized. The chambers are in series flow relationship proceeding from an inlet chamber to a heating chamber and then to an outlet chamber. The intermediate heating and cooling chambers have a common wall between them so that heat is transferred from the sand to be cooled to the incoming sand. The heat input to the heating chamber is controlled in response to the sand temperature in that chamber.

The fluidizing air to the inlet and outlet chambers may be separated from the remainder of the fluidized air and controlled in response to the outlet sand temperature by increasing the activity in the inlet and outlet chambers to reduce sand temperature and by decreasing the activity to reduce the outlet sand temperature. The inlet chamber is segmented to provide for the removal of heavy particles introduced to that chamber on an intermittent basis.

A process developed by *H. Jacob; U.S. Patent 3,857,201; December 31, 1974* is one in which spent contaminant coated casting sand is mixed with fresh uncontaminated casting sand, and the thus obtained mixture is subjected to mechanical agitation, during which the fresh sand scours the contaminant coating off the spent sand and thus regenerates the sand.

A process developed by *E.A. Sears; U.S. Patent 4,008,856; February 22, 1977* provides a system for reclaiming used foundry sand containing sodium silicate binders in a continuous economical process which makes possible wide-scale usage of sodium silicate binders for general foundry use. Such usage will permit a foundry to capitalize on the dimensional stability of the sand-binder mixture now felt justifiable only for more precise casting requirements and for cores.

Conventional wet sand-reclamation systems are expensive and somewhat inefficient. They often are not economically practical except in areas where the freight cost of new sand overshadows the cost of reclaiming used sand. Only relatively small percentages of reclaimed sand can be used, requiring substantial addition of new sand for reuse.

The process, however, is one in which chunks of sand and binder are disintegrated by mechanical crushing in a water wash. A primary slurry containing discrete particles of sand and both dissolved and suspended binder is conveyed to a settling tank where the sand settles from the slurry. The heavier particles of sand are removed by a ladder conveyer, which allows liquid to drain from the sand as it is being elevated. The sand is dried in rotary drum dryers, being subjected to a constant flow of heated air.

The remaining secondary fluid, containing particles of binder, is directed to a centrifugal separator, where the suspended materials such as binder particles are removed from the liquid. The liquid is then recirculated in the system, eliminating discharge of liquid or the need for substantial replenishing of liquid during the process.

An apparatus developed by *E.A. Mueller, T.H. Painter and M.K. Harrod; U.S. Patent 4,050,635; September 27, 1977; assigned to Whirl-Air-Flow Corporation* is one for reclaiming sand from foundry molds and forms. A drum is mounted in a frame for rotation about its longitudinal axis. The frame is hinged at its rearward end so that the longitudinal axis of the drum may be adjustably inclined from the horizontal. The drum is provided with first and second breaking zones for reducing large pieces of material in the drum as the drum rotates.

The first breaking zone contains longitudinal rib members and spikes attached thereto while the second breaking zone contains longitudinal rib members only. A perforate partition separates the first and second breaking zones and prevents the movement of material from the first breaking zone to the second breaking zone until it has been sufficiently reduced.

There is also provided a grinding zone for further reducing particles having a predetermined maximum size by means of grinding balls. Various sizing screens are provided for classifying the material being reclaimed into a finished product having a predetermined maximum size and oversized particles.

A technique developed by *G.Y. Richard; U.S. Patent 4,113,510; September 12, 1978; assigned to Societe d'Applications de Procedes Industriels et Chimiques SAPIC, France* utilizes chemical means rather than mechanical means (as used in many other processes) to remove binders from foundry sand so the sand can be recycled.

Foundry sand particles are regenerated by contact with a liquid containing either a peroxide oxidant or an oxidizing salt to remove resinous binder material accumulated around each of the particles.

A process developed by *K.V. Hauser and F. Mencenberg; U.S. Patent 4,130,436; December 19, 1978; assigned to Edw. C. Levy Co.* is one for treating waste foundry core and/or molding sand in a manner which upgrades these scrap streams to new sand quality for entry at the core making point. The process consists of incrementally heating the sand as required up to temperatures in the 1800° to 2500°F range, followed by cooling. The degree of heating is chosen to achieve a desired lowering of acid demand value (ADV) from that which would result from heating only to conventional high temperatures, about 1500°F. Lowering of ADV reduces the variability which would otherwise be introduced into a mixture of new and recycled sand for cores.

Heating waste molding sand to the temperatures contemplated by this process also causes clay additives to fuse and become nonabsorptive, thus eliminating the need for mechanical attrition which would otherwise be needed to remove the clay so that it would not absorb high quantities of resin in the core making process. The process also burns off resins from the sand being treated in a manner which eliminates environmental pollution.

The process will be described first with respect to waste foundry core sand, then waste molding sand and finally with respect to a mixture of these two sands which is the nature of most foundry waste sand.

Waste Foundry Core Sand: This sand is bonded with either sodium silicate or organic resin binders. Binders of the former type are not amenable to treatment by this method. Sand with organic resin binders is first incinerated to about 1500°F to burn off the binder. However, during this process calcium and magnesium carbonates, found in small amounts in natural sand from which the cores were made, decompose and are calcined to calcium oxide and magnesium oxide which are more chemically active than the original carbonates. This produces a product with an abnormally high acid demand. In foundry terminology this is called ADV.

If the heating stops at this point, the decomposition products will cause problems in catalyst control of the resin setting times during reuse of the sand. The catalyst used in making cores from the recycled sand will react in a different way than with new sand. Since cores are normally made from a mixture of new and recycled sand, the variability thus produced into the mixture is intolerable.

An unacceptable increase in ADV of recycled sand can be reduced by further heating of the core sand to the 2200° to 2500°F range. At these increased temperatures, the calcium oxide and magnesium oxide formed at the lower temperatures will first melt, assuming hard burned (a term meaning collapse of the microcrystalline structure) or less reactive forms which have lower solubility and thus markedly lower ADV.

As the temperature increases toward 2500°F these compounds react chemically with the sand grain surface to form aluminates and silicates in a manner similar to the reaction which occurs in the manufacture of cement. This will reduce ADV to zero or near zero and the aluminates and silicates will form an integral coating on the sand grains. The sand would then be cooled.

It should be noted with respect to core sand treated in this manner that because this sand has no clays which would react with calcium oxide and magnesium oxide at lower temperatures than will the sand surface, ADV cannot be lowered without achieving relatively a high temperature level. It would therefore appear desirable to process core sand jointly with mold sand as described below, unless sand with a low ADV is used as the original raw material.

Waste Foundry Molding Sand: This sand is a product which includes both clay binders and a fine carbon additive, both of which must be removed or deactivated if the recycled sand is to be placed in a condition suitable for reuse in core making. Otherwise, the carbon and clay additives would cause weakness in a core due to poor adhesion of the resin to the sand. The clay binder would also absorb an excessive amount of resin in the core making process, which results in weakening of the core and is a very costly problem. In addition, this waste foundry molding sand has fine sand resulting from cracking of sand grains during multiple reuse in the molding operations.

Waste foundry molding sand can be upgraded for recycling by first heating the sand with air available to burn off the carbon additives at a temperature of 1400° to 1600°F in a manner already well-known in the industry. This heating step can also weaken the bond between the clay and sand grains. However, if the recycling process is stopped at this point the result is inadequate in several ways. The acid demand problem described with respect to the core sand would still arise. Secondly, the clay would still have to be removed, for example, by abrasion, in order that it not absorb high quantities of resin in the core making process and thus weaken the core.

Thus, as a second step, the process contemplates continuing to heat the waste foundry molding sand to a temperature range from 1800° to 2200°F. In this temperature range, varying somewhat with the type of clay used, the clay additives fuse onto the surface of the sand, becoming an integral part of the sand grain and thus become nonabsorptive with respect to the resin which will be added later during the core making process.

As a third step, if the ADV after the second step is still too high, heating of the sand will be continued into the range of 2200° to 2500°F. In this range, calcium oxide and magnesium oxide which had been formed at the lower temperatures, in the manner described above, will first melt, becoming hard burned, less reactive forms which have markedly lower ADV, and will then chemically react with the fused clay and the sand itself, forming aluminates and silicates to further lower the ADV, and forming an integral coating on the sand grains. The final temperature can be selected to achieve the described lowering of ADV. The sand would then be cooled.

Overheating could cause stress cracking and fracture of the sand grains themselves. This could happen at 2400°F or above, depending on sand quality and rates of heating and cooling.

Mixed Core Sand and Molding Sand: Most foundries have waste sand of both types in varying proportions. A preferred method for mixed sands is as follows:

(1) Proceed with treatment of the waste molding sand by heating with air available to burn off carbon additives in the range of 1400° to 1600°F, as described above.

(2) The next step is to inject core sand, crushed if necessary to achieve a fast reaction, into the hot molding sand, with an excess of air. This burns out new resins in a manner which recovers the fuel values of the burned resins, resulting in energy saving. Secondly, the resin will be burned off at a temperature high enough to insure complete incineration; that is, no unburned organics will be emitted into the atmosphere which could pollute the environment. It should be noted that this is a potential problem with conventional core sand in incineration systems such as that described in U.S. Patent 2,478,461.

(3) As a third step in the treatment of the mixed core and molding sand, the heating would be completed at incremental temperatures as described above with respect to the treatment of waste foundry molding sand, that is, first to temperature range of 1800° to 2200°F and then, if desired, to a range of 2200° to 2500°F. The sand mixture would then be cooled.

The equipment for performing the above described method, either with respect to molding sand, core sand, or mixtures, could be a rotary kiln, fluid bed calciner or hearth type roaster, all substantially as manufactured for use in other fields of mineralogical processing.

Final product preparation of recycled core sand, molding sand or a mixture as set forth above would be similar to that for new natural sand for foundry use. The cooled product would be screened and classified to remove oversize and undersize grains and achieve the correct gradation for satisfactory core manufacture.

If some melting together of particles has taken place, this will require crushing of agglomerates.

An apparatus developed by *R.A. Cina and A.D. Kluge; U.S. Patent 4,137,675; February 6, 1979; assigned to Roberts Corporation* is a cylindrical sand reclaimer drum for removing spent bonding contaminants and refuse from chemically bonded foundry molding sand and the like. The cylindrical drum is rotatably mounted along its longitudinal axis and is supported in a substantially horizontal inclined position along its longitudinal axis. The cylindrical drum has an intake opening at the elevated end thereof to receive the contaminated sand and process it through a plurality of internal compartments containing spherical or cube shaped abrasion media.

The compartments are formed by a plurality of spaced-apart circular divider panels mounted on the central spindle or shaft of the cylindrical drum and which are perpendicular to the longitudinal axis thereof and are axially adjustable therealong.

A plurality of spaced-apart openings are provided along the circumferential edge of each of the divider panels. Each opening is provided with a selectively adjustable gate so that the size of the opening can be adjusted as desired so that the size of the openings in each succeeding divider panel is smaller than those in the preceding divider panel. Thus, each divider panel removes refuse of a particular size from the sand as it passes therethrough.

The lowermost final compartment has screen panels provided in the circumferential wall thereof which form the discharge opening for the clean reclaimed sand. The lower end wall of the cylindrical drum is provided with discharge openings through which the contaminants and refuse are discarded. Each compartment is provided with an access panel in the wall of the cylindrical drum so that spherical or cube shaped abrasion media can be selectively added or removed therefrom.

Motor drive means are provided in association with the cylindrical sand reclaimer drum so as to rotate the drum so as to cause the contaminated sand to advance therethrough. Cyclone air removal suction means are provided in association with the sand reclaimer drum which are adapted to remove free bonding materials such as porous burnt resin particles which have been abraded from the sand during passage of the contaminated sand mass through the rotary sand reclaimer.

A process developed by *W.L. Adams; U.S. Patent 4,144,088; March 13, 1979; assigned to Kenzler Engineering Company, Inc.* is one for reclaiming used foundry sand and combining it with new sand to meet specifications. The resulting sand mixture may then be coated with a bonding material such as resin. During the process, the reclaimed sand is stabilized as to its size so that upon reuse in the foundry it does not appreciably change its grain size and consequently will not cause casting defects which are generally referred to as expansion defects.

Each time the silica sand passes through the phasing process where it is expanded and quickly contracted, the subsequent amount of expansion and the severity thereof decreases. The process provides an economical and efficient process for reclaiming used foundry sand and provides a finished, blended sand which is particularly stable for subsequent foundry uses. In addition, the reclaimed sand which has been thermally reclaimed provides sand grains of sufficient similar size and screen analysis which requires considerably less bonding material to achieve proper molding strengths. Such a process is shown in Figure 61.

New sand is stored in conventional hoppers or bins **1** and conveyed by the transporter **T** via a line to the new sand tanks **4** which deposit new sand in the blender **6**.

The system utilizes several transporters **T** for transporting the sand from one station to the other via suitable conduits. These transporters are of themselves conventional and it is believed sufficient to say that they are of the dense phase air type so that the sand is conveyed pneumatically at low velocity by air under pressure which is introduced into the transporter.

Reducing: Used sand which is to be reclaimed is deposited and held in the storage bin or tanks **8** and is then discharged as required onto the conveyor means **9** for delivery into a reducer lump crusher **10**. The reducer lump crusher is of the orbital rotating type and acts to break up or reduce the clumps of used sand and then deposits the loose sand on a conventional conveyor **12**.

In the systems, several conventional vibratory, spring mounted conveyors **9, 9a, 12, 89, 99** and **100** are used and they all are of the same type, and are not believed to require any detailed explanation.

Magnetic Separating: The conveyor **12** conveys the reduced sand beneath a magnetic separator **14**. The magnetic separator may be of the endless belt type which contains magnets that pass over the reduced sand and pick up tramp metal, rods, nut, bolts, and other metallic materials from the sand and then discharges the metal, in the known manner, to one side of and separate from the conveyor **12**.

Nonmetallic Separating: The iron-free, crushed sand then passes through a fluid bed (air) separator **20** which removes the nonmetallic foreign matter from the sand by floating the foreign matter and separating it, by a series of inclined screens, from the heavier sand grains, and discharging the foreign matter to the side, via outlet **21**. It is believed sufficient to say that the fluid bed separator is of the multiscreen type that separates the nonmetallic material by weight. That is the solids are suspended by the screens and within an upward moving stream of gas and the solids/gaseous mixture behaves like a liquid which can be handled, measured and controlled as though it were a liquid.

Heating: Upon leaving the fluid bed separator, the sand enters a furnace or thermal conveyor **24** which may be fired by coal, gas, oil, or the like. The furnace heats the sand and its foreign material such as clay, carbonaceous material, and other nonmetallic foreign matter to a temperature preferably of about 1400°F. As the material is being conveyed through the furnace, the carbonaceous material and some of the other foreign material is burned off. The heating of the sand grains causes them to change size thereby further causing the foreign material, to a certain extent, to be cracked or removed from the grain of sand.

Air Sweep Classifier: After leaving the furnace **24**, the hot sand is quickly cooled to about 400° to 500°F, in a water quencher **Q**. This further cooling, also helps break up coatings of foreign material remaining on the sand grains.

Figure 61: Kenzler Engineering Co. Inc. Process of Reclaiming Used Foundry Sand

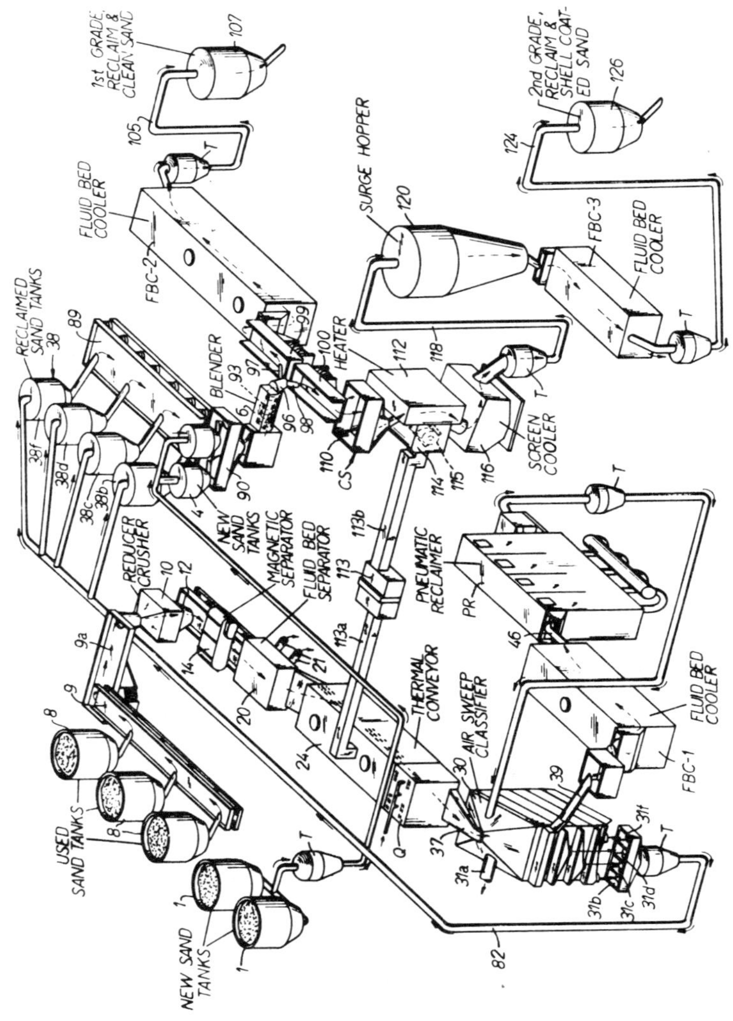

Source: U.S. Patent 4,144,088

The sand is then delivered via conduit **37** to the air sweep classifier **30**. The air sweep classifier **30** has an air inlet which directs the air upwardly through grading screens and the dust ladened air is discharged through outlet **31a**. The air sweeps the sand, causing it to be tossed about for good contact with the screens. The heated sand is conveyed on the inclined, first coarse screen and the large sand grains are delivered to the conduit **39** where it is conducted to the fluid bed cooler **FBC-1.**

A portion of the sand which passes through the first screen is small enough and has had its impurities and coatings sufficiently removed by the previous crushing, magnetic separation, fluid bed separation (nonmetallic separation), and heating processes, so that it is acceptable and consequently passes into the conduit **82** where it is conducted to the reclaimed sand tanks **38**.

Various receiving compartments **31b, 31c, 31d** and **31f** are provided for receiving the various grades of sand from the screens of the classifier. These various grades are then selectively conveyed by conduit **82** to their respective tanks **38b, 38c, 38d,** and **38f,** where the desired grades are drawn off for blending with new sand as may be required.

As the sand passes through the air sweep classifier, it is further cooled somewhat and upon leaving the classifier via conduit **39**, it is at a temperature of about 450°F.

That portion of the sand, which because of its large size passes directly to the fluid bed cooler **FBC-1**, is treated in a cooler to further reduce the sand and cool it to about 300°F.

The system utilizes a number of fluid bed coolers **FBC-1, FBC-2,** and **FBC-3,** through which the sand is conveyed from one end to the other. These fluid bed coolers are of the type having a system of water coils which act as heat exchangers and provide multiple weirs over which the sand mixed in the fluid bed (air bed) is forced over. The water coil receives recirculating water from a cooling tower (not shown). Fluidized air is provided by a turbo-compressor (not shown). The cooled sand then passes to the outlet **46**.

Pneumatic Reclaimer (Sand Scrubber): After the sand has been reconditioned to some extent by the previously described reducer crusher, the magnetic separator, fluid bed separator, thermal conveyor, and the screening process provided by the air sweep classifier, all of which treat the sand as a mass, the compound grains and the accumulated coatings of burned clay and carbonaceous material respond to only more intensive treatment of the individual sand grains. Wet processing and thermal reclamation methods, when combined, are effective but economically impractical except for very large high volume operations.

In other words, during the thermal destruction of resins, pitches, wood products, coal and oils in the sands, certain of these residues combined with clays will form a coke-like matrix in which several individual sand grains may be trapped. In a normal screening, these agglomerates could easily retain their identity and present a false specification. Of equal importance, the coke-like material is a sponge thirsty material which would greatly decrease the value of the product.

A grain of sand may be coated with layers of clay and carbonaceous material, for example, the original grain may have a first layer of vitrified clay and carbonaceous material and also a second layer of similar material, both layers of which in addition also contain fines and live clays. Thus, sand of this character which is referred to as a waste shake-out sand, even after the previously mentioned reconditioning cannot continuously be reused for facing and cores. The grains become roughened, fused and enlarged. Excessive fines and size variations of the grains reduce response to ramming, making sand hard to control and thermally unstable.

Consequently, after the partially reconditioned sand leaves the air classifier **30**, it is passed through a fluid bed cooler **FBC-1** and into a pneumatic reclaimer **PR**. The pneumatic reclaimer (or sand scrubber) combines the performance of the thermal and wet systems of reconditioning into a single compact unit. The pneumatic reclaimer **PR** utilizes air as the motivating force and uses the sand grains themselves as the abrasive body.

Blending: The reclaimed sand from tanks **38** is delivered by a vibratory conveyor **89** to a hopper **90** located above the blender **6**. The new sand from tanks **1** is also delivered to the hopper **90** as previously mentioned. The blender **6** provides continuous high speed blending of the sands and employs a mix-folding action provided by a slowly rotating paddle and it also provides controlled vibration to fluidize solids. The result is a rapid dispersion from one solid phase into another. Good mixing efficiency is provided by the gentle action and the discharged mix contains the same ratio of materials as supplied at the inlet. The paddle employed in the mixer is shown and includes a rotatable auger and paddle assembly **93**.

The mixed sand moves through the blender by gravity and consequently retention time and mixing time are maximum for any given feed rate. The mixed material then passes through the outlet of the blender **96** which includes a direction delivery spout **97, 98** which can convey the mixed material to either the vibratory conveyor **99** or **100**, respectively, for delivery to a fluid bed cooler **FBC-2** or to the coating station **CS**, respectively.

The mixed sand that passes through the fluid bed cooler **FBC-2** is conveyed by the transporter **T** through conduit **105** and into a storage bin **107** which will then contain first grades of reclaimed and clean sand.

Heating-Coating-Cooling: The mixed or blended sand which is diverted by diverter **98** is conveyed to the coating station **CS.** More specifically, it is delivered by conveyor **100** to the hopper **110** and then into a heater **112** of the coating station.

The heater **112** receives its heat from heat that has been reclaimed from the furnace **24.** This is accomplished by a heat exchanger **113** that is connected by air conduit **113a** with the furnace and by conduit **113b** which connects the heat exchanger with the hot coater **114.** Thus, the heat from the furnace is utilized without permitting the contaminated air from entering and contaminating the hot coater **114.** Heat exchangers of the air-to-air type or water jacket types, among others may be used, provided they deliver clean heat to the coater. The sand which is to be coated is then heated to the proper temperature of about 450°F for the subsequent bond coating process. The sand then passes to the hot coater **114** which applies or mixes a hot coating shell of bond to the sand. Slowly revolving augers **115** having mixing paddles are used for this coating process.

The shell coating sand then passes to a screen cooler **116** where the coated product is completely cooled and is then delivered via the transporter **T** and conduit **118** to the surge hopper **120** which accommodates varying rates of flow in the systems. The sand then passes through the fluid bed cooler **FBC-3** of the type previously described and the transporter **T** then conveys it via conduit **124** to the storage bin **126** where it is availble for reuse as a reclaimed and shell coated sand.

Foundry operation has long been recognized as presenting potential hazards for foundry workers, in part due to dust and the like arising from molding material handling processes including shake out of rough castings. For these reasons, it has become conventional to provide pollution control means for gathering, from a foundry, airborne fine particles otherwise presenting hazards to workers.

Heretofore, the gathering of airborne fine particles has presented further difficulties in that increasingly stringent requirements for air quality have prohibited direct emission of the airborne fine particles into an ambient atmosphere. One approach to disposal of such fine particles which has achieved wide acceptance has been to wash such particles out of an air flow and discard the particles as a wet waste. It has been assumed that the fine particles and wet waste have not been useful and, in fact, disposal of such wet waste has presented an increasingly difficult problem. Foundry sludge is difficult to handle and transport, and is deemed unacceptable by many operators of land fills and other normal disposal areas.

Thus, it is the object of a process developed by *B.W. Adkison and T.E. Burden; U.S. Patent 4,149,581; April 17, 1979; assigned to Cole Manufacturing Company* to provide, in a foundry having certain conventional apparatus, an improvement which recycles fine particles removed from the foundry operation as wet waste. In realizing this object wet waste removed from the foundry is delivered to a muller for mixing with sand and water in preparing molding material for subsequent use. Figure 62 is a schematic representation of the process.

It has been conventional to provide, in the operation of a foundry, for the preparation of molding material by mixing sand and water within a muller, here generally indicated at **10**. The specific details of the muller means **10** have not been here disclosed, as detailed information concerning such apparatus is available to interested readers from manufacturers of commercially available foundry equipment. Molding material prepared in the muller by mixing sand, appropriate additives and water, is delivered to appropriate personnel and/or mechanism for forming of molds and casting of metal. From the pouring location, molds containing rough castings are delivered to a shake out location here generally indicated at **11**, at which the rough castings and molding material are separated.

Due to the quantities of airborne particles created in the foundry operation, and frequently particularly at a shake out location **11**, it is common to provide pollution control means within a foundry for removing airborne fine particles. In the form illustrated, the pollution control means includes means for inducing airflow in the form of an exhaust fan **12**. The exhaust fan **12** operatively communicates with suitable enclosures and/or hood means within the foundry, such as a hood generally indicated at **14** which may be located over or adjacent a shake out location **11**. By means of an airflow induced through the hood **14** by the exhaust fan **12**, airborne fine particles of molding material are entrained and removed from within the foundry.

Figure 62: Cole Mfg. Co. Apparatus for Recycling Fine Particles from Foundry Sand Reclamation Processes

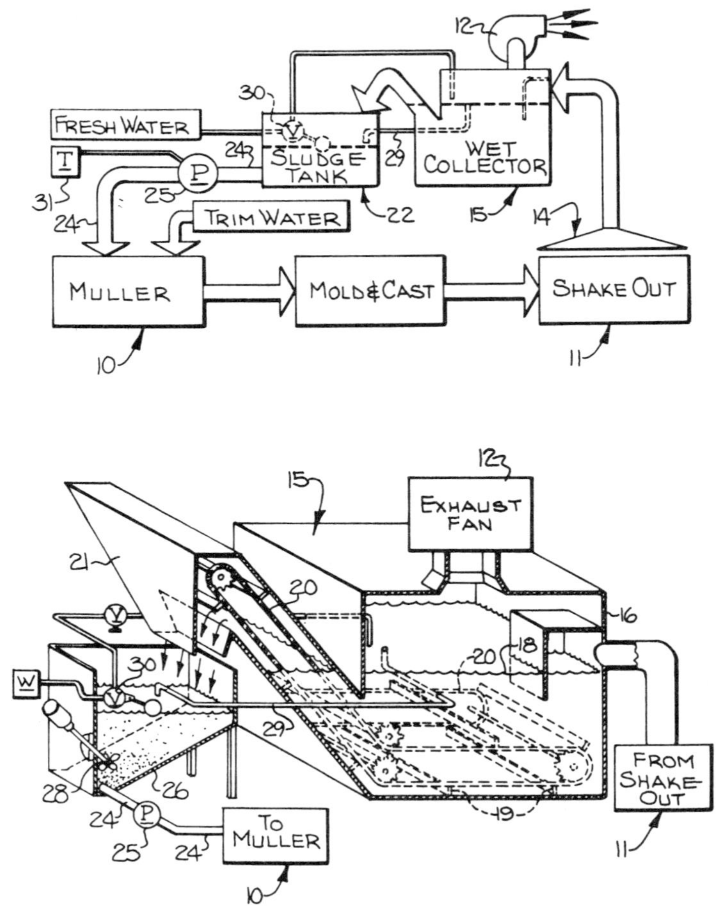

Source: U.S. Patent 4,149,581

Air and entrained fine particles thus removed from a foundry are conventionally delivered to a wet collector, here generally indicated at **15**. Wet collector means such as the apparatus **15** are generally well-known to persons operating foundries and other establishments subject to air pollution control regulations. One form of such a wet collector means **15**, as indicated in the view at the lower part of the figure, may include a housing **16** constructed as a substantially closed tank containing a body of water **18.** Air drawn from within the foundry is forced to pass through the water **18** within the housing **16,** for removal of entrained, air-borne fine particles therefrom. Such removed fine particles eventually settle toward the bottom of the housing **16** to form a sludge deposit thereupon. The sludge deposit is removed from the wet collector **15** in the form shown, by a series of scraper bars **19** mounted upon driven chains (one of which is visible at **20**) and extending transversely of the housing **16.** By means of the scraper bars **19,** sludge is withdrawn from the bottom of the housing **16,** raised to a discharge chute **21,** and discharged from the wet collector **15.**

Recirculating means are provided which operatively communicate with the scrubber means **15** for receiving wet waste therefrom and operatively communicate with the muller means **10** for delivering wet waste thereto for mixing with sand. By such recycling of fine particles, the disposal problems heretofore encountered are obviated without degrading the quality of the molding material used or the castings made. More particularly, a sludge tank **22** is, in the illustrated embodiment, disposed beneath the discharge chute **21,** for receiving wet waste from the collector **15.**

By means of a discharge conduit **24** and pump means **25** such as an air-driven diaphragm pump, liquid and wet waste from the sludge tank **22** is delivered into the muller means **10.** Preferably, and in the form shown, the sludge tank **22** has a bottom wall **26** which slopes toward the location from which the conduit **24** leads. Additionally, it is preferred to provide agitating means in the form of a motor driven propeller **28** or the like disposed within the sludge tank **22** for maintaining suspension of fine particles in liquid contained therewithin. Sludge discharged from the wet collector **15** is diluted within the sludge tank **22** and maintained in suspension prior to delivery to the muller means **10.**

In order to achieve a desired dilution of wet waste from the wet collector **15,** and in the particular form shown, water is delivered from the wet collector **15** to the sludge tank **22.** More particularly, the wet collector **15,** in the form shown, is of a type wherein the desired water level within the wet collector is maintained by a constant overflow from within the housing **16** through an overflow conduit **29.**

Fresh water (from any suitable source such as a city water supply) is supplied to the wet collector **15** by a float valve **30** responsive to water levels within the sludge tank **22.** That is, should pumping of diluted wet waste from the sludge tank **22** lower the water level of that tank below the level maintained by the normal overflow from the wet collector **15,** then additional water is added through the wet collector to restore the sludge tank water level.

In operating a foundry having the apparatus described, wet waste is delivered from the sludge tank **22** by the pump **25** upon preparation of a batch of molding material in the muller **10.** Preferably, the quantity of diluted wet waste supplied to the muller **10** is controlled by a timer **31** operatively connected with the pump **25** and causing operation of the pump **25** for an interval of time sufficient to deliver a quantity of liquid less than the entire quantity necessary for preparation of a batch of molding material. Typically, the liquid provided from the sludge tank **22** may be on the order of 75 to 90% of the liquid required for preparation of a batch of molding material. Provision is made for the addition of trim water from a suitable source such as a city water supply under the manual control of an operator or otherwise as may be necessary or appropriate to make up the exact quantity of water required for preparation of the batch of molding material.

Experience with an operating installation of equipment operated in accordance with the methods of this process has demonstrated the advisibility of some provision for back flushing the conduit **24** periodically, in order to assure that the pump **25** and conduit **24** are not clogged by fine particles which have settled out therewithin. In that installation, the casting quality of the molding material is not degraded by recirculation of fine particles, as corresponding quantities of more coarse particles (or sand) are removed from the molding material which is recycled through normal foundry operation.

G

GADOLINIUM FROM BUBBLE MEMORY UNIT SCRAP

Gadolinium-gallium-garnet (GGG) which has the formula $Gd_3Ga_5O_{12}$ is used mainly in the manufacture of components for bubble domain memory units. In the manufacturing stage about 80% of the starting material ends up as rejected material or scrap. This is waste material from a relatively expensive compound, and its recovery is of considerable economic interest.

If the recovery and processing of the scrap leads to sufficiently pure products, then these can be recycled and used again as starting material. As a result, the economics of the production of GGG wafers can be improved considerably. Under the term scrap is to be understood particulate material (remainder of charges used to grow crystals, fragments of crystals from all stages of processing to wafer form) and also fines or powders which are formed during cutting, grinding, polishing, etc. the garnet $Gd_3Ga_5O_{12}$ or similar materials.

The digestion of such scrap is a problem which has existed for only a few years and one which has not been solved up to now. All previous attempts have failed because of the slow rate of dissolution of these complex oxides.

An improved process has therefore been developed by *E. Gusset; U.S. Patent 4,198,231; April 15, 1980; assigned to Swiss Aluminium Ltd., Switzerland* in which gallium and gadolinium are recovered and separated from scrap material of substances which contain both of these elements in the form of oxides or compounds which can be transformed to the oxides.

The scrap is finely ground and then dissolved in a strong, aqueous mineral acid. The gadolinium is precipitated from the purified solution as the oxalate; then the gallium is deposited in the metallic form by means of electrolysis. The electrolytic deposition of the gallium can also take place before the gadolinium is precipitated out as the oxalate.

The following is a specific example of the conduct of this process. The scrap materials which can be used as the starting material for the process are particulate residues of charges for crystal growing, broken parts of crystals from all stages of preparation to wafers, and also fines or powder arising from grinding, polishing and lapping the garnet $Gd_3Ga_5O_{12}$. Round grinding fines from the preparation of the GGG crystal were dried at 120°C and then heated at 600°C for a few hours to decompose the volatile impurities.

1,000 g of fine particulate dry scrap all of which were $<40\ \mu m$ in size and containing 34% gallium and 46% gadolinium were boiled for 2 hours in 2,100 ml of 35% hydrochloric acid with reflux. This corresponds to a charge of 99%. After digestion of the scrap the undissolved part was filtered off and rinsed. This residue weighed 20 g when dry. This material was collected and then digested once more. The filtrate, including washing and rinsing water, 2,300 ml in total, was treated with 4,000 g metallic gallium at 50°C for 45 minutes. In doing so the metal must be dispersed as finely as possible.

In the cementation process the more noble elements present in the aqueous solution are separated out and partly dissolved in the gallium until saturated and then precipitated as intermetallic impurities or in an elemental form. The cementation can also be carried out with a much smaller amount of gallium. The metal can be used repeatedly until the necessary purification has been achieved. The result of the purification process was a solution of 2,400 ml containing 140 g/ℓ gallium and 190 g/ℓ gadolinium.

The pH was adjusted to a value of 1 by addition of 900 ml of 4 N potassium hydroxide solution, at the same time stirring well, and then approximately 20 ml of 30% hydrogen peroxide added to oxidize the iron present. The precipitation of the gadolinium was carried out at 50°C by addition of 1,500 g of crystallized, technical grade oxalic acid ($C_2O_4H_2 \cdot 2H_2O$) and the suspension stirred gently overnight to achieve complete precipitation of the gadolinium oxalate.

The gadolinium oxalate $Gd_2(C_2O_4)_3 \cdot 10H_2O$ was separated out by centrifuging, rinsed with about 200 ml dilute oxalic acid (6 g/ℓ) and dried at 130°C before converting to gadolinium oxide (520 g) by heating at 800°C. The gadolinium can be purified by precipitating in acid and renewed precipitation as the oxalate.

The liquid left over after centrifuging and the rinsing water from the first separation of gadolinium oxalate (3,300 ml) was brought to a pH of 12 by adding 2,300 g potassium hydroxide and stirring strongly. The 6,000 ml of solution containing 55 g/ℓ of gallium and 1 g/ℓ of gadolinium was then electrolyzed at 60°C using a stainless steel cathode and graphite anode, and using a current density of approximately 0.1 A/cm^2. After 48 hours, 325 g of gallium had been deposited, and the electrolyte containing 0.4 g/ℓ of gallium was thrown away.

The precipitated metal represented scrap of 99.99% purity and can be converted directly to the oxide. The gadolinium content of the gallium lay below the detection limit of 0.3 ppm, by wt, of the mass spectrometer.

GALLIUM FROM ALUMINA PLANT LIQUORS

Gallium, a by-product of the aluminum system, is produced by tapping a portion of the green liquor generated in the production of alumina (Al_2O_3). Gallia (Ga_2O_3), present in bauxite, dissolves along with the alumina during the digestion of the bauxite ore. The soluble gallia builds up in recycled plant liquors by dissolving to saturation (about 0.2 g/ℓ) and partial precipitation. After the solution is saturated with gallia, its concentration in the liquor remains relatively constant with the excess exiting the process via the red mud waste.

Gallia is produced by withdrawing a small portion of the plant liquor and subjecting it to a separate precipitation. In this process, a small amount of alumina is lost equal to about 100 parts alumina for each part of gallia produced. Once the alumina-gallia precipitate is obtained, the gallium is separated from it electrolytically and is refined by subjecting it to as many as 16 steps of processing on a laboratory scale.

The sand removed in classification prior to filtration of the green liquor is of interest because it is an example of the variety of minerals contained in the plant liquor. The black sand is considered waste, though it might some day be considered worthy of by-product processing. If the sand removed by the classifier is subjected to magnetic separation, a black (magnetic) portion can be separated from the total. A sample of this material at Alcoa's Bauxite, Arkansas plant assayed 40% iron oxide and 3.8% titania with smaller amounts of zircon, manganese, magnesium, vanadium, niobium, gallium, chromium, molybdenum, nickel, lead, and copper oxides. The system depicted in Figure 63 is based on the alumina plant of the Aluminum Company of America at Bauxite, Arkansas.

Figure 63: Alumina System Showing Gallium By-Product

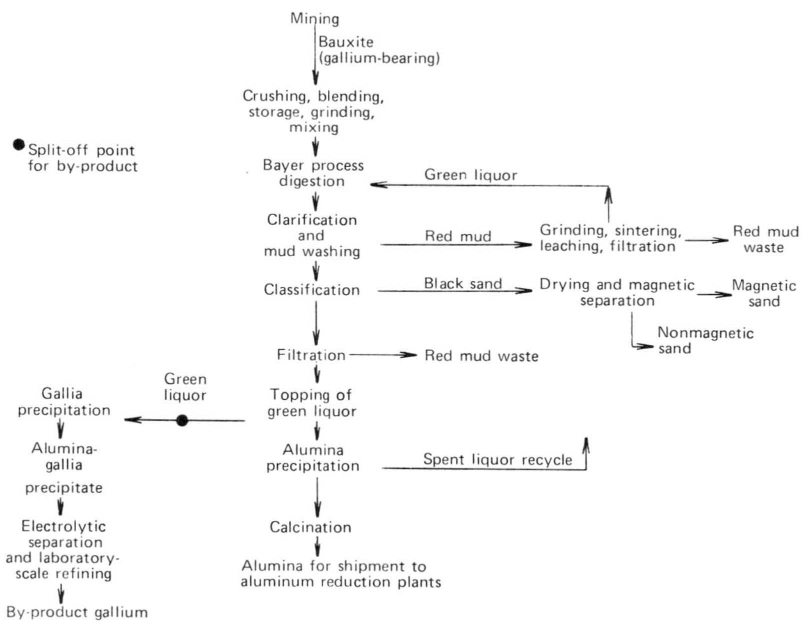

Source: Reference (17)

Gallia is produced by extracting a very small portion of plant liquors which flow through the plant at 300,000 gpm. Labor is by far the major item of cost in the production of gallium. Even the methods of packing and shipping have significant impact on incremental costs. There is no full-time labor devoted to gallium production. The labor force that extracts gallia is also concerned with the production of several other specialty products. Because there is a market for only small quantities of gallium, large quantities are discarded in the red mud wastes. Bauxites are known to contain up to 0.01% gallium or 0.22 pound per long ton. 50% recovery of contained gallium would yield 0.11 lb/ton, and 90% recovery would yield 0.20 lb/ton.

GALLIUM FROM BUBBLE MEMORY UNIT SCRAP

See "Gadolinium From Bubble Memory Unit Scrap."

GALLIUM FROM PHOSPHORUS FURNACE FLUE DUST

Gallium is found in association with zinc, germanium, aluminum and copper, and occurs in coals. Although the element gallium is relatively abundant, it is widely disseminated and, except in rare instances, these sources often contain small concentrations only in the order of 0.001 to 0.02% gallium.

Gallium is also found in minuscule amounts in phosphate minerals and these amounts are somewhat concentrated during the processes for the production of elemental phosphorus. In these processes, for example, the gases leaving the electric furnaces are treated in precipitators for the collection of the flue dusts which contain gallium, as well as a large number of other elements, the most important of which are zinc, silver, cadmium, aluminum, phosphorus, sodium, potassium, calcium, fluorine, chlorine, silicon and carbon.

The extraction of gallium and germanium from coal fly ash and from phosphorus furnace flue dust have been described by R.F. Waters and H. Kenworthy in Report of Investigation No. 6940, Washington, D.C., U.S. Bureau of Mines (April 1967).

Laboratory-scale selective volatilization methods developed by the Bureau of Mines to recover germanium and gallium concentrates from coal fly ash and phosphorus furnace flue dust are summarized. In the better experiments between 85 and 98% of the germanium was recovered as the lower oxide and as sulfides, and between 75 and 97% of the gallium was recovered as the lower oxide and as the trichloride. Grades of the condensates were inconsistent and varied between a few tenths of a percent and 8%, depending on the method used.

In the primary extractions of germanium alone the degree of enrichment surpassed 100 to 1, and for gallium the best ratio was 30 to 1. Upgrading of condensates by vaporization retreatment and by leaching and precipitation is discussed. An appreciable degree of mechanical preconcentration by attrition scrubbing and elutriation was possible on two fly ash samples.

A process developed by *T.L. Charlton, R.F. Redden, H.E. Hirsch and S.C. Liang; U.S. Patent 4,071,422; January 31, 1978; assigned to Cominco Ltd., Canada* is one in which gallium and other values are recovered from phosphorus-furnace flue-dust by treating the dust with sulfuric acid to form a solution and a residue. The residue may be treated for recovery of values. Zinc is precipitated from the solution as zinc ammonium sulfate hexahydrate and an alkaline material is added to precipitate a gallium concentrate and to form a gallium-free solution, which may be treated for recovery of sulfate and phosphate values.

The gallium concentrate may be upgraded by mixing concentrate with lime to precipitate calcium phosphate and by subsequently adding sodium hydroxide to redissolve coprecipitated gallium and aluminum compounds. Residual precipitate is recycled and the solution is neutralized to precipitate an upgraded concentrate. Aluminum may be removed as calcium aluminate. Upgraded concentrate may be dissolved in alkaline solution and the solution electrolyzed for deposition and recovery of gallium. Such a process is illustrated in Figure 64.

Figure 64: Cominco Ltd. Process for Concentrating and Recovering Gallium from Phosphorus Furnace Flue Dust

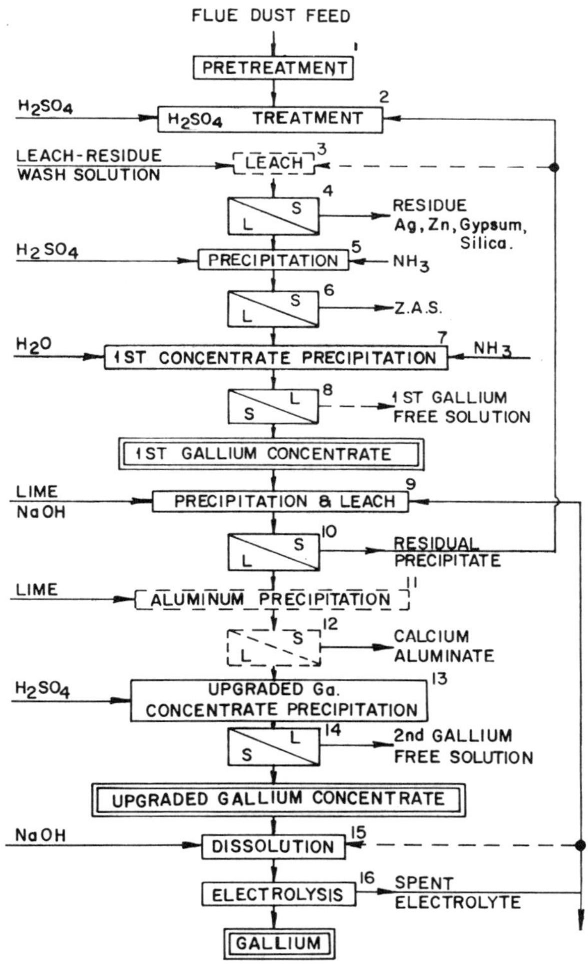

Source: U.S. Patent 4,071,422

Flue dusts treated according to this process are obtained from precipitators in the electric furnace production of elemental phosphorus and vary in composition depending on the composition of the phosphate minerals used. Such dusts may contain, for example, 0.02 to 0.05% Ga, 5 to 15% Zn, 0.5% Cd, 10 to 25 ounces per ton Ag, 1 to 4% Al_2O_3, 1 to 3% Na_2O, 5 to 25% K_2O, 7 to 12% CaO, 25 to 35% P_2O_5, 0.3 to 1.0% Cl, 13 to 19% SiO_2, 1 to 5% F and 5 to 20% H_2O. When removed from the precipitators and exposed to air, the dusts burn and partly fuse causing the formation of agglomerates with a very broad range of sizes. Agglomerates may reach sizes up to 12 to 18 inches. Such agglomerates are not suitable for direct processing.

Flue dust is subjected to pretreatment **1** (Figure 64) where the particle size of the dust may be reduced to a convenient particle size, which is suitable for feeding to the process, such as, for example, a size of 1 cm or less. The pretreatment to the desired particle size may be accomplished by feeding dust to size-reduction equipment, such as a hammermill, screening out the undersize material, returning oversize material to the mill and feeding the pretreated dust to sulfuric acid treatment **2**. Simple size reduction to particle sizes smaller than about 6 M (mesh, Tyler Series) is unnecessary as this process can treat coarse material without difficulties.

Alternatively, pretreatment may be accomplished by wet-grinding flue dust in an aqueous slurry followed by separating the ground dust from the liquid. Wet-grinding has advantages when the flue dust has a high content of water-soluble constituents. Wet-grinding not only reduces the particle size of the dust but solubilizes significant quantities of the soluble constituents. The separated liquid contains amounts of potassium and phosphate values which make the recovery of these values, for example, as fertilizer, economically feasible. After separation from the liquid, the wet-ground dust may be partially dried, if so desired, before being fed to sulfuric acid treatment **2**.

In sulfuric acid treatment **2**, the pretreated dust is treated with sulfuric acid for the conversion of metal values in the dust to sulfates and for the elimination of at least a portion of fluorine. Pretreated dust is mixed with sulfuric acid, heated to an elevated temperature and agitated for a period of time of sufficient duration to effect the desired degree of leaching.

In one method of the acid treatment **2**, dust is mixed with sulfuric acid in an amount sufficient to provide in the resulting leach solution 15 to 25% free acid found necessary to ensure high extractions of metal values.

The mixture of dust and acid is heated to a temperature of from 60° to 100°C and is agitated for a period of from one-half hour to 6 hours. During the reaction, metal values are converted to sulfates, and fluoride in the dust reacts with the acid and silica to form silicon tetrafluoride which evolves from the reaction mixture. If so desired, the pressure over the reaction mixture may be reduced below atmospheric to enhance the removal of fluorine. The leach solution is separated from the leach residue in liquid-solids separation **4**.

The high free acid concentration in the leach solution can cause corrosion in the acid treatment **2** and in subsequent process steps. This may be alleviated by using corrosion-resistant equipment, which, however, is expensive.

High extractions and good fluorine removal can also be obtained, while maintaining the free-acid concentration in the leach solution low to minimize corrosion, preferably by subjecting the dust to a two-stage treatment comprising a sulfation with a limited amount of concentrated sulfuric acid followed by an aqueous leach. According to this preferred method, dust is mixed with concentrated sulfuric acid of, for example, 93% strength in a limited amount sufficient to convert metal values to sulfates and to provide 1 to 4% free acid in the subsequent aqueous leach. This low amount of free acid alleviates corrosion problems and obviates the need for expensive corrosion-resistant equipment in subsequent processing.

The variables in temperature and mixing time must be controlled to ensure adequate mixing and conversions as well as adequate fluorine removal. These variables also control the fluidity of the acid-dust mixture. To maintain the necessary fluidity of the mixture, the temperature should be at least 100°C and should preferably be maintained in the range of 100° to 250°C. A mixing time of at least 15 minutes is necessary to ensure adequate mixing and sulfation. A mixing time of from 1 to 2 hours is preferred.

The sulfuric acid treatment according to the preferred method is carried out in a suitable mixing device, such as, for example, a pug mill or a blunger. The mixing device may be externally heated to maintain the mixture at the desired temperature. The evolved gases are scrubbed and, upon completion of the sulfation, the reaction mixture is discharged from the mixing device and fed to leach **3**.

In leach **3**, the sulfated mixture is slurried with acid wash-solution from the subsequent washing of the leach residue and the mixture is leached to dissolve the soluble constituents and to crystallize gypsum in readily-separable form. The leach is preferably carried out in a single-compartmented leach vessel, or in two leach vessels connected in series. Sulfated mixture and wash-solution are fed to and slurried in the first vessel and leach residue and leach solution are discharged from the second vessel.

The leach is conducted at a temperature of from 60° to 100°C. The temperature may be maintained by suitable means, such as, for example, by the use of steam. To maintain solubility of the leached products, and to prevent corrosion of stainless steel equipment, the concentration of free sulfuric acid in the slurry is maintained in the range of 1 to 4%. Concentration of free acid is mainly controlled by the amount of sulfuric acid added in sulfuric acid treatment **2**, as described above, and is further controlled by the acid content of the wash-solution. The free acid concentration may be adjusted by evaporation of wash-solution or by addition of water.

The leach residue is separated from the leach solution in liquid-solids separation **4**. Liquid-solids separations and washing of the solids in the process are carried out using standard practices, methods and equipment. Adequate washing ensures improved separation of values.

To minimize loss of gallium, zinc and phosphate with the leach residue, the residue is washed in a countercurrent system comprising at least two stages (not shown). After liquid-solids separation and following the washing, the acid wash solution is returned to the leach **3**. If so desired, the wash solution may be evaporated or diluted to adjust its specific gravity, prior to being fed to leach **3**.

The washed leach residue which contains silver, zinc, gypsum and silica may be further treated for the recovery of metal values. The leach solution is treated in precipitation **5** for removal of a major portion of the zinc. Zinc can be effectively precipitated from the leach solution as zinc ammonium sulfate hexahydrate (ZAS), represented by the formula $(NH_4)_2Zn(SO_4)_2 \cdot 6H_2O$, which is only slightly soluble in solutions containing at least 100 g/ℓ ammonium sulfate and preferably containing ammonium sulfate in the range of 100 to 200 g/ℓ.

The required ammonium sulfate concentration may be provided in precipitation **5** by adding ammonium sulfate in the required amount, or by adding ammonia and/or sulfuric acid in controlled amounts to the leach solution. The pH of the solution is preferably monitored and maintained at values of from 1 to 2, and preferably at a value of 1.5, by controlled addition of ammonia and/or acid. The addition of ammonia and/or acid is continued until the solution contains ammonium sulfate in the range of 100 to 200 g/ℓ.

To ensure maximum precipitation of ZAS, the temperature of the solution should be lowered as much as is practical, for example, to 25°C or lower. In case of addition of ammonium sulfate, the solution may be heated to, for example, 60° to 100°C and then cooled, to facilitate crystal growth. Depending on the composition of the flue dust, the specific gravity of the solution may be adjusted to enable easy separation of the precipitated ZAS crystals from solution and to prevent other salts from precipitating. The specific gravity of leach solution should be in the range of 1.2 to 1.5.

The ZAS crystals, which are essentially free of gallium, are separated from solution in liquid-solids separation **6**, washed with a solution containing at least 100 g/ℓ ammonium sulfate and are recovered and may be further processed, e.g., into zinc-containing fertilizer.

In cases wherein the flue dust contains cadmium, cadmium will coprecipitate with the ZAS. Cadmium may be recovered by redissolving the ZAS crystals, cementing the cadmium with zinc dust and recrystallizing ZAS from the cadmium-free solution.

Using this method of zinc precipitation, the concentration of zinc in the solution may be reduced to about 1 g/ℓ. The solution is subsequently treated by first concentrate precipitation **7** in which solution from liquid-solids separation **6** is treated with an alkaline material to precipitate the first gallium-containing concentrate. This concentrate is a complex product of variable composition and comprises substantially all gallium, aluminum and fluorine originally present in the solution, a substantial amount of phosphate, as well as amounts of sodium, potassium, ammonium and sulfate.

Prior to adding alkaline material, the specific gravity of the solution is adjusted, depending on the composition of the flue dust, to a value sufficient to prevent precipitation of phosphates. The specific gravity is adjusted to within the range of 1.1 to 1.3. The solution is then heated to a temperature of from 60° to 100°C and the pH of the solution is raised to a value of 3.5 to 5.0 by the addition of an alkaline material such as, for example, ammonia or ammonium hydroxide. Precipitation of the concentrate is complete after agitating the resulting slurry for about one hour.

As it appears that for a given weight of aluminum the weight of phosphate contained in the first concentrate is directly related to the weight of fluorine in solution, it is important that as much fluorine as feasible be removed from the process prior to the precipitation of the first concentrate. The prior removal of a major portion of fluorine can be effectively accomplished in the preferred method of sulfuric acid treatment **2**, described above, by carrying out the sulfuric acid treatment at temperatures and with retention times in the upper region of the defined range.

The first gallium concentrate is separated from the first gallium-free solution in liquid-solids separation **8** and the concentrate is washed with water to remove soluble compounds. The first gallium-free solution and wash solution, which contain mainly sodium, potassium and ammonium sulfates and phosphates, are treated for the recovery of these compounds, for example, as fertilizer. The solutions are substantially free of gallium, i.e., contain typically 1 to 2 mg/ℓ gallium or less. The washed first gallium-containing concentrate, which may contain about 1% gallium, a 20 to 50 times concentration of the gallium content of the dust, may be further upgraded as will be described under the description of the second embodiment of the process.

According to the second embodiment of this process, flue dust is treated according to the process as has been described above, to give a first gallium concentrate which is then upgraded as will be described hereinbelow.

The first concentrate is fed to an alkaline precipitation-and-leach step **9** in which the concentrate is slurried with water and lime is added to precipitate and remove phosphate as calcium phosphate. Any sulfate present is precipitated as gypsum. The lime is preferably first slurried with water and then added to the slurry of the concentrate.

The amount of lime is determined by the phosphate content of the concentrate, which may be as high as 50% P_2O_5, and is added in an amount at least stoichiometric to the phosphate content. Phosphate is preferentially precipitated while some gallium and aluminum compounds may coprecipitate. The slurry is agitated for a first period of at least one hour and is maintained at a temperature of from 60° to 100°C.

After the period required for the formation of calcium phosphate, an amount of a suitable alkaline material such as, for example, potassium hydroxide or sodium hydroxide, is added to leach and redissolve coprecipitated gallium and aluminum compounds from the precipitate. Alkaline material is added in an amount sufficient to raise the pH to and maintain the pH at a value of at least 11.

The precipitation-and-leach step **9** can be carried out in the same vessel, but in order to obtain a good separation of phosphate from gallium, it is essential that the phosphate is first precipitated with lime before the alkaline material is added to dissolve gallium. Agitating the slurry with the required amount of alkaline material for a second period of at least one hour dissolves substantially all gallium and aluminum, and a residual precipitate remains.

The slurry is discharged from step **9** and is subjected to liquid-solids separation **10**. The solids fraction, comprising the residual precipitate, is washed and is treated for the recovery of values. Preferably, the residual precipitate is recycled to sulfuric acid treatment **2**. If so desired, at least a portion of the residual precipitate may be recycled to leach **3**, in which case the acid content in the leach must be adjusted.

The liquid fraction from liquid-solids separation **10**, which is a gallate-aluminate-containing solution, may be directly treated in upgraded gallium-concentrate precipitation **13** wherein the liquid is neutralized with acid for the formation of the upgraded gallium concentrate. Preferably, sulfuric acid, for example, of 93% strength, is added to lower the pH to a value of about 6. A pH of about 6 represents the point at which the gallium precipitate reaches its minimum solubility.

The precipitation is carried out at a temperature of from 60° to 100°C and for a time of sufficient duration to obtain a precipitate which can be easily separated from solution. The upgraded gallium concentrate is subsequently separated from solution in liquid-solids separation **14** to yield the second gallium-free solution. The solids are washed and recovered.

If desired, aluminum may be removed from upgraded concentrate by dissolving the concentrate, adding a suitable calcium compound to precipitate calcium aluminate, removing the precipitate (not shown) and treating the solution for the reprecipitation of upgraded gallium concentrate as described.

Alternatively, in the preferred method, the liquid fraction from liquid-solids separation **10** is first treated in aluminum precipitation **11**, followed by liquid-solids separation **12**. In aluminum precipitation **11**, aluminum is precipitated as calcium aluminate. A suitable calcium compound is added to the liquid fraction from separation **10**, preferably in the form of a slurry of lime or calcium hydroxide, and in an amount which is at least stoichiometric and preferably one to two times the stoichiometric amount of aluminum in the liquid.

The pH should be maintained at a value of at least 10, and preferably of at least 11, to prevent gallium from precipitating. The temperature of the liquid is maintained in the range of 60° to 100°C. The mixture of liquid and calcium slurry is agitated for a period of time to complete the reaction, usually at least two hours, and is then subjected to liquid-solids separation **12**. The separated solids are washed and removed from the process, and the gallate solution is then treated in upgraded gallium-concentrate precipitation **13**, followed by liquid-solids separation **14**, as described above.

The upgraded gallium concentrate which may contain, for example, 20% gallium, represents 70 to 90% of the gallium originally present in the flue dust. Upgraded gallium concentrate may be further processed for the recovery of gallium metal.

According to the third embodiment of this process, flue dust is treated according to the process as described above to obtain upgraded gallium concentrate which is treated in dissolution step **15** wherein the concentrate is dissolved by mixing the concentrate with an alkaline solution such as a sodium hydroxide solution. Alternatively, the concentrate may be repulped with water and solid sodium hydroxide may be added. The dissolved concentrate is then subjected to electrolysis **16** which is carried out according to conventional practice.

Gallium metal is recovered from the electrolysis and spent electrolyte may be removed from the process or may be recycled to precipitation-and-leach step **9**. If so desired, a portion of the spent electrolyte may be recycled to dissolution step **15**, as indicated by the broken line in Figure 64.

The gallium metal, recovered from this process, may have a purity of about 99%, the main impurity being zinc. If so desired, this gallium metal can be further purified for the production of high-purity gallium for electronic purposes.

GERMANIUM FROM ZINC SMELTER FLUE DUSTS

The production of by-product germanium has been described by A. Petrick, Jr., H.J. Bennett, K.E. Starch and R.C. Weisner (17). Particular reference to germanium has been drawn from A.P. Thompson and J.R. Musgrave in *J. Metals* 4, 1132-37 (Nov. 1952).

The most relevant by-product subsystem for germanium includes all processing of germanium-bearing residues after their precipitation with zinc dust and separation from cadmium-bearing solution. The production of germanium can be viewed as a by-product cadmium system if the cadmium system is profitable in itself.

The flow diagram shown in Figure 65 is based primarily on the article by Thompson and Musgrave printed in 1952. The processing is essentially unchanged. Some additional technical material was reviewed to assure consistency with modern practice. The process shown in Figure 65 begins with the collection of the cadmium-germanium-bearing fume from the roasting and sintering operation in processing zinc sulfide concentrates.

The split-off point for germanium separates the collection, leaching and filtration of the fume from the first partial precipitation which was done expressly to separate the germanium-bearing precipitate from the cadmium-bearing solution. This partial precipitation is achieved by adding zinc dust sufficient to precipitate copper and germanium without precipitating the cadmium.

Some other impurities, principally arsenic, are also thrown down with the copper-germanium precipitate. From this point on, the process is similar whether the 1% germanium product comes from zinc smelter residues, zinc flue dusts, or coal ashes. The germanium-bearing sludge is redissolved in sulfuric acid, and the germanium is concentrated by a second precipitation.

The germanium concentrate is dried and roasted and is then dissolved in strong hydrochloric acid. The resulting germanium tetrachloride and arsenic chloride are distilled from the solution and are then ready for purification. Purification involves multiple distillation of the tetrachloride, hydrolyzing with water, and calcination to produce a pure germanium dioxide. The metal is produced by hydrogen reduction of the germanium dioxide in a tube furnace.

The resulting germanium metal powder is melted in an inert atmosphere to prevent oxidation and is cast into bars. The product is not pure enough for electronic applications and must be further purified by such techniques as zone refining.

Figure 65: By-Product Germanium Subsystem

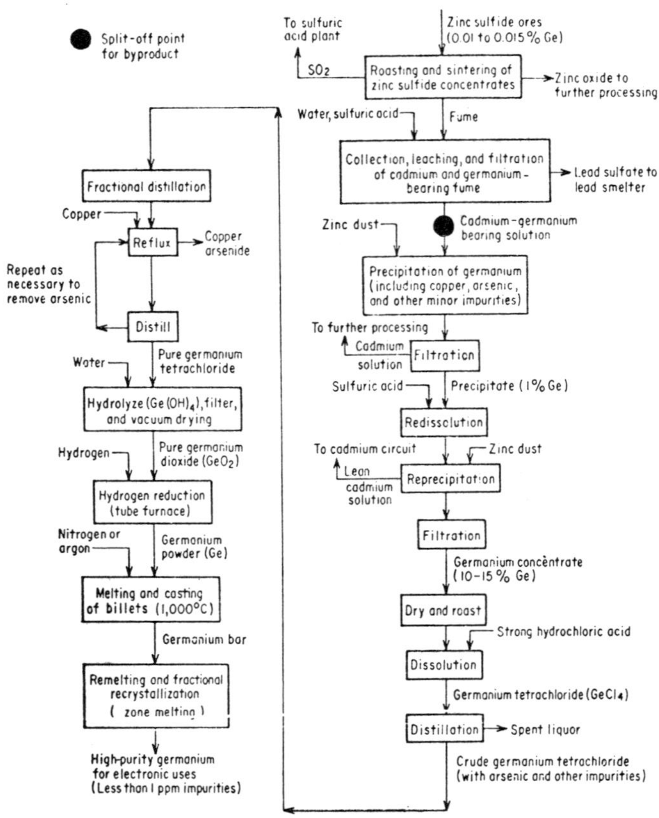

Source: Reference (17)

GLASS FROM GLASS CONTAINER WASTE

Collected waste glass contains caps and the like of glass bottles, synthetic resin lids or covers for cup-shaped glass containers, aluminum caps for liquor bottles and aluminum rings separated from the caps. Waste glass may further contain empty cans for beverages. In recent years, many of the caps have been made of aluminum, while some are made of iron, synthetic resin, cork, etc. Empty cans are generally aluminum and iron cans.

When waste glass containing such extraneous matter is used as cullet, the glass product obtained contains bubbles and unmelted substances, i.e., so-called stones, and involves changes in color or transparency due to the presence of the extraneous matter. These deficiencies reduce the strength and appearance of the product, seriously impairing the commercial value of the product.

Accordingly, the extraneous matter must be removed to the greatest possible extent. However, the extraneous matter varies greatly in shape and properties. Moreover, regardless of whether the waste glass has been fractured or not, some kinds of extraneous matter may not differ from the waste glass in shape, size or specific gravity. Thus, with the exception of magnetic materials which are magnetically separable and empty cans which can easily be separated, the undesirable matter is not readily separable by the conventional methods, such as screening or sink and float separation, which separate the extraneous matter from the waste glass on the basis of very specific differences such as shape, size and specific gravity and are applicable to only one type of material at a time. In fact, the conventional methods give very low yields.

For this reason, very cumbersome procedures are usually followed which involve various kinds of separation methods including the final step of manual separation. However, the manual separation is not fully efficient in removing the extraneous matter which remains to be separated off. Consequently, the waste glass resulting from the separating operation still contains a considerable amount of extraneous matter.

In a process developed by *J. Shimoiizaka, A. Konosu, K. Morita, A. Sugie and K. Isioka; U.S. Patent 4,083,776; April 11, 1978; assigned to Yamamura Glass KK, Japan* water is supplied to an inclined drum having a helical wall on its inner peripheral surface to cause the water to flow over the helical wall and also on the inner peripheral surface of the drum and to thereby form a flow of water running from the upper end of the drum to its lower end.

Waste glass is fed to the drum while the drum is rotated at a high speed. During falling of the waste glass onto the flow of water, during settling of the waste glass in the water flow while it is being sent toward the drum upper end by the helical wall, and while the settled waste glass is maintained by the high-speed rotation of the drum in a suspension-resembling state in which pieces of the waste glass are dispersible and in rocking motion circumferentially of the drum, extraneous matter is forced toward the drum lower end and run off therefrom due to its resistance per unit weight thereof to the flow of water which resistance is greater than that of the waste glass. The waste glass separated from the extraneous matter is discharged from the drum upper end.

An apparatus comprises the abovementioned drum rotatable at a high speed, means for supplying water and means for feeding waste glass. The helical wall of the drum has a height and a pitch which are suitable for producing the flow of water.

Such an apparatus is shown in Figure 66. It includes a drum **1** installed in its inclined position and rotatably supported by four rollers **2**. The drum **1** has a smooth inner peripheral surface on which a helical wall **3** resembling a ribbon screw is formed. When the drum **1** is rotated by the rollers **2**, the waste glass fed by a feeder **4** to the lower end **1a** of the drum **1** is conveyed by the helical wall **3** from the lower end toward the upper end **1b** of the drum. The opposite ends **1a** and **1b** of the drum are open.

Figure 66: Yamamura Glass Company Apparatus for Removing Extraneous Matter from Waste Glass

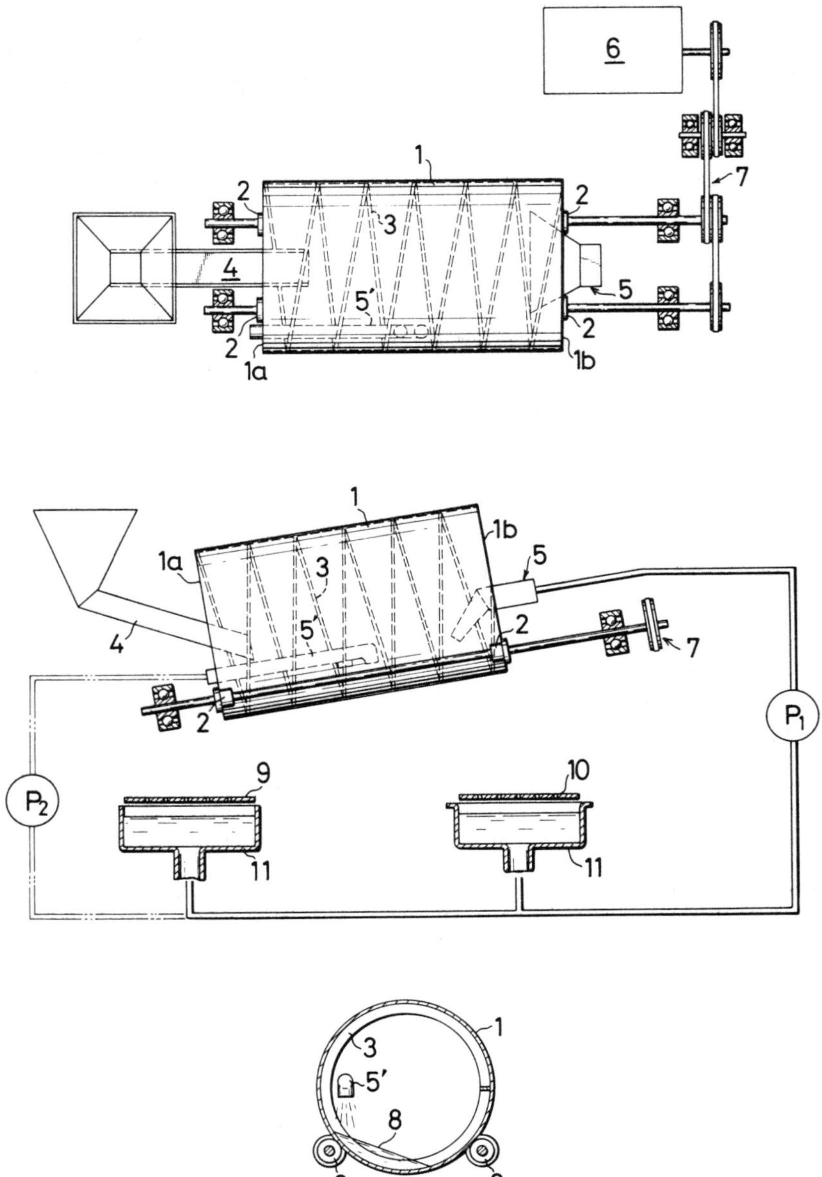

Source: U.S. Patent 4,083,776

The drum **1** is provided with means **5** for supplying water to the drum **1** to form a flow of water running from the upper end **1b** to the lower end **1a**. The drum **1** is further provided in its interior with a tube **5'** for introducing dynamic pressure water by which extraneous matter, which tends to settle on the bottom of the drum **1**, is maintained in suspension. The water tube **5'** has an outlet which is disposed at a position close to one side of the drum **1** toward which the water surface is raised by the rotation of the drum **1**. A drive system **7** for the rollers **2** has a speed change gear **6**.

The helical wall **3** in the drum **1** has a height, for example, of 15 to 40 mm. The water supplying means **5** supplies water at a rate capable of forming a flow of water required to force extraneous matter toward the drum lower end **1a** over the coils of the helical wall **3**. The drum **1** is rotated at a circumferential speed, for example, of 80 to 300 m/min. Broken waste glass is fed to the drum **1** by a feeder **4**.

The waste glass falls onto the water flow **8** along with the extraneous matter contained therein. The extraneous matter, which is not brittle, remains unbroken and almost free of deformation even if the waste glass is broken up under gravity during its passage through or from the feeder **4** (by its own energy of impact on falling). Accordingly, pieces of extraneous matter retain their original characteristic shapes; caps and like pieces retain three-dimensional shapes and are very bulky relative to their weight, so that they are readily floatable on the water flow **8** and easy to wash away. Synthetic resin cap covers and like extraneous pieces having a low specific gravity can also be easily washed away with the water flow **8**.

On the other hand, flat extraneous pieces, such as iron caps, aluminum rings, aluminum pull-top pieces having an opening knob and a cutout portion integral with the knob, have a slightly greater tendency to settle than the abovementioned three-dimensional pieces but are somewhat less likely to settle than broken glass pieces. Furthermore, such extraneous pieces tend to be caught between glass pieces, possibly settling onto the inner peripheral surface of the drum **1** along with the glass pieces.

The extraneous matter in the waste glass is washed away by the water flow **8** toward the drum lower end **1a** and is run off from that end, while the waste glass is transported by the helical wall **3** toward the drum upper end **1b** against the flow of water **8** and is discharged from the upper end **1b**.

Conveyors **9** and **10** receive the extraneous matter and waste glass thus discharged. The conveyors are in the form of a liquid permeable belt and are provided with water receptacles **11** thereunder. The water supplying means **5** and water tube **5'** are provided with pumps $\mathbf{P}_1$ and $\mathbf{P}_2$ for circulating water.

A process developed by *J. Shimoiizaka, A. Konosu, Y. Hayashi, K. Morita, A. Sugie and K. Ishioka; U.S. Patent 4,084,755; April 18, 1978; assigned to Yamamura Glass KK, Japan* is one in which waste glass containing three-dimensional extraneous matter to be removed is dropped onto a grizzly and is thereby broken up by gravity and separated into an undersize portion containing planar fragments of waste glass and of extraneous matter and an oversize portion containing three-dimensional large pieces of waste glass and unbreakable extraneous matter, the grizzly comprising parallel bars arranged at a spacing smaller than the minimum width of projection of the three-dimensional extraneous matter.

The oversize portion is repeatedly subjected to the same procedure as above to break up the waste glass to an undersize material within a specified range of sizes and to remove the three-dimensional extraneous matter from the waste glass. A breaking and separating apparatus comprises grizzlies arranged in multistage fashion within a passage for dropping the waste glass or a rotatable drum-shaped grizzly having scraping plates.

An apparatus developed by *K. Morita, A. Sugie and K. Ishioka; U.S. Patents 4,069,979; January 24, 1978 and 4,136,832; January 30, 1979; both assigned to Yamamura Glass Kabushiki Kaisha, Japan* is one involving breaking up and separating waste glass with the use of a lattice comprising a grizzly made up of parallel bars arranged at a spacing smaller than the minimum width of projection of three-dimensional extraneous matter and auxiliary parallel bars disposed over one surface of the grizzly and arranged at right angles to the bars. The diagonal dimension of the openings of the lattice is smaller than the maximum width of projection of planar extraneous matter.

The method is characterized by dropping the waste glass onto the grizzly surface of the lattice to break up the waste glass under gravity and to separate the waste glass into an undersize portion passing through the lattice and an oversize portion, and repeatedly subjecting the oversize portion of the abovementioned operation to obtain cullet as an undersize material and to remove the extraneous matter from the waste glass as an oversize material.

The lattice is in the form of a rotatable drum with the grizzly positioned on the inner side of its peripheral portion. The drum-shaped lattice has lifter plates attached to its inner periphery and an inlet for the waste glass and an outlet for the extraneous matter at its opposite ends respectively.

Such an apparatus is shown in Figure 67. The apparatus comprises a lattice including a grizzly **2** made up of parallel bars **1** arranged at a spacing smaller than the minimum width of projection of three-dimensional extraneous matter such as empty cans, and auxiliary parallel bars **3** disposed over the grizzly **2** and arranged at right angles to the bars **1**. The diagonal dimension of the openings of the lattice is smaller than the maximum width of projection of planar extraneous matter such as pull-top pieces, rings, flat caps, covers, etc.

As shown, the lattice is in the form of a rotatable drum with the grizzly **2** defining the inner periphery of the drum. Lifter plates **4** are attached to the inner periphery of the drum. The drum has an inlet **5** for waste glass and an outlet **6** for extraneous matter at the opposite ends respectively.

With the embodiment shown, the bars **1** of the grizzly **2** extend in the direction of rotation of the drum, while with other embodiments not shown, the bars **1** extend at right angles to the direction of rotation of the drum. The figure shows guide rollers **7**, guide rails **8**, a drive gear **9** and a driven gear **10**.

When the drum-shaped lattice is rotated with waste glass placed therein, the glass is repeatedly raised by the lifter plates **4** and then dropped onto the grizzly **2**.

Figure 67: Yamamura Glass Company Apparatus for Breaking Up Waste Glass Containers to Obtain Cullet

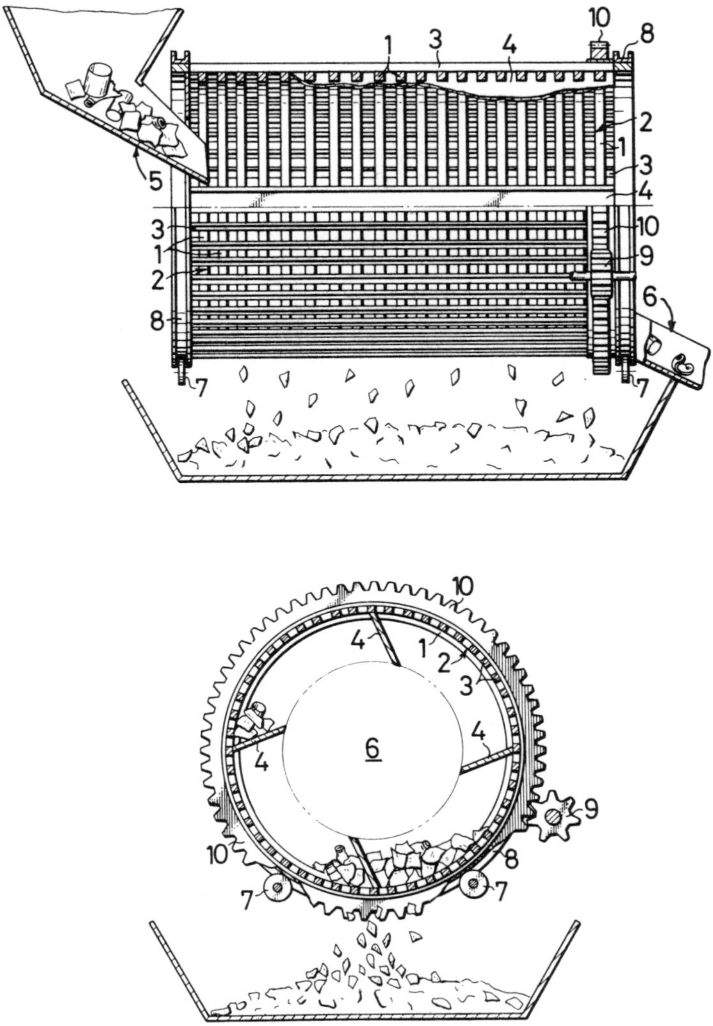

Source: U.S. Patent 4,069,979

The larger the falling glass piece, the greater is the energy of impact involved and the more likely is the glass piece to be subjected to concentrated stress on striking the bar **1**. Thus, large glass pieces are readily breakable, whereas small glass pieces are less breakable. The planar glass pieces which have such shape originally or which have been so shaped by being broken, even if the maximum width of projection thereof is larger than the spacing between the grizzly bars **1**, pass through the grizzly **2** easily and fall onto the auxiliary bars **3** because the minimum width of projection thereof is smaller enough than the spacing between the grizzly bars **1**. Among these glass pieces, those having a maximum width of projection which is smaller than the spacing between the auxiliary bars **3** rapidly pass between the bars **3**, for example, through the openings of the lattice.

On the other hand, pieces of extraneous matter fall at a relatively small velocity and involve smaller energy of falling impact. Because of their elastic nature, light weight and characteristic shapes, they will not be broken nor will collapse upon striking the lattice, while three-dimensional articles such as empty cans will not pass through the grizzly **2**.

Planar pieces such as rings, pull-top pieces, flat caps or covers, etc., tend to readily pass through the grizzly **2** and are positioned on the auxiliary bars **3**. However, their posture is governed by the parallel bars **1** of the grizzly **2** so that the maximum width of projection thereof is at right angles to the auxiliary bars **3**. Consequently, they are retained on the bars **3** and unable to pass through the openings of the lattice.

Thus, extraneous pieces of three-dimensional shape as well as of planar shape remain within the lattice as an oversize portion and are repeatedly raised and dropped by the lifter plates **4** along with coarse pieces of waste glass. During this process, brittle waste glass alone is broken up to smaller pieces, which pass through the lattice openings. The waste glass therefore progressively reduces in amount. Finally the extraneous matter alone remains. As a result, the waste glass is made into pieces of cullet not larger than a specified size, with the extraneous matter removed therefrom.

Because the remaining extraneous matter involves small energy of impact when falling, has an elastic nature and is protected from the impact exerted by falling glass pieces by being positioned between the parallel bars **1** of the grizzly **2**, it remains in the form of separate pieces without being collapsed. Thus a very high proportion of the extraneous matter remains within the lattice. Moreover, extraneous pieces will not engage and retain glass pieces.

Small pieces of glass broken by the first fall onto the bars **1** of the grizzly fall onto the auxiliary bars **3** secondly. At this second fall the small pieces of glass will not be further broken because of less falling impact. Glass pieces which are of such sizes that they need not be further broken strike the auxiliary bars **3** with their planar direction positioned vertically by being guided by the grizzly bars **1** and readily pass through the bars **3**, so that they are less likely to be broken by the bars **3**. Accordingly, glass pieces are broken up to sizes within a narrow range without entailing noticeable formation of fines or very small pieces as is the case with use of the grizzly alone.

The auxiliary parallel bars **3** will not break the glass pieces other than those of large sizes falling directly against them. The combination of the bars **3** and the grizzly in the form of a lattice having two different surface levels achieves the synergistic effect of removing the extraneous matter with greatly improved efficiency, without impairing the function of the grizzly which gives broken pieces in a narrow size range, while protecting the extraneous matter from falling glass pieces. The effect to remove the extraneous matter can be ensured to the greatest extent by minimizing the spacing between the bars **3** within the range allowable in view of the size of the pieces of cullet to be obtained.

One avenue to glass recovery from glass container waste is to design glass containers which lend themselves to recovery as described by *L.J. Taylor and H.N. Troy; U.S. Patents 4,053,666; October 11, 1977 and 4,055,441; October 25, 1977; both assigned to Owens-Illinois, Inc.*

The container in this case is a glass-based container having at least a portion of its external surface coated after the annealing of the glass with a protective, decorable, water-removable, organic polymeric film which can be removed by an aqueous medium such that the annealed glass container base can be recovered, recycled and reused.

GLASS FROM LAMINATED GLASS SCRAP

In making safety glass windshields, pairs of glass sheets are laminated together with an interposed laminating interlayer. In making such windshields, there is some spoilage because of processing conditions. Such laminated glass has heretofore been disposed of as a solid waste material. In view of the increase in costs of the raw materials used to manufacture glass and reduction of generated cullet, it is desirable to recover the glass in any spoiled laminated glass product. Usable by-product produced in recovering the cullet may also be recovered in the form of heat and condensed vapors.

A process developed by *D.J. Gurta; U.S. Patent 3,912,534; October 14, 1975; assigned to Ford Motor Company* is one for recovering cullet from laminated glass scrap. The laminated scrap is crushed to fragment the glass but the glass generally remains in bonding contact with the laminating interlayer.

The crushed scrap is mixed as its temperature is increased so that the laminating interlayer softens to permit the interlayer to be torn thus reducing the size of the scrap. The finely divided scrap is heated to a temperature at which the interlayer burns away. Any carbon residue left on the crushed scrap is also burned away. The crushed scrap is cooled and recovered as cullet.

Figure 68 shows a suitable form of apparatus **10** for the conduct of this process. A conveyor **12** delivers laminated scrap **14** from a source such as a windshield production line. The laminated scrap is in the form of a defectively manufactured windshield in which a pair of glass sheets are laminated together by an interposed laminating layer such as one of polyvinyl butyral. The method for forming such a laminated windshield is well known in the art.

The conveyor **12** delivers the laminated scrap **14** to a crusher device **16** at which the scrap is fragmented. Although the glass is fragmented, the glass generally remains attached to the laminating interlayer. The crusher crushes the scrap so that the individual pieces of glass have a size of about 2 in^2.

Figure 68: Ford Motor Company Apparatus for Cullet Recovery from Laminated Glass Scrap

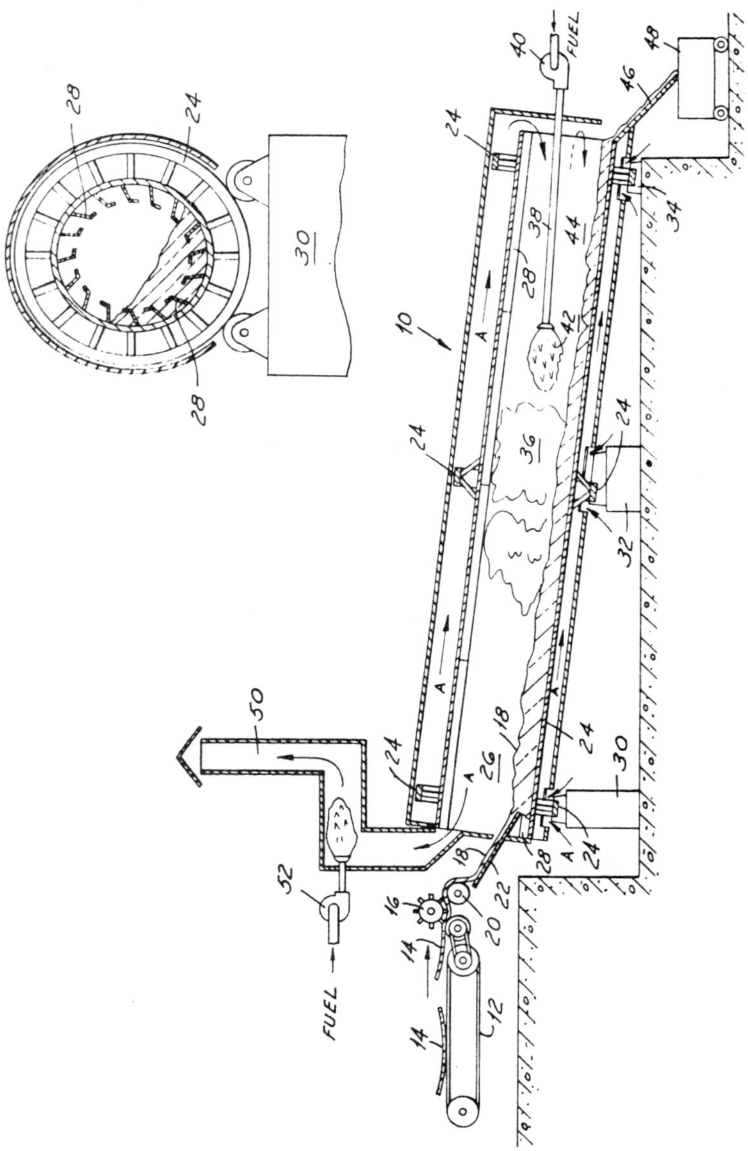

Source: U.S. Patent 3,912,534

Crushed laminated scrap **18** is delivered over a roll **20** onto a charging device **22** of a rotating incinerator **24**. The incinerator is rotated by known devices not shown. The rotational rate of the incinerator is about 7 rpm. The incinerator has a diameter of about 4 feet. The incinerator is inclined at a slope of ½ inch/foot so that upon rotation thereof the scrap is slowly mixed as it moves along the length thereof.

In an initial zone **26** of the incinerator the crushed scrap is preheated. During the preheating stage the interlayer softens so that as the scrap is mixed, the individual scrap pieces tend to separate from other scrap pieces to form small portions wherein two small pieces of glass are held together with the interlayer material. As shown, individual lifting and mixing blades **28** are disposed about the internal circumference of the incinerator **24** to lift and mix the scrap as the incinerator rotates. The incinerator is mounted for rotation on support blocks **30, 32** and **34**.

Scrap moves along the length of the incinerator **24** from the initial zone **26** to a combustion zone **28.** In passing between the two zones, the scrap is heated slowly which results in a reduction of the size of individual portions of laminated scrap to a finely divided state as previously described. By the time the scrap reaches the combustion zone **28**, sufficient heat has been provided that the interlayer burns. The heat is provided to the combustion zone by a burner **38** at start up, after which the combustion is self-supporting. The burner is fed from an air/fuel mixture which enters the burner by way of a pipe **40**.

The laminating interlayer thus is burned away separating the individual pieces of glass which have been laminated in a back-to-back manner. In being burned away, the interlayer may leave some residual carbon on the glass pieces. In a flame propagation zone **42** of the incinerator, the residual carbon is burned away. The pieces of glass then pass beyond the flame propagation zone into a cooling zone **44**. After cooling the individual glass pieces **46** are discharged into a cullet collector **48**.

Combustion air for the process is drawn into the incinerator **24** in a countercurrent manner so it may be preheated prior to entry into the burning zone of the incinerator **24**. In particular, arrows **A** indicate the flow of the air to and through the incinerator. The gases produced in the combustion process are discharged to the atmosphere through a stack **50**. The stack may have an afterburner **52** mounted therein for cleaning up and burning off any residual materials so that the discharge complies with local ordinances. The gases may also be condensed to recover that portion of the interlayer which might be vaporized rather than burned.

GLASS FROM MUNICIPAL WASTE

See "Ferrous Metals from Municipal Waste."

See also processes described under "Paper from Municipal Waste."

A process developed by *P.G. Marsh; U.S. Patent 3,945,575; March 23, 1976; assigned to Black Clawson Fibreclaim Inc.* is one in which mixed solid waste, particularly such as municipal waste, is subjected to a pulping action which results in fragmenting the glass and metal constituents into particles of similar size, with the cast and foil aluminum being converted into the form of chunks and pellets, respectively, while the aluminum which is present as alloy, such as is used in cans and other food containers, is cut into flakelike particles, after which the mixed particles are dried and then subjected to agitation causing the flakes of aluminum to be concentrated at the top of the mix for separate removal by skimming or scalping.

The remaining mixture of glass, aluminum and other materials is then subjected to high tension electrostatic treatment effecting separation of the glass from the other materials.

Such a process is shown in Figure 69. Heterogeneous municipal waste is deposited in a treatment vessel **10** with sufficient water to dilute it to a solids content of 4 to 6%. In the vessel **10**, it is subjected to forces of mechanical and hydraulic sheer and mechanical impact of sufficient intensity to reduce the frangible portions of the waste to a predictable maximum particle size. This is accomplished by means of a rotor **11** rotating above a bed plate **12** having perforations therethrough of such diameter that as the solids in the mixture are reduced to the desired particle size range, they are extracted continuously from the vessel through the outlet line **13** to pump **14** with a portion of the water or other liquid in slurry form for further treatment.

U.S. Patent 3,595,488 shows a treatment vessel particularly suited for the practice of this process by reason of the provision therein of flails or hammer members **15** pivoted on the rotor and cooperating with stator means **16** in the form of a series of attrition bar members to effect chopping and shearing of solid materials which are engaged therebetween, such particularly as the plate aluminum components of municipal waste. The same patent shows the relationship to the rotor of an appropriate extraction plate, and for the purposes of this process, satisfactory results are obtained with cylindrical extraction holes in the range of 1 to 2 inches in diameter.

The treatment vessel **10** is also provided with an outlet **20** located radially outwardly of the stator **16** and connecting downwardly with a junk remover **21** for removing relatively large infrangible solids, which usually include the majority of the ferrous metal pieces. A suitable junk remover is shown in detail in U.S. Patent 3,549,092, and as indicated in the drawing, it includes conveyor means **22** for carrying large solids upwardly from below outlet **20** for gravity discharge to a separator **23**. The separator **23** should include magnetic means for separating ferrous metals from other materials, and also means for screening liquid and fines for recirculation to the treatment vessel as indicated at **24**.

Preferred results have been obtained when the separator **23** is adapted to deliver only pieces having a maximum dimension of 4 to 6 inches to the outlet lines **25** and **26**. The outlet line **26** for nonferrous materials will include the structural aluminum, and this line may be connected at **27** back to the pulper **10** for further size reduction facilitating further treatment of this scrap aluminum in the balance of the system shown.

The drawing shows a connection **28** for continuously resupplying liquid to the treatment vessel **10** to maintain the desired level of liquid therein.

Figure 69: Black Clawson Fibreclaim Inc. Process for Recovery of Glass and Other Valuable Materials from Municipal Waste

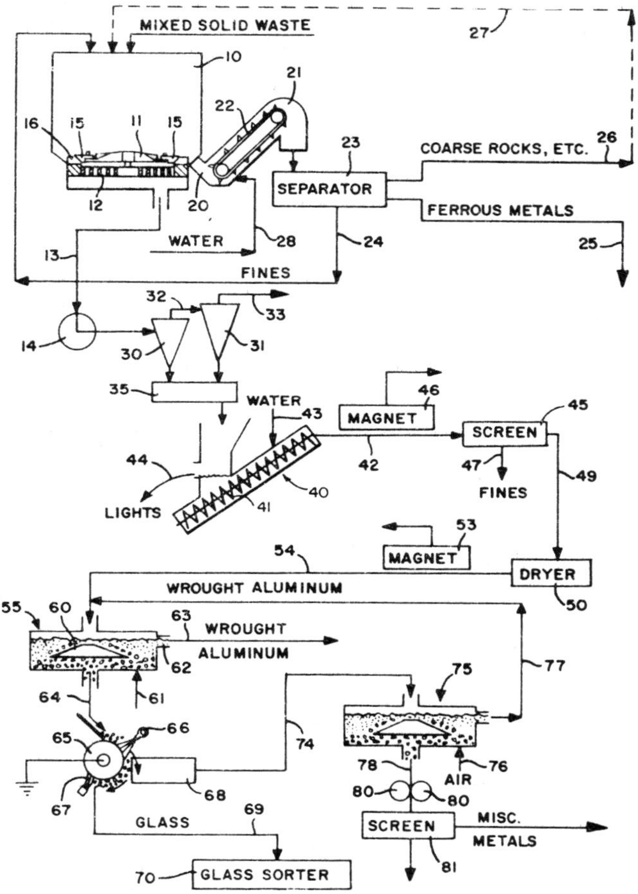

Source: U.S. Patent 3,945,575

Preferably the liquid is supplied directly to the junk remover **21** as shown, to create a flow of liquid upwardly through the outlet **20** in counterflow relation with the heavy materials moving downwardly through the outlet **20** under the influence of gravity.

This liquid counterflow will not only wash fibrous and other small particles from the solids discharging through the outlet **20**, but it has the important function of carrying back into the treatment vessel the pieces of relatively light metal which would otherwise tend to remain with the heavier ferrous pieces. This is especially true of aluminum present in the form of cans, trays and other food containers, which are thus caused to remain within the treatment vessel for comminution and extraction through the plate **12**.

The ultimate effect of these conditions in a treatment vessel as described is that the aluminum tends to remain in the vessel until there has been substantial separation through the outlet **20** of the heavier metals which were present in the charge of waste materials, and the action of the rotor **11** and flails **15** on the aluminum is to reduce it to particles of sufficiently small size for extraction through the plate **12**. It appears that this size reduction is effected by wadding or crumpling pieces of foil into the form of pellets of roughly spherical shape, and the cast aluminum pieces are broken and chopped into chunks similar to the pellets in size, along with the other nonferrous metals.

The wrought aluminum, however, due to the fact that it is an alloy, and thicker than foil, is stiffer and substantially more resistant to balling or wadding than is the foil. It therefore tends primarily to be cut by the flails **15** and stator **16** into predominately two-dimensional flakes which are comparable in length and width to the pellets and chunks but of a thickness comparable with the original thickness of the trays, can tops and bodies, and other aluminum elements present in the charge of waste.

Preferred results have been obtained in the practice of the process, as noted above, by utilizing an extraction plate **12** wherein the holes are approximately 1 inch in diameter. As a practical matter with such hole sizes, the maximum dimension of the particles passing therethrough will usually not exceed 1½ inches, and the average dimensions of the wrought aluminum pieces will be about ½ inch. The glass and other materials which also remain in the vessel for comminution and extraction through the plate **12** will be of similar dimensions when they reach the outlet line **13**.

In the next stage of the system, the organic and other light materials are removed as completely as possible from the glass, aluminum and other inorganic materials extracted through the outlet line **13**. This stage is represented in the drawing by a pair of liquid cyclones **30** and **31** connected in series so that the light fraction from the first unit **30** is delivered by the line **32** to the inlet of the second cyclone **31**. The light fraction from the second cyclone will comprise the great majority of the fibrous materials in the initial charge, along with other organic materials such as particularly garbage and vegetation, and this material is transported by a line **33** for whatever further treatment may be desired, such as, for example, recovery of fiber therefrom as disclosed in U.S. Patent 3,736,223.

The glass, aluminum and other heavy metal and nonmetallic particles discharged from the cyclones **30** and **31** are collected by a conveyor **35** which delivers them to an elutriation tower **40** for removal of as much as possible of any remaining organic and other light materials. As indicated diagrammatically in the drawing the material discharged from the conveyor **35** passes downwardly in the tower and is then carried upwardly by conveyor means such as a screw **41** to the discharge line **42**.

Water or other elutriation liquid is added at **43** for upward counterflow against the heavy materials on the inlet side of the tower in order to wash light particles back out of the tower at **44**, and with the water inlet **43** located, as shown, at a level below discharge line **42** but above the overflow outlet **44**, the heavy materials will be washed and discharged relatively dry.

The outlet line **42** for accepted material from tower **40** leads to a screen **45** where the total bulk of solid materials is reduced by removal of fines, e.g., particles with a maximum dimension of ⅛ to $^3/_{16}$ inch, in order to reduce the load on the subsequent processing equipment. Preferably a magnetic drive **46** is used in conjunction with the line **42** to remove ferrous metal particles from the feed ahead of the screen **45**. Satisfactory results have been obtained if 15 to 25% of the total solids present in the line **42** are removed by the discharge line **47** from the screen **45**.

The material retained at the screen **45** is conducted by a line **49** to a dryer **50**. To remove whatever ferrous metal particles remain in the mass, a second magnetic separator **53** is shown as located for action in conjunction with the discharge line **54** from the dryer **50**. The purpose of the dryer **50** is to prepare the solid particles for subjection to a dry separating action at the next station **55**, which comprises means for agitating the mixed particles in such manner as to effect the removal of the plate aluminum pieces.

Satisfactory results have been obtained when the station **55** comprises a gravity separator of the shaker screen type which includes means for providing an upward flow of air through the particles while they are being subjected to a shaking action. This is represented in the drawing by an oscillating element **60** into which the line **54** delivers the particulate mixture, and the air supply is represented by the arrow **61**.

The various particles in the mixture interact with each other in such manner as to cause the aluminum flakes to be concentrated in the top layer from which they are readily removed by a mechanical scalping action or by a skimming action such as may be effected by causing the top layer to flow over a weir **62** to the outlet line **63**. Apparently this result is in large measure due to the physical characteristics of the wrought aluminum flakes with relation to the other particles in the mixture. More specifically, these flakes have relatively large surface areas in comparison with their bulk and weight, and agitation causes them to migrate upwardly in the mixture.

This concentrating action can be carried out by agitation alone, but it is accelerated in the presence of an upward flow of a fluid, such as air, through the mixture, as indicated at **61**. An example of a commercially available apparatus which is suitable for this purpose is a vibrating Bauer Specific Gravity Separator.

The wrought aluminum flakes recovered in the manner just described have been found to be relatively free not only of glass and other materials, but also of the other forms of aluminum which were present in the original waste, i.e., the pellets and chunks into which the cast and foil aluminum is reduced, as well as the other nonferrous metal particles. The wrought aluminum recovered in this manner is accordingly of such high quality that it can be prepared for reuse for the same purposes as before with minimum further processing.

The material remaining in the mixture from which the wrought aluminum pieces have been removed has been found to comprise primarily glass, aluminum in pellet and chunk form, other metals including brass, bronze, copper, zinc and lead, and a residue of nonmetallic materials such as particularly ceramics and stones. These materials are delivered by a line **64** to the next stage of the system where the mixture is subjected to high tension electrostatic separating forces with the aid of a grounded drum **65** and high voltage charging electrode **66** which cooperate to pin the glass particles and other nonconductors to the drum for removal by wiper **67**, while the particles of conducting materials are thrown off the drum surface to a collecting bin **68**.

The process contributes significantly to the efficiency of the electrostatic sorting procedure, because the majority of the particles of relatively large area are removed at station **55**. The mass delivered to the drum **65** therefore comprises predominately particles of small area which are correspondingly more responsive to electrostatic forms.

The glass removed from drum **65** is conducted away at **69** for separation in accordance with color by optical sorting means **70**, such as are shown in U.S. Patent 3,650,396. The materials which collect in the bin **68** are preferably delivered by line **74** to a second shaker screen station **75**, which may be a duplicate of station **55** having an air inlet **76**, for separation of whatever wrought aluminum was not removed at station **55**. The accepted material from station **75** will usually include more than wrought aluminum, and it is therefore recycled at **77** to the inlet side of station **55** to assure both that the maximum feasible amount of the plate aluminum is recovered separately and also that it is as uncontaminated as possible.

The materials rejected at the station **75** are delivered by a line **78** to a crushing station, represented by a pair of rolls **80**, for the purpose of crushing the ceramic, stone and other hard nonmetallic materials of minimum value to a sufficiently fine particle size for ready elimination by screening as indicated at **81**. The remaining metal particles can be sold as a mixture, or separated by conventional chemical or other treatments into the individual metal components. Since the majority of this material will usually be aluminum, it can readily be recovered for reuse for purposes requiring a lower grade of metal than the plate aluminum removed at station **55**.

Another process developed by *P.G. Marsh; U.S. Patent 3,970,254; July 20, 1976; assigned to Black Clawson Fibreclaim Inc.* is one in which glass is recovered for use as cullet from mixtures of broken glass with pieces of metal and other heat resistant materials by a method wherein the mixture is thermally shocked by subjecting it to a rapid increase or decrease in temperature to induce cracks in and/or break-up of the glass particles, while the nonglass particles remain unchanged because of their nature, and the mixture is mechanically treated, by agitating or the like, so that the glass particles are broken up by further propagation of the induced cracks.

Prior to the mechanical treatment, and preferably prior to the thermal shock treatment, the mixture is screened or otherwise treated to eliminate fines of all kinds. After the mechanical treatment, the mixture is further screened so that the glass particles, which have been reduced in size, can be recovered as the undersized fraction, while the nonglass particles, which have remained generally the same size, are retained on the screen.

Such a process is shown in Figure 70. It shows the separation of glass in conjunction with means for optically sorting glass particles of substantial size from contaminant particles.

Figure 70: Black Clawson Fibreclaim Inc. Method for Separating Glass from Heat Resistant Materials

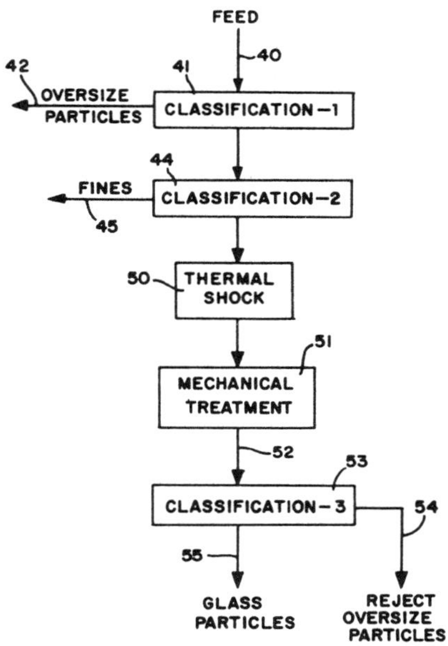

Source: U.S. Patent 3,970,254

The feed **40** of mixed glass and heat resistant contaminants is subjected first to two classification stages, the station **41** being illustrated as having a screen of relatively large perforations, e.g., ¼ inch, for the removal of oversized particles at **42**, followed by a second station **44** having a relatively fine screen, e.g., ⅛ inch perforations, for the removal of fines at **45**. The resulting intermediate fraction is then advanced successively through the thermal shock station **50**, mechanical treatment station **51**, conveyor **52** and classification station **53**, where a screen of the same size perforations as at station **44**, or slightly smaller, separates the contaminant particles **54** from the glass particles **55**.

In the system shown, and with perforations of the stated sizes, the oversized particles removed at **42** can be optically sorted, but they could also be passed through a series of stations similar to the stations **50, 51** and **53** wherein the screen at the final classifying station has perforations of the same size as, or slightly smaller than, those at the station **41**, and will therefore accept the shattered glass particles but again reject the contaminant particles. It will also be apparent that the station **41** may represent a series of two or more screens of progressively smaller hole sizes, for the purpose of separating a plurality of fractions for further treatment.

A process developed by *C.B. Webb; U.S. Patent 4,116,822; September 26, 1978; assigned to Carpco, Inc.* is one for separation of glass from waste material by high tension techniques. The waste is sized to a range from ⅛ to 1½ inches, dried to a moisture content less than 2%, air classified to remove low bulk density material, and subjected to magnetic separation to remove ferrous metal. The waste is then fed to the top of a rotating grounded drum and subjected successively to ion bombardment from corona discharge electrodes and to an electrostatic field from a static electrode. The glass remains pinned to the roll while other materials come off in the static field.

Figure 71 is a schematic representation of such a process.

Figure 71: Carpco, Inc. High Tension Method of Selectively Separating Glass from Waste Material

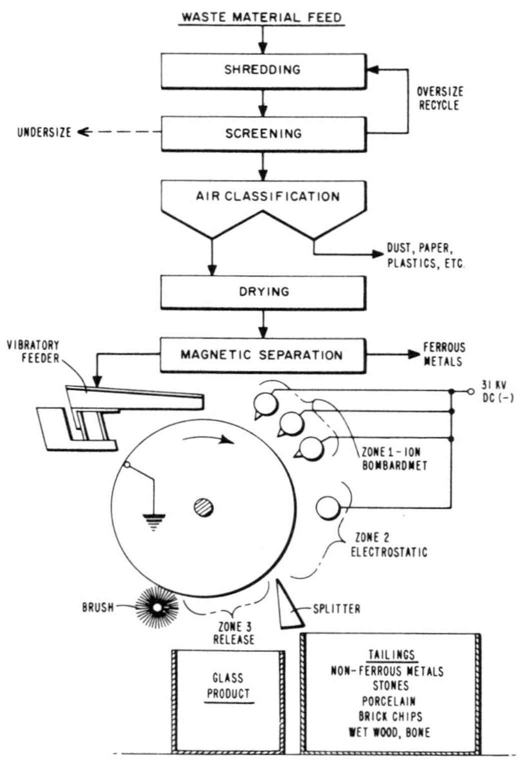

Source: U.S. Patent 4,116,822

A process developed by *B.W. Morey; U.S. Patent 4,070,273; January 24, 1978; assigned to Occidental Petroleum Corporation* is one for recovering purified glass from solid waste which includes a predominantly organic fraction and a predominantly inorganic fraction both of which contain glass particles. The predominantly inorganic fraction is further subdivided into a glass-rich fraction containing glass, inorganic, and organic constituents. The organic constituents are initially separated from the glass-rich fraction.

The organics-free fraction is comminuted and screened to remove coarse particles greater than a first predetermined mesh size. The residue is classified to remove fine particles less than a second mesh size. The resultant mixture, after removal of coarse and fine particles, is subjected to froth flotation to form a float containing predominantly glass which is recovered for further use.

Details of the flotation step in such a glass recovery process have also been published with particular reference to specific flotation reagents. Thus, *B.W. Morey and W.R. White; U.S. Patent 4,067,502; January 10, 1978; assigned to Occidental Petroleum Corporation* describe a process in which glass is separated from the mass of inorganic matter of a similar particle size by a flotation process using as the flotation reagent at least one amine containing a hydrocarbon having from 8 to 22 carbon atoms, in at least one hydrocarbon group attached to a nitrogen atom as the selective flotation reagent for the glass particles. The preferred amines are primary amines and secondary amines with primary amines being particularly preferred.

Another flotation separation technique for glass recovery has been described by *W.R. White and G.P. Van Tighem; U.S. Patent 4,122,950; October 31, 1978; assigned to Occidental Petroleum Corporation*. It is one in which particulate glass values contained in the final inorganic fraction from comminuted municipal wastes and having the particle size between 200 and 20 mesh are recovered by froth flotation by activating the glass surface with a bi- or trivalent metal ion and using as the beneficiation reagent a water compatible sulfonated hydrocarbon.

Another process for glass recovery from municipal waste is described by *B.W. Morey; U.S. Patent 4,065,282; December 27, 1977; assigned to Occidental Petroleum Corporation*. It involves shredding the waste, subjecting the shredded waste containing the glass to treatment in an air classifier, and recovering the overhead including glass from the air classifier, the latter being operated under conditions such that the major portion of the glass particles contained in such overhead are of a fineness smaller than 6 mesh, and usually of a size such that a substantial portion of such glass particles are less than 20 mesh.

It then involves screening such overhead fraction and recovering an enriched glass fraction of the aforementioned fineness, subjecting the screened glass-enriched material to treatment on an air table, and recovering a fine glass product of substantially reduced organic content, and having a particle size less than 6 mesh, with a substantial portion less than 20 mesh.

Such fine glass product containing a substantial portion of glass particles of a size less than 20 mesh, can be employed in construction applications, e.g., in admixture with asphalt to form slurry seal, for production of glass and resin binder sheets, and the like, and may be further beneficiated to sufficient purity for recycle to glass manufacture.

Figure 72 is a block diagram of the essential steps in this process. Referring to Figure 72, municipal waste or trash, indicated at **10**, and containing 7 to 10% glass, is fed to a shredder to shred the material as indicated at **12**.

Figure 72: Occidental Petroleum Corporation Process Using Air Classification for Glass Recovery from Municipal Waste

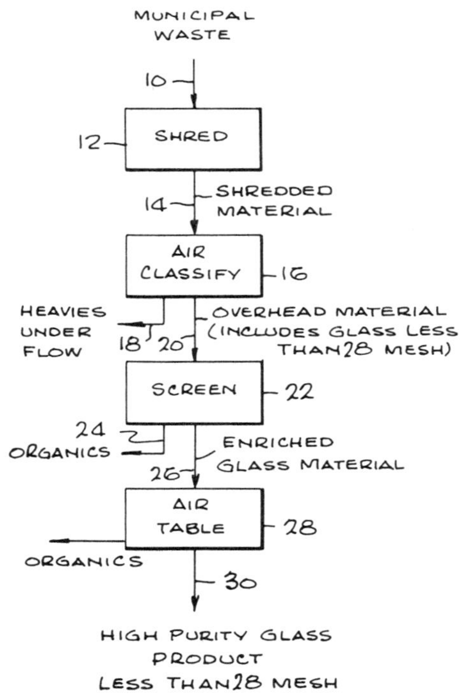

Source: U.S. Patent 4,065,282

In the shredding operation the glass is broken into various size pieces generally less than ½ inch while 80% or more of the other components are broken into pieces ranging in size from dust to 12 inches. Such other components include rock, metals, paper, cardboard, wood, plastics and the like.

The shredded material at **14** is then fed to an air classifier, as indicated at **16**, and an overhead material **20** is recovered. Such overhead material includes 4 to 6% of fine glass particles, 2 to 4% other inorganics including rock chips, brick, etc., and 60 to 70% of combustible organics including paper, cardboard, plastics, etc. The remainder is mostly water. The air classifier is operated so that the glass particles have a size less than 28 mesh.

The underflow or heavies from the air classifier, indicated at **18**, is composed of about 70% of the glass contained in the waste material **10**, but such glass particles are of a size larger than 28 mesh. Such heavies also include brick and rock chips, metals and the like.

The overhead material at **20**, including the fine glass particles comprising about 30% of the glass contained in the waste material **10** is then placed on a screen as indicated at **22**, such screen being of 28 mesh size.

The screened product thus produced is essentially in the form of an enriched glass material consisting of about 50% glass of particle size less than 28 mesh, 20% other inorganics, including brick and rock chips, and 30% combustible organics, including paper, plastics, etc. The material retained on the screen and of a particle size larger than 28 mesh, is essentially composed of organics and is separated as indicated at **24**.

The enriched glass material **26** thus separated in the screening operation is fed to an air table as indicated at **28**. The heavier glass particles are separated on the air table into a high purity glass product of particle size less than 28 mesh, as indicated at **30**. There is also separated on the air table a dust product, a light inorganic product and an organic product. The high purity glass product at **30** consists of about 90% glass, the dust product containing less than 1% glass and the light inorganic product containing about 13% glass. The recovery of glass in the glass product is about 80% of the glass in the enriched glass material fed to the air table.

GLASS FROM SCRAP AUTOMOBILES

See "Steel from Scrap Automobiles."

GLASS FROM TELEVISION PICTURE TUBES

The phosphor screen of a black matrix color picture tube is ordinarily formed in the following way. A uniform coating of photoresist is applied on the interior surface of a glass panel and subjected to exposure corresponding to three primary colors, G (green), B (blue) and R (red) through a shadow mask. On developing, respective relief images of photoresist film corresponding to three primary colors, G, B and R, remain on the panel. The panel is then coated all over the surface with a nonluminescent light-absorptive substance.

Subsequently, the panel is treated with an aqueous solution of oxidizing agents or the like, to remove the remaining photoresist film corresponding to three primary colors, G, B and R, and the nonluminescent light-absorptive substance coated on the photoresist film, to form so-called matrix holes corresponding to three primary colors, G, B and R. The matrix holes are coated with respectively predetermined phosphors corresponding to three primary colors, G, B and R, to form the phosphor screen.

For the photoresist, there have been used conventionally a polyvinyl alcohol (PVA)—ammonium bichromate type, and more recently a polyvinylpyrrolidone (PVP)—water-soluble azide compound type. For the nonluminescent light-absorptive substances, black graphite is generally used.

When any panel is found not having the desired characteristics in any step of the abovementioned manufacturing operation of a phosphor screen, such as formation of the relief image of photoresist film or formation of the black matrix, it becomes necessary to remove photoresist, etc., adhered on the defective panel by cleaning to reclaim the panel.

Conventionally, in reclaiming the panels, an aqueous solution containing several to ten plus several percent, or in some cases, even higher percent of hydrofluoric acid and ammonium fluoride have been used as a cleaning preparation to treat the panels at room temperature for several minutes. However, such a cleaning composition is unsatisfactory for completely removing the photoresist and carbon. This led to a serious problem because it is impossible to obtain a color picture tube with a uniform phosphor screen when such an imperfectly cleaned panel is used.

A process developed by *Y. Oba, S. Nonogaki, M. Oikawa, K. Miura and Y. Tomita; U.S. Patent 3,876,468; April 8, 1975; assigned to Hitachi, Ltd., Japan* is designed to overcome the problems posed by the prior art and give a perfectly reclaimed panel. Another object is to provide a cleaning composition which is less hazardous in handling in the aforementioned panel cleaning operation, yet can manifest full effect in a short period of time.

In this process, an aqueous solution containing hydrofluoric acid and tannic acid was found useful as a glass cleaning composition in reclaiming clean glass panels from defective panels rejected during manufacture of a phosphor screen for color picture tubes. The composition is superior to conventional compositions not only in cleaning efficiency but also in operational safety because of lower hydrofluoric acid concentration and lower treating temperature.

GLASS FIBERS FROM GLASS FIBER SCRAP

It is known to form glass fiber or filaments into insulating mats or boards which are bonded together by a binding material, for example, a phenolic binder. The binder is applied as the filaments are directed down onto a moving conveyor. Glass fiber is also manufactured as continuous filament, by causing glass streams to issue through orifices in the base of a vessel known as a bushing. The glass streams are attenuated into filaments by mechanical means, i.e., the filaments are drawn down by being wound onto a winding drum or collet. Tissue is made in a similar bushing to continuous filament but attenuation is caused by a gaseous stream to form staple fiber.

In all these processes, relatively large amounts of scrap material are produced, e.g., in the centrifugal process, once the mat has been formed, the edges are trimmed to give a standard size, and this selvedge may be scrap. In the continuous filament process scrap is produced, when starting to wind on, or when cakes are badly wound, or breaks occur before a suitable size has been attained.

It has previously been considered that the recovery and return of such material as cullet to the melting process was uneconomical due to both the bulk of the material and the presence of organic polymers and resins. However, with the rising cost of raw materials, and increased production of glass wool for insulation purposes, pressure has been increased to reduce the quantity of raw materials going to waste, and the quantity of scrap to be disposed of.

In a process developed by *J.S. Thomas and A.F. Mason; U.S. Patent 4,054,434; October 18, 1977; assigned to Fibreglass Limited, England* scrap material in both wool staple and continuous filament form can, if reduced to a milled form of a preferred mesh size and bulk density, be used directly as a batch material, and fed to a glass tank to feed a fiberizing process without any harmful effects.

This is a process in which scrap glass fibers containing carbonaceous material to provide a carbon content of 0.4 to 0.04% in a batch is combined with batch materials in which the scrap amounts to 5 to 25% by the weight of the batch. The batch is melted under oxidizing conditions and the resultant melt is used to form glass fibers.

Figure 73 shows the essential features of such a process.

Figure 73: Fibreglass Limited Process for Recovering Reusable Glass Feed for Fiber Production from Glass Fiber Scrap

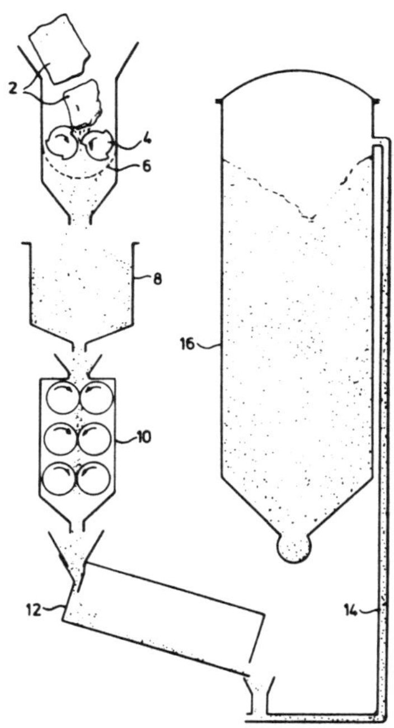

Source: U.S. Patent 4,054,434

Referring to Figure 73, scrap fiber **2** is fed into the top of a knife mill in which the scrap fiber is cut down into particles which are sufficiently small to pass through a screen **6**. The graded particles are then passed out from the knife mill **4** by way of a surge hopper **8** to a roll crusher **10**. The roll crusher breaks down the particles into a finely divided form which are then passed through an incinerator **12** to burn off excess binder and other nonmineral substances. The resulting product is fed from the outlet of the incinerator by way of a feed pipe **14** to a storage hopper **16** from which it can be dispensed for mixing with the batch material.

A process developed by *S.F. Brzozowski; U.S. Patent 4,188,228; February 12, 1980; assigned to PPG Industries, Inc.* is one in which glass pellets for use in producing glass for the manufacture of glass fibers are formulated in which a significant portion of the pellets are made up of fiber glass cullet. The pellets of glass fiber batch contain quantities of fiber glass cullet from 5% by weight to 35% or more by weight. The glass fiber cullet employed is in the form of ground glass or short glass fibers, 0.0625 to 0.5" (0.158 to 1.27 cm).

GOLD FROM MINE WASTE DUMPS

The experimental leaching of gold from mine wastes has been described by H.J. Heinen and B. Porter in Report of Investigation No. 7250, U.S. Bureau of Mines, Washington, D.C., (April 1969). Excellent gold recovery was demonstrated in a laboratory-scale leaching test on waste material from an open-pit mine in northeastern Nevada.

An injection system for leaching was applied in which the sodium cyanide solution was introduced into the interior of the dump at selected points. Dissolution of gold from the waste material, which assayed 0.06 ounce gold per ton, was readily achieved because the gold occurred as submicroscopic particles in a porous host rock. Gold recovery of 80% was obtained in 11 days from material crushed to 1 inch in size, and recovery was 90% after 21 days.

A process developed by *R.M. Darrah; U.S. Patent 4,056,261; November 1, 1977* provides a system for recovering at least one of the metals consisting of gold and silver from mine run dumps or crushed ores containing at least gold or silver. The equipment used consists of an inclined surface for supporting a heap of the mine run dumps or crushed ores, means for distributing an alkali and cyanide containing aqueous solution onto the heap, a storage zone for the resulting pregnant solution arranged to collect the run-off of the solution from the inclined surface, a plurality of activated carbon loaded ion exchange vessels mounted on mobile support means, conduit means for feeding the pregnant solution serially through the vessels thereby to strip the gold and silver values from the solution and conduit means for recycling resulting barren solution for distribution onto the heap.

Such a system is shown in Figure 74. The illustrated embodiment includes spraying **1** a dilute cyanide and caustic solution on the surface of the mine run dumps or a crushed ores heap **2** and allowing the solution to slowly percolate through the heap **2** until it encounters the impervious pad **3**, constituted of compacted soil or mill fallings for example, under the ore.

The natural slope then directs the solution into a collection ditch **4** lined with a suitable membrane such as butyl rubber **5** and the solution flows by gravity to an impervious holding pond **6**, also lined with an impervious membrane such as compacted soil or polyvinyl chloride (PVC) plastic.

Figure 74: Darrah Process for Recovery of Gold and Silver from Mine Run Dumps or Crushed Ores Using a Portable Ion-Exchange Carbon Plant

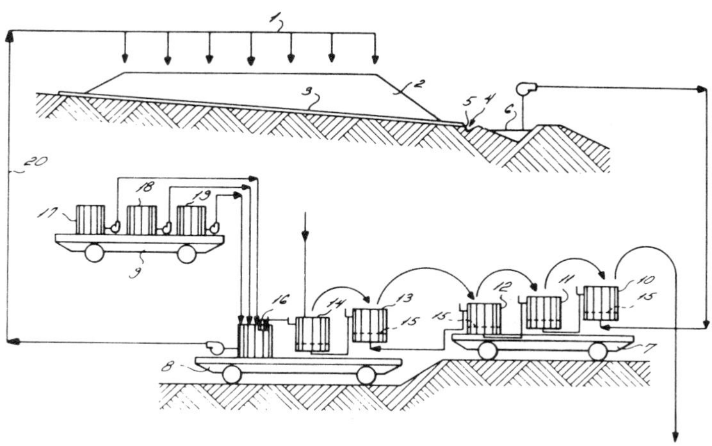

Source: U.S. Patent 4,056,261

The dissolved gold and silver values (the solution being called pregnant solution) are pumped to the portable carbon ion-exchange plant which is mounted on three trailers **7, 8** and **9**. Each of tanks **10, 11, 12, 13** and **14**, mounted on trailers **7** and **8**, includes a punch-plate distributor **15** near the bottom of the tank for distributing the solution which is fed to the bottom of the tank and a charge of granular activated carbon for the ion-exchange supported in the tank by the distributor **15**.

The solution flows upwardly through each tank through the punch-plate distributor **15**. The upward velocity of the solution keeps the granular carbon in a teetered or expanded state. The activated carbon adsorbs the aqueous gold and silver values. This step is repeated in the series of identical tanks **10, 11, 12, 13** and **14** until the required amount of values has been recovered.

To achieve maximum efficiency, the carbon is periodically transferred in the opposite direction of the solution flow. This countercurrent flow effects contact of the highest values of carbon (i.e., carbon having the lesser amount of unused adsorptive capacity) with the solution containing the largest amount of gold and silver. The transfer of carbon can be effected by pumps, screens or water-eductors. The eductor is preferable, since it may be connected to a double-sequence timer to cycle the carbon transfer automatically on a schedule determined by the operator.

The solution overflowing the launder on the last tank flows into a safety screen, which removes any floating carbon, and enters a surge tank **16**. Liquid caustic and cyanide are added from respective tanks **17** and **18**, using calibrated reagent meters and the barren solution is pumped back to the heap through conduit **20**. Fresh, make-up water is usually added from a tank **19** mounted on the trailer **9**. However, if serious plant scaling is encountered, it may be preferable to add make-up water to the pond to allow the resultant carbonate scale to settle in the pond.

A typical run for the abovedescribed portable carbon plant will vary considerably dependant upon the ore grade and the leach rate of the gold and silver values.

GOLD FROM PLATING WASTE SOLUTIONS

Substantially large amounts of precious metals such as gold, palladium, rhodium and platinum, as well as copper and zinc, are employed in certain plating baths, drag-out solutions and rinses. With the increasing shortages of these metals and their concomitant increases in price, efficient, inexpensive procedures for recovering the metals are desirable.

In order for a procedure to be satisfactory, it must have a high efficiency in removing substantially all of the metal ion from the solution, reducing it to metal form. The metal must be produced in a form substantially free of contaminants and readily isolable.

Furthermore, it is desirable to be able to reuse the resulting solution, rather than discarding the relatively expensive chemicals which are employed in the plating, drag-out or rinse baths.

Therefore, in reducing the metal, it is preferable that the oxidized form of reducing agents that may be employed in the various solutions should also be reduced. Not only should they be reduced, but they should be returned to the solution in a form useful for reuse of the solution.

A process designed to answer these requirements has been developed by *R.A. Wallace; U.S. Patent 4,082,546; April 4, 1978.* The process is one for recovering precious metals such as gold, palladium, rhodium and platinum as well as copper and zinc from respective waste solutions laden with such metals, such as stop baths, stabilizers, solution tailings and waste plating drag-out and rinse solutions.

The metal ion-laden solutions are treated with mechanically worked narrow aluminum foil strips, threads or chaff, substantially completely covered with a few microns thick imperfect layer of aluminum oxide. The resulting metal-free solution may be repeatedly used and restored.

The method provides high metal recovery efficiency, and little maintenance, while providing the capability for repetitive use of various solutions employed in plating and other processes.

GOLD FROM SCRAP GOLD

Refiners receive scrap in a variety of forms and must determine processing steps depending on total gold values present, average content, and a wide range of impurities to be separated. Scrap dealers may partly process the scrap and then ship the upgraded product elsewhere for intermediate treatment and refining. Scrap is generally either purchased or treated on a toll basis. The bulk of the scrap processed is generated in manufacturing operations, and because of tight controls over waste materials handling in precious metals plants, nearly all "home generated" scrap can be recovered. Some gold is lost, however, because in practice some manufactured products cannot be efficiently treated for complete recovery (14).

Two-thirds of the secondary production of gold has been home-scrap, recycled by fabricators. Probably the greatest loss of gold in fabrication occurs in gold plating plants where fouled or depleted solutions may be discarded. Military scrap has become a significant source of reclamation, and government agencies maintain an extensive collection system (14).

GOLD FROM SILVER REFINERY MUDS

The block flow diagram for by-product gold recovery is shown in Figure 75. The Wohlwill electrolytic process is used to recover gold from the residue of the silver cells after the excess of silver is removed by leaching with hot sulfuric acid. The Wohlwill process consists of electrolyzing gold anodes in a hot solution of silver chloride. The cathode consists of a thin ribbon of gold. The deposit which adheres to the cathode is exceedingly pure, averaging 99.975% purity with minute impurities of platinum and palladium.

In the copper and silver refining preceding gold refining, the more noble metals are insoluble and form mud and slime residues. The process is repeated again in gold electrolysis so that the platinum and palladium appear in the slime and become the by-products of the by-products of the by-products.

The scale of operations in the gold subsystem is very small compared to the silver circuit. In the hypothetical circuit shown, only 1 ounce of gold is handled for each 50 ounces of silver. Equipment and labor required are correspondingly small. Rates of operation are kept high to keep the high value in processed metal at a minimum level.

GOLD FROM WASTEWATERS

A technique developed by *M. Markels, Jr.; U.S. Patent 4,033,763; July 5, 1977; assigned to World Resources Company* is one for economically recovering metal values, especially precious metal values, from wastewaters comprising a mixture of natural surface water, industrial wastewater and organic waste, where the metal values appear in concentrations below levels previously commercially of interest.

Figure 75: Gold from Silver Refinery Muds

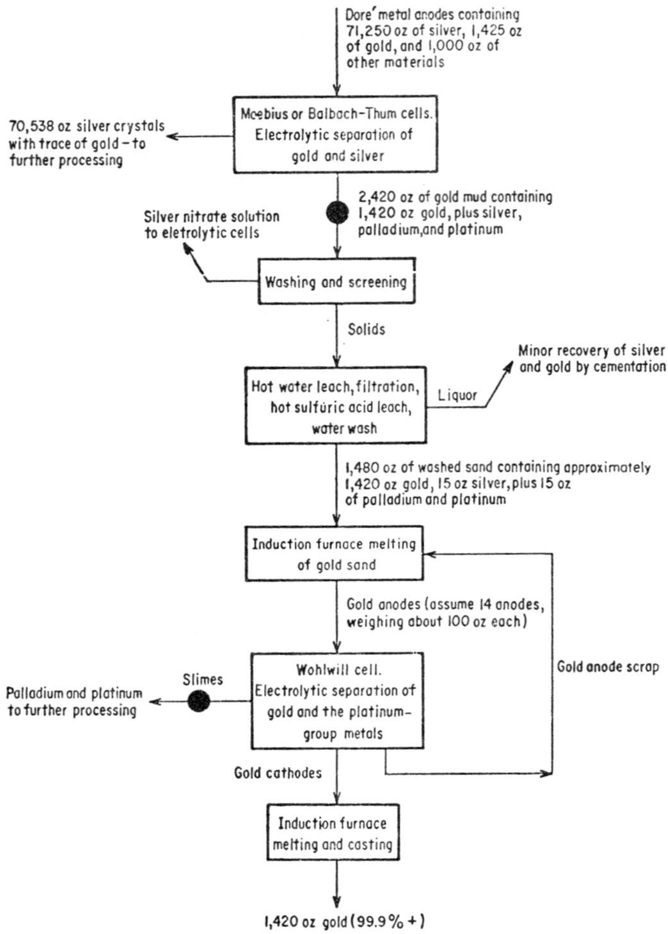

Source: Reference (17)

The process includes the steps of concentrating the metal values of the wastewaters by bacterial treatment thereof to imbibe and capture the metal values from the waters and immobilize them in a sludge, coagulating and separating the sludge from the water, dewatering the sludge and burning off the organic matter, and recovering selected metal values by inorganic metal-winning process steps.

Figure 76 is a block flow diagram of such a process.

Figure 76: World Resources Company Process for Recovery of Selected Precious Metal Values from Wastewaters

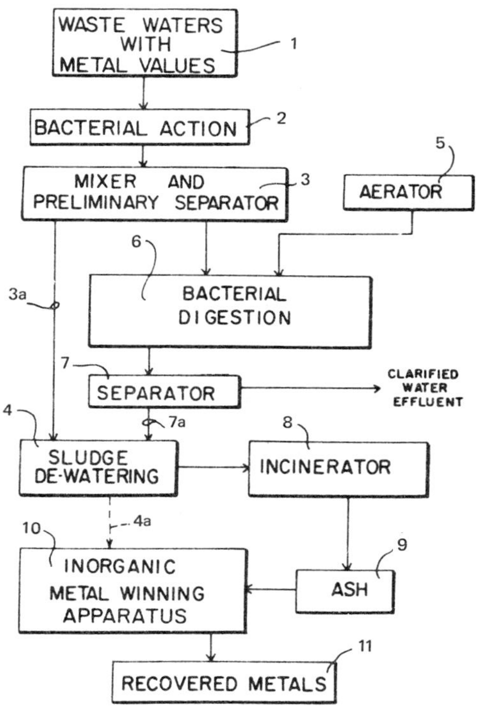

Source: U.S. Patent 4,033,763

The wastewaters contain metal values, the recovery of which is the purpose of this process. These metals appear in wastewaters which are represented by the block **1**, these waters comprising, for instance, industrial wastewater and natural surface water containing very dilute quantities of the various metals appearing as dissolved compounds. These waters enter the system having entrained therein organic waste of a type and quantity suitable to support the growth of bacteria. The source is generally a city sewage system.

In a typical illustrative system, the influent including the organic waste, undergoes bacterial action represented by the block **2** before passing through a mixer and preliminary separator **3** designed to remove solids which can be made to precipitate immediately, and these solids are then delivered directly to a sludge dewatering device **4** which comprises any one of the several types which are well known in the sewage treatment art.

In a more advanced type of sewage system the main stream of the influent passes from the preliminary separator **3** into a bacterial digestion chamber **6**.

These bacterial steps can be either aerobic or anaerobic. In the digestion chamber **6**, in the case of aerobic bacterial treatment the mixture is aerated by a source of air **5**. The activated sludge is circulated in the bacterial digestion chamber **6** for a protracted period of time. The mixture is then separated in a separator **7** which delivers clarified water and concentrated sludge through separate outlets. This sludge which may typically be about 5% solids is also introduced into a sludge dewatering device **4** where it is mixed with the sludge from the preliminary separator **3**. When dewatered, the sludge is about 16% solids by weight.

The diagram is a composite showing two flow routes which can be followed concurrently by the sewage, or alternatively. One route involves the mere mechanical separation of the sludge via the flow route **3a**, and the other flow route involves the inclusion of bacterial digestion in the digester **6**, either aerobic or anaerobic, followed by the delivery to the dewatering device **4** by way of the route **7a**.

After the solid waste material has been dewatered in the sludge dewatering device **4**, it can be delivered to an incinerator **8** where it is incinerated to an ash which as discharged is essentially devoid of organic material. The ash comprises compounds of the various metals, mostly oxides. The ash is represented by the block **9** on the diagram, the ash comprising about 15 to 50% by weight of the sludge entering the incinerator if this sludge were dried. The ash is then subjected to selected inorganic metal-winning process steps represented by the block **10**, illustrative examples of which are given hereinafter, which steps recover selected metals comprising the output of the process as represented by the block **11**.

As an alternative process, the dewatered sludge from the block **4** can be mixed as indicated by the dashed line **4a** with a base metal ore and smelted in the metal-winning step **10**. In this case the incineration step **8** is omitted, and the organic matter in the sludge is burned off incidental to smelting of the ore. For purposes of this disclosure the word ore is defined to include not only a substantially untreated raw ore, but includes suitable intermediates resulting from prior treatment steps of the raw ore.

In an illustrative example, the influent mixed wastewaters comprising natural surface water, industrial wastewater, and organic waste, for instance sanitary waste, enter at a daily flow rate of about 28,000,000 gallons. The influent temperature is about 70°F with a pH of about 7. The influent contains a total solid content of about 500 to 1,500 ppm, the precious metal content being, for silver, about 20 to 100 ppb, for gold about 1 to 5 ppb, and for platinum about 0.03 to 0.16 ppb. These precious metal values appear in the presence of very much higher relative quantities of base metals, which for purposes of the subsequent metal-winning steps appear as contaminants.

In the bacterial treatment steps of the process, there is a residence time of from 3 to 24 hours during which the bacteria feed on the organic matter and imbibe the metal values, thereby immobilizing and concentrating them in a sludge, the temperature remaining at about 70°F and the pH remaining at about 7.

The sludge is then coagulated and separated to recover clarified water and wet sludge in which the total solids content has been increased to about 5% by weight.

This wet sludge is then dewatered so that its total solids content is further increased to about 16% before incineration. However, where the sludge is to be added to an ore in a smelting step, the sludge should be further dewatered or dried. In either event, the sludge is burned to drive off the organic matter.

Where incineration is used to recover an ash, this ash comprises inorganic contaminants and metal values appearing about as follows: 20% P_2O_5; 5% NO_3; 7% SiO_2; 3.4% Ca; 0.5% Mg; 48,800 ppm Fe; 31,600 ppm Al; 11,400 ppm Zn; 10,900 ppm Cu; 2,180 ppm Cr; 1,900 ppm Ni; 1,450 ppm Pb; 472 ppm Mn; 300 ppm Ga; 190 ppm Cd; 600 ppm Ag; 30 ppm Au; 4 ppm Pd; 1 ppm Pt; and 10 ppm Tl.

From the point of view of physical characteristics the ash is a finely divided reddish powder with a density of about 40 lb/ft^3, wherein the above metallic values appear mostly as oxides which are physically dispersed and undifferentiated from each other in the mass. The ash is produced in quantities of about 8,000 pounds per day, so that its concentration when compared with total influent is about 30,000:1. The effluent from the organic steps of the process comprises about 27,000,000 gallons of clarified water per day. The incineration is carried out with excess air during a residence time of about one-half hour and the ash leaves the incinerator at about 650°F.

When using one of the inorganic smelting processes the following results are obtainable: percentage of the gold and silver recovered, 90%; gold production per day, 0.36 pounds (5.25 troy ounces); and silver production per day, 3.6 pounds (52.5 troy ounces).

The content of the incinerated ash will, of course, vary with the geographic area from which the wastewaters are taken. The mineral content of the surface waters of that area constitute one variable, and the type of industry contributing to the industrial wastewaters is another important variable.

There are various different metal-winning processes which can comprise the block **10** representing the inorganic steps required to recover one or more particular metals from the ash shown in the block **9**. Among the commonest of these are various pyrometallurgical processes including smelting and/or sintering, and various hydrometallurgy processes, which represent two different approaches which can be used alternatively.

As mentioned above, the incinerated ash containing the precious metal values also contains large percentages of base metals which are undesirable in cyanide extraction because they consume large amounts of the cyanide. Moreover, the cyanide forms complexes with the base metals as well as with the precious metals and these complexes tend also to be removed along with the selected precious metal values from the cyanide bath, whereby the desired isolation of precious metal values is defeated. Therefore, acid leaching of the base metal values using concentrated H_2SO_4 is employed as a pretreatment step to leach out active base metal values before cyanidation of the remaining ash.

As a practical process, three or more acid baths are prepared in different vessels. The pH of the baths is maintained below 2.5, and the ash is removed from bath to bath at about 24 hour intervals in such a way that the weakest acid bath receives the untreated ash, and then the ash and/or the acid is moved to the next vessel for another 24 hours, so that with each move, the increasingly leached ash encounters a more concentrated acid. At the end of three days, this pretreatment is concluded and the remaining ash is ready for cyanidation, the base metal values having been removed in solution, with the spent acid, and the precious metal values remaining in the ash since they are not soluble to a significant extent in the sulfuric acid.

The acid bath can advantageously contain about 400 pounds of concentrated H_2SO_4 per ton of ash. About 2 tons of water are used to dilute each ton of concentrated H_2SO_4, the proportion being noncritical so long as there is sufficient water present to facilitate pumping of the bath and filtering out of the ash.

The leached ash, still containing the precious metal values, is washed. Its residual acid content is neutralized by the addition of 2 to 4% of the weight of the ash of hydrated lime to raise the pH of the bath and prevent the formation of hydrogen cyanide when the ash is subsequently moved to a cyanidation bath which is continuously stirred and aerated by bubbling air therethrough. The bath contains from 1 to 6% of the weight of the ash of sodium cyanide, together with the ash from the pretreatment step, and washed coconut charcoal to the extent of about one ton per thousand ounces of precious metal.

For a cyanide concentration of about one pound NaCN per one hundred pounds of ash, over a period of 24 to 72 hours the silver recovery from the ash by adsorption onto the charcoal is between 85 and 90%, and the gold recovery is about 95%. The recovery is not much increased by further increasing the cyanide concentration.

The charcoal with the precious metal complexes adsorbed thereon is recovered from the bath and the metal values are separated, for instance by burning off the charcoal. The precious metals can then be purified by well known steps, for instance by electrolytic refining.

H

HYDROFLUORIC ACID FROM SPENT PICKLE LIQUORS

See "Zirconium from Waste Pickle Liquors".

HYDROGEN CHLORIDE FROM CHLORINATION OF ORGANICS

Hydrogen chloride as a gas and in aqueous solution as hydrochloric acid are commercially valuable products which may have a variety of uses including the production of chlorinated organics, e.g., hydrochlorination of methanol to form methyl chloride, and other chlorination, polymerization, isomerization, alkylation and nitration reactions; production of food products, e.g., corn syrup and sodium glutamate; in the metals industry, e.g., pickling and metal cleaning; in the petroleum industry, e.g., acidizing of formations; and other miscellaneous end uses such as industrial cleaning of equipment.

Hydrogen chloride and hydrochloric acid may be produced by several different routes, including as a by-product in organic syntheses such as the thermal chlorination of methyl chloride to form more highly chlorinated methanes, e.g., methylene chloride, chloroform and carbon tetrachloride.

A process developed by *S. Schäfer, A. Ohorodnik, K. Gehrmann and A. Mainski; U.S. Patent 4,003,723; January 18, 1977; assigned to Hoechst AG, Germany* is one in which crude hydrogen chloride gas obtained as a by-product in the production of chloroacetic acids by the catalytic chlorination of acetic acid with chlorine gas in contact with acetic anhydride and/or acetyl chloride, and contaminated with about 0.6 up to 3% by volume of acetyl chloride, is purified. To this end, crude hydrogen chloride gas is introduced into the base portion of a scrubbing zone and scrubbed countercurrently therein with 0.5 up to 20 ℓ, per normal cubic meter (STP) of hydrogen chloride gas, of a cooled scrubbing liquid under circulation, consisting substantially of 20 to 80 wt % of concentrated H_2SO_4, 15 to 60 wt % of acetic acid, and 5 to 50 wt % of water.

The resulting purified moist hydrogen chloride gas issuing at the head of the scrubbing zone is introduced into the base portion of a drying zone and scrubbed countercurrently with precooled sulfuric acid under circulation. A suitable arrangement of apparatus for the conduct of such a process is shown in Figure 77.

Crude hydrogen chloride gas, which is to undergo purification is introduced through a conduit **1** into the base portion of a scrubbing column **2** and scrubbed countercurrently therein with a scrubbing liquid delivered to the head of the scrubbing column **2** and consisting, e.g., of 30 to 60 wt % of H_2SO_4, 20 to 50 wt % of acetic acid and 10 to 40 wt % of water.

Figure 77: Hoechst AG Process for Purification of Crude Hydrogen Chloride from Chloroacetic Acid Manufacture

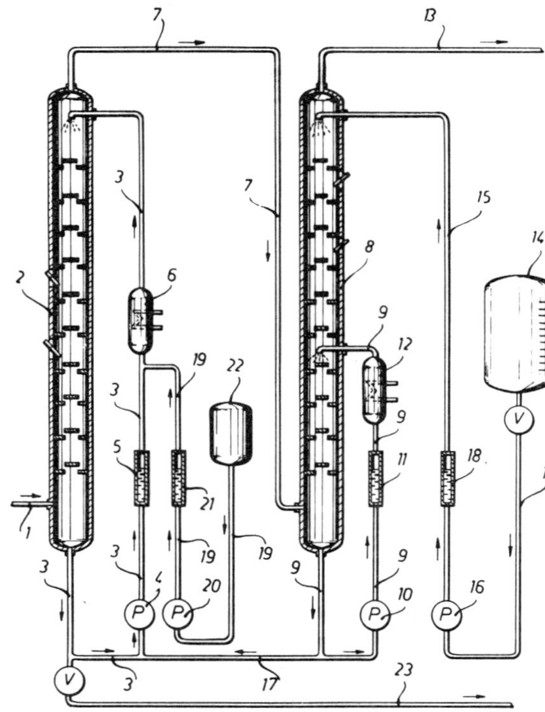

Source: U.S. Patent 4,003,723

The scrubbing column **2** may comprise, e.g., a bubble tray or packed column, with an efficiency corresponding to the separating power of 10 to 15 theoretical trays. Scrubbing liquid coming from the base portion of the scrubbing column **2** is recycled to the head of the column **2** by means of a pump **4** and through a conduit **3**. The quantity of scrubbing liquid under circulation is metered by means of a quantity recording meter **5**. Also disposed in the conduit **3** is a heat exchanger **6** enabling the reaction heat set free during scrubbing to be dissipated.

Hydrogen chloride freed from carboxylic acid chlorides is taken from the head of the scrubbing column **2** through a conduit **7** and delivered to the base portion of a drying column **8** which may equally comprise a bubble tray column with a separating power corresponding to that of at least 10 to 15 theoretical trays. Gas which ascends in the drying column **8** is scrubbed countercurrently with concentrated sulfuric acid and dried in this manner. Sulfuric acid which accumulates in the base portion of the drying column **8** is continuously taken therefrom and recycled to the drying column through a conduit **9**. Disposed in the conduit **9** are a pump **10**, a quantity recording meter **11**, and a heat exchanger **12** for delivering, metering and cooling the sulfuric acid. Purified dry hydrogen chloride gas leaves the drying column **8** through a conduit **13**.

The sulfuric acid under circulation in the drying column **8** is continuously replaced by a metered quantity (metered by means of a quantity recording meter **18**) of fresh sulfuric acid coming from a reservoir **14** and conveyed through a conduit **15** by means of a pump **16**. At the same time, an adequate proportion of circulated sulfuric acid is taken from the base portion of the drying column **8** through conduit **9**, and recycled to the scrubbing column **2** through the conduit **3**. The sulfuric acid fed to the cycle line **3** through the conduits **9** and **17** is diluted with a metered quantity (metered by means of a quantity recording meter **21**) of water. The water comes from a reservoir **22**, travels through a conduit **19** and is pumped into the cycle line **3** by means of the pump **20**.

The volume of liquid conveyed through the cycle line **3** and kept under circulation in the scrubbing column **2** is maintained constant. To this end, the process provides for fresh sulfuric acid and water to be introduced continuously into the cycle line **3**, and for an adequate proportion of scrubbing liquid to be removed simultaneously from the base portion of the scrubbing column **2** through the conduits **3** and **23**.

The process enables the crude hydrogen chloride gas by-product, which is invariably obtained in the production of chloroacetic acids by the chlorination of acetic acid with chlorine gas at elevated temperature and in contact with an acetic anhydride and/or acetyl chloride catalyst, to be purified under commercially attractive conditions and compares very favorably in this respect with the prior art methods. The resulting purified hydrogen chloride gas contains very minor residual proportions of water and acetic acid, i.e., less than 10 ppm of water and less than 200 ppm of acetic acid, so that it is possible for it to be used, e.g., as a reactant in oxychlorination reactions.

The composition of the scrubbing liquid is a factor which critically determines successful operation. More particularly, the scrubbing liquid should be so composed that it is possible for the carboxylic acid chlorides, which are contained in crude hydrogen chloride gas, to be dissolved in the scrubbing liquid. In this connection, it is of interest to note that carboxylic acid chlorides contained in crude hydrogen chloride are insufficiently soluble only in concentrated sulfuric acid. In other words, concentrated sulfuric acid cannot be used alone as the scrubbing liquid.

In HCl recovery from hydrocarbon chlorination processes, common practice is to separate the hydrogen chloride from the organics by partially condensing the gaseous stream at essentially atmospheric pressure, which commonly requires large amounts of refrigeration involving the use of liquefied gases such as liquid ethylene at low temperatures, for example, of from -50° to -85°C.

The need to use such low temperature refrigeration necessarily entails correspondingly high energy consumption. Further, the use of such low temperature refrigeration may necessitate defrosting, insofar as ice and frozen hydrogen chloride hydrates accumulate on the equipment which interferes with its normal operation and which may also cause increased corrosion problems due to accumulation of such ice along with some hydrogen chloride. Such defrosting operations typically entail a complete shutdown of the condenser equipment and removal of the frozen material from the unit, and may also lead to certain losses in product yields.

A process developed by *J.E. Loyless; U.S. Patent 4,010,017; March 1, 1977; assigned to Vulcan Materials Company* is one in which high purity hydrogen chloride is recovered from a gaseous product stream consisting essentially of hydrogen chloride and at least one chlorinated hydrocarbon of from one to two carbon atoms, such as the chloromethanes, by a process which has significantly lowered refrigeration requirements, thus resulting in savings of energy and materials. The process involves:

(a) condensing essentially all of the gaseous stream to form a liquid phase feed stream wherein the liquid phase consists essentially of the hydrogen chloride and the chlorinated hydrocarbon(s);

(b) passing the condensed liquid phase feed stream to a stripping zone;

(c) maintaining the stripping zone at a pressure and temperature profile sufficient to retain essentially all of the chlorinated hydrocarbon(s) in a liquid phase and to allow essentially all of the hydrogen chloride to separate in a vapor phase;

(d) removing a first recovery stream consisting of the high purity hydrogen chloride vapor phase from the stripping zone; and

(e) removing a second recovery stream consisting of the chlorinated hydrocarbon(s) liquid phase from the stripping zone.

In a preferred embodiment, the gaseous product stream is from a methyl chloride thermal chlorinator and consists essentially of hydrogen chloride and a chloromethane mixture; this gaseous product stream is completely condensed and cooled to a temperature between about $-30°$ and $-10°C$ and then fed to the stripping zone, which is maintained at a pressure between about 200 and 250 psia. The equipment arrangement for such a process is shown in Figure 78 and an example of process operation will be given with reference to that figure.

This example illustrates the recovery of high purity hydrogen chloride from a methyl chloride thermal chlorinator gaseous product stream consisting essentially of hydrogen chloride and a chloromethane mixture consisting essentially of methyl chloride, methylene chloride, chloroform and carbon tetrachloride.

Referring to the figure, a gaseous product stream from a methyl chloride thermal chlorinator (not shown) and subsequent quench zone (not shown) is passed through a line or conduit **1**. This gaseous product stream is at a pressure of about 78 psia and a temperature of about 73°C and contains a vapor which is composed as follows on a parts by weight basis.

Hydrogen chloride	~6,270
Methyl chloride	~35,000
Methylene chloride	~29,500
Chloroform	~10,400
Carbon tetrachloride	~1,370
Other higher molecular weight chlorinated organics as a heavy ends portion	~50

Figure 78: Vulcan Materials Co. Process for Recovery of Hydrogen Chloride from Chlorinated Hydrocarbons Such as Chloromethanes

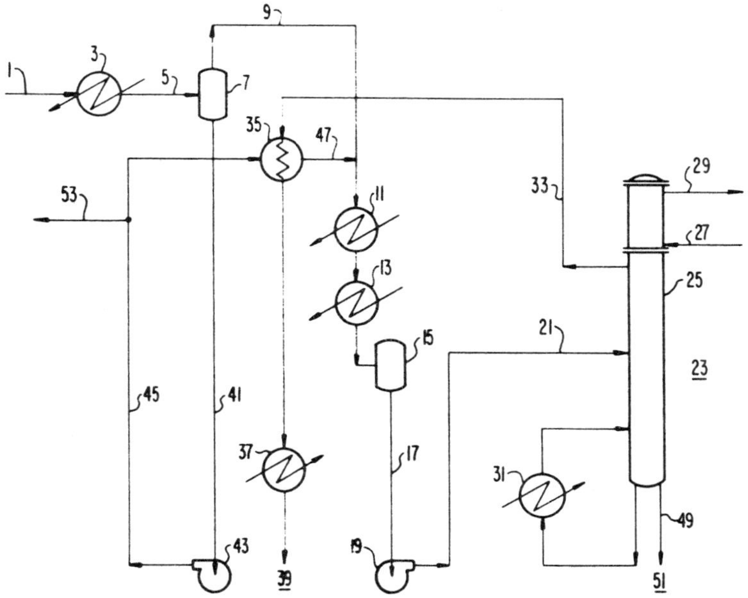

Source: U.S. Patent 4,010,017

The gaseous stream is passed through line **1** to a water-cooled heat exchanger **3** and then through a line **5** to a reflux drum **7**. From the reflux drum the gaseous material at a pressure of 78 psia and a temperature of about 43°C is passed through a line **9** to a heat exchanger **11** and then to another heat exchanger **13** wherein the coolant is R500 fluorocarbon mixture entering at a temperature of -30°C. After passing through the heat exchanger **13**, the gaseous stream has been condensed by cooling to a temperature of about -18°C, and the pressure is 78 psia, and the stream at this point is entirely in the liquid phase which consists essentially of the hydrogen chloride and the chloromethanes. After leaving the heat exchanger **13**, the liquid phase feed stream is passed to a feed drum **15** and then through a line **17** to a pump **19**.

The condensed liquid phase feed stream is then pumped at a pressure of about 235 psia to a stripping zone **23**. The stripping zone **23** is composed of a stripper **25** which is at a pressure of about 235 psia and is cooled at its upper end by a R500 fluorocarbon mixture coolant which enters the stripping zone **23** via a line **27** at a temperature of -20°C and is withdrawn from the stripping zone **23** via a line **29**. The stripper **25** is a conventional Monel shell unit having a 3 ft i.d. and a 47 ft height with 30 Monel trays and is fitted with a steam reboiler unit **31**, which is adjusted to give a temperature of about 105°C in the lower portion and about 25°C in the middle portion of the stripper unit.

A recovery stream consisting of 5,950 pbw of hydrogen chloride in the vapor phase is removed from the stripper zone **23** via an off-take line **33** at a temperature of about -15°C and is then passed through a heat exchanger **35** and yet another water-cooled heat exchanger **37** and is sent to a recovery zone **39** at a temperature of about 10°C. Also from the reflux drum **7**, a liquid phase portion of the feed stream which portion contains about 65 pbw hydrogen chloride; 1,970 pbw methyl chloride; 3,980 pbw methylene chloride; 1,540 pbw chloroform and 225 pbw carbon tetrachloride are passed via line **41** to a pump **43** and is thereafter passed via a line **45** to the heat exchanger **35** and into indirect heat exchange contact with the hydrogen chloride product in line **33** and thereafter is passed via a line **47** into admixture with the gaseous stream in line **9**.

A second recovery stream consisting of the chloromethanes liquid phase is removed from the stripping zone **23** via a line **49** at a temperature of about 105°C and is passed to a recovery zone **51**. This chloromethanes liquid phase is essentially hydrogen-chloride free and contains about 25,000 pbw methyl chloride; 9,500 pbw methylene chloride; 2,700 pbw chloroform; 270 pbw carbon tetrachloride and 10 pbw of the heavy ends residue. Additional chloromethanes mixture comprised of about 300 pbw hydrogen chloride; 20,000 pbw methylene chloride; 7,700 pbw chloroform; 1,100 pbw carbon tetrachloride and 40 pbw heavy ends residue is removed from line **45** via a line **53** and is passed to the thermal chlorinator quench zone (not shown).

HYDROGEN CHLORIDE FROM DILUTE HCl SOLUTIONS

It is known that many industrial processes produce effluents consisting of hydrochloric acid either in dilute aqueous solution or in the state of mixtures with other gases. The industrial manufacturer is faced with a choice between two possible courses: either the solutions or the gases containing hydrochloric acid are not recovered, in which case it is necessary to have recourse to costly neutralization treatments; or else an attempt is made to recover the acid. Recovery of the acid is a difficult operation since the hydrogen chloride gas produces in aqueous solution an azeotropic mixture having a concentration of 20% of hydrochloric acid. In the case of lower initial concentrations of hydrochloric acid, conventional distillation produces steam and the azeotropic mixture, the concentration of which is insufficient in the majority of industrial applications.

There do exist, however, a number of different methods which result in the production of hydrogen chloride gas or of concentrated aqueous solutions. The first of these methods consists in distilling the solution after having displaced the point of azeotropy by addition of chloride. This method involves distillation of the entire quantity of water contained in the initial hydrochloric acid solution at a temperature of 140° to 150°C and entails high heat consumption at considerable cost.

In the second method of known type, hydrochloric acid is extracted by an organic solvent; since a part of the water is extracted at the same time, there is added a third solvent which gives rise to separation of the loaded phase into a concentrated aqueous solution of hydrochloric acid and an organic phase made up of two solvents practically without either hydrochloric acid or water. This method results in solutions which have a concentration of 25% of hydrochloric acid and which are distilled so as to obtain dry HCl gas. It is also necessary to distill the mixture of the two solvents before recycling the latter.

The prior art processes have the disadvantages of high power consumption and an improved process is claimed by *J. Duhayon, M.J. Jacques, P. Michel, M. Tarnero and E. Zellner; U.S. Patent 4,036,939; July 19, 1977; assigned to Commissariat a l'Energie Atomique, France*. It is a process for the recovery of hydrogen chloride in gaseous form from an organic solvent loaded with hydrochloric acid in a dilute aqueous solution which consists in carrying out prior to distillation a dehydration of the loaded organic solvent in a dehydration extractor in which the solvent is circulated countercurrent to an aqueous solution having a high chloride concentration.

A process developed by *A. Coenen, K. Kosswig and G. Prominski; U.S. Patent 4,115,530; September 19, 1978; assigned to Chemische Werke Hüls AG, Germany* is a process for the production of gaseous hydrogen chloride by extraction of dilute aqueous hydrochloric acid with amines, consisting of the following steps:

(a) The aqueous hydrochloric acid is extracted with an amine or with a mixture of an amine and an inert organic solvent which is immiscible with water and has a lower boiling point than the amine used, the amines utilized being tertiary alkylamines, tertiary aryldialkylamines, secondary arylalkylamines, primary alkylarylamines, or mixtures thereof, which contain 14 to 32 carbon atoms in the nitrogen-bound side chains, and among these at most 1 nitrogen-bound methyl group and at least 1 aliphatic residue containing at least 6 carbon atoms, and wherein the acid constant K_a of the amine is smaller than 10^{-3};

(b) An inert, water-immiscible organic solvent which has a lower boiling point than the amine used is added to the extract, unless the solvent has already been added in stage (a);

(c) The extract is distilled, the vapors thus formed are condensed; the water is continuously separated from the two-phase condensate; and the organic phase is returned into the distillation process; and

(d) After removal of the water, the extract is distilled under reflux at sump temperatures of between 100° and 250°C and the gaseous hydrogen chloride thus liberated at the head of the column is withdrawn.

HYDROGEN CHLORIDE FROM INCINERATION GASES

The chlorine consuming industry produces large quantities of chlorine-containing organic by-products, referred to herein as Cl-waste. Disposal of Cl-waste poses environmental, technical and economic problems. The conventional and environmentally least objectionable method of Cl-waste disposal is its reaction with oxygen at elevated temperatures, referred to herein as incineration or combustion.

In the combustion of Cl-waste its carbon content is converted essentially to carbon dioxide, and its hydrogen content to water, both environmentally unobjectionable. The chlorine content of Cl-waste is converted by combustion partly to hydrogen chloride and partly to chlorine. It is the presence of these two compounds in the combustion gas of Cl-waste that is mainly responsible for the environmental, technical and economic problems.

The conventional way of dealing with chlorine and hydrogen chloride in the combustion gases is to reduce the former to hydrogen chloride or dispose of it in some other way and then to absorb the hydrogen chloride in water, forming aqueous hydrochloric acid, mostly of low concentration. Although aqueous hydrochloric acid is more easily handled and less objectionable than gaseous hydrogen chloride diluted with huge quantities of the other components of the combustion gas, it is still essentially an undesired waste product, the disposal of which poses further problems. A better way of separating the chlorine and hydrogen chloride from the other components of the combustion gas of Cl-waste than absorption in water is desired.

A process developed by *W. Opitz and H. Hennen; U.S. Patent 4,115,531; September 19, 1978; assigned to Hoechst AG, Germany* is one in which hydrochloric acid having a substantially constant HCl content of 20 to 36% is produced continuously by removing hydrogen chloride from hot combustion gas. To this end, hot combustion gas containing 1 to 10% by volume of HCl and steam, this gas being obtained by the joint combustion of off-gases containing chlorinated hydrocarbons and liquid residues of chlorinated hydrocarbons, is introduced into a quenching zone and quenched therein down to a temperature lower than its dew point by means of cooled hydrochloric acid containing hydrogen chloride in a concentration of 20 to 36%, which concentration corresponds to the concentration of the hydrochloric acid which is to be produced.

The resulting gas-liquid mixture is separated in an absorption cooling zone into cooled hydrochloric acid and cooled combustion gas containing 0.1 to 1% by volume of HCl. The cooled hydrochloric acid is introduced into a reservoir zone capable of receiving a multiple of the quantity of hydrochloric acid produced per hour and recycled from the reservoir zone to the quenching zone. Additional 20 to 36% hydrochloric acid produced is taken out of the quenching cycle. Such a process is shown in Figure 79 and the process will be described in more detail with reference to that figure.

Hot combustion gas at about 1000°C, containing hydrogen chloride is introduced into a cooling device providing a quenching zone, shown at **1**, in which it is quenched down to 60°C, i.e., down to a temperature well below its dew point, by means of a quenching liquid (hydrochloric acid) which is circulated by means of a pump **4** through an absorption cooler **2** and a reservoir **3**, the quenching liquid being admitted to the quenching zone **1** through at least three pipes **5** (only one of which is shown), providing inlets which are disposed in a single plane.

The combustion gas is further cooled in the cooler **2**, whereby most of the hydrogen chloride is absorbed with the resultant formation of additional hydrochloric acid. The circulated quantity of hydrochloric acid, which has the concentration desired for the final product, and the cooling surface of the cooler **2** are so related that the liquid leaves the cooler **2** with a temperature of 15° to 20°C, e.g., 17°C. The acid is reintroduced into the quenching zone **1** by way of the pipes **5**. In this manner, it is possible to obtain hydrochloric acid of 30%

strength, for example, according to the partial pressure prevailing. The reservoir **3** serves as temporary storage for hydrochloric acid of 20 to 36% strength.

Figure 79: Hoechst AG Process for Production of Hydrochloric Acid by Removing Hydrogen Chloride from Combustion Gas

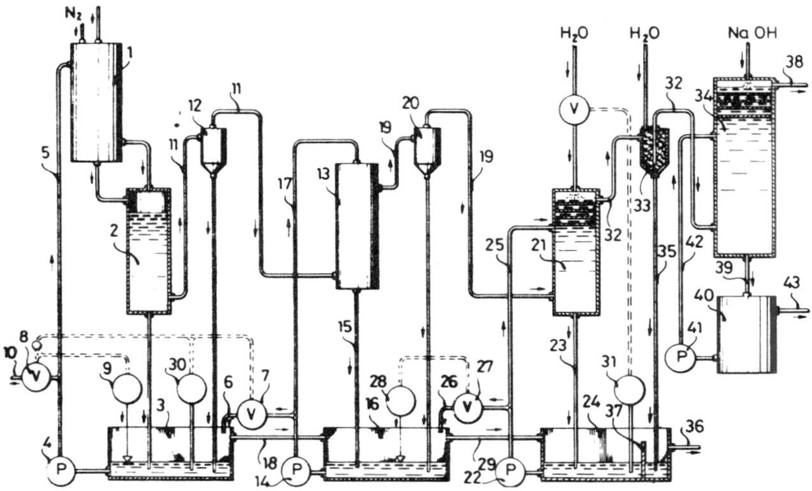

Source: U.S. Patent 4,115,531

The HCl concentration range desired in the reservoir **3**, e.g., 29 to 30% HCl, is established by means of a density recorder **30**, operating in association with the quenching cycle. The recorder **30** controls a valve **7**, the arrangement being such as to prevent this desired concentration range from being exceeded by providing for the admission, as necessary, of hydrochloric acid having a strength of 3 to 19%, preferably 5 to 15%, this acid coming from a first scrubbing cycle, as described below, via a conduit **6**. While maintaining the desired concentration range, the recorder **30** regulates the opening of a valve **8**, the latter being additionally controlled, however, by an automatic level indicator **9**.

As a result, continually-formed fresh hydrochloric acid in concentrated form, e.g., 29 to 30% HCl, can be taken from the quenching cycle for any desired use, through a conduit **10**, as long as the hydrochloric acid in the reservoir **3** does not fall below a certain minimum level, this being controlled by means of the level indicator **9**.

Those constituents of the combustion gas which are not condensed in the cooler **2**, which usually consist predominantly of nitrogen and carbon dioxide with minor proportions of hydrogen, oxygen and hydrogen chloride, are conveyed at approximately 30°C through a conduit **11** and a droplet separator **12** to a washing or scrubbing tower **13**, forming part of the first scrubbing cycle. In the tower **13**, the gas is further cooled down to about 20°C by scrubbing it with cooled hydrochloric acid having a strength of 3 to 19%, preferably 5 to 15%, whereby its hydrogen chloride content is further reduced.

The scrubbing liquid is circulated by means of a pump **14** through a conduit **15**; a second reservoir **16**, in which the abovementioned 3 to 19% (preferably 5 to 15%) hydrochloric acid is temporarily stored; a conduit **17**; and the tower **13**. The conduit **17** branches off to the abovementioned conduit **6** and valve **7**, whereby dilute hydrochloric acid from the first scrubbing cycle can be admitted to the reservoir **3** as necessary. An overflow line **18** provides a connection from an upper level of the reservoir **3** to the reservoir **16**, this overflow line being used in the event of excessively large quantities of hydrochloric acid of lower concentration than desirable being admitted to the reservoir **3**. The overflow of hydrochloric acid of insufficiently high concentration or density, e.g., having a strength of less than 30 wt % of HCl, is temporarily stored in the reservoir **16** of the first scrubbing cycle.

The gas scrubbed in the tower **13** is conveyed through a conduit **19** and a second droplet separator **20** to a scrubbing tower **21** which forms part of a second scrubbing cycle, and of which the upper third is provided with bubble trays. In the tower **21**, the gas is scrubbed with cooled hydrochloric acid having a strength of 0.5 to 2%, and its hydrogen chloride content is thus further reduced. In this second scrubbing cycle, the scrubbing liquid is circulated by means of a pump **22** through a conduit **23**, a third reservoir **24** in which the 0.5 to 2% acid is temporarily stored, a conduit **25** and the tower **21**. To ensure constant coverage with water of the bubble trays in the tower **21**, and to ensure a maximum hydrogen chloride content of 2% for the scrubbing liquid used in the second scrubbing cycle, fresh water is continually admitted to the head of the tower **21**. The admission of fresh water to the bubble trays of the tower **21** is controlled by means of a second density recorder **31**.

The conduit **25** branches off to a pipe **26** and a valve **27**, whereby the acid in the second reservoir **16** can be further diluted; a level indicator **28** causes the valve **27** to be opened when the level of acid in the reservoir **16** falls below a predetermined minimum level. A second overflow line **29** provides a connection from an upper level of the reservoir **16** to the reservoir **24**, this overflow line being used in the event of unduly large quantities of dilute acid being collected in the reservoir **16**. The overflow of this dilute acid is conveyed for further temporary storage to the reservoir **24** of the second scrubbing cycle. Excess quantities of 0.5 to 2% hydrochloric acid leave the reservoir **24** by way of a conduit **36** provided beyond a partition **37**, as shown.

The gas scrubbed in the tower **21** is introduced through a conduit **32** via a separating vessel **33** into a spray tower **34**. In the separating vessel **33**, which is filled with metal turnings or chips (e.g., iron or copper) which are water-sprayed from above, the very minor proportions of chlorine and hydrogen chloride still present in the gas are chemically bound, apart from the traces thereof to which reference is made below. Dissolved metal chlorides run down through a conduit **35** and leave through the conduit **36**.

The spray tower **34** is supplied from above with dilute sodium hydroxide solution having a strength of approximately 1%, which enables the last traces of chlorine and hydrogen chloride to be removed from the gas. The purified off-gas (N_2, CO_2, O_2, H_2O) leaves the apparatus through a conduit **38**. The sodium hydroxide solution is recirculated by means of a pump **41** through a bottom outlet **39**, a container **40**, a conduit **42** and the spray tower **34**, but is finally removed as spent liquor through a conduit **43** and may then be utilized for neutralizing the hydrochloric acid flowing out through the conduit **36**.

A process developed by *W.H. Prahl; U.S. Patent 4,157,380; June 5, 1979* is one for removing chlorine and hydrogen chloride from a combustion gas, such as a combustion gas formed from incinerating chlorine-containing organic materials, which comprises:

(a) Lowering the temperature of the combustion gas below the melting point of cupric chloride or mixture thereof with other salts;

(b) Contacting the cooled combustion gas of (a), in the presence of oxygen, with copper of lower than a divalent oxidation state, such as a cuprous compound or its equivalent in a quantity sufficient to absorb chlorine and hydrogen chloride present in the gas, thereby converting the cuprous compound to cupric chloride (in order to absorb substantially all HCl and Cl_2, it is desirable to use a stoichiometric excess of cuprous compound); and

(c) Contacting cupric chloride with a reducing agent, thereby converting the cupric chloride to cuprous chloride or an equivalent compound of lower than divalent oxidation state, and the reducing agent to a chlorinated product.

The products formed from the reducing agent in (c) can be removed and cuprous chloride formed in (c) can be reused in (b).

One such reducing agent is hydrogen. Its reaction product is hydrogen chloride, according to the equation: $2CuCl_2 + H_2 = 2CuCl + 2HCl$. In that case, the economic advantage of this process is the recovery and upgrading of the chlorine and hydrogen chloride from the combustion gas in form of concentrated gaseous hydrogen chloride, rather than as dilute aqueous hydrochloric acid.

Other reducing agents are aliphatic or aromatic hydrocarbons. They are converted to the corresponding chlorinated hydrocarbons, as represented, for instance, in the case of methane and benzene by the equations:

$$2CuCl_2 + CH_4 = 2CuCl + HCl + CH_3Cl$$

$$2CuCl_2 + C_6H_6 = 2CuCl + C_6H_5Cl + HCl.$$

In such reactions, which are substitution reactions, only one-half of the chlorine and hydrogen chloride extracted from the combustion gas goes into the chlorinated product. The other half is still in form of hydrogen chloride, although in concentrated gaseous form, rather than in form of dilute aqueous acid.

Reducing agents capable of utilizing the full quantity of chlorine and hydrogen chloride in the combustion gas are compounds capable of addition reactions. An example of such compounds is ethylene, which reduces cupric chloride according to the equation: $2CuCl_2 + C_2H_4 = 2CuCl + C_2H_4Cl_2$, forming essentially pure ethylene dichloride without by-products. Ethylene dichloride is an important intermediate in the production of vinyl chloride.

HYDROGEN CHLORIDE FROM OXYCHLORINATION PROCESS EXIT GASES

A molten mixture containing the higher and lower valent forms of a multivalent metal chloride (in particular, cuprous and cupric chloride) is contacted with oxygen and aqueous hydrogen chloride recycle to recover the chlorine values by generation of the higher valent metal chloride and also effect oxidation of the melt by production of the oxychloride. The gas withdrawn from the oxidation contains hydrogen chloride, chlorine and water-vapor, with the hydrogen chloride being separated from the gas as aqueous hydrogen chloride and recycled to the oxidation. The chlorine and water-vapor in the gas are contacted with activated carbon to produce hydrogen chloride, which is separated as aqueous hydrogen chloride and also recycled to the oxidation.

The process flow diagram of this process is shown in Figure 80 as described by *H.D. Riegel, H.D. Schindler and V.A. Strangio; U.S. Patent 4,036,776; July 19, 1977; assigned to The Lummus Company.*

Figure 80: Lummus Co. Process for Recovery of Chlorine Values from a Melt Oxidation Gas as Hydrogen Chloride

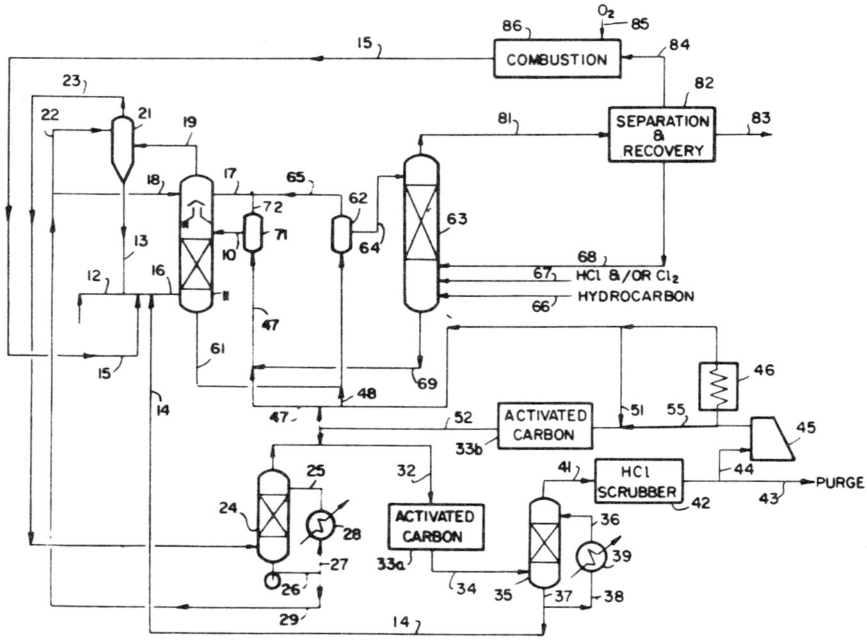

Source: U.S. Patent 4,036,776

Referring to the figure, a molten chloride salt, such as a mixture of potassium chloride, cupric chloride and cuprous chloride in line **10** is introduced into the top of an oxidation vessel **11**, maintained at 750° to 870°F and 1 to 20 atmospheres. A compressed molecular oxygen-containing gas, such as air, in line **12**; an aqueous solution of hydrogen chloride, in lines **13** and **14**, obtained as described below; and a by-product combustion effluent, in line **15**, obtained as described below, and comprising chlorine, hydrogen chloride, as well as carbon oxides, water-vapor, nitrogen and, perhaps, unreacted oxygen; are combined in line **16** and introduced into reactor **11**. It is to be understood that the feed stream introduction may be effected otherwise than as particularly described.

As a result of the countercurrent contact between the feed introduced through line **16** and the descending molten salt mixture, introduced through line **10**, the salt is oxidized to produce copper oxychloride, and the hydrogen chloride and chlorine introduced as aqueous hydrogen chloride, are selectively reacted with the molten salt to produce cupric chloride. In addition, the water introduced with the aqueous hydrogen chloride is vaporized. Furthermore, the amount of oxygen present in the gas introduced through line **12** is in an amount in excess of that required to react with the hydrogen chloride, whereby there is net production of copper oxychloride.

An effluent gas, including water-vapor, nitrogen, carbon oxides, chlorine and hydrogen chloride, rises into the top vessel **11** wherein the gas is combined with lift gas, as described below, and introduced through line **17**. The combined gas is directly contacted in the top of vessel **11** with a spray of aqueous hydrogen chloride quench liquid introduced through line **18** to cool the combined gas and eliminate any vaporized and entrained salts therefrom. The effluent gas is cooled to a temperature at which the molten salt is still in the form of a melt to permit the molten salt to flow back into the reactor **11**.

The cooled gaseous stream withdrawn from reactor **11** through line **19** is introduced into quench vessel **21**, wherein the gaseous stream is directly contacted with an aqueous hydrogen chloride stream introduced through line **22**. The quenching in vessel **21** is effected in a manner such that the aqueous hydrogen chloride quench liquid introduced through line **22** is partially vaporized which produces a remaining aqueous hydrogen chloride quench liquid having a higher concentration of hydrogen chloride. The quenching in vessel **21** also functions to separate any remaining entrained salt from the gaseous effluent. In accordance with a preferred embodiment, quenching in vessel **21** is effected in a manner to cool the effluent to a temperature of 200° to 250°F.

The remaining liquid aqueous hydrogen chloride quench liquid, containing any remaining salt, is withdrawn from vessel **21** through line **13** and introduced, as recycle, into reactor **11**.

The effluent gas, also containing vaporized quench liquid, withdrawn from vessel **21** through line **23**, is introduced into a direct contact quench tower **24** of a type known in the art, wherein the gas is cooled by direct contact with aqueous hydrogen chloride quench liquid introduced through line **25**. The quenching in tower **24** is controlled in a manner such that not all the hydrogen chloride present in the off-gas is recovered therefrom in that such complete recovery will be accompanied by an unacceptable corresponding amount of water condensation.

In accordance with a preferred embodiment, condensation is preferably controlled to condense an aqueous hydrogen chloride solution having a hydrogen chloride concentration of 8 to 20%, preferably from 10 to 18%, all by weight. In general, such a result can be achieved by effecting cooling of the gas introduced into tower **24** to a temperature of from 120° to 250°F.

An aqueous hydrogen chloride solution is withdrawn from tower **24** through line **26** and a first portion thereof passed through line **27** and cooler **28** for introduction into the quench tower through line **25** to meet the cooling requirements. A second portion of the aqueous hydrogen chloride is passed through line **29** for meeting the quenching requirements of the reactor **11** through line **18** and the requirements of vessel **21** through line **22**.

The gas withdrawn from tower **24** through line **32** contains inerts, water-vapor, some hydrogen chloride and chlorine, and is introduced into reactor **33a** of a pair of reactors **33a** and **33b**, each containing a bed of activated carbon. For purposes of illustration, reactor **33a** is on-stream and reactor **33b** has been taken off-stream for replenishment of activated carbon. As a result of the contact between the gas introduced through line **32** and the activated carbon in reactor **33a**, the chlorine is converted to hydrogen chloride. The gas withdrawn from reactor **33a** through line **34** is essentially free of chlorine and contains increased amounts of hydrogen chloride, water-vapor, nitrogen and carbon oxides. The gas may also include some oxygen. The gas in line **34** is introduced into a direct contact quench tower **35**, of a type known in the art, wherein the gas is cooled by direct contact with aqueous hydrogen chloride quench liquid introduced through line **36**.

The quenching in tower **35** is controlled in a manner such that not all of the hydrogen chloride present in the gas is recovered therefrom, as described with reference to the operation of quench tower **24**. Condensation is preferably controlled to provide a condensed aqueous hydrogen chloride solution having a hydrogen chloride concentration from 8 to 20%, preferably from 10 to 18%, all by weight. In general, such a result can be achieved by effecting cooling of the gas introduced into tower **35** to a temperature of from 100° to 250°F.

An aqueous hydrogen chloride solution is withdrawn from tower **35** through line **37** and a first portion thereof passed through line **38**, including cooler **39**, for introduction into the quench tower through **36**. A second portion of the aqueous hydrogen chloride is passed through line **14** for introduction into the oxidation vessel **11** to recover the chlorine values thereof. The aqueous hydrogen chloride in line **14** can be concentrated, e.g., by stripping of water prior to introduction into oxidation vessel **11**.

The gas withdrawn from tower **35** through line **41** is caustic- and water-washed in zone 42 to remove remaining hydrogen chloride, and a portion thereof released to the atmosphere through line **43**. The remaining portion of the gas in line **44** is compressed in compressor **45** and the temperature thereof regulated in heat exchanger **46** prior to passage through lines **47** and **48** for use as a lift gas for transporting molten salt, as described below.

A portion of the compressed lift gas, in line **51** or line **55** (depending on temperature conditions), is passed through reactor **33b** containing activated carbon previously used for removing chlorine from the gas in line **32**, in order to strip chlorine and hydrogen chloride therefrom prior to replenishing the activated carbon. The lift gas, withdrawn from reactor **33b** in line **52**, containing components desorbed from the activated carbon bed (in particular, chlorine and/or hydrogen chloride), is combined with the gas in line **32**, with the chlorine values ultimately being recovered therefrom, as described above.

Alternatively, the lift gas containing chlorine and/or hydrogen chloride can be combined with lift gas in line **47** for ultimate reusing of chlorine values. After replenishment of the activated carbon, bed **33b** can be heated to operation conditions by a portion of the lift gas in line **51** or **55**.

Referring back to the oxidation vessel **11**, a molten salt containing copper oxychloride and enriched in the cupric chloride content thereof, is withdrawn from the bottom of vessel **11** through line **61** and lifted by the lift gas in line **48** into a separation vessel **62**, positioned adjacent to the top of chlorination vessel **63**. In separator **62**, the molten salt is separated from the lift gas, with the molten salt being introduced into the top of chlorination vessel **63** through line **64**. The lift gas is withdrawn from vessel **62** through line **65** and combined with lift gas used for transporting salt to the oxidation reactor **11**, for introduction into the quenching portion of vessel **11** through line **17** to thereby separate any entrained and vaporized salt therefrom.

Fresh feed methane in line **66**, fresh feed hydrogen chloride and/or chlorine in line **67** and recycle components in line **68** are introduced into the bottom of reactor **63**, wherein they are contacted with the descending molten salt to effect chlorination of the methane and recycle components.

Molten salt withdrawn from reactor **63** through line **69** is lifted by lift gas into a separation vessel **71**, positioned adjacent to the top of reactor **11**. In separator **71**, the molten salt is separated from lift gas, and introduced through line **10** into vessel **11**. Lift gas is withdrawn from separator **71** through line **72** and combined with the lift gas in line **65** for introduction through line **17** into the top of the quenching section of the vessel **11**.

A chlorination effluent is withdrawn from reactor **63** through line **81** and introduced into a separation and recovery section, schematically indicated as **82**. In separation and recovery section **82**, recycle components are recovered and recycled to reactor **63** through line **68**. Chlorinated methane reaction product is recovered from separation section **82** through line **83**.

Chlorinated hydrocarbons which are not recovered as reaction product and not suitable, as recycle, for the production of chlorinated methanes, are withdrawn from the separation and recovery section **82** through line **84** and introduced along with molecular oxygen in line **85** into a combustion zone **86**, wherein the chlorinated hydrocarbons are burned to recover the chlorine values thereof. A combustion effluent, including hydrogen chloride, chlorine, carbon oxides, water-vapor and nitrogen, and optionally oxygen, is withdrawn from combustion chamber **86** through line **15** for introduction into the reactor **11**, as described above.

HYDROGEN CHLORIDE FROM SPENT PICKLE LIQUOR

A process developed by *L.J. Hansen; U.S. Patent 4,107,267; August 15, 1978; assigned to Toledo Pickling & Steel Service, Inc.* is one for regenerating waste acid, such as waste hydrochloric acid pickling liquor, by producing aqueous hydrochloric acid of pickling concentration and generating iron oxide including that of pigment grade. The process comprises the following steps: (a) concentrating the waste liquor; (b) reacting the concentrated liquor at a temperature below 1000°F in the presence of oxygen and hydrocarbon combustion gases and on a recycled moving bed of particles without vibration of the reactor bottom; (c) separating the iron oxide dust from the other gases from the reactor; (d) cooling these gases using the same concentrator; (e) absorbing the hydrogen chloride from the cooled gas in an aqueous solution; (f) scrubbing the unabsorbed gases in water to absorb the hydrogen chloride gas remaining in the gases from the absorber; and (g) condensing any liquids and any residual HCl remaining in the exhaust gases before exhausting them as nonpolluting gases into the atmosphere and passing the condensate back into the waste acid concentrator.

See also "Iron from Spent Pickle Liquor" and specifically U.S. Patents 4,113,158 and 4,177,119.

HYDROGEN FLUORIDE FROM ALKYLATION PROCESS

It is common practice in the petroleum industry to produce a high octane motor fuel by alkylating isoparaffins with olefins in the presence of a catalyst which preferably is hydrofluoric acid or hydrogen fluoride (HF). The effluent from the alkylation reactor is usually passed to a settling vessel, wherein a hydrocarbon phase is separated from an HF acid phase. The hydrocarbon phase is fractionated to separate low boiling hydrocarbons from the alkylate product. The acid phase is cooled and recycled to the alkylation reactor for use in the alkylation process. However, it is necessary to purify a portion of the recycled acid phase in order to prevent a buildup of acid-soluble oils (ASO) and water in the system.

Generally, this purification is accomplished by passing an HF-containing stream from the settler to an HF purification or rerun fractionation column, wherein HF is stripped from the ASO by means of a vaporous hydrocarbon which is preferably isobutane. While the HF purification systems used heretofore have generally been effective to produce an acid stream having the required purity, the recovered ASO stream from the purification unit usually has contained undesirable amounts of HF, resulting in the loss of valuable HF and in ecology problems. This has required subsequent purification steps to remove HF before the ASO can be processed further, such as by recycling to the refinery as fuel or passed to a disposal unit.

Thus, a process has been developed by *F.W. Skraba; U.S. Patent 4,199,409; April 22, 1980; assigned to Phillips Petroleum Company* for recovering HF catalyst from an alkylation unit acid stream containing acid-soluble oil, wherein a bottoms stream, which contains HF, acid-soluble oils and a minor quantity of water from an HF rerun column, is processed. The bottoms mixture is cooled in a heat exchanger to a temperature sufficiently low so as to effect separation of HF from the acid-soluble oil with the mixture then being passed to a separating vessel to allow the separation of the HF from the acid-soluble oil due to the reduced mutual solubility.

A gradient of HF concentration is effected within the separation vessel, wherein an HF-rich acid-soluble oil mixture is withdrawn as a stream separate from a substantially HF-free acid-soluble oil portion, which is also recovered as a separate stream. Such a process is shown in Figure 81.

Figure 81: Phillips Petroleum Co. Process for Recovery of HF from an Alkylation Unit Acid Stream Containing Acid-Soluble Oil

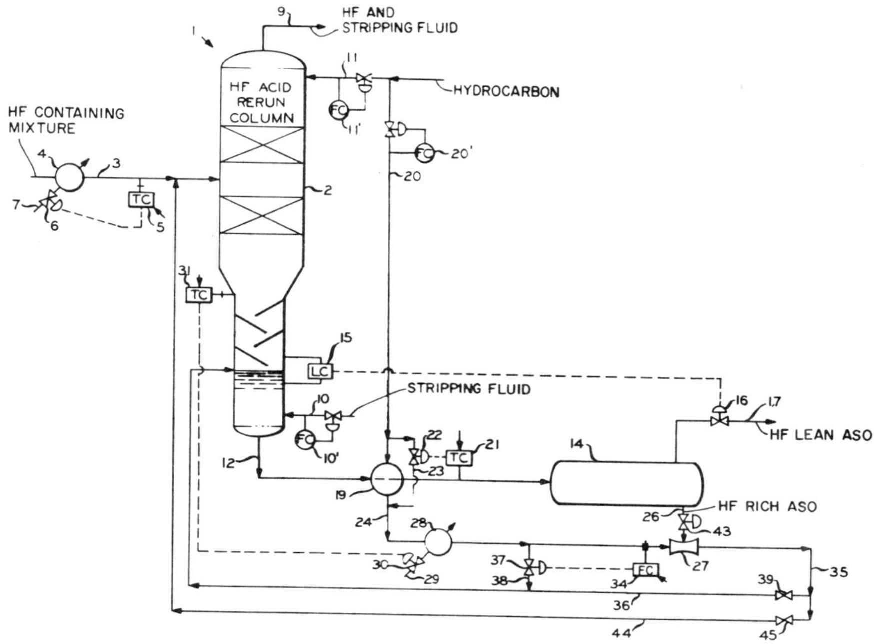

Source: U.S. Patent 4,199,409

The reference numeral **1** designates generally an apparatus used to recover HF from an alkylation HF acid catalyst stream. The apparatus **1** includes what is commonly referred to as a rerun fractionation or contacting column **2**, which has a construction well-known in the art and, therefore, need not be further described.

A line **3** connects the rerun column **2** to portions of an alkylation unit (not shown) which has a construction and operation that is also well-known in the art. The line **3** is operable to conduct an alkylation HF catalyst stream to the rerun column **2**. Typically, the alkylation HF catalyst stream contains HF, ASO, water and isobutane. It should be noted that there is a very low water content contained in the stream and it is generally about 3 wt % or less and the low water content is due to drying of the olefin and isobutane feedstocks. Preferably, an indirect heat exchanger **4** is connected in the line **3** and is operable for controlling the temperature, by means of a suitable temperature controller **5**, of the impure acid stream supplied to the rerun column **2**.

The temperature controller **5** is operably connected to the line **3** and a valve **6**, whereby the temperature controller senses the temperature of the HF acid catalyst stream in the line **3** and controls the valve **6** in response to the temperature.

The valve **6** is operable to control the amount of heat exchange medium supplied to the heat exchanger **4** via line **7** which is connected to a suitable source of heat exchange medium. Preferably, the line **3** is connected to rerun column **2** adjacent an upper portion thereof, and used HF catalyst is added as substantially all-vapor feed.

A conduit **9** communicates with an upper portion of the rerun column **2** and is provided as an overhead outlet for vaporous HF and a stripping hydrocarbon and reflux hydrocarbon such as isobutane, which is introduced into the rerun column **2**. The conduit **9** returns the overhead products from the rerun column **2** to the alkylation unit. Preferably, the stripping hydrocarbon is introduced into the rerun column **2** at a plurality of points to effect optimum separation of HF from the HF acid stream mixture supplied to the rerun column **2**. The mixture or HF acid stream contains, among other things, HF, acid-soluble oil and water, as described above.

As shown, vaporous stripping hydrocarbon is introduced into a lower portion of the rerun column **2** through a line **10** which communicates with a lower portion of the rerun column **2**. Preferably, the line **10** communicates with the rerun column **2** at a position below the level of accumulated liquid bottoms retained in a lower portion of the rerun column. Preferably, the stripping hydrocarbon is introduced at a temperature of approximately 400° to 450°F. The bottoms herein referred to are collected portions of the mixture which has a major portion of the HF removed therefrom by contact with the stripping hydrocarbon wherein the stripping hydrocarbon and HF are taken off through the overhead conduit **9**. Preferably, reflux liquid hydrocarbon, such as isobutane, is introduced into an upper portion of the rerun column **2** through a conduit **11**.

The conduit **11** communicates with an upper portion of the rerun column and since the reflux hydrocarbon is liquid and at a lower temperature than the mixture of gases and vaporized bottoms and vaporized mixture introduced into the column **2** through the line **3**, there is effected condensation of heavier or higher boiling portions of the vaporized materials. The abovedescribed operation is well-known in the art and therefore need not be more specifically disclosed herein.

Bottoms are removed from the lower portion of the rerun column **2**, preferably through a line **12** which connects the bottom portion of the rerun column **2** to a settler vessel **14**, which is of generally standard construction. Preferably, means are provided to control the level of the bottoms in the rerun column **2** and, as shown, a level controller **15** is operably associated with the rerun column **2** and is also operably connected to a valve **16**. The valve **16** is connected to a discharge line **17** which, in turn, communicates with an upper portion of the settler **14** whereby a change in the level of the bottoms is sensed by the level controller **15**, which controls the valve **16** so that when the level of the bottoms becomes excessive, the valve **16** will open further and allow discharge of a portion of the contents of the settler and thereby lower or maintain the desired level of bottoms in the rerun column **2**.

The bottoms removed from the rerun column **2** are at a temperature of approximately 300° to 400°F and the higher the temperature of the bottoms, the more the mutual solubilities of the constituents of the bottoms, i.e., the higher the temperature the more ASO that will be dissolved in the HF, and the more HF that will be dissolved in the ASO.

It is to be noted that occasionally light acid-soluble oils are made in HF alkylation. When this occurs, the rerun bottoms temperature is lowered to become approximately 275° to 300°F. This allows the light ASO to go out with the rest of the bottoms rather than as an overhead product through the conduit **9**. When the light ASO is no longer present, the bottoms temperature can be raised to the normal temperature as described above. This portion of the process is well-known in the art.

Indirect heat exchange means **19** is connected in the line **12** between the rerun column **2** and the settler **14** and is operable to cool the bottoms as same are being conducted from the rerun column **2** to the settler **14**. The heat exchange means **19** is connected to a suitable source of cooling medium to provide indirect cooling for the bottoms. Preferably, and as shown, the heat exchange means **19** is connected to the line **11** by a line **20**, whereby cooled liquid hydrocarbon from the line **11** is conducted to the heat exchange means **19** to effect the cooling of the discharged bottoms. As shown, flow controllers **10', 11'** and **20'** are provided in lines **10, 11** and **20**, respectively, for regulation of flow, therethrough. Preferably, temperature control means is provided to control the temperature of the bottoms entering the settler **14**. As shown, a temperature controller **21** is operably connected to the line **12** to sense the temperature of the fluid therein and is also operably connected to the flow control valve **22**, which is connected in a bypass line **23**.

The bypass line **23** is connected between the line **20** and a discharge line **24** to form a bypass around the heat exchange means **19**, whereby a change in temperature is sensed by the temperature controller **21** to thereby control the valve **22** as, for example, when the bottoms entering the settler **14** are too hot, the valve **22** will be further closed in response to operation of the temperature controller **21** to allow more cooling medium to flow through the heat exchanger **19** to cool the bottoms to the desired temperature. Preferably, the bottoms entering the settler **14** are maintained at a temperature of up to approximately 125°F and, preferably, in the range of 75° to 125°F and, more preferably, in a temperature range of 90° to 115°F.

The liquid hydrocarbon which is used as a cooling medium is heated in the heat exchange process conducted in the heat exchange means **19** and leaves the heat exchange means **19** in a heated condition as, for example, at a temperature in the range of about 130° to 175°F, and preferably, about 145°F. However, it is to be noted that this condition may be varied in the process.

The material in settler **14** consists of a continuous ASO phase in which is dispersed small droplets of HF. Most of the HF droplets, in time, will settle near the bottom of the settler. Thus, by withdrawing material from the bottom of the settler, there is obtained an HF-rich ASO phase portion and from the top, there is obtained an HF-lean ASO portion.

A line **26** communicates with a lower portion of the settler **14** and is operable for discharge of an HF-rich ASO phase. The ASO phase withdrawn from the top of the settler **14** through the line **17** is in a substantially HF-free condition. The recovered heavy phase from the settler **14** preferably is returned to the rerun column **2** for recycling and reuse as is desired. To accomplish this, a preferred means is to provide an eductor **27** which has the line **26** communicating therewith and the educting fluid preferably is the vaporized hydrocarbon discharged through the line **24**, which is connected in flow communication with the eductor **27** whereby the stripping hydrocarbon is used to educt the heavier phase from the settler **14** back to the rerun column **2**. However, it is to be understood that other methods of recycling the heavier phase containing the HF, which is discharged through line **26**, can be provided. In the form shown, the line **24** has an indirect heat exchanger **28** connected thereto. The heat exchanger **28** is connected to a suitable source (not shown) of heat exchange medium by a line **29** which has a control valve **30** connected thereto. Preferably, the valve **30** is operably connected to a temperature controller **31** which, in turn, is operably associated with the rerun column **2** to sense the temperature therein.

The vaporous hydrocarbon from the line **24**, downstream of the heat exchanger **28**, is introduced into the rerun column **2** in a manner later described with the temperature of the hydrocarbon being controlled to maintain a predetermined temperature in the rerun column **2** by use of the heat exchanger **28** and control means associated therewith.

Preferably, a flow controller **34** is operably connected to the line **24** to sense the flow of the vaporous hydrocarbon therethrough. A line **35** communicates with the eductor **27** and, in turn, is connected to a line **36** which communicates with the rerun column **2**, whereby vaporous hydrocarbon and educted heavy phase from the settler **14** is returned to the rerun column **2**. The flow controller **34** is operably connected to a flow control valve **37** which, in turn, is connected in a bypass line **38**, which communicates between the line **24** and the line **36** to form a flow control bypass.

The flow of vaporous hydrocarbon through the eductor **27** is controlled by the flow controller **34** which, in turn, controls the amount of bypass stripping hydrocarbon flowing through the line **38** to bypass the eductor **27**. Preferably, a valve **39** is connected in the line **36** between the eductor and the connection of the line **36** with the line **38** to control the amount of educting fluid and educted heavier phase flowing back to a preselected locus in the rerun column **2**. This can be either a manually controlled valve or can also be an automatically controlled valve. It is to be noted that the educted heavier phase is heated by the educting fluid so that the mixture of same is in a heated condition in a preferred embodiment, before introduction into the rerun column **2**.

The process is more fully understood by a description of the operation thereof. The alkylation HF-acid stream from an alkylation unit which is to be rerun is introduced into the rerun column **2** through the line **3** with the temperature of same being controlled by the use of the heat exchanger **4** and a temperature controller combination **5**. The acid waste stream is contacted by stripping vaporous hydrocarbon, which is introduced through the line **10**, whereby a major portion of the HF contained in the acid waste stream is removed as a vaporous overhead product through the line **9** along with the stripping hydrocarbon and the vaporized reflux hydrocarbon which is introduced through the line **11**.

Liquid reflux hydrocarbon preferably is introduced into an upper portion of the rerun column to contact any vaporous heavier materials and to condense same so that the condensed materials will collect in the bottom of the rerun column.

The level controller **15** is operable to control the level of the bottoms in the bottom of the rerun column **2** by allowing discharge of ASO, which is substantially free of HF through the line **17** from the settler **14**. Preferably, the amount of ASO discharged through the line **17** will be about 25 vol % or less of the amount of the material added via conduit **12**. The collected bottoms are conducted from the rerun column **2** to the settler **14** via line **12** and before introduction into the settler **14**, the bottoms are cooled to a temperature of up to approximately 125°F as described above.

Upon entering the settler, the cooled bottoms remain a sufficient length of time to allow the ASO and the HF to sufficiently separate and because of the difference in densities, the HF will concentrate in the lower portion of the settler **14** for discharge as an HF-rich oil phase through the line **26**. This separation is due to the reduced mutual solubilities because of the lowering of the temperature of the bottoms entering the settler **14**.

The ASO discharged through the line **17** will be substantially free of HF, having about 1% by weight, and preferably less than about 0.5%, and most preferably less than about 0.1%. A small amount of water will also be discharged through the line **26** with the HF-rich ASO phase. This water will be in the amount of approximately 0.5% or less and can be considered a trace material.

Up to about 90% of the ASO-containing HF entering the settler **14** can be recycled back to the rerun column by being discharged with the second or heavy phase containing the HF and water through the line **26**.

It is to be understood that anywhere from 0 to about 90% of the ASO entering the rerun column **2** can be recycled and preferably about 75 to 90% of the ASO is recycled to the rerun column **2**.

This importantly eliminates the need for complete phase separation of the ASO and HF (and water). However, it is to be noted that complete phase separation could be accomplished in the settler **14** whereby little or no ASO would be discharged through the second phase discharge **26**. However, this would require prohibitively long separation times and a prohibitively large settler vessel **14**.

A line **44** can be connected to the line **35** and the line **3** so that a portion of the recovered second phase can be returned to an upper portion of the rerun column **2** with the line **44** preferably having a flow control valve **45** connected thereto to control the amount of materials flowing through the line **44**. Since this is an alternate mode of operation of the process, it is to be understood that the line **44** and valve **45** can be eliminated, if desired. The ASO recovered through the line **17** is conducted to other apparatus (not shown) for further processing of the ASO. The second phase recovered through the line **26** is educted and recycled back to the rerun column for use in the HF stripping operation. It is seen, therefore, that a portion of the ASO is continually recycled to the rerun column **2** and is heated before recycling to the rerun column **2** to further enhance the operating efficiency of the column **2**.

Pollution of the air with HF acid vapor and light hydrocarbons occurs when noncondensible gas is vented from a depropanizer accumulator as known in the art. The venting is continuous. The vent gas heretofore has been neutralized in a closed system with sodium or potassium hydroxide or otherwise chemically contacted with an agent of which a suitable disposal must be made. In any event, however the HF contaminating the vent gas has been recovered, there has been a disposal problem as in disposing of spent caustic combined with fluoride and a concomitant waste of the HF.

A process developed by *C.O. Carter; U.S. Patent 3,886,220; May 27, 1975; assigned to Phillips Petroleum Company* permits the recovery of HF from such vent gases and also involves generating, in situ, in the alkylation operation, in which HF catalyst is rerun or regenerated, a reaction medium which can be used as absorbent for recovering from the vent gas the HF normally contained therein.

By operating the conventional HF-acid rerun operation in two stages, a first in which acid-soluble oil is removed by subjecting the used HF to stripping with a hydrocarbon, e.g., hot isobutane; and a second in which the recovered HF and isobutane stream is refluxed to obtain a substantially purified HF-isobutane stream, a water-stream containing HF can be obtained which can be used for a reaction with an olefin, e.g., as herein named, to generate an alkyl fluoride-containing stream or phase which is then passed to the original alkylation or to a subsequent alkylation reaction or otherwise utilized in the process.

Thus, the water containing HF is used as absorbent to scrub vent gas emanating from the process thus enriching still further the water with HF, which has been removed from the vent gases which can, therefore, be vented safely or can be used as a fluoride-free fuel gas following which this enriched water is reacted with an olefin to produce a stream containing alkyl fluoride, which, by its further utilization, retains in situ in the system, the HF. The process flow diagram is shown in Figure 82, which follows.

For sake of convenience, three sections of the figure have been framed and these are termed **A, B** and **C**. Generally, **A** represents a conventional isobutane-olefin alkylation in which are present a conventional isostripper and HF or propane recovery column; **B** represents an HF-acid rerun unit; and **C** represents a vent gas-scrubbing operation which is operated in conjunction with **A** and **B**.

Isobutane and olefin are fed to the system by **1** and **2**. Recycled isobutane from **3** is combined with the isobutane feed and passed through **4**, together with olefin from **2** by **5** into riser-reactor **6**, wherein it is intimately contacted with a stream of HF catalyst as known in the art. Catalyst is normally cycled from settler **7** by **8, 9,** cooler **10**, and **6** back to the settler. Reaction for the most part takes place in riser-reactor **6**. In settler **7**, there is formed a supernatant hydrocarbon phase passed by **11** to an isostripper **12**, while the lower phase, consisting of used HF acid, is cycled as described.

The operation of isostripper **12** is conventional in this description in the sense that there are removed as bottoms by **13**, an alkylate product-containing stream; by **14**, a normal butane-containing stream; and overhead at **15** to cooler-condenser **16**, a vaporous stream containing principally HF vapor, propane, and noncondensibles. Stream **15**, cooled as far as possible with ordinary cooling water available in the plant, is passed by **17** to accumulator **18** from which a hydrocarbon phase formed is passed in part as reflux to the isostripper by **19** and in yield quantities by **20** to HF column **21**. An acid phase **22** is recycled by passage by **23** into the upper portion of settler **7**. The isostripper and HF column are reboiled as shown and as conventional in the art.

Figure 82: Phillips Petroleum Co. Process for HF Recovery from Alkylation Process

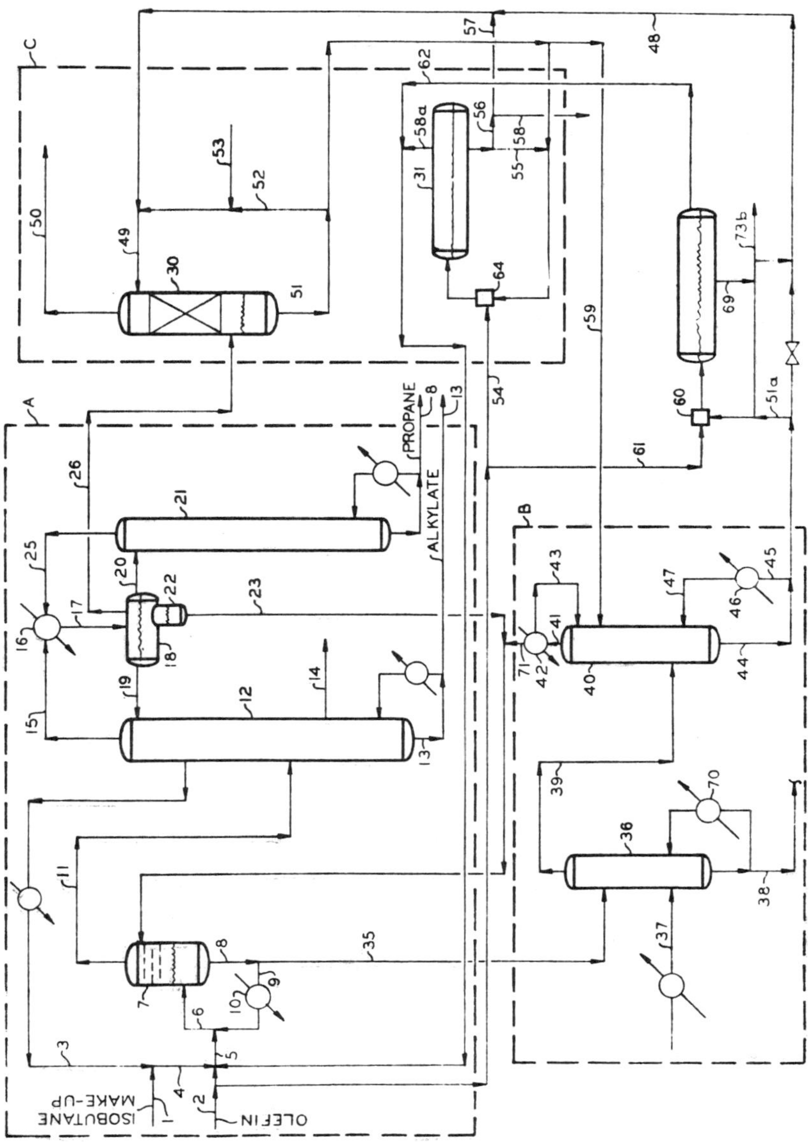

Source: U.S. Patent 3,886,220

Bottoms stream **8** from the tower **21** constitutes essentially propane, substantially free from HF. Overhead **25** is passed from column **21** to the cooler-condenser **16** and thus recycled to accumulator **18**. Gases, not condensed, pass by **26** into zone **C**. In zone **C** there are provided a scrubber **30**, a reactor **64** and a settler **31**.

According to the process, scrubber **30** is supplied with a scrubbing or absorption medium from zone **B**. The scrubbing medium is enriched in HF in zone **C** and the enriched medium is used in reactor **64** to produce alkyl fluoride.

A portion of the acid passes from the foot of settler **7** by **35** to stripper **36**. The stripper is heated and reboiled by hot isobutane vapors entering at **37**, as well as by conventional reboiler **70**. The used HF catalyst is stripped by countercurrent contact, while it passes downwardly against the rising isobutane vapors in a manner and under conditions such that substantially only an acid-soluble oil residue is withdrawn at **38**. HF, hydrocarbon vapor and water-vapor, according to the process, pass overhead at **39** into a second stripper **40** which is refluxed with some of its own condensed overhead passed by **41** through condenser **42** and by **43** into a top portion of the column **40**. Column **40** is reboiled by heating a portion of the bottoms stream, comprising water and HF withdrawn at **44** and passed by **45** to heater **46** and by **47** into the bottom of column **40**.

Yield quantities of water and HF are taken by **48** into the top of scrubber **30** in section **C**. The point of entry of the HF-containing water stream is selected to obtain the best or optimum desired scrubbing conditions so that gases vented from the system at **50** will be substantially free from HF. As shown, the absorbing medium entering at **49** is withdrawn at **51** and can be cycled by **52** to **49** and thence into the top portion of scrubber **30**. There can be added directly to scrubber **30**, as desired, additional scrubbing medium, for example, additional water; or, additional water can be introduced at **53** into the cycling stream **52**.

The bottoms **51** are passed to reactor **64** to which also is passed, in this embodiment, a portion of the olefin stream **2** fed to the process. Thus, olefin is fed by **54** into reactor **64**, wherein a reaction between the olefin and HF in the water containing the same takes place. Phases are formed in settler **31** and a lower water-phase removed at **55** is cycled to reactor **64** and/or by **56** and **57** to **48**, and thence to the scrubber. Conditions are chosen in reactor **64** such as to substantially deplete the water-phase of HF. To this end, a series of countercurrently-operated contactors can be used in lieu of a single reactor as shown via **64**.

Liquid water accumulating in the system is removed at **58** and **73b** from the process. This water will be only that water which together with the amount of water-vapor emanating at **50** equals the water introduced into the system via the feedstocks. This relatively small quantity of water will be treated as may be desired before it is discarded. Alkyl fluoride from **31** is passed to the alkylation reactor **6** by **58a**, **62** and **5**.

HYDROGEN FLUORIDE FROM FURNACE POT LININGS

Cathode pots of electrolytic furnaces used in the production of aluminum are lined with side carbon and bottom carbon compositions which are electrically conductive. The bottom carbon is generally of graded anthracite coal and coke bonded together with pitch. During electrolytic operation, the cryolite components, rich in sodium and fluoride values, are slowly absorbed into the lining. Eventually the cathode liners foul to the point where they must be replaced. The spent cathode liners contain substantial amounts of fluoride values, aluminum in the form of aluminates, sodium fluoride, and absorbed sodium metal which, on exposure to atmospheric moisture, are converted to a caustic. Other materials, such as the anthracite carbon contained in the monolith spent liner, are valuable.

The monolith liner is composed of sidewalls and a bottom wall, with the bottom wall containing the anthracite carbon and constituting about two-thirds of the total weight of the cathode lining. Disposal of the spent pot liner has posed a problem due to the leaching of fluorides and other contaminants into ground water.

Methods for recovering cryolite from spent cathode liners are known. Such methods have generally used caustic (sodium hydroxide), sodium carbonate or water to extract the fluoride values from used cathode liners.

A process developed by *E.Q. Dahl; U.S. Patent 4,065,551; December 27, 1977; assigned to Elkem-Spigerverket A/S, Norway* is a process for recovering fluorine from solid carbonaceous waste material from furnaces for melt electrolytic production of aluminum. Such waste carbonaceous material is recovered as used furnace linings and bottoms, soot or scum, and dust from cyclones and electrofilters. Used furnace linings and bottoms from an aluminum furnace will usually contain about 10 to 15% by weight of fluorine, and the scum and cyclone dust about 30% fluorine.

The process is one in which particles of solid carbonaceous waste material which contain fluorine are suspended in a stream of gas consisting of a mixture of steam and air to establish a fluidized bed in a retort. The particles are heated to a temperature of not less than 1000°C to cause pyrohydrolysis and the recovery of fluorine as gaseous hydrogen fluoride at a very efficient rate. A process flow diagram is shown in Figure 83.

Referring to the figure, a reactor **10** is provided with a plate **12** having small openings distributed over its entire surface. The plate **12** is a ceramic grid, but it may be made of any porous material which will withstand the temperature of the pyrohydrolysis reaction. The reactor shown in the figure has a diameter of 25 cm at the grid and a diameter of 65 cm in the top cylindrical portion. The reactor measures 200 cm from top to bottom.

The charge of particles of waste carbonaceous material containing fluorine is fed to the reactor by means of feed pipe **14**, from a hopper (not shown) having a bed of charge of sufficient depth to seal the reactor or the charge may be blown into the reactor under a slight pressure. The rate at which the charge is fed is controlled by a conventional valve (not shown).

In the apparatus illustrated in the figure, a mixture of steam and air for fluidizing the charge is supplied to the bottom of the reactor by pipe **16** in sufficient volume and velocity to maintain a fluidized bed of particles in the conical portion of the reactor below feed pipe **14** where the pyrohydrolysis reaction takes place.

Figure 83: Elkem-Spigerverket A/S Process for Fluorine Recovery from Carbonaceous Waste Furnace Linings and Dust from Aluminum Manufacture

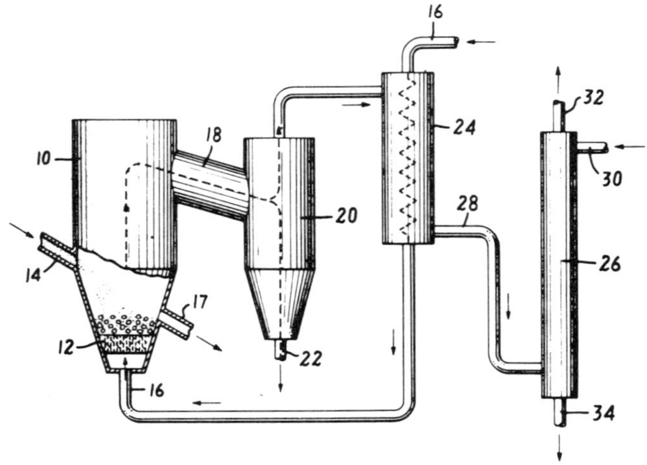

Source: U.S. Patent 4,065,551

As the reaction continues, ash will collect in the bottom of the reactor **10**. This ash may be continuously removed in batches by means of an overflow **17** which is suitably a conventional lock device. To remove a batch of ash, the pressure within the reactor is dropped. However, there is no need to completely shut down the reactor.

The gas containing hydrogen fluoride and dust and some of the lighter ash passes from the top of the reactor through pipe **18** into accumulator **20** where the dust and lighter ash settle and are removed through pipe **22**. The gas passes overhead, preferably into the heat exchanger **24**, to preheat the mixture of steam and air supplied to the reactor through pipe **16**. Thereafter, the gas is fed to the bottom of conventional wash tower **26** by means of pipe **28** where the gas moves up through wash water supplied by pipe **30**. The inert gases leave the top of the wash tower at **32** and the water which has absorbed the hydrogen fluoride gas is recovered from the bottom of the tower at **34** as hydrofluoric acid.

A process developed by *G.O. Orth, Jr. and R.D. Orth; U.S. Patent 4,113,831; September 12, 1978; assigned to National Fluoride and Chemical Corporation* is one in which fluoride and aluminum values, as well as carbon, are recovered from waste cathode liner material from aluminum electrolytic cells by leaching of the liner at ambient temperature with a caustic solution, followed by precipitation of sodium fluoride by saturating the leach liquor with a compound which suppresses the solubility of sodium fluoride in the leach liquor. Ammonia is a preferred compound.

Aluminum compounds, as well as the carbon values, are also recovered. Treating chemicals used in the process are recycled. The process is essentially a closed-cycle process with substantially no discharge of effluent. The process is shown schematically in Figure 84.

Figure 84: National Fluoride and Chemical Corp. Process for Recovery of Sodium Fluoride and Other Chemicals from Spent Carbon Liners

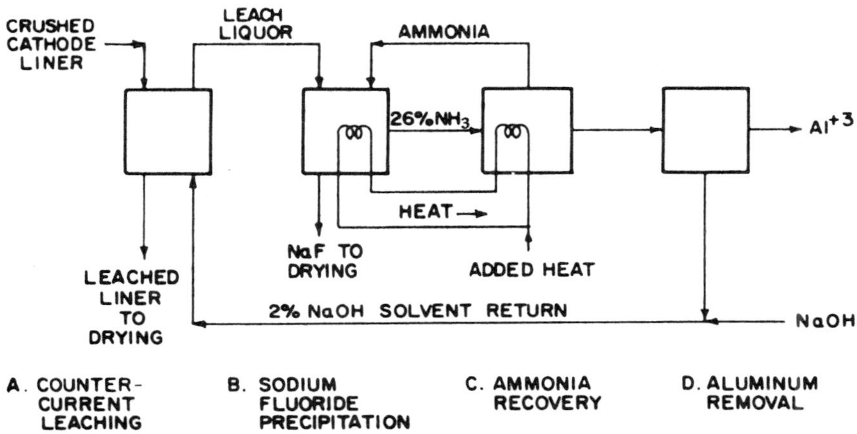

Source: U.S. Patent 4,113,831

A process developed by *J.N. Andersen and N. Bell; U.S. Patent 4,158,701; June 19, 1979; assigned to Kaiser Aluminum & Chemical Corporation* is one in which a pyrohydrolysis system is provided for the recovery of valuable components from waste and spent materials generated in electrolytic aluminum reduction facilities.

The pyrohydrolysis system uses a dense phase fluidized bed reaction zone for the pyrohydrolysis of coarse feed and a dilute phase fluidized reaction zone for pyrohydrolyzing fine feed, this zone being superimposed and interconnected with the dense phase zone.

The off-gases, after removal of the solids, are contacted in a dilute phase fluidized zone with a source of Al_2O_3 to remove residual Na values and to produce Na-free HF. The solids from the first dilute zone, having a desired high Na:Al atom ratio, can be combined with the product clinker from the dense bed zone. The off-gas, containing the source of Al_2O_3, is subjected to solids separation, the solids-free and Na-free HF is utilized, while the solids of low Na:Al mol ratio are recycled to the residual Na conversion step. Such a process is shown in Figure 85.

A very similar process is described by *J.N. Andersen and N. Bell; U.S. Patent 4,160,808; July 10, 1979.*

Figure 85: Kaiser Aluminum & Chemical Corp. Pyrohydrolysis System for Processing Fluorine-Containing Spent and Waste Materials

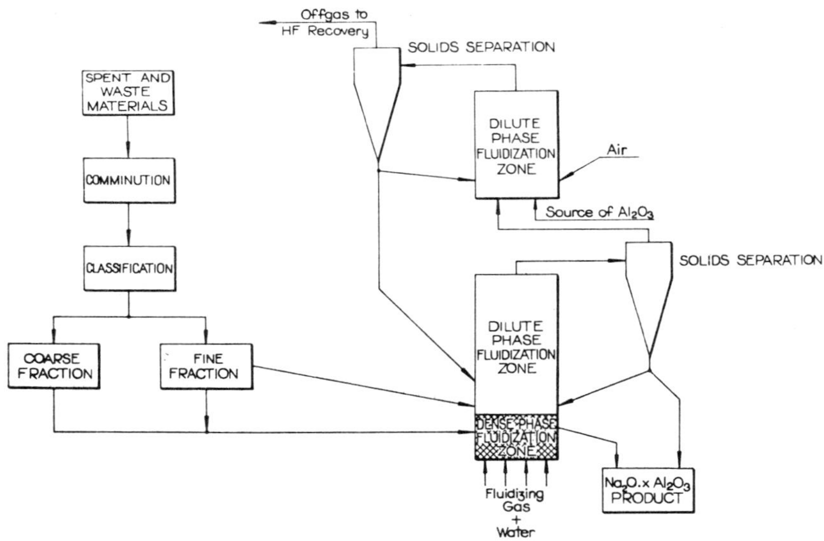

Source: U.S. Patent 4,158,701

I

IODINE FROM ACETOXYLATION PROCESS WASTES

Recent technical literature indicates the usefulness of vicinal glycol esters as precursors for certain important chemicals. For example, U.S. Patent No. 3,586,716 teaches the hydrolysis of esters such as ethylene glycol mono- and diacetate to ethylene glycol, and German Offenlegungsschrift No. 2,412,136 discloses a method of cracking propylene glycol monoacetate to propylene oxide.

These esters are produced by the acetoxylation of an olefin with oxygen and a carboxylic acid using a homogeneous liquid-phase catalyst system, usually one containing a halogen, most desirably iodine. When iodine is employed during the reaction, it reacts to form a variety of compounds the majority of which are organic. For reasons explained below, substantially all of the iodine must be removed from the esters before they are further utilized, and the iodine must be recycled to the reactor in an active form with as little loss as possible.

A large plant producing these esters requires 5,000 to 30,000 pounds of iodine to be fed to the reactor per hour. If the iodine recovery recycle efficiency is 95%, that is, 5% of the iodine charged to the reactor is not recovered from the effluent and recycled, the iodine make-up is 2.0 to 12.0 million pounds per year. With a 99% recovery and recycle efficiency, the annual make-up of iodine is 0.4 to 2.4 million pounds/year. However, iodine is in short supply (the 1974 U.S. consumption was 9 million pounds, 80% imported) and is expensive.

Furthermore, if iodine is not removed from the crude vicinal glycol esters, the iodine may contaminate the various end-products produced from these esters. For example, ethylene glycol produced from crude ethylene glycol acetates made with iodine catalyst but which are not treated contains too much iodine to be used as synthetic fiber feed-stock. Thus for an iodine-catalyzed acetoxyltion process to be technically and economically feasible, the esters must be highly purified, and the iodine catalyst recovery-recycle efficiency must be very high.

A process developed by *M.B. Sherwin and R. Hansen, U.S. Patent 4,087,623; May 2, 1978; assigned to Chem Systems, Inc.* for treating the acetoxylation reactor effluent has been developed which produces highly purified esters and concomitantly recovers and recycles the iodine catalyst at very high efficiency.

Briefly, the method consists of five steps.

> First, the acetoxylation reactor effluent is fractionated to produce several cuts, one of which is a crude ester cut containing some (approximately one-third to one-twentieth) of the iodine fed to the acetoxylation reactor. Additional fractionation cannot completely remove this remaining iodine (primarily organic iodine compounds) because of the close-

ness of the respective boiling points. The other cuts from fractionation, containing the rest of the iodine, are recycled to the reactor.

Second, the crude esters from fractionation are contacted in the liquid phase with certain basic materials to convert the contained iodine compounds to inorganic iodine salts.

Third, the esters are separated from the salts and are ready for further processing (e.g., cracking and hydrolysis).

Fourth, elemental iodine is liberated from these salts by a peroxidation reaction which also produces a base.

Fifth, the elemental iodine is separated from the peroxidation reaction mixture and is recycled to the acetoxylation reactor. The base from step 4 may be recycled to the second step.

IODINE FROM DEHYDROGENATION PROCESS WASTES

Processes wherein certain hydrocarbons are converted to less saturated hydrocarbons by dehydrogenation in which iodine is employed are well known in the art. These dehydrogenation reactions include, for example, the conversion of paraffins to olefins and diolefins, of olefins to diolefins, of certain paraffins to aromatics (dehydrocyclization) and of alkyl aromatics to vinyl aromatics including dehydrocoupling as in the conversion of toluene to stilbene. Such iodinative dehydrogenation processes employ iodine as such or iodine in conjunction with oxygen whereby the hydrogen iodide simultaneously formed in the reaction is converted under the reaction conditions to elemental iodine.

In some processes, there is also a concomitant production of organic and inorganic iodides as by-products. In order for such processes to be economically feasible, iodine utilization must be maintained at a high level which requires efficient recovery of the iodine from process streams containing it for re-use in the process.

Also, the removal of iodine from reaction effluents has a salutary effect in some instances. In the production of silbene, for example, by dehydrocoupling of toluene in the presence of iodine, the removal of inorganic iodine from the reaction effluent facilitates purification of the silbene by distillation to a product having >99% purity which is essentially free of iodine.

Moreover, it is well known that iodine, when absorbed into the human system, may cause various toxic symptoms, and that iodine compounds, such as methyliodide, ethyliodide, which are usually generated during various chemical reactions in the presence of iodine, are poisonous to the human body.

Therefore, at chemical plants which use iodine as a catalyst for reaction of organic substances, there exist problems which require immediate attention. The iodine should be recovered on an economical basis. Such recovery from plant waste will aid and control pollution problems.

A process developed by *M.M. McLane and R.A. Newson; U.S. Patent 4,036,940; July 19, 1977; assigned to Monsanto Company* is one in which iodine is recovered from a process stream containing it or iodine-containing compounds by passing the stream while in the vapor state in contact with a solid treating agent or adsorbent comprising alumina impregnated with the oxides of copper and chromium and thereafter subjecting the solid treating agent to heating to a temperature from about 400° to 600°C, while passing a stream containing air or oxygen over it. The resulting effluent gas containing iodine can be fed directly back to the process from which it was derived or otherwise treated as by sublimation, for example, if desired.

A process developed by *M. Ishigami, K. Arimoto, Y. Inoue and R. Fujii; U.S. Patent 3,992,510; November 16, 1976; assigned to Harima Chemicals, Inc., Japan* is a process for recovery of iodine from waste containing iodine or iodine compounds, comprising the steps of burning the waste in a combustion chamber, and scrubbing out the iodine or iodine compound in the resulting combustion gas with a basic aqueous solution of sodium thiosulfate.

Such a process is shown in Figure 86. Waste oil **5** (or waste water **6**, or waste gas **7**) containing iodine or iodine compound is fed into combustion chamber **1** to be burned. When the waste oil is less combustible or incombustible, some fuel oil may be admixed therewith and burned, keeping the temperature of the combustion chamber **1** at more than 500°C, and preferably more than 800°C. The waste oil **5** may be burned in the combustion chamber admixed with a fixed quantity of waste water **6** and/or waste gas **7**, all of which may contain iodine or iodine compound. The iodine compound contained in waste oil **5**, waste water **6**, or waste gas **7**, becomes free iodine by oxidation decomposition resulting from burning.

The resultant combustion gas **8** containing the free iodine is led to absorption chamber **2**, which is suitably filled with absorption liquor **12**, consisting essentially of a basic aqueous solution of sodium thiosulfate. The free iodine in the combustion gas **8** is absorbed and converted to sodium iodide by reacting with sodium thiosulfate. If desired, absorption liquor in absorption chamber **2** may be cooled by cooler **4** to an optimum temperature. After absorption treatment in the absorption chamber **2**, unabsorbed combustion gas **8** is further led to absorption tower **3** to absorb and react iodine remaining in the gas and thereafter the remaining gas may be discharged as exhaust gas **9** from tower **3** to the external air.

Absorption liquor **12** is continuously fed into absorption tower **3** from an external source not shown, and after circulation in tower **3**, is successively fed into the absorption chamber **2**. Thus, the absorption liquor is used to absorb any unreacted gas from chamber **2** before being supplied to chamber **2** for the primary reacting. After absorbing iodine in absorption chamber **2**, the liquor is discharged continuously from chamber **2** into stock system **11**, to recover sodium iodide. The concentration of sodium iodide in the recovered liquor is maintained at a substantially constant value.

Since iodine is irreversibly absorbed into the absorption liquor of sodium thiosulfate solution, both in the absorption chamber **2** and the absorption tower **3**, there is very little free iodine that is lost through exhaust gas **9**. It has been found that exhaust gas **9** contains iodine in an amount of less than 1 mg/m^3, and advantageously, the recovery rate is excellent.

Figure 86: Harima Chemicals Incorporated Process

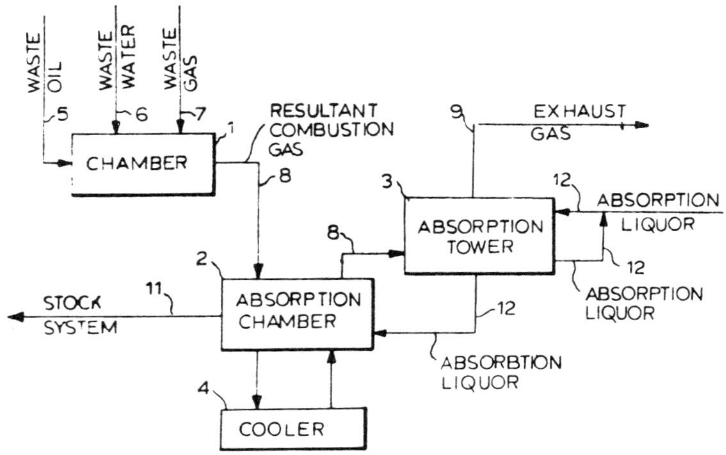

Source: U.S. Patent 3,992,510

The resultant liquor after reaction with iodine essentially contains sodium sulfate and sodium iodide with a small quantity of sodium carbonate, sodium bicarbonate and unreacted sodium thiosulfate. Iodine may be recovered from this resultant liquor by any conventionally known process. For example, after making the liquor acidic, chlorine gas may be blown into the liquor to precipitate free iodine; the process is simple and inexpensive.

IRIDIUM FROM SPENT CATALYSTS

It is well known that certain catalysts, such as reforming catalysts, may include one or several catalytically active metals, and particularly metals of the platinum series, platinum and iridium being among those used most frequently, and that the supports of these catalysts very often are aluminous, and most frequently formed of active alumina.

It goes without saying that the rejects from the fabrication of such catalysts as well as the used catalyst must be treated to recover therefrom at least the precious metals, and this must be as completely as possible.

Generally the methods recommended by prior art for treatment of these catalysts consist of either solubilizing the supports, leaving the unattacked precious metals to be found in the slurries, and the slurries then are treated with acids, or to solubilize all of these catalysts as completely as possible, then to proceed with the separations on the obtained solutions. The solubilization agents may be caustic soda lyes, used with or without pressure, and the current acids used alone or in mixture. The different metals than are recovered by treating the solutions of the metal salts, obtained directly or those corresponding to their recovery in the slurries, in case of a selective attack of the support.

Generally, these different methods include too large a number of operations which despite the precautions taken, cause substantial losses in precious metals. However, the precious metals frequently present are incompletely separated from each other, which requires further purification treatments which result in a further reduction of the recovery yield.

A process developed by *M. Thomas and J. Grosbois; U.S. Patent 4,069,040; January 17, 1978; assigned to Rhone-Poulenc Industries, France* is one in which the recovery of the constituent elements of catalysts containing precious metals, such as platinum and iridium deposited on aluminous supports is accomplished by a substantially total solubilization of the catalysts by means of hydrochloric acid and by extraction, for example by means of ion exchange resins. This system allows for a good separation of the platinum and the iridium.

The following is one specific example of the conduct of such a process. A quantity of 5 kilograms of catalyst is calcined for 2 hours at 400°C to destroy the organic substances, then attacked under reflux by 50 liters of 20% by weight (6N) hydrochloric acid and 0.1 liter of 58% by weight nitric acid in an externally heated glass lined reactor. After three hours of attack at a temperature of about 110°C, 0.25 liter formic acid is added and the heating is continued for 1 hour. The solution then is drawn and filtered.

A dry residue of 10.5 grams is obtained which essentially contains silica and alpha alumina, with traces of platinum and iridium being present in the residue. Considering its weight which is very small, this corresponds to a substantially complete placement in solution of the precious metals.

The clear solution obtained is maintained at a temperature of 70°C and then passed over a bed of a strong Duolite A 101D anion exchange resin, on which the platinum becomes strongly fixed, while the effluent solution contains aluminum and almost all of the iridium.

The resin containing the platinum is rinsed with an aqueous solution of diluted 0.1 M hydrochloric acid, then dried and calcined, by heating progressively to 800°C in an oxidizing atmosphere. That way, a metallic mass of 17.6 grams is recovered composed particularly of platinum and containing only very small amounts of iridium.

One liter of 58% nitric acid is added to the effluent solution previously collected and kept under reflux for 1 hour. This solution, in which iridium is thus brought to a valence of four, is passed over a second anion resin bed under the same conditions as for the operation of fixing the platinum. After calcination of the resin at 800°C, in a successively oxidizing and reducing atmosphere, a metallic mass of 2.35 grams iridium free of platinum is obtained.

The recovered quantities of precious metals substantially correspond with the total amount present in the treated catalyst.

A process developed by *R.F. Belt, R.C. Puttbach and C.T. Schreiber; U.S. Patent 4,022,443; May 10, 1977; assigned to Litton Systems, Incorporated* is one for the concentration and recovery of platinum family metals from ceramic materials. The metal desired is first transported from the ceramic material through the utilization of a temperature gradient in a specially adapted furnace utilizing the volatile characteristics of the oxide of the metal desired. The volatile oxide is deposited upon a nucleation surface where it is reduced by means of hydrogen as the furace cools to recover the metal.

A suitable form of apparatus for the conduct of such a process is shown in Figure 87.

Figure 87: Litton Systems Inc. Apparatus for the Concentration and Recovery of Platinum Family Metals from Ceramic Materials

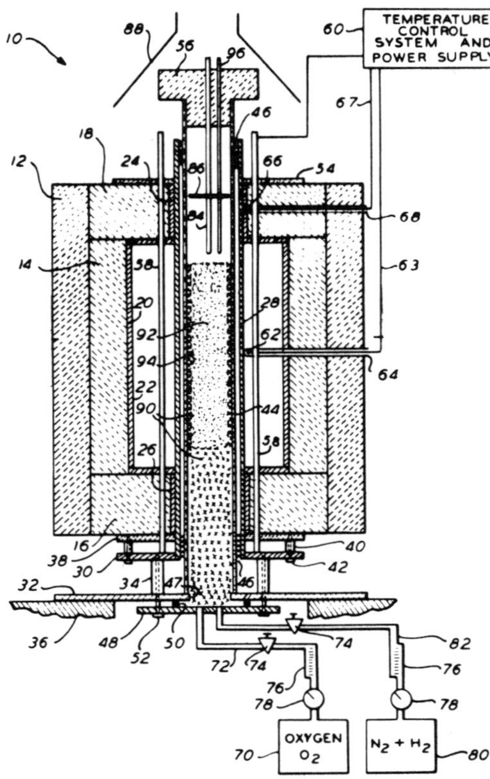

Source: U.S. Patent 4,022,443

Referring now to the drawing, a furnace for the concentration and recovery of platinum family metals, including pure iridium metal (Ir) and/or iridium oxide (IrO_2) is shown generally at **10** having an outer insulating cover **12** constructed from ceramic fiber. The outer cover **12** is typically 30 inches in diameter and 40 inches long, constructed tubularly to surround an inner insulating cover **14** of firebrick that is provided with a bottom flange **16** and a separate top flange **18** which forms an inner chamber **20**.

Inner chamber **20** receives a cylindrically shaped alumina casting **22** which is closed at each end to line the chamber **20** and provide the inner support structure for the furnace **10**. Aligned with the longitudinal axis of the inner support structure **22** are upper and lower tubular extensions **24** and **26**, respectively, which are aligned to receive a tubular core member **28** that passes through the full vertical height of the furnace outer and inner insulating covers **12** and **14** to close the inner chamber **20**. The upper and lower extensions **24** and **26** and the core **28** are all formed from alumina.

The alumina core **28** is attached to the inner opening of a washer shaped steel support plate **30** which, in turn, is supported on a steel base plate **32** by standoffs **34**. The base plate **32** rests upon a suitable frame **36**. A second washer shaped support plate **38**, constructed from a refractory material known as maronite, is separated from the base plate **32** by standoffs **40** also constructed from maronite. The standoffs **40** are held in place by a ring of steel bolts **42** which pass from the lower surface of the plate **30**, through the standoffs **40** and into plate **38**. The inner insulating cover **14** and inner support structure **22** rest upon the support plate **38**.

Passing through the center of the core **28** and extending beyond each end thereof is a tube **44** typically constructed from a refractory material, such as mullite, with a length of 60 inches and an inside diameter which may vary from 2.75 to 6.25 inches. The mullite tube **44** is separated from the core **28** by ceramic fiber **46** inserted at the ends of the core **28** between the inner surface of the core and the outer surface of the tube **44**.

The refractory tube **44** is attached to the base plate **32** and aligned with a centrally located opening **47** within the plate **32** which is closed by a sealing plate **48**. The sealing plate **48** seals the base plate opening **47** by compressing a rubber O-ring **50** as a ring of bolts **52** is tightened into the base plate **32**.

A top washer shaped sealing plate **54**, constructed from maronite, rests tightly against the core member **28** on the top flange **18** of the inner insulating cover **14**. The mullite tube **44** is closed at its upper end by a cap **56** constructed from firebrick to complete the structure of the furnace **10**.

Supported upon support plate **30** and passing through the support plate **38**, lower flange **16**, inner support structure **22**, top flange **18** and top sealing plate **54**, are six heating elements **58**, constructed from silicon carbide. These elements parallel the core **28** and refractory tube **44** and pass through the inner chamber **20** formed by the inner support structure **22**. The heaters are energized from a power supply **60** which provides power thereto under the control of a control thermocouple **62**. The control thermocouple **62** mounts in a thermocouple tube **64** that passes through the outer insulating cover **12**, inner insulating cover **14** and inner support structure **22** to rest against the furnace core tube **28**. Wires **63** connect the thermocouple **62** to the power supply **60**.

A second control thermocouple **66** may be mounted within a second thermocouple tube **68** which passes through the outer insulating cover **12** and the top flange **18** of inner cover **14** in the upper areas of the furnace. The second control thermocouple **66** connects to the power supply **60** via wires **67** and is used to monitor the temperature gradient between the hot zone and the cold zone.

The so-called hot zone is found within the inner support structure **22** generally in the area of the thermocouple **62**, while the cold zone is found in the general area of the refractory tube **44** which passes through the top flange **18**. The temperature gradient is used to encourage the transportation of the iridium oxide or other oxides of platinum group metals which are volatile.

Another method of testing the temperature gradient during the heating of the furnace is to simply remove the firebrick cap **56** and slowly insert a ceramic protected thermocouple along the longitudinal axis of the refractory tube **44**.

A pressure tank **70** supplies oxygen to the chamber formed by the refractory tube **44**, sealing plate **48** and cap **56** through a suitable connecting tube **72** which includes a control valve **74**, flow meter **76** and pressure gauge **78**. A second pressure tank **80** supplies a mixture of nitrogen and hydrogen through the sealing plate **48** via a tube **82** which also incorporates a control valve **74**, flow meter **76** and pressure gauge **78**. The mixture of nitrogen and hydrogen may vary, but in the preferred embodiment, a mixture of 85% nitrogen to 15% hydrogen by volume has been satisfactorily used.

The flow rate of the oxygen or the mixture of nitrogen and hydrogen is adjusted by valves **74** and monitored by flow meters **76**. As each gas passes up the refractory tube **44**, it passes around a nucleation tube **84** which is mounted along the longitudinal axis of the refractory tube **44** within the firebrick cap **56**. The nucleation tube **84** mounts a nucleation disc **86** which substantially closes the cyclindrical cross section of the tube **44**. After the gas flows around the nucleation disc **86**, it passes around and through the firebrick cap **56** to be exhausted through an exhaust hood **88**.

The refractory tube **44** is loaded with clear zirconium oxide (ZrO_2) to a depth of approximately two inches inside the chamber formed by the inner support structure **22**, as shown at **90**. The ceramic material **92** which bears metals from the platinum family is then placed within a paper tube **94**, shown by dashed lines, and inserted along the longitudinal axis of refractory tube **44**. Clear zirconium oxide **90** is then placed between the outer surface of the paper tube and the inner surface of the refractory tube to complete the charging of the furnace.

The furnace is then closed by placing the nucleation tube **84** and disc **86** in the position shown as the cap **56** is placed in the top of tube **44**. The nucleation tube **84** and disc **86** are constructed from mullite. An alumina rod **96** may be inserted through the firebrick cap **56** and the nucleation disc **86** to monitor the transportation of the platinum family metal from within the ceramic bearing material **92** in the hot zone of the furnace toward the cold zone.

The ceramic material **92** may be in a powder or grog form and may be aluminum oxide, zirconium oxide, a combination of both or any other ceramic material which contains platinum family metals whose oxides are volatile. The latter include: iridium (Ir), ruthenium (Ru), rhodium (Rh), and osmium (Os).

These platinum family metals share one common feature in that their oxides are volatile below 1200°C. The grog or powder **92** bearing the platinum family metal is located within the high temperature zone created by the resistance heating elements **58**. In this zone the temperature is increased until it reaches 1200° to 1475°C, at which point the platinum family metal within the ceramic powder, for example, iridium metal within aluminum oxide, is oxidized to form an oxide which is vaporized and carried by the flowing oxygen from the bottom to the top of the refractory tube **44**.

As the flowing iridium oxide, for example, passes into the area of the nucleation tube **84** and disc **86**, the reduced temperatures within this relatively cold zone, which is beyond the heating elements **58**, causes the iridium oxide to collect upon the nucleation tube and disc. The disc **86** may be moved up or down within the cold zone to additionally decrease or increase the cold zone temperature. It should be apparent that the hot zone configuration cannot be easily adjusted once its maximum temperature is established. However, hot zone configuration can be affected to some degree by the tube size, loading of the charge or the shape of the charge.

In operation, the furnance **10** is charged as described above. Once the nucleation tube **84**, disc **86** and cap **56** are replaced, the furnace is turned on and heated at a rate of 30° to 40°C per hour to a temperature of about 1350°C. It will be understood that the paper tube **94** will burn off at a temperature well below the temperature required to combine the flowing oxygen with the platinum family metals. The temperature build-up within the furnace **10** is monitored through the utilization of the thermocouples **62** and **66**.

As the temperature within the hot zone reaches approximately 800°C, the flow of oxygen is initiated by opening the control valve **74** to allow the oxygen to flow at a rate which may vary from two to twenty cubic feet per hour. At this time the temperature gradient within the furnace is also checked.

After the temperature of 1200° to 1475° is reached in the hot zone, the furnace temperature is monitored on a daily basis during operation. The alumina rod **96** and/or the nucleation tube **84** and disc **86** are removed daily to monitor the transportation progress of the volatile oxides. As the rods and disc are removed, any excess iridium or iridium oxide, for example, which has been transported onto the surfaces in the cold zone may be recovered by scraping the iridium into a container.

When the transport of the iridium, in the form of iridium oxide, from the grog **92** has been completed, the flow of oxygen is terminated; and the forming gas comprising nitrogen and hydrogen is turned on through the manipulation of the second control valve **74**. The forming gas may flow at a rate which varies between one to four cubic feet per hour. The furnace is then allowed to cool at a rate of about 30° to 40°C per hour. Once the furnace has cooled, the nucleation tube **84** and disc **86** are removed; and the iridium deposited thereon recovered. The iridium is now in the form of a pure iridium as the oxygen associated therewith has been carried off in the form of water vapor.

The bottom seal plate **48** is then removed by removing bolts **52** which allows the charge to drop from the furnace. A flat metal plate (not shown) replaces the sealing plate **48** at the bottom of the mullite tube **44**. Using an alumina rod (also not shown), the iridium is then scraped off the walls of the mullite tube **44** and collected on the plate just placed in the position of the sealing plate **48.** After scraping, the plate is removed and the iridium recovered.

IRON FROM OXYGEN CONVERTER SLAGS

In certain areas often large amounts of slag are found, originating, i.e., from the method of the oxygen steel production, the so-called L.D.-slag. This slag, contrary to blast furnace slag, after cooling (quenching) is not hydraulic and so not very valuable. The possibility of recycling the slag in a blast furnace to extract the iron is limited as a result of accumulation of undesired elements. The waste piles of this slag can so be considered as being detrimental to the environment.

The main components of the L.D.-slag are usually within the following limits:

	Percent by Weight
SiO_2	7-18
Al_2O_3	0.3-2
CaO	35-55
MgO	2.6-6
Fe + FeO_x	15-40
MnO_x	2-6.5
P_2O_5	1-2.5

The iron content of the slag is such that extraction of iron from the slag appears useful. Extraction of the iron from the slag by a simple reduction treatment is difficult and economically little remunerative. In a reducing atmosphere at a temperature of about 1200°C, it is true, metallic iron is formed, but this iron remains finely divided in the pores of the remaining slag because the temperature at which this slag becomes liquid is considerably increased by reducing the iron compounds. The iron can hardly be separated from the remaining slag.

A method developed by *A.G.J. Wolfs and L.H. Goessens; U.S. Patent 4,174,961; November 20, 1979; assigned to Eerste Nederlandse Cement Industrie (ENCI) NV, Netherlands and C.B.R. Cementbedrijven NV, Belgium* is one for recovering elemental iron and manganese present in oxide form in waste slag obtained from oxygen steel production. The metal oxides are reduced in the presence of carbon or carbon combined with a coal mine waste material and an Al_2O_3 containing additive present in the form of a silicate.

Such a process is shown in Figure 88. In the flow-sheet at **11** L.D.-slag and at **12** schist (e.g. the waste stone of a coal mine) are supplied to a crusher **13** crushing the raw materials. At **14** the broken product is supplied to a rotary kiln **15** wherein the iron compounds of the L.D.-slag are reduced by the carbon being present with the schist, optionally supplemented with another reducing agent. At **16** heating gas is supplied to bring the mass in the furnace at the desired temperature. In the furnace the mass finally melts at about 1400°C and at **17** from the melt separated pig iron can be drained off. By means of a drain **18** finally the remaining new slag is extracted. The slag is brought to a mixer **110** via route **19** in liquid form or after cooling in solid form.

Figure 88: Dutch-Belgian Process for Simultaneously Recovering Iron and Cement Clinker from Oxygen Converter Slags

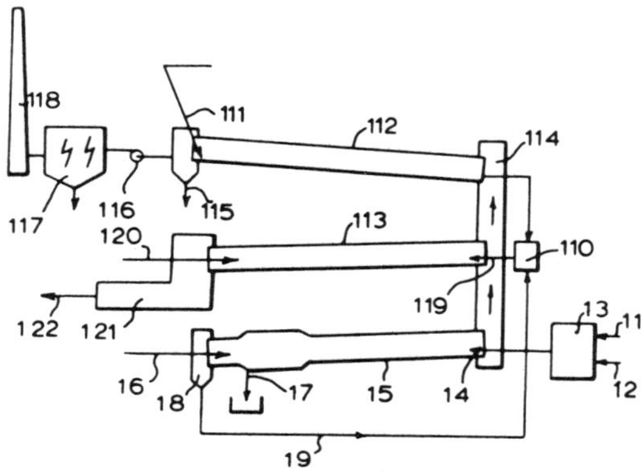

Source: U.S. Patent 4,174,961

At **111** wet or dry finely divided limestone or chalk is supplied to a heat exchanger **112** (e.g. a drum) and is preheated. For the benefit of this preheating flue gases of the rotary kiln **15** and the sintering furnace **113** can be used, which gases are supplied via a flue gas channel **114**. The heat exchanger **112** is of course provided with a dust discharge **115**, a flue gas fan **116** and an electrofilter **117** for catching the finest dust while the flue gases are released in the atmosphere via chimney **118**.

The preheated meal from the heat exchanger **112** is also introduced into the mixer **110** (optionally with dust separated in **115** and **117**) and after mixing with the slag which was extracted at **18** is supplied to the sintering furnace **113**. The supplied hot mixture **119** of slag and limestone meal is heated in the sintering furnace **113** up to about 1400°C, for which purpose at **120** heating gas is introduced. The sintered product is cooled in a cooler **121** and at **122** is discharged as portland cement clinker for further processing to cement.

A process developed by *P.R. Pack; U.S. Patent 4,177,062; December 4, 1979; assigned to Harsco Corporation* involves contacting oil-coated mill scale with hot slag derived from the steel making process whereby the oil is ignited and flashed away and the scale commingled with the slag for subsequent processing, recovery of metallic values, and recycle to the steelmaking operation.

IRON FROM RED MUD FROM BAYER PROCESS

One process, universally used today, for refining bauxite is the Bayer process. By this process bauxite, as mined, is finely ground and charged to digesters with a hot caustic soda solution to put the aluminum in solution as sodium aluminate.

After digestion, the caustic and soluble sodium aluminate are separated by settling and filtration from insoluble residue. The residue is red mud about 3 tons of which are formed for each ton of aluminum product. The mud appears as fine particles (less than a micron size).

The red mud, depleted of alumina, ordinarily imposes severe disposal problems inasmuch as the caustic content is detrimental to plant and animal life. Also large unsightly ponds are needed as settling ponds for storage of the waste material. Red mud generally comprises oxides of aluminum, iron, silicon, titanium, calcium and sodium. Attempts have been made to utilize the red mud, generally without much success.

A process developed by *G. Dobos, Z. Felföldi, G. Horváth, G. Kaptay, Z. Osvald and K. Solymár; U.S. Patent 3,989,513; November 2, 1976; assigned to Magyar Aluminiumipari Troszt, Hungary* is one in which limestone, hydrated lime or other CaO containing material is added to the red mud in certain molar proportions to the ingredients of the red mud, then the mixture is further mixed with a reducing agent, and the entire mixture is subjected to a reducing heat treatment during the course of which the mix becomes softened or molten. The iron and the slag are separated from each other while the mix is still molten, or after it has been cooled, and then the Na and Al content of the slag is extracted with a sodium carbonate solution containing a caustic, or with a sodium carbonate solution and a solution of a caustic in two steps.

IRON FROM ROLLING MILL SCALE

See "Iron from Oxygen Converter Slags."

IRON FROM SCRAP IRON

About 5% of the total tonnage of iron ore and agglomerates consumed in U.S. iron- and steelmaking furnaces is derived from secondary materials such as blast furnace dust, mill scale, and steel-furnace slag. Flue dust is usually recycled through sintering plants. Mill scale and slag may be recycled through a sintering plant or charged directly to the furnaces. Slag scrap and other scrap are also charged to the blast furnace. The flow of materials in 1976 as indicated from consumers' reports is diagramed in Figure 89. Approximate average iron contents of materials recycled in 1976 were as follows: Flue dust, 50%; mill scale, 65.5%; slag, 23%; and slag scrap, 53%.

IRON FROM SCRAP TIRES

See "Blast Furnace Fuel from Scrap Tires" in reference (1).

Figure 89: Flow of U.S. Primary and Secondary Iron-Bearing Materials at Iron and Steelmaking Plants, 1976
(Million Short Tons of Contained Iron)

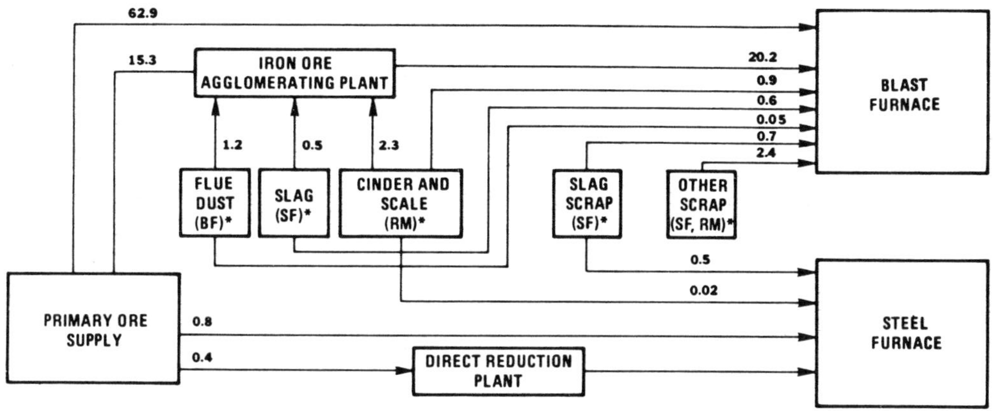

*PRINCIPAL SOURCES OF SECONDARY MATERIALS:

(BF) BLAST FURNACE
(SF) STEEL FURNACE
(RM) ROLLING MILLS

Source: Reference (16)

IRON FROM SPENT PICKLE LIQUOR

See also "Sulfuric Acid from Spent Pickle Liquor."

A process has been developed by *M. Watanabe and S. Nishimura; U.S. Patent 4,113,588; September 12, 1978 and U.S. Patent 4,177,119; December 4, 1979; both assigned to Solex Research Corporation of Japan* for recovering iron and H_2SO_4 or HCl from waste pickling liquors which contain large amounts of iron ions.

The process involves treating waste H_2SO_4 or HCl containing large amounts of Fe ions and particularly obtaining reusable H_2SO_4 or HCl and high purity electrolytic iron or iron hydroxide using both solvent extraction and diaphragm electrolysis techniques. In general, since most of Fe ions contained in the waste acids are present as Fe^{2+} ions, the oxidation method of Fe^{2+} ions to Fe^{3+} ions with high pressured air or oxygen, or H_2O is often used.

It involves in the case of the recovery of waste HCl:

(1) converting the Fe^{2+} ions in the waste to Fe^{3+} ions in an oxidation-reduction cell wherein the waste solution is the anolyte and the catholyte is an HCl solution containing Fe^{3+} ions, the used catholyte thus containing Fe^{2+} ions;

(2) extracting the Fe^{3+} ions from the anolyte from step (1) with an organic solvent **C** to form an $FeCl_3$-complex therein and to produce an aqueous raffinate;

(3) regenerating solvent **C** by contacting it with an aqueous extractant to extract the Fe^{3+} therefrom and using the Fe^{3+} containing extractant as the catholyte in step (1);

(4) recovering the iron values in the used catholyte from step (1) by diaphragm electrolysis; and

(5) subjecting the raffinate from step (2) to the diaphragm electrolysis of step (4) to recover HCl therefrom.

Figure 90 shows such a process for reusable HCl recovery from waste hydrochloric acid containing a relatively small amount of free acid; and Figure 91 shows such a process for reusable HCl recovery from waste hydrochloric acid containing large amounts of free acid and total Cl^- ions.

The organic solvent **A** is a solution of a compound selected from the group consisting of dialkyl phosphoric acid, mixtures of dialkyl phosphoric acid and naphthenic acid or an oxime together with 3 to 15% of a higher alcohol having from 5 to 15 carbon atoms as a modifier and a petroleum hydrocarbon as a diluent.

The organic solvent **B** is a phosphoric acid ester or amine together with a petroleum hydrocarbon as a diluent.

The organic solvent **C** is a phosphoric acid ester or amine together with a petroleum hydrocarbon as a diluent.

Figure 92 shows a flow sheet of the reusable H_2SO_4 recovery from waste sulfuric acid.

Figure 90: Solex Research Corporation of Japan Process for Iron and HCl Recovery from Pickling Liquors Containing Little Free Acid

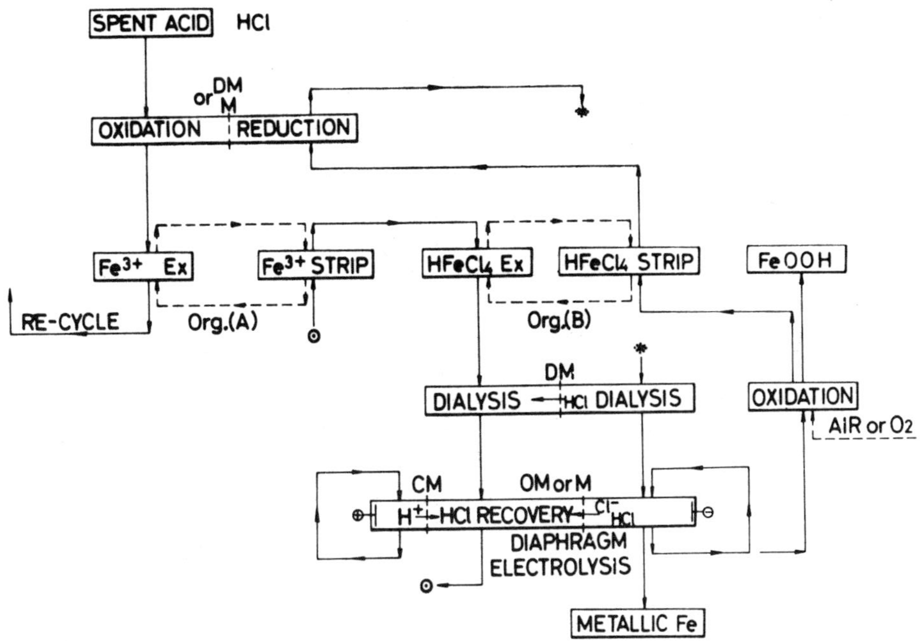

Source: U.S. Patent 4,113,588

Figure 91: Solex Research Corporation of Japan Process for Iron and HCl Recovery from Pickle Liquors Containing Large Amounts of Free Acid

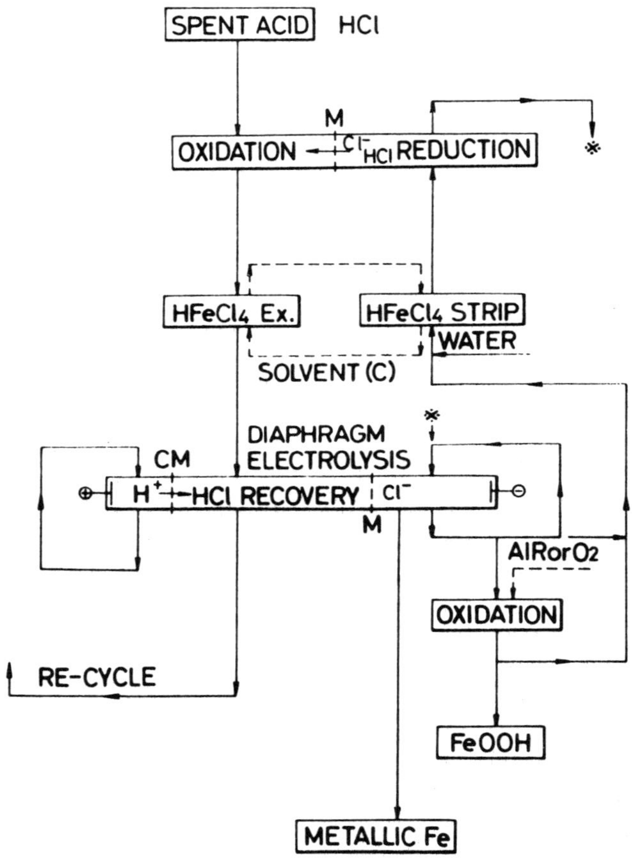

Source: U.S. Patent 4,113,588

Figure 92: Solex Research Corporation of Japan Process for Recovery of Iron and H_2SO_4 from Pickling Liquors

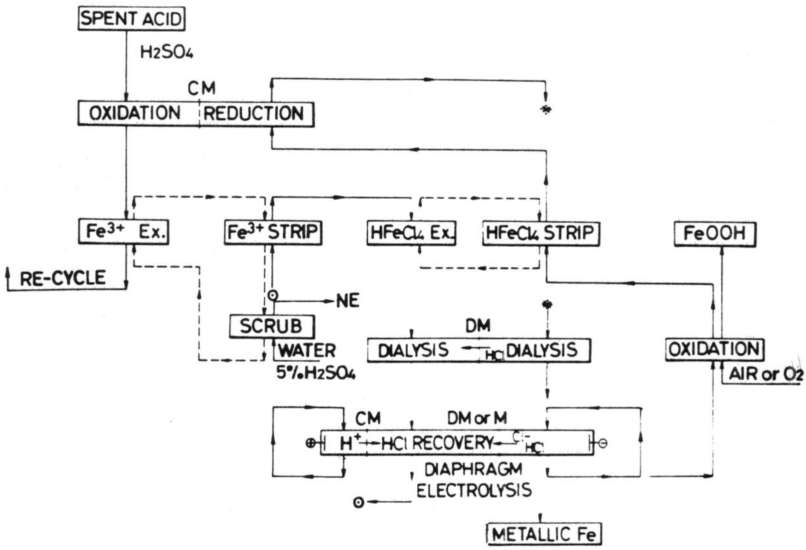

Source: U.S. Patent 4,113,588

IRON FROM STEEL MILL FURNACE DUSTS

Steel mill operations generally involve the production of liquid pig iron from ore with coke (ordinarily produced on site) and flux, the conversion of this liquid iron to steel, and the formation of the steel into shapes. Each one of these operations produces particulate solid wastes which must be disposed of by recovery of the metallics for best economics. In the past, the disposal, generally by the sintering process, was a source of pollution and much of the carbon in the waste was consumed.

In a typical iron producing operation, iron oxide ore, flux and the reductant coke are charged into a blast furnace, where the charge is smelted and the iron oxide is reduced to iron, resulting in liquid pig iron and a slag which are recovered as molten products. The high temperature and the rush of air (at velocities in excess of 10 feet per second), carry off large quantities of fines, which are removed in dust recovery systems either dry or as a sludge.

This blast furnace dust is a mixture of iron oxides, carbon, lime dust, and ash components. In either form, it presents a serious disposal problem, since about 2.5 tons of dry weight dust and sludge are produced for each 100 tons of iron.

When the pig iron is converted into steel, in the open hearth or basic oxygen furnace, high iron oxide dusts are produced. However, they too present a serious disposal problem, as a major proportion is submicron in particle size and since about 1.5 tons are produced for each 100 tons of raw steel.

Another source of particulate waste is that obtained from the steel-forming processes. The steel ingots, billets and other semifinished forms are reheated in furnaces and converted into shapes such as sheets, bars and structural forms in continuous rolling mills, which force hot oxidized particles, known as scale, from the billets or other shapes during the operation.

This mill scale is largely particulate magnetic iron oxide (Fe_3O_4) and generally contains varying quantities of oil and grease; hence, it too presents a disposal problem.

Dumping or sale of these wastes is not economical, as the metallic oxide and carbon units contained therein are valuable. Moreover, it is gradually being prohibited by statute. Sintering has been the most common means of agglomeration heretofore practiced, but because it involves high temperatures and high gas volume has been a very serious air pollution source. The correction of this problem has resulted in ever increasing cost for construction and operation of pollution control facilities.

The sintering process further burns (oxidizes) away much of the carbon in order to produce a semiplastic condition, thus resulting in loss of substantial heat values. Hot briquetting, though widely discussed, has not proven commercially practical, as it is very expensive and cannot successfully agglomerate carbon fines.

Processes using varying types and amounts of cement have added slag-forming ingredients, and all have displayed a loss of bonding strength at high temperatures. It is therefore evident that the most desirable agglomeration should be one which can be operated at low temperatures and low gas volumes in order to provide a low level of pollution.

The process should also be capable of retaining all valuable ingredients while producing an agglomerate that is stable at high temperatures with good load-bearing strength and of uniform size.

A process developed by *H.W. Maurice and J.K. Pearce; U.S. Patent 3,948,644; April 6, 1976; assigned to Pennsylvania Engineering Corporation* is one in which the flue gases from a steel refining furnace or vessel are delivered to a gas cleaning system in which the dust is removed and collected in slurry form. The slurry is dried and reduced to dust again after which it is delivered to a storage vessel.

The dust is transported pneumatically from the storage vessel to tuyeres in the steel refining vessel and thereby injected back into the molten metal within the vessel. The dust is rich in iron oxide and contributes to the steel yield from the refining vessel.

Such a process is shown in Figure 93.

Figure 93: Pennsylvania Engineering Corporation Process for Recycle of Steel Refining Furnace Dust

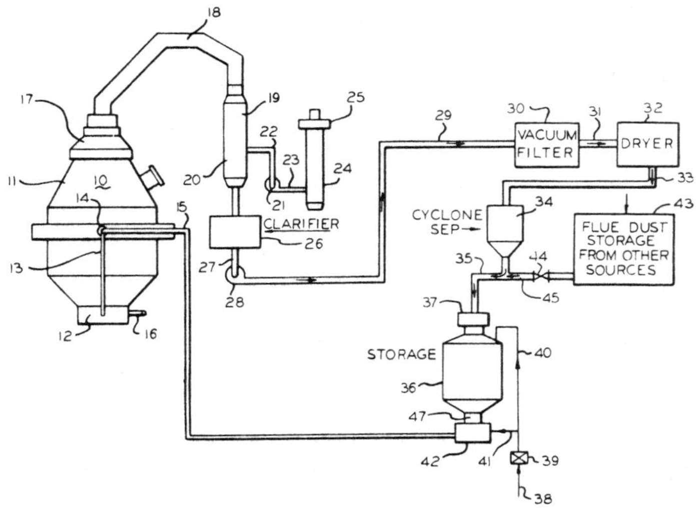

U.S. Patent 3,948,644

Although the new dust reclaiming method and apparatus may be used in connection with various steel converter vessels and furnaces such as bottom and top blown basic oxygen converters and open hearth furnaces, for the sake of brevity the process will be illustrated in connection with a bottom blown steel converter vessel which is designated generally in the drawing by the reference numeral **10**. Vessel **10** comprises a metal shell **11** which is lined with refractory material in a known manner.

There is a chamber **12** at the bottom of vessel **10** and this chamber contains two different headers, not shown, which are for distributing gases and gases containing finely divided materials to a set of tuyeres, not shown, which project through the bottom of the vessel. The bottom blown tuyere arrangement, which is known to those who are currently involved in the steel making arts requires no further elaboration. Suffice is to say that the tuyeres each have two independent passageways for injecting gases and gases in which finely divided materials are entrained into the molten metal within vessel **10**.

Each tuyere is comprised of two concentric pipes, the center one of which delivers oxygen and finely divided materials entrained therein into the molten metal within the vessel. The other passageway which results from the second pipe being concentric with the first is used to inject a hydrocarbon gas into the melt concurrently with injection of oxygen.

The hydrocarbon cracks into its hydrogen and carbon constituents where the gas evolves from the tuyere at its interface with the molten metal. This cracking process is endothermic and results in cooling the tuyeres so that they do not deteriorate under the intense heat that results from the reaction between oxygen and constituents of the melt which is being refined.

Gases such as oxygen containing finely divided material are delivered to the tuyeres by means of a pipe **13** which runs down the side of the vessel and connects with the header within chamber **12**. Pipe **13** is the outlet of a rotary joint **14**. The rotary joint connects to a stationary pipe **15** which is in turn connected to one or more sources of oxygen or other gases and gases in which finely divided materials are entrained, as will be explained more fully hereinafter. The hydrocarbon gas may be delivered to the tuyeres through a pipe **16** which may lead to a rotary joint, not shown, on the rear side of vessel **10**. This rotary joint connects to a pressure regulated source, not shown, of hydrocarbon gas such as propane.

Vessel **10** is supported in a trunnion ring which is adapted to tilt on a horizontal axis to thereby tilt the vessel with it. The trunnion pins, their journals, and the tilt drive mechanism are not shown because they are conventional.

Pig iron is converted to steel in vessel **10** by injecting oxygen into the hot molten metal along with other gases and solids which are needed for effectuating the necessary metallurgical reactions. Thus, gases and materials are injected into the molten metal and large quantities of reaction gases and smoke evolve therefrom. These gases and smoke are collected by a hood **17** which is situated over the mouth of vessel **10**. Hood **17** is coupled with a duct **18** which delivers the dirty gases to a wet scrubber **19**.

Scrubbers suitable for this duty are well known to those involved in the steel making art. It is sufficient for present purposes to observe that the scrubber precipitates the finely divided solids from the gases whereupon the solids settle in slurry form in the bottom region **20** of the scrubber. The separated gases are withdrawn from the scrubber by a blower **21** whose inlet pipe **22** leads to the scrubber and whose outlet pipe **23** leads to a stack or chimney **24**. This stack is equipped with a burner **25** at which the carbon monoxide and other combustible gases are completely oxidized or burned and discharged to the atmosphere. In some installations, the gases are delivered by blower **21** to suitable storage vessels for ultimate use as a fuel which can be done because of the high combustible carbon monoxide content of the gases.

The slurry which is collected in the bottom region **20** of scrubber **19** and which includes an appreciable portion of finely divided iron oxide is delivered to a clarifier **26** which is shown in block form because it is a known device. In clarifier **26** much of the water is separated from the slurry and more concentrated slurry is withdrawn through a pipe **27** which connects with the input of a pump **28**. Pump **28** conveys the heavy slurry through a conduit **29** to a vacuum filter **30**.

Vacuum filters are known to those involved in the steel making arts and need not be described in greater detail. It is sufficient to say that additional water is removed from the mass of finely dividied solids in the vacuum filter **30**. From filter **30** the semidried material is transported through a conduit **31** to a dryer **32** wherein substantially all of the residual water is driven from the mass of finely divided material. Various fuel fired dryers are commercially available for performing the duty required by dryer **32** so it need not be described in any greater detail.

After drying, the coalesced solids are delivered by means of a conduit **33** to a separator **34** which may be of the well known cyclone type. Separators of this type further dry the particulate material which is passed through them and are capable of separating particles of different masses. In the present example, however, separation is inconsequential and the finely divided materials are delivered from the separator through a conduit **35** to a storage vessel **36**. The storage vessel **36** is the ultimate destination of the finely divided materials for that part of the system which has been described up to this point, but it will be understood that there may be intermediate bins or other suitable vessels, not shown, interposed between separator **34** and pressure vessel **36** so as to be able to accommodate and store all of the finely divided material that is produced by the system.

The fine materials containing a high percentage of iron oxide are admitted into storage vessel **36** through a gas-lock type valve **37** at the top of the vessel. Vessel **36**, is of course, depressurized while finely divided material is being admitted to it.

Vessel **36** may be pressurized from a pipe **38** which leads to a high pressure source of gas, not shown, of any type that is useful in connection with refining in vessel **10**, but most generally oxygen is used. The gas is delivered through a regulating valve **39** to a line **40**, which communicates with the interior of the storage vessel to pressurize the volume above the finely divided materials which are in its bottom region.

A branch line **41** leads to a chamber **42** in which there is suitable known apparatus for entraining finely divided solids coming down throat **47** from the storage vessel in gas which is delivered from pipe **41**. The gas entrained finely divided material which in this case is rich in iron oxide is delivered from chamber **42** by means of pipe **15** to the rotary joint **14** and ultimately to the tuyeres in the bottom of vessel **10**. The finely divided solids are thereby injected into the molten metal.

Of course, the iron oxide so injected is, to a large extent, reduced to pure iron which contributes to the yield of the batch of steel which is being produced in the converter vessel. Thus, iron oxide which left vessel **10** from some preceding heat in the form of smoke is returned and reclaimed to perform a useful purpose in a subsequent heat.

Although various kinds of apparatus were demonstrated as being useful for effecting drying of the slurry between the wet gas scrubber **19** and storage vessel **36** it will be understood that the apparatus is merely suggestive and that the primary object is to deliver the finely divided solids derived from the smoke to a storage vessel in dry form so as to facilitate injection into a converter vessel for reuse. It should also be understood that the method hereinabove described is applicable to any type of converter vessel or furnace which is adapted for direct injection of powdered materials to the interior thereof.

Furthermore, the finely divided particulate matter may be obtained from one kind of refining system such as an open hearth furnace, a blast furnace, or a top-blown basic oxygen furnace and then delivered through valve **44** and pipe **45** to a storage bin **43** for further delivery to pressure vessel **36** from which the dust or particulate matter is transported to the bottom tuyeres for use in another vessel such as vessel **10**.

Thus, the important aspect of the process is realized which is to dispose of dust from any metallurgical process and to abate pollution by injecting the dust directly into molten metal through submerged tuyeres irrespective of the type of metallurgical vessel to which such tuyeres may be adapted.

Those skilled in the art will appreciate that the bottom blown converter vessel system herein described is normally associated with other storage vessels similar to storage vessel **36** and that their outputs are connected to pipe **15** which serves as a common header and enables delivery of various finely divided materials to the converter vessel on demand.

A process developed by *J.W. Kreiger and C.E. Jablonski; U.S. Patent 4,003,736; January 18, 1977; assigned to Bethlehem Steel Corporation* is one for preparing strong, stable green balls of dry collected iron-bearing fume, which contains not less than about 8% combined amount of lime (CaO) and magnesia (MgO). The fume is charged to a balling device where there is added to the fume a quantity of water equal to both the stoichiometric amount required to completely hydrate such lime and magnesia and to produce in the green balls a water content between about 7 to 12%. The residence time of the fume on the balling device is not less than about 8 minutes.

The product consists of strong stable balls suitable for charging into steelmaking furnaces.

A suitable form of apparatus for the conduct of such a process is shown in Figure 94. The drawings show a typical balling machine **10** supported on base **11** and having drive **12** and disc **13**, which has bottom **14** and side **15** and rotates in a clockwise manner, as shown by the arrow. Iron-bearing oxygen steelmaking fume particles **16** are fed on conveyor **17** and discharged from its end onto a lower portion of disc **13**. Some of the fume particles fall directly into the trough formed by bottom **14** and side **15**, while the remainder of the particles fall onto and adhere to small balls formed of particles which have previously fallen on such disc.

As disc **13** rotates, it carries with it the fume particles and the small balls with their adherent fume particles which are sprayed with water from nozzles **18** that are positioned at various locations above disc **13**. By continuously spraying water onto the fume and the balls, layers of fume are successively deposited one on top of another until the fume particles agglomerate and form balls. In addition the water causes the lime (CaO) and magnesia (MgO) to hydrate. The residence time of the fume on the disc is dependent upon several factors, as is well known to those skilled in the art, including the disc angle, its rotational speed, the rate of feed of fume to the disc, the size of the balls desired, etc.

Any suitable prior art method for balling may be used provided there is added to the fume the stoichiometric amount of water necessary to hydrate the lime (CaO) and magnesia (MgO) and an additional amount of water sufficient to provide between about 7% and 12% water in the green balls.

Figure 94: Elevation and End View of Bethlehem Steel Corporation Balling Device for Handling Steelmaking Dusts for Re-Use in Furnaces

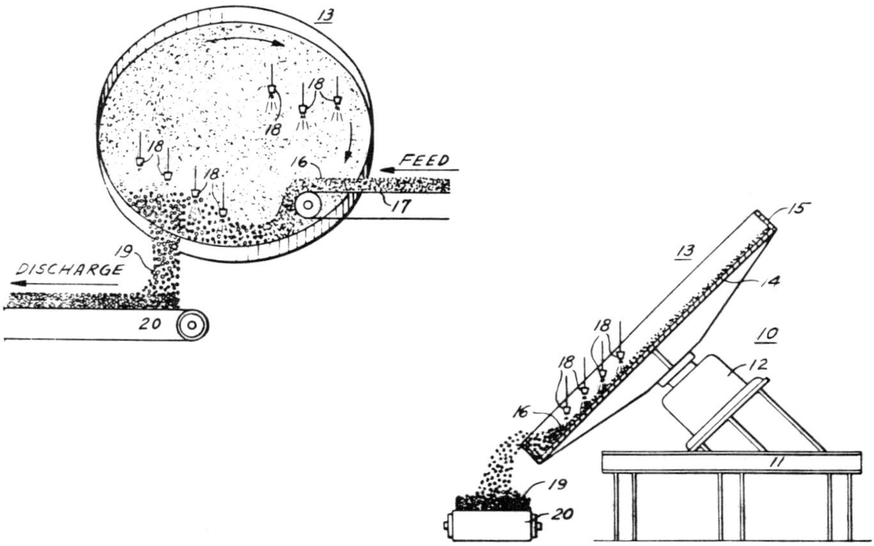

Source: U.S. Patent 4,003,736

In addition, the fume must be retained on the balling disc for a minimum period of not less than about 8 minutes. By following balling practices well known to those skilled in the art and by using the proper amount of water and the proper retention time on the disc as mentioned above, there will be formed strong, stable green balls **19** which will be discharged from disc **13** onto conveyor **20**.

A process developed by *J.W. Kreiger; U.S. Patent 4,004,916; January 25, 1977; assigned to Bethlehem Steel Corporation* is one for preparing moist iron-bearing fume for use in steelmaking furnaces to recover the iron contained in the fume. The fume is thickened, substantially instantaneously dried, impact-fractured and balled. Moisture is added to the fume on the balling device to bind the fume to produce green balls which resist degradation during handling and transport. The green balls have sufficient strength to resist degradation during handling prior to and during charging to steelmaking furnaces.

Such a process is shown in Figure 95. Turning now to the drawing, iron-bearing wet-collected fume from various steelmaking furnaces is thickened in individual thickeners **13** and **15** and dry-collected fume from an open hearth precipitator is water leached to reduce the sulfur content and is thickened in a thickener **11** as shown. Some water leached fume from a basic oxygen furnace is also passed to thickener **15**. The underflows from the individual thickeners **11, 13** and **15** are mixed together in a common thickener **17**, which can also be considered as a mixing tank, to form a single slurry composed of water and fume and containing between 35 to 55% solids.

Figure 95: Bethlehem Steel Corporation Method for Agglomerating Wet-Collected Fume for Use in Metallurgical Furnaces

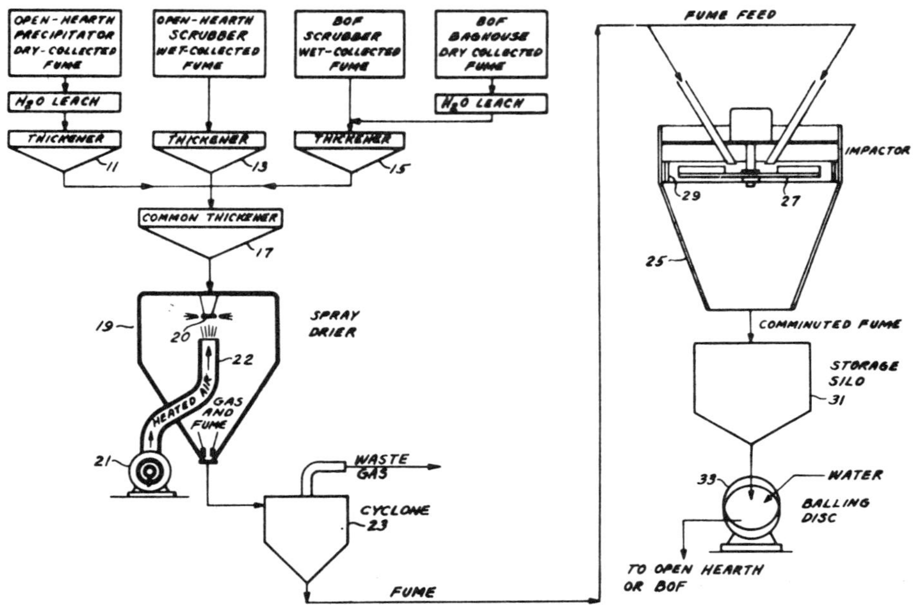

Source: U.S. Patent 4,004,916

The slurry of fume from the mixing tank, or common thickener **17**, is next dried in a spray dryer **19**. The slurry is in the example fed into the spray dryer **19** at a rate of between 80 gallons per minute to 100 gallons per minute. The slurry is atomized by feeding it into a rapidly rotating atomizing head **20** located near the top of the dryer and is flung from the head in very fine droplets. The size of the droplets is controlled by the diameter of small orifices in the rotating head and by the speed of rotation. Heated air at a temperature of about 1200°F is blown upwardly toward the atomized slurry by a blower **21** through a duct **22** positioned under the atomizing head **20**.

The water in the slurry is substantially instantaneously vaporized as it contacts the hot gases and small dense hardened spheroids having a more or less smooth surface and containing a plurality of fume particles are formed. The fume is usually raised to a temperature of about 280°F (138°C) and the temperature of the hot air is reduced to about 280°F (138°C) by heat exchange between the particles and the hot gases and vaporization of the water. The small dense spheroids formed by the vaporization of water retain their shape identity and are discharged from the bottom of the spray dryer with the remainder of the fume which may not have been included in spheroids at a temperature of about 280°F (138°C) into any suitable cyclone device or the like **23** which serves to separate the fume particles and spheroids from the gas stream which exists from the spray dryer. The formation of the spheroids reduces the surface area of the fume to below about 8,000 square centimeters per gram.

The dried fume is next conveyed to and impacted in an impactor **25**. In the impactor **25**, the fume is fed to a horizontal rotating member **27**. The speed of the rotating member, for example about 3,500 revolutions per minute, imparts a velocity to the fume such that the small dense spheroids and a portion of the coarsest particles of fume itself are hurled from the rotating member against a liner or target **29**, which is positioned a predetermined distance from the periphery of the rotating member, at a velocity sufficient to cause the spheroids and particles to be fractured.

New, dry, fresh, clean, angular fracture surfaces are formed on these particles of fume and fractured spheroids. A portion of the intermediate size particles of fume may also be fractured. The smallest particles of fume are hurled from the rotating member, but are not fractured.

The formation of the new, dry, fresh, clean angular fractures on the surface of a portion of the fume increases the Blaine surface area of the fume from less than about 8,000 square centimeters per gram to more than about 8,000 square centimeters per gram and preferably to not less than 12,000 square centimeters per gram.

The impacted fume is next passed to storage in silo **31** or other suitable storage facility for use as needed. The impacted fume is removed as needed from storage and fed to a balling device **33**, such as a cone or disc type balling device. Moisture which is needed for balling is added to the fume on the balling device. Green balls within a size range of about ⅜ of an inch to 1¼ inches in diameter for use in steelmaking furnaces are produced on the balling device. All the green balls are substantially spheroidal in shape.

A process developed by *B.E. Cass, D.W. Coate and J.R. Quigley; U.S. Patent 4,119,455; October 10, 1978; assigned to Carad, Incorporated* is one for recovering iron-bearing flue dust collected as a by-product in wet sludge or dry form from metallurgical processes for recycling. The moisture content of the collected flue dust is adjusted to a level at which the wet dust is of a plastic consistency such that it will extrude into cohesive agglomerates (generally 8 to 16% moisture content).

If the dust is collected in a dry state, moisture is added; if collected in a wet state, the moisture content is adjusted by the addition of a complementary dry material. Hydraulic cement is added to the mixture in the range of approximately 4 to 15% by weight and the mixture is extruded into cohesive agglomerates and thereafter cured for subsequent charging to metallurgical furnaces.

A process developed by *I. Arai, Y. Yamada, A. Mutsuta and K. Kadota; U.S. Patent 4,133,756; January 9, 1979; assigned to Sumitomo Heavy Industries, Ltd., and Sumitomo Metal Industries, Ltd. both of Japan* is one for feeding a constant amount of filter cake to a dryer thereby producing dried powder having a predetermined moisture content so that materials of uniform quality are fed at a constant rate to the stages subsequent to the dryer in a pelletizing system of the iron and steel industry.

There is shown a schematic illustration of the system for treating the dust discharged from several types of furnaces used in the iron and steel industry in Figure 96.

A blast furnace dust slurry and a converter dust slurry are fed through conduits **11** and **12** to a mixing bath **13**. Also, in order to maintain the concentration of the mixture of the slurries (mixed slurry) approximately in the range of 35 to 45% (by weight), the dry dust collected by dry type collectors is fed through a conveyor **15** into the bath **13** and the whole mixture is agitated by a mixing agitator **16** associated with the bath **13**. All of the mixed slurry is then directed to a stirring basin **18** through a mixed slurry conduit **17**.

The basin **18** is equipped with a suitable stirring means such as a rotatable stirrer **19** to maintain the uniformity of the mixed slurry. The mixed slurry is then fed through a conduit **20** to a receiving vessel **21** where a drum type belt filter **22** is disposed.

The rotational speed of the belt filter is variable and the belt filter is equipped with a variable speed drive **23**. Rotation of the belt filter **22** produces filter cake **24** having a fixed or predetermined moisture content if the concentration of the mixed slurry is within a predetermined range.

The filter cake **24** is discharged onto a conveyor **25** which delivers the filter cake **24** through a chute **27** to a rotary dryer **26**. A suitable weighing device **28** is associated with the conveyor to sense the variation in amount of the filter cake discharged onto the conveyor and to regulate the variable drive **23** to control the rotational speed of the belt filter **22** so that the filter cake is supplied at a constant rate to the dryer **26**.

A hot air generator **29** is associated with the dryer **26** which dries the filter cake and discharges it as powder **30**. The dried powder is conveyed to the next stage by a conveyor **31**.

Also, there is provided an exhaust duct **32** to which a temperature detector **33** is coupled to sense the temperature of the exhaust gas. According to the temperature of the exhaust gas thus sensed by the detector **33** and the value obtained by the weighing device **28**, it is possible to control the fuel supplied to the hot air generator through a fuel supply conduit **34** to set and maintain the temperature at the detector **33** at a desirable value in order to produce dried powder having a fixed or predetermined moisture content at a constant rate. To such end, a flow regulator valve **35** is disposed in the fuel supply conduit **34** with which an actuator **36** is coupled.

The actuator **36** is adapted to receive a signal from a control unit **38** which receives temperature information from the detector to generate the signal. The signal controls the actuator **36** to determine the opening degree of the valve **35**. Also, two air intake pipes **39** and **40** are connected to the hot air generator, the pipe **39** is adapted to introduce air necessary for combustion of fuel while the pipe **40** introduces fresh air for controlling the temperature of hot air.

Figure 96: Sumitomo Process for Processing Dry Dust and Slurried Dust from Steelmaking Processes to Yield a Dry Powder Suitable for Pelletizing and Recycle to Steel Manufacture

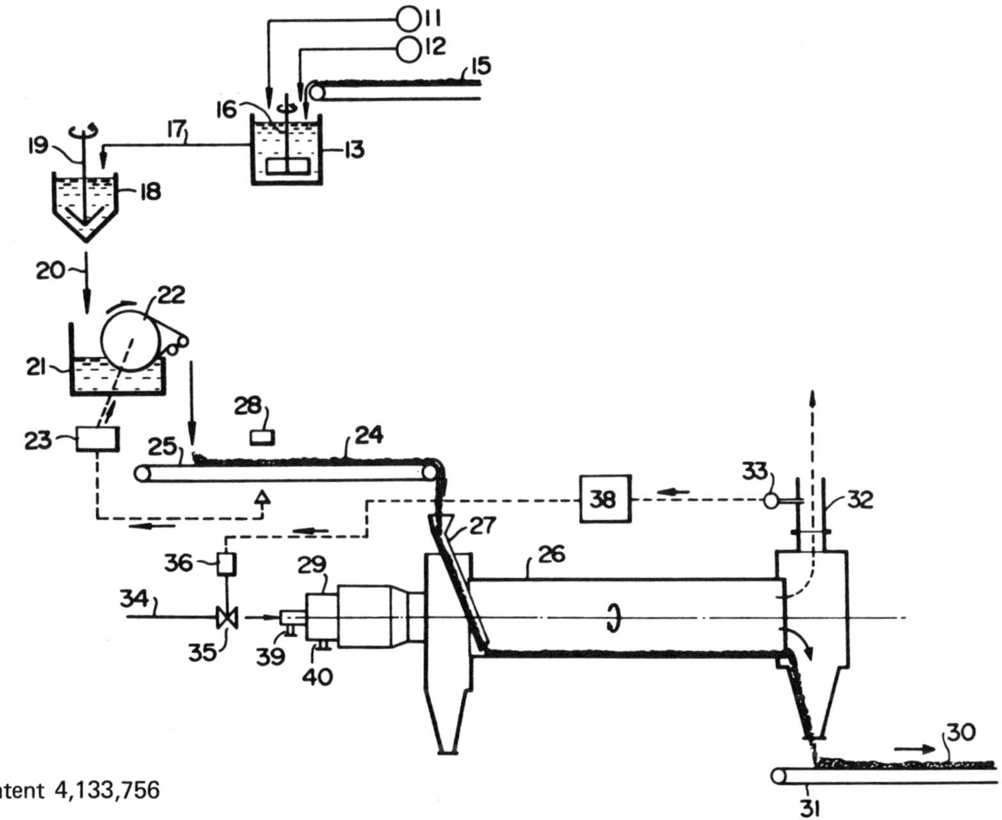

Source: U.S. Patent 4,133,756

A process developed by *K. Maeda; U.S. Patent 4,134,755; January 16, 1979* is one for recycling iron material components of dusts discharged from iron manufacturing furnaces such as blast furnaces and converters and removing zinc type metals and compounds from such dusts to avoid their deposition and accumulation on the furnace walls.

A wet method is employed for separating the iron dust components from the zinc and removing these zinc type metals and compounds. The zinc material is made identifiable or earmarked from the iron material by being selectively ground or otherwise reduced to finer particles than the iron material, the zinc material then being removed from the iron material by wet separation under negative pressures for reclamation or disposal. The iron material remaining is processed for recycling into the furnace. Such a process is shown in Figure 97.

Figure 97: Maeda Process for Removing Zinc Type Metals and Compounds in Dust Recovery Process in Iron Manufacture

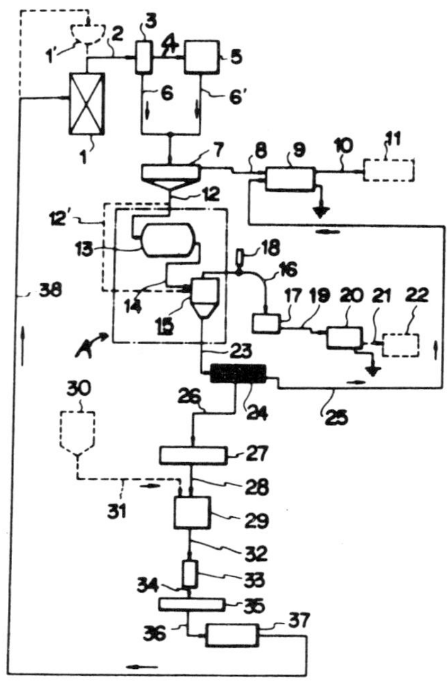

Source: U. S. Patent 4,134,755

An ordinary blast furnace **1** as the iron manufacturing furnace is connected to a known electric dust precipitator **3** as the primary dust collector through an exhaust gas duct **2**. When the iron manufacturing furnace is a converter or electric furnace **1'**, it also is connected in a similar manner.

The electric dust precipitator **3** is connected to a wet venturi scrubber **5** as the secondary dust collector through a dust collecting duct **4**, and the electric dust precipitator **3** and the wet venturi scrubber **5** are appropriately connected to a thickener **7** through dust collecting ducts **6** and **6'**.

An overflow pipe **8** from the thickener **7** is appropriately connected to a waste water treatment apparatus **9** which is then connected to a re-treatment apparatus **11** through a transporting and connecting pipe **10**.

A mechanism **A** for removing zinc type metals and compounds is connected to an underflow pipe **12** of the thickener **7**. Mechanism **A** includes a grinding mill **13** such as a vortex mixer which receives the output of underflow pipe **12** and operates to effect selective pulverization of concentrated dusts, namely the iron and zinc-containing components, which are treated according to the work index of their toughness and brittleness. The softer zinc components are thus ground into finer particles than the iron. A discharge opening (not shown) of grinding mill **13** is connected to a feed inlet of a negative pressure type wet cyclone **15** through a discharge pipe **14**.

When the iron and zinc particles of the collected dusts in the thickener already made to have different particle sizes, as described hereinafter, by pipe flowing or pumping during the transportation process, the underflow pipe **12** is directly connected to the negative pressure cyclone **15** through a second underflow pipe **12'**.

In situations where the dusts are slime-coated in the thickener **7**, a line mixer or the like is employed in place of the grinding mill **13** so as to prevent recohesion of dust particles, the output of the line mixer being connected to the feed opening of the negative pressure cyclone **15**.

An overflow sucking finder pipe **16** of the negative pressure type wet cyclone **15** is run to a seal tank **17** of a water tank and is downwardly extended so that a water column is formed therein, whereby a negative pressure siphon tube is formed. A siphon regulator **18** capable of pressure adjustment is disposed on the top of the finder pipe **16**.

The seal tank **17** is connected to an appropriate waste water treatment apparatus **20** through a connecting pipe **19** and the waste treatment apparatus **20** is connected to a retreatment apparatus **22** for zinc type metals and compounds through a connecting pipe **21**.

An underflow pipe **23** of the negative pressure type cyclone **15** is connected to a filter press **24** for solid-liquid separation through an appropriate regulator (not shown), and a waste water pipe **25** from the filter press **24** is used to connect the waste water treatment apparatus **9** to the thickener **7**.

A transportation conveyor **26** connected to filter press **24** is connected to an appropriate drying apparatus **27**, and a transportation conveyor **28** from the drying apparatus **27** runs to an appropriate agitating and mixing apparatus **29** such as a pan type mixer.

A conveyor **31** from a dry fine powder hopper **30** for starting ores or the like may be connected to the agitating and mixing apparatus **29** as may be necessary.

An appropriate granulating apparatus **33** is connected to the agitating and mixing apparatus **29** through a transportation conveyor **32**, and the granulation apparatus **33** is connected to a calcination apparatus **35** through a transportation conveyor **34** and to a sintering apparatus **37** as the final treatment apparatus through a transportation conveyor **36**. The sintering apparatus **37** is connected by means of a transportation conveyor **38** to blast furnace **1** or converter **1'**.

In the system having the above structure, when the blast furnace **1** or the converter or electric furnace **1'** is operated and the electric dust precipitator **3** and the wet venturi scrubber **5** are operated, exhaust dusts from the blast furnace **1** or the converter or electric furnace **1'**, which contain ZnO, Zn, Fe_2O_3 and the like are first collected in the form of coarse particles by the electric dust precipitator **3**, and then collected in the form of fine particles by the wet venturi scrubber **5**.

The solid dust portion is gathered in the thickener **7** by means of the dust collecting ducts **6** and **6'** and undergoes solid-liquid separation and sedimentation actions, and the overflowed portion is transferred to the waste water treatment apparatus **9** via overflow pipe **8**. Any remaining solid dust portion is fed by means of pipe **10** to the retreatment apparatus **11** for an appropriate subsequent treatment.

A spigot of the thickener **7** for the precipitated and concentrated underflowed (solid dust) portion is connected by means of an underflow pipe **12** to the grinding mill **13** of the mechanism **A** for removal of zinc type metals and compounds, which is operated in a relay manner. Selective pulverization is conducted by the grinding mill **13** based on the work index of the preliminarily analyzed toughness, brittleness and the like of Pb, Zn, ZnO, and Fe_2O_3 of the precipitate so that particle size identification or earmarking is accomplished.

This selective pulverization of grinding mill **13** is such that Zn, Pb and Zn components are pulverized as finely as possible, but the Fe_2O_3 component is not pulverized. Accordingly, particle size identification of the zinc components from the iron components is clearly achieved by this wet pulverization, and these size-earmarked particles are discharged and fed in a mingled condition to a feeder at the feed opening of the negative pressure type wet separator **15**.

When the particle size sorting is preliminarily attained in the transportation line to the thickener **7**, the dust particles are directly fed to the feed opening of the negative pressure type cyclone **15**.

When the dust particles are slime-coated, recohesion is appropriately prevented by a line mixer or the like as pointed out hereinbefore, and they are fed to the feeder in the particle size-sorted state.

Since in the negative pressure type wet separator **15** the finder pipe **16** is connected to the seal tank **17** located below to form a siphon pipe, a negative pressure is maintained in the body portion, and by means of the siphon regulator **18**, the finder pressure, namely the operation negative pressure may be controlled to an optimum level.

Accordingly, the solid component fed to the negative pressure wet separator **15** is maintained under a negative pressure condition by suction. Namely, finer particles of Pb, Zn and ZnO are positively sucked more strongly and separated from the coarser particles of Fe_2O_3 and the like, and they are sedimented and gathered in the water tank **17** of the seal tank through the overflow pipe **16** and substantially completely together by the effect of the above particle size earmarking.

Coarse particles of the iron-containing components such as Fe_2O_3 are discharged intermittently by predetermined amounts from the underflow pipe **23** through the action of the above-mentioned regulator.

Accordingly, in the mechanism **A** for removing zinc type metals and compounds, not only the wet selective pulverization in the first stage and the pulverizing particle size earmarking, but also the positive sucking separation of fine particles by the negative pressure type wet separation in the second state are accomplished. Therefore, the solid portion discharged from the underflow pipe **23** has a much reduced content of zinc type metals and compounds and the iron-containing components occupy a substantial portion of this solid portion.

Zinc type metals and compounds thus agglomerated and gathered in the seal tank **17** through the overflow pipe **16** are transferred to the waste water treatment apparatus **20** through the transfer pipe **19** and subjected to the solid-liquid separation. Solid zinc metals and compounds are treated in the treatment apparatus **22** so that they can be utilized again.

Zinc-free iron-containing components discharged from the underflow pipe **23** are fed to the filter press **24** where solid-liquid separation is performed. The liquid is fed to the waste water treatment apparatus **9** through the waste water pipe **25**, where the liquid is appropriately treated. The solid material in treatment apparatus **9** is fed to the retreatment apparatus **11** so that it can be retreated, as pointed out hereinbefore.

The iron-containing components which have been subjected to solid-liquid separation in filter press **24** are transferred to the drying apparatus **27** by the transportation conveyor **26** after opening of the filter press **24**, and they are appropriately dried.

Then, they are fed to the agitating and mixing apparatus **29** by the transportation conveyor **28** where they are subjected to the mixing and kneading operation. In some cases, they are mixed and kneaded with fine powder of a raw material to form a homogeneous mixture. Then, the mixture is fed to the granulation apparatus **33** by the conveyor **32** and formed into granules having an appropriate size. The granules are then fed to the calcining apparatus **35** by the transportation conveyor **34** where they are calcined and shaped. The calcined and shaped product is then fed to the sintering apparatus **37** through the transportation conveyor **36** where it is pelletized. The pellets are recycled for reduction in the blast furnace **1** or converter or electric furnace **1'**.

A process developed by *L. Fiege, H.W. Höhle, W.J.P. Janssen and K.-H. Ulrich; U.S. Patent 4,189,314; February 19, 1980; assigned to Fried. Krupp Huttenwerke AG, Germany* is one for utilizing waste dust and waste sludge from blast furnaces and steel works for subsequent steel production.

It involves forming the waste dust and waste sludge into pellets containing basic gangue and high sulfur contents and having a metallization degree of over 80%.

Up to 10% by weight of the pellets are intimately mixed with liquid pig iron relative to the weight of the pig iron to form a pig iron melt. The pellets are placed into a pig iron charging ladle prior to and/or during the reladling of the liquid pig iron from a torpedo ladle into a charging ladle, and the pig iron is deslagged prior to the pouring thereof into a refining vessel.

IRON OXIDE FROM ILMENITE CHLORINATOR DUST

In the chlorination of titaniferous materials, as for example, titanium-containing ores such as ilmenite, by reacting chlorine and carbon with the ore at elevated temperatures to produce titanium tetrachloride, large amounts of dust are formed which comprise essentially particulate ferrous chloride contaminated with TiO_2, coke and other metal oxides and chlorides, as for example, magnesium chloride and manganese chloride.

The nature of this dust, hereinafter referred to as chlorinator dust, is such that it would constitute an environmental nuisance if stored or disposed of as land fill. It is desirable therefore to use the chlorinator dust as a source material for production of useful products such as iron oxide and gaseous chlorine, the latter, for example, for use in chlorinating titaniferous ores.

There are several known processes for recovering iron oxide and chlorine by oxidation of iron chlorides.

An improved process for chlorine and iron oxide recovery has been developed by *A. Hartmann, A. Kulling and H. Thumm; U.S. Patent 4,060,584; Nov. 29, 1977; assigned to Kronos Titan GmbH, Germany*. It is one in which the dust produced in the chlorination of titaniferous ores which comprises essentially particulate ferrous chloride plus solid contaminants including coke and various metal chlorides and oxides is oxidized in successive stages at relatively low temperatures to recover particulate iron oxide and gaseous chlorine.

In particular, the process is characterized by the following steps:

a. The chlorinator dust comprising essentially ferrous iron chloride is first reacted at temperatures between 500° and 800°C with only so much oxygen that solid Fe_2O_3 and gaseous ferric iron chloride are formed. Essentially no chlorine is formed;

b. the solid ferric iron oxide so produced, may contain coke and other contaminants, and these solids are separated from the gaseous products of the reaction at temperatures from 500° to 800°C;

c. the gaseous products of the reaction, i.e., ferric iron chloride and CO_2 if present, are reacted with additional oxygen to form gaseous chlorine and additional solid Fe_2O_3, the reaction being started at temperatures between 600° and 800°C, the temperature being gradually lowered to below 600°C; and

d. the solid Fe_2O_3 formed in step c is separated from the gases produced in stage c.

This procedure avoids to a large extent, the disadvantages of the prior art processes.

The process is shown schematically in Figure 98.

Figure 98: Kronos Titan GmbH Process for Oxidation of Chlorinator Dust

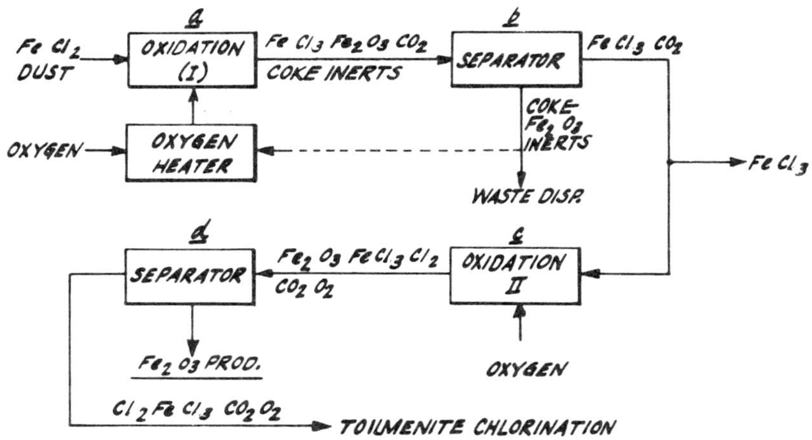

Source: U.S. Patent 4,060,584

Referring to the drawing, step a of the oxidation process occurs essentially according to equation (I):

$$6\ FeCl_2 + 1.5\ O_2 \longrightarrow Fe_2O_3 + 4\ FeCl_3 \qquad (I)$$

Since the reaction is carried out at relatively low temperatures, i.e., 500° to 800°C, combustible secondary constituents of the chlorinator dust, such as carbon, are definitely not oxidized. For this reason no overheating or sintering of the iron oxide occurs and the reaction can be easily controlled. The solid iron oxide formed is relatively coarse and has substantially no tendency to adhere to the reactor walls and hence the iron oxide together with the secondary solid constituents of the chlorinator dust, such as coke, are easily separated from the gas mixture, the essential constituent of which is gaseous ferric iron chloride.

The oxygen used in step a is substantially stoichiometric and hence no chlorine is formed, the amount of oxygen used being calculated from the composition of the chlorinator dust employed with the aid of the above stated equation. The separation of this solid iron oxide and other solids from the gas mixture (step b) is carried out at the prevailing temperatures of 500° to 800°C, using conventional separating means; and the ferric iron chloride containing gas mixture can be worked up further without loss of energy.

Thus the gaseous ferric iron chloride of step b may be condensed and separated from the other constituents of the gas mixture; and because of its purity can be used in various fields, for example, in water purification. However, it is preferred to react the gas mixture containing the ferric iron chloride with additional oxygen (step c) to form additional solid Fe_2O_3 and gaseous chlorine according to the equation:

$$2\ FeCl_3 + 1.5\ O_2 \longrightarrow Fe_2O_3 + 3\ Cl_2 \qquad (II)$$

and subsequently in step d to separate the substantially pure Fe_2O_3 so formed from the chlorine-containing gas mixture.

A great advantage of the process lies in the fact that the solid iron oxide produced in successive steps c and d is substantially uncontaminated and may be worked up further directly, for example, in the smelter industry or for the purpose of manufacturing iron oxide pigment. If the reaction according to equation (II) runs to completion in step c then twice as much iron oxide is formed as in step a; hence the greater part of the iron values of the chlorinator dust is here obtained in usable form.

The reaction of step c occurs, to a greater or lesser extent, depending on the temperature at which oxidation of the gaseous ferric iron chloride is carried out. Preferably the reaction of step c is carried out in such a way that it is started initially at temperatures between 600°C and 800°C and thereafter is gradually lowered to below 600°C. It is particularly advantageous to lower the temperature to as low as 350°C. In this connection the gaseous ferric chloride mixture is introduced from the preceding step, i.e., step b, into the oxidation reactor of step c without heating or cooling which procedure effects relatively simple reaction control and the use of simply constructed devices.

As the temperature is lowered the equilibrium of equation (II) is shifted to an increasing extent in the direction of the Fe_2O_3 formation so that an extensive almost complete iron oxide formation is achieved. Any further lowering of the temperature, that is to say, below 350°C is not advisable as a rule in order to avoid condensation of any unreacted gaseous ferric iron chloride.

Since the reaction in step a is weakly exothermic, it is necessary to introduce energy in order to heat the chlorinator dust to reaction temperature and to compensate for the heat losses occuring due to radiation, and so forth. The heat addition can be carried out by a preliminary heating of one or both reaction constituents. A preferred form of carrying out the process comprises preheating the oxygen before step a by combustion of a suitable fuel material; and can be carried out directly in that, for example, the fuel is burned in the oxygen stream before being fed into the reactor; or the oxygen stream may be mixed with hot combustion products or heated indirectly by passing over heat exchange surfaces.

As fuel materials, any desired substances may be employed in so far as they do not form products which have an unfavorable influence on the oxidation of the ferrous or ferric iron chlorides. Thus, they should be essentially free of hydrogen or hydrogen compounds. Suitable fuel materials, are, for example, carbon monoxide and carbon. In case the ferrous iron chloride dust includes coke, this too may be employed as fuel material.

Moreover, if the solid mixture separated in step b contains coke, this solid mixture may, at least in part, be reacted with excess oxygen, or a gas mixture containing oxygen, at a temperature above the ignition point of the coke and the preheated oxygen containing gas mixture separated from the solid materials and introduced into step a.

It is possible that either the entire solid mixture separated in step b or only a part of it is reacted with oxygen. The operation depends in detail on how high the preheating temperature of the oxygen should go, and on the size of the coke content of the solids mixture as well as on the permissible dilution of the preheated oxygen containing gas mixture with the combustion products. Too great a dilution of the gas mixture is disadvantageous because, in that case, the chlorine containing gas recovered in step d is also too highly diluted.

Any unreacted ferric iron chloride present in the chlorine containing gas leaving step d may be separated therefrom in a manner known as such, prior to further use of the chlorine containing gas mixture.

The process can be carried out at normal pressure. However, it is advantageous to use elevated pressure. Apart from other advantages (smaller apparatus, higher through-put, and so forth) an additional advantage in this form of operation is that following step d any unreacted ferric iron chloride does not have to be removed from the chlorine containing gas mixture and accordingly the gas mixture, freed of solid Fe_2O_3 may be introduced directly into a chlorinator without separation of unreacted ferric iron chloride.

IRON OXIDE FROM ROLLING MILL SLUDGE

Scale or cutting scrap in a small quantity is conventionally mixed with iron ore fines to serve as a sinter raw material. Along with the recent drastic increase in pig iron production, there is a rapidly increasing demand for sinter as an iron-making raw material, thus requiring installation of large-scale sintering machines. The resultant increase in the consumption of iron sources other than iron ore is leading to the utilization of scales and cutting scraps having a high content of grease deposit in larger quantities. From the point of view of pollution control, on the other hand, it is now a general practice to use an electrostatic precipitator for collecting the dust contained in the exhaust gas from a sintering machine.

In this practice, however, if a sinter raw material containing a grease deposit such as a scale or a cutting scrap is mixed in a large quantity into the raw sinter mix, the combustion of the raw sinter mix would cause evaporation of deposited grease, and the grease vapor would be directed, together with the exhaust gas, into the electrostatic precipitator.

The dust collected in the electrostatic precipitator has therefore a grease content of from about 0.1% to about 10%. This grease, if deposited in a large quantity on the electrodes of the electrostatic precipitator would be burnt by the spark between the electrodes or by the high temperature exhaust gas, and cause troubles such as a fire in the electrostatic precipitator.

Because a rolling mill sludge deposited in a waste water pit of a steel rolling mill has a moisture (water) content of from about 20% to about 30% and a grease content of about 2% to about 3% even after dehydration through a filter press, the following measures are taken for the treatment thereof:

(1) To use the rolling mill sludge for reclamation or to dump the same for ocean disposal;

(2) To utilize the rolling mill sludge as an iron source after removing the grease deposit thereof by a chemical washing (i.e., washing with a chemical);

(3) To utilize the rolling mill sludge as an iron source after removing the grease deposit thereof by combustion.

However, in view of the Fe content of from about 60% to about 70% of the rolling mill sludge, method (1) above is not preferable from the standpoint of the effective use of iron resources. Furthermore, the increasing requirement for pollution control is more and more restricting places for reclamation or dumping of the same, and the disposal of rolling mill scale by this method will be impossible in the near future.

Because of the small particle diameters and the high moisture content of from about 20% to about 30%, even after the dehydration through a filter press, the rolling mill sludge has a high viscosity. Treatment of the rolling mill sludge by method (2) or (3) above, therefore, is not easy. In method (2) above, furthermore, the very difficult separation of the grease limits the use of this method.

Method (3) above requires high equipment costs for a fluidized bed furnace, a rotary kiln, a multi-stage combustion furnace and other facilities and high operation costs. In any of these furnaces, the treatment of the rolling mill scale at a high temperature of from about 700° to about 800°C causes agglomeration of the rolling mill sludge, thus forming another disadvantage of this method. There are many other problems in this method, such as the necessity to install an electrostatic precipitator for collecting the dust contained in the combustion waste gas.

For these reasons, no satisfactory results are being obtained by any of methods (1) to (3) described above.

Thus, as improved process has been developed by *T. Izawa, S. Kajikawa, K. Satomi, K. Nakano, S. Kurosawa, K. Hirai and Y. Tsukada; U.S. Patent 4,091,545; May 30, 1978; assigned to Nippon Kokan Kabushiki Kaisha, Japan*.

For the purpose of utilizing as a sinter raw material a rolling mill sludge with a grease deposit having a high moisture content, which is deposited in a waste water pit of a steel rolling mill, the moisture content of the rolling mill sludge is reduced to from about 1% to 10% by drying the rolling mill sludge at a temperature of from about 120° to about 200°C, preferably of about 140°C to about 150°C, with the use of the waste heat of a sinter cooler of a sintering machine, or by mixing the rolling mill sludge with a mill scale of a low moisture content collected without the use of roll cooling water.

The rolling mill sludge, the moisture content of which has thus been reduced, is continuously fed, by cutting into substantially equal batches, to at least one of a sinter discharge section, an entry section of the sinter cooler, belt-conveyor for fines return provided adjacent to a hot screen, and a return hopper of the sintering machine. By this process the grease deposit is substantially completely removed through combustion or evaporation with the use of the heat of the sinter produced by the sintering machine.

The apparatus which may be used in this process is shown in Figure 99.

Figure 99: Nippon Kokan Kabushiki Kaiha Process for Removing Water and Grease from Rolling Mill Sludge to Aid in Iron Oxide Recovery

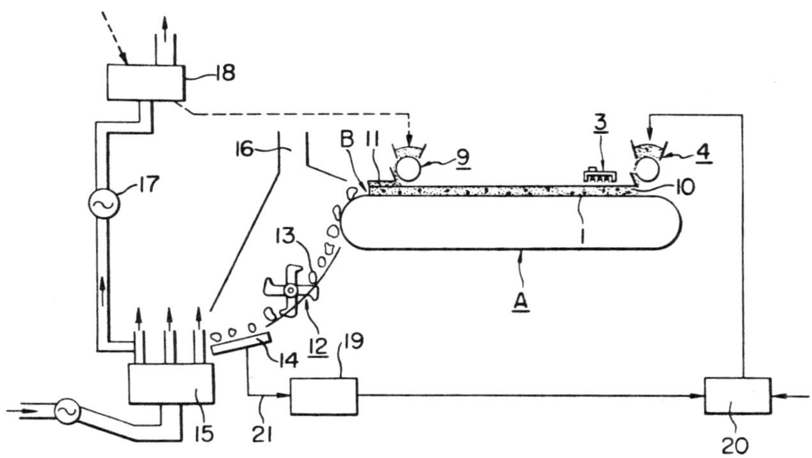

Source: U.S. Patent 4,091,545

As shown there, a raw sinter mix **10** having a known conventional composition, fed from a raw material feeder **4** onto a pallet **1** of a sintering machine **A** is ignited by an ignition furnace **3** and sintered as it travels to the left in the figure. A sludge feeder **9** provided at a sinter discharge section **B** of the sintering machine **A** feeds a dried rolling mill sludge **11**, the moisture content of which has been reduced as described below, and which contains a grease deposit, in substantially equal batches, and feeds the batches onto the sinter on pallet **1**. The dried rolling mill sludge **11** fed onto the sinter on pallet **1** is heated by the heat of the sinter **13** while it travels to a hot screen **14** through a sinter breaker **12**, and the grease deposit is substantially completely removed through combustion or evaporation.

The rolling mill sludge **11**, the grease deposit of which has thus been removed, is sieved through the hot screen **14**, and is sent, together with the under-screen return fines, to a return hopper **19** through a belt-conveyor for fines return **21** for temporary storage. The portion of the sinter remaining on the screen **14** is sent to a sinter cooler **15**, where it is cooled.

On the other hand, part of the cooling gas (air) heated by heat exchange with the sinter at the sinter cooler **15** is directed by a blower **17** to a sludge dryer **18** (for example, a fluidized bed dryer), where the hot gas dries a rolling mill sludge charged therein. The sludge which contains a grease deposit and which has a high moisture content is one which is deposited in a waste water pit of a steel rolling mill. For example, in the case where a rolling mill sludge containing 28.0% moisture and 2.0% grease, charged into the dryer **18**, was dried at a drying temperature of 140° to 150°C for 15 to 20 minutes, a dried rolling mill sludge containing 1.13% moisture and 2.5% grease was obtained. A drying temperature of the rolling mill sludge exceeding 200°C in the dryer **18** causes evaporation of the grease deposit, and this requires installation of equipment to collect the grease vapor thus produced in the dryer **18**.

If the drying temperature is under 120°C, drying of the rolling mill sludge would take too much time. Test results reveal that the drying temperature should be from about 120° to about 200°C, preferably about 140° to about 150°C. In the present invention, therefore the rolling mill sludge is dried with the use of the gas fraction having a temperature of about 120° to about 200°C, preferably of from about 140° to about 150°C, of the cooling gas heated by the heat exchange.

It is desirable to reduce the viscosity of the rolling mill sludge by drying the rolling mill sludge to a moisture (water) content not exceeding 10%, thereby ensuring a smooth supply of the rolling mill sludge by the sludge feeder **9**; however, it is not necessary to reduce the moisture content to under 1%. The rolling mill sludge **11** thus dried but still containing a grease deposit is fed by the sludge feeder **9** onto the sinter as described above to remove the grease deposit thereof.

The exhaust gas containing dust, grease combustion gas and grease vapor generated in the process between the sinter discharge section **B** and the hot screen **14**, is collected through a hood **16**, sent to a wet dust collector (not shown) where it is cleaned, and then discharged. The dehydrated and degreased rolling mill sludge and return fines sent from the return hopper **19** and the sinter raw materials sent from a raw material hooper (not shown) are mixed into a known conventional composition by a mixer **20**, and fed to the raw material feeder **4** in the form of pellets produced by a pelletizer (not shown) as required.

IRON POWDER FROM MACHINING SCRAP

There are strong economic as well as ecological incentives favoring development of a process for the direct converion of machine scrap or swarf into a powder which can be utilized by the ferrous powder metallurgy industry. At Ford Motor Company alone an estimated 105,000 tons of low alloy steel machine turnings are generated by the various manufacturing plants and sold as scrap on the open market, destined to be part of a furnace charge in some ferrous melting operation. Its utility, even as a furnace charge material, is limited by high bulk-to-weight ratios and residual machine oil content.

The alloying elements present in most of this kind of scrap represent a valuable resource if they could be economically recovered. In melting operations much of this alloy content is oxidized and lost to slag. The ecological advantages of direct conversion of swarf to powder with no intermediate melting operation stem from use of a cleaner, less polluting process and the theoretical 100% recovery of valuable alloying elements. Economically, direct conversion is also very attractive. Based on current prices for iron powders or prealloyed iron powders and the price of swarf on the open market, a significant differential is available for conversion costs and profit. In spite of these incentives, technological problems exist which have thwarted previous attempts to employ powder made from swarf in standard powder metallurgy sintering operations.

A process which is designed to overcome these problems has been developed by *S.M. Kaufman; U.S. Patent 4,129,443; December 12, 1978; assigned to Ford Motor Company*.

It is a process which impacts the scrap metal at a temperature below its ductile-brittle transition temperature to cryogenically provide scrap metal powder. The scrap metal is subjected to two impacting operations (such as by use of a ball milling machine), the first operation being carried out simultaneously with the cryogenic medium and the second impacting operation is carried out at ambient temperature conditions utilizing milling elements which impact cold work to at least a portion of the cryogenic powder.

An anti-oxidation agent, such as copper or iron, is mechanically transferred to substantially each particle of the cryogenic powder during the second operation to form a continuous copper or iron envelope thereabout. The coated cryogenic powder is then compacted and sintered; the sintered product may be subjected to hardening or tempering treatments if desired.

The steps in the process are shown in the diagrams in Figure 100. Scrap metal and particularly machine turnings **10** are selected as the starting material. Machine turnings is defined herein to mean segments of ribbons of low alloy steel. They typically are shavings cut from alloy bar (produced by melting to a conventional SAE designation). But machine turnings, preferably ferrous based, include alloying ingredients such as manganese, silicon, chromium, nickel and molybdenum.

The turnings should be selected to have a surface-to-volume ratio of at least 60:1, which is characteristic of machine turnings. The scrap pieces will have a size characterized by a width 0.1 to 1.0 inch, thickness of 0.005 to 0.03 inch, and a length of 1 to 100 inches. Machine turnings are usually not suitable for melting in an electric furnace because they prevent efficient melt down due to such surface-to-volume ratio.

This process can be performed with other types or larger pieces of scrap metal, although capital investment costs may increase due to the difficulty of impacting scrap metal sized in particle pieces beyond 0.03 inch thick. The scrap pieces should be selected to be generally compatible in chemistry as combined in the final product; this is achieved optimally when the scrap is selected from a common machining operation where the same metal stock was utilized in forming the turnings.

The selected scrap pieces **10** are then put into a suitable charging passage **11** leading to a ball milling machine **12** or equivalent impacting device. Within the passage, means **13** for freezing such metal pieces is introduced such as liquid nitrogen; it is sprayed directly onto the metal pieces. Mere contact of the liquid nitrogen with the scrap pieces will freeze them instantly. The application of the liquid nitrogen should be applied uniformly throughout its path to the point of impaction.

The iron ball milling elements **14** are motivated preferably by rotation of the housing **17**, to contact and impact the frozen pieces **15** of scrap metal causing them to fracture and be comminuted. Such impaction is carried out to apply sufficient fracturing force (defined to mean less than 1 ft.-lb) and a sufficient period of time and rate to reduce the scrap pieces to a powder form.

Figure 100: Ford Motor Co. Process for Improving the Sinterability of Iron Powder Derived from Comminuted Scrap Metal

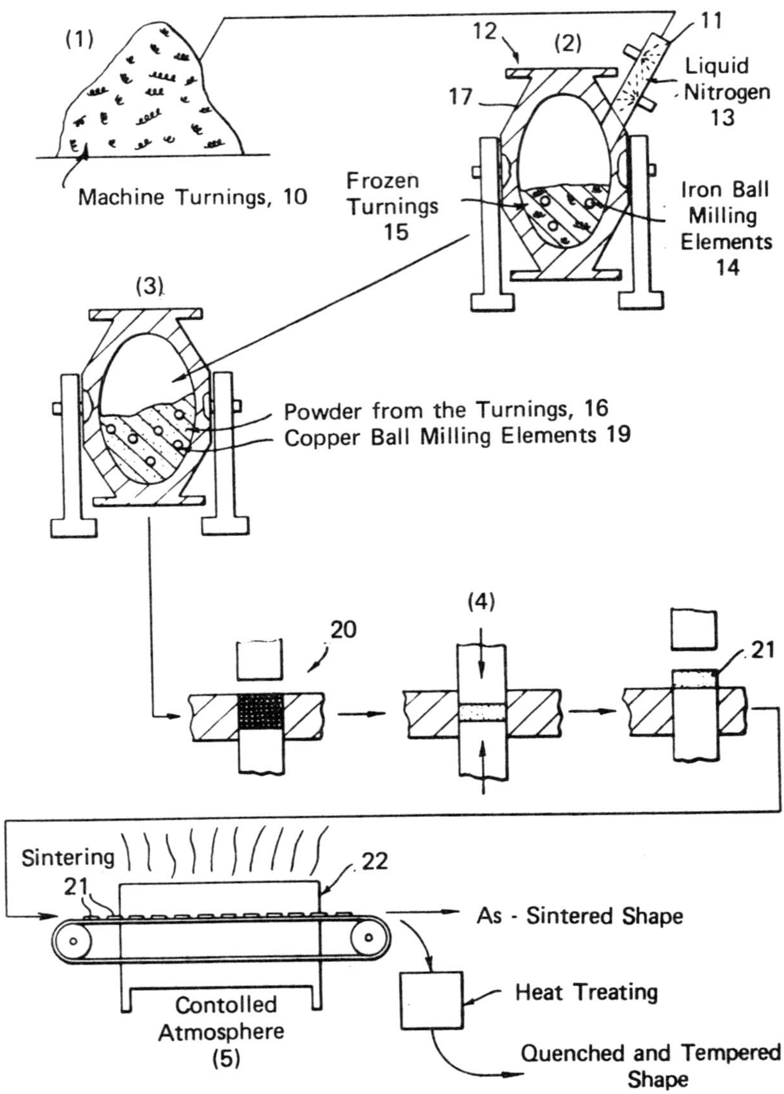

Source: U.S. Patent 4,129,443

The powder **16** will typically have both a coarse and a fine powder proportion. Both proportions will be comprised of particles which are flake or layered in configuration; each particle will be highly irregular in shape and dimension, none being spherical in shape.

The comminuted cryogenic powder **16** is then subjected to another impacting step, but this time at ambient temperature conditions. The powder is placed preferably in another ball milling machine, the machine having elements **19** laden with a special anti-oxidation coatable agent, such as copper or iron. The elements are preferably in the form of solid balls of about 0.5 inch in diameter and consisting of iron or copper. The coatable protective agent should be characterized by (a) a hardness less than that of the coated powder to promote transfer to the particles upon impact between the elements and particles, (b) being completely soluble in the metal of the particles, (c) being easy to abrade, and (d) acting as an oxidation barrier.

In trials, the interior chamber was a 3 inch x 6 inch cylinder, powder charge was 10 in^3, and the milling time was about 48 hours. Milling time and rate depend on mill volume, mill diameter size of copper or iron balls and the speed of rotation. The function of this second impacting step is two-fold: to provide an oxidation barrier on each powder particle and to cold work each coarse particle. The balls transfer, by impact, a portion of the copper or iron ingredient carried by the ball milling elements **19** so as to form a copper or iron shell about substantially each particle of the powder **16**. The finer powder will obtain a copper or iron or a combination coating by true abrasion or scratching with the surface of the ball milling elements **19**. Ball milling elements **19** should have a diameter at least 50 times the largest dimension of any of the particle shapes of the cryogenic powder **16**.

The ball milling operation also must generate non-natural defect sites (resulting from cold working) in substantially all powder particles above 124 microns; the ball milling operation herein should be carried out so that substantially each coarse particle has at least one defect site therein. This can be accomplished by rotating the housing **20** to impart a predetermined abrading force from the balls **19**.

A quantity of powder conditioned from the above impacting is compacted by a conventional press **20** to a predetermined density, preferably, about 6.6 g/cc. This is brought about by the application of forces in the range of 30 to 35 tsi. The presence of the copper or iron envelope about the powder particles improves compressibility. With prior uncoated powders, a density of about 6.4 g/cc is typically obtained using a compressive force of 85,000 psi; with the powder herein, densities of about 6.6 g/cc are now obtained at the same force level.

The shape **21** into which such powder is compacted is designed to have an outer configuration larger than that desired for the final part. A significant and highly improved shrinkage takes place as a result of the next step; the shrinkage can be a predetermined known factor and allowance can be made in the compacted shape **21** of this step. Shrinkage will be in the controlled limits of 0 to 15%.

The compacted shape **21** is subjected to a sintering treatment within a furnace **22** wherein it is heated to a temperature preferably in the range of 2000° to 2100°F, for ferrous based cryogenic powder. The temperature to which the compact is heated should be at least the plastic region (typically sintering temperature) for the metal constituting the powder. A controlled or protective atmosphere is maintained in the furnace, preferably consisting of inert or reducing gases.

At the sintering temperatures, atomic diffusion takes place between particles of the powder particularly at solid contact points therebetween; certain atoms of one particle are supplied to fill the defect sites or absence of certain atoms in the crystal structure of the contacting particle, the defect sites being present as a result of cold working in the above step. Diffusion is accelerated to such an extent, that an increase of more than 100 times is obtained.

It is theorized that at least 60% of the improvement in physical properties of the resulting sintered shape is due to the controlled cold working of the coarse powder particles. The increased diffusion is responsible for the increase in shrinkage.

The copper or iron envelope on the particles serves to essentially prevent oxidation of certain elements or ingredients within the powder particles, particularly manganese and silicon. With typical ball milling parameters (physical size of mill, speed change and ball size) sufficient to the job, it can be statistically calculated that substantially each particle of the cryogenic powder will possess an impervious copper or iron shell. However, a totally impervious shell is not absolutely essential to obtaining an improvement of some of the properties herein.

L

LEAD FROM GUN CLUB SOILS

A substantial market exists for scrap lead and one largely unexploited source of such lead is the ranges of gun clubs, particularly the ranges of those clubs which specialize in trap shooting. Over the course of years considerable lead is expended at these clubs; it being in the form of shot discharged from shotgun shells. This shot lies in a band located between 200 and 300 yd from the firing position, and may be buried as deep as 6".

Heretofore, lead shot has been successfully recovered from relatively loose or sandy soils, but only at great expense and considerable difficulty. Moreover, the lead is relatively dirty and must be washed. Efforts to recover lead shot from soils containing a substantial amount of clay or highly compacted top soil have met with little success.

Therefore, a process has been developed by *F.C. Peterson; U.S. Patent 4,039,433; August 2, 1977; assigned to C. Hager & Sons Hinge Manufacturing Company* which is applicable to soils and particularly to highly compacted soils such as clay, which is economic and which produces a washed lead product.

The process involves mixing soil containing metal with water so as to convert the soil to mud, maintaining the mud in suspension within the water while allowing the metal to gravitate through the water, and thereafter removing the metal from the water. It is also embodied in an apparatus including a tank, means for introducing water into the tank such that turbulence is created, and means at the bottom of the tank for conveying the metal which gravitates downwardly away from the bottom of the tank. Figure 101 is a perspective view of the apparatus for performing the process.

The first step in this process is collecting the soil containing the lead shot. This may be achieved by removing the top 6" of soil with a conventional scraper which is a large piece of earth moving equipment. The scraper, of course, is operated only over the 100 yd band in which the shot lies. The soil stripped by the scraper is delivered to an earth shredder **A** which resembles a conventional shredding machine. The rotating knives or hammers of the shredder **A** reduce the soil to a relatively small size and break up any large clumps. The shredder **A** discharges the shredded soil downwardly where it falls onto a conveyor **B** which transports the soil to a lead separator **C**.

Within the separator **C** the soil is mixed with relatively turbulent water so that the soil is quickly converted into mud which is maintained in suspension. Clean water is continually forced into the separator **C** from the bottom thereof and this flow of clean water buoys the suspended mud upwardly and indeed causes it to overflow the lead separator **C**. The lead shot, being considerably denser than the soil, gravitates downwardly through the incoming clean water and is collected. As the shot passes through clean water it is thoroughly washed so that the shot which is collected is entirely free of the soil. The recovered lead shot may be reused in shot-gun shells or sold as lead scrap.

Figure 101: C. Hager & Sons Apparatus for Lead Recovery from Target Range Soils

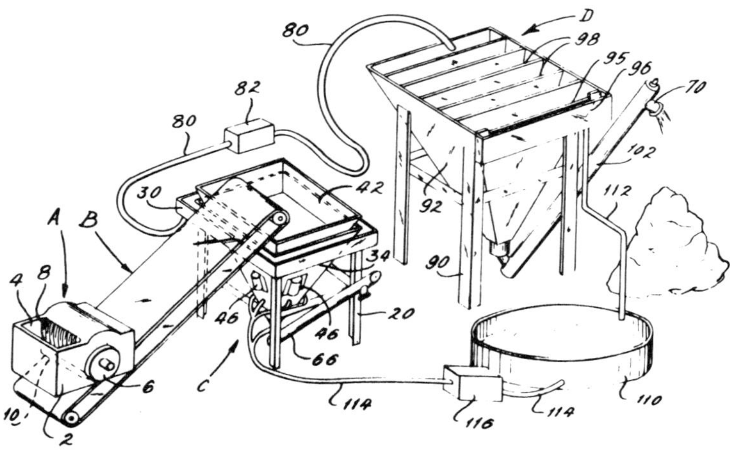

Source: U.S. Patent 4,039,433

The soil or mud suspension which overflows the lead separator is directed to a settling basin **D** where the mud settles out and is thereafter pumped to a mud pile. This mud may be returned to the range to replace the soil previously stripped away, or it may be used for other purposes such as land fill. The water overflows the settling basin **D** and is pumped back to the lead separator **C** where it is utilized as the clean water supply. In other words, the same water is circulated between the settling basin **D** and the lead separator **C** so a continuous supply of fresh water is not necessary.

The earth shredder **A** is conventional and includes a housing **2** provided with an upwardly opening inlet **4**. The housing **2** contains a rotor **6** provided with knives or hammers **8** which move past the inner end of the inlet **4** and thereafter past a grate **10** in the bottom of the housing **2**. The shot-bearing soil which is collected by the scraper is dumped into the inlet **4** of the shredder **A** and thereupon falls into the paths of the rapidly revolving knives **8** which reduce the soil to a relatively fine composition. All of the soil discharged from the shredder **A** should pass a ½" screen. The reduced or shredded soil drops into the conveyor **B** which delivers it to the lead separator **C** mounted on frame **20.**

Attached to the top members is collecting trough **30,** the bottom of which is somewhat below the top edge of a wier so that any water which flows over the wier will cascade into the trough **30** where it will collect.

The frame **20** supports an agitating tank **34** which possesses the shape of an inverted pyramid and is formed by four generally triangular walls joined together at their sides. Each wall has a gusset welded to it midway between its sides.

The gussets support an inlet funnel **42** which, like the agitating tank **34**, is in the shape of a truncated pyramid.

Vibrator units **46** induce vibrations in the four walls at a force of between 0 and 1,000 lb. The frequency at which the vibrator units operate whould be between 0 and 2,500 cp. The four vibrator units **46** need not be synchronized nor need they even operate at the same frequency.

The discharge pipe leads to a screw-type conveyor **66** which extends upwardly at an oblique angle and projects from the frame **20** between two of the legs thereof. The conveyor **66** includes a barrel having an inlet port at its lower end and an outlet spout at its upper end.

The discharge spout in the trough **30** surrounding the upper end of the agitating tank **34** is connected with a drain line **80** which leads to the settling basin **D**. The drain line **80** passes through a pump **82** which lifts the suspension upwardly to the upper end of the settling basin **D**, inasmuch as the settling basin **D** is somewhat higher than the lead separator **C**.

The settling basin **D** includes a frame **90** which supports a settling tank **92** having the shape of a pyramid, the base or larger end of which is open and presented upwardly. Being a pyramid, the tank **92** has four generally triangular walls **96** which are welded together along their sides. The upper margin of one of the walls has an elongated depression **95** so that it is slightly lower than the upper margins of the remaining walls and adjacent to the depression **95** is an overflow trough **96** which is likewise mounted on the frame **90**.

The drain line **80** opens into the settling tank **92** adjacent to the wall opposite from the wall along which the trough extends. Between these two walls are a plurality of baffles **98** which traverse the interior of the tank **92** generally parallel to the trough **96**. The baffles **98** are oriented generally vertically and have their upper margins located at the upper edges of the tank **92**. Their lower margins, however, are spaced upwardly from the bottom of the tank **92** so that water may pass beneath the baffles **98**. At the very bottom of the tank **92** where the four walls converge, the tank **92** is provided with a discharge spout which leads to a screw-type conveyor **102**.

The overflow trough **96** is connected with a storage tank through a drain line **112** so that water which flows over the depression **95** on the sidewall adjacent to the overflow trough **96** will collect in the storage tank **110**. The storage tank **110** is in turn connected through a supply line **114** to the several supply pipes of the lead separator **C**. The supply line **114** has a pump **116** in it.

LEAD FROM LEAD REFINERY DROSS

Most lead ores contain significant amounts of arsenic and antimony in addition to other impurities. Lead bullion is produced by roasting the ores and then smelting the roasted ore in a blast furnace. Molten lead bullion from the furnace is charged to primary drossing kettles in which oxides and other insolubles float to the surface of the molten lead at a later stage. Sulfur is then added to the partially drossed molten lead to form sulfides of copper, and other impurities.

During this so-called decopperizing stage the molten lead is stirred to insure complete reaction between the sulfur and the copper. Once the decopperizing reactions are completed the sulfides formed are allowed to float to the surface of the bath, and the refined lead bullion and the sulfur dross are then separated. The sulfide dross obtained in this stage can be recycled to either the blast furnace or to the primary drossing kettles but in either case the dross from the primary drossing kettles is conventionally smelted in a reverberatory furnace to recover additional amounts of lead and to recover copper and other metals in the form of matte and/or speiss which can be treated by copper smelters.

Lead ores mined in the southeastern portion of Missouri are unusual in that they contain significant quantities of nickel and cobalt as well as arsenic and copper. These characteristics of southeastern Missouri lead ores cause problems in the treatment of the lead refining kettle drosses resulting therefrom. The presence of significant quantities of nickel and cobalt results in the formation and the precipitation on the furnace sidewalls of high melting point nickel- and cobalt-containing speisses. The continual buildup of these deposits in a conventional dross smelting furnace, such as a reverberatory furnace, gradually reduces the capacity of the furnace and ultimately forces furnace shutdown for removal of the deposit.

Thus, a process has been developed by *D.C. Cusanelli, W.R. Opie and M. Yavorsky; U.S. Patent 4,168,155; September 18, 1979; assigned to Amax, Inc.* in which lead refining drosses are fed to a furnace containing molten slag that is heated to at least 1200°C by the resistance to electric current flowing between at least two electrodes immersed in the slag. The drosses are smelted to separate matte and metal phases which are tapped from the furnace.

The process is particularly advantageous for smelting drosses containing significant amounts of nickel or cobalt and arsenic which react to form speisses. The speisses have densities between that of the matte and the metal and melting points that are high compared to the matte and metal. Resistance heating of the slag insures that any speiss formed is maintained in the molten state thereby minimizing precipitation of the speiss on the furnace walls.

LEAD FROM LEAD SCRAP

Secondary lead is lead recovered from scrap, wastes, drosses, and residues. Most secondary lead is derived from wornout, damaged, or obsolete fabricated products such as battery plates, cable covering, pipe, and sheet. Such material is collected, remelted, and refined in secondary smelters to produce refined lead or soft lead and various lead-base alloys. Additional secondary lead is recovered from process scrap, largely drosses and residues generated during the fabrication of lead products, and recycled to secondary smelters for production of refined lead. Some secondary lead materials are reused after remelting without refining, but an increasing proportion is processed in refineries because of the need, in most uses, to meet specifications. About 35 to 40% of the total world use of lead is estimated to be met from secondary materials.

The chief source of secondary lead is from automobile storage batteries that have been scrapped after use. In the United States, about 80% of the lead used in the manufacture of storage batteries is recycled; in some foreign industrial countries, a somewhat higher percentage is recycled. The importance of secondary lead is illustrated by production in the United States, which amounted to 580,000 tons from old scrap in 1976 compared with mine production of 610,000 tons. Over the 10-year period 1967 to 1976, secondary lead from old scrap accounted for 37% of the total amount of lead consumed in the United States.

Much of the secondary lead produced is sold as lead alloy; this is particularly true of lead from reclaimed batteries and cable covering, which contains small percentages of antimony and other metals. The bulk of the secondary lead containing antimony is resold to battery manufacturers. Secondary lead containing tin is most often reused in the manufacture of solder, bearing metals, and other lead-tin alloys. Through proper refining techniques, soft pig lead meeting primary lead specifications is also recovered from lead-base scrap. Secondary lead is mainly recovered as antimonial lead, as soft lead, and in other lead- and copper base alloys.

During 1970 to 1975, antimonial lead averaged 55% of the total secondary lead recovered; soft lead, 30%; and lead in other lead and copper base alloys, 15%. Since 1973 lead recovered as antimonial lead has declined from about 57 to 49% of the total recovered in 1976. Soft lead recovery increased from about 28% in 1973 to about 44% in 1976, largely reflecting the growing use of maintenance-free batteries using lead alloys containing little or no antimony. Gasoline and pigments are the major use categories in which lead cannot be reclaimed. Lead in ammunition, foil, solder, annealing, and galvanizing is to a large degree unreclaimable. Details of lead scrap recovery processes and their environmental impacts are reviewed by Nack et al (15).

LEAD FROM SOLDER SKIMMINGS

Solder baths used in can manufacturing are covered with a layer of ammonium chloride flux to prevent oxidation of the solder. Zinc chloride is added to the flux to reduce fuming of the molten ammonium chloride. Lead oxide present reacts with the chlorides to form lead chloride. In skimming off the slag which is formed as the cans are soldered some of the flux and solder is removed because of the inefficiency of the skimming procedure. The skimmings after they harden contain solder chunks and granular lead chloride in a matrix of solid ammonium chloride and zinc chloride. These skimmings are referred to herein as solder skimmings or solder bath skimmings.

The existing process for recovering solder from solder bath skimmings consists of melting the skimmings in a gas-fired furnace, tapping the molten solder and discarding the top layer of zinc chloride, ammonium chloride, and lead chloride. In the process of furnacing, an ammonium chloride fume is emitted from the stack at a level far in excess of current air pollution control limits. Air pollution correction equipment for the off-gas is costly. High energy wet venturi scrubbers will not meet air pollution codes on plume opacity. Electrostatic precipitators are prohibitively expensive since the elements must resist corrosive attack by the acidic fumes.

In view of the disadvantages of prior processes for recovering solder from scrap solder skimmings, there is a need for a method for economically and practically recovering solder from the solder skimmings which is within acceptable air pollution control limitations.

A process which was designed to overcome these problems has been developed by *J.E. Reynolds; U.S. Patent 4,108,635; August 22, 1978; assigned to Hazen Research Inc.* involves grinding the skimmings, water leaching the skimmings to dissolve the zinc chloride and ammonium chloride, screening the leach residue to recover solder from the leach slurry, separating the ammonium and zinc chloride solutions from the lead chloride in the leach slurry, and contacting the lead chloride with an alkaline flux, a reductant, and an oxygen-containing gas under heat to recover lead from the lead chloride. Such a process is shown in Figure 102.

Figure 102: Hazen Research Inc. Solder Skimmings Recovery Process

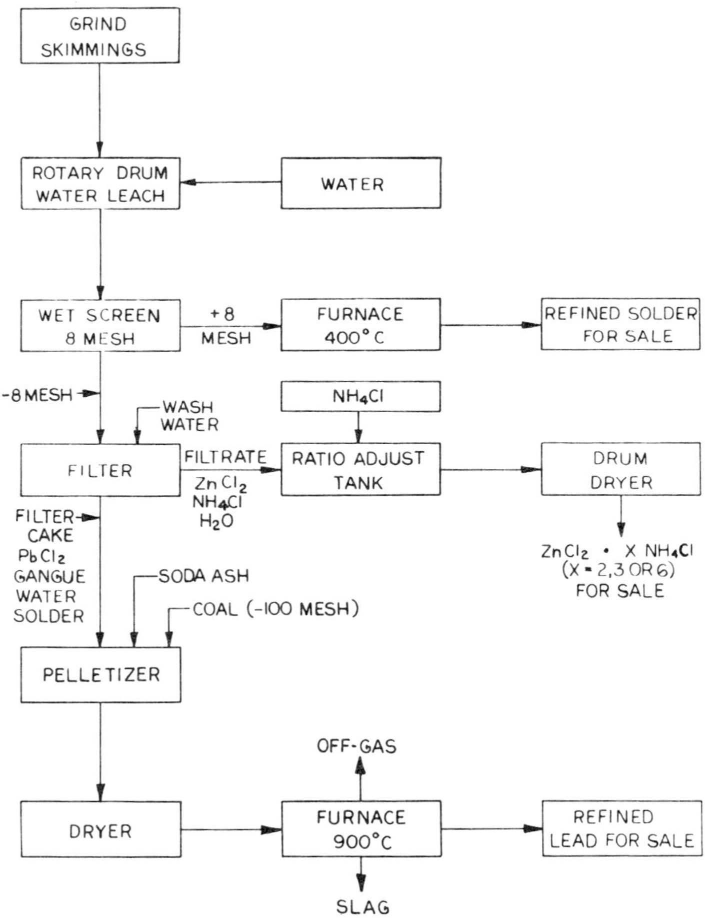

Source: U.S. Patent 4,108,635

LEAD FROM STORAGE BATTERY SCRAP

See also "Polypropylene from Scrap Storage Batteries" in reference (1).

Conventional processes for recovering lead from lead battery scrap depend on the use of a reverberatory or lead blast furnace.

These prior art processes result in production of copious quantities of SO_2 and, therefore, present severe air pollution problems. Attempts to solve these problems by use of improved emission control processes and equipment have resulted in greatly increased costs of operation.

It has been found that the pollution problems can be largely overcome by means of a process in which a hydrometallurgical step is employed to convert $PbSO_4$ to an easily reduced form of PbO, without emission of significant amounts of SO_2. This process employs an initial step in which the battery scrap, the essential component of which is $PbSO_4$, is reacted with a slurry of calcium hydroxide.

The resulting product consists of an aqueous slurry of PbO and $CaSO_4 \cdot 2H_2O$, which is then filtered and dried in the second step of the process.

The third step of the process comprises reduction of the PbO to metallic lead by means of carbon, which may be readily accomplished at temperatures of about 600° to 650°C. However, it was found that the presence of calcium sulfate particles from the first step of the process prevents coalescense of the lead into a molten pool that is readily separated as a metallic lead product of high purity.

However, it has been found by *D.A. Wilson; U.S. Patent 3,940,265; February 24, 1976; assigned to U.S. Secretary of the Interior* that the calcium sulfate may be fluxed by means of a mixture of KCl and NaCl, thereby fluidizing the calcium sulfate and permitting formation of a molten pool of the reduced lead that is readily separated from the reaction mixture.

The process is shown in the block flow diagram in Figure 103. Feed material **1** consists of conventional lead battery scrap, which typically consists of about 61% paste material and 39% metallic lead-antimony grid material. The paste material consists predominantly, i.e., about 41%, of $PbSO_4$, the balance being PbO_2 and fine metallic lead.

The scrap is initially ball milled, wet or dry, in ball mill **2** to liberate the paste and reduce it to a particle size of about -100 mesh. The metallic grid material is not affected but remains as pieces of chunks of metal. The ball-milled product is then fed to reactor **3** and admixed with $Ca(OH)_2$ slurry **4**. The reactor may be any suitable reaction or mixing vessel such as a revolving drum, twin shell, or double cone mixer. The $Ca(OH)_2$ slurry will consist of 15 to 20% CaO in water, and is admixed with the battery scrap in an amount sufficient to supply approximately the stoichiometric amount of $Ca(OH)_2$ as required by the equation:

$$PbSO_4 + Ca(OH)_2 + H_2O \longrightarrow PbO + CaSO_4 \cdot 2H_2O$$

The resulting mixture is allowed to react for a time sufficient to substantially complete the reaction, a reaction period of 2 to 5 min usually being sufficient. Room temperature is usually sufficient; however, temperatures up to 50°C may be employed if shorter reaction times are desired.

Figure 103: U.S. Dept. of Interior Process for Recovery of Lead from Battery Scrap

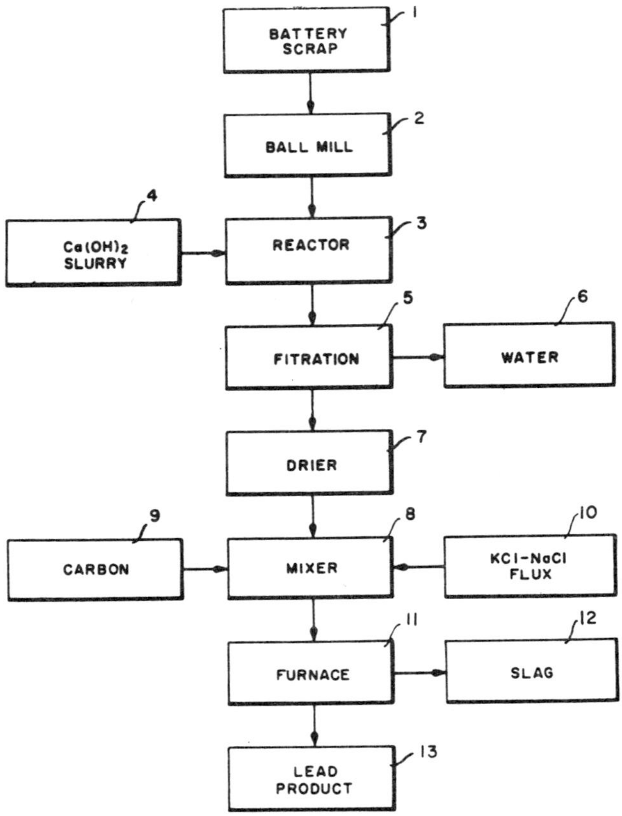

Source: U.S. Patent 3,940,265

The reaction mixture is then filtered, reference **5**, or otherwise treated for removal of water **6**, and the residue passed to drier **7** where a temperature of 150° to 200°C is employed to substantially completely remove any remaining free water. The dried product is then fed to mixer **8**, which may be any conventional mixing apparatus such as those mentioned above, where it is dry mixed with the carbon reductant **9** and the KCl-NaCl flux **10**.

This mixture is then fed to furnace **11** where the combined fluxing and reducing reactions take place. The carbon may be in any conventional form suitable for use as a reductant, e.g., charcoal, coke, carbon black, lampblack, etc. It is preferably employed in the form of a fine powder, but granular forms may also be used. It is used in approximately the stoichiometric amount required to reduce the lead oxide formed.

The KCl-NaCl flux mixture is added in an amount sufficient to flux the calcium sulfate formed, thereby forming a fluid mixture from which the molten lead product is readily recovered. Preferably, the amounts of KCl and NaCl are sufficient to form a ternary eutectic mixture with the calcium sulfate. This mixture consists of 38.5 mol % of KCl, 19.0 mol % of $CaSO_4$ and 42.5 mol % of NaCl, and melts at 605°C. However, mixtures comprising 32 to 54 mol % KCl, 21 to 19 mol % $CaSO_4$ and 48 to 25 mol % NaCl are generally sufficient to fluidize the calcium sulfate and permit efficient recovery of the lead.

Furnace **11** may be any conventional reactor designed to provide the required reaction temperature and to be inert to the reagents employed. Reaction temperatures of 605° to 700°C are satisfactory, with a temperature of about 650°C generally being preferred. Optimum time of the reaction will depend on the specific types and amounts of reactants, as well as temperature, but a period of 1 to 2 hr is usually sufficient for substantially complete conversion to the PbO to molten metallic lead. Since the carbon has a lower density than the flux, an up and down stirring motion of the mixture during the reaction is generally necessary to ensure efficient reduction of the PbO.

Lead product **13** collects as a molten pool at the bottom of the furnace and is readily separated from slag **12** by means of conventional methods such as tapping, pouring, or use of a molten metal pump.

Inevitably in the battery manufacturing process there are defective or damaged finished dry plates or elements which must be scrapped. It is, of course, desirable to reclaim as much of the scrap material as possible in order to bottom manufacturing costs.

One current method for processing defective or damaged plates or elements for scrap reclamation comprises placing the plates or elements in a melt-down kettle having a temperature about the melting point of lead. Approximately 50% of the grid metal or approximately 20% of the entire plate or element weight can be recovered as metallic lead in this melt-down process. The remaining 50% of grid metal together with the nonmetallic components of the plates or elements are collected as dross in the melt-down process which must be further processed by an expensive smelting process for further metallic lead recovery.

The smelting process yields about 85% metallic lead from the dross with approximately 12% of the entire weight of the original scrap not recovered by these combined processes, i.e., melt-down and smelting.

In addition to the value of the nonrecovered portions of the original scrap, the smelting operation costs the battery manufacturer about 50% of the entire value of the original plates or elements. It can thus be seen that a substantial savings can be obtained by avoiding the expensive smelting process.

A process developed by *T.D. Litt, R.C. Shelton and E.J. Colwell; U.S. Patent 4,009,833; March 1, 1977; assigned to ESB Incorporated* is one for separating lead grids from battery active material without the need for melting and smelting. The process has the added features that it requires little labor and the separated active materials of the battery plates are reduced to small grains or powders suitable for reuse by direct addition to paste mixing apparatus.

More specifically, the process is one in which lead-acid storage battery plates damaged during battery manufacture are placed in a trough having an upward slope. The trough is vibrated causing the nonmetallic components of the plates to be shaken free from the metallic components of the plates with the nonmetallic components being reduced in size during the vibration process. The nonmetallic components of the plates are discharged from the trough as a fine powder through suitable openings in the bottom of the trough while the metallic components climb the trough and are discharged at the upper end of the trough. Such an apparatus is shown in plan and elevation in Figure 104.

Figure 104: Plan and Elevation of ESB Inc. Apparatus for Separating Metallic from Nonmetallic Components in Scrap Storage Battery Plates

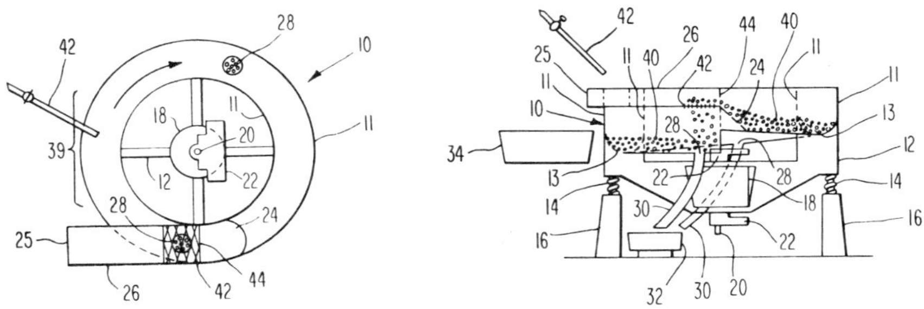

Source: U.S. Patent 4,009,833

A helical trough is shown generally at **10** having upstanding sides **11** and bottom **13**. The trough **10** is mounted on a frame **12**. The frame is supported by springs **14** and foundations **16**. A motor **18** having a vertical shaft **20** is also mounted on frame **12**. Two eccentric weights **22** are attached to shaft **20**. When the motor **18** is energized, the motion of weights **22** imparts a circular vibrating motion to the trough **10**. The slope of the bottom of trough **10** is uniform except for a short steep ramp or dam **24** at the upper end of the trough. The ramp feeds a discharge chute **26**. Two openings **28** having a screen with a series of small holes therein allow active material in wet or dry form to flow out of the trough. Chutes **30** direct the active material to a pan **32** where it is collected for reuse. A second pan **34** collects the grids dropping from the discharge end **25** of chute **26**.

In operation, damaged battery plates and other scrap material are fed to the trough **10** at the general area indicated by **39**. The plates slowly ascend the slope of the trough in the direction of the arrow and as they do so, all nonmetallic material is shaken from them. There is a build-up of grid parts just before ramp **24**, forcing a stream of such parts up the ramp **24** out chute **26** and into the pan **34**. Nonmetallic materials tend to drift vertically down through the grinding media in the trough **10**, pass out through openings **28** into chutes **30** and into pan **32**. During the passage of the nonmetallic particles down the trough, they rub together and rub on the ascending lead parts with the result that they are broken up into fine powder, or grains. The largest has been measured to be no greater than 500 μ in diameter.

It has been found that separation occurs whether the plates are wet or dry, formed or unformed. However, with dry plates, there is an undesirable emission of dust from the apparatus. It has been found that a small trickle of water fed into the trough with the battery plates as at **42** eliminates the dust problem and aids in the efficiency of the process.

The separating trough need not be circular. In certain installations, it may be more convenient to use a straight trough. Also, while openings **28** have been illustrated for the discharge from the trough of the separated active materials of the battery plates, it should be understood that the bottom **13** of the trough may be partially or wholly formed of a screen having an appropriate mesh.

In order to improve the efficiency of the separation process by decreasing the time required for treating the scrap battery plates, a grinding media may be provided in the trough **10**. Such a grinding media is illustrated by the reference numeral **40**. For ease of illustration and clarity, the grinding media **40** is not illustrated in the planview. The grinding media provided may comprise any suitable material having a hardness greater than the materials introduced into the trough **10**, as for example, steel or alumina balls or cylinders. Alumina balls or cylinders are preferred for the reason that they are noncontaminating. The grinding media serves to speed the pulverizing of the active materials of the scrap battery plates as well as aiding in the separation of the active materials from the lead grids.

The grinding media **40** tends to build up and work itself up the ramp **24** just as the grid parts do. The grinding media passes through a screened area **42** provided in the forward end **44** of chute **26** and returns to the lowermost part of the trough **10**.

The metallics collected in pan **34** may be fed to a grid casting pot or other lead melting facility while the nonmetallics collected in pan **32** may be fed to a paste mixing operation.

A suitable slope for the trough is a rise of 8" in 142" or 2.4°. A helicoid trough 14" wide, 16" or more deep and having a mean diameter of 61" vibrated with an excursion of about ⅛" with a 20 hp motor turning at approximately 1,750 rpm will separate up to 2,000 lb of scrap plates into bare metal and nonmetallics per hour.

A process developed by *A.E. La Point; U.S. Patent 4,018,567; April 19, 1977* is one in which whole or shredded lead-acid batteries, together with a quantity of sodium carbonate and water, are fed continuously into a rotating drum separator containing a charge of grinding balls. Agitation of the mixture, aided by the internal drum construction features, further breaks up and degrades the battery fragments, neutralizes any contaminating electrolyte, transposes the finer particles of lead sulfate into lead carbonate, and forms a heavy-medium suspension of the active material on which organic battery fragments float.

The suspension of active material constantly overflows from the drum at one end and carries the organic fragments with it into a first trommel, while sinkable fragments of grid metal and other battery parts of lower grade antimonial lead alloy are mechanically removed at the opposite end and deposited in a second trommel. A portion of the overflowing suspension is pumped back to the drum together with a regulated quantity of water, while the remainder is delivered to a thickener for further processing.

A process developed by *G. Tremolada; U.S. Patent 4,026,477; May 31, 1977; assigned to A. Tonolli & Co. SpA Italy* is one for separation of the components of scrap lead storage batteries including crushing, and then heating the pieces while tumbling, dry screening the paste from the residual materials which are fed to a sink-float apparatus whose fluidized pulp comprises a liquid suspension of battery paste. The pH and density of the pulp are regulated in the pH range of 7 to 10 by addition of sulfuric acid. After sink-float separation the drained and washed materials are recovered and the pulp and wash liquid are recycled back into the system. Such a system is shown in Figure 105.

Figure 105: A. Tonolli & Co. SpA Process for the Separation of Components of Scrap Storage Batteries

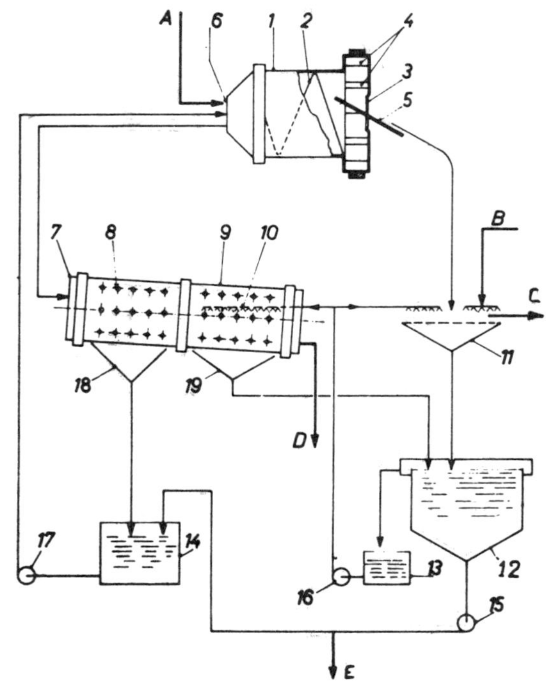

Source: U.S. Patent 4,026,477

The plant comprises a conventional wet separator **1** having a revolving drum provided with a spiral **2**, a discharge mouth **3** furnished with paddles **4**, a chute **5** and a charging mouth **6**; a revolving drum **7** the axis of which is inclined slightly to the horizontal with a plurality of openings in the walls and divided in two sections **8** and **9**, drum **7** is also provided with bored pipes **10** for washing water; two hoppers **18** and **19** for collecting and conveying the drain liquids from sections **8** and **9** of drum **7**; a screen **11**; a thickener **12**; a tank **13** for collecting the decanted water from thickener **12**; a tank **14** for collecting the pulp, provided with a nonrepresented stirrer; and pumps **15**, **16** and **17**. The process is as follows. Lead storage batteries are loaded in bulk in a drum crusher, together with a stream of counter-current hot gas. The crushed batteries, in sizes between 2 and 15 cm, are freed of paste by means of the screening section of the drum crusher, provided with openings between 2 and 30 mm, are fed at **A** together with the pulp into the mouth **6** of separator **1**. The pulp comes out from mouth **6** of separator **1** together with the floating inactive materials, while the metallic components are conveyed through spiral **2** towards mouth **3** and thereto discharged by paddles **4** and chute **5**.

The inactive materials and the pulp discharged from **1** are conveyed in drum **7**, the pulp is recovered in the tank **14** and the inactive materials after being washed with water, by means of pipes **10** and pump **16**, are discharged at **D**. The metallic components are sent onto screen **11** and after being washed with water recycled from **13** and with fresh water **B**, are discharged at **C**. Washing waters from section **9** and drum **7** and from screen **11** are sent into the continuous thickener **12**, the overflow of which is sent into tank **13**, the thickened pulp is recycled, through pump **15**, into the tank **14**; any possible excess of paste is removed from the cycle at **E**, e.g., by filtering part of the thickened pulp discharged from thickener **12**.

It is evident that by adopting a sufficiently powerful thickener **12**, it is possible, by controlling the flow of pump **16** to regulate within wide limits the amount of washing water and thus to reduce at will the amount of active substance adhering to the sink and to the float products.

A process developed by *M. Liniger; U.S. Patent 4,030,916; June 21, 1977* is one in which lead is recovered from the lead-containing sludge from used batteries by first heating the sludge at 100° to 150°C to drive off the water. The sludge is subsequently reacted at about 1000°C with carbon powder in a closed electric furnace. The relatively small volume of water-free waste gas produced leads to a marked saving in energy. Not only dust but also the SO_2 can be removed simply from the waste gases.

A process developed by *J.R. Birk; U.S. Patent 4,058,396; November 15, 1977; assigned to Electric Power Research Institute, Inc.* is one in which lead is recovered from batteries by contacting comminuted batteries with a molten alkali metal salt, reducing lead compounds to lead, separating molten lead from the molten salt, and transferring the molten salt and char to a combustion zone, where the salt is reheated to the desired temperature and recycled for contacting with the batteries. A portion of the alkali metal salt is withdrawn and purified, so as to maintain a relatively constant composition of the alkali metal salt.

A process developed by *H. Ahonen, J. Lindroos, V. Sarkkinen and R. Seppanen; U.S. Patent 4,096,045; June 20, 1978; assigned to Lyijyvalkoistehdas Gronberg Oy-Blyvittfabriken Gronberg AB, Finland* is one for the recovery of lead from battery scraps by mixing the lead scraps at elevated temperature with an aqueous solution of an alkali metal chloride and hydrochloric acid in order to dissolve, in the form of lead chloride, the lead compounds present in the lead scraps. The process then involves separating an aqueous solution of this lead chloride from insoluble hard lead, which is recovered, crystallizing lead chloride from the cooling aqueous solution, which is separated from the lead chloride crystals, and finally subjecting the obtained lead chloride to electrolysis for the recovery of metallic lead. Such a process is shown in Figure 106.

Figure 106: Finnish Process for the Recovery of Lead from Lead Scraps

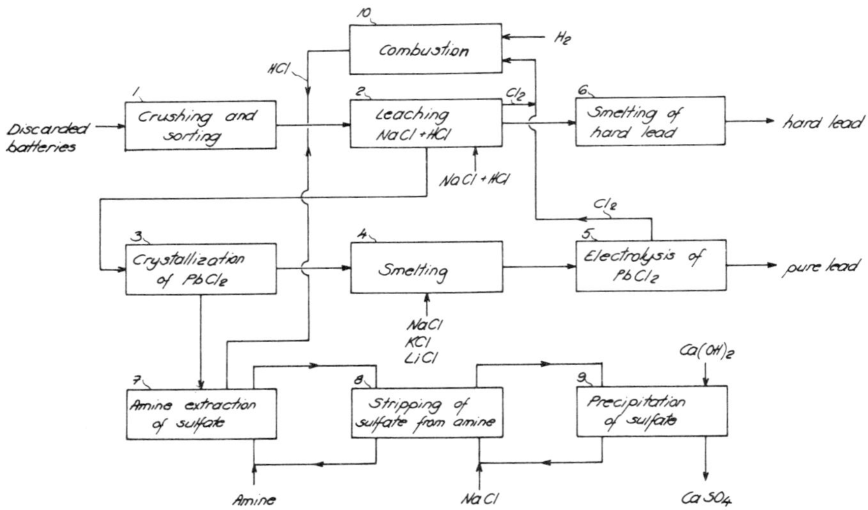

Source: U.S. Patent 4,096,045

The discarded batteries are crushed **1**, and the accumulator boxes, separators and other such materials with a lower density are separated from the lead parts in a water suspension with the aid of their different settling rates, which per se is conventional. The separated and crushed material containing lead, lead oxides, and lead sulfates is broken in an impact crusher or the like into suitably sized pieces for the leaching of the pulp and for the reactions occurring during the leach. The pieces are fed into a horizontal leaching reactor **2** provided with rotating mixer arms. Into the same reactor a solution consisting of aqueous solutions of NaCl and HCl is fed from the opposite direction.

In the experiments performed, it has been observed that not nearly as good a leaching result is obtained with pure HCl or pure NaCl as with a mixture of the two. Furthermore, it has been found that sulfuric acid greatly increases the leaching capacity. It has been observed that the best leaching solution is an aqueous solution containing hydrochloric acid approximately 80 g/ℓ, sodium chloride approximately 160 g/ℓ, and sulfuric acid approximately 20 to 70 g/ℓ. Approximately 70° to 80°C has been found to be the best leaching temperature.

The PbO_2 present in the pulp is reduced in PbO, which dissolves as a chloro complex, and the $PbSO_4$ present in the pulp dissolves as a chloro complex and sulfuric acid. When it is reduced, PbO_2 releases chlorine. In this particular leaching process a large portion of the released chlorine can be used for oxidizing the surface of the metallic lead which is continuously exposed owing to the effective mechanical stirring; thus, more lead chloride is obtained.

The chlorine which has not reacted is directed from reactor **2** for burning **10** with hydrogen into hydrochloric acid, which is returned to the leach **2**.

In a well performed leach **2** a concentration of at least 37 g/ℓ is obtained for the lead chloride, and 99% of the lead sulfate and lead oxide present in the pulp is dissolved. Moreover, part of the metallic lead present in the original pulp dissolves as lead chloride.

It has also been found in the leach **2** described above that some cementation of the impurities dissolved on the surface of the exposed lead, i.e., metals nobler than lead, occurs in spite of the acid solution. Thereby a double advantage is simultaneously gained. The products thus obtained are a very pure lead chloride and an already metallic hard lead which is so rich in antimony, for example, that after melting **6** it is suitable for alloying.

Since the obtained lead chloride soution has a high concentration and the solubility of $PbCl_2$ is greatly dependent on the temperature most of the $PbCl_2$ can be precipitated **3** as very pure crystals by lowering the temperature to approximately 18°C, in which case approximately 84% of the lead present in it is in the precipitated crystals. The mother liquor from the crystallization is filtered and returned to the process circulation for further treatment.

The obtained $PbCl_2$ crystals are dried and smelted **4** with NaCl, KCl, or LiCl. Thereby a salt mixture with a low melting point is obtained, from which Pb can be electrolyzed **5** by a known method. The lead obtained from the molten salt electrolysis is very pure, 99.99%.

The mother liquor which is to be returned to the circulation of the process from the crystallization **3** and into which impurities, such as sulfuric acid and metallic impurities, have accumulated from the battery scraps is fed, before its return to the process, to a purifying treatment **7**.

Bisulfate ions and sulfate ions can be extracted **7** very advantageously from the circulating solution by using, for example, some water-insoluble tertiary amine, such as trioctyl amine, for the extraction. In particular, it has been observed in experiments that an extraction **7**, performed by means of a rapid, effective contact, gives a better yield than a long contact terminating in a final equilibrium. For a reduction of the viscosity, and for a better clarification of the extraction dispersion, the reagent has been diluted with, for example, kerosene or xylene. Amines other than tertiary can also be used.

The obtained bisulfate solution of a tertiary or respective amine can be regenerated **8**, **9** for a new extraction cycle by means of, for example, $Ca(OH)_2$, NaOH, or NH_4OH plus $CaCl_2$, using the lastmentioned at the rate at which make-up chlorides are required in the process.

The metallic impurities in the mother liquor after the crystallization of $PbCl_2$ include Fe^{3+}, Cu^{2+}, Sb^{3+} and Sb^{5+}, Sn^{2+} and Sn^{4+}, As^{3+} and As^{5+}, Bi^{3+} and Bi^{5+}, Ag^{+} and Ni^{2+}. Out of these, at least Fe^{3+}, Ag^{+} and Ni^{2+} pass to a high degree into the organic phase. The Cu, Sb and Bi ions cement on the surface of metallic lead during the leaching of the lead salts. During the cementation, more pure lead is simultaneously dissolved in the solution.

The circulating solution purified by the methods described above is returned, after the make-up additions, to the battery scrap leach **2**.

Chlorine is released both from the leach **2** and especially from the lead chloride electrolysis **5**. This chlorine must be burned **10** back into hydrochloric acid so as to be again available in the process cycle.

A process developed by *E.V. Akerlow and K.M. Buchholz; U.S. Patent 4,098,685; July 4, 1978; assigned to Akerlow Industries, Inc.* is one in which lead battery components such as metallic grids and paste are fed into a tumbling drum which has inlet and outlet ends and an intermediate region in which there are screens for the fine particles of paste to discharge to the exterior of the drum. Water is sprayed onto the plates in the drum. A shroud or housing surrounds the region of the drum containing the screens.

Means are provided to create subatmospheric air pressure in the shroud and, consequently, in the drum. The slurry of paste particles is collected in a tank which is integral with the bottom of the shroud. The slurry is passed through a cyclone separator and the water extracted by the separator is recirculated through the lines which supply the spray nozzles. All openings in the drum are subjected to subatmospheric pressure.

Known methods for reclaiming the components of battery plates are detrimental to the comfort, safety and health of those who are required to perform the method. There is always water containing lead oxide particles in the working environment due to spilling and splashing that is incidental to the method. Lead compounds are known to be hazardous to health. Even though the workers wear waterproof clothing, lead compounds deposit on them and expose them to the danger of lead poisoning.

An even greater health hazard results from lead oxide particles inevitably becoming airborne even though the environment and the materials are kept wet. The size of the lead oxide particles is believed to range between 0.1 and 0.5 μ with most particles being below 0.3 μ although some are even smaller than 0.1 μ. These small particles are readily entrained by gas and vapor molecules in the atmosphere. Inhalation of lead compounds is even more detrimental to health than skin contact. Some workers exhibit the symptoms of lead poisoning after a very short exposure to such environment. The Akerlow process is designed to overcome these problems and provide a safe operation for the workers involved.

An apparatus developed by *L.H. Jaquay; U.S. Patents 4,102,676; July 25, 1978; and 4,180,251; December 25, 1979; both assigned to Dravo Corporation* is one in which finely divided dried battery mud is injected into a stream of air through the veni contracta of a venturi in an air conduit. A reducing gas and oxygen are successively introduced into the stream through apertures in the walls of the conduit communicating with annular chambers in fuel and oxygen manifolds surrounding the conduit in sufficient quantities to maintain a stable reducing flame. The droplets of metallic lead and slag which fall from the reducing flame are collected in a refractory lined chamber and separated in a conventional manner.

A suitable form of apparatus for the conduct of such a process is shown in Figure 107. The air is introduced through a conduit **1** having a venturi **3** located at an intermediate point thereof. A feed tube **5** having a flared end **7** for receiving finely divided dried battery mud communicates with the conduit **1** at the veni contracta **9** of the venturi **3**.

The finely divided dried battery mud which is fed through the feed tube **5** becomes entrained in the air introduced in the direction of the arrow **11** into the conduit **1**.

Figure 107: Dravo Corporation Apparatus for Recovering Lead from Battery Mud by an Oxygen Smelting Process

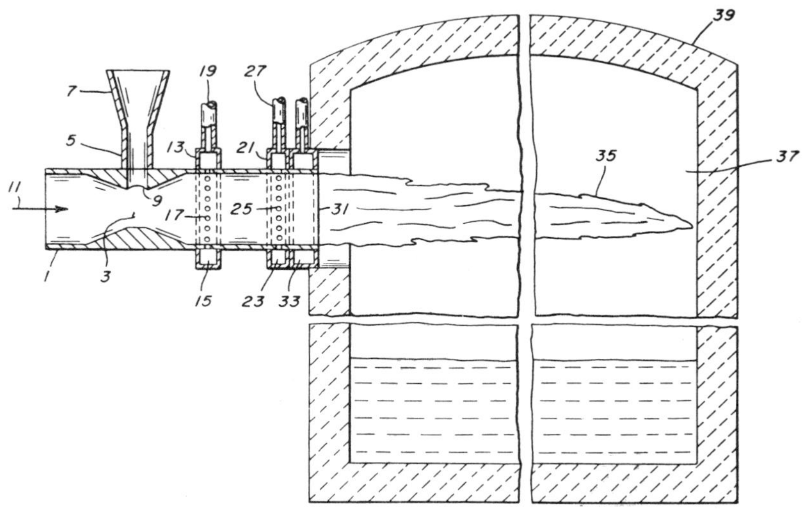

Source: U.S. Patent 4,102,676

Downstream of the venturi **3** is a fuel manifold **13** having an annular chamber **15** which surrounds the conduit and communicates therewith through a plurality of angularly distributed apertures **17** in the wall of the conduit **1**. Natural gas, or any other suitable reducing gas, is introduced into the annular chamber **15** of fuel manifold **13** through supply line **19**.

Downstream of the fuel manifold **13** is an oxygen manifold **21** which also has an annular chamber **23** surrounding the conduit and communicating therewith through a plurality of second apertures **25** in the walls of the conduit. Oxygen of 90%-plus purity is supplied to the annular chamber **23** of the oxygen manifold through supply line **27** in sufficient quantity to ensure a stable flame and to maintain the proper flame temperature for reduction of the mud (about 2100° to 2500°F).

The distance between the apertures in the fuel manifold and the oxygen manifold and the number and diameter of holes in each set are selected to give a fuel rich layer against the burner wall by the time that the mixture reaches the oxygen manifold apertures. The injection of oxygen into this fuel rich layer provides a layer which burns with a high propagation rate and a high temperature which maintains a stable flame envelope for conducting the reduction reactions in the central part of the flame. As an example, these conditions should be met in a burner with a two inch diameter conduit by locating the fuel apertures **17** about two inches downstream of the veni contracta **9** of the venturi **3** and the oxygen apertures **25** about another two inches downstream of the fuel apertures.

In order to protect the tip **31** of the conduit **1** from the high temperature of the reducing flame, cooling water is circulated through an annular chamber **33** surrounding the tip **31**. Even with water cooling, the distance between the oxygen injection apertures **25** and the tip **31** of the conduit must be short, on the order of ½ inch, to prevent destruction of the burner tip.

The burner flame **35**, carrying with it the finely divided dried battery mud, projects into a chamber **37** of a refractory lined vessel **39**. The droplets of metallic lead and slag fall into a pool at the bottom of the chamber **37** where they separate into layers which can be separately tapped in a conventional manner.

Suitable parameters for a pilot plant test model for carrying out the process would be 75 to 120 lb/min of dried and screened battery mud, 650 to 750 scfm of air, 75 to 120 scfm of natural gas and 15 to 25 scfm of 90%-plus oxygen. These proportions when introduced into a two inch diameter conduit produce a stable flame with sufficient temperature to completely reduce the lead compounds while they are still suspended in the flame.

A technique developed by *A.F. Gäumann and V. Dancilovic-Matulas; U.S. Patent 4,107,007; August 15, 1978; assigned to Gesellschaft zur Förderung der Forschung an der Eidgenössischen Technischen Hochschule, Switzerland* is one for the recovery of lead from scrapped lead batteries including mechanically comminuting them and freeing them from battery acid. Then the battery scrap is treated with an alkaline reagent solution to convert lead oxide and sulfate to dissolved lead compounds, pure lead is recovered by electrolyzing the solution, and metallic lead components of the scrap are recovered by a physical separation from the scrap remaining after the dissolution treatment.

A process developed by *K.D. Libsch and M.V. Rao; U.S. Patent 4,115,109; September 19, 1978; assigned to NL Industries, Inc.* is one for producing metallic lead from lead battery plates and other types of lead scrap in a secondary lead smelting operation. This process comprises adding from 1 to 8% moisture and 0.5 to 5% oxygen to the air which is used to react with carbonaceous material such as coke to maintain the temperature in the melting region from 1100° to 1300°C. The moisture is added to the air to react with the coke to form carbon monoxide and hydrogen, and the oxygen is added to maintain the temperature in the melting zone. Increased amounts of lead are produced using smaller quantities of coke and iron when the moisture and the oxygen are added to the air.

An integrated process for the recovery of lead and other components from junk batteries has been described by *M.E. Elmore and J.K. Klang; U.S. Patent 4,118,219; October 3, 1978; assigned to Gould Inc.* This hydrometallurgical process for recycling the paste and electrolyte of junk lead-acid batteries begins by separating the paste and electrolyte from the junk batteries. The paste and electrolyte are then reacted to increase the lead sulfate content of the paste and reduce the acid content of the electrolyte. Excess liquid is then removed from the mixture of paste and electrolyte, after which the paste is mixed with an aqueous ammoniacal ammonium sulfate solution to dissolve a substantial portion of the lead sulfate from the paste.

The remaining solid materials are then separated from the resulting pregnant lead solution, after which lead carbonate is precipitated from the pregnant lead solution by the addition of ammonium carbonate, ammonium bicarbonate or carbon dioxide. The precipitated lead carbonate is removed from the spent liquid and either converted to lead oxide by calcining, or mixed with sulfuric acid and then calcined or reacted with a chemical reducing agent to convert lead dioxide therein to lead oxide, and the lead oxide-containing material is then mixed with sulfuric acid in a second sulfating reaction to increase the lead sulfate content thereof. The resulting product is mixed with an aqueous ammoniacal ammonium sulfate solution to dissolve further lead sulfate therefrom. The resulting pregnant lead solution is separated from any solid materials remaining therein, after which lead carbonate is precipitated from the solution.

Figure 108: Gould, Inc. Process for Recycling Junk Lead-Acid Batteries

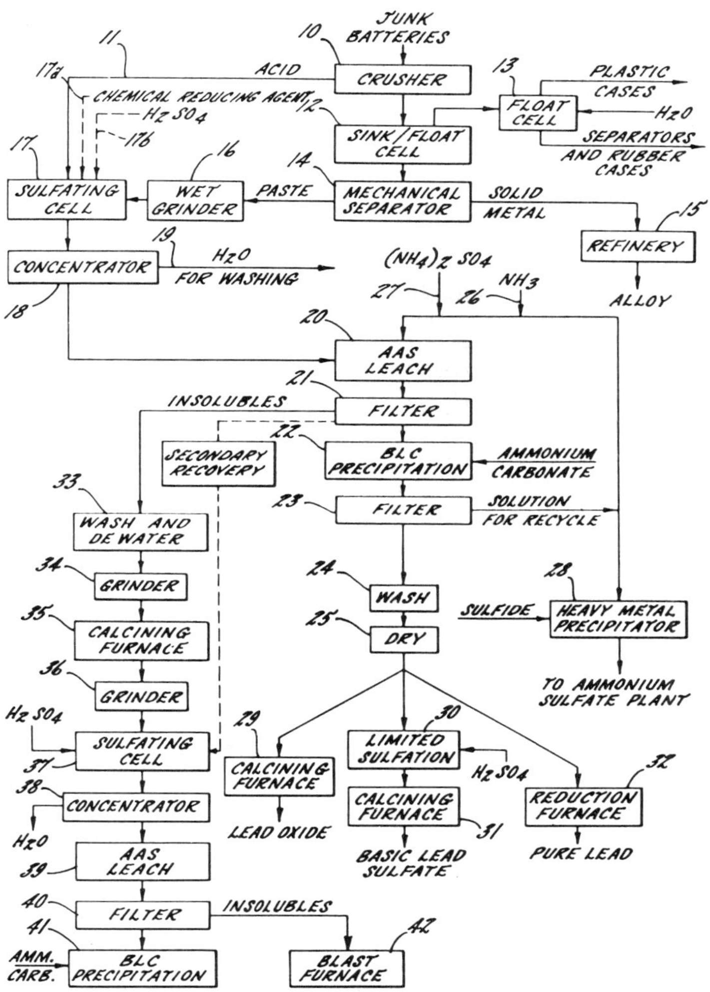

Source: U.S. Patent 4,118,219

Such a process is shown in Figure 108 and the process will be described in detail with reference to that figure. Referring now to the drawing, the scrap batteries are broken in a mechanical crusher **10** which breaks the solid parts of the batteries into fragments of about 2 to 6 cm while allowing the major portion of the liquid electrolyte to drain from the crusher through a line **11**. The mechanical crusher **10** may be a conventional hogger or other crushing device known in the art. The solid components, including the battery paste, are conveyed from the crusher **10** to a sink/float cell **12** where the wet leady materials form a high density media on which the lower density materials from the battery cases and separators float and, therefore, can be separated from the leady materials.

The lower density materials are skimmed off the top of the cell **12** and conveyed to a water float cell **13** where the materials which float in water, such as polypropylene from plastic battery cases, are separated from the materials which sink in water, such as fragments of hard rubber battery cases and battery separator materials. The plastic material removed from the top of the cell **13** is generally reprocessed, while the rubber and separator material removed from the bottom of the cell **13** are usually discarded as solid waste material.

The heavy density media which sinks in the cell **12** includes both lead paste and structural metal from the original batteries. These two materials are separated from each other in a mechanical separator **14**, such as wet screening or vibratory milling equipment, which receives the heavy density media from the bottom of the cell **12**. The resulting solid metal fraction is transported to a refinery **15** where the metal is dried, melted and/or smelted and refined to produce lead alloys which can be reused in new batteries. The wet paste from the mechanical separator **14** is fed to a wet grinder **16** which increases the surface area and leachability of the wet paste before it is fed into a sulfating cell **17**.

The paste and electrolyte separated from the junk batteries are reacted to increase the lead sulfate content of the paste and reduce the acid content of the electrolyte, and the resulting paste is then mixed with an aqueous ammoniacal ammonium sulfate solution to dissolve a substantial portion of the lead sulfate from the paste. Thus, in the illustrative process the wet paste from the grinder **16** and the liquid electrolyte from the crusher **10** are mixed in the sulfating cell **17**. In this cell **17**, the paste is allowed to react, preferably under mild agitation, with the sulfuric acid contained in the electrolyte so as to substantially increase the content of lead sulfate in the paste and reduce the acid concentration of the electrolyte.

More specifically, the paste consists primarily of lead sulfate ($PbSO_4$), lead dioxide (PbO_2), lead oxide (PbO) and a minor amount of lead metal (Pb), the lead oxide reacting directly with the sulfuric acid contained in the electrolyte to produce lead sulfate and water. In addition, as in a discharging battery described previously, lead dioxide and lead metal react in the presence of sulfuric acid to produce lead sulfate and water. It will be appreciated that the utilization of the electrolyte in the sulfating step avoids the otherwise troublesome problem of disposing of the waste electrolyte.

Unlike $PbSO_4$ and PbO, PbO_2 is not soluble in the aqueous ammoniacal ammonium sulfate solution, and does not react with the sulfuric acid to form $PbSO_4$ until it is reduced to PbO. A certain amount of the PbO_2 is reduced by the Pb in the paste, but there is generally a stoichiometric deficiency of Pb (produced by the negative battery plates) relative to the PbO_2 (produced by the positive battery plates). Thus, the excess PbO_2 must be reduced by a supplementary treatment, e.g., by calcining or by the addition of a chemical reducing agent such as hydrogen peroxide, formaldehyde, additional Pb or the like. If desired, the chemical reducing agent may be added to the sulfating cell **17** via line **17a**, or it may be added in a secondary recovery process farther downstream. In any event, it is preferred to grind the paste before the addition of the reducing agent because the $PbSO_4$ and PbO make it difficult to reduce the PbO_2.

When calcining is used to reduce the PbO_2 (by dissociation), the calcining may be carried out between the grinder **16** and the sulfating cell **17** or in a secondary recovery process farther downstream. After the PbO_2 is reduced, it can be easily sulfated so that it will dissolve in the aqueous ammoniacal ammonium sulfate solution.

The sulfating reaction in the cell **17** requires several days to achieve completion under mild agitation at room temperature, but complete reaction is not required for purposes of this process. In general, it is sufficient for the reaction to proceed for only a few hours, e.g., 3 to 5 hr. If desired, the reaction rate may be increased by elevating the temperature within the cell **17**, up to a maximum of about 100°C. Additional sulfuric acid may be added to the sulfating cell **17**, as indicated at **17b**, to compensate for any deficiency in the amount of electrolyte required to react with all the PbO in the paste. As an alternative, sulfuric acid may be used in place of all the electrolyte, but this is normally not preferred because of the problem of disposing of the electrolyte from the junk batteries.

In order to remove excess liquid from the reaction product formed in the sulfating cell **17**, the reaction product is fed into a concentrator **18** where the sulfated paste is allowed to settle to separate it from the major portion of the liquid fraction, which is primarily water. This settled portion is further expressed to reduce the liquid content to less than about 20 wt %, preferably less than 10 wt %. The water or dilute electrolyte that is removed from the sulfated paste in the concentrator **18** is discharged through a line **19** and may be recycled for use in washing operations such as **33** to be described below, or in separation operations such as the float cell **13**, or in other operations where high purity water is not required.

To dissolve the lead sulfate out of the concentrated sulfated paste, the paste is transformed from the concentrator **18** to a leaching chamber **20** containing an aqueous ammoniacal ammonium sulfate leaching solution. The leaching solution contains about 2 to 25 wt % ammonia and about 10 to 45 wt % ammonium sulfate, with the preferred composition falling within the ranges of 10 to 15 wt % ammonia and 20 to 35 wt % ammonium sulfate. Relatively high concentrations of both ammonia and ammonium sulfate are preferred in order to leach the maximum amount of lead sulfate from the paste. The leaching solution rapidly dissolves significant portions of the lead sulfate contained in the paste, but it does not dissolve lead dioxide or lead metal components of the paste. Nor is there any significant dissolving of antimony, barium, bismuth, arsenic, tin or iron, all of which are metals typically associated with lead-acid storage batteries.

Thus, the leaching operation effects a significant improvement in the purity of the lead sulfate derived from the battery paste. Metals which dissolve in the leaching solution include copper, silver, cadmium and the alkali metals.

The leaching operation proceeds at a relatively rapid rate and generally requires less than an hour, typically less than about 5 min, at ambient temperatures under mild agitation. The leaching solution is capable of dissolving up to 10 wt % lead at ambient temperatures, but in actual practice the solution is generally composed of only about 5 wt % lead. Increasing the temperature of the solution increases its dissolving power, but also produces higher ammonia vapor pressures and greater instability of the lead complex in the solution. The optimum reaction time within the leaching chamber **20** is that required to achieve a lead concentration within the range of 5 to 10 wt % in the solution.

For the purpose of separating the undissolved materials, including both impurities and insoluble lead compounds, from the pregnant lead solution produced in the leaching chamber **20**, the output of this leaching chamber is passed through a filter **21**. The resulting filtrate, which is the pregnant lead solution, enters a precipitation vessel **22** where it is treated to produce lead carbonate, e.g., basic lead carbonate, which is insoluble in the solution and precipitates out as a fine crystalline material.

The preferred reactant for effecting the precipitation is ammonium carbonate, ammonium bicarbonate or carbon dioxide. The lead carbonate precipitation step effects a further refinement in the lead being recovered, because the carbonates of other metals which are frequently found in junk batteries such as copper, silver and cadmium remain soluble in the solution to which the ammonium carbonate, ammonium bicarbonate or carbon dioxide is added, and thus such metals are separated from the lead carbonate precipitate.

To separate the lead carbonate from the barren solution from which it is precipitated, the slurry formed in the precipitation vessel is passed through a filter **23**, with the recovered solids then being washed and dried as indicated at **24** and **25**. The filtrate or spent liquor from the filter **23** is recycled to the leaching chamber **20**, with additional ammonia and ammonium sulfate being added via lines **26** and **27** to maintain the desired composition in the ammoniacal ammonium sulfate leaching solution.

The ammonia addition is adjusted to maintain the desired free ammonia content in the leaching solution, and the ammonium sulfate is added if water dilution of the leaching solution becomes excessive. These additions can be readily effected by adding the desired amount of ammonia and then forcing the solution through a bed of ammonium sulfate which can be renewed as needed. This technique ensures a saturated leaching solution of ammonia and ammonium sulfate which has the requisite leaching power.

In order to maintain a relatively constant volume of leaching solution in the leaching chamber **20**, a portion of the solution being recycled from the filter **23** to the leaching chamber **20** is continually removed from the system to compensate for the added ammonia and ammonium sulfate and dilution due to the liquid content of the paste entering the leaching chamber. Thus, in the illustrative process a portion of the filtrate from the filter **23** is diverted to a heavy metal precipitator **28** where any dissolved heavy metals, such as copper, silver, cadmium and lead, are precipitated as sulfides by the addition of hydrogen sulfide, ammonium sulfide or the like. The resulting metal sulfides precipitates out of the solution and can be removed by filtering. If the solution contains excess lead, it may be forced through a bed of ammonium carbonate to form lead carbonate which is then removed by filtering. The remaining ammonium sulfate solution, which typically contains 20 to 30 wt % ammonium sulfate and 5 to 15 wt % ammonia, can then be mixed with a high concentration of sulfuric acid to neutralize the free ammonia content and to increase the ammonium sulfate. This is in contrast to the ammonium sulfate solutions that are produced as by-products in other lead recovery processes, such as in smelting operations, where the ammonium sulfate solution is generally too dilute and impure to be useful as a feed stock for ammonium sulfate plants.

Returning to the lead carbonate that is recovered in the filter **23** and then subsequently washed and dried to remove residual leaching solution and water soluble impurities, this material may be further processed in several different ways. Three optional processes are generally illustrated in the drawing for converting the lead carbonate to lead oxide, basic lead sulfates or pure lead, respectively. To produce lead oxide, the lead carbonate is heated in a calcining furance **29** at a temperature within the range of 400° to 800°C to dissociate the lead carbonate into lead oxide, carbon dioxide and water vapor, in addition to any ammonia released from residual ammonium compounds.

The lead oxide produced by this calcining operation is a fine powder with a large specific surface. Small amounts of lead sulfate, generally less than 10 wt % are usually associated with the lead oxide, normally in a tetrabasic lead sulfate crystal structure. The lead sulfate presumably comes from sulfate substitution in the basic lead carbonate crystal structure. Calcining the lead carbonate to the oxide typically requires several hours, depending upon the particular equipment and operating temperature employed.

To convert the lead carbonate to basic lead sulfates, which is the second option illustrated in the drawing, the lead carbonate is sulfated with sulfuric acid in a sulfating cell **30**, and then calcined in a furnace **31** at a temperature of 400° to 800°C for several hours. The resulting products are basic lead sulfates, e.g., tetrabasic lead sulfate, monobasic lead sulfate, normal lead sulfate or mixtures thereof. The amount of sulfuric acid added will determine the type of lead sulfate produced.

If it is desired to convert the lead carbonate to pure lead, the lead carbonate is smelted in a reduction furnace **32** at a temperature of 800° to 1200°C to produce pure lead metal. It has been found that the resulting lead metal has an extremely high degree of purity. The carbonate is preferably mixed with carbonaceous reducing agents and simple fluxes when fed to the furnace.

In accordance with a further aspect of this process, additional lead is recovered from the insoluble materials removed from the leaching solution by converting lead dioxide therein to lead oxide, then mixing the lead oxide-containing material with sulfuric acid to convert the lead oxide to lead sulfate, and finally mixing the resulting product with an aqueous ammoniacal ammonium sulfate solution to dissolve a substantial portion of the lead sulfate therefrom. Thus, in the illustrative process the insoluble materials removed from the leaching solution by the filter **21** are washed and dewatered as indicated at **33** to remove excess liquid.

The resulting material is then ground in a wet grinder **34**, dried, and calcined in a furnace **35** at a temperature of about 300° to 800°C to dissociate any lead dioxide present to lead monoxide, or to basic lead sulfates if sulfate is present, and to oxidize any residual lead metal to lead monoxide or basic lead sulfates. Any pieces of battery case or separator material or other organic materials present in the charge to the calcining furnace **35** are consumed in the calcining operation.

To convert the lead oxide in the calcined product from the furnace **35** to lead sulfate, the calcined product is ground in a grinder **36** and then fed into a secondary sulfating cell **37** where it is reacted with a sulfuric acid solution and/or spent electrolyte. Sulfation of this material is relatively slow, depending upon the composition and specific surface characteristics of the calcined material. Usually several days of sulfation at ambient temperatures are required to complete the reaction, although the reaction can be substantially completed in a matter of hours by increasing the temperature in the sulfating cell **37** to at least 60°C.

To remove excess liquid from the sulfated solution, the output of the cell **37** is fed into a concentrator **38** where the paste is allowed to settle so that excess liquid can be removed to reduce the liquid content of the paste to less than 20 wt %, preferably less than 10 wt %. The resulting paste is then transferred to a secondary leaching chamber **39** containing an aqueous ammoniacal ammonium sulfate solution having essentially the same composition as that described for the primary leaching chamber **20**. Insoluble materials are removed from the resulting solution by passing it through a filter **40**, and the resulting filtrate, which is the pregnant lead solution, is passed into a precipitator **41** where ammonium carbonate, ammonium bicarbonate or carbon dioxide is added to precipitate the lead in the form of lead carbonate.

The insoluble materials removed by the filter **40** may be simply melted in a reduction operation carried out in a blast furnace **42**, or further processed by other means to recover their separate metallic contents.

As an alternative to the illustrative secondary recovery process utilizing calcining, the lead dioxide in the insoluble materials removed from the filter **21** can be converted to lead oxide by grinding the insolubles and then adding a chemical reducing agent thereto. The resulting lead oxide can then be fed directly into the sulfating cell **37**.

If the original paste from the mechanical separator **14** is treated by calcining or a chemical reducing agent to reduce PbO_2 to PbO before the paste is fed to the initial sulfating cell **17**, the amount of lead compounds contained in the insolubles removed from the filter **21** should be so small that the secondary lead recovery process can be eliminated completely. In this case, the insolubles are merely treated to recover any other by-products of value therein.

Regardless of whether the reduction of the PbO_2 to PbO is effected in the primary or secondary recovery stage of the process, the use of a chemical reducing agent is generally preferred over calcining because the use of the reducing agent provides a continuous wet process. This not only eliminates unnecessary drying and grinding steps, but also avoids the production of dry dusts which can present pollution, disposal and safety problems.

LEAD FROM TEL MANUFACTURING WASTES

Tetraethyllead has been manufactured commercially for many years by reacting an excess of ethyl chloride with lead-monosodium alloy. In such reaction, the sodium in the alloy is mostly converted to sodium chloride, about 25% of the lead in the alloy is converted to tetraethyllead, and most of the rest of the lead in the alloy is converted to metallic lead in finely divided form. After the reaction has been completed, the excess ethyl chloride is distilled off, and the reaction mass is then drowned in water and the tetraethyllead removed by steam distillation in the presence of a still aid which largely prevents agglomeration of the finely-divided metallic lead particles.

The steam still residue comprises a suspension of the by-product lead particles in a dilute solution of sodium chloride. This suspension is conveyed to a sludge pit where it is allowed to settle, forming an upper layer of aqueous salt solution and a lower layer of wet sludge which is composed mainly of the finely-divided lead mixed with about 8 to 20 wt % of the aqueous salt solution. The water layer is drawn off, the wet sludge is washed with water to remove most of the salt and then the sludge is dried to remove most of the water.

The resulting dried by-product lead sludge is impure, containing sodium chloride, lead chloride, sodium hydroxide and, in some cases, lead sulfate, lead sulfide and lead chromate (formed by reaction of the lead with the still aid), usually in a proportion of less than 1 wt %. In addition, the lead particles are coated with lead oxide. Usually, the lead oxide is present in a proportion of from 2 to 10 wt %, frequently as high as about 20 wt % and, in extreme cases, as high as about 30 wt %. It is possible, in some cases, to keep the amount of lead oxide down to 0.1 to 0.2 wt %. The by-product lead sludge will sometimes contain iron salts, such as ferric chloride, ferric sulfate and iron sulfides, when an iron compound (usually ferrous sulfate) is employed as an ingredient of the still aid.

The recovery of the lead in refined form from such impure by-product lead sludge has been difficult and has required a series of treatments which are inconvenient and costly. Usually, the impure by-product lead sludge has been refined by melting it in a reverberatory furnace at a temperature of from 700° to 900°C to form an upper layer of slag and a lower layer of molten lead consisting of most of the metallic lead originally present in the impure by-product lead sludge. A temperature of at least 700°C is required in the furnace to release molten metallic lead therefrom. The lower layer of molten lead is drawn off into a pump pot, a dross usually being formed during the pot filling and being recovered as skimmings, and returned to the alloying operation. The slag and the dross are then further treated to recover their lead content.

Another waste product produced in the manufacture of tetraalkyllead is an alkali metal in admixture with an alkaline earth metal, specifically, sodium metal mixed with calcium metal and their salts. Sodium is used as described above in an alloy with lead in the manufacture of tetraalkyllead. Sodium for alloying with lead is produced in one method by an electrolytic process, utilizing as the electrolyte a fused metal salt mixture. For example, in the production of sodium a fused mixture of sodium chloride and calcium chloride is employed as the electrolyte. The purpose of adding the calcium chloride is to reduce the melting point of sodium. In this way, the sodium produced at the cathode will be in the molten state and will not be vaporized at the temperature of the electrolytic mixture. While the use of the fluxing agent such as calcium chloride has this advantage, it does introduce additional difficulties into the overall process. The main difficulty is that calcium metal is produced at the cathode along with the sodium. This crude product is collected, cooled and subsequently filtered so as to recover most of the sodium in substantially pure form for use in alloying with lead. The residue or filter sludge which remains after this purification contains appreciable quantities of both sodium and calcium and it is the production of this sludge as a by-product which has presented a disposal problem to the industry for a number of years.

This sludge contains on the average 90 to 95 wt % sodium and calcium, the remainder consisting of the various salts and oxides of these metals and other impurities. The sodium content averages about 70 wt % and the calcium content varies between 5 and 30 wt %. The sludge is in the form of crystals of calcium metal and electrolyte embedded in a matrix of sodium.

A process has been developed by *L.T. Crasto, R.H. Jones and J.C. Klock; U.S. Patent 4,153,451; May 8, 1979; assigned to Ethyl Corporation* which provides for recovery of by-product lead from reverberatory furnace lead slag and the use of waste sodium/calcium filter sludge as the major reductant in the process.

More specifically, this waste-plus-waste recovery process involves the steps of:

(a) Charging the lead by-product materials to a rotary furnace;

(b) Drying and melting the lead-containing materials at 330° to about 650°C until any residual moisture is driven off;

(c) Adding to the molten furnace charge about 0.05 to 0.25 lb/lb of the charge of an active alkali metal mixture;

(d) Smelting the furnace charge in the presence of the alkali metal mixture for a time sufficient to form a separate metallic lead phase and a residue; and

(e) Removing the metallic lead from the furnace.

LEAD FROM ZINC REFINERY FLUE DUSTS

See also "Cadmium from Zinc Refinery Flue Dusts."

A process developed by *I.G. Matthew, R.W. Pickering, T. Suzuki and H. Uchida; U.S. Patent 4,135,912; January 23, 1979; assigned to Electrolytic Zinc Company of Australia Limited, Australia and to Mitsubishi Metal Corporation, Japan* is one applicable to the treatment of lead residues that remain undissolved after converter dusts and other dusts, produced during the pyrometallurgical production of copper, have been leached in dilute sulfuric acid and is particularly suitable for the treatment of residues such as those that remain undissolved when solids that contain previously undissolved zinc, copper and cadmium ferrites are subjected to a hot sulfuric acid leaching treatment as part of the process for the production of electrolytic zinc.

This process provides a method for the treatment of lead residues in which the lead is present at relatively low concentrations and is mostly present in the form of lead sulfate. The process has the advantages of treating the lead residues without the need for prior sintering, and the process is essentially continuous. The main products from the process are a high grade lead bullion containing major proportions of the lead, silver and gold, and a slag containing major proportions of the other constituents, excepting sulfur, of the lead residues. After cooling, such slags are very inert, and they are eminently suitable for disposal in environmentally acceptable ways.

Gaseous emissions from the process consist essentially of carbon oxides and sulfur oxides admixed with air. These gaseous emissions may be subjected to standard procedures for the removal of contained or entrained dust and fume and to standard procedures for absorption and treatment of the sulfur oxides.

The process is suitably carried out in an electric furnace, with the electrodes partly immersed in the slag phase. Furnaces of the Heroult type are very suitable, with three or more electrodes.

The furnace is operated at a feed rate and with an electric power input such that the molten slag layer is maintained at a temperature of 1000° to 1500°C, but preferably in the range of 1100° to 1350°C.

LEAD CARBONATE FROM STORAGE BATTERY MANUFACTURE

See "Ammonium Sulfate from Storage Battery Plate Wash Water."

LEAD OXIDE FROM FLUE DUSTS

A process developed by *J.E. Allen; U.S. Patent 3,850,613; November 26, 1974; assigned to Ferro-Carb Agglomeration Ltd., Inc.* is one in which steel mill and foundry waste dusts containing zinc and/or lead oxides in addition to iron oxides are briquetted by means of a hydrocarbonaceous binder, together with sufficient carbon to reduce at least the lead and zinc oxides to the metal.

The briquettes are preferably preferentially treated with an oxygen-containing gas at 350° to 600°F to selectively react with the binder to obtain strong briquettes; and the briquettes are then heated to a temperature of between 1800° and 2500°F to cause any zinc or lead present to first be reduced to metal and be volatilized as metal, and then be oxidized to ZnO and PbO respectively in the gaseous phase. The zinc and lead oxides are recovered from the exhaust gases in an appropriate dust collector while the iron-bearing briquettes are cooled in a nonoxidizing atmosphere, and are then used as ferrous charge in the iron or steelmaking processes.

When scrap materials containing lead, e.g., from storage batteries, are smelted in metallurgical furnaces for the purpose of recovering lead and other metals contained therein, some metalliferous dust is formed as a by-product. This so-called flue dust is carried with the hot gases leaving the furnace through the vent pipes arranged for this purpose, and it is known after cooling of the smoke to collect the dust in electrostatic filters, bag filters or other separation devices and to return it to the metallurgical furnaces in order to recover its contents of metal.

The dust consists mainly of lead oxide and small quantities of oxides of other metals, which have been present as alloying components or impurities in the original furnace charges. However, the dust also contains other chemical compounds of the metals present, especially chlorides, sulfides and sulfates. When it is attempted to smelt and reduce the flue dust in metallurgical furnaces, either separately or mixed with other materials, some of the dust will again be carried off with the hot gases, and formation of new dust will also take place by evaporation and subsequent condensation of metalliferous compounds.

Even if the necessary reducing and slagging agents are present in the furnace charges, only a partial reduction of the returned flue dust will be obtained, and continued operation will increase the quantity of circulating flue dust, and at the same time its composition will be adversely changed, as the quantity of chlorides, sulfides and sulfates, which are not readily reduced, is increased relative to that of oxides so that the content of metal is reduced.

Thus, a process has been developed by *S. Bergsoe; U.S. Patent 4,013,456; March 22, 1977; assigned to Paul Bergsoe & Son A/S, Denmark,* in which the aforementioned disadvantages are eliminated and a material is produced which is more suitable for treatment in a metallurgical furnace, so that a large increase of the yield of metal is obtained by the method as compared with the yield obtainable by the known technique.

In this process, the flue dust is fused at a relatively low temperature without any substantial reduction taking place. The slag thus formed is cooled until solidification, reagents for the improvement of the process being added to one or more of the process steps before, during or after fusion of the flue dust, the reagents comprising slag-forming chemicals, a material for raising the melting temperatures, and a ferrous reducing agent.

A suitable form of apparatus for carrying out this process is shown in Figure 109. In the apparatus shown, the flue dust **1** containing lead is collected in a separation device **2** in a smoke-cleaning plant. The flue dust **1** is taken to a fusing furnace **3** by means of a conveyor **4**, shown in the drawing as a worm conveyor.

An additive **5**, such as sodium carbonate or borax, can be added to the flue dust **1** in the conveyor **4** from a dispensing apparatus **15**, which in known manner can be arranged to dispense in proportion to the quantity of flue dust conveyed at any time by the conveyor **4**. During such an addition in the conveyor **4**, the conveyor also operates as a mixing apparatus for the flue dust **1** and the additive **5**. In the fusing furnace **3** the mixture of dust **6** thus produced falls upon an inclined hearth **7**, where it is heated by the flame **8** of an oil burner **9**, placed opposite to and directed towards the hearth **7**.

The molten mass **10** flows down the hearth **7** to an outlet **11**, where an additive **12**, e.g., in the form of a fine-grained, ferrous material, can be supplied from a dispensing apparatus **13**, which may be controlled in the same way as the dispensing apparatus **15** referred to in the foregoing. The slag **16** thus produced flows down into a collecting vessel, where it is kept heated by means of an oil burner **19** while being constantly stirred by means of a stirring machine **17**. When the vessel **14** is full, the stirring machine **17** is removed, and the vessel may either be emptied into another vessel or be exchanged for a corresponding vessel, the slag **16** being in any case cooled till it solidifies, subsequently to be divided, if so desired, into suitable blocks for charging in a metallurgical melting furnace.

Figure 109: Bergsoe Process for Treating Flue Dust Containing Lead

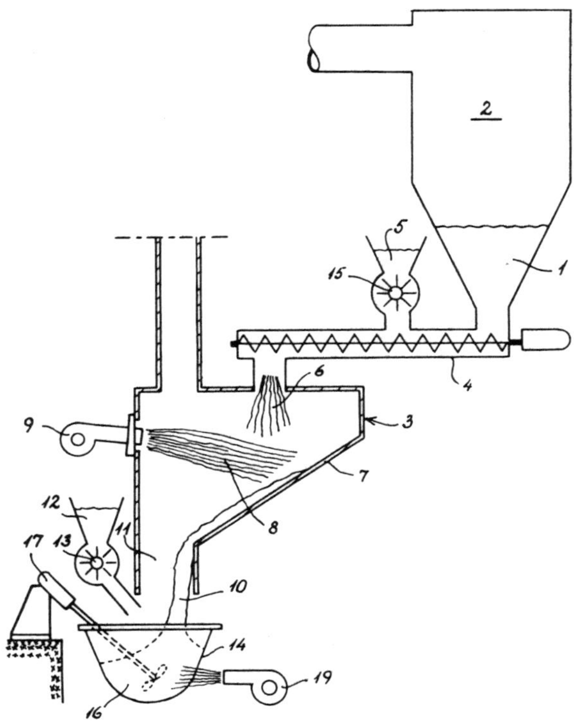

Source: U.S. Patent 4,013,456

LIME FROM SEWAGE SLUDGE INCINERATION RESIDUES

In water and wastewater treatment processes, sludges are often produced that contain calcium-bearing compounds in substantial quantities. The calcium-bearing sludges are formed, for example, in tertiary sewage treatment processes wherein phosphorus is removed from wastewater by lime precipitation. In such sludges, the phosphorus values appear as solid compounds which are generally classified as apatites by workers in the art and which can be generally designated $Ca_x(PO_4)_y$ where x and y are integers. As another example, calcium-bearing sludges are produced when lime and/or slaked lime are added to raw sewage to serve as a flocculating agent or to water to produce chemical softening; in such cases, the sludges usually include substantial quantities of calcium carbonate. Calcium-bearing sludges are found also in some sugar beet refining operations and in some high-turbidity water treatment systems.

It is well known that the organic materials in such sludges can be disposed of by incineration. Incineration reduces the organic materials to inert ash and concomitantly produces gases and vapors, principally carbon dioxide and steam. It is also well known that sewage sludges can be calcined under controlled conditions to form calcium oxide solids from the calcium carbonate values. In the case of phosphorus-bearing sewage sludges, the apatites which remain after incineration are intimately mixed with the ash and calcium oxide values.

Workers in the field of water and wastewater treatment have long felt a need for an economical system for reclaiming lime values from the solid residuals of incinerated calcium-bearing sludges. If the lime values could be inexpensively reclaimed and recycled in a relatively concentrated form, plant operating costs could be reduced insofar as expenditures for the purchase of lime or limestone are concerned.

A process developed by *O.E. Albertson; U.S. Patent 3,996,133; December 7, 1976; assigned to Envirotech Corporation* is one for reclaiming calcium values from the solid residual of incineration or calcination of calcium-bearing sewage and water sludges which includes the following steps.

(a) Classifying the residuals on the basis of their size and density to form two classified streams, the first of which contains relatively larger sized solid particles and has a relatively decreased concentration of calcium and the second of which contains relatively increased concentration of calcium for reuse.

(b) Then mixing the particles in the first stream with water to form an aqueous solution containing calcium hydroxide.

(c) Then clarifying the aqueous solution in a settling zone to form a sediment relatively rich in the insoluble inert solids and a decanted stream which is relatively rich in calcium values.

Such a process is shown in Figure 110. The calcium-bearing sludge is carried by a conduit **11** to an incineration device **13** which is illustrated as a conventional multiple hearth furnace but which could also comprise a rotary kiln or a fluidized bed furnace. The gases of incineration (principally nitrogen, oxygen, carbon-dioxide, and steam) are vented from the incineration device via a conduit **15**. The residual solids of incineration are carried from the incineration device via a conduit **17**.

If the sludge feed to the incineration device is from a tertiary sewage treatment process, for example, the sludge will typically contain apatites, organics, inerts, and calcium carbonates. By controlling the combustion of such sludges, the organics can be oxidized to ash and, at the same time, the calcium carbonates can be calcined to produce calcium oxide. Such a process is taught, for example, in U.S. Patent 3,623,975.

Since some of the residual solids of incineration are readily entrained in the exhaust gases, it is usually desirable to interpose a conventional particulate collection device **19** in the vent conduit **15**. In the drawings the exhaust gases enter an inlet **21** in the collection device, the cleaned gases are discharged via a conduit **23**, and the collected particulates are discharged via a conduit **25**. A conventional fan or blower **26** can be provided to enhance the flow of the gases.

Figure 110: Envirotech, Inc. Calcium Reclamation Process for Sludge Treatment

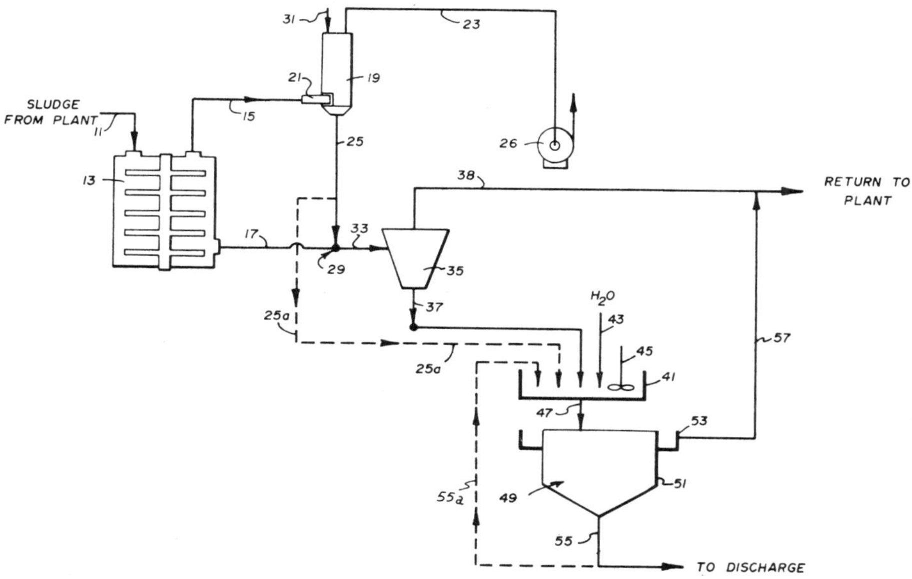

Source: U.S. Patent 3,996,133

The particulate collector **19** can be either of the dry type such as a cyclonic collector, or of the wet type such as a venturi scrubber. Where the former type of device is used, the dry collected particles are carried by the conduit **25** to a juncture **29** for addition to the solids stream **17**. In the case of a wet scrubbing device, water or other scrubbing liquid is added to the scrubber through a line **31** and the collected wet particulates are carried via a bypass line **25a** for processing as will hereinafter be described.

If the incineration device **13** is a conventional fluidized bed furnace, all of the solid products of incineration are carried out of the furnace with the exhaust gases by elutriation and, therefore, the conduit **17** would be omitted. In such an instance, it is mandatory that the particulate recovery device **19** be provided and that it be of the dry type to prevent reversion of CaO to $CaCO_3$, which reaction would occur if water and carbon dioxide gas contacted the calcium oxide solids.

If a dry-type particulate collection device is used with a fluidized bed furnace or if a kiln or multiple hearth incinerator is utilized, the dry residual solids from the incinerated sludge are carried from juncture **29** by a conduit **33** to a classification device **35**. A conventional gaseous-type classifier, a cyclone separator, for example, could be used here to classify the particulate solid materials in an air current according to their size and density. Two or more classifiers can be utilized here in series to provide improved separation.

It is also contemplated that a conventional hydraulic-type classification unit could be employed to accomplish classification by particle size and density. In the case of hydraulic separation, the dissolved calcium values in the classified streams appear principally as calcium hydroxide.

Two conduits **37** and **38** lead from the classifier **35**. The first conduit **37** carries the relatively coarser particles (rejects) while the second conduit **38** carries a stream of the finer particulates (accepts). Typically, the accepts stream will comprise particles less than 50 μ in size while the rejects stream includes larger particles. (Normally, the maximum particle size is 250 μ.) It has been found in practice that the accepts stream is rich in calcium oxide and relatively poorer in ash (i.e., inerts) than the rejects stream and usually includes between 40 and 75% of the total available calcium oxide.

The rejects stream solids are carried, via the conduit **37**, to a conventional mixing or solubilizing unit **41** into which a stream of water is added at **43** to dissolve or slake the calcium oxide values to produce calcium hydroxide $Ca(OH)_2$. (If a wet-type particulate recovery unit **19** is utilized, the aforementioned conduit **25a** is arranged to discharge directly into the mixing unit **41**.)

The mixed solution is conveyed from the mixing unit **41** via a conduit **47** to a conventional sedimentation device **49** which generally includes a large liquid-holding tank **51**, an overflow launder **53** for the removal of clarified liquid from the tank, and a bottom discharge **55** for the withdrawal of settled solids. In the tank, the dense insoluble ash and calcium-phosphate components settle rapidly and, accordingly, the underflow stream **55** from the tank includes mostly ash and apatites. The calcium values, on the other hand, are dissolved and are removed in the overflow in the form of calcium hydroxide, $Ca(OH)_2$, and are returned to the treatment plant via a conduit **57**.

In an optional modification designated by dashed line **55a**, a fraction of the underflow stream **55** is recycled to the mixing device **41** to further solubilize calcium values that slake slowly and which have settled in tank **51**.

M

MAGNESIA FROM SCRAP FURNACE LINING BRICK

Magnesia-based refractory lining materials have only a limited length of life in those furnaces in which they are used. The materials are deleteriously affected, inter alia, by molten slag and iron oxide, by high temperatures and by the variations in temperature to which they are subjected in the furnaces, the linings becoming pitted and their mechanical strength impaired to such an extent that they must finally be replaced by a new lining. The material removed when changing a lining, this material, for example, being in the form of scrapped bricks, comprises to a large extent MgO, of which a part is contaminated with slag, etc. The recovery of the MgO content of this lining material would enable valuable raw materials to be returned for reuse in the manufacture of, for example, magnesite brick or for the manufacture of other qualified products, particularly if the recovered magnesia material is of good quality and at least substantially free from impurities, primarily impurities in the form of SiO_2 and iron oxides.

Previous experiments have shown that certain magnesium-rich minerals, e.g., brucite ($MgO \cdot H_2O$), which is not soluble in pure water, can be dissolved to some extent in water containing CO_2 at a CO_2-pressure of a few atmospheres and without substantially elevated temperature. Experiments have also been made to leach under CO_2-pressure impure magnesite materials which had been fired at approximately 600° to 650°C. These experiments, however, have not resulted in a commercially usable method, since in both cases the leaching solutions were too weak. The content of MgO per liter of leaching solution has, in general, reached only 3 to 9 g/l.

As opposed to the MgO products calcined at the lower temperatures, commercial magnesite brick fired at higher temperatures and corresponding molten magnesite material have been considered more or less inert to the influence of, inter alia, water, and leaching with water under carbon dioxide pressure has not produced any worthwhile results either.

A process developed by *H.E.B. Hassler and P.G. Kihlstedt; U.S. Patent 4,140,745; February 20, 1979; assigned to Advanced Mineral Research AB, Sweden* is one in which magnesia is recovered from scrap lining material rich in magnesia by leach treatment of the material with water in the presence of carbon dioxide under pressure, separating the leaching solution from solid leaching residues, precipitating out magnesium hydrocarbonates from the leaching solution and calcining the precipitate to form magnesia. To make the scrap lining material more readily leachable and enable the production of a rich leaching solution, the lining material, prior to the leaching treatment, is treated with steam at a temperature within the range of 100° to 300°C and corresponding saturation pressures for a period of 1 to 50 hours.

MAGNESIUM FROM CELL SLUDGES

The separation of solid Mg metal spheroids from entrapment in a solid contiguous matrix of a friable salt or mixture of salts presents particular problems to one who may desire to recover the Mg in its spheroidal form and also retain on each spheroid a thin protective coating of the matrix material. Whereas it has been known for many years that such a Mg-containing matrix is removed as cell sludge from the electrolysis of molten $MgCl_2$ and as a slag material from Mg or Mg-alloy casting operations, attempts to recover the Mg or Mg-alloy particles by grinding or intensive ball-milling have generally resulted in smashing, breaking, or flattening a large portion of the Mg particles. Such deformed particles may be acceptable if the primary purpose of recovering the Mg metal is that of remelting the Mg for coalescence or for recasting.

In many cases, however, what is of special interest is the recovery, from the solid matrix, of Mg spheroids which each have a thin protective coating of the matrix remaining thereon. Such spheroidal Mg particles are of particular interest for use in inoculating molten ferrous metals, e.g., the desulfurization of steel. The thin protective coating of matrix on the Mg spheroids helps avoid the hydrolysis of Mg by moisture or the oxidation of Mg by air. Mg particles which are substantially flattened or elongated or which do not have a high degree of rotundity are not as readily useful in operations where the particles are injected through a lance beneath the surface of molten iron or steel. Ideally, the operators of such lances would prefer that the Mg particles be of consistent size, consistent Mg content, and consistent rotundity in order to avoid unwelcome variances during the inoculation process.

A process developed by *W.M. Goldberger, B.K. Parekh, and B.P. Faulkner; U.S. Patent 4,182,498; January 8, 1980; assigned to The Dow Chemical Company* is one in which magnesium beads, or Mg-alloy beads, having a high degree of rotundity, and having a thin coating of sludge or slag ingredients, are recovered from entrapment in a friable contiguous material of sludge or slag material by: (a) a primary milling of the friable matrix in a hammer mill or impact mill, (b) screening the material to collect the desired particle sizes, (c) attriting the material in a secondary milling operation to further, and gently, grind sludge material from around the beads, and (d) separating the rotund beads from the pulverized matrix material by using a shape-classifier, such as a slanted shaker table. The material may be treated beforehand by melting the metal-containing sludge or slag, adding a flux or emulsifier material with stirring to cause dispersion of the metal into a relatively narrow particle size range and then freezing the molten mixture.

MAGNESIUM FROM SCRAP MAGNESIUM

Magnesium has been used in very substantial quantities for many years, both in the nearly pure form and in high magnesium content alloys. Though magnesium and high magnesium content alloys (hereinafter simply magnesium) present a very substantial fire hazard, machine shop and other processing operations for controlling the likelihood of occurrence of fires are well-known. Consequently, magnesium castings, machined parts and parts formed of other processes are commonly used in applications requiring the special properties of low density and a high strength to weight ratio. However, in spite of the routine use of magnesium for many years, no really satisfactory method of disposing of magnesium fines, chips and scraps and of recovering the magnesium has been found.

However, a process has been developed by *G.T. Roberts, D.V. Owens, and R.F. Goodspeed; U.S. Patent 4,065,299; December 27, 1977; assigned to Teledyne Industries, Inc.* which is one whereby magnesium and magnesium alloy fines, chips and scraps may be recovered and remelted for reuse. The method is comprised of the steps of pelletizing the magnesium by first compacting the magnesium with a low pressure press so as to initially expel the majority of the air, followed by a high pressure press under a controlled environment to form a compact pellet. The pellets are then melted either by stacking on the cold heel of a pot and heating under a controlled environment or preheating in a basket over a melt, and then subsequently immersing the load. Melting can also be accomplished by controlled melting, utilizing an induction melting furnace whereby temperature can be slowly increased. Pelletizing pressures range from approximately the yield pressure to the cold welding pressure of the material being reclaimed. A typical pelletizing environment is 20% SF_6 and 80% CO_2 and a suitable melting environment is 98% CO_2 and 2% SF_6.

The apparatus, which may be used in the conduct of such a process, is shown in Figure 111.

Figure 111: Teledyne Industries, Inc., Magnesium Reclamation Apparatus

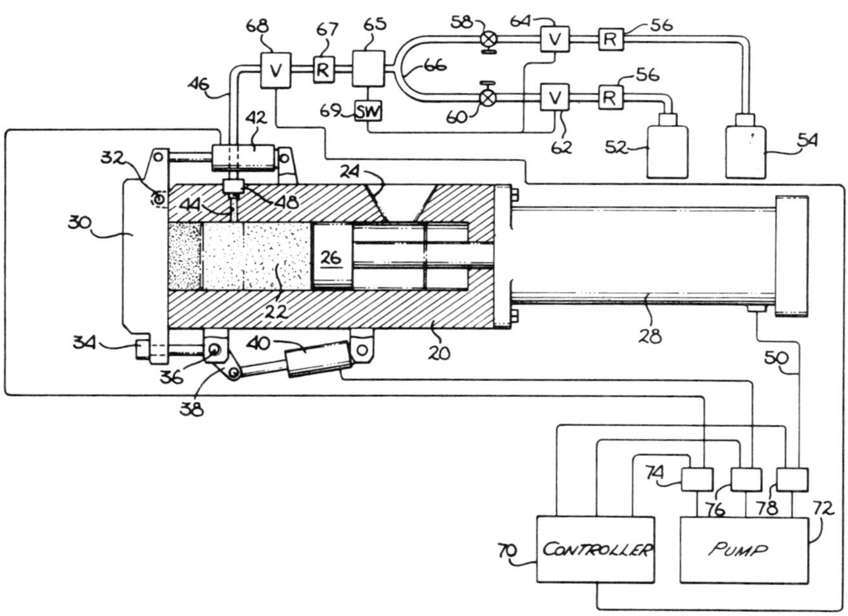

Source: U.S. Patent 4,065,299

The compactor has a housing **20** defining a cylindrical chamber **22** which may receive the magnesium fines, chips and scraps through an opening **24** at one end thereof when the piston **26** is withdrawn by the hydraulic cylinder **28**. A door **30** supported on pivots **32** closes the opposite end of chamber **22**, with the door being retained in the closed position by a lock **34** pivotally supported from the housing **20** by pivot **36**. The assembly comprising lock **34** contains an arm **38** driven by a small hydraulic cylinder **40** so as to controllably rotate the lock **34** from the locked position as shown in the figure to a lower unlocked position. Also, the door itself may be driven in rotation by a third hydraulic cylinder **42** coupled between the housing **20** and the door **30** at a position above the pivot **32**.

At a position adjacent to door **30** is a vent **44** extending to a line **46** for providing a controlled environment to the cavity **22**. Preferably, a one-way valve **48** (a ball valve or the like) is disposed adjacent to vent **44** to prevent reverse flow through the line.

The pelletizing in the preferred embodiment is actually a three-step process. After the pelletizer is loaded with a charge through opening **24**, hydraulic cylinder **28** is activated with a low pressure so as to force the magnesium particles to the end of the cylinder adjacent door **30**, providing some minor initial compaction of the more fragile chips such as turnings and the like, and expelling the majority of the air from the chamber or cylinder. In this regard, it should be noted that while door **30** provides a closure of one end of the cylinder, it is not an airtight closure so that the air in the cylinder may pass between the door and the end of the cylinder as the piston advances. If desired, some vents may be made in the door **30** to further relieve any air pressure which would tend to build up, though it has been found that such additional vents typically are not required.

The purpose of this first compression is to expel the majority of the air from the cylinder and to collect the magnesium particles at one end of the cylinder. Since the atmosphere surrounding the magnesium during this first compression is air and, therefore, an oxidizing environment, this first compression should be carried out with negligible heating and friction with the particles. Thus, the pressure of this first compression should not exceed 15,000 psi though a pressure as low as 5,000 psi or even lower is adequate.

When the pressure in the hydraulic line **50** supplying hydraulic cylinder **28** reaches a pressure indicative of the desired low pressure on the magnesium particles, the motion of piston **26** is reversed and an inert atmosphere is provided to the then expanding volume of chamber **22** through line **46**. In the preferred embodiment, a mixture of SF_6 and CO_2 is used, with a mixture by volume of 20% SF_6 and 80% CO_2 being most preferred. Thus, for this purpose, standard gas cylinders **52** and **54** containing each of the gases may be used, with a regulator **56** on each tank dropping the output pressure thereof to a few psi over atmospheric to insure suitable flow.

The flow rates in each line may be adjusted with adjustable valves **58** and **60**, and turned on and off with solenoid control valves **62** and **64** in each of the gas delivery lines. The two gases are fed to a mixer tank **65**, and through a regulator **67** and main solenoid valve **68**, to assure proper mixing of the two gases prior to delivery into the chamber **22**. In the preferred embodiment, the pressure at the mixer tank is controlled by the pressure switch **69**, and should cycle between 85 to 110 psi. The regulator **67** should be set at approximately 30 psi. The main solenoid valve **68** is coupled to the controller **70**.

After the initial compression, the piston is withdrawn and the gas provided into the increasing chamber volume as described above, with the piston retracting to a position not quite exposing the opening **24**, thus avoiding the opportunity of contamination of the new environment in the chamber with the surrounding air (since the CO_2-SF_6 mixture is heavier than air, little contamination would occur anyway unless there were substantial drafts in the region of the compactor). At this point in the piston travel, the piston motion is again reversed to extend toward the door **30**, and the flow of gas is shut off by shutting off the main solenoid valve **68**. While the piston motion tends to increase the pressure of the gas in chamber **22**, the one-way valve **48** prevents back-flow of the gas into line **46** so that the gas must necessarily penetrate and permeate the magnesium against door **30**, and incidently leak out around the door. Consequently, when the piston **26** starts the high pressure compaction of the magnesium, the initial air atmosphere around the magnesium particles has been replaced by the CO_2-SF_6 mixture. Accordingly, a high pressure compaction may now be carried out without risk of fire from the heat and friction of the compaction.

In order to insure suitable compaction and the desired resulting self-supporting pellet, a compaction pressure which is higher than the yield strength of the magnesium should be used. By way of specific example, a compaction pressure of approximately 40,000 psi has been found to provide the desired result. Higher pressures could also be used, although there is little advantage in attempting to exceed the cold welding pressure for the magnesium, as the process is intended to be readily used for the recovery of magnesium fines, chips and scraps of various vintages and histories, and accordingly the extent of cold welding could be expected to be random and poor because of the presence of oxides, cutting oils, and other contaminants in the mix.

At the end of the high pressure compaction, the force on piston **26** is relieved, lock **34** is released and door **30** opened. Piston **26** is then further extended to expel the resulting pellet, and withdrawn for loading of another charge.

In the preferred embodiment, a cylinder having an inner diameter of approximately 3" is used, and pellets of good density uniformity have been achieved having lengths of up to 3". Above this length, however, resulting pellets have poor self-supporting characteristics and more nonuniformity in density than desired. Consequently, it is believed that, at least for the pressure stated, best results will be achieved if the resulting pellet length is kept equal to or less than the diameter of the pellet. Also, while the pellets are self-supporting, they are not particularly structurally sound.

Therefore, when being ejected from the cylinder, friction between the pellet and the cylinder walls may cause some chattering of the pellet and dislodging of the surface particles from the exposed pellet face. This may be avoided or at least minimized by providing a slight tapering of the cylinder to a larger diameter adjacent the door and/or the use of a noncontaminating lubricant on the cylinder walls, such as a silicon oil. Inasmuch as the best location for port **44** and the best location for the start of the taper in the chamber adjacent the door depend upon the length of the pellet being formed, best results are obtained if the amount of magnesium per pellet is reasonably well controlled, such as by a rough weighing of the charge or by holding the volume of the charge constant.

In the preferred embodiment, the cycle is automatically controlled by a controller **70** controlling the output of a hydraulic pump **72** through valves **74**, **76** and **78**, which control the main cylinder **28** and the door actuators **42** and **40**. The controller **70** also provides a signal to control the main solenoid valve **68** to provide automatic control of the gas mixture supply. In this regard, SF_6 is an expensive gas, and while it should be used in sufficient quantities to minimize the fire hazards, the automatic control of the gas will avoid any unnecessary excess use thereof. The complete cycle, once initiated, is completed in approximately 20 seconds so that an output of approximately two pellets, 3" diameter by 3" long per minute, may be achieved. Thus, the system and process is not only effective in solving the previously existing disposal problems, but in addition is cost effective purely as a reclamation process.

MAGNESIUM FROM TRACER AMMUNITION

Most tracer ammunition compositions are made from varying percentages of the same basic materials and these are, in order of quantity, strontium nitrate, magnesium, strontium peroxide, polyvinyl chloride, calcium resinate, barium peroxide, oxamide, zinc stearate, polyethylene, strontium oxalate, and lead dioxide. The strontium nitrate and magnesium account for about 60% of the total. Waste treatment methods used for disposing of tracer ammunition involve burning or chemical degradation of the pyrotechnic material and produce air and water pollution.

The method used for disposing of tracer material depends on the stage of the manufacturing process when the material is scrapped. Rejected finished tracer cartridges are burned in metal containers at a burning ground. If a batch of tracer material must be discarded, it is placed in oil and burned. Dry wastes spilled during the assembly process are collected by a vacuum system, under water, and the vacuum collectors are periodically dumped and the contents are treated with caustic, water and steam in a system of sumps. The waste from the sumps are taken by truck to basins on a high hill and the degraded material is leached to the soil. As the methods used put degraded pyrotechnic material into the air and water, these methods are ecologically unsound and no longer desirable.

A process developed by *C.F. Parrish, J.E. Short, Jr., and K.S. Whorrall; U.S. Patent 3,930,844; January 6, 1976; assigned to The U.S. Secretary of the Navy* is one for disposing of tracer ammunition pyrotechnic material containing strontium nitrate, magnesium, strontium peroxide, polyvinyl chloride, calcium resinate, barium peroxide, oxamide, zinc stearate, polyethylene, strontium oxalate and lead dioxide, with strontium nitrate and magnesium accounting for about 60% of the total material. Strontium nitrate is first removed from the material by dissolving in cold water, and the water solution of strontium nitrate is filtered and evaporated to reclaim the strontium nitrate. The remaining materials are given successive washes in hot water, ethyl alcohol and methylene chloride to remove all the other materials except magnesium. The magnesium is dried and reclaimed. Such a process is shown schematically in Figure 112.

Figure 112: U.S. Navy Process for Magnesium Recovery from Tracer Ammunition on a Commercial Scale

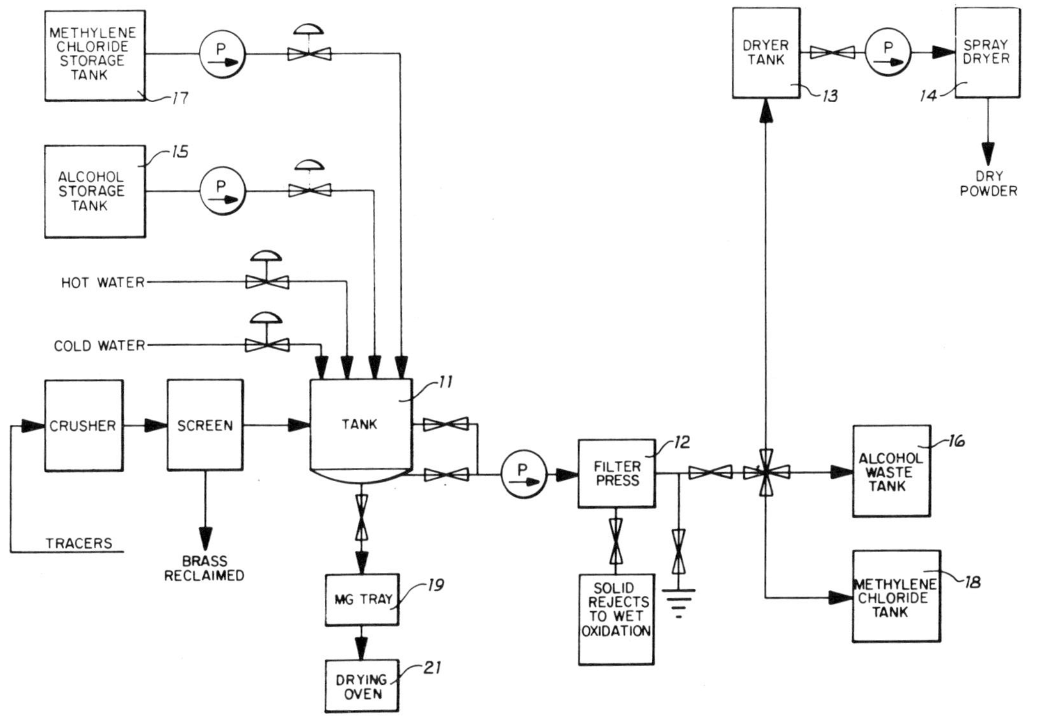

Source: U.S. Patent 3,930,844

Waste tracers are crushed and screened to remove the pyrotechnic material which is introduced into an agitated tank **11**. Pyrotechnic material left over from batch mixes and that material collected by collectors can be added directly to the tank. First, cold water is added to the agitated tank **11**. Agitation will wet down the solids and speed dissolution of the strontium nitrate. This solution is removed through an outlet above the bottom of tank **11**, leaving behind the undissolved, sinking solids. The solution is pumped through filter press **12** where floating and suspended lightweight impurities, such as calcium resinate and decomposed oxides are removed. The clarified strontium nitrate solution is collected in tank **13** and then spray dried in spray dryer **14** to give a dry powder that can be used to produce more tracers.

Next, hot water is introduced into tank **11** and the hot water washes remove more of the material other than magnesium from the remaining solids. The hot water is decanted, filtered and the liquid discarded. Then, ethyl alcohol is pumped from tank **15** into tank **11** to dissolve additional materials. The alcohol is then decanted, filtered and collected in tank **16** for recycling or reprocessing. In addition to dissolving additional materials, ethyl alcohol prevents clumping of the material that occurs when a water wash is followed by a wash with methylene chloride.

Methylene chloride is then pumped from tank **17** into tank **11** to dissolve the last of the impurities from the magnesium. The methylene chloride is decanted, filtered and stored in tank **18** for recycling or reprocessing. The bottom valve of tank **11** is then opened and magnesium flows onto trays **19** to be dried in a vented oven **21**. The sludge collected by filter press **12** is washed to a storage drum to await shipment to a wet oxidation facility.

By using this method, tracer mixes from various types of ammunition can be handled together thus significantly reducing labor costs. Also, the likelihood of failures is decreased when operators do not have to make decisions as to which process to use or where to place scrap.

A similar magnesium recovery process is described by *W.E. Doades, J.E. Short, Jr., and K.S. Whorrall; U.S. Patent 3,982,930; September 28, 1976; assigned to The U.S. Secretary of the Navy.* It, however, is devoted to the disposal of tracer bullets having in a steel bullet a pyrotechnic material containing strontium nitrate, magnesium, etc.

The pyrotechnic material is removed from the steel bullet by drilling and the steel is used as scrap. The balance of the recovery process is identical to that described in connection with U.S. Patent 3,930,844 above.

MERCURY FROM BRINE IN CHLOR-ALKALI PLANTS

The mercury amalgam electrolytic cell for the production of chlorine and caustic has been used primarily because of the high grade caustic soda which is produced. However, the loss of mercury from the effluent streams issuing from these electrolytic cells creates serious ecological problems. Furthermore, the loss of mercury is a costly expenditure in the caustic and chlorine producing plants.

Accordingly, it is imperative that means are found for the removal and recovery of mercury and/or other heavy metal ions from both solid and liquid plant effluents.

A process developed by *J.H. Entiwisle and R.W. Griffiths; U.S. Patent 3,718,457; February 27, 1973; assigned to Imperial Chemical Industries, Ltd., England* is a process for recovering mercury from waste brine from mercury cells comprising adding sulfide or hydrosulfide ions to the brine to precipitate mercury sulfide, trapping the mercury sulfide on a mixed media filter in which the average size of particles in successive superimposed layers decreases and the specific gravity of the particles increases from the top of the filter downwards, removing the mercury sulfide from the filter by backwashing with an aqueous liquor, separating the mercury sulfide from the backwash liquor and treating the sulfide to produce metallic mercury.

A process developed by *E.N. Balko, W.F. Schmitt, and S.D. Argade; U.S. Patent 4,012,297; March 15, 1977; assigned to BASF Wyandotte Corporation* provides for an integrated system in which mercury is extracted from plant waste streams, the waste streams are discharged, and the mercury recovered by electroreduction in the plant electrolytic cells.

Such a system will be described in more detail with reference to Figure 113. As shown, a brine stream **2**, which has been depleted in the mercury cell **12** and has been dechlorinated at **14**, is brought to saturation by passing brine stream 2 through a bed of solid rock salt in a saturator tank **4**. A sodium carbonate solution **6** is added to brine stream **2**, which stream is then pumped into clarifier **8**, wherein a solution of sodium hydroxide **10** is added whereupon the insoluble impurities are settled and stored in collection tank **16** together with a 5 to 10 gpm brine purge stream from clarifier **8**. Brine stream **2** is then electrolyzed in mercury cell **12**, stripped of chlorine at dechlorinator tower **14** and depleted brine stream **2** is returned to saturator tank **4**.

The slurry from collection tank **16** is transferred to mixing tank **18** where it is reacted with spent sulfuric acid **20** from chlorine drying towers (not shown). This provides for including the mercury in the acid in the recovery cycle and simultaneously provides for neutralization of the acid. Any carbon dioxide **22** produced during this neutralization is vented from the tank. The resulting slurry is transferred to reactor tank **24**, the pH is adjusted from about 6.0 to 9.0 and sufficient oxidant **26** is added to oxidize all of the metallic mercury and insoluble mercury salts to the soluble Hg^{+2} ion form. The oxidant may be sodium hypochlorite which has free NaOH present or if chlorine gas is employed, then additional NaOH may be required. The slurry is then filtered on filter wheel **28** employing a polypropylene filter cloth.

The resulting filter cake is washed with water and the solids **30**, which are discharged, are removed to a suitable waste disposal such as a landfill. The filtrate from filter wheel **28**, including the wash water, is collected in tank **32** where other liquid plant waste stream(s) **34**, such as sump and wash streams containing dissolved mercury are combined with the filtrate. Residual oxidant is destroyed by the addition of either sodium thiosulfate or sodium bisulfite **36**. The pH of the solution is then adjusted from about 2.5 to 6, preferably a pH of 3, employing a solution of hydrochloric acid **37** and an excess of NaHS **38** is added to form insoluble mercury sulfide. The mercury sulfide is filtered through a paper filter **40** which contains a cake of cellulose filter aid built upon it.

Figure 113: BASF Wyandotte Corp. Mercury Recovery and Recycle Process for Handling Mercury Cell Effluents

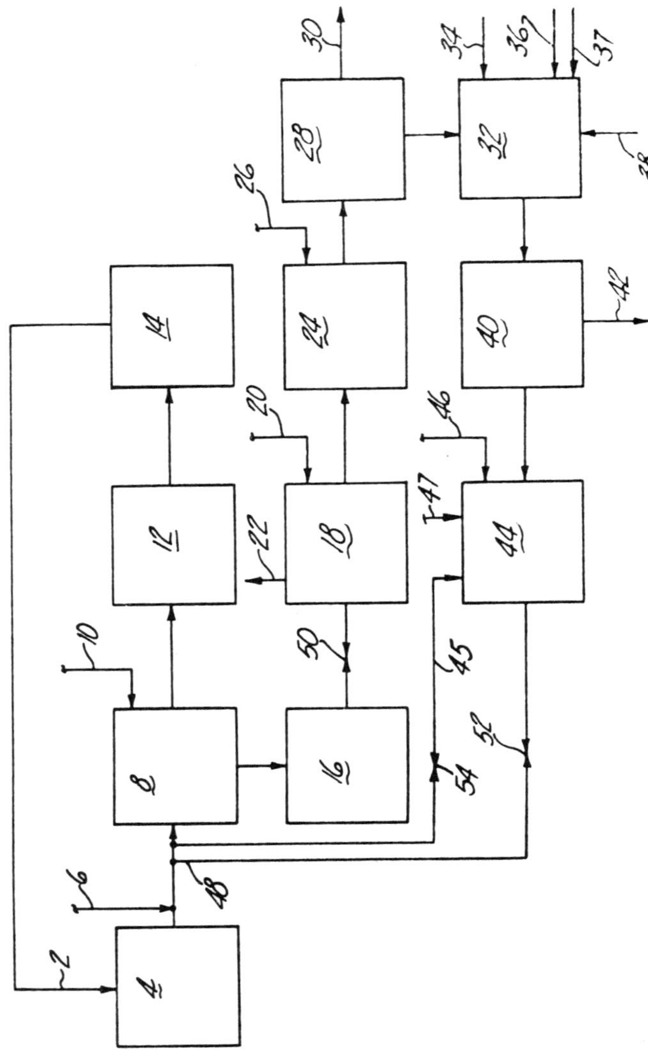

Source: U.S. Patent 4,012,297

The filtrate **42** essentially devoid of mercury is sewered. The filter cake from **40** is then slurried in plant brine **45** employing mechanical agitation in tank **44** to cause a fine dispersion of the cellulose filter medium and the mercury sulfide. A conventional agitator may be employed but in the preferred embodiment, it is desirable to use a special high shear dispersion mixer. The dispersion is then treated with oxidant **46** after the pH of the slurry is adjusted with sodium hydroxide **47** to a value from about 8 to 10, preferably a pH of 9.

The quantity of oxidant added, assuming sodium hypochlorite is employed, is within the range of about 3 to 6 mols of hypochlorite added per mol of mercury sulfide present and preferably about 5 mols. If the pH is appreciably below 8, some oxidant may be lost due to a reaction with the cellulose filter matrix while a pH appreciably above 10 may precipitate some of the mercury as mercuric oxide. The resulting mixture containing cellulose fibers and particles suspended in a mercuric chloride-sodium chloride solution is then injected into the brine stream between the saturator **4** and clarifier **8** at point **48**. The entire mercury recovery process may be kept isolated from the caustic chloride process by closing off valves **50**, **52** and **54**.

The oxidant employed in the process may be either sodium hypochlorite or hypochlorous acid which may be produced by adding chlorine gas to either an alkaline or an acidified brine solution until an amount of chlorine gas sufficient to oxidize all of the mercury is solubilized.

The pH range employed for the extraction of the mercury from the brine muds in collection tank **16** is from about 6.0 to 9.0, preferably from about 6.0 to 6.5.

This process in a single processing system treats all mercury-chlorine plant solid and liquid wastes. This process has substantial advantages over the prior art in that it:

(1) Provides a purge to the plant brine system for the control of undesirable impurities, i.e., sulfate ion.

(2) Performs well with a salt of relatively high impurity levels containing from 0.5 to 5% impurities.

(3) Provides a mercury removal and disposal system for spent sulfuric acid from the plant chlorine drying towers.

(4) Provides a mercury removal process for all waste liquid streams in the plant and is not restricted to tailbox wash water. More particularly, sump and wash streams are treated without the need for separate processing.

(5) Does not require separate purification of the soluble mercury recycled to the brine system.

(6) Provides for an efficient extraction of mercury from brine muds and produces a filter cake which is more easily filtered and handled.

MERCURY FROM CAUSTIC/CHLORINE PROCESS SLUDGES

During the production of chlorine and alkali metal hydroxides by electrolysis of alkali metal chloride brines in a cell using a mercury cathode, waste materials are accumulated which contain high concentrations of the alkali metal hydroxide and metallic mercury in a semisolid or sludge form.

Previously, these waste materials have been disposed of, for example, in landfill operations. This method of waste treatment results in a considerable economic loss of both product and raw material, in addition to causing pollution of natural resources. Furthermore, there is a danger of injury to personnel handling these wastes, both from caustic burns and from possible contamination by mercury. In addition, sludges containing high caustic concentrations have high freezing points, for example, about 50°F for a sludge containing 50% NaOH, posing a problem for equipment used in processing or storing the sludge.

In the treatment of mercury-containing waste materials, one widely used method is to heat the mercury-containing waste in a retort to recover the metallic mercury. However, caustic sludges, such as those obtained from a filter for concentrated caustic solutions or from an amalgam decomposer, cannot be fed to a retort without causing corrosive damage to the equipment. Therefore, there is need for a process of treating mercury-containing sludges which permits recovery of economically valuable materials while reducing the pollution of natural resources.

A process developed by *G. Generini; U.S. Patent 4,087,276; May 2, 1978; assigned to ANIC SpA, Italy* is one for stripping mercury from mercury-containing sludges, the improvement consisting in that the sludges are first dried and then roasted, all these operations being carried out under a subatmospherical pressure. A final condensation of the vapor phases takes place under superatmospherical pressure. A high efficiency of recovery is thus warranted. A suitable form of apparatus for use in such a process is shown in Figure 114.

The method can be regarded as being carried out in two stages, which correspond to two discrete sections, that is, an oven for the continuous roasting of the sludges and a condensation system under subatmospherical pressures of the barometric type.

The oven is of the continuous type, which is indirectly heated by means of electric resistors. It is composed of a horizontal cylindrical body **1**, which is heated from the outside by a set of resistors **2**, independent from each other, so as to have several heat degrees, according to the load of the oven and the concentration of water and mercury.

At one end of the cylindrical body, there is formed the horizontal sludge inlet mouth **3**; on the other there has been formed a dome **4**, for drawing vapors and, in the underlying portion, a mouth 5, the diameter of which is equal to that of the oven, for dumping the exhausted ashes. The sludge loading mechanism, as shown in the cross-sectional view A–A, is of the pressing screw **6** type (the pitch of the screw grows shorter and shorter) with an extrusion head equipped with cutting blades; the system is driven by an electric motor with reduction gear having a speed variator **7**.

Such an expedient is responsible, during processing, for the perfectly tight seal against the losses of vacuum in the system. The automatic variation of the rpm of the loading screw is a function of the temperature of the vapors in the terminal section of the oven concerned. The product charged into the oven must be a sludge having at the most 45 to 50% moisture and which has been neutralized (if and when acidic) to a pH which is certainly alkaline (8 to 9). In the interior of the cylindrical body, a shaft **8** is rotated, which is equipped with a screw **9** for the feeding of material. This is equipped with special scraping blades **10**, the latter provided to stir the sludges as they are being roasted while ensuring, moreover, that no danger of jamming is possible for the screw.

Figure 114: ANIC SpA Apparatus for Removal of Mercury from Sludge by Heating and Condensing

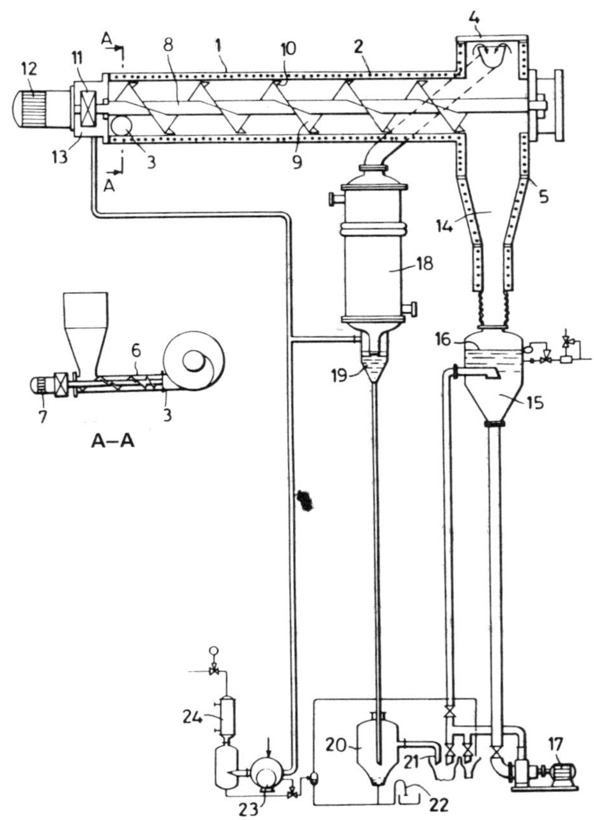

Source: U.S. Patent 4,087,276

The rotation of the screw is furnished by a speed-reducing gear **11**, arranged at either end of the cylindrical body of the oven, the electric motor of which, **12**, (two-speed) is of the kind which transfers the torque by magnetic expanders so as to guarantee the vacuum tightness in an absolute manner. The reducing unit **11**, of necessity, will be arranged in a vacuum chamber **13** and separated from the oven body by a gland seal. The run of the oven is very simple and wholly automated.

The sludges, as they come from the loading extruder, are fed forward by the screw so that, after having been first dehydrated, start to become roasted, the result being the distillation of the mercury contents.

The working conditions are 0.06 effective atmosphere at about 250°C, these conditions being such as to ensure the vaporization of all the mercury as contained in the sludges concerned. These conditions as specified above are automatically maintained by the feeding rate of flow of the sludges and by the following run parameters: (a) speed of rotation of the feeding screw (2 speeds), and (b) insertions of the several sets of electric resistors. There is thus the possibility of ensuring that the mercury contents in the ashes emerging from the screw is virtually zero.

Inasmuch as in the terminal portion of the oven, it is imperative to ensure that all the vapors evolved from the sludges are conveyed towards the condensation system while it is concurrently required that the exhausted ashes be continually removed from a system which is under a negative pressure, the following expedient has been envisaged.

The ashes emerging from the screw fall into a duct **14**, which is also heated by electric resistors, which feeds them to a container **15**, the latter being placed at a barometric level, a so-called dissolver in which a certain level of water **16** is present.

The temperature of such water is maintained, by an automatic system, at a constant value (about 37°C) which is slightly above the temperature corresponding to the vapor pressure of water under the same conditions of normal run of 0.06 effective atmosphere (35.8°C). This expedient acts in such a way that, when the hot ashes fall in water, some steam is generated (a direct function of the ashes' temperature and their specific heat) which, superheated by the hot walls of the duct to 250°C, acts in counterflow relative to the ashes and prevents a condensation of mercury and thus the presence of the metal in the exhausted ashes.

In order then to encourage the dissolution of the ashes in water, there is a pump **17** which provides a vigorous recirculation through the dissolver.

Lastly, with conventional level-checking systems, thermostatically controlled water is provided and fed in, as the dissolved ashes are being withdrawn. The vapors emerging from the dome **4** of the oven at the temperature of 250°C contain, almost entirely, superheated steam coming, predominantly, from the moisture contained in the fed-in sludges plus the counterflow washing steam of the exhausted ashes plus mercury vapors.

All these steam and vapors are conveyed to a vertical tube bundle condenser **18**, which is water-cooled (processing in the tubes on account of the possibility of dusts being entrained).

The condensate is collected in the bottom section of an accumulator **19**, having a hydraulic seal, which is placed at a barometric level and the outlet of which directly communicates, at the zero level, with an atmospherical collecting vessel **20**. Such a condensate is composed of water and mercury: water overflows through an overflow outlet **21**, whereas the mercury is automatically dumped by a siphon **22** from the bottom of the vessel.

The vapor phase of the accumulator **19** is composed of steam plus traces of mercury vapors plus the unavoidable noncondensed fractions; these latter are composed almost exclusively of air, coming from the air dissolved in ash-dissolving water and that occluded in the neutralized loaded sludges.

These vapors are drawn by a liquid ring vacuum pump **23**, which is intended to keep the system at a pressure of 0.06 effective atmosphere.

Obviously, the percentage of mercury vapors at the intake side of the pump is a direct function of the temperature of the cooling water in the exchanger **18**, and thus to ensure the value of 0.1 mg per normal cubic meter of Hg in the atmospherical effluent, the delivery of the pump is sent to a second water condenser **24**, wherein at the pressure of 1 to 3 atmospheres, as controlled by a pressure adjuster, an additional condensation is effected.

As an alternative, if the specifications of the vacuum pump do not permit a very high pressure drop, the coolant for the condenser **24** can also be water-cooled at least as low as 4° to 6°C, coming from a refrigeration system which will have very reduced dimensions on account of the small amounts of vapor involved.

Summing up, irrespective of whichever of the two abovementioned systems is adopted, the quantity of the atmospheric effluent is at any rate in the order of a few tens of liters an hour with a content of Hg which is always less than 0.1 mg per normal cubic meter.

One treatment process has been developed by *W.M. Loo; U.S. Patent 4,149,879; April 17, 1979; assigned to Olin Corporation*. It involves intermixing the caustic sludge with a liquid under pressure or agitation to produce a caustic slurry which is separated into a caustic liquor, metallic mercury and a remaining sludge which is suitable for roasting in a retort to recover better than 99% of the mercury in the sludge.

The process recovers valuable mercury and caustic materials and eliminates pollution of natural resources which was formerly caused when these sludges were disposed of in landfill operations. In addition, the process reduces potential hazards to personnel handling the caustic sludges and requires no addition of chemicals which produce undesirable by-products and add to the cost of the recovery process. A schematic of this process is presented in Figure 115.

Figure 115: Olin Corporation Process for Recovery of Mercury and Caustic Values from Caustic Sludges

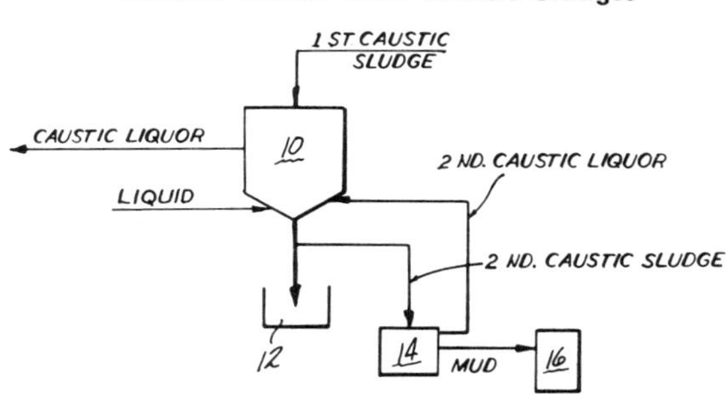

Source: U.S. Patent 4,149,879

As illustrated, a first caustic sludge is introduced into vessel **10**. Liquid is fed through an opening in the bottom of vessel **10** to flow up through the caustic sludge to form a slurry containing metallic mercury, caustic liquor, and solids. After separation, metallic mercury is withdrawn and introduced into vessel **12**. Caustic liquor is removed through an opening in the upper portion of vessel **10**. The remaining sludge is fed as second caustic sludge to vessel **14**. After separation, additional caustic liquor is removed through an opening in the upper portion of vessel **14** and recycled to vessel **10**. The mud remaining in vessel **14** is fed to retort **16** for the recovery of any remaining mercury.

MERCURY FROM SCRAP MERCURY

Virtually all metal can be reclaimed from mercury-cell plants, electrical apparatus, and control instruments, when the plant or equipment is dismantled or scrapped. Reported high-volume sources of scrap for reprocessing include instrument and electrical manufacturers, research laboratories, and mercury batteries. The most common single source was dental amalgams. During the past several years, the drastic decrease in the price of mercury made recovery from low-grade scrap uneconomical, but this decrease in supply was more than offset by increased emphasis on eliminating mercury discharges into the environment. In-house scrap is processed by the consumer and reintroduced into the manufacturing process.

MERCURY FROM SPENT CATALYSTS

It is known that vinyl chloride can be produced from acetylene and hydrogen chloride in contact with the active carbon-catalyst generally impregnated with 10% by weight mercuric chloride. The reaction of the components is initiated at a temperature of about 80°C which, inside a contact furnace, increases to 120° to 150°C, as a result of the exothermal reaction, and is maintained at that value by cooling. The fact that the mercury chloride is subject to sublimation gradually results in fatigue phenomena for the catalyst. These are even favored by the fact that the active carbon surface area becomes reduced by condensation and polymerization products that penetrate into the pores of the active carbon.

The catalyst will often be found to be exhausted and inactive after an operation period of 8 to 15 months, depending on the catalyst load. Such exhausted catalyst still contains between about 2.2 and 5.8% by weight mercuric chloride. This depends on the operation period and reaction conditions. The inactive catalyst is required to be removed from the contact furnace which has previously been banked out in a prepared bed of lime. The step of rejecting a mercury catalyst is disadvantageous for reasons of economy and involves the danger of ground water poisoning.

A process developed by *A. Jacobowsky; U.S. Patent 3,537,843; November 3, 1970; assigned to Knapsack AG, Germany* involves the recovery of mercury from a contaminated inactive mercuric chloride/active carbon-catalyst by burning the active carbon with a deficiency of oxygen or air, expelling or subliming off, together with the combustion gases, a mixture of gaseous mercury, mercurous chloride and mercuric chloride, condensing the gaseous mixture by cooling it, and, by adding a suitable reducing agent, reducing to mercury the mercurous chloride and mercuric chloride contained in the condensate.

A process developed by *H. Maiwald, G. Hockele, and H. Sauer; U.S. Patent 3,981,967; September 21, 1976; assigned to Chemische Werke Huls AG, Germany* is one in which bound mercury, especially mercury(II) chloride, is recovered from mercury-containing exhausted catalysts, in a process in which the catalysts are heated at 270° to 600°C while passing a carrier gas therethrough and the mixture of carrier and desorption gases is then withdrawn and cooled. The process particularly involves pretreating the inactive catalyst with chlorine at temperatures of 100° to 300°C, preferably at 150° to 270°C. In this process, the deposition of carbon black and organic solid products and the formation and precipitation of mercury and insoluble mercury(I) chloride is avoided. Figure 116 is a flow diagram of such a process.

Figure 116: Chemische Werke Huls AG Process for the Recovery of Bound Mercury from Mercury-Containing Catalysts

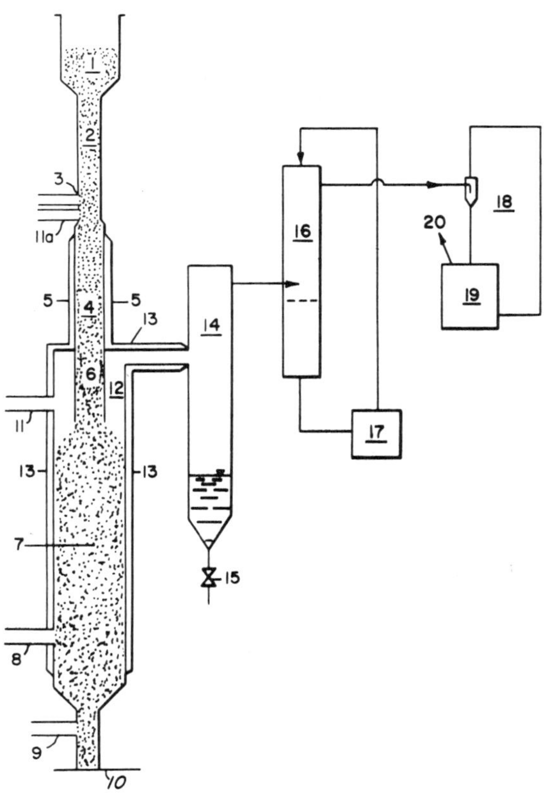

Source: U.S. Patent 3,981,967

The inactive catalyst flows from a storage vessel **1** for inactive catalyst via a feed pipe **2** into a pipe **4** which is externally heated to 300° to 500°C, preferably 350° to 450°C, by a heating unit **5**. A nitrogen inlet pipe **3** (nitrogen barrier) is disposed in the lower zone of feed pipe **2**. Between the nitrogen inlet pipe **3** and the preheating zone provided by pipe **4** and external outer wall heating unit **5**, chlorine is introduced through a feed line **11a** for pretreating the inactive catalyst at the point at which a temperature gradient exists, as a result of heating unit **5** of the preheating zone **4**, so that the spent catalyst is pretreated with chlorine at from 100° to 300°C, preferably from 150° to 270°C. These temperatures can also be maintained in the chlorination pretreatment reaction zone by diluting the chlorine with an inert gas, preferably nitrogen. By diluting the chlorine with an inert gas, the strongly exothermic reaction of the chlorination by the impurities, in the form of various organic compounds present on the inactive catalyst, is throttled and the heat of reaction is removed in a more advantageous manner.

The figure also shows an extension pipe **6** for feeding pretreated catalyst to chlorinating chamber **12** as well as a second chlorine feed line **11** and an external wall heating unit **13**.

The desorption vessel **7** is provided with a gas supply **8** for air or oxygen, a gas supply **9** for nitrogen, and a metering unit **10**. Connected to the desorption vessel **7** is a condensation tower **14** provided with emptying means **15** for removing condensed $HgCl_2$. For removing residual $HgCl_2$ there is provided scrubber **16** provided with hydrochloric acid recycle solution **17**. Nitrogen is withdrawn from scrubber **16** by jet pump **18** which is operated with dilute caustic soda solution in the cycle **19**, and is ventilated via ventilation means **20**.

A process developed by *H. Müller; U.S. Patent 4,059,438; November 22, 1977; assigned to Hoechst AG, Germany* is one in which the active carbon is burnt with the quantity of air or oxygen or mixtures thereof necessary to produce combustion gases containing metallic mercury in vapor form together with between 0.5 and 5% by volume of oxygen and between 2 and 15% by volume of carbon monoxide. The combustion gases are then contacted in a condensation zone with water so as to effect separation of the mercury therein. The gases issuing from the condensation zone are conveyed through a scrubbing zone having an alkaline liquor therein; and the gases leaving the scrubbing zone are delivered to the atmosphere.

The apparatus for carrying out the process is comprised of a shaft furnace being provided in its base portion with a grate and with an ignition flap opening thereinto, and of a scrubbing tower positioned downstream, in the direction of flow, of the furnace and receiving an alkaline liquor. The shaft furnace is provided in its base portion with a catalyst-supplying downpipe opening thereinto above the ignition flap. A condensation zone receiving water and terminating with its lower end in an outlet valve for the removal of mercury therethrough is positioned between the shaft furnace and the scrubbing tower; and the upper end of the scrubbing tower terminates in a conduit running to the suction side of a fan.

Such an arrangement of apparatus is shown in Figure 117. As can be seen, contaminated inactive catalyst is delivered from a reservoir **1** through a conveyor belt **2** and a downpipe **3** to a grate **5** disposed in the bottom portion of a shaft furnace **4**. It is generally good practice to provide for the grate **5** to be charged with a relatively little quantity of catalyst. Arranged between the grate **5** and the point where the downpipe **3** opens into the shaft furnace **4** is an ignition flap **6**. The shaft furnace **4** is furthermore provided with an air inlet **7** having a stop valve **8** positioned therein and with a combustion gas outlet **9**. The air inlet **7** opens into the bottom portion, and the gas outlet **9** extends from the head of, furnace **4**.

Figure 117: Hoechst AG Process for the Workup of a Contaminated Inactive Mercury(II) Chloride/Active Carbon Catalyst

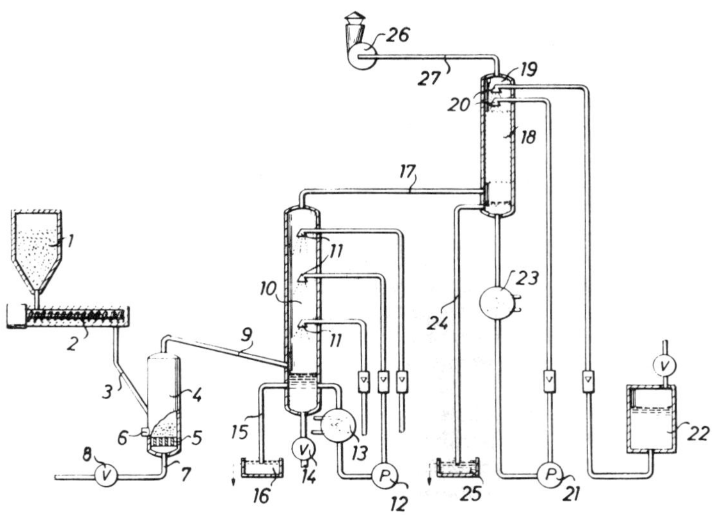

Source: U.S. Patent 4,059,438

Combustion outlet **9** opens into the base portion of a condensation tower **10** having a plurality of nozzles **11** centrally disposed therein with one downstream of another. The nozzles **11** are connected to a pump **12** receiving liquid coming from the base portion of the condensation tower **10** and travelling through a heat exchanger **13** positioned between the tower **10** and the pump **12**. Positioned downstream of the tower **10** is an outlet valve **14** permitting condensed metallic mercury to be removed therethrough. Placed below the point where the combustion gas outlet **9** opens into the condensation tower **10** is a goosenecked outlet **15**, which runs from the tower **10** to, and terminates in, a neutralization basin **16**. The head of tower **10** terminates in a further gas conduit **17**.

The gas conduit **17** opens into the base portion of a scrubbing tower **19** packed with filler material **18**. A plurality of nozzle structures **20** are centrally disposed, one downstream of another in the upper portion of tower **19**. The nozzle structures **20** are connected to a pump **21** or a reservoir **22** having sodium hydroxide solution therein. The pump **21** receives and conveys an alkaline liquor coming from the base portion of the scrubbing tower **19** and travelling through a heat exchanger positioned between the tower **19** and the pump **21**. Placed below the point where the gas outlet **17** opens into the scrubbing water **19** is a goose-necked outlet **24** which runs from the tower **19** to, and terminates in, a basin **25**.

The head of scrubbing tower **19** terminates in a conduit **27** which is connected to the suction side of a fan **26** delivering off-gas having between 1 and 2 mg of mercury per cubic meter therein to the atmosphere.

METALS FROM MIXED METAL SLIMES

Such metal slime is obtained, for example, as metal oxide or hydroxide from plants for pickling metal with acid, from plants for coating metals electrolytically or from other surface-treating plants. It is a particular object to provide a process for recovering iron and zinc, and for separating the zinc, thus recovered, from the iron. It is another particular object to provide an inexpensive method for recovering metals from metal-containing waste, so as to provide an economic alternative to the disposal of metal wastes by releasing them into lakes or rivers.

A process developed by *H. Reinhardt and H.D. Ottertun; U.S. Patent 3,966,569; June 29, 1976; assigned to MX Processor Reinhardt & Co. AB, Sweden* is one in which metal can be recovered from metal-containing waste by means of liquid extraction processes. The waste is leached with sulfuric acid. The resulting metal sulfate solution is contacted with an organic solution of a reagent, so as to extract iron and zinc to the organic solution. The organic solution is washed with sulfuric acid in two steps, viz, with weaker acid in the first step, so as to transfer zinc to the washing solution, and with stronger acid in the second step, so as to transfer iron to the washing solution. Zinc and iron have now been separated, and are recovered from the washing solutions in known ways, for example, by crystallization. If the waste contains other metals, in addition to iron and zinc, additional selective liquid extraction processes are added, before or after the iron-zinc-extraction.

Figure 118 shows schematically a plant for recovering metals from slime containing not only iron and zinc but also copper, nickel and chromium.

Figure 118: MX Processor Reinhardt & Co. AB Method of Recovering Metal from Metalliferous Waste

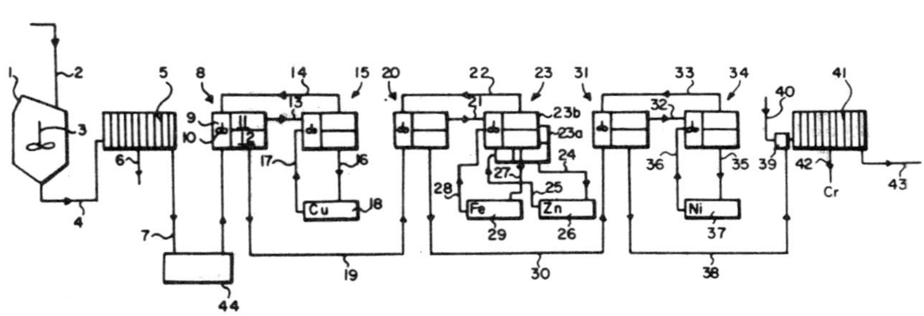

Source: U.S. Patent 3,966,569

The plant comprises a leaching vessel **1** with an inlet **2** for slime, sulfuric acid, water, a stirrer **3** and an outlet **4** connected to a filter **5**. This filter is provided with an outlet **6** for leaching remains and an outlet **7** which conducts the filtered leaching solution through a tank **44** to a first extraction unit **8**. Each stage in the extraction unit is of the mixer-settler type and comprises a mixing chamber **9** in which the leaching solution is mixed with an organic solution with the help of a stirrer **10**, resulting in copper being extracted over to the organic solution, and a separation chamber **11, 12** in which the two liquid phases, which should be substantially insoluble in each other, are allowed to separate due to their different densities. The lighter phase (usually the organic solution) is then collected in the upper part **11** and the heavier phase in the lower part **12**.

An extraction unit consists of one or more mixer-settler units. If multistep extraction is used, this is performed in countercurrent flow. Pipes **13, 14** connect the extraction unit **8** to a washing (stripping) unit **15** where each step is of the same mixer-settler type as the steps in the extraction unit. The washing unit may also consist of one or more mixer-settler units. Copper is removed in the washing unit **15** from the organic solution to an aqueous washing solution. Pipes **16, 17** connect the washing unit **15** to an apparatus **18** for removing the copper from the washing solution.

The lower part **12** of the separation chamber **11, 12** is connected by a pipe **19** to a second extraction unit **20** for the extraction of iron and zinc. The extraction unit **20** is connected by pipes **21, 22** to a double washing unit **23** comprising two units **23a, 23b** where the various steps in each unit are of mixer-settler type. Unit **23a** is connected by pipes **24, 25** to an apparatus **26** for removing zinc. Unit **23b** is connected by pipes **27, 28** to an apparatus **29** for removing iron.

The lower part of the separation chamber in the extraction unit **20** is connected by a pipe **30** to a third extraction unit **31** for the extraction of nickel. The extraction unit **31** is connected by pipes **32, 33** to a washing unit **34** which, in turn, is connected by pipes **35, 36** to an apparatus **37** for the removal of nickel from the washing solution. The lower part of the separation chamber in the extraction chamber **31** is connected by a pipe **38** to a tank **39** to which sodium hydroxide is supplied through a pipe **40** to neutralize the acid solution. The chromium hydroxide thus precipitated is filtered off in a filter **41** and the filter remains are removed through an outlet **42** and the filtrate through a pipe **43**.

The following is one specific example of the conduct of such a process. Metal hydroxide slime was collected from a plurality of zinc coating industries. After mixing the average metal content was in percent by weight of the solids content: 6.6 Zn, 15.4 Fe, 1.8 Cu, 2.4 Ni and 4.4 Cr. 300 kg of the slime and 9 liters of water were added to a leaching vessel. An aqueous solution containing 40% by weight sulfuric acid was added at a rate adequate for maintaining a pH value of 0.5. The leaching process was continued at 20° to 25°C while stirring for 20 hours. The solution was filtered and analyzed. It contained, in grams per liter, 4.17 Zn, 9.33 Fe, 0.90 Cu, 2.67 Ni and 0.34 Cr.

The solution was extracted in countercurrent flow in a device containing four mixer-settlers with an organic solution consisting of 10% by volume of LIX 64N (General Mills, Inc.) and 90% by volume kerosene. The volume ratio of organic solution to aqueous solution was 2.5:1. The resulting organic solution contained

0.36 g/ℓ copper, whereas the contents of zinc and iron were below 0.001 g/ℓ. The resulting aqueous solution contained, in grams per liter, 4.16 Zn, 9.3 Fe, <0.001 Cu, 1.9 Ni and 2.7 Cr.

The organic solution was washed in countercurrent flow in a stripping device containing three mixer-settler apparatuses with an aqueous solution from the electrolysis of copper containing 30% by weight sulfuric acid and 15 g/ℓ copper. The volume ratio organic solution to sulfuric acid solution was 30:1. The copper content of the organic solution was transferred to the sulfuric acid solution, which contained 28 g/ℓ copper after the stripping operation. Copper was recovered by electrolysis from the solution in a quantity of 15 g/ℓ in each cycle. The organic solution from the stripping operation was returned for another extraction process.

The aqueous solution from the copper extraction process was extracted in countercurrent flow in a device containing three mixer-settler apparatuses with an organic solution containing 25% by volume di(2-ethylhexyl)-phosphoric acid, 5% by volume dodecanol and 70% by volume kerosene. The volume ratio organic solution to aqueous solution was 10:1. The resulting aqueous solution contained, in grams per liter, 0.04 Zn, 1.3 Fe, 1.9 Ni and 2.7 Cu.

The organic solution was washed in countercurrent flow in a stripping device containing two mixer-stirrer apparatuses with an aqueous solution containing 14% by weight sulfuric acid, resulting in the zinc being washed out from the organic solution into the sulfuric acid solution. The volume ratio organic solution to sulfuric acid solution was 20:1. The resulting organic solution contained 0.01 g/ℓ zinc and 0.05 g/ℓ iron. The resulting sulfuric acid solution contained 2.8 g/ℓ zinc and 0.001 g/ℓ iron. Zinc sulfate was recovered by crystallization from the sulfuric acid solution in a quantity of 5 g/ℓ zinc. Sulfuric acid was added to restore the desired concentration of 14% by weight and the solution was returned for another stripping operation. The organic solution was returned for extracting zinc and iron in another extraction process.

The aqueous solution coming from the extraction of zinc and iron was passed to a third extraction process in a device containing two mixer-settlers. The solution was extracted in countercurrent flow with an organic solution consisting of 25% by volume dioctyldithiophosphoric acid and 75% by volume kerosene. The volume ratio organic solution to aqueous solution was 2:1. The resulting organic solution contained 0.95 g/ℓ nickel. The resulting aqueous solution contained 0.02 g/ℓ nickel and 2.7 g/ℓ chromium.

The organic solution was stripped in countercurrent flow in a device containing two mixer-settler apparatuses with an aqueous solution containing 50% by weight sulfuric acid. The dwelling time in the apparatuses was longer than that in the apparatuses referred to earlier. The volume ratio organic solution to sulfuric acid solution was 90:1. This stripping operation resulted in nickel and small quantities of zinc and iron being washed out from the organic solution to the sulfuric acid solution. The resulting sulfuric acid solution contained 86 g/ℓ nickel. Nickel sulfate was recovered by crystallization from this solution in a quantity of 50 g/ℓ nickel. Sulfuric acid was added to restore the concentration to 50% by weight. The solution was returned for another stripping operation. The organic solution was returned for another extraction operation.

The aqueous solution from the nickel extraction process contained chromium and small amounts of other metals. It was neutralized with a 45% by weight solution of sodium hydroxide to a pH value of 8 to 9. Chromium hydroxide precipitated and was removed by filtration.

METALS FROM OILY SCRAP METALS

The melting of scrap metals for reuse is a well-known procedure, but, quite often, the scrap metals are contaminated with oils, such as cutting oils, and the like, and the contaminating material will be volatized when the metal is heated and will pass off from the metal as an undesirable atmospheric pollutant. Such emissions, due to the fact that the principal contaminant of scrap metal is oil, are essentially hydrocarbon in nature and are quite objectionable when discharged directly into the atmosphere.

A number of different schemes have been proposed for collecting such emissions to prevent them from entering the atmosphere, but theretofore, arrangements for accomplishing this have been expensive and imperfect in operation and have represented no economy whatsoever in respect to the melting of the scrap metal for reuse.

Two different alternative schemes have been proposed and will be described here. One involves an improved pollution-free incineration process and the second is a solvent-washing process which avoids many of the problems posed by the incineration process.

An improved incineration apparatus has been described by *K.M. Habayeb; U.S. Patent 3,869,112; March 4, 1975; assigned to Wabash Alloys Inc.* The apparatus collects the hydrocarbon emissions and supplies the emissions to a place of combustion which may be, for example, the burners which are supplying heat to the respective furnace, to the base of the stack of the furnace, to the furnace downstream of the burners, or to burners for another furnace, or to a burner for a heated ladle.

The furnace is so constructed that the chamber in which the metal is melted extends beyond the furnace enclosure to form a charging well and an enclosure is erected around the charging well which, at the top, forms a hood in which the emissions are collected and from which the emissions are withdrawn at least as rapidly as the emissions are generated from the metal introduced into the charging well. Advantageously, the furnace is heated by gas burners and the emissions are supplied to the gas burners as combustion air.

During periods of hydrocarbon emissions from the metal being melted which exceed the requirements of the burners for combustion air, at least some of the emissions can be conveyed from the enclosure surrounding the charging well to another burner, for example, a burner for another furnace, or for a ladle in which molten metal is transported to a place of use. The emission can also be burned by injection thereof into the furnace or the base of the stack of the furnace.

The result of collecting the emissions and conveying the emissions to a place of combustion, is a great reduction in the emissions from a melting furnace of the nature referred to, to a level well below the allowable level required by pollution control codes while, at the same time, a substantial economy can be realized if the emissions are burned where the heat therefrom is useful.

In addition to the potential pollution problems noted above, a substantial quantity of fuel is required to incinerate the combustibles, a typical unit capable of handling 20 tons of scrap per hour requiring 54,000,000 Btu to incinerate the contaminants.

It has also been a standard practice to clean contaminated metallic parts by passing them through a drum-type washer wherein they are selectively subjected to various chemical washing and rinsing operations, with or without subsequent drying of the parts, depending upon how and when they are to be recycled. By way of comparison, a drum-type washer capable of processing 20 tons of scrap per hour only consumes about 8,000,000 Btu per hour to heat the washing solution, and at the same time does not pollute the air. Furthermore, the parts cleaned by a washing operation contain less residue than parts cleaned by incineration, and where the parts are melted for reuse, it has been found that washed parts produce from 20 to 25% less slag in the melting furnace than do parts cleaned by incineration.

There are, however, a number of disadvantages to drum-type washers, particularly insofar as fines recovery and sludge removal are concerned. For example, a substantial amount of scrap fines are discharged from the washing drum along with the washing solution, and heretofore the washing solution has been diverted into a tank or trough extending along one side of the drum in which a series of perforated catch baskets are suspended to collect and retain the fines. In normal operation, these baskets must be removed and emptied every few hours. In addition, a portion of the removed contaminants will settle in the bottom of the trough as sludge, and must be removed by dredging at regular intervals.

A scrap washing process which has been designed to overcome these latter problems has been developed by *P.J. Seiwert; U.S. Patent 3,973,573; August 10, 1976; assigned to The MacLeod Company*. It is a scrap washing system wherein scrap metal, such as crushed engine blocks or other oil and grease contaminated metal scrap, is cleaned by a washing operation to remove the oil, grease or other foreign materials prior to the melting of the scrap for reuse.

The system includes a drum-type tumble and hot wash machine wherein the scrap parts are first subjected to a tumbling or impacting step to remove loose residue. The scrap is then soaked in a bath of hot cleaning solution and subjected to a spray cleaning step prior to discharge from the drum. The system includes a conveyor mechanism for continuously recovering scrap fines discharged from the drum as an incident of the recovery and recycling of the washing solution, together with means for continuously collecting and removing both oil and sludge from the system as the washing solution is collected for reuse.

METALS FROM PLASTIC/METAL COMPOSITE PARTS

There are many plastic/metal composites such as wire scrap and reinforced hose scrap which may be susceptible to both metal and polymer recovery. One system for plastic/metal separation has been described by *C.G. Bradbury, S.T. Gotham and J.P. Wudyka; U.S. Patent 3,941,684; November 2, 1976; assigned to Leesona Corp.*

This system includes a granulator, followed by a fiber/dust separator with these components preferably being repeated, followed by a screening separator and, finally, a high-tension separator for separating plastic and metal. The system particularly incorporates an impact-counterflow separator for removing fiber/dust from heavier plastic and metallic particles.

Also read "ABS Plastics from Plated ABS Scrap" in reference (1).

MOLYBDENUM FROM COPPER SULFIDE ORES

The split-off point for by-product molybdenum occurs at the flotation stage in the milling of copper sulfide ores according to A. Petrick, Jr., H.J. Bennett, K.E. Starch and R.C. Weisner in Information Circular 8569, Washington, D.C., U.S. Bureau of Mines (1973). From a combination of a literature review and plant visits, the above authors prepared the hypothetical flow sheet shown in Figure 119.

The plant has a 6 ton per day throughput of by-product molybdenum. The tonnage of copper concentrates handled per ton of molybdenite production varies widely. The ratio of copper sulfide production to molybdenum sulfide production varies from less than 10 to 1 to thousands to 1.

There are several approaches used in producing separate molybdenum and copper concentrates. The one chosen for the hypothetical subsystem is differential flotation. First a collective float of both copper and molybdenum sulfides is made; then a separation is made by selective flotation in a second circuit. This is the approach used when the molybdenum content of the ores is so low that selective flotation in the primary circuit is not justified. This occurs in the great majority of cases.

The collective flotation of molybdenite along with copper sulfide minerals presents no major problems. The process used for copper-molybdenum ores preparatory to flotation is similar to that used for copper only ores. The ores are crushed and ground to a particle size sufficient to free the copper minerals for flotation. They are then treated with reagents that promote flotation of both molybdenum and copper sulfides. The primary collective concentrate in the hypothetical flow sheet is assumed to be at a rate of 430 tons per day. The collective concentrate is thickened, reground, and conditioned for molybdenum disulfide flotation.

The major separation of copper and molybdenum sulfides is made in the first and second rougher cells. The nonfloatable material from the second rougher cells becomes the final copper concentrate and the float material which is molybdenum disulfide continues through several cleaning operations. It was decided to include all operations beyond primary flotation in the by-product subsystems.

Figure 119: By-Product Molybdenum Subsystem

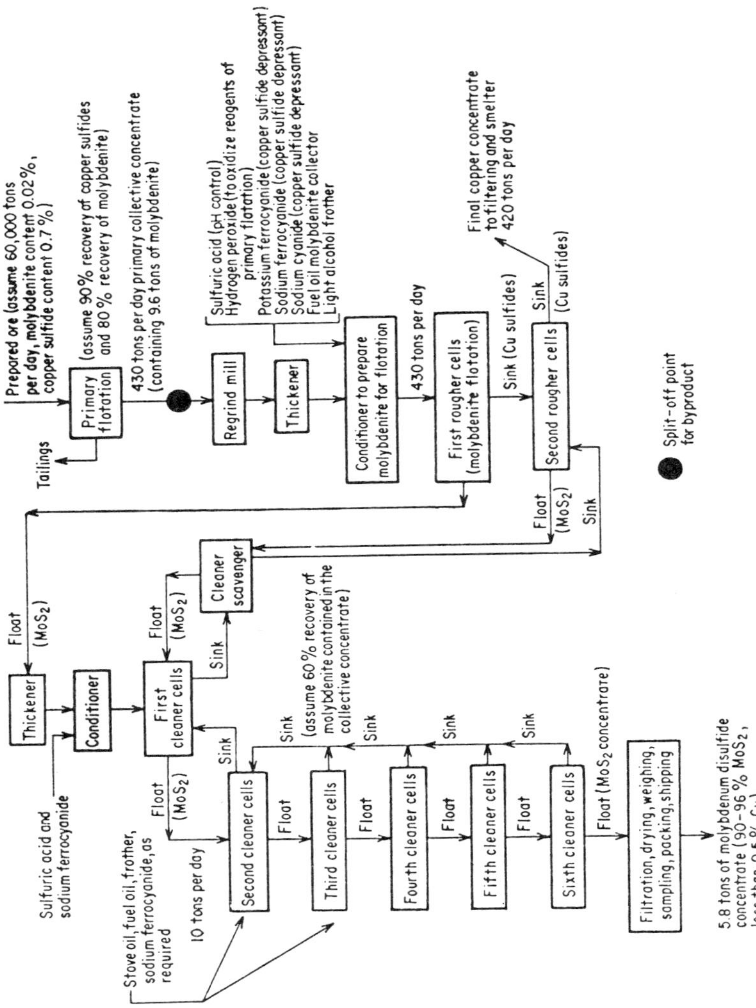

Source: Reference (17)

The by-product molybdenum subsystem consists of all processing of molybdenum disulfide subsequent to the production of the collective concentrate. There is considerable variation in the approach. The hypothetical flow sheet is confined to the operations shown in Figure 119.

These consist of thickening, grinding and conditioning prior to the first flotation as well as six-stage cleaning of the molybdenite concentrate. Recovery of molybdenum disulfide in primary flotation is assumed to be 80% and recovery in the molybdenum subsystem beyond that point is assumed to be 60%. The result is a total of 5.8 tons of 11,600 lb of molybdenum disulfide daily derived from the original processing of 60,000 tons per day of copper ore averaging 0.02% molybdenum disulfide.

MOLYBDENUM FROM SCRAP MOLYBDENUM

Some secondary molybdenum is recovered in the production of alloy steels, superalloys and metal, but data on quantities are incomplete and inconclusive. Most of this recycled material is generated and reused directly at metal-forming or fabricating plants (runaround scrap) or shipped to plants that then reclaim it (prompt industrial scrap). Very little molybdenum-containing obsolete scrap is processed for its molybdenum content.

Although some molybdenum is recycled as a minor constituent of scrap alloy steels, the use of such scrap does not generally depend on its molybdenum content. An increasing, but indeterminate quantity of molybdenum is being reclaimed from spent catalysts and chemical residues (16).

MOLYBDENUM FROM SMELTER FLUE DUST

See "Copper from Smelter Flue Dust".

MOLYBDENUM FROM SPENT CATALYSTS

See also "Cobalt from Spent Catalysts", especially U.S. Patent 4,145,397.

A process developed by *H. Castagna, G. Gravey and A. Roth; U.S. Patent 4,075,277; February 21, 1978; assigned to Metaux Speciaux SA, France* is one for recovering molybdenum in the form of molybdic acid from waste products and particularly from spent catalysts containing a support based on γ-alumina wherein the molybdenum is in the form of molybdenum oxide or molybdenum sulfide. The process involves treating the waste products with sodium carbonate and heating to convert the molybdenum present to sodium molybdate and converting the sodium molydate to molybdic acid by treatment with concentrated nitric acid to produce molybdic acid high in purity that is suitable for the production of molybdenum compounds as well as for the recovery of the pure metal. Such a process is shown in block flow diagram form in Figure 120.

Figure 120: Metaux Speciaux SA Process for Recovering Molybdenum Values from Spent Catalysts

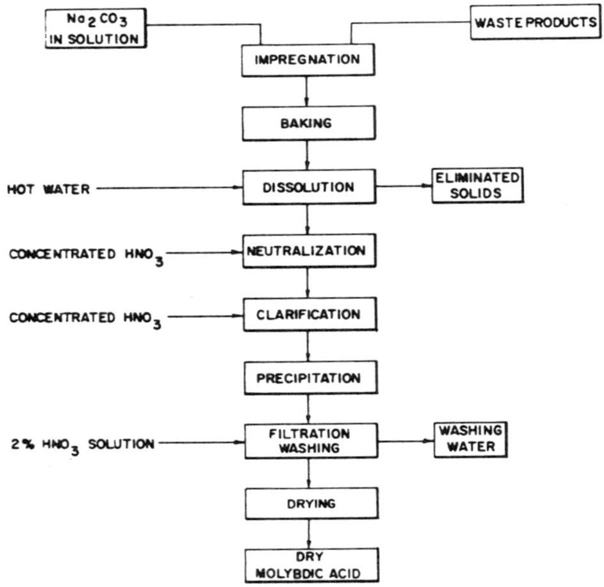

Source: U.S. Patent 4,075,277

The waste catalyst products to be treated have undergone preliminary oxidizing roasting at a temperature in the order of 500°C, which has substantially freed them from volatile materials, carbon and part of the sulfur. They are in the form of small cylinders or small balls. These waste products contain 8% of Mo, 1.5% of S and approximately 2% of Co; the support is based on γ-alumina. These waste products are treated in a rotary mixer with an aqueous solution heated to approximately 70°C and containing 400 g/ℓ of Na_2CO_3. Mixing is carried out in batches of 100 kg, which are mixed for about 30 minutes with 37.5 ℓ of solution. The quantity of Na_2CO_3 amounts to 150 g/kg of waste products.

Calculation shows that the quantity of Na_2CO_3 theoretically necessary for converting the 8% of Mo into sodium molybdate amounts to approximately 87 g/kg of waste products. Similarly, the quantity of Na_2CO_3 required for converting the 1.5% of S into sodium sulfate amounts to 44 g/kg of waste products. Accordingly, this leaves an excess of 19 g of Na_2CO_3 per kilogram of waste products, i.e., 1.9%.

The waste products are alternately impregnated in two mixers arranged in parallel so as to enable a batch which has just been impregnated to be progressively transferred to the baking furnace while a second batch is being impregnated.

The baking furnace is a rotary furnace approximately 4.2 m long with an internal diameter of 630 mm which is heated to between 650° and 750°C by a propane burner. At one end, it is continuously fed with impregnated product at a rate of approximately 100 kg/hr. The residence time of the product in the hot zone of the furnace is fixed at about 1 hour by means well-known to the expert. At the output end of the furnace, the temperature of the product is lowered to between 70° and 80°C by passage through a cooler of which the walls are cooled by the circulation of water. Approximately 95% of the molybdenum contained in the product is then present in the form of sodium molybdate.

The sodium molybdate is dissolved by washing the product in a layer approximately 10 cm thick in countercurrent on a continuous belt filter with a filtering surface of 1 m^2.

The washing unit comprises six stages; the sixth and last stage is fed with hot water at 80°C at a rate of approximately 120 ℓ/hr. The concentrated solution of sodium molybdate is removed at the output end of the first stage at a rate of about 104 ℓ/hr with a content of from 45 to 50 g/ℓ of Mo in dissolved form. The washed solid product which is removed from the belt after washing with pure water in the last stage contains approximately 0.27% of Mo in soluble form. Accordingly, the washing yield is of the order of 97%.

The solution containing the Mo is introduced into a first neutralization reactor at a constant rate adjusted to 104 ℓ/hr. At the same time, a first metering pump introduces an HNO_3 solution with an HNO_3 concentration of 53% into the reactor. The reactor is equipped with cooling means in the form of a double jacket filled with circulating water so as to keep the temperature below 20°C. The volume of this reactor is approximately 150 ℓ and the residence time of the solution is slightly more than 1 hour. The operation of this first metering pump is controlled by a probe for continuously measuring pH which is placed in the reactor so as to maintain a pH in the range from 5.2 to 5.5. Under these conditions, the average output of the pump is 8 ℓ/hr.

The solution, which is slightly cloudy because of the incipient precipitation of alumina, enters a second clarification reactor equal in volume to the first reactor where it receives another addition of nitric acid by means of a second metering pump which is controlled in dependence upon the first metering pump so that it delivers exactly the same volume of a nitric acid which has the same composition because it is taken from the same reservoir. The temperature of this reactor is kept below 30°C again by the circulation of water through a double jacket.

Under these conditions, the alumina is redissolved and clarified. The solution then enters a third precipitation reactor which has the same volume as the first two reactors and which is heated to 100°C by the circulation of steam through a double jacket.

Under these conditions, the molybdic acid precipitates while most of the aluminum remains in solution. An agitator enables the precipitate to be kept in suspension. The suspension is then delivered to a rotary filter on which the precipitate is collected and continuously washed with demineralized water containing 2% by volume of concentrated HNO_3. The precipitate then enters a hot air dryer where it is heated to approximately 100°C. This precipitate, based on molybdic acid, has an average Mo content of 61.2%. Its aluminum content is only 0.004%. Its mean apparent density is 2. The yield of Mo recovered in the form of molybdic acid, based on the Mo contained in the waste products, amounts to approximately 85%.

A somewhat similar process has been described by *G. Gravey, P. Grolla and A. Roth; U.S. Patent 4,075,278; February 21, 1978; assigned to Metaux Speciaux SA, France*. This process variation involves mixing the spent catalyst with sodium carbonate powder, adding water and remixing to absorb the water in the catalyst particles, baking the mixture to convert the molybdenum present to sodium molybdate, treating the mixture at ambient temperature with carbon dioxide gas, treating the resultant mixture with hot water to dissolve the sodium molybdate, neutralizing and clarifying with concentrated nitric acid, and precipitating to produce molybdic acid high in purity that is suitable for the production of molybdenum compounds as well as for the recovery of the pure metal.

MOLYBDENUM FROM TUNGSTEN LAMP MANUFACTURE

In the processing of tungsten filament coils and especially so-called coiled coils, a very fine tungsten wire is wrapped about a small molybdenum mandrel to form a first coil. The coiled tungsten and molybdenum mandrel are then wrapped about a second mandrel so that the tungsten is in the form of a coil which, in turn, is formed into a coil. The formed coiled-coil is then slipped off from the largest mandrel, leaving the first molybdenum mandrel in place within the turns of the smaller tungsten coil. After cutting to size, this member is immersed in a solution of mixed sulfuric acid and nitric acid which dissolves the retained molybdenum mandrel, leaving the coiled-coil tungsten intact.

The resulting coiled-coil of tungsten is then in a form suitable for use, such as a filament for an electric lamp. A molybdenum mandrel is also used in the manufacture of single coil lamp filaments.

The relative concentrations of the mixed sulfuric acid and nitric acid are subject to some variation and a mixed acid solution of sulfuric acid (normality of 13) and nitric acid (normality of 7) has been found to be very suitable. When this acid solution has sufficient molybdenum dissolved therein so that the molybdenum is present in amount of from about 40 to 75 g/ℓ of acid solution, the rate of molybdenum dissolution becomes sufficiently slow so that the acid is regarded as spent. In the usual practices of the prior art, this spent acid solution has suitably been treated and then discarded, since it has not been economical to recover the dissolved molybdenum.

A process for molybdenum recovery has been developed by *A.D. Kulkarni; U.S. Patent 3,963,823; June 15, 1976; assigned to Westinghouse Electric Corporation.* In this process, as shown in Figure 121, dissolved molybdenum is recovered from spent sulfuric acid-nitric acid solution by adding ammonium hydroxide to spent acid solution to raise pH thereof to about 1.5 to 3, preferably from 2 to 3.

The solution is then agitated and simultaneously heated to a temperature less than about 95°C until substantially all dissolved molybdenum is precipitated as ammonium molybdate complex, from which the precipitate is then recovered.

Figure 121: Westinghouse Electric Corp. Process for Molybdenum Recovery from Spent Acid Solution in Tungsten Lamp Filament Manufacture

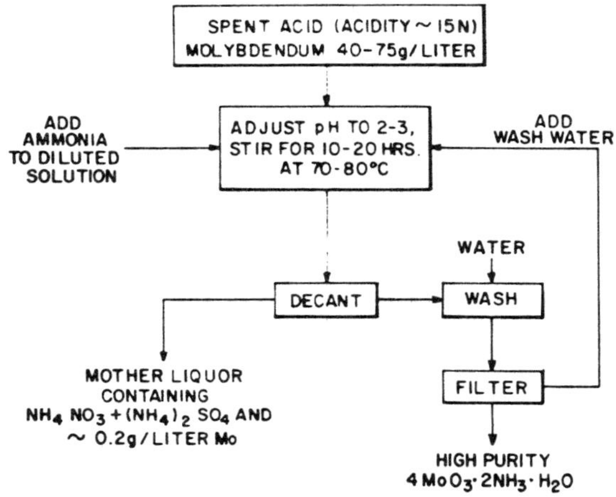

Source: U.S. Patent 3,963,823

N

NICKEL FROM BATTERY SCRAP

See "Cadmium from Battery Scrap."

NICKEL FROM BRIGHT NICKEL PLATING BATHS

In conventional decorative chrome plating the workpiece is protected from corrosion by a first layer of nickel deposited from a so-called semibright bath of an aqueous solution of nickel salts. To provide a smooth and lustrous surface on which to deposit the chromium layer an additional overlying layer or layers of nickel are deposited from so-called bright baths of an aqueous solution of nickel salts and various brightener additives of organic compounds containing sulfur. After the bright nickel layer or layers are applied to the workpiece it is rinsed in an aqueous mixture to wash away the nickel plating solutions and then a layer of chrome is deposited from a chromium bath on the workpiece over the bright nickel layer to provide a decorative exterior chrome surface.

The nickel salts, such as nickel chloride and nickel sulfate, in the baths of conventional electroplating processes must be periodically replenished and particularly the nickel salts in the semibright baths, because as the workpieces are transferred from one bath to another and into the aqueous rinse before deposition of the chromium layer they drag out or carry away with them a portion of the nickel baths. The portion of the nickel baths carried away by the workpieces and remaining in the aqueous rinses is not normally reused in conventional electroplating processes and cannot be reused in a semibright nickel bath because it contains so much sulfur from the brightener additives that it would seriously impair the corrosion resistance of the semibright nickel layer. Thus, the aqueous rinses of conventional electroplating processes are disposed of and usually must be treated to permit safe disposal thereof. Both such treatment of the aqueous rinses and the replacement of the nickel salts in the plating baths are costly.

Thus, a process has been developed by *L.S. Wright, Jr.; U.S. Patent 4,197,167; April 8, 1980* which permits recovery of nickel plating solutions from aqueous rinses and reuse thereof in both semibright and bright nickel baths and substantially decreases and sometimes eliminates the need for treatment and disposal of aqueous rinses containing nickel bath mixtures washed from workpieces. The quantity of brightener additives needed for nickel electroplating and the amount of power per workpiece required for nickel electroplating are decreased, thus providing a plating process which is more economical, easily carried out, and reliable in use and resulting in chrome plating of improved quality.

The process involves adding acid to the rinse mixture, passing the rinse and acid mixture through a carbon filter to remove the brighteners and decrease the sulfur content of the mixture, and adding the filtered mixture to and reusing it in the semibright and bright nickel baths of the nickel electroplating process.

Such a process is illustrated in Figure 122. The drawing shows in diagrammatic form a suitable apparatus **10** in conjunction with a nickel plating line **12** for carrying out this method. Plating line **12** has a semibright nickel tank **14**, bright nickel tank **16** and first, second and third rinse tanks **18, 20** and **22** through which each workpiece is successively transferred. The anodes in tanks **14** and **16** are not shown. The nickel plating bath mixture carried away from tanks **14** and **16** by the workpieces is rinsed from the workpieces by aqueous mixtures in rinse tanks **18, 20** and **22** which tanks are arranged so that any overflow of tank **22** flows back into tank **20** and any overflow of tank **20** flows back into tank **18.**

Figure 122: Wright/Ladney Process for Recovery and Reuse of Bright Nickel Electroplating Baths Carried Away by Workpieces

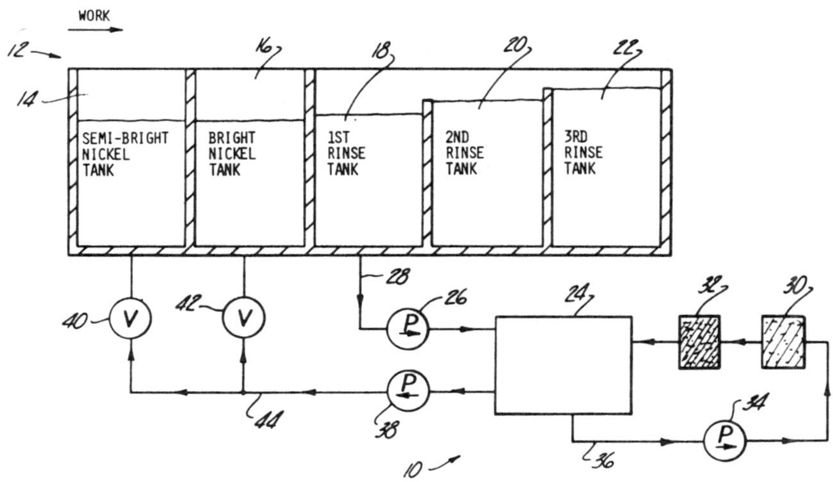

Source: U.S. Patent 4,197,167

The apparatus **10** has a reservoir **24** which is supplied with a portion of the aqueous rinse mixture from tank **18** by a pump **26** and suitable conduits **28.** Acid is added to the rinse mixture in reservoir **24** and the rinse and acid mixture is recirculated through a particle filter **30** and a carbon filter **32** by a pump **34** and suitable conduits **36.** The filtered rinse and acid mixture in reservoir **24** is supplied to nickel tanks **14** and **16** by a pump **38**, control valves **40** and **42**, and suitable conduit **44.**

Preferably the rinse mixture from tank **18** is supplied to reservoir **24** by pump **26** either continuously or at frequent intervals and at a rate which is equal to the rate of additions to tank **18** of the nickel plating baths from tanks **14** and **16** by the workpieces and the back flow from tank **20** so that the level of the aqueous rinse mixture in tank **18** remains substantially constant.

Preferably the filtered rinse and acid mixture from reservoir **24** is supplied to nickel tanks **14** and **16** by pump **38** and valves **40** and **42** either continuously or at frequent intervals and at a rate which is equal to the rate at which the nickel plating baths are removed from tanks **14** and **16** by the workpieces so that the level of the baths within tanks **14** and **16** remains substantially constant and the chemical composition and pH value of the baths is substantially uniform throughout the plating process.

To assure that the rinse and acid mixture in reservoir **24** is adequately filtered when it is returned to the baths in tanks **14** and **16**, pump **34** preferably recirculates such mixture through filters **30** and **32** at a much higher rate than the rate at which the mixture is removed from the reservoir by pump **38** so that the mixture in the reservoir in effect passes through filters **30** and **32** several times before it is added to the nickel plating baths in tanks **14** and **16**.

If desired, suitable apparatus for automatically maintaining the pH value of the mixture within reservoir **24** substantially constant by adding small quantities of acid as needed can be utilized. Particle filter **30** may be a paper filter or depth cartridge from Summit Scientific of Rutherford, N.J. and carbon filter **32** is preferably of activated carbon such as the Karbo Klear activated carbon impregnated depth filter with 4 ounces of carbon per 10" section available from Summit Scientific of Rutherford, N.J.

In using plating line **12** tanks **14** and **16** are filled with conventional nickel plating baths having an aqueous solution of about 300 to 550 g/ℓ of nickel chloride and nickel sulfate salts and a buffer of about 30 to 40 g/ℓ of boric acid. The baths may have a pH value in the range of about 2 to 5 and preferably a pH value of about 3 to 4 and are normally operated at a temperature in the range of about 70° to 180°F and preferably about 130° to 150°F. In order to provide plated workpieces with good corrosion resistance the semibright bath has a low sulfur content and usually does not contain any organic brightener compounds.

The bright nickel bath contains various brightener additives which are organic compounds and at least some of which contain sulfur which results in the bright nickel bath having a substantially higher sulfur content than that of the semibright bath. Several suitable brighteners are disclosed in U.S. Patent 3,288,574 and the patents referred to therein. Rinse tanks **18, 20** and **22** contain an aqueous acidic mixture having a pH value in the range of about 1.5 to 2.5 and are normally operated at a temperature in the range of 60° to 80°F.

Commercially available nickel plating bath mixtures with which this process has been successfully used are Perflow and No. 701 semibright bath mixtures and Zodiac and Galaxie bright bath mixtures produced by the Harshaw Chemical Company of Cleveland, Ohio, and the N2E semibright bath and No. 66 and 724 bright bath mixtures produced by the Udylite Company of Detroit, Mich.

In utilizing the apparatus **10** with a plating line **12** a portion of the aqueous mixture in rinse tank **18** is withdrawn at a slow rate by pump **26** and supplied to reservoir **24** where it is mixed with a sufficient quantity of hydrochloric and/or sulfuric acid to maintain the pH value of the mixture within the reservoir at not greater than 3.0 and preferably in the range of 2.0 to 2.5. The rinse and acid mixture in in reservoir **24** is recirculated by pump **34** so that it passes through filters **30** and **32** and is returned to tank **24** several times thereby assuring that the mixture in reservoir **24** is adequately filtered. Filter **30** removes particulate particles from the mixture and filter **32** removes the organic brighteners from the mixture by absorption so that it can be returned to and used in both the semibright and bright nickel baths in tanks **14** and **16.** Additional nickel salts are added as needed to the semibright bath in tank **14** and additional brightener additives are added as needed to the bright nickel bath in tank **16** just as in conventional electroplating processes.

Using the apparatus **10** to carry out the process with a conventional nickel plating line to plate metal or plastic workpieces with a semibright nickel layer 0.0005" thick and bright nickel layer 0.003" thick with a surface area of about 20 ft^2 at the rate of about 100 workpieces per hour has proved to be highly satisfactory. The semibright bath contained about 32 oz per gal $NiSO_4$, 6 oz per gal $NiCl_2$, and 7 oz per gal boric acid with the mixture having a pH value of about 3.2, and the bright plating bath contained about 32 oz per gal $NiSO_4$, 10 oz per gal $NiCl_2$, and 7 oz per gal boric acid with the mixture having a pH value of about 3.6. The nickel plating baths were at a temperature of about 145°F and the first aqueous rinse was at a temperature of about 80°F.

In the semibright bath the workpieces were subjected to a potential of about 7 volts and a current of about 40 amp/ft^2, and in the bright nickel tank the workpieces were subjected to a potential of about 7 volts and a current of about 50 amp/ft^2. Apparatus **10** supplied mixture from the first rinse tank to the reservoir **24** substantially continuously at a rate of about 200 gal/hr, recirculated the mixture in reservoir **24** through filters **30** and **32** at the rate of about 200 gal/hr, and supplied filtered mixture from reservoir **24** to the semibright nickel bath in tank **40** at about 75 gal/hr and to the bright nickel bath in tank **42** at about 30 gal/hr. Reservoir **24** had a capacity of about 3,000 gal and sufficient acid was added to the mixture in the reservoir as needed to maintain the pH value of the mixture in the range of about 1.5 to 2.5 with an average value of about 2.0.

Such use of the process has resulted in substantial cost savings due to the greatly decreased amount of waste nickel plating baths and rinse water which must be treated for disposal and the decrease in the amount of nickel salts and brightener additives consumed by the plating process. With the use of this process it has become economically feasible to increase the concentration of nickel salts in the plating baths which has resulted in improved quality of nickel plating, a decrease in the quantity of brightener additives, and a decrease in the amount of power required to produce a given thickness of nickel plating on a workpiece.

By returning the filtered mixture to the nickel plating bath in small quantities and in at least frequent intervals the variation in the chemical composition and the pH value of the baths during plating operations is believed to have been substantially decreased which is believed to result in more consistent and reliable plating operations which produce plated workpieces of improved quality.

NICKEL FROM NONFERROUS SLUDGES

See also "Metals from Mixed Metal Slimes."

A process developed by *L.E. Lancy; U.S. Patent 3,953,306; April 27, 1976; assigned to Dart Environment and Services Company* is an improved process for preparing highly concentrated salt solutions of metals of high purity from waste metal sludges at superior dissolution rates.

The process is carried out by preparing, concentrating and washing the sludge containing insoluble metal hydroxide under specific conditions to safeguard against the solids being exposed to air or dried out. The wet sludge is then dissolved in a dilute acid. The resultant solution can be used directly, e.g., as makeup in many electrolytic metal finishing processes to compensate for dragout losses occurred, or as the primary metal source in electrolytic processes employing insoluble anodes. Other specific uses for the process include the preparation of aluminum sulfate coagulants.

A process developed by *M. Kawasaki, S. Mizumoto and H. Nawafune; U.S. Patent 4,064,022; December 20, 1977; assigned to Ebara-Udylite Co., Ltd., Japan* is one in which a sludge containing a sparingly soluble metal compound such as nickel carbonate or the like is supplied to an anode compartment or an intermediate compartment of an electrolytic cell provided with an anode compartment and a cathode compartment or with an anode compartment, an intermediate compartment and a cathode compartment each sectioned by a filter membrane or an ion-exchange membrane.

Thereafter an electrolytic treatment is conducted by using an insoluble electrode as anode and a usual electrode as cathode to precipitate a metal such as nickel on the cathode. Effective recovery of a metal is attained directly from a sludge by way of electrolysis.

The following is a specific example of such a process, described with respect to the apparatus shown in Figure 123.

Figure 123: Kawasaki/Ebara-Udylite Co., Ltd. Apparatus for Nickel Recovery Directly from Sludge by Electrolysis

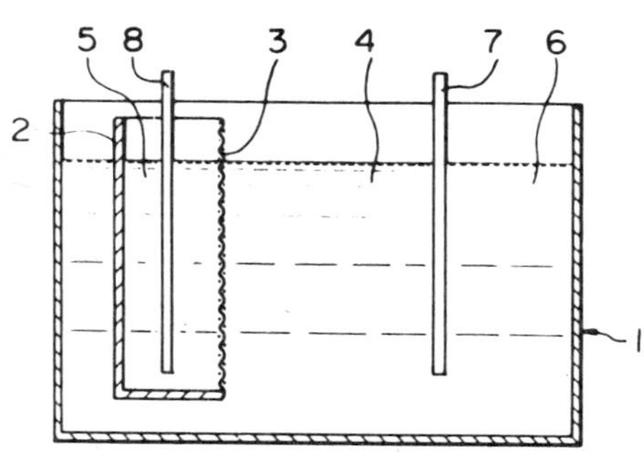

Source: U.S. Patent 4,064,022

In an electrolytic cell **1**, a cubic container **2** made up of polyvinyl chloride of which one of the perpendicular walls was replaced by a felt cloth **3** was dipped into an electrolytic solution **4** in the electrolytic cell **1** to prepare an anode compartment **5**. The remaining part in the electrolytic tank **1** was used as a cathode compartment **6**. 1 liter of an aqueous solution of nickel sulfate with a concentration of 1 mol/liter having dissolved therein 30 g of boric acid was used as the electrolytic solution **4**.

A calendered copper plate one side of which was coated with an insulating material and a platinum plate were used as a cathode **7** and an anode **8**, respectively. While keeping the temperature of the solution at 45°C, a nickel carbonate sludge having an average moisture content of 85% was added to the anode compartment **5** at a rate of 100 to 150 g/hr. An electrolytic treatment of the sludge was carried out under the following conditions. Electric current was 6.25 A and total quantity of electricity, 600,000 coul.

In this electrolytic treatment, 8.90 g of precipitated nickel were obtained at every 30,000 coul.

NICKEL FROM PLATING RINSE WATERS

Recycle treatment of nickel plating wastewater by means of ion exchange resins has been previously known. Thus, in the processes practiced to date, nickel plating wastewater is allowed to flow over cation resins so as to adsorb nickel ions, and when the cation resins become saturated, the nickel ions adsorbed on the resins are eluted by passing through mineral acids, particularly sulfuric acid, so as to regenerate the cation resins and, at the same time, to recover nickel salt solutions; in these cases, due to the presence of an excessive amount of sulfuric acid, the eluted solutions, namely the nickel salt solutions, cannot be reused as a nickel plating solution as such, and therefore, they are further treated over anion resins to remove the excessive acid by adsorption so as to recover reusable solutions.

However, in the course of removing the acid, by means of treatment with anion resins, from the solution eluted from the cation resins, the nickel ion concentration in the solution drops by about 20% and thus a solution with a desired concentration cannot be recovered; furthermore, a considerable amount of nickel is present in the wash water and has to be removed from the washings of the anion resins after adsorption by passing the wash water through cation resin beds, thus causing an increased load on the cation resins, and in addition, the amount of alkali required for the regeneration of anion resins is several times greater than the theoretical, thus causing increased secondary pollution of the treatment water, to mention some of the drawbacks.

A process which overcomes these problems has been developed by *T. Hayashi; U.S. Patent 4,009,101; February 22, 1977; assigned to Kayabakogyo KK, Japan.*

In order to efficiently recover the nickel components from wastewater which has been used for the washing of nickel plated products in a form reusable as a nickel plating solution as such, the nickel plating wastewater is mixed with an aqueous alkaline component so as to remove, by precipitation in the form of hydroxides, contaminating metal ions other than nickel. The filtrate so obtained is treated through an ion exchange process so as to obtain an aqueous solution of nickel salts containing excess free sulfuric acid, a portion of which is then treated with alkali so as to precipitate nickel hydroxide which is separated in the form of nickel hydroxide through filtration by means of a centrifuge.

The nickel hydroxide so obtained is then added to the rest of the aqueous nickel salt solution for neutralization with the free sulfuric acid contained therein to form nickel sulfate; and the free sulfuric acid is then removed so as to obtain a highly concentrated nickel sulfate solution.

Figure 124 is a schematic of such a process.

Figure 124: Kayabakogyo KK Process for Recycle Ion-Exchange Treatment of Wastewater from Nickel Plating

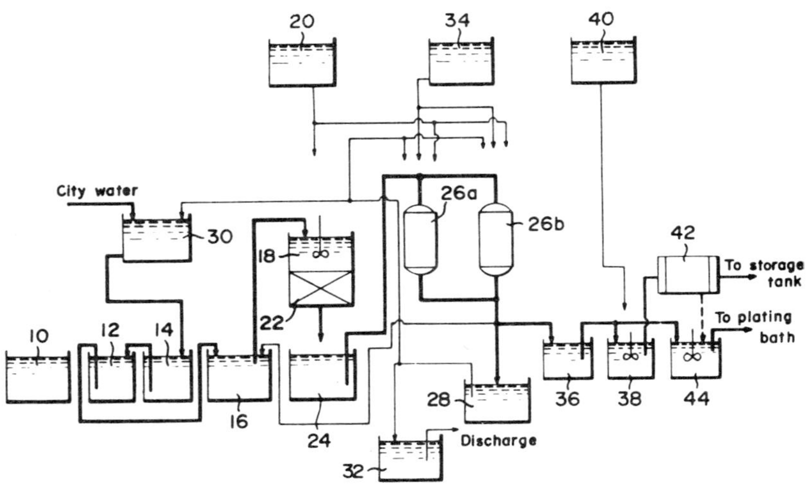

Source: U.S. Patent 4,009,101

The products nickel plated in the plating bath **10** are washed first in the first washing bath **12** and then in the second washing bath **14**. The wash water used for the washing of the nickel plated products is first supplied into the second washing bath **14**, the wash water enriched with nickel salts in the second washing bath **14** is transferred into the first washing bath **12**, and the wash water further enriched with nickel salts in the first washing bath **12** is sent to the storage tank **16** for storage. In general, the nickel plating wastewater in the storage tank **16** has an acidity of pH 3.5 to 4.5.

The nickel plating wastewater stored in the storage tank **16** is transferred into the adjustment tank **18**, where aqueous sodium hydroxide supplied from the sodium hydroxide tank **20** is added for adjustment to make the pH of the solution about 7.5 so as to precipitate out the metallic components other than nickel, such as iron and copper, etc., in the form of hydroxides, and the resultant mixture is filtered through the filter **22** installed below the adjustment tank **18** to obtain the filtrate, which is then passed to the filtrate tank **24**.

Then, the filtrate in the filtrate tank as such is allowed to pass through adsorption towers **26a** and **26b** which are packed with cation resins so as to adsorb nickel ions, and the treated water eluted from these adsoprtion towers **26a** and **26b** is stored in the first water recovery tank **28**. Through this treatment, the treated water stored in the first water recovery tank **28** becomes innocuous, and thus it may be used again as wash water by admitting it into the second water recovery tank **30** or it may be discarded after adjustment of the pH in the pH adjustment tank **32**. When the adsorption capacity of the cation resins in the adsorption towers **26a** and **26b** has reached saturation after repeating such treatments, the flow of filtrate is stopped, and the adsorption towers **26a** and **26b** are subjected to

(1) Purging: First, air is blown through so as to purge filtrate remaining among the cation resin particles.

(2) Prewashing: Then about 20 volumes of water per volume of resins are used for upward and downward washing of the cation resins. The treated water in the first water recovery tank **28** is utilized as wash water for this treatment.

(3) Purging: Air is again blown through so as to purge wash water remaining among the resin particles, and the purged filtrate as well as the purged wash water is returned to the storage tank **16**.

(4) Reagent injection: An equivalent volume of aqueous sulfuric acid from the sulfuric acid tank **34** is introduced per volume of the resin so as to effect desorption of the nickel ions and regeneration of the resins, and is eluted in the form of a nickel sulfate solution.

(5) Purging: Air is then blown in so as to purge the nickel sulfate solution remaining among the resin particles, and all of these solutions are sent to the primary solution recovery tank **36** for storage.

Through the treatments described above, the nickel ions adsorbed over the cation resins are removed in the form of a nickel sulfate solution and, at the same time, the adsorption towers **26a** and **26b** are regenerated. In this case, it is recommended to regenerate one of the adsorption towers **26a** and **26b** while the other is being used for adsorption, so as to make continuous operation possible.

Thus, the eluted solution stored in the primary solution recovery tank **36** becomes a nickel sulfate solution containing nickel ions plus free sulfuric acid plus a very small amount of sodium. A portion of this eluted solution in the primary solution recovery tank **36** is then transferred into the reaction tank **38** for treatment with aqueous sodium hydroxide supplied from the sodium hydroxide tank **40** in order to precipitate the nickel ions in the form of hydroxide, which is to be removed by filtration through the filter press **42**. The filtrate is again returned to the storage tank **16**, whereas the filter press is water-washed in order to recover the nickel hydroxide. On the other hand, the rest of the eluted solution in the primary solution recovery tank **36** is sent to the secondary solution recovery tank **44** for mixing with the nickel hydroxide separated by filtration through the filter press **42**, the nickel hydroxide is allowed to react with the excess free sulfuric acid contained in the eluted solution so as to reduce the amount of free sulfuric acid by the formation of nickel sulfate; and the resulting nickel salts solution of desired concentration is recycled back into the plating bath **10** for reuse as a plating solution.

NICKEL FROM SCRAP NICKEL

Scrap is a significant source of nickel supply. Nickel scrap is made in forming and shaping operations in primary processing plants, equivalent to the material considered as home scrap in steel mills. Nickel scrap is generated in fabricating plants that use nickel-bearing materials such as stainless steels and superalloys. It also is generated when consumer goods made from nickel-bearing materials become obsolete.

Approximately one third of the nickel scrap consumed follows the normal pattern of primary consumption after passing through scrap collectors, brokers, smelters, refiners, and foundries. The exception is Monel alloy scrap, which has unique use in steel mills producing specialty copper-nickel, corrosion-resistant, high-strength steels.

On the other hand, stainless steel, high-strength steel, and superalloy scraps are consumed in plants producing these materials. Almost invariably stainless steel scrap is used to make stainless steel, nickel-bearing alloy scrap is used to make nickel-bearing alloys, and nickel-base superalloy scrap is used to make superalloys. It is significant that nickel consumption can be correlated with stainless steel production only by taking into account the large quantity of runaround stainless steel scrap in the steel mills.

Nickel home scrap is generated in integrated steel mills, in nonferrous smelter and refining plants, and in foundries. It does not normally reach an outside market. Prompt industrial scrap nickel is sold to steel mills, smelters, and refiners directly by manufacturing companies, or indirectly through scrap brokers. Essentially all nickel scrap obtained from obsolete equipment is returned through scrap brokers to the steel mills, smelters, refineries, and foundries. Normally, brokers sell stainless and alloy steel to integrated steel mills and other nickel-bearing scrap to nonferrous smelters and refiners. However, in an active market, brokers may sell segregated nickel scrap directly to foundries.

The quantity of scrap metal generated in making many nickel alloys is unusually high compared with that generated in making steel and nonferrous alloys of copper, lead, zinc, and silver. In making and working with stainless and alloy steels, yields average less than 60%; in making and working high-nickel alloys, yield is often as low as 20%. Inhouse loss in recycling high-nickel alloy is significant but not critical, but once the nickel-bearing material leaves the primary plant, scrap loss is 30% or more. The refractory nature of most high-nickel alloys makes processing them to yield separate elements both technically difficult and expensive. In the United States, high-nickel alloy scrap is normally not utilized unless its composition is known within close limits so that it can be reused as is.

Otherwise, it usually is exported to Japan or the Federal Republic of Germany where it is processed to separate the contained elements in a form suitable for reuse (16).

NICKEL FROM SPENT CATALYSTS

As is well known to those skilled in the art, nickel-containing catalysts are used in many reactions such as hydrogenation reactions, alkylation reactions, hydroalkylation reactions, cracking processes, etc. Initially, these catalysts perform at a high level, but as the reaction proceeds, the catalyst becomes less active. Eventually, the activity of the catalyst decreases to a point where it is not sufficiently effective to be used in a commercial process.

A wide variety of nickel catalysts and modifications thereof have been described in the art in which they are utilized. Nickel catalysts are used extensively in hydrogenation reactions such as in the hydrogenation of unsaturated organic compounds. Usually, catalysts which are used in hydroalkylation reactions will contain in addition to nickel, other metals such as tungsten. Nickel catalysts used in cracking operations often contain molybdenum and other elements.

Considerable research has been conducted on methods for regenerating spent catalysts and/or recovering nickel from spent catalysts and other nickel-containing waste materials since nickel is an expensive metal to be discarded and, moreover, the safe disposal of waste nickel requires consideration of environmental hazards.

See also "Cobalt from Spent Catalysts," especially U.S. Patent 4,145,397.

See also "Triarylboranes from Spent Catalysts" in Reference (1).

A process developed by *Y. Hirayama; U.S. Patent 4,029,495; June 14, 1977* is one in which a supported catalyst contaminated with organic impurities or a mixture of such a catalyst and a flux is introduced into a rotary furnace, rocking furnace or gradient furnace and agitated and heated therein until sintered or transformed into a semimelted state. After sintered or transformed into a semimelted state, the mix is then solidified, cooled and pulverized. The heavy metal is recovered by gravity concentration or magnetic separation. Alternatively, the sintered or semimelted mixture may be fused at a higher temperature to separate the components by difference in their specific gravities.

The process is shown schematically in Figure 125 on the following page.

The following is one specific example of the conduct of the process. 100 parts of a waste catalyst containing 50% oils and/or fats, 35% kieselguhr, and 15% of Ni was mixed with 5 parts of caustic soda flakes. The mixture was then kneaded to form pellets having a diameter of 30 mm and a length of 50 mm. These pellets were introduced into a preheated rotary furnace and reductively melted at 1000° to 1300°C while the furnace was rotated. The mixture was discharged from the furnace, cooled and crushed. The slag was finely pulverized and then Ni powder was recovered by means of a magnetic separator (Ni recovery 95%). Analysis of the recovered Ni powder gave the following: C, 0.3%; P, 0.024%; S, 0.502%; SiO_2 (free), 1.8%; Al_2O_3, 0.5%; and Ni, 97%.

Figure 125: Hirayama Process for Recovering a Heavy Metal Catalyst Component from a Spent Catalyst

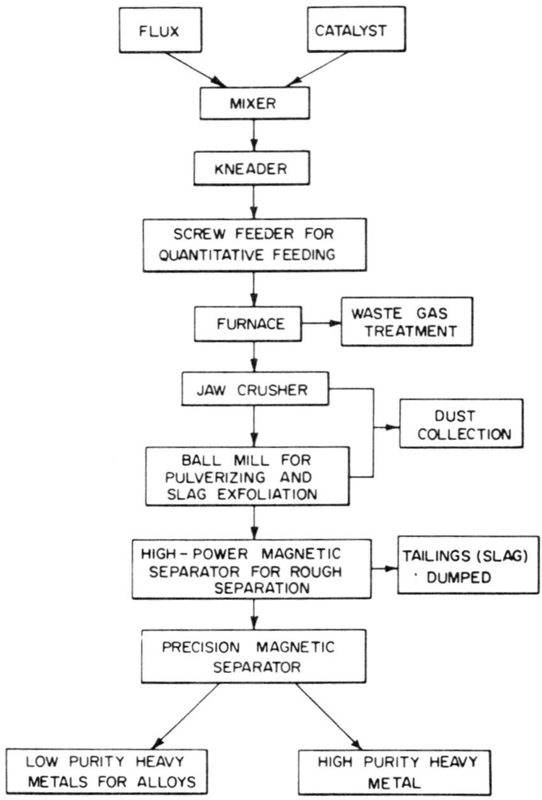

Source: U.S. Patent 4,029,495

A process has been developed by *W.D. Atchison, A. English and D. Halter; U.S. Patent 4,120,698; October 17, 1978; assigned to The Hanna Mining Company* which involves (a) subjecting a mixture of the waste catalyst material and a nickel-containing ore to oxidizing conditions at an elevated temperature; (b) melting the oxidized mixture; (c) subjecting the melt to reducing conditions; and (d) recovering the desired nickel.

More particularly, an inexpensive method is described for increasing the nickel content and/or production of ferronickel obtained from nickel-containing ores using waste nickel materials such as spent nickel catalysts. The process provides a method for recovering nickel from spent catalyst systems which otherwise often have to be discarded because of contaminants and for using the heat values of the contaminants in the process. Such a process is shown in Figure 126 on the following page.

Figure 126: Hanna Mining Co. Process for Recovery of Nickel from Catalysts by Integration with Ferronickel Production

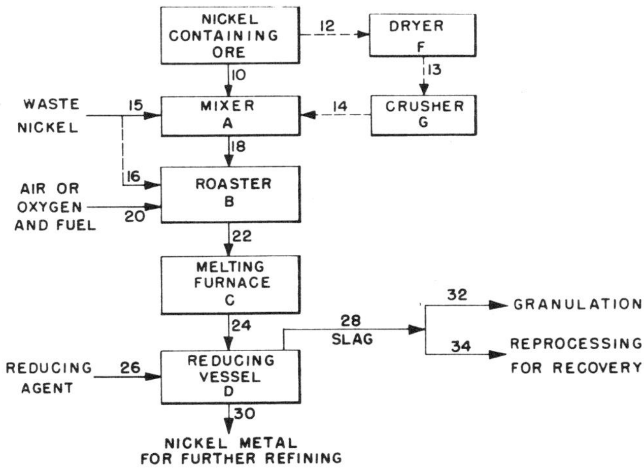

Source: U.S. Patent 4,120,698

The nickel-containing ore and the waste nickel are fed to mixer **A** as indicated by arrows **10** and **15** respectively. The ores generally are wet and must be reduced to a suitable particle size range. The nickel-containing ore is advanced to a dryer **F** via **12** then to a crusher **G** via **13** and to mixer **A** via **14**. The mixer may be any type for mixing dry powder.

The nickel-containing ore and the spent nickel catalysts are mixed in mixer **A** and then advanced via **18** to roaster **B** which contains an oxidizing atmosphere. Sawdust generally is added to the roaster either with the ore or separately. The amount of sawdust added is determined by monitoring the temperature within the furnace. Multiple hearth roasters (Herreschaft) are useful as oxidizing furnaces. An oxygen source such as air or pure oxygen and fuel are fed to roaster **B** via **20** to maintain the operation of the roaster under oxidizing conditions and at the desired temperature of up to 925° to 955°C.

Alternatively, the waste nickel is fed directly into the roaster via **16** at a prescribed rate where it mixes with the ore and is oxidized. This procedure is preferred when the waste nickel material has heat values resulting in a spontaneous increase of the temperature within the oxidizing furnace. Fewer heat units must be supplied from an external fuel source thereby reducing costs and minimizing the added energy requirements of recovering nickel. Various amounts of the waste nickel material can be mixed with the nickel-containing ore either in the mixer or the roaster. Generally, the waste material is added to the ore to provide a mixture containing about 1 to 10% of waste material.

The oxidized mixture then is advanced to a melting furnace **C** via **22** where it is heated to a temperature of up to 1600° to 1700°C. Electric furnaces are useful as melting furnaces. The molten ore is advanced via **24** to a reducing container **D**, which may be one or more large ladles. Reduction of the nickel is accomplished by adding a reducing agent via **26** with vigorous mixing action to provide good contact between the reducing agent and the molten ore. Examples of preferred reducing agents include silicon and ferrosilicons. Carbon also can be used as a reducing agent particularly when a submerged ore furnace is used. Ferrosilicons containing from about 45 to 55% silicon particularly are useful. When the vigorous mixing is completed, the nickel is allowed to settle to the bottom of the container and slag is skimmed off the top via **28**. The slag is either granulated via **32** or reprocessed for recovery via **34**.

As the reducing reactions continue, the nickel accumulates in the ladle and is removed via **30**. Impurities present in the recovered nickel, such as phosphorus, can be removed subsequently by suitable refining slags. When the nickel-containing ore is a nickeliferrous ore, such as a Riddle laterite ore, the nickel recovered as a product of the method is a ferronickel. The introduction of waste nickel into mixer **A** and/or oxidizing furnace **B** increases production of ferronickel and/or ferronickel having an increased nickel content, the increase depending upon the nature and amount of waste nickel used in the method. In general, the ferronickel will contain from about 45 to 55% nickel. By incorporating waste nickel, the nickel content of the ferronickel prepared from a nickeliferrous ore can be increased up to about 4 to 5%, and/or increased production of suitable ferronickel can be obtained from lower grade ores.

Industrial processes, which use nickel compounds, often generate wastes containing low but recoverable levels of nickel. A typical example is the catalytic process for the manufacture of acrylates using nickel carbonyl. Waste streams from this process contain about 4% of recoverable nickel, which economics no longer allow to be discarded. Similarly, other processes giving rise to nickel-containing wastes can no longer afford to discard them in spite of the low level of nickel contained therein.

The problems associated with recovery of nickel from such industrial wastes require a carefully defined recovery process. First, the amount of nickel present is low. This then requires that recovery be as complete as possible to obtain the maximum benefit from the waste. Second, the nickel waste contains other elements, such as iron, copper and sulfur, along with various organic compounds which must be dealt with so as to permit maximum nickel recovery. Third, the nickel must be recovered in a form which allows it to be used as recovered or at least with a minimum of further processing. Fourth, any components used in the recovery process must themselves be recoverable if not completely used in the recovery. Fifth, the cost of recovery must be at least competitive with the cost of comparable nickel available commercially. Sixth, the recovery of nickel from waste must not itself generate any atmospheric or water polluting waste, and ideally is such that it can be run continuously with component recycle.

These requirements have been met by a process developed by *M.S. Brown, R.M. Burch and G.M. Warth and described in U.S. Patent 4,131,641; December 26, 1978; assigned to Rohm and Haas Company.*

Such a process is shown in Figure 127 on the following page.

Figure 127: Rohm and Haas Co. Process for Recovery of Nickel from Low-Level Nickel Carbonyl Catalyst Wastes

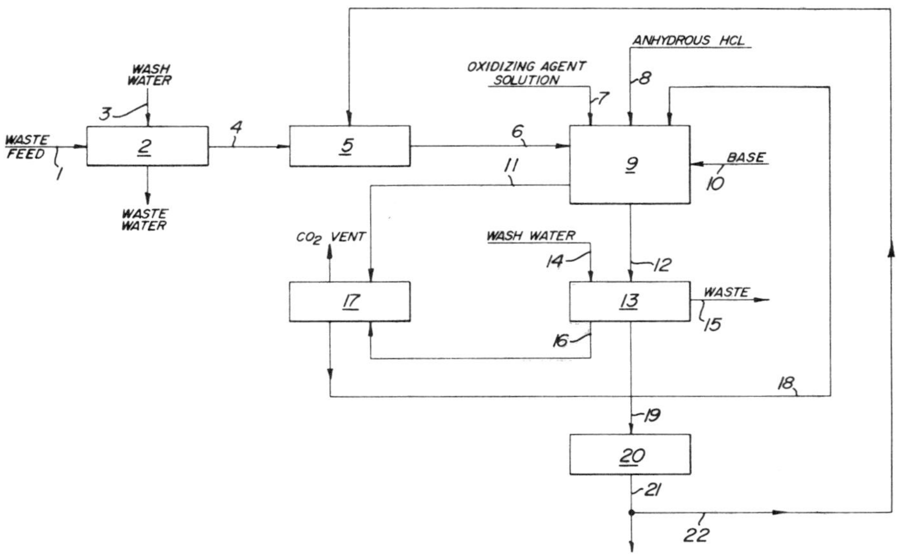

Source: U.S. Patent 4,131,641

A nickel-containing waste stream **1** is fed to filtration apparatus **2** wherein it is filtered and washed with a wash water stream **3**. The filter cake is washed until the filter cake contains a constant maximum nickel concentration and the wastewater contains a constant minimum concentration of soluble components, such as copper, iron and so forth. The filter cake is then transferred by **4** to slurry tank **5**, wherein the filter cake is slurried with nickel chloride brine solution. The latter is a partial recycle, via line **22**, of the nickel chloride product stream **21**.

The slurried filter cake is then fed to reactor **9** via line **6**. The slurry is then reacted by the sequential addition of an oxidizing agent solution via **7**, absorber off-gas solution containing absorbed hydrogen chloride via **18** and anhydrous hydrogen chloride via **8**. The exothermic reaction is complete when the reaction mix changes to a green color, the color change being preceded by an increase in the rate of temperature increase. By maximizing the hydrogen chloride feed rate and minimizing hydrogen chloride loss to the off-gas absorber **17**, the time of reaction can be controlled. Carbon dioxide and excess hydrogen chloride from the reaction are scrubbed via **11** in off-gas absorber **17**, the carbon dioxide being vented off.

The reaction mix after cooling is adjusted from a pH of about 0.3 to about 2.5 with base added via **10**. The cooled, pH-stabilized reaction mix is fed via **12** to a filtration unit **13**. Iron and sulfur are removed as precipitates, and the filtrate is sent via **19** to unit **20** for final filtration and pH adjustment.

At this point, copper can be removed from the product nickel chloride brine solution by use of suitable complexing agents. The final product is recovered via **21**, a portion of this nickel chloride brine solution being recycled via **22** to slurry tank **5**. The filtered reaction mix cake is then washed with water wash **14**, the filtrate being sent via **16** to the off-gas absorber **17**, wherein this wash is used to scrub excess hydrogen chloride from reactor **9**.

The process, as outlined, provides an extremely effective way of recovering nickel from industrial wastes containing low levels of nickel. When run continuously, the only wastes from the process itself are the wash water from the initial waste wash and the spent filter cake.

NICKEL FROM SUPERALLOY SCRAP

In recent years the use of so-called superalloys has burgeoned due to the need for wares having the special properties afforded by these materials, such as high temperature strength, high temperature oxidation resistance, corrosion resistance and the like. As the use of these superalloys grows so does the amount of scrap and other wastes containing them. The usual superalloy scraps include such physically diverse items as spent and off-quality wares, turnings, grindings, pouring skulls, mold gates, flashings and waste products such as slimes, sludges and slags.

In view of the fact that scraps of the foregoing types contain metals of substantial commercial value, there presently exists a clear need for suitable methods by which such scraps may be treated so as to recover at least certain of the metal values therefrom in commercially usable forms. Those scraps comprising substantial quantities of the superalloys usually contain significant quantities of nickel and/or cobalt and various schemes have been proposed by which to recover these metal values from such scraps.

A process developed by *B.H. Rosof; U.S. Patents 4,138,249; February 6, 1979; and 4,173,467; November 6, 1979; both assigned to Cabot Corporation* is one for recovering valuable metals from superalloy scrap.

The process described in the first patent cited uses an ammoniacal leaching step whereas the process described in the second patent includes the steps of melting the initial scrap feed, carburizing the melt, solidifying the carburized melt, and recovering the valuable metals from the resultant material by an appropriate acid leaching step.

The latter process variant is shown schematically in Figure 128 on the following page.

In a typical example of the operation of this process, an initial scrap metal feed was prepared and melted down in an electric arc furnace. The initial scrap metal feed consisted of 1,000 lb of miscellaneous hard metal scrap, 11,000 lb of grindings, 909 lb of other miscellaneous materials, 480 lb of burnt lime (CaO) and 300 lb of fluorospar (CaF_2). The grindings forming part of the initial scrap metal feed were obtained from grinding operations and consisted of a mixture of metal particles and grinding wheel debris.

Figure 128: Cabot Corp. Process for Recovering Valuable Metals from Superalloy Scrap

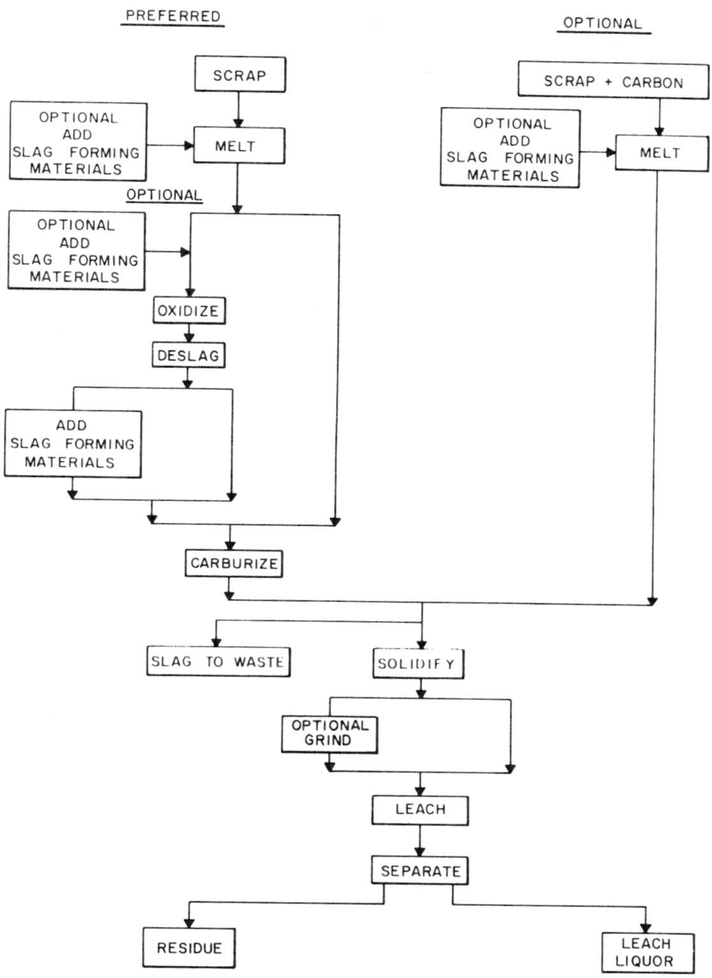

Source: U.S. Patent 4,173,467

In the preparation and melting steps, all of the hard scrap and a portion of the grindings were charged into the furnace prior to startup of the heating operation. The balance of the grindings, the lime and the fluorospar were added to the furnace during the melt down of this preliminary charge. A sample of the resulting molten metal bath was taken which analyzed 0.63 wt % carbon. Most, but not all of the slag was then removed. Some oxidation of highly oxidizable elements contained in the scrap usually occurs during melt down and the resulting oxides generally report to and are thus removed with the slag. The temperature of the bath was adjusted to about 3050°F and oxygen was injected into the bath at a pressure of about 70 psi for five minutes by means of a refractory covered, one-half inch diameter, iron pipe lance. A sample of the molten metal bath was again taken which analyzed 0.33 wt % carbon. The balance of the slag was then removed and replaced with a new slag consisting of 200 lb of fluorospar and 80 lb of burnt lime.

Next, about 350 lb of graphite powder was injected into the scrap metal bath by means of a ceramic covered iron pipe lance and employing argon as the carrier fluid. A small amount of fluorospar was then added in order to aid in fluidizing the slag.

A sample of the scrap metal bath was then taken which analyzed 2.78 wt % C. An additional 75 lb of graphite was then injected into the bath in the manner previously described and the bath resampled. The carbon content of this sample was 3.21 wt %. The temperature of the bath was then adjusted to about 2650°F and the heat tapped into the teeming ladle, an additional 20 lb of graphite being added to the teeming ladle during the tapping step.

Water jets were used to break up the teem stream as the heat was teemed into a water filled pit. The resulting shotted carburized scrap metal product was washed, dried and screened. 7,698 lb of retained shot and 2,365 lb of undersize were recovered from the screening operation. Of the original scrap metal charge, spills amounted to about 90 lb and skulls were roughly estimated to be about 200 lb.

Exclusive of the burnt lime and fluorospar components, a total of 13,354 lb of hard metal scrap, grindings, miscellaneous other materials and graphite were added to the bath. About 10,353 lb of scrap metal as retained shot product, undersize, spills and skulls were recovered. The difference between the total original charge weight and that of the total recovered metal is due to grinding wheel debris which is included in the original charge weights but which debris reports to the slag, losses of carbon due to oxidation, losses of readily oxidizable scrap metal elements such as aluminum and silicon to the slag, metal entrained in the slag, as well as other usual losses and measurement errors.

The composition of the shot product was from a teem stream sample and is given below. The low value reported for the silicon is a good indication that the preparation and treatment of the initial scrap metal feed was effective in removing readily oxidizable elements such as silicon and aluminum which are usually present.

Composition of Screened Shot

Element	Weight Percent
C	3.37
Cb (Nb)	$<$0.10
Co	5.15
Cr	11.35
Cu	1.19
Fe	23.69
Mn	0.30
Mo	10.00

(continued)

Element	Weight Percent
Ni	42.47
P	0.023
S	0.115
Si	0.01
W	1.50

100 g of the screened shot and 400 ml of 20 wt % aqueous HCl leach liquor were placed in a flask fitted with a condenser. The liquor was boiled for 24 hours, the contents of the flask filtered, and distilled water added to the liquor filtrate in order to bring the volume thereof up to 500 ml. This diluted leach liquor was then analyzed, the results of which analysis are given below.

Composition of Leach Liquor

Element	Grams/Liter
Fe	21.80
Ni	72.60
Co	8.58
Cr	4.35
Mo	0.24
W	0.10

From the composition of the shot and leach liquor the percentages of various elements of the initial scrap metal feed which reported to and dissolved in the leach liquor are calculated. These results are given below.

Initial Scrap Metal Feed Elements Dissolved in Leach Liquor

Element	Grams per 100 Grams Shot	Leach Liquor (g/500 ml)	Element in Leach Liquor (wt %)
Fe	23.69	10.9	46.0
Ni	42.47	36.3	85.5
Co	5.15	4.29	83.3
Cr	11.35	2.18	19.2
Mo	10.00	0.12	1.2
W	1.50	0.05	3.3

It is apparent from examination of this data that dissolution of nickel and cobalt in the leach liquor proceeds to a substantially greater extent than does the dissolution of chromium, molybdenum or tungsten.

NITRIC ACID FROM USED ACRYLIC FIBER SPINNING SOLUTIONS

See "Sodium Thiocyanate from Used Acrylic Fiber Spinning Solutions."

NITROGEN TETROXIDE FROM CELLULOSE SPIN BATH

See "Dimethylformamide from Cellulose Spin Bath" in reference (1).

NONFERROUS METALS FROM MUNICIPAL WASTE

See also processes under "Paper from Municipal Waste" in reference (1).

NONFERROUS METALS FROM SCRAP AUTOMOBILES

A process developed by *F.J. Fontein and H.H. Dreissen; U.S. Patent 4,034,861; July 12, 1977; assigned to Stamicarbon BV, Netherlands* is one for recovering usable materials from discarded scrap material, such as junked automobiles, in which nonferrous metals are separated from nonmetals in a hydrocyclone using water as the aqueous separating medium. A particulate scrap/water suspension is injected into the cylindrical area of the hydrocyclone and a predominantly nonmetal-containing portion is removed from an overflow area of the cyclone while the nonferrous-metal-containing portion is removed from the apex area of the cyclone. If required, the scrap material is preferably pulverized prior to treatment and is subjected to a ferrometallic separation operation if such metals are present.

A somewhat similar process has been developed by *A.T. Basten and H. H. Dreissen; U.S. Patent 4,036,441; July 19, 1977; assigned to Stamicarbon BV, Netherlands.* It is a process for separating scrap essentially consisting of nonferromagnetic metals into several fractions by specific gravity in which only one heavy separating medium is required.

The scrap has been recovered from waste products composed of metals and nonmetals by reducing the particle size of the waste through grinding to a size less than 50 mm including particles as small as 0.5 mm. Thereafter, the ferromagnetic components are magnetically separated from the ground waste and the ground waste is fed to a primary specific-gravity separator where nonmetallic components are separated from the nonferromagnetic portion of the ground waste. Thereafter, further separation of the nonferromagnetic portion in a secondary and in some cases, tertiary separators, is accomplished.

Such a process is shown in Figure 129 on the following page. A discarded car **1** is flattened between two rollers **2** and **3** into a bundle **4** in such a way that the car's engine block may, mainly, pass without deformation. The bundle **4** is supplied to a shredder **5** and the resulting scrap is transferred onto a conveyor **6** which carries the ground scrap past a magnetic separator **7**, which separates out the ferromagnetic components from the rest of the scrap. The ferromagnetic components are then discharged via a chute **8** to a suitable collection point (not shown). The ferromagnetic component consists principally of ferroalloys and may be used as raw material in the fabrication of steel.

The remaining scrap composed of nonferromagnetic metals, plastics, etc., is then deposited upon a screen sieve **9**; parts which are too coarse for further processing, for instance if sized over 70 mm, are returned to the shredder **5** by means of a suitable conveyor, generally indicated at **10**. The portion passing through the sieve **9** is sent to a supply vessel **11**. Water may be supplied to the supply vessel **11** through a line **12**. By means of a line **13**, a mixture of solids and water can be carried from the supply vessel **11** to the primary hydrocyclone **14**.

Figure 129: Stamicarbon Process for Recovering Usable Nonferrous Materials from Scrap Automobiles

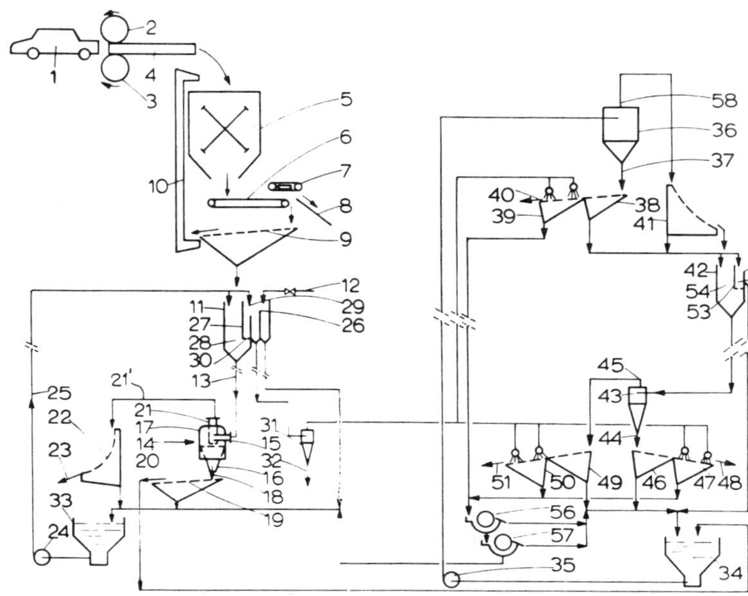

Source: U.S. Patent 4,036,441

The supply vessel **11** is installed at a level which is sufficiently, e.g., about 10 m, higher than that of the hydrocyclone **14** such that the mixture of solids and water enters the hydrocyclone through a tangential feed **15** with sufficient speed.

In the hydrocyclone separation is effected between specifically heavier and specifically lighter parts. The first fraction, which consists of specifically heavier parts (chiefly metals, like copper- and aluminum-alloys) leaves the hydrocyclone through the discharge opening at the tip **18** thereof and is dewatered on a dewatering screen **19**. The second fraction, which consists of specifically lighter parts (for instance plastics and rubber) leaves the hydrocyclone through the overflow tube **21** and line **21'** and is transported to a sieve bend **22** where it is dewatered and finally discharged at **23**. The water separated off on the screens **19** and **22** is collected in a pump sump **33** and sent back to the supply vessel with the aid of a pump **24** through a line **25**.

The supply vessel has an overflow **26** for the purpose of maintaining a constant water pressure at the feed of the cyclone. In this case a vertical baffle **27** is installed in the supply vessel, which baffle divides the supply vessel into two compartments **28** and **29**. Through the aperture **30** the compartment **29** is connected with the compartment **28**. Part of the water recycled via the line **25** is, with the solids, fed to compartment **28**, the remainder of the recycled water being sent to compartment **29**, in such a ratio that the downward velocity of the water through the aperture **30** is so large that no light materials can reach the overflow **26** which connects with the compartment **29**.

Fine components will accumulate in the recycled process water. In order to avoid too great a concentration of fine particles, part of the recycling process water is passed through a cyclone-thickener **31**; the purified water is used again, the thickened fraction being discharged at **32**.

Preferably, the primary hydrocyclone **14** is of the type in which the cone-shaped part **16** may be axially adjusted relative to the cylindrical part **17**.

The first fraction leaving the draining screen **19** at **20** is supplied to a mixing vessel **34** in which it is taken up by a heavy suspension medium of finely ground ferrosilicon in water, having a specific gravity of about 2.45. The mixture of the heavy suspension medium and metal parts from the first fraction is passed by a pump **35** to a secondary hydrocyclone **36** whose top angle of the cone-shaped part amounts to 60°. In one embodiment the diameter of the cyclone **36** is 600 mm, the diameter of the feed pipe is 150 mm and the diameter of the vortex finder is 210 mm.

At the discharge opening at the tip **37** of the cyclone, a specifically heavy fraction, comprised of parts with a specific gravity of over about 3.15, is discharged along with the suspension medium (mainly copper and copper alloys), which fraction is drained of the suspension medium on a draining screen **38** and a rinsing screen **39** and is discharged at **40**. Through the overflow line **58** of the secondary hydrocyclone **36**, a specifically lighter fraction is discharged along with the suspension medium and largely drained of the suspension medium on a sieve bend **41**. The lighter fraction, together with the suspension medium from the draining screen **38** and the sieve bend, is sent to a supply vessel **42** from which the mixture is passed to a tertiary hydrocyclone **43**.

In this tertiary hydrocylone **43**, whose top angle of the cone-shaped tip amounts to 20°, the lighter fraction is separated into a middle metal fraction, which mainly consists of parts having a specific gravity of over 2.65, and which is discharged along with the suspension medium at the discharge opening at the tip **44** thereof, and a lightest metal fraction, which is mainly comprised of parts having a specific gravity below 2.65, which is discharged at the overflow **45** of the hydrocyclone **43**. The middle fraction comprises chiefly of aluminum; it is drained of the suspension medium on a draining screen **46** and a rinsing screen **47** and discharged at **48**. The lightest fraction, which contains the lighter metals and alloys, is drained of the suspension medium on a draining screen **49** and a rinsing screen **50** and discharged at **51**.

The supply vessel **42** is installed at a sufficient height (e.g., about 10 m) above the hydrocyclone **43** which together with an overflow **52** provide the required constant liquid pressure at the feed of the tertiary hydrocyclone **43**. Supply vessels **11** and **42** are of the same type as the vessel described in Netherlands Patent Specification 96,726, and the action of supply vessel **42** is similar to vessel **11** which has already been discussed above. The supply vessel **42** is divided by a partitioning baffle **53** into two compartments **54** and **55**. The lighter fraction coming from the sieve bend **41** is supplied to the compartment **54**, while the suspension medium separated off by the sieve bend **41** and the draining screen **38** is distributed over both compartments **54** and **55** in such a ratio that no floating parts can reach the overflow **52**. In order that the suspension medium can be distributed over the compartments **54** and **55** in the proper ratio, it must be separated from the solid parts, which is the function carried out by the sieve bend **41**. The suspension medium coming from the draining screens **46** and **49** and the overflow **52** is returned to the mixing vessel **35**. On the rinsing screens **39, 47,** and **50** the purified water obtained at the overflow of the cyclone-thickener **31** is used as spraying water.

The diluted suspension medium from the rinsing screens **39, 47,** and **50** is regenerated with two series-connected magnetic separators **56** and **57**. The regenerated magnetic fraction is passed to the mixing vessel **34**, while the water is separated off and directed to the pump sump **33** for the process water to be used in the primary hydrocyclone **14**. In this way part of the water is continusouly withdrawn from the suspension circuit and led through the process water circuit of the primary cyclone, part of which, in turn, is led through the cyclone-thickener. In this manner, therefore, an accumulation of fine parts is also avoided in the suspension medium cycle.

The following tabulation shows the specific gravity distribution of the nonferromagnetic part of a type of motorcar scrap which is to be considered representative of the type of waste scrap being processed.

Specific Gravity Fraction	Specific Gravity	Percent by Weight	Fraction Designation
A	<1.78	20	nonmetal fraction
B	1.78-2.65	23	lightest fraction
C	2.65-2.79	14	middle fraction
D	>3.32	39	heavy fraction

The fraction A mainly contains the nonmetals, fraction B is comprised of the light metals and alloys, while fraction C is comprised practically completely of aluminum. Fraction D contains the heavier metals, chiefly copper and copper alloys. Material having a specific gravity of between 2.79 and 3.32 is present in only a very small quantity.

See also "Steel from Scrap Automobiles."

O

OLIVE PROCESSING LIQUOR FROM OLIVE PROCESSING WASTES

Ripe olives in their natural (or fresh) state are extremely bitter and thus must be processed for human consumption. To remove the bitter principle olives are soaked in aqueous sodium hydroxide (lye) for extended periods of time. After soaking, the olives are removed from the lye bath, washed with water, neutralized with acetic acid, and then soaked in aqueous sodium chloride (brine) for flavoring.

The lye (or alkaline) solution employed in the initial soaking of the fresh olives becomes contaminated with organic impurities. These impurities impart a deep brown, almost black, color and an extremely unpleasant odor to the lye solution. Because the biological oxygen demand (BOD) of this waste is extremely high, it cannot be released into municipal sewage treatment systems without first being diluted with enormous quantities of water.

Currently, disposal of the olive processing fluid is accomplished by forming ponds into which the liquid is placed. The water is allowed to evaporate from these ponds, leaving a sludge of lye and the organic contaminants. Because of its unpleasant aroma the used fluid must be piped or trucked many miles to uninhabited areas where pollution consequences hopefully are minimized. The lye, of course, poisons the soil in and around the ponds. This disposal operation involves considerable expense both for handling the waste and for land needed to house the ponds. Also the cost of lye has recently increased tremendously. Olive processors fear that expenses will become prohibitive and that they will be forced to discontinue part of their operations.

Thus, a process has been developed and described by *R. Teranishi and D.J. Stern; U.S. Patent 3,975,270; August 17, 1976; assigned to U.S. Secretary of Agriculture* in which useful olive processing liquor is recovered from olive processing waste solution by a process wherein lime, charcoal, and calcium carbonate are successively added to the olive processing waste solution to form a mixture and the mixture is allowed to settle. The treated liquor is separated from the settled contaminants and is then immediately recyclable to process fresh olives.

P

PALLADIUM FROM PLATING WASTE SOLUTIONS

See "Gold from Plating Waste Solutions."

PALLADIUM FROM SPENT CATALYSTS

Spent palladium catalysts typically comprise palladium in an amount ranging up to 2%, the remainder thereof comprising activated carbon, alumina and diverse impurities such as superfine sand, metal soaps and high molecular weight substances.

Among the more popular procedures for recovery of palladium from spent catalysts are those which involve incineration of the catalyst in an oven to yield an ash having a palladium concentration ranging from 15 to 25% and a small percentage of carbon, frequently in the form of graphite. Unfortunately, the presence of carbon impairs the dissolution of the ash in acids and, consequently, must first be removed. This end is typically attained by annealing the ash at a temperature within the range of 600° to 1000°C. However, some of the carbon remains in the ash so treated, thereby preventing quantitative dissolution of palladium in aqua regia or hydrochloric acid admixed with hydrogen peroxide.

Nonetheless, after stirring the resultant solution, it is permitted to clarify and is then decanted. The sediment is next subjected to repeated dissolution and decantation and the extracts united. The residue is then filtered, washed, dried and treated with sulfuric acid. Thereafter it is diluted with water, decanted and filtered. The filtrates so obtained are then reduced with zinc to yield raw palladium which is dissolved in an acid to form a concentrated palladium solution which is finally processed in a two-stage process to a palladium sponge. The sponge may then be washed and dissolved to yield a solution which is diluted to 120 g/ℓ palladium. In this manner, about 99% palladium is obtained from the ash.

As noted from a review of the conventional process described above, it is a time-consuming operation which places heavy demands upon labor and is objectionable from the standpoint of safety and hygiene. Additionally, the process does not permit the entire palladium content in the spent catalyst to be reclaimed.

See "Copper from Spent Catalysts" and specifically U.S. Patent 3,947,543.

Thus, an improved process has been developed by *L. Vodicka, I. Peka and V. Suchy; U.S. Patent 4,072,508; February 7, 1978; assigned to Vysoka skola chemicko-technologicka, Czechoslovakia* which involves fluorinating the spent catalyst or ashes thereof with elemental fluorine or a mixture of elemental fluorine with hydrogen fluoride at a temperature which increases gradually from 200° to 500°C, thereby forming palladium fluoride and decomposing carbides and silicates present in the catalyst. Finally, the palladium fluoride is decomposed with a mineral acid at a temperature within the range of 90° to 100°C.

The prime thrust of the instant process resides in the fact that substantially all of the palladium may be reclaimed from the spent catalyst, but for trace amounts, in a short period of time and in a simplified process as compared with those of the prior art.

A process developed by *G.J.K. Acres and A.J. Bird; U.S. Patent 4,123,500; October 31, 1978; assigned to Johnson, Matthey & Co., Limited, England* is a process for the recovery of a catalytic metal from a carbonaceous residue or support comprising subjecting an aqueous slurry of material containing catalytic metal to be recovered to a temperature of not less than 200°C and a pressure not less than 75 atmospheres in an atmosphere comprising a quantity of oxygen at least sufficient to oxidize substantially all the carbon present, the oxidation being carried out in the presence of a liquid aqueous phase.

The following illustrates the recovery of palladium using such a process. The palladium from 10 g of a Pd on C catalyst containing 5.0% Pd (w/w) was recovered using a process as described. A temperature of 200°C and a pressure of 75 atmospheres oxygen were maintained from 4 hours over an aqueous slurry containing 250 ml water. Recovery of Pd: 97.7% by weight.

PAVING MATERIAL FROM FLY ASH AND WATER TREATMENT SLUDGE

A composition developed by *D.I. Angelbeck; U.S. Patent 4,105,463; August 8, 1978; assigned to The University of Toledo* is an improved lime-fly ash-aggregate (LFA) pozzolanic composition which is useful as a load-carrying base for roads, airport runways, parking lots and the like.

As the cost of raw materials has increased, greater interest in utilization of waste materials as a means of providing cost benefits has also increased. This interest in waste utilization has also been spurred by the increasing cost and difficulty of disposal of waste materials in view of increasingly stringent pollution standards. In addition, utilization of waste materials also helps to conserve natural resources.

An outstanding example of utilization of waste material is the use of fly ash from the combustion of powdered coal, recovered from power plant flue gases by mechanical and/or electrostatic precipitation. One economical use of fly ash has been its use in LFA mixtures.

The composition is prepared by mixing together lime, fly ash, graded aggregate, adding water, and from 0.5 to 4%, based on total dry weight, of sludge from a water treatment plant. The mixture is placed on top of a prepared subgrade, spread uniformly, and compacted to a high degree of relative density to provide the main load-carrying component of a road, airport runway, parking lot or the like.

PLATINUM FROM PLATING WASTE SOLUTIONS

See "Gold from Plating Waste Solutions."

PLATINUM FROM PLATINUM METALS REFINERY EFFLUENT

In the course of work to recover the values from the effluent from a platinum metals refinery, it was found that all the possible methods that gave excellent results on synthetic solutions failed to do so when actual refinery effluent was treated. The treated effluent appeared barren on chemical analysis, but on evaporation to dryness spectrographic examination of the residue showed contents of up to 100 mg/ℓ of various metals in the solution. Since there is no way to isolate these compounds, their structure cannot be determined. However, their slow degradation is accompanied by the release of ammonia. Other evidence suggests the possibility of stable hydroxyl groups being involved.

Therefore, an improved process was developed by *J.J. MacGregor; U.S. Patents 4,127,458; November 28, 1978 and 4,162,950; July 31, 1979; both assigned to Matthey Rustenburg Refiners (Proprietary) Limited, South Africa.*

The process involves the steps of recovering platinum group metal present as a stable complex dissolved in an aqueous effluent from a platinum group metals refinery: (1) Adjusting the pH of the effluent as necessary to an alkaline pH of at least 10; (2) heating the effluent to a temperature above 60°C; (3) electrolyzing the effluent at the temperature and alkaline pH using an electrode comprising at least one metal selected from the group consisting of Ru, Rh, Pd, Ir, Pt and alloys thereof at an anode potential having a half-cell voltage of at least 7.5 V and a current density of at least 0.2 A/cm^2 so as to break down the complex by anodic oxidation and precipitate metal present in the complex; and (4) filtering off the thus precipitated metal.

PLATINUM FROM SPENT CATALYSTS

See also "Iridium from Spent Catalysts."

Among the more prominent industries employing the catalytic characteristics of platinum is the petrochemical industry. Thus, platinum catalysts, typically deposited on a carrier such as alumina and the like, are widely used in petrochemical applications for hydrogenation, dehydrogenation, cracking, refining, detoxification of waste gases, etc.

It has been found that with each succeeding application, the platinum catalysts evidence a reduction in efficiency, such being attributed either to the accumulation of sediments formed by secondary reactions during the processing sequence or to the presence of deleterious materials entering the reaction system with raw materials. Heretofore, it has been common practice to effect regeneration of the catalysts by annealing at elevated temperatures to remove organic materials, by saturating the catalyst with an ammonium nitrate or chloride solution and finally heating to a temperature ranging from 400° to 500°C.

Nonetheless, studies have revealed that catalysts regenerated in this fashion several times lose their effectiveness. At that juncture, it is advantageous to recover the precious metal and this end is normally attained by a processing procedure categorized either as a chlorination or leaching method.

In the chlorination procedure, the spent catalyst is roasted in air at temperatures ranging up to 870°C and exposed at such temperatures to chlorinating agents such as aluminum chloride, silicon tetrachloride, carbonyl chloride and the like. The resultant platinum compounds are absorbed by water and reclamation of platinum therefrom is effected from solution by reduction with aluminum, zinc and the like.

The leaching technique may be based either upon an acid or alkaline reaction. The acid process also involves roasting the catalyst at a temperature in excess of 800°C to remove organic substances and, subsequently, leaching for several hours with a 20 to 60% sulfuric acid solution or a 20 to 50% hydrochloric acid solution at temperatures ranging from 100° to 140°C. Oftimes, this reaction is effected at superatmospheric pressures ranging up to 5.0 MPa (50 kp/cm^2). Platinum may then be recovered from a separated sludge.

In the alkaline process, the spent catalyst, either annealed or nonannealed, is melted with an excess of sodium hydroxide, the resultant melt being leached with water and the platinum concentrate converted to platinum. Dissolution of nonannealed catalysts in such processes may typically be attained by means of a 40% sodium hydroxide solution at temperatures ranging from 120° to 145°C.

Unfortunately, each of these prior art processes has inherent limitations. The platinum catalysts regenerated in the foregoing manner are freed from organic compounds at temperatures ranging from 800° to 1200°C, thereby increasing power costs and introducing complexities in processing. Additionally, the plurality of steps in the regeneration process, namely, roasting, melting, sedimentation, decantation, filtering, etc., results in platinum losses ranging from 3 to 5%.

Similarly, the alkali and acid processes are limited by aluminate liquor processing and high acid volume requirements, respectively. Accordingly, efforts to overcome these difficulties have continued.

It is well known in catalysis to apply to a substrate a metallic value which constitutes the catalyst material. The metallic value may be a precious or platinum group metal and may be applied to a catalyst support in the form of a so-called washcoat. In catalysis the quantity of metal used as the catalyst material (the metallic value) is, generally speaking, a relatively small proportion of the total weight of the catalyst. For example, catalysts intended primarily for use in the purification of vehicle exhaust gases contain metallic values as low as 0.1 to 0.4% of the weight of the total catalyst. Chemical recovery of metallic values from supports bearing such relatively low quantities of the metallic values by, for example, acid leaching, would be very costly due to the size of the plant and to the quantities of reagents required.

One washcoat method of laying down or applying a metallic value catalyst to a substrate involves applying to a cordierite support an alumina washcoat which is impregnated or subsequently coated with a platinum group metal or alloy, for example, platinum or rhodium or alloys thereof.

A process developed by *A. Gibbon; U.S. Patent 4,074,865; February 21, 1978; assigned to Johnson, Matthey & Co., Limited, England* is one for the recovery of metallic values from a washcoated catalyst. The following are specific examples of the application of such a process.

Example 1: A sample (6 g) of used car exhaust catalyst containing nominally 0.35% Pt and 0.03% Rh was ground in a Tema mill until 50% was –400 mesh, (<36 μm). The material was treated on a Superpanner (wet gravity separator, Infrasizers Ltd.). The concentrate obtained weighed 27.1% of the weight of the input material and the tails 72.9%. The concentrate assayed at 1.054% Pt and 0.044% Rh and the tails at 0.041% Pt, 0.002% Rh. Thus, 90% of the Pt and 89.1% of the Rh was recovered in the concentrate.

Example 2: A sample (30.2 g) of used car exhaust catalyst containing nominally 0.22% Pt + Rh was passed through a Glen Creston Laboratory Hammer Mill fitted with a 1 mm 16 mesh discharge screen. On the Superpanner a 12.5% concentrate was obtained containing 1.11% Pt, 0.082% Rh. The tails were ground for 20 seconds in a Tema mill and retreated on the Superpanner to give a second concentrate referred to here as the middling fraction (14.5%), containing 0.20% Pt, 0.013% Rh and a tails (73.0%) at 0.03% Pt, 0.002% Rh. Thus, 88.5% of the Pt and 89.3% of the Rh was recovered in the concentrate and middling which together represented 27% of the original bulk.

A process developed by *I. Peka, L. Vodicka, J. Ulrich and J. Hajek; U.S. Patent 4,077,800; March 7, 1978; assigned to Vysoka skola chemicko-technologicka, Czechoslovakia* is one for the recovery of platinum from a spent catalyst which involves subjecting the catalyst to fluorination with elemental fluorine or a mixture thereof with hydrogen fluoride at a temperature ranging from 100° to 600°C. The resultant platinum fluoride is then converted to platinum in elemental form.

PLATINUM METALS FROM PRECIOUS METALS SCRAP

Consumption of the platinum group metals is largely nondissipative. In normal uses, the platinum group metals are virtually indestructible. Moreover, their high cost insures that every effort will be made to conserve and recover them. The relative ease with which they can be separated from other less chemically stable materials allows nearly complete recovery. In addition, at least 97% of the platinum group metals used each year in the United States is purchased by industrial organizations, rather than individuals, which tends to assure systematic conservation and recovery of the metals. However, platinum group metals used in electronic equipment are dissipated, since recovery of precious metals from used electronic equipment is generally uneconomic even when physically possible.

There are other minor uses that are essentially dissipative, and it appears that some, as yet unknown, part of the platinum and palladium in automobile exhaust catalysts will not be recovered.

Recycling of the platinum group metals can entail remelting and refabrication of a worn metal part or coating, or complete re-refining of the constituent platinum group metals. The consumer of the metal seldom performs the reclaiming operation. Certain industries, such as the chemical, petroleum refining, and glass industries, usually retain ownership of the metal and pay the refiner or refiner/fabricator a conversion fee. The quantity of platinum group metals toll-refined in this way each year was large—more than 1 million ounces in 1974. The quantity refined from purchased scrap was considerably smaller, only 325,000 ounces in 1974 (14).

PLUTONIUM OXIDE FROM NUCLEAR FUEL PROCESS SCRAP

See "Uranium Oxide from Nuclear Fuel Process Scrap."

POROUS CERAMICS FROM FLY DUSTS

The finely divided particles of material entrained in waste gases emanating from different industrial processes and discharged through chimneys to the surrounding countryside create serious environmental problems. Consequently, it is now generally required that filter devices be provided for the purpose of capturing these particles to the benefit of the environment. Particles originating from processes such as steel and other metal manufacturing processes and collected in electrofilters or the like also create serious environmental problems. Previously, no useful purpose has been found for these waste products. The following process enables some of these products to be used.

Characteristic of these products which may be used in accordance with the process and which will hereinafter be referred to as fly dust is that firstly they have an extremely small particle size with an enormously large specific surface area, often in excess of 20 m^2/g, and secondly they have a high proportion (75 to 92%) of silica (SiO_2) and contain different metal oxides, such as one or more of the oxides MgO, Cr_2O_3, FeO, Fe_2O_3, Al_2O_3, ZnO, Ni_2O_3, TiO_2, CuO, PbO and often also a minor amount of C or S, each small particle having exactly the same composition. By way of example of fly dusts can be mentioned:

(a) Fly dust (silica fumes) obtained from ferrosilicon manufacturing processes, containing between 75 and 92% SiO_2 and varying percentages of MgO, Cr_2O_3, FeO, together with minor quantities of C and possibly also S. Such dust normally has a specific surface area in excess of 20 m^2/g, and is amorphous;

(b) Fly dust obtained from silicon-chrome manufacturing processes and having a similar composition and properties to the silica fumes above;

(c) Fly dust obtained as waste from perlite expanding processes and containing about 70 to 75% SiO_2, 12 to 15% Al_2O_3, 6 to 10% alkali oxide, the remainder being various metal oxides; and

(d) Fly dust obtained from crushing quartz, quartzite or other stones (in many countries the extra-fine waste dust of stone-crushing, e.g., for road constructions has to be collected and is not allowed to be spread into the air).

A process developed by *F.W. Kurz and H. Rudmark; U.S. Patent 4,071,369; January 31, 1978* is one for manufacturing porous ceramic products by mixing ceramic material with a fly dust containing silica and metal oxides and having a large specific surface area of at least 20 m^2/g. The mixture is heated to at least 800°C to convert it into a porous ceramic material. The process is carried out so as to result in either products having mainly closed pores by heating at sintering temperatures or products having mainly open pores by heating at consolidation temperatures. The admixture may be wrapped in a sheet of material which upon heating the admixture to form pores in the material encloses the gases formed at low temperatures but which ruptures at high temperatures to expose the clay to an oxidizing atmosphere.

The following is one specific example of the application of this process. 100 pbw clay were mixed with 10 pbw of a fly dust obtained from a ferrosilicon manufacturing process. Admixed therewith were 10 pbw of a 20% sulfite waste liquor and sufficient to homogenize the dough. The homogenized dough was then divided into small pieces, which after being dried were either placed in molds or onto a belt with side barriers and heated.

The pieces can be introduced into the kiln at room temperature or at temperatures as high as 300° to 400°C or higher. The time taken to reach the top temperature—which varies between 1100°C and 1200°C according to the composition of the clay, and even higher with refractory clays—is from 1 to 5 hours, depending on the type of the clay. The pieces are maintained at this temperature, at which expansion takes place as a result of a sinter skin being formed, for roughly 30 minutes, again depending on the type of clay used. Further, according to the composition of the clay and the siliceous fly dust, the amount of silica fumes recited in the example may, if desired, be reduced to 6 pbw and supplemented with 3 pbw of filter dust obtained from a steel manufacturing process.

PORTLAND CEMENT FROM COAL ASH SLAG

Generally speaking, coal gasification involves the reaction of coal, at high temperatures, with a gas containing oxygen and steam to produce a gas substantially comprising CO and H_2, which is suitable for use as a fuel. As a by-product of gasification, a char or slag component is produced which must be disposed of. This disposal problem is compounded by the fact that the slag has limited value as a structural material and yet a substantial quantity is produced which must somehow be discarded. Depending upon the type and source of coal, the ash or slag content may vary from 5 to 50% by wt. This percentage becomes significant for a gasifier which handles 100,000 lb/hr.

In the past, slag has mainly been disposed of by merely dumping it with little consideration being given to its use for other purposes, mainly because of its limited uses. Accordingly, there is a need and there will be a greater future need for large volume uses of coal ash slag.

A process developed by *F.F. Fondriest; U.S. Patent 4,174,974; November 20, 1979; assigned to Standard Oil Company (Indiana)* is a manufacturing process for converting coal ash slag from a slagging coal gasifier into a marketable cement product having the characteristics and qualities of portland cement.

This process comprises the steps of transferring molten slag from a slagging coal gasifier to a melt chamber and reacting it with a mineral containing lime, for example: calcium oxide, calcium hydroxide or calcium carbonate, to form a homogeneous cement product. This cement product is then transferred to a quench chamber where it is allowed to cool and solidify in the form of clinkers, which are later reduced to powder form.

This process also provides an efficient means to conserve energy when producing a portland cement and provides an effective and economical way to dispose of a waste product while increasing the cost efficiency of a slagging coal gasifier.

Figure 130 shows such a manufacturing process **10** whereby coal is fed into a slagging coal gasifier reactor vessel **13** by feed input mechanism **11**.

Figure 130: Standard Oil Co. (Indiana) Process for Converting Coal Gasifier Slag to Portland Cement

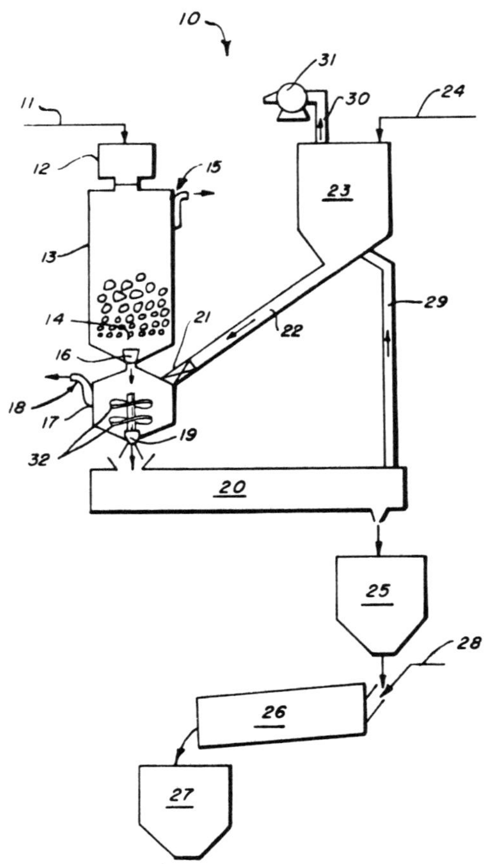

Source: U.S. Patent 4,174,974

Most commonly, bituminous, semibituminous and lignitic noncoking coals are burned or pyrolyzed in synthesis gas plants. The reason for this is that coking coals such as those used as a reducing agent in the production of iron ore are expensive, low in volatility and are in relatively short supply. Basically, coal of the same grade as that used for steam raising or the generation of electricity are used in slagging coal gasifiers.

The coal feedstocks are usually prepared by cleaning and washing in order to lower the ash content, reduce caking, and to remove contaminants and impurities, such as pyrites. The pretreated coal arriving through feed input mechanism **11** accumulates in lock hopper **12** which is a pressurized feed mechanism and is retained there until needed by reactor vessel **13**. When needed, the coal is supplied to reactor vessel **13** wherein the coal is burned or pyrolyzed under pressure in order to convert it to synthesis gas. During this process, the heavy hydrocarbons, those having a carbon to hydrogen atomic ratio of 1 or 2 or higher, are converted to methane which has a carbon to hydrogen ratio of 1 to 4.

Inside reactor vessel **13**, the synthesis gas containing hydrogen, carbon monoxide and carbon dioxide is adjusted in composition so that the hydrogen/carbon monoxide ratio desired for a particular synthesis is obtained. This gas is then drawn off through vent **15** and can be used for the production of ammonia, methanol, oxoalcohols and/or synthetic hydrocarbons.

The coal ash slag **14** gradually descends to the bottom portion of reactor vessel **13** and funnels downwards toward slag tap **16**. The slag tap is a specially designed valve which permits the controlled withdrawal of the liquid slag from the pressurized reactor vessel. This is to insure that the synthesis process is carried out in the desired environment and for the purpose of insuring that the coal is completely burned or pyrolyzed to slag. The hot molten slag, having a temperature range between approximately 2800°F and 3600°F, passes through slag tap **16** to melt tank **17** wherein the slag is reacted with a mineral containing lime, for example: calcium oxide, calcium hydroxide or calcium carbonate, to form a homogeneous cement product having a temperature range between approximately 2200°F and 2800°F.

The mineral containing lime that enters melt tank **17** is first accumulated in hopper **23** through input line **24**. Here in hopper **23** the mineral containing lime is preheated with heated air from quench chamber **20** which is being recirculated through return line **29**. The heated air is drawn up through the material in hopper **23** and out of the hopper through vent line **30** and into suction pump **31**. The preheated mineral containing lime, assisted by gravity, discharges from hopper **23** through feeder tube **22** to slide valve **21**. Slide valve **21** restricts the flow rate to melt tank **17** so that the slag and mineral containing lime can be properly admixed by impellers **32** within the melt tank.

During the reacting process, carbon dioxide gas is given off and is vented from melt tank **17** through vent line **18**. The slag and mineral containing lime is reacted in the ratio of approximately 1 part of slag to every 1.2 to 4 parts of the mineral containing lime and preferably in the ratio of about 1 part of slag to 2.5 parts of mineral containing lime to form a conglomerate. The slag and mineral containing lime or conglomerate is stirred until a homogeneous cement product is formed at which time it is withdrawn from melt tank **17** through tap **19** to quench chamber **20**.

In quench chamber **20** the cement product cools and solidifies in the form of clinkers which can range in size from ¼" to 2" in diameter, preferably less than 1" in diameter. It is advantageous to construct quench chamber **20** so that the clinkers can be carried toward a storage facility while they are in the process of being cooled. This can be accomplished by employing a conveyor belt or a rotary screw arrangement with air or water being used as the cooling medium, preferably air. Air from the atmosphere enters the quench chamber along with the cement and acts as the cooling medium. The heated air is drawn off through return line **29** and is directed back to hopper **23** so it can be utilized to preheat the mineral containing lime.

The clinkers are removed from quench chamber **20** and are transferred to storage bin **25** where they are retained until needed. At that time, the clinkers are combined with a set regulating additive, for example, anhydrite or gypsum, which is added through input line **28** and together the mixture is transported to grinding mill **26**. In addition to the anhydrite or gypsum, other compounds such as alumina, aluminum sulfate, sodium/potassium sulfate, potassium carbonate or borax can be added in the exact quantity necessary to produce other desired properties in the cement. In grinding mill **26**, the clinkers, set regulating additive and any other desirable additive are reduced to powder form. The powdered cement is then transferred to storage hopper **27** and is ready for present or future use.

PORTLAND CEMENT FROM OXYGEN CONVERTER SLAGS

See also "Iron from Oxygen Converter Slags."

POZZOLAN FROM FLY ASH

See also "Paving Material from Fly Ash and Water Treatment Sludge."

A process developed by *V.J. Hurst and R.W. Styron; U.S. Patent 4,121,945; October 24, 1978; assigned to Amax Resource Recovery Systems, Inc.* is a wet process for the beneficiation of electrostatic fly ash to provide therefrom a series of products which have higher commercial value than the raw fly ash, including a carbon concentrate, cenospheres, an iron concentrate, an enhanced pozzolan, and an inert mineral filler.

The process is shown in Figure 131. As shown there, electrostatic fly ash is applied, either wet or dry, to a sieve **10** in which the material exceeding 50 mesh is scalped off for impoundment. By removing the 50+ mesh material, the larger pieces of carbon and other agglomerated particles are removed. The material which passes the 50 mesh screen is applied to a blunger **12** in which water is added to provide a consistency of about 55% solids or higher. The material is now passed through a series of conditioners or mixing tanks **13a** through **13d.** The tanks **13** are essentially conditioning and holding tanks. Kerosene at the rate of 2 lb/ton of fly ash is added to the slurry in conditioner **13a**. This conditions the surface of the carbon particles in the fly ash and allows interaction with the air bubbles for subsequent froth flotation.

Figure 131: Amax Resource Recovery Systems, Inc.
Fly Ash Beneficiation Process

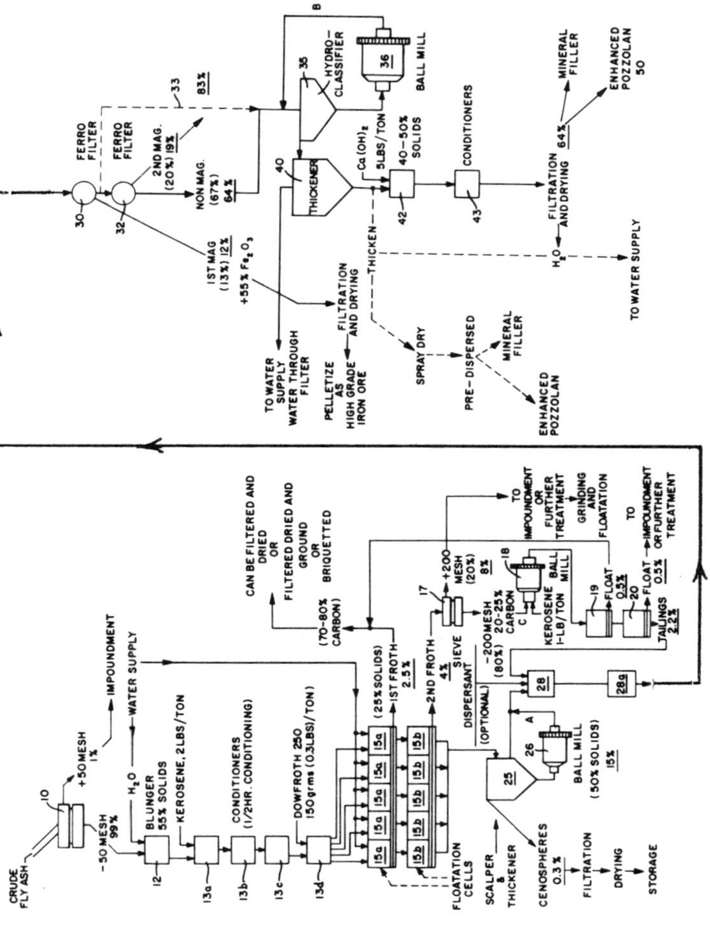

Source: U.S. Patent 4,121,945

The fly ash slurry with the kerosene is conditioned for at least one half hour of total in the series-connected conditioners **13a** to **13d**. It has been found that recovery of carbon can be improved by the addition of a small amount of frothing agent in the final conditioning tank **13d**. This agent may be added at the rate of 150 g/ton of fly ash input.

The conditioned material which now contains kerosene and the particle frothing agent is fed off to a series of parallel-connected flotation cells **15a** and the carbon is floated off the top using the froth flotation technique. The flotation cells **15a** are banked in parallel, depending on the total size and tonnage required, and employ air bubbles to separate the carbon material. Dilution water is added to the flotation cells to bring the solids content down to 25%. The particles which are coated thus become attached to the air bubbles which rise to the surface.

The first froth or overflow comes out from 75 to 80% carbon and may consist of about 2.5% of the total input. This product can be utilized directly as a carbon black and can be dried or filtered, ground and briquetted to produce a commercial product.

Further, carbon removal is effected by a second series or bank of flotation cells **15b** which are connected in series with the cells **15a**. A second froth is removed which may comprise about 4% to the total input. The output of the second froth may be discarded or may be further processed. As frothing of the sample continues, the carbon concentrate becomes finer grained and lower in carbon. These relations are illustrated by analyzing the carbon concentrate obtained from Bowen fly ash by frothing until all carbon is removed. Sieving and splitting each size fraction into a magnetic and nonmagnetic portion (with a hand magnet) yield the following distribution of carbon:

Size Fraction	Carbon (%)
+30 mesh	74.6
-30+100 mesh	
magnetic	45.5
nonmagnetic	62.0
-100+200 mesh	
magnetic	46.1
nonmagnetic	61.6
-200 mesh	24.6

The addition of sodium hexametaphosphate, a dispersant, during preflotation conditioning, gives a cleaner carbon separation. Some common dispersants, however, adversely affect carbon separation. Orzan S, for example, cannot be used because it is not only a dispersant but is also a carbon depressant.

The carbon that is removable from fly ash is mostly vesicular. It crushes readily. When ground to submicron size particles, it can be used as a filler and reinforcing agent in rubber.

The purpose of the cells **15b** is to effect a substantial removal of the initial carbon content of the fly ash to produce an end item; either an enhanced pozzolan or an industrial extender, which is substantially free of carbon, that is, it contains less than 1% carbon by weight.

By applying the output to a 200 mesh sieve **17**, about 20% of the input is removed as 20 to 25% pure carbon. The -200 mesh material from the second froth which passes through the sieve **17** may be further conditioned with the addition of kerosene and applied to a ball mill **18** for grinding in order to provide an extremely fine particle size to liberate the trapped carbon and thus upgrade the fraction of this carbon for blending with the overflow with the first flotation position. Thus by the milling at ball mill **18** and by applying the output to additional flotation tanks **19** and **20**, a first flotation of 75% pure carbon consisting of about 0.5% of the input, may be added to the froth output of the tanks **15a** as a relatively pure carbon black material.

The froth output from the tank **20** may go to impoundment for further treatment, or may be discarded. Therefore, by means of the flotation cells **15a** and **15b**, a tailing is provided which is relatively free of carbon, and a concentrated commercial carbon product consisting of about 3% of the input is also obtained.

The underflow from the second tank of flotation cells **15b** is fed to a thickener or scalper **25** in which the cenospheres are floated off. After removal of the carbon by flotation, the cenospheres are the lightest component of the fly ash, and are readily separable by allowing the fly ash slurry to settle a few minutes and scalping off the cenospheres that have floated to the surface. These may consist of about 0.3% of the input and are subjected to filtration, drying and storage. The yield of cenospheres can be increased by a brief agitation of the slurry followed by a second settling. Cenospheres continue to rise to the surface during several stirring-settling-scalping cycles. The cenospheres are, themselves, commercially valuable as acoustical insulation and as a lightweight ballast material. Microwave drying may be used to dry the cenospheres.

It is important to remove the cenospheres from the product, particularly where an enhanced pozzolan is produced, as these hollow spheres are weak and crush rather easily, and degrade the strength of the concrete product.

At this point, the underflow from the scalper **25** is fed partially to a ball mill **26**. The ball mill further reduces the particle size and is used to control consistency so that the combined output from the thickener **25** is about 50% solids. This output is applied to a further set of series connected conditioners **28** and **28a**. Also, at the conditioner **28**, the underflow from the second flotation tank **20** is reapplied as an input to the main stream flow. At this point, a suitable dispersing agent may optionally be added, particularly if the end product is to be spray dried. However, the dispersing agent may be omitted.

The output of the conditioner **28a** is applied to a wet magnetic separator, otherwise known as ferro filter **30**, as shown. Any number of suitable wet magnetic filter devices may be used, but it has been found that relatively low intensity wet magnetic separation may be used. A suitable magnetic separator is described in U.S. Patent 2,074,085.

The magnetic separator **30** is adjusted so that the first pass removes about 13% of the input thereto for approximately 12% of the total fly ash input. This first separation removes about 12% of the iron rich material and may run between 50 and 80% Fe_2O_3. Thus, the output of the ferro filter **30** represents a high concentration iron product which may be pelletized or may be fed to a plasma arc furnace for the production of pit iron. It may also be used in heavy media separation.

Preferably, the nonmagnetic output of the first ferro filter is applied to a second ferro filter **32** where a second magnetic fraction of approximately 19% of the total input is removed. The second filter **32** may be eliminated depending upon the necessity to have a relatively iron-free ultimate product. This is shown by the broken lines **33** where 83% total of the output is applied directly to a hydroclassifier **35** and ball mill **36**, and such products may be then used as an industrial filler.

However, to provide an enhanced pozzolan it is preferred to employ the second ferro filter **32** which removes an additional 20% of a lower grade magnetic iron product, and the nonmagnetic output of the filter **32** is then applied to the hydroclassifier **35** and ball mill **36**. At this point, the enhanced pozzolan product is ground down further in fineness and minimizes the need for a dispersing agent. A dispersing agent may be added at this point in the form of 0.018 to 0.025% of calcium lignin sulfate or others, as commonly known in the cement and concrete mix industry as water reducing agents.

The output from the hydroclassifer **35** is fed into a thickener **40** where substantial quantities of water are removed so that the output is about 40 to 50% solids. At this point, lime at the rate of about 5 lb/ton of solids may be added and blended in conditioners **42** and **43**, the output of which is again filtered and dried, and provides the enhanced pozzolan **50** of the process.

The pozzolan is essentially free of carbon in that it has less than 1% carbon by weight and which is substantially free of magnetic iron in that it preferably contains no more than 6% Fe_2O_3 by weight and may vary from source to source. However, the output of the thickener **40** may be used directly without the addition of lime as an enhanced pozzolan or as an inert mineral filler. In either case, the dried output may be fed to storage silos and is a very fine product similar to kaolin clay or cement.

R

RHENIUM FROM MOLYBDENUM SULFIDE ROASTER GASES

Rhenium is a relatively scarce material and, in recent years, it has achieved attention as a technological material in a variety of fields. It is employed in electrical contacts, thermocouples, filaments for electronic devices, as a ductilizing addition agent in molybdenum and tungsten for electronic uses and has also been proposed for use as a catalyst in the processing of petroleum.

Figure 132: By-Product Rhenium Subsystem

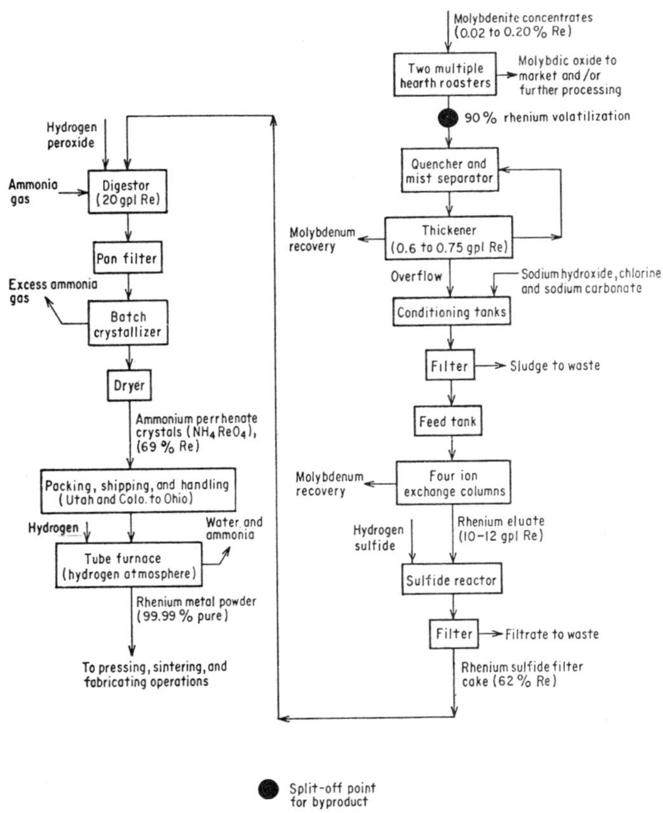

Source: Reference (17)

The presence of rhenium in ores and minerals does not generally exceed a fraction of 1% (e.g., up to 0.5%) and usually rhenium is found in the greatest quantities in certain sulfide ores. During the oxidation roasting of sulfide ores, such as molybdenum and copper sulfides, the rhenium is converted to a large extent to rhenium heptoxide (Re_2O_7) which is volatile and passes off with the gaseous effluents resulting from roasting. The roasting gases may include N_2, O_2, H_2O, SO_2, SO_3, and the like. The roaster gas stream also contains solid particles in suspension, such as molybdenum sulfide, molybdenum oxide, silica and other extraneous materials which are generally removed by precipitators, scrubbers, etc.

The essential features of Kennecott and Chase Brass processes are summarized in Figure 132. The flow sheet with the exception of the hydrogen reduction step is based on the Kennecott Copper Corp. molybdic oxide plant near Salt Lake City, Utah.

The process involves roasting of molybdenite concentrates to molybdic oxide in multiple hearth roasters. A portion of the contained rhenium is volatilized and appears in the effluent gas stream along with entrained molybdenite particles. If sulfur in the molybdenite is reduced to 0.2 to 0.4% somewhat better than 50% of the rhenium is released. The rhenium oxide is then dissolved in a liquid solvent, separated from the solvent by an ion exchange technique and subsequently recovered in the form of rhenium salts from ion exchange resins.

RHENIUM FROM WASTEWATERS

A process developed by *K.J. Richards and C.N. Wright; U.S. Patent 3,894,866; July 15, 1975; assigned to Kennecott Copper Corp.* involves recovering rhenium values from dilute solutions containing from 0.00001 to 1.0 g/ℓ of rhenium. The process consists of:

> Adding particles of a base metal which is more electropositive than copper to such a solution to react with and to precipitate the dissolved rhenium values therein;
>
> Introducing into the solution, simultaneously with or prior to the addition of the base metal thereto, particles of a collector metal selected from the group consisting of copper and silver to collect the precipitated rhenium values, the particles of the collector metal having sufficient surface area to retain and carry the precipitated rhenium values;
>
> Separating the solids and the liquids; and
>
> Treating the solids to recover the rhenium values therefrom.

A process developed by *C.N. Wright and K.J. Richards; U.S. Patent 3,870,779; March 11, 1975; assigned to Kennecott Copper Corp.* is one in which rhenium values are recovered from dilute solutions thereof by introducing a soluble sulfide into the solution and reacting the rhenium values with a base metal in the presence of the sulfide to form a precipitate which may be easily separated from the solution. The rhenium is recovered from the precipitate by conventional procedures.

Such a process is shown in Figure 133.

Figure 133: Kennecott Copper Corp. Process for Recovering Rhenium Values from Complex Industrial Solutions

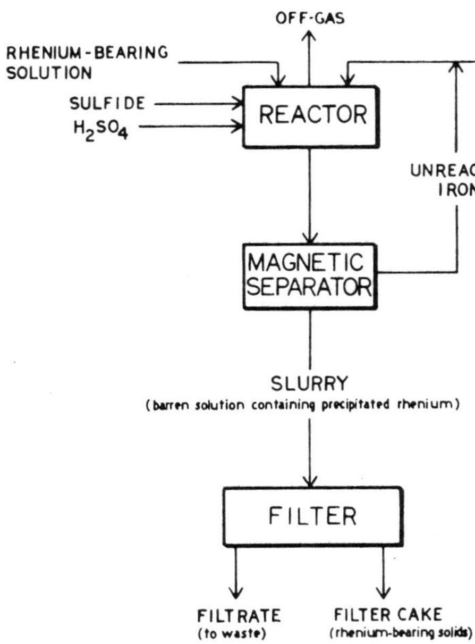

Source: U.S. Patent 3,870,779

RHODIUM FROM PLATING WASTE SOLUTIONS

See "Gold from Plating Waste Solutions."

RHODIUM FROM SPENT CATALYSTS

See also "Platinum from Spent Catalysts" and especially U.S. Patent 4,074,865 which describes rhodium recovery from spent automotive emission catalysts.

Another major use of rhodium is in hydroformylation or oxo process catalysts. Such a process involves the reaction of olefins with carbon monoxide and hydrogen at elevated temperature and superatmospheric pressure, in the presence of certain catalytic metal carbonyl complexes, to give aldehydes.

$$R{-}CH{=}CH_2 \xrightarrow[CO\ +\ H_2]{\text{catalyst}} \underset{\text{n-aldehyde}}{R{-}CH_2{-}CH_2{-}CHO} + \underset{\text{isoaldehyde}}{R{-}\underset{CHO}{\underset{|}{CH}}{-}CH_3}$$

In these equations, R is an organic radical.

If cobalt-containing catalysts are used for this reaction, as is frequently the case, the reaction temperatures required are relatively high, and this favors the formation of the isoaldehydes, which as a rule are undesired products. It is true that rhodium-containing catalysts permit the use of substantially milder reaction conditions so that a higher proportion of n-aldehydes is produced, but hitherto these catalysts have only found hesitant acceptance in industrial hydroformylation practice, because the recovery and regeneration of the expensive noble metal presents substantial difficulties.

Both in batch and in continuous operation, the more volatile constituents of the reaction mixture, including the process products, are separated off by distillation, while the catalysts accumulate in the higher-boiling distillation residue. It is true that this catalyst-containing residue can be recycled to the hydroformylation, but ultimately not all of it can be recycled because the amount of residue progressively increases and because the activity of the catalyst progressively decreases.

The recovery and regeneration of the expensive rhodium catalysts are therefore of vital economic importance but the conventional processes for this purpose have proved rather unsatisfactory.

Therefore, an improved process has been developed by *R. Kummer, H.W. Schneider and K. Schwirten; U.S. Patent 4,113,754; September 12, 1978; assigned to BASF AG, Germany.* It involves the isolation and regeneration of catalysts of type I $[ClRh(CO)(PR_3)_2]$ or type II $[HRh(CO)(PR_3)_3$, where the Rs are identical or different hydrocarbon radicals, to give the catalysts in a pure form.

This process is carried out by regenerating aqueous rhodium salt solutions, as obtained on treating distillation residues of hydroformylation mixtures with oxygen-containing mineral acids and peroxides. The aqueous rhodium salt solutions are treated with a cation exchanger, the latter is then separated from the solution, the absorbed rhodium ions are desorbed with hydrochloric acid. The hexachlororhodate solutions, containing hydrochloric acid, are reacted, in the presence of a water-soluble organic solvent and a tertiary phosphine PR_3, with carbon monoxide, or with compounds which eliminate carbon monoxide, at from 0° to 150°C and from 1 to 5 bars, and the resulting complexes I or, if the process is carried out under hydrogenating conditions, the resulting complexes II, are separated off.

A process developed by *N. von Kutepow, F.J. Mueller and P. Reuter; U.S. Patent 4,131,640; December 26, 1978; assigned to BASF AG, Germany* is one for the separation of rhodium from a solution containing a rhodium carbonyl complex such as is obtained in the carbonylation of an alkanol or of an olefinically or acetylenically unsaturated compound with carbon monoxide and water. The process involves conversion of the separated rhodium into a soluble rhodium carbonyl complex, in which the solution containing the rhodium carbonyl complex is treated with hydrogen in the presence of a carrier substance at a temperature of from 25° to 300°C and at a pressure of from 1 to 700 atmospheres.

The solution freed from rhodium is separated and the rhodium precipitated on the carrier is converted into a soluble rhodium carbonyl complex by treatment with carbon monoxide in the presence of a solvent and a compound which can replace the carbon monoxide as a ligand at a temperature of from 20° to 300°C and at a pressure of from 1 to 700 atmospheres.

A process developed by *J.L. Balmat; U.S. Patent 4,135,911; January 23, 1979; assigned to E.I. Du Pont de Nemours and Company* is one for recovering rhodium present in a still heel from the hydroformylation of an unsaturated hydrocarbon wherein rhodium and a triaryl phosphite ligand were used. The recovery process involves:

(a) pretreating the still heel with a compatible organic solvent or a mixture thereof, at least 5 ml of water per 100 g of still heel, oxygen gas or a precursor thereof and sufficient base to produce a pH of 2 to 7 in the reaction mixture after rhodium precipitation;

(b) heating to a temperature of from 0° to 80°C to oxidize the triaryl phosphite ligand to the corresponding phosphate compound;

(c) heating the mixture from (b) for from about 15 minutes to about 120 minutes at about 115°C to about 175°C to precipitate zero valent rhodium, and optionally purifying the precipitated rhodium from (c) by

(d) separating the rhodium precipitate;

(e) washing the rhodium precipitate with a solvent-acid solution at a pH of 3 to 4;

(f) treating the rhodium precipitate with an alkaline reducing solution to reduce the trivalent rhodium impurity to zero valent rhodium;

(g) quenching the rhodium with glacial acetic acid;

(h) separating the rhodium;

(i) washing the rhodium with an acid solution at a pH of 3 to 4;

(j) drying the rhodium in an inert atmosphere at from 250° to 400°C; and

(k) oxidizing the rhodium at a temperature of from 300° to 900°C to convert the rhodium to Rh_2O_3.

A process developed by *B. Fell and W. Dolkemeyer; U.S. Patent 4,188,363; February 12, 1980; assigned to Bayer AG, Germany* is one for the separation of rhodium catalysts from high-boiling and tarry distillation residues obtained in the processing of reaction products from organic, catalytic reactions utilizing dissolved, homogeneous rhodium complex catalysts, such as hydroformylation reactions, hydrocarboxylation reactions, isomerization, dimerization and oligomerization reactions. The recovery process involves passing a stream of a nonoxidizing, and preferably reducing, carrier gas, such as hydrogen or a hydrogen-containing gas, through the residue at an elevated temperature to selectively remove the organic constituents, leaving a heterogenous residue containing the rhodium, and dissolving this heterogenous residue in an inorganic acid to, thereby, form a water-soluble rhodium compound, which is suitable for the production of a complex rhodium catalyst in the known manner. Oleum is preferably used as the inorganic acid, and after mixing with the heterogenous residue, a portion is preferably fumed off. Figure 134 shows a laboratory apparatus which may be used for the conduct of this process.

Figure 134: Bayer AG Laboratory Apparatus for Rhodium Catalyst Recovery

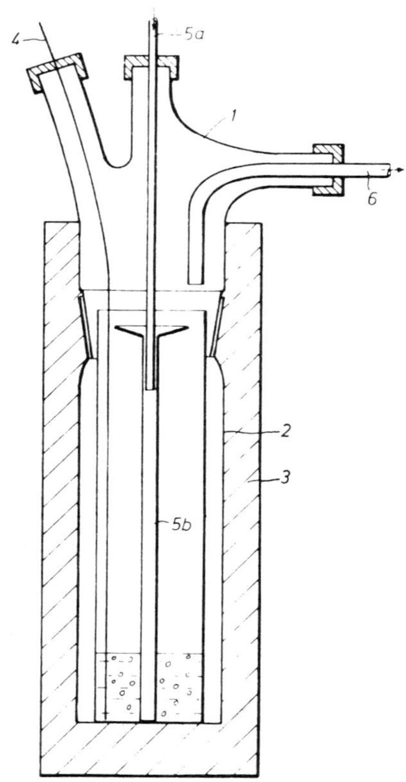

Source: U.S. Patent 4,188,363

A homogeneous solution consisting of a rhodium complex and a high-boiling distillation residue is heated in a reaction vessel through which a stream of carrier gas continuously flows. The vessel is described with the aid of the diagram, the numbers of which have the following meaning: **(1)** upper part of glass reactor; **(2)** lower part of glass reactor; **(3)** heating; **(4)** thermocouple; **(5a)** metal capillary, glass inlet; **(5b)** glass capillary; and **(6)** product outlet. All the constituents of the heavy oil as well as all the products formed by the pyrolysis reaction are carried off with the gas. The noble metal remains in the reaction chamber in a high concentration.

The following is one specific example of the conduct of the present process. 500 g of n-valeraldehyde are heated to 300°C for 72 hours in a piston-stirred autoclave. The low-boiling polycondensation products are then distilled off under a high vacuum. 300 mg of $HRhCO(PPh_3)_3$ together with 1,200 mg of PPh_3 are dissolved in 15 g of the residue, which cannot be distilled, at about 100°C, while adding hydrogen. A continuous stream of H_2 of 0.4 ℓ/minute is passed through the mixture in the reaction vessel in the diagram.

After its exit from the reaction chamber, the gas stream is passed through two wash bottles, connected in series, containing the above polycondensation products and toluene as the absorption liquid. The reactor is heated to 300°C for 90 minutes and subsequently cooled. 169 mg of solid residue remain in the reaction chamber. No rhodium can be detected in either of the two wash bottles. Thus the entire noble metal constituent remains in the solid residue.

15 g of oleum (10% strength) are filled into the reaction vessel and 25% of this oleum is evaporated off as fumes at about 300°C. The off-gases are withdrawn under a slight vacuum and passed through a wash bottle containing H_2O as the absorption liquid. It is also not possible to detect rhodium in the water. The residual acid, containing noble metal, is diluted with 162 ml of H_2O to give a 7% strength sulfuric acid. The solution is colored dark brown. The entire amount of rhodium employed is detected in the solution.

RUTHENIUM FROM SPENT CATALYSTS AND/OR SCRAP ELECTRODES

Various salts of ruthenium have long been known to have utility in a wide variety of catalytic applications. Most commonly, a suitable salt of ruthenium, either alone or with salts of one or more additional metals, was deposited on a catalyst support such as alumina, titania, zirconia, silica, or silica alumina.

More recently, ruthenium has found significant commercial application as a coating for electrodes, particularly so-called dimensionally stable anodes. In such applications, the ruthenium may be deposited by electroplating or by other deposition means, either as the metal or in the form of a suitable salt, on an electrode substrate such as, for example, titanium, tantalum, or the like.

The wide variety of advantages obtained by the use of such electrodes almost immediately established them as commercially significant. Since their first introduction, these electrodes have not only found almost universal acceptance in their original commercial applications, but are continually being modified for an ever-widening variety of additional commercial applications. This, in turn, made it more and more economically desirable to recover the ruthenium from used or defective anodes.

A recovery process has been developed by *R.S. DePablo, D.E. Harrington, and W.R. Bramstedt; U.S. Patent 4,132,569; January 2, 1979; assigned to Diamond Shamrock Corporation.* In this process, ruthenium is stripped from a catalyst or electrode substrate by immersion in a fluoboric acid solution, converted to ruthenium oxide, and the ruthenium oxide is then converted to the alpha ruthenium trichloride for use in the preparation of fresh catalyst and/or electrodes.

S

SALT CAKE FROM ALUMINUM REFINING

See "Aluminum from Furnace Skim or Dross," especially U.S. Patent 4,073,644.

As pointed out there, in a number of prior art processes for recovering aluminum metal value from aluminum dross, the dross is heated under a flux comprised of various salts or combinations of salts, usually comprising chloride salts of alkali or alkaline earth metals or mixtures of these salts. In the fluxing step aluminum metal, protected by the flux, is removed from the dross by draining or other gravity or mechanical separating techniques leaving a product sometimes referred to as salt cake which is composed of the fluxing salt mixture, aluminum dross which is primarily oxides of aluminum, entrained aluminum metal particles or chunks, and, in general, all impurities which are less dense than aluminum or which are formed into particles with sufficient porosity to be less dense than aluminum.

Traditionally, the salt cake is allowed to cool in large rocks or chunks resembling mine slag. These chunks are extremely hard, quite large, and, in general, are extremely difficult to handle.

These chunks or rocks of salt cake usually are broken up by heavy, specially designed machinery, or by laborious manual or mechanical operations to recover the aluminum metal values which are entrained in the salt cake. The aluminum metal may appear as chunks ranging from the very fine to pieces several inches in thickness. Following removal of the valuable aluminum metal, the salt cake has, in the past, usually been discarded since it has no value in the form in which it exists and it has been too expensive and difficult to recover the flux salt components therefrom.

Disposal of the salt cake presents particular problems and improper disposal often has resulted in serious ecological disruptions. Any disposal which presents the immediate or long term potential of exposure to moisture presents the hazardous possibility that salts may be leached from the salt cake and carried into agricultural, industrial or potable water supplies. In the past, salt cake has been disposed of by simply depositing it in marble quarries, etc., in isolated locations. This type of disposal is unacceptable from the safety and ecological point of view since it presents the serious possibility that salts leached from the salt cake may find their way into underground or surface water streams or reservoirs.

Thus, a process has been developed by *P.N. Papafingos and R.T. Lance; U.S. Patent 4,060,118; November 29, 1977; assigned to Alumax Mill Products, Inc.* for recovering flux salt and recycling or reusing the same in an aluminum refining method by introducing salt brine into a cylindrical evaporator with hot evaporator gas, rotating the evaporator to expose the brine to the hot gas and removing the slurry from the evaporator.

A suitable form of apparatus for the conduct of such a process is shown in Figure 135. The evaporator, indicated generally at **200**, comprises a cylindrical drum **202** which is supported for rotation by a pair of tires **204** and **206**. The tire **204** rides on a pair of trunions **208** and **210** which are mounted on journaled axles **212** and **214** by brackets on a vertical support stand **216** in the manner conventionally used for supporting rotating processing drums. The tire **206** is likewise supported by a pair of trunions, one of which is shown at **218**. The drum is supported at an angle, as illustrated.

The drum is equipped with a sprocket **220** which is driven by a chain **222** and a smaller sprocket **224** which is, in turn, powered by a geared motor system **226** and a pair of pulleys and belts indicated generally at **228**. The motor **226** is a variable speed, reversible motor, or, equivalently, may be a constant speed motor equipped with a variable speed drive connector and reversing gears, to permit the drum to be rotated at the desired speed in either direction.

The drum is equipped at the input end, shown at the left of the figure, with an opening **230**. At the other end, the output end, a large diameter cylindrical well is mounted to the end of the drum to receive the brine-salt crystal slurry. In the center of the well **232** is an output opening **234**. The output opening is surrounded by a cowling **236** which, in turn, is connected to a stack conduit **238** for dumping the gaseous products of the evaporation, principally water vapor, into the atmosphere, or conducting the gaseous output to a gas clean up system or to a vapor recovery system, as may be desired.

A slurry recovery line, **240**, extends down into the well **232** and out through the opening **234** to permit removal by pumping of the brine-salt crystal slurry. The pipe **240** is L-shaped with the end which extends into the well being offset from the vertical slightly to compensate for the displacement of fluid slurry in the well resulting from rotation of the drum. The amount of offset can be varied to obtain maximum depth in the salt in the well.

A gas fired blower furnace **242**, provided with a nozzle **244**, which can extend into the opening **230** in the input end of the evaporator drum and with an input conduit **246** provides a source of heat for evaporation and a source of air to carry out the evaporated moisture. In the preferred embodiment, the conduit **246** is connected to the stack of one or more aluminum melting furnaces or one or more aluminum recovery furnaces.

The interior of the evaporator furnace is of a unique design, especially adapted to accomplish the evaporation by the introduction of hot evaporator gas into the evaporator. Welded or otherwise secured to the interior wall of the drum, around the entire circumference, are a multiplicity of fins, indicated by way of example as fin **250**. The fins are not, however, secured directly to the wall. Instead, two or more tabs **252** are secured to the fins at one end and are secured to the wall at the other end so as to space the proximal edge of the fin some distance away from the wall.

The fins are arranged, as shown, in sets, each of which extends around the interior circumference of the drum, along the length of the drum, with wider fins being used at the output end of the drum, as shown at **254**, intermediate height fins being used in the central section of the drum, as shown at **256**, and narrower fins being used at the input end of the drum, as shown at **258**.

Figure 135: Alumax Mill Products, Inc. Apparatus for Recovering Chloride Flux Salt Cake from Aluminum Production

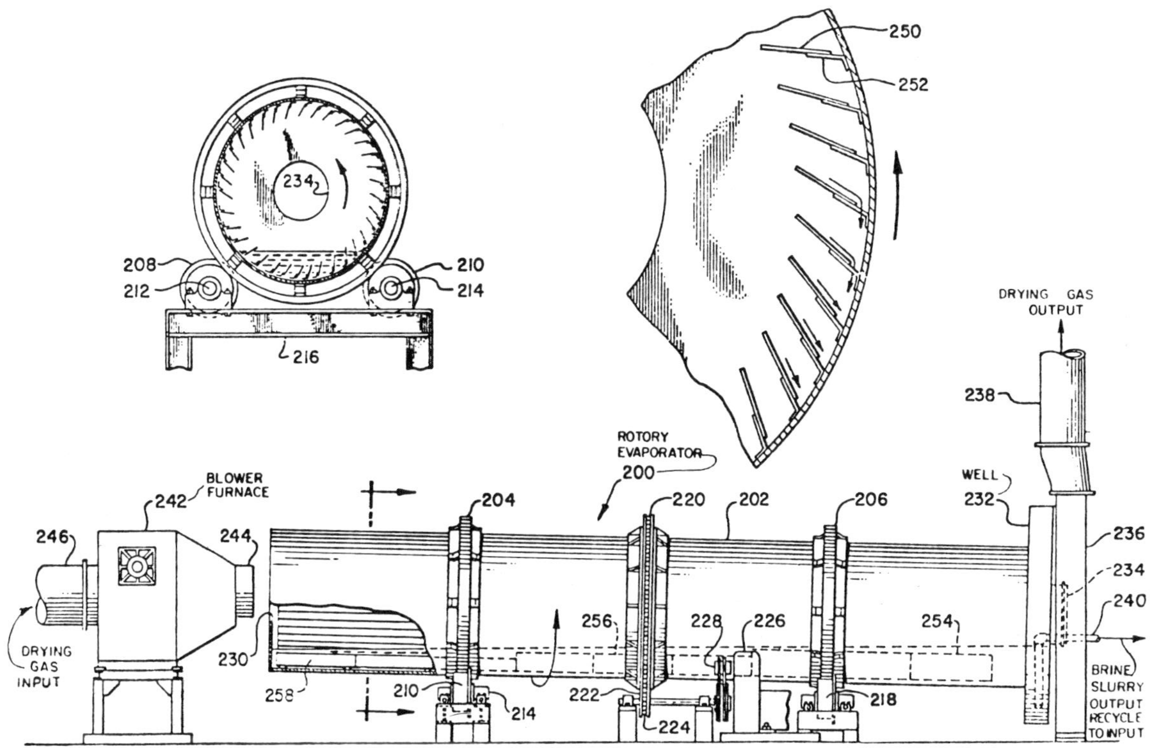

Source: U.S. Patent 4,060,118

In an exemplary embodiment, the drum is approximately 38 to 40 feet long and has a diameter of about 7 feet. The wall thickness of the drum is three-eighths inch. Of course, the drum can be constructed in any convenient length or diameter.

The evaporator is mounted at an incline of from 1° to 10°, preferably from 2° to 5°. The actual amount of incline is subject to considerable variation, depending upon the type of salt used and the desired rate of travel of the salt-brine slurry through the evaporator.

It has been found, using the drum as described above, that an angle from 2° to 5° is quite satisfactory to provide efficient flow through the evaporator and, at the same time, permit optimum evaporation.

The brine, which forms the brine-salt crystal slurry upon partial evaporation, lies in the drum at a variable depth, the depth being greater at the output end than at the input end.

The width of the fins is such that the fins are completely submerged in the brine when the fins are at the bottom of the drum during the rotational cycle of the drum, as the evaporator is in operation.

The rotation of the fins into and out of the liquid causes a disturbance of the liquid surface and constant mixing of the liquid, thus increasing the rate of evaporation because of a constantly changing and disturbed surface area in the brine which has the net effect of reducing surface tension and preventing the formation of a film of salt on the surface.

Thus, the flow of salt and brine is from the input end, where the brine is introduced by simply extending a pipe into the evaporator or by any other desired means, to the output end where the remaining liquid and the crystallized salt are removed from the well and the vapor is removed from the central opening in the drum and is vented to the stack or to a recovery or clean up system as may be desired.

SCANDIUM FROM URANIUM PROCESS SLUDGES

As described by A. Petrick, Jr., H.J. Bennett, K.E. Starch and R.C. Weisner (17), scandium can be recovered from uranium process sludges. The relationship of the scandium by-product subsystem to the uranium production system is illustrated in Figure 136 based on the Vitro process. Two split-off points for by-product scandium are shown. In the Vitro Chemical Co. process, thorium and titanium and scandium building up in the organic stream of the uranium solvent extraction system are selectively tapped and subjected to a special stripping with hydrofluoric acid.

The white cake precipitated contains 10% scandium oxide and is further refined to produce high-purity scandium oxide. The system is based on processing of an iron sludge containing scandium, titanium, thorium, uranium and iron. This sludge is filtered from the loaded strip liquor and is either recycled to recover uranium, discarded, or is used as a raw material for scandium recovery as shown.

Figure 136: By-Product Scandium Subsystem

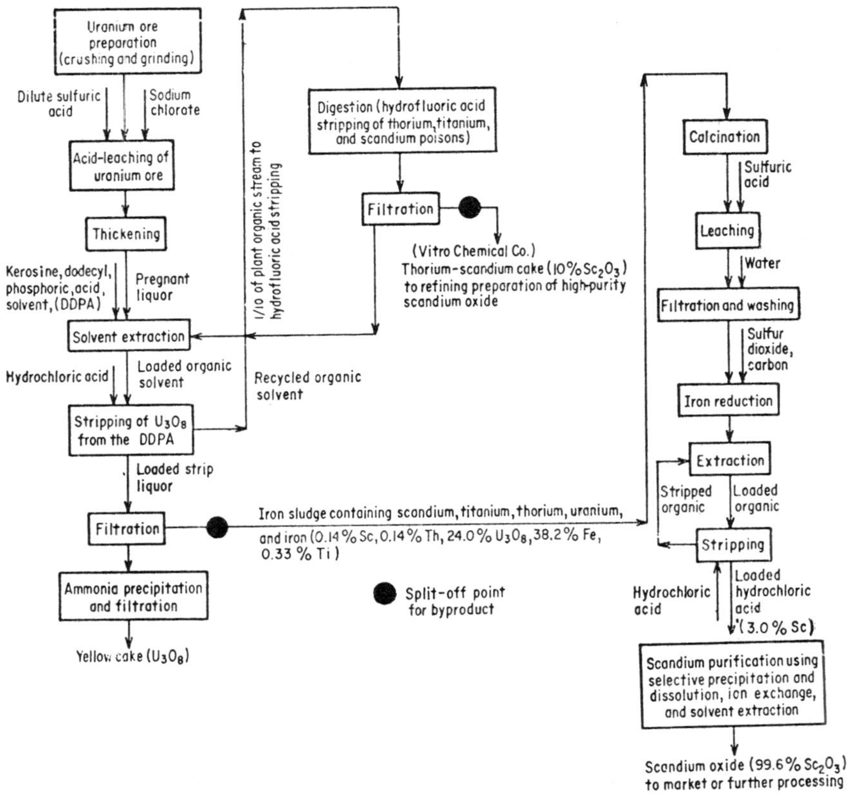

Source: Reference (17)

SELENIUM FROM COPPER REFINERY ANODE SLIMES

See also "Copper from Copper Refinery Anode Slimes."

As described by A. Petrick, Jr., H.J. Bennett, K.E. Starch and R.C. Weisner (17), the three main methods of recovering selenium from copper muds and slimes are: roasting with soda; roasting with an acid; and smelting with soda and niter; the last is illustrated in Figure 137. The metallurgy and the economics of Se and Te are closely interlocked.

At a precious metals refinery the volatile Se in gases from the Cupel furnace is collected by water scrubbing to yield an acid solution. The addition of alkaline fluxes to the Cupel yields a soda slag containing both Se and Te. When the high-acid selenium solution is contacted with the soda slag in a neutralizing tank TeO_2 is precipitated and the neutralized solution can then be treated to precipitate the selenium. The metallurgy and economics of selenium alone can best be described by following the processing of the volatile gases and soda slag from the Cupel.

Figure 137: By-Product Selenium Subsystem

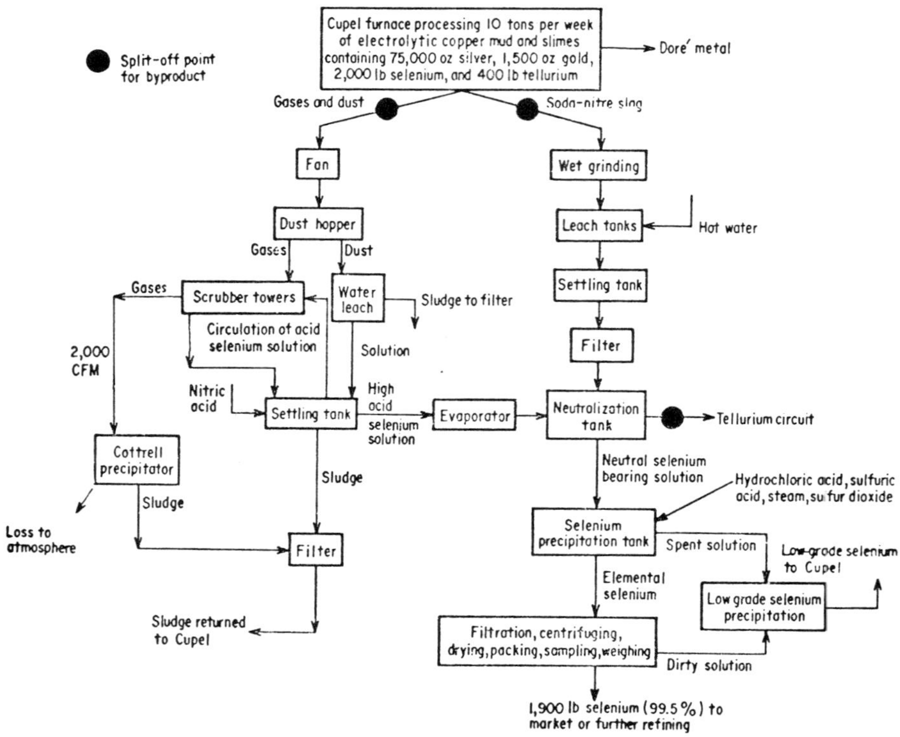

Source: Reference (17)

Figure 137 indicates a production of 1,900 pounds of selenium from the 2,000 pounds of selenium in the charge. This is based on an assumed recovery of 95% overall in selenium processing. Such a recovery assumes recirculation of part of the selenium content via precipitation of a low-grade selenium product which is returned to the Cupel furnace. Actual recovery is likely to be higher than 95% but the 100 pound loss is assumed to account for a portion of the recirculated selenium.

A process developed by *K.N. Subramanian, A. Illis and N.C. Nissen; U.S. Patent 4,163,046; July 31, 1979; assigned to The International Nickel Company, Inc.* is a hydrometallurgical process for treating copper refinery anode slimes containing selenium, lead and precious metals in which a selenium product of better than 99.7% purity can be produced and the final residue is suitable for smelting to produce anodes for silver refining. Such a process is shown in the flow diagram in Figure 138.

Figure 138: International Nickel Company Process for Recovery of Selenium from Copper Refinery Anode Slimes

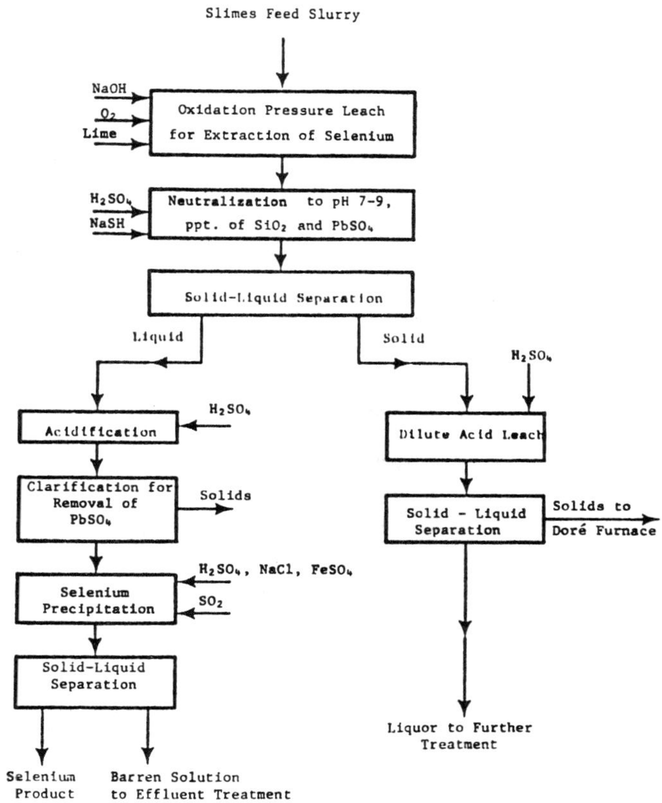

Source: U.S. Patent 4,163,046

Referring to Figure 138, detellurized and decopperized slimes containing the following metal values (approximate), 35% selenium, 1% tellurium, 3% copper, 2% nickel, 30% silver, 0.5% gold, 0.5% platinum, 1.0% palladium and 30% gangue including SiO_2, are slurried with 1 kilogram of caustic soda per kilogram of slimes to provide a pulp density of 15% by weight solids and with about 5% lime based on the weight of the slimes.

The slimes are subjected to a caustic leach at about 180° to 220°C, e.g., 200°C, for about 3 to 7 hours, e.g., 5 hours, under 0.3 to 1.0 MPa, e.g., 0.7 MPa (100 psi) oxygen partial pressure. A temperature of at least 180°C is required for complete oxidation. Caustic soda is preferred because it is less expensive than other alkali metal hydroxides such as potassium or lithium hydroxide. The amount of alkali metal hydroxide required will vary according to the composition of the slimes treated. In general, the amount is calculated to be about 30 to 100% in excess of the reactions taking place. The oxygen can be provided, for example, by commercial oxygen at a total pressure of, e.g., 2.2 MPa. Air may be used.

Oxidation of selenium to the hexavalent state is advantageous in that, for example, an essentially tellurium-free caustic solution can be obtained and there is no substantial loss of selenium with the lead precipitate as occurs when lead is removed from a solution containing mainly selenite. There is no requirement for controlling the oxidation of selenium to mainly Se^{+4} since the problem of difficulty of reduction of Se^{+6} to Se^0 is taken care of in this process, as explained below.

The lime which is added to the caustic leach medium serves to aid subsequent filtration of the leach product slurry. The caustic oxidative leach treatment extracts in excess of 98% of the selenium and less than 5 mg/ℓ of tellurium into the leach liquor. When the caustic leach is carried out at a temperature of about 200°C, an O_2 partial pressure of about 0.7 MPa, and a solids content of about 15%, it has been found that about 50% lead, 60% silica, about 2 to 10% of the platinum, and about 1 to 3% of the palladium are also extracted into the caustic leach liquor in about 4 hours.

The reactions which occur during the various stages of this process are very complex. Generally, it is believed that the overall reactions which occur during the caustic oxidation leach step are typified by the following equations:

(A) $$2Se + 4NaOH + 3O_2 \rightarrow 2Na_2SeO_4 + 2H_2O$$

(B) $$2Ag_2Se + 4NaOH + 3O_2 \rightarrow 4Ag + 2Na_2SeO_4 + 2H_2O$$

(C) $$2Te + 4NaOH + 3O_2 \rightarrow 2Na_2TeO_4 + 2H_2O$$

(D) $$2Cu + NaOH + O_2 + 2H_2O \rightarrow 2Cu(OH)_2 + NaOH$$

(E) $$2Cu + NaOH + O_2 \rightarrow 2CuO + NaOH$$

(F) $$PbSO_4 + 4NaOH \rightarrow Na_2PbO_2 + Na_2SO_4 + 2H_2O$$

(G) $$SiO_2 + 2NaOH \rightarrow Na_2O{\cdot}SiO_2 + H_2O$$

(H) $$C + 2NaOH + O_2 \rightarrow Na_2CO_3 + H_2O$$

In reaction (H) C represents organic compounds. The Na_2TeO_4 (reaction C) precipitates substantially quantitatively.

The caustic leach slurry is neutralized at a temperature of 30° to 70°C, e.g., about 50°C, with sulfuric acid to a pH of between 7 and 9, e.g., about 7.5, and sodium hydrosulfide (NaSH) is added in the amount of from 1 to 7% by weight of slimes. At a pH of about 7 to 9, lead values and silica precipitate. The extent of platinum and palladium precipitation varies, but the NaSH lowers the platinum and palladium levels in the purified liquor to less than about 5 mg each per liter.

Instead of NaSH, a sulfide such as an alkali metal sulfide or alkaline earth metal sulfide or ammonium sulfide or a polysulfide may be used. The use of sulfide is particularly advantageous since it aids in the purification of the Se as well as in the separation and greater recovery of valuable precious metals including silver. Also, it will dispense with the requirements for the use of other means such as an expensive ion exchange step to recover precious metals, particularly to lower the precious metal levels in the selenium product. For example, in the absence of NaSH addition, the selenium product may analyze up to 500 ppm platinum plus palladium and with a suitable amount of NaSH addition the platinum and palladium level may be decreased to as low as 5 ppm. The resultant product of the neutralized caustic leach is very complex.

Consequently, for convenience the sulfide addition is expressed relative to the weight of the slimes. The amount of sulfide added is in effect a compromise to maximize the precipitation of precious metals from the neutralized caustic leach liquor without sacrificing the selenium purity or tellurium extraction during the subsequent acid leach.

With respect to reactions which occur in the neutralization step, it is believed that the overall reactions can be represented by the following equations:

(I) $$2NaOH + H_2SO_4 \rightarrow Na_2SO_4 + 2H_2O$$

(J) $$Na_2PbO_2 + 2H_2SO_4 \rightarrow PbSO_4 + Na_2SO_4 + 2H_2O$$

(K) $$Na_2O{\cdot}SiO_2 + H_2SO_4 \rightarrow SiO_2 + Na_2SO_4 + H_2O$$

After neutralization with sulfuric acid, the resultant liquor is then separated from the residue, e.g., by filtration. As indicated above, filtration of the slurry is aided by the presence of the lime added to the caustic pressure leach slurry. A filtration rate of 300 to 600 liters per square meter per hour, e.g., about 400, has been achieved when filtered under vacuum at about 75°C. The leach residue is then washed thoroughly to ensure that soluble selenium does not remain with and contaminate the tellurium, precious metals or other components of the tellurium-copper-precious metal recovery system.

The separated neutralized caustic leach liquor, containing about 2% of the lead, less than about 5 mg/ℓ tellurium and less than about 70 mg/ℓ copper, is further acidified with sulfuric acid directly to about 100 g/ℓ acid at a temperature of about 50° to 60°C. Additional lead sulfate precipitates and the liquor is clarified to remove precipitated lead sulfate. It has been found that the clarified liquor should contain at least 100 g/ℓ H_2SO_4 to achieve acceptable rates and conversion of the selenate to elemental selenium.

The acidified leach liquor contains selenium in solution substantially as sodium selenate (Se^{+6}) and minor amounts, if any, of sodium selenite (Se^{+4}). Reduction to elemental selenium is carried out with SO_2 in the presence of a halide and ferrous ion. Typically, halide is added, e.g., as a chloride in an amount equivalent to 30 to 70 g/ℓ of sodium chloride, e.g., 50 g/ℓ of NaCl, and 5 to 20 g/ℓ of ferrous ion in the form of sulfate, e.g., about 10 g/ℓ of ferrous ion in the form of ferrous sulfate, are added to the leach liquor, the mixture is heated to about 90° to 100°C, e.g., 95°C, and SO_2 is fed to the heated mixture at the rate of 0.1 to 0.3 ℓ/min/ℓ of solution to precipitate elemental selenium.

Almost 100% of the selenium values in solution precipitate in about 4 hours. The reactions which occur are typified by the following equations:

(L) $Na_2SeO_4 + 2HCl \rightarrow Na_2SeO_3 + H_2O + Cl_2$

(M) $H_2SeO_3 + 2SO_2 + H_2O \rightarrow Se + 2H_2SO_4$

(N) $3H_2SeO_4 + 6HCl + 6FeSO_4 \rightarrow 3H_2SeO_3 + 2FeCl_3 + 2Fe_2(SO_4)_3 + 3H_2O$

(O) $H_2SeO_4 + 2HCl \rightarrow H_2SeO_3 + Cl_2 + H_2O$

(P) $6FeSO_4 + 3Cl_2 \rightarrow 2FeCl_3 + 2Fe_2(SO_4)_3$

(Q) $Fe_2(SO_4)_3 + SO_2 + 2H_2O \rightarrow 2FeSO_4 + 2H_2SO_4$

(R) $2FeCl_3 + SO_2 + H_2SO_4 + 2H_2O \rightarrow 2FeSO_4 + 6HCl$

In the reactions iron acts as a cyclic intermediate. The elemental selenium is separated, e.g., by filtration, and the barren solution is neutralized to a pH of 8 to 9 with lime. A selenium product can be produced having a purity of better than 99.7%.

The neutralized and sulfided caustic leach residue contains essentially all of the copper and tellurium in the feed to caustic pressure leaching. Preponderant parts of these elements are extracted by repulping the residue in sulfuric acid at a temperature of from 40° to 70°C, e.g., about 50°C, and a pH of from 1.5 to 3.0, e.g., about 2.0. Approximately 50 to 80% of the tellurium and copper values are extracted. This tellurium-bearing liquor is separated from the precious metals containing residue, and it may be treated elsewhere in the plant for Te recovery.

As shown in Figure 138, the separated residue resulting from the neutralized oxidative caustic leach product, after subjection to a dilute acid leach, and separation from the tellurium-bearing acid leach liquor, is fed to a Doré furnace for recovery of precious metals. With feeds of the composition used to illustrate the process, such residues analyze about 30 to 45% silver, about 1 to 2% copper, less than about 1% tellurium, e.g., 0.5 or 0.6% tellurium, up to about 4% selenium, e.g., 1% selenium, and about 5% nickel.

SELENIUM FROM URETHANE REACTION PRODUCTS

Urethane compounds may be made by reacting an organic compound containing at least one hydroxyl group, e.g., ethyl alcohol, with carbon monoxide and a nitrogenous organic compound at elevated temperature and pressure in the presence of a selenium catalyst and a base and/or water.

Because of the high reactivity of selenium, when used as a catalyst as in the above reaction, its combination with organic compounds very frequently results in the formation of organoselenium compounds, such as benzoselenazol and diethyl diselenide, which remain in solution with the urethane reaction product. In addition, the inorganic selenium compounds used as catalysts or formed in the reaction may also remain in solution with the urethane product.

Because of the cost and toxicity of selenium, it is essential that as much of the selenium be recovered from the urethane reaction product as is possible and from the inorganic or organoselenium compounds in a form suitable for reuse as a catalyst. Prior art processes have been proposed for the recovery of selenium from certain organic reaction products and aqueous or acidic solutions. However, such prior art processes are generally narrow in scope and application and have proven to be of little or no value to the recovery of selenium from urethane solutions containing same.

Therefore, a process has been developed by *J.H. Vorachek and T.A. Washall; U.S. Patent 4,038,375; July 26, 1977; assigned to Atlantic Richfield Company* providing for the recovery of selenium from a urethane solution containing selenium compounds by contacting the selenium-containing urethane solution at a suitable temperature with an aqueous solution of a mercuric salt compound or mixtures thereof to form an insoluble precipitate of mercury-selenium compounds or complexes. The selenium and mercury may be recovered from the resulting precipitate in a form suitable for reuse in the preparation of urethanes and selenium removal respectively.

A process developed by *J.J. McCoy and J.G. Zajacek; U.S. Patent 4,038,376; July 26, 1977; assigned to Atlantic Richfield Company* involves contacting the selenium-containing urethane solution at suitable temperatures with a supported metal or a mixture of metals and/or metal oxide compounds, the metal selectively chosen from Group B metals, i.e., Group IB through VIIIB, metals of the Periodic Table and particularly supported copper and mixtures of copper and other metal oxide compounds of the group.

The selenium retained by the supported metal or mixture of metals and/or metal oxides is then recovered in a form suitable for recycle and reuse as a catalyst, for example, by reaction with a suitable oxidant such as air at an elevated temperature and the selenium-free supported metal or mixture of metals and/or metal oxides is regenerated in a reducing atmosphere to render it suitable for reuse in the selenium removal step.

Essentially similar ground is covered by *J.J. McCoy and J.G. Zajacek; U.S. Patent 4,055,630; October 25, 1977; assigned to Atlantic Richfield Company.*

A process developed by *T.A. Washall and H. Morgan; U.S. Patent 4,038,377; July 26, 1977; assigned to Atlantic Richfield Company* involves contacting a hydrocarbon solvent solution of the selenium-containing urethane at a suitable temperature with an aqueous ammonium hydroxide or alkali metal hydroxide solution and recovering the selenium from the resulting selenium compound enriched aqueous hydroxide phase; the deselenized urethane remaining in the hydrocarbon phase.

A process developed by *T.A. Washall; U.S. Patent 4,041,139; August 9, 1977; assigned to Atlantic Richfield Company* involves contacting the selenium-containing urethane solution at a suitable temperature with water and a saturated aliphatic or alicyclic paraffin hydrocarbon or mixtures thereof. The selenium may be recovered from the resulting hydrocarbon phase in the form of selenium per se or selenium dioxide for reuse in the preparation of urethane.

A process developed by *R. Rosenthal and E.T. Shawl; U.S. Patent 4,055,629; October 25, 1977; assigned to Atlantic Richfield Company* involves contacting the selenium-containing urethane solution at a suitable temperature with copper(1) chloride, oxygen or an oxygen-containing gas such as air, and a complexing solvent such as an amine or nitrile to form an insoluble precipitate of copper-selenium compounds or complexes. The selenium and copper may be recovered from the resulting precipitate in a form suitable for reuse in the preparation of urethanes and selenium removal respectively.

A process developed by *E.T. Shawl, M.N. Sheng and J.G. Zajacek; U.S. Patent 4,130,633; December 19, 1978; assigned to Atlantic Richfield Company* involves contacting the selenium-containing urethane solution at a suitable temperature with an aqueous hydrogen peroxide solution and recovering the selenium from the resulting selenium compound enriched aqueous phase in a form suitable for reuse in the preparation of urethanes, the deselenized urethane remaining in the hydrocarbon and/or organic phase.

SELENIUM FROM XEROGRAPHIC EQUIPMENT DISCARDS

The art of xerography involves the use of a photoconductive element, i.e., drum or plate, which is uniformly electrostatically charged in order to sensitize its surface. The plate is then exposed in an imagewise manner to activating electromagnetic radiation which selectively dissipates the charge in the exposed areas of the photoconductive material while leaving behind a latent electrostatic image in the nonexposed areas. This latent electrostatic image may then be developed by depositing a finely divided electroscopic marking material or toner on the surface of the photoconductive material. This concept was originally disclosed in U.S. Patent 2,297,691 and is further amplified and described in many related patents.

The photoconductive insulating material is characterized in that it has a high electrical resistance in the dark which resistance decreases significantly upon exposure to activating radiation. Amorphous selenium was found to be an efficient inorganic photoconductor and has been widely used in certain plain paper copiers.

More recently, arsenic/selenium alloys have come into their own as commercially acceptable inorganic photoconductors. Arsenic alloys of selenium are preferable to selenium for several reasons. First of all, the photodischarge speed of such alloys is faster than that of pure selenium. In addition, these alloys are photoresponsive to longer wavelengths of light than is pure selenium and, therefore, can be used to copy a wide spectrum of material.

Furthermore, arsenic/selenium alloys are generally harder than pure selenium and are, therefore, more resistant to physical abrasion. In spite of the increased resistance to physical abrasion, photosensitive devices employing arsenic/selenium as the photoconductor do tend to wear out and must be replaced.

It is apparent that sound economics dictate the recovery and reuse of the arsenic/selenium alloy remaining on the discarded photosensitive devices. Such reclamation involves removing the alloy from its substrate and purifying it to a state of purity suitable for reuse as a photoconductive alloy. Several methods have been used to remove the alloy from the substrate. One such method involves a thermal shock treatment. This method is effective, but its efficiency decreases as further research improves the adhesion between the alloy layer and the substrate. Another method involves hydraulic lathing which works well when the substrate is flexible as in the case of an endless flexible belt, but is not readily adaptable to those devices wherein the substrate is of a thickness too great to permit flexibility.

Mechanical shaving may be used to remove the layer of arsenic/selenium from the substrate in this situation. Mechanical shaving effectively removes the arsenic/selenium from the substrate but tends to produce an alloy contaminated by metal, especially aluminum. In these cases, the final separation from the metal becomes very difficult. Distillation gives a residue of metal selenides; attempted metal dissolution in acid gives toxic arsine and hydrogen selenide by-products. Dissolution of the alloys in aqueous sodium hydroxide introduces its own set of bulk contaminants.

A process developed by *K.A. Waehner and A.T. Giammarise; U.S. Patent 3,992,511; November 16, 1976; assigned to Xerox Corporation* is one for recovering pure selenium from metal substrates having a selenium coating thereon without impairment of the substrate for reuse. This process comprises substantially converting the selenium in the coating to a water-soluble selenocyanate by immersing the coated substrate in an aqueous solution of an alkali metal cyanide, removing the substrate from resultant solution, and treating the resultant solution with an acid to precipitate the selenium from the remaining liquid.

More recently, the endless xerographic belt has been used in xerographic duplication. The endless xerographic belt must be flexible and is preferably seamless. Suitable belts are quite thin and have a surface with a high degree of smoothness due to the need for the production of high quality images on the image retention side of the belt.

A further requirement is that the belt have a relatively high tensile strength. Satisfactory belts can be prepared by electroplating a ductile metal, e.g., stainless steel, brass, aluminum or nickel, onto a mandrel to form a thin, uniform layer of the metal. Removal of the metal layer from the mandrel provides the substrate upon which the photoconductive material can be deposited to form the xerographic belt.

Xerographic photoreceptors, whether they be in the form of a drum or belt, normally have a barrier layer between the conductive substrate and the layer of photoconductive material to prevent charge injection from the conductive substrate in the dark. When the substrate used is an aluminum drum, a thin layer of aluminum oxide, formed by oxidizing the surface of the drum, serves well as the blocking layer. When the substrate is an endless belt, it has been found that a thin, normally submicron layer of an insulating organic resin is desirable for use as the blocking layer.

Typically, the organic barrier layer used is a polymer blend of polycarbonate and polyurethane resins in a ratio of about 7 parts by weight polycarbonate to 1 part by weight polyurethane. After application of the organic blocking material to the substrate, amorphous selenium or an alloy thereof is applied by vacuum deposition.

Belts made in this manner serve admirably in high speed xerographic duplicators but, of course, wear out over extended periods of use. When a belt wears out, sound economics dictate that the selenium or selenium alloy be recovered and recycled. Various methods are available for stripping selenium from the substrate. These include heat stripping, water quenching, ultrasonics and bead blasting. These processes, however, are not readily adaptable to the previously described xerographic belt because of the presence of the organic interface on the flexible metal substrate. Stripping selenium by these processes also strips part of the organic material thereby contaminating the recovered selenium.

Cold shocking the belt, such as by dipping it in liquid nitrogen, removes the selenium without affecting the organic interface, but is inherently a two-step process; the first step being a liquid nitrogen dip, and the second step consisting of mechanical vibration or shaking.

It would be desirable to provide a method for recovering selenium or selenium alloy from the surface of a xerographic belt comprising a ductile metal substrate having a thin layer of an organic resinous material on its surface which is, in turn, overcoated with selenium.

Such a process has been developed by *J.K. Williams; U.S. Patent 4,047,973; September 13, 1977; assigned to Xerox Corporation*. This process comprises the consecutive steps of:

(1) mounting the belt on a rigid mandrel having a ribbed outer surface;

(2) subjecting the surface of the belt to one or more jets of high pressure water, the jets being expelled from one or more nozzles at a pressure of from 8,000 to 11,000 psi to strip the selenium or selenium alloy from the belt without substantially affecting the layer of organic resinous material to thereby form a water slurry of selenium or selenium alloy; and

(3) recovering the water slurry and removing the water to obtain substantially pure selenium or selenium alloy.

A process developed by *W.H.H. Gunther; U.S. Patent 4,097,273; June 27, 1978; assigned to Xerox Corporation* is one for the recovery and purification of arsenic/selenium alloys which are contaminated with organic and inorganic materials. The method comprises:

(1) dissolving the alloy in an aqueous solution of morpholine, piperidine or a lower aliphatic primary or secondary amine which is a solvent for the alloy but a nonsolvent for the contaminants therein;

(2) physically separating the resulting alloy solution from the undissolved contaminants; and

(3) evaporating the alloy solution to dryness to recover the alloy.

As an example, a series of xerographic photoreceptor drums containing a photoelectrically active layer of arsenic/selenium alloy is immersed in a bath of 40% methylamine in water. Dissolution of the alloy begins immediately and is complete in 1 to 2 hours depending on the degree of agitation, freshness of the stripping bath, temperature, etc. The stripped drums are removed from the bath, rinsed in turn with 5% methylamine in water and deionized water and dried. After drying, the drums are reusable in alloy vapor deposition processes.

The stripping solution, after a sufficient number of drums have been treated, is filtered to remove insoluble contaminants contained therein and then distilled to recover the arsenic/selenium alloy. The recovered alloy is recycled into production whereas the distillate of aqueous methylamine is recovered for reuse in stripping baths.

A process developed by *G.P. Baccaro and J.F. Seitz; U.S. Patent 4,097,267; June 27, 1978; assigned to Xerox Corporation* is one for the reclamation of arsenic containing selenium scrap which comprises separating the scrap from various contaminants therein by fractional distillation to provide selenium of high purity. The recovered selenium is realloyed by contacting it with a master arsenic/selenium alloy containing 10 to 20% arsenic which master alloy has been prepared by the vapor/liquid combination of arsenic and selenium.

Figure 139 shows a suitable form of apparatus for the conduct of such a process with the realloying **41** and shotting apparatus **46** shown in more detail in the lower view. As shown in the figure, arsenic/selenium scrap is fed from scrap bin **11** by gravimetric feeder **13** to induction premelter **15**.

In the premelter, the alloy is liquified by bringing it to a temperature of approximately 450°C. The liquid alloy leaves the premelter via vessel **17** and enters the distillation column **19** at a temperature of greater than 450°C. It should at least be a free flowing liquid and can be heated to above its boiling point and introduced as a gas.

The distillation column is designed to vaporize the liquid mixture of arsenic, selenium and impurities and to recover selenium in a nearly pure state. The column can consist of a series of sieve plates designed to achieve equilibrium by intimate contact of the falling liquid stream with the rising vapor stream. A certain degree of separation is thereby obtained at each plate or stage. The distillation can also be performed in packed columns. Whereas in plate columns each stage is an equilibrium stage, in packed columns the vapor and liquid streams never reach equilibrium and point-to-point relationships of the contacting phases are exceedingly important.

A column packed with Rashig rings is generally preferred because of ease of construction and greater range of operating throughputs. The column also incorporates a reboiler **21** and a condenser **23**. A high arsenic collector **22** is connected to the reboiler by line **24**. The purpose of the high arsenic collector is to remove excess arsenic so that substantial amounts of As_2Se_3 will not form.

In order to achieve the desired distillation, the concentration of arsenic in the reboiler and column is kept below 25 weight percent. Volatile constituents, consisting of selenium, oxides of arsenic (As_2O_3), chlorides ($AsCl_3/SeCl_x$), metallic chlorides, e.g., $FeCl_3$, and hydrocarbons move upward in the column, whereas heavier materials, consisting primarily of arsenic (as arsenic triselenide) and metallic impurities, move to the bottom of the column.

Figure 139: Xerox Corporation Process for Purification and Realloying of Arsenic and Selenium Alloys

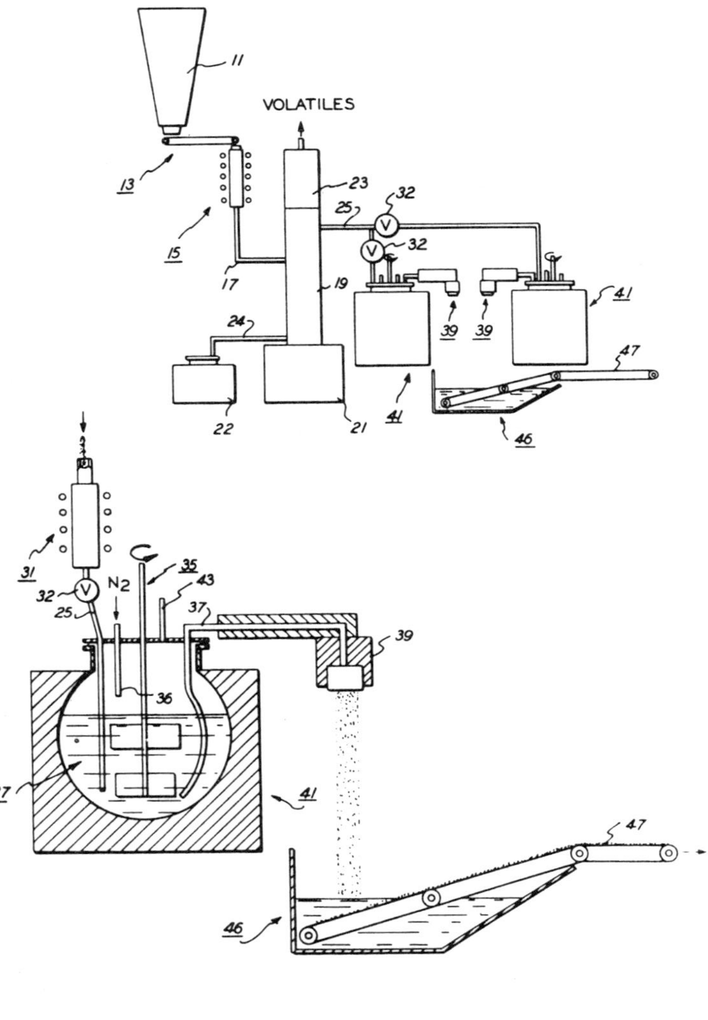

Source: U.S. Patent 4,097,267

Selenium is condensed in the condenser at approximately 680°C. At this temperature the other volatiles are not condensed and are withdrawn as overhead impurities. Pure selenium is partially withdrawn from the top of the column through line **25** with the remainder recycled into the column as reflux. Metallic impurities collect in the reboiler **21** and a high arsenic-selenium waste stream is withdrawn from the bottom plate of the reboiler **21**. Withdrawal can be either on a continuous or periodic basis. The distillation column is normally quartz lined although an alumina or mullite surface may be used.

The selenium condensate (99.999% purity) is collected in two pyrex reactors **41** maintained at 300°C. In operation, the reactors will alternately collect the condensate. Upon completion of one charge of purified selenium, the selenium is realloyed by introducing an arsenic/selenium master alloy (a high arsenic/selenium alloy) and alternatively, chlorine gas. The realloying apparatus is illustrated in the lower view in the figure.

The master alloy which contains from 10 to 20 weight percent arsenic is conveyed from a feed hopper (not shown) to the induction premelter **31**. The molten alloy flows to the pyrex reaction vessel **27** through line **25** with the rate of flow being controlled by valve **32**. The molten arsenic/selenium master alloy and the molten selenium from the distillation process are combined in the proportions necessary to form the final alloy desired with stirring being supplied by agitator **35**.

Alternatively, nitrogen gas is sparged into the molten arsenic/selenium through inlet **36** to provide agitation. The nitrogen gas is also sparged into the vessel to blanket the molten material and prevent oxidation. Chlorine is introduced as a gas through a gas dispersion tube **43**. This step is normally carried out at a temperature of about 300°C for a period of about 20 minutes depending on the final chlorine composition desired.

Upon formation of the desired alloy, it is pumped through line **37** by pressurizing the vessel to the shotter **39**. Heating mantles **41** are employed to maintain the alloy in the molten stage during this operation. The shotter is designed to disperse the molten alloy into a fine dispersion which is dropped into shotting tank **46** which is filled with distilled, deionized water. The droplets solidify into discrete particles in the shotting tank and are transported for further processing by conveyor **47**.

SILVER FROM COPPER REFINERY ANODE SLIMES

See also "Copper from Copper Refinery Anode Slimes."

Silver production involves processing of anode mud and slime from electrolytic copper refining. As described by A. Petrick, Jr., H.J. Bennett, K.E. Starch and R.C. Weisner (17), the silver production scheme starts from the point where the silver-bearing material is stripped of copper to the point of production of silver crystals or bricks as shown in Figure 140.

Where silver production is obviously justified economically by comparative revenue and incremental costs, as is usually the case, one may view the production of the other by-products, selenium, tellurium, gold, and platinum metals as by-products rather than as coproducts of the silver subsystem. This further highlights the economic dependence of one by-product upon another.

Figure 140: By-Product Silver Subsystem

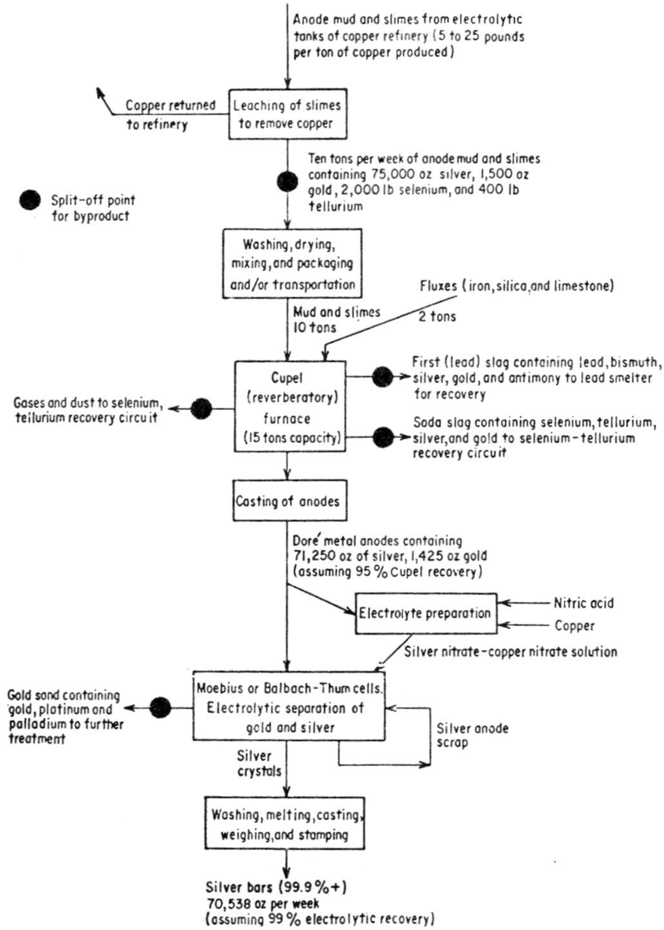

Source: Reference (17)

The hypothetical system shown in Figure 140 assumes as its raw material 10 tons of leached slimes per week from copper electrolytic refining. The assumed mud-slimes analysis is 7,500 ounces of silver and 150 ounces of gold per ton of slimes, 10% selenium, and 2% tellurium. The slimes are first washed, dried, mixed, packaged and transported from an electrolytic copper refinery to the facility for recovering precious metals. The metal content of the mud and slime are hypothetical based on review of available data on treated slime analyses from several electrolytic copper refineries.

When the mud-slimes are received at the precious metal refinery, they are loaded into a 15-ton Cupel furnace, and fluxes are added. The Cupel is a reverberatory-type furnace with wall construction of a steel shell, lined with basic refractories and having a roof of alumina brick. The Cupel furnace is fired with oil or gas. The operation of the Cupel is a batch process, consisting of 5 to 8 days smelting and refining in the same furnace. The length of time in the furnace depends upon the impurities to be removed. Selenium and tellurium recovered in gases, dust and slag can be regarded as separate subsystems involving economic decisions, essentially separate from the production of the silver.

The product of the Cupel furnace is cast into Doré (gold-silver alloy) anodes and moved to the Balbach-Thum electrolytic cells for separation of the silver and gold. Recovery of silver in the Cupel furnace is estimated at 95% of that in the anode-mud slimes. The Balbach-Thum cells look like washtubs made of fiber glass and plastic, 36 by 24 inches wide and 9 inches deep. The flat Doré metal anodes, about 20 inches square and 1 inch thick, weighing about 1,000 ounces each, are suspended 3 to 4 inches above and parallel to the bottom of the cell. The bottom of the cell is the cathode. The electrolyte is a mixture of copper nitrate and silver nitrate. There is no circulation or external heating of the electrolyte.

The anode basket is a filter cloth bag resting on glass rods that support the weight of the anodes. Silver crystals precipitate on the cathode and are scraped out every 3 to 4 hours. Silver crystals recovered from the cells are then washed, melted in an induction furnace and cast into silver bricks. The gold mud collecting in the filter bag during the silver electrolysis contains gold and platinum metal. This is removed and subjected to further treatment. Recovery in the Balbach-Thum electrolytic cells is estimated at 99%.

SILVER FROM ELECTROPLATING WASTES

A process developed by *R.A. Wallace; U.S. Patent 4,008,077; February 15, 1977* is one for recovering metallic silver from used silver-laden solutions, such as spent photographic fix, stop baths, stabilizers, electrolytic fix tailings, and waste silver-plating drag out and rinse solutions.

The silver-laden solutions are treated with mechanically worked narrow aluminum foil strips, threads or chaff, substantially completely covered with a few microns thick imperfect layer of aluminum oxide. The resulting silver-free solution may then be chelated with a suitable chelating agent to provide a rejuvenated solution, which may be repeatedly used and restored.

The method provides high silver recovery efficiency, and little maintenance, while providing the capability of repetitive use of various silver-containing solutions.

A process developed by *A.E.H.J. Cremers, A.P.A. Maes and J.L. Pleysier; U.S. Patent 4,051,026; September 27, 1977; assigned to Leuven Research & Development VZW, Belgium* is one in which silver is recovered from silver-containing solutions such as wastewaters from the photographic and electroplating industries by treating such solutions with a cation exchanger in the presence of thiourea or thiourea derivatives. The cation exchanger may be selected from tectosilicates, phyllosilicates and organic cation exchangers.

Experiments have shown that the presence of thiourea in silver-containing water has a significantly synergistic effect on the adsorption of silver by cationic exchange materials. This fact resides probably on the formation of a silver complex of thiourea, such complex being adsorbed easily then by the cation exchanger. In particular, it has been shown that the distribution coefficient of the silver-thiourea complex between the ion exchange material and the water is several orders of magnitude higher than the distribution coefficient of the silver ion alone.

Therefore, silver may be removed quite easily from aqueous solutions at a yield of 99% or more, independently from the initial silver content of the solution, i.e., independent from the question whether this silver content was several grams or several milligrams per liter.

SILVER FROM MINE WASTE DUMPS

See "Gold from Mine Waste Dumps."

SILVER FROM SCRAP PHOTOGRAPHIC FILM

See also "Terephthalic Acid from Photographic Film Waste."

A process developed by *W.J. Van Slyke; U.S. Patent 3,899,323; August 12, 1975; assigned to Battelle Development Corporation* is one in which silver-bearing manufactured organic products such as photographic film, photographic paper, printed circuit boards, etc., are loaded into a reduction zone of a retort and pyrolyzed by hot nonoxidizing gases produced by the burning of the carbon residue of previously pyrolyzed organic products in an adjacent oxidation zone in the same retort.

The carbon residue is oxidized by subjecting the residue to a heated air and water vapor mixture. The rate of pyrolysis is controlled so that the film entering the retort is contacted by nonoxidizing gases and vapors having a temperature less than 200°C. The off-gases of the pyrolysis are burned to heat the air and water vapor mixture.

A suitable form of apparatus for use in this process is shown in Figure 141. The retort **80** is substantially upright having a lined chamber **81** with a grate **82** located near the lower end thereof for supporting solid materials. The side walls **83** of the retort are substantially vertical to direct the film downward toward the grate **82** and the lower end.

A solid particle removal means **84** is located at the lower end of the retort for removing solid particles without admitting air. Such a device is frequently referred to as a rotary air-lock valve. A similar means **85** is located at the upper end of the retort for feeding the silver-bearing film into the chamber without admitting air. A conveyor **87** is shown for feeding the film to the retort on a continuous or intermittent basis.

Figure 141: Battelle Development Corporation Process for Silver Ash Recovery from Combustible Silver-Bearing Waste Products

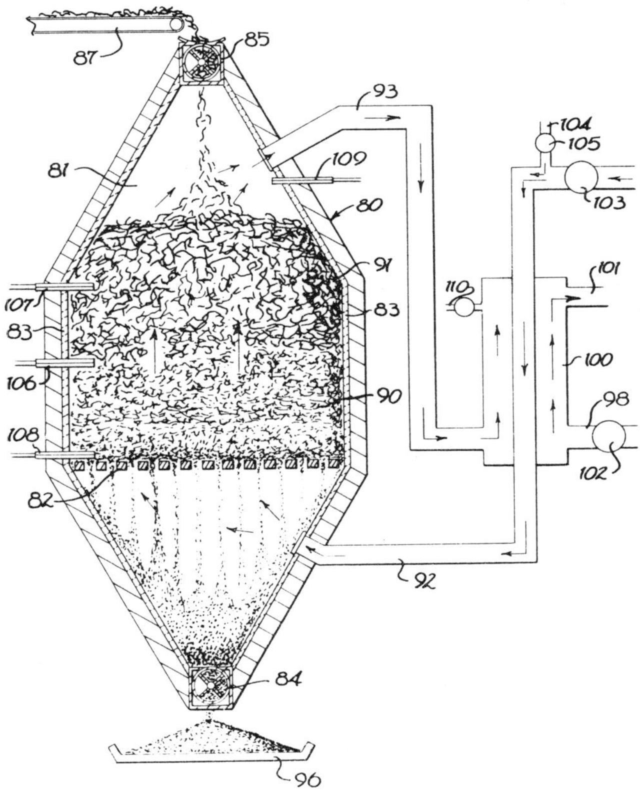

Source: U.S. Patent 3,899,323

The chamber **81** has two reaction zones above the grate–an oxidation zone **90** and a reduction zone **91**–with the reduction zone **91** being contiguous and immediately above the oxidation zone **90**. A gas input line **92** extends into the retort at the oxidation zone or below the grate. In the figure, line **92** is shown extending into the retort below the grate. An off-gas line **93** extends from the retort above the reduction zone for removing the product off-gases produced by pyrolytic decomposition of the film.

The burning is accomplished by directing a preheated air (containing free oxygen) and water vapor mixture through the line **92** and into the retort of a sufficient temperature and mixture ratio to burn the silver-bearing carbon residue. The resulting silver-bearing ash drops through the grate and is removed from the retort through the removal means **84**. A receiving container **96** is positioned below the retort to catch the ash. The ash is subsequently processed to separate the silver from the ash by a conventional flux or flotation process.

The pyrolytic off-gas mixture is utilized to preheat the air and water vapor (steam) to a temperature between 300° and 800°C before it reaches the retort. The preheating is accomplished by adding oxygen or air through line **98** to the pyrolytic or process off-gas mixture to burn the combustible gases present to provide a very hot gaseous product of nitrogen, carbon dioxide and water vapor of approximately 2500°F.

The hot gaseous product is passed through a heat exchanger or chamber **100** that transfers the heat to line **92** and to the air and water vapor that flows therethrough. The hot combustion gaseous product is then emitted to the atmosphere through line **101** with no accompanying soot or smoke. The amount of air combined with the pyrolytic off-gases is controlled by valve **102**. Excess air may be admitted to regulate the temperature of the gaseous product.

Temperature sensing devices are mounted in the retort to monitor the progress of the process. The sensors are operatively connected to valves **103** and **105** to adjust and control the flow rates and ratios of oxygen and water vapor. A sensing device **109** is mounted in the retort adjacent to the off-gas line **93** to sense and monitor the temperature of the pyrolytic off-gases as they leave the retort. The process is controlled and regulated to maintain the temperature sensed by device **109** less than 200°C.

Thermocouples **106, 107** and **108**, extend into the chamber **81** to measure the temperature of the processed material during its flow through the retort. Thermocouple **108** is located adjacent the grate to measure the bed temperature of the carbon residue. Thermocouple **106** is located in the interface between the oxidizing zone and the reduction zone. Thermocouple **107** extends into the reduction zone near the top thereof.

Between the measurements of the thermocouple **106** and **107**, a good indication can be obtained as to the rate of pyrolysis taking place in the reduction zone. It is desirable to have the temperature vary in the bed from 700°C at the bottom of zone **91** to less than 200°C near the top of the reduction zone to provide a smooth pyrolytic reaction without loss of silver in the off-gases.

Although there are several methods of initially starting the process, one of the preferable methods is to initially place a carbonaceous material in the oxidizing zone and then load the film on top in the reduction zone. A combustible fuel such as natural gas is injected into the chamber **100** through line **110** and ignited with air from line **98** to heat the air and water vapor mixture sufficiently to burn the carbonaceous material.

Hot gases from the carbonaceous material pyrolyze the film to reduce the film to a silver-bearing carbon residue and produce the desired off-gases. The valve **110** is then closed enabling the process to be self-sustaining with the addition of the air through line **98**. As the process proceeds, the grate **82** may be vibrated or opened and closed periodically or continuously to assist in the flow of the solids through the zones.

A process developed by *W.M. Fassel and D.W. Bridges; U.S. Patent 3,956,088; May 11, 1976; assigned to Barber-Colman Company* is one in which film is subjected to wet oxidation at high temperature and high pressure in the presence of a wet oxidation catalyst and a silver complexing compound. The liquor from the wet oxidation step is then subjected to processing to separate the solubilized silver which is freed from the film and to separate components derived from the film material in the liquor that remains. Such a process is shown schematically in Figure 142.

Figure 142: Barber-Colman Company Process for the Destruction of Developed and Undeveloped Photosensitive Film and the Recovery of Products Therefrom

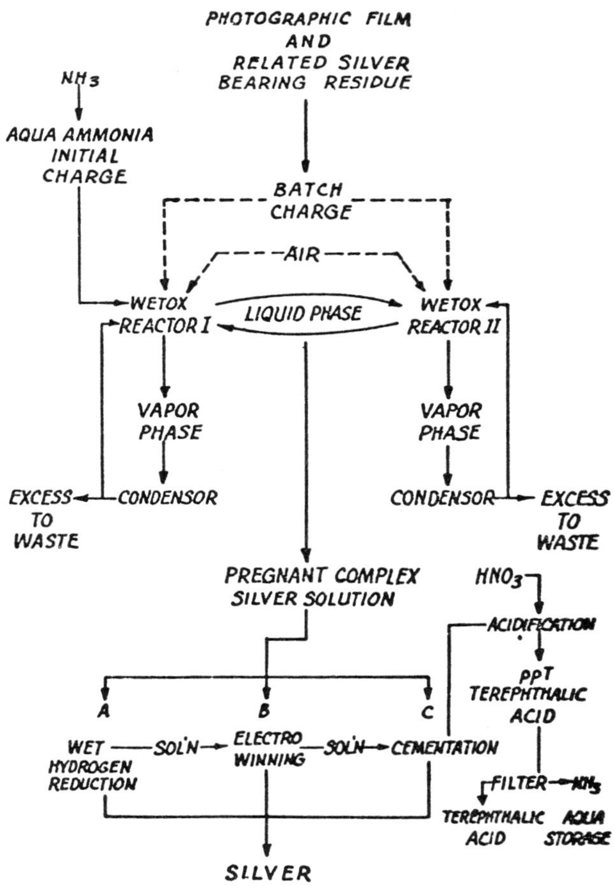

Source: U.S. Patent 3,956,088

The following is a specific example of the conduct of the process. 30.8 g of undeveloped photographic film, containing 1.14% silver concentrated in the silver emulsion, on a Mylar base, was chopped into small pieces and introduced along with 150 g of ammonium sulfate, 10^{-3} mol of copper sulfate and water in an amount to make up 1.5 liters, into a wet oxidation reactor.

The wet oxidation reactor, in the form of an autoclave, for high pressure operation, is provided with a high speed stirrer and inlets for the introduction of air at or near the vortex of the stirrer for immediate and uniform distribution of the air as fine particles throughout the liquor for maximum dissolution of oxygen therein.

The wet oxidation was carried out with the constant introduction of air, and constant stirring, with the liquor maintained at a temperature of 460°F and under 1,000 psig until dissolution of the material was substantially complete, or for about 1 to 2 hours. The COD of the effluent was calculated as 812 mg/ℓ oxygen.

The effluent was recycled to the wet oxidation reactor with 32.8 g of chopped film introduced into the reactor as a second charge. The reactor was operated in the same manner as before, except that a pressure of 970 psig was used. The effluent from the wet oxidation of the second batch had a COD of 685 mg/ℓ oxygen.

The liquor effluent from the second batch was recycled for a second time for wet oxidation of an additional 32.5 g of chopped film at 461°F and 1,035 psig. The effluent liquor, having a COD of 869 mg/ℓ oxygen was recycled for the third time for wet oxidation with an additional increment of 34.9 g film, at 460°F and 695 psig. The final liquid phase, identified in the flow diagram as pregnant complex silver solution, had a COD of 897 mg/ℓ oxygen. The final effluent contained 1.246 g/ℓ of silver.

The silver can be recovered from the pregnant complex silver solution or leach liquor by a number of ways, including:

(1) the addition of sulfuric acid to produce silver sulfate which precipitates and can be separated by conventional water-solid separation techniques;

(2) the addition of nitric acid to convert the silver amine complex to silver nitrate, followed by electrowinning as by the Moebius process;

(3) by hydrogen reduction of silver chloride precipitated from the silver amine solution by the introduction of chloride ion;

(4) by direct hydrogen reduction of the silver amine complex solution with gaseous hydrogen under high temperature and pressure; and

(5) by direct reduction of the silver from solution by such reducing compounds as glucose and formaldehyde.

In the example described, direct reduction with hydrogen was used by introduction of hydrogen under high pressure in the same reactor or autoclave used to carry out the high pressure, high temperature wet oxidation. Reduction with hydrogen was carried out at a temperature of 420° to 500°F under a pressure within the range of 700 to 1,000 psig. The freed silver precipitated and was recovered from the liquor by filtration, centrifugation, decantation, or the like liquid-solid separation techniques.

After removal of the silver, the remaining leach liquor contains the terephthalate derived from the Mylar film base, in the form of the amine or amide. The liquor is acidified, as with a mineral acid such as nitric or sulfuric acid to a pH within the range of 2 to 5 to form terephthalic acid which separates out as a precipitate. It is preferred to make use of nitric acid for acidification, since any trace quantities of silver will then be retained in solution for recycling with a new batch of pregnant complex silver solution.

A process developed by *W.L. Dusenberry; U.S. Patent 3,960,550; June 1, 1976* is one for reclaiming silver from photographic substrates and particularly a method wherein silver halide coated substrates, such as plastic film, are subjected to a bleach solution for removing the silver-bearing emulsion from the substrate. The emulsion is then placed in a settling tank and subjected to treatment with sodium hydroxide and a flocculating agent, such that the sodium hydroxide acts to precipitate the silver, while the flocculating agent tends to flocculate it.

The silver sludge is then removed from the lower strata of the settling tank and placed in canvas bags to allow liquids to drain from the sludge while the flocculating agent retains the sludge in the bags. The sludge then dries to a substantially mudlike consistency and then is placed in an oven and heated, until the moisture thereof is released, thereby rendering the mudlike sludge dry and in a substantially hard cakelike form.

The hard cakelike silver material is then pulverized and placed in a crucible with a flux on top, the flux consisting of sodium carbonate and boric acid granules. The crucible is then subjected to heat sufficient to melt the silver out of the pulverized cake. Then the molten silver is poured in a first crucible and thereby forms a silver casting which is a primary casting.

These primary castings may be then positioned in a second crucible with charcoal on the upper surface thereof, whereupon the primary castings are remelted, and whereupon the charcoal on top of the molten silver absorbs impurities, and then the molten silver is poured into a second casting of high purity.

A process developed by *A.H. Gerber and E. Wainer; U.S. Patent 4,078,916; March 14, 1978; assigned to Horizons Research Incorporated* is one in which scrap photographic film comprising at least a four component system involving a polyalkyleneterephthalate film base, an adhesive or subbing layer utilizing a terpolymer consisting chiefly of polyvinylidene chloride and a gelatin layer containing metallic silver or a compound thereof and a variety of contaminants such as causal dirt, dyes and the like is broken down for recovery of the valuable constituents therein including the silver, the polyvinylidene chloride terpolymer, and the polyesters.

The process involves treating the chopped photographic film at an elevated temperature with a solvent for both the polyester and polyvinylidene chloride values in which solvent both the gelatin and silver values are insoluble, under conditions chosen to insure the quantitative recovery of the silver values.

A process developed by *N.G. Branibar; U.S. Patent 4,093,532; June 6, 1978; assigned to Anken Industries* is one in which photographic film is transported through a plurality of compartments within which fluids act upon the constituents in the film coating. One of the compartments contains fluid that includes a silver complex dissolved therein, and a silver precipitating unit is positioned within that compartment. The precipitating unit includes anode-cathode elements connected to a source of electrical potential for depositing silver upon the elements.

A suitable form of apparatus for the conduct of such a process is shown in Figure 143.

Figure 143: Anken Industries Apparatus for Recovery of Silver from Photographic Film

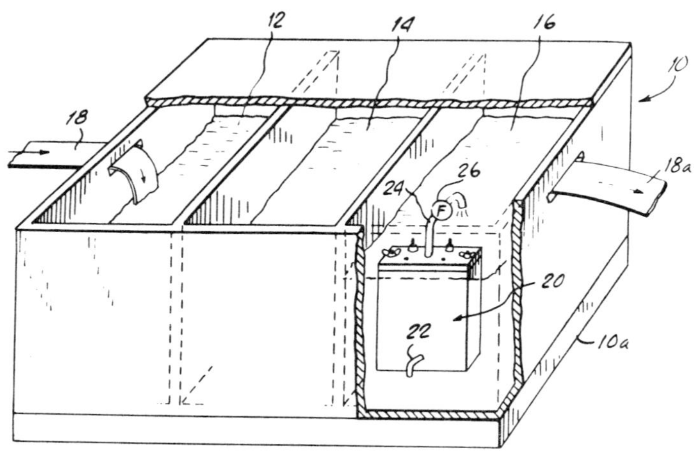

Source: U.S. Patent 4,093,532

The silver reclaiming unit **10** may be a conventional film processing unit of the type used for developing film in automatic typesetting operations. The processor includes three compartments **12, 14** and **16** through which a film **18** is transported for processing. As shown, the film enters the processor from the left and exits the processor from the right. The movement through the compartments is automatic.

Each of the compartments contains fluid which acts upon the constituents of the film coating. For automatic typesetting operations, the fluids provide for the development of light-exposed film. In the case of reclaiming silver from spent or already developed photographic film, the compartments would contain fluids for first converting silver in the film to silver halide form and then dissolving that silver halide so that it may be electrodeposited out of solution.

Specifically, the solution in the first compartment **12** may be a solution of ferric chloride and water. The solution may include some hydrochloric acid, or it may be one of ferric chloride and hydrochloric acid. The solution in this compartment converts the silver in the spent film to silver chloride form. The solution in the compartment **14** may comprise a sodium sulfite solution to neutralize the ferric chloride. The final solution in the compartment **16** may be a typical fixing solution, e.g., a thiosulfate solution which converts the silver halide to a soluble silver complex, namely, silver thiosulfate.

Positioned within the final compartment **16** is a silver precipitating unit **20**. The silver precipitating unit receives solution from the compartment **16** via an inlet **22**. This solution is pumped into the unit, and exits from the unit via an outlet **24** which may include a filter **26**. Within the silver precipitating unit, silver is electrodeposited. The photographic film emerges from the processing unit **10** as at **18a** with all of the silver removed and precipitated therefrom within the unit **20**.

A process developed by *D.L. Higgins; U.S. Patent 4,127,465; November 28, 1978; assigned to Ag-Met, Inc.* is one in which silver is recovered from used film plates by passing them into a solution into which their silver material is dissolved and by electroplating the dissolved silver out of the solution. Electrical circuits are provided to increase the magnitude of a continuous plating current in proportion to the rate of film and silver input and to decrease the current gradually to coincide with the decrease in silver concentration as plating continues.

A process developed by *P.A. Phillips; U.S. Patent 4,150,977; April 24, 1979; assigned to Keltek Processing, Inc.* is one for treating photographic film to recover silver and the plastic base layer for reuse. A wash tank containing warm water at about neutral pH contains at least one enzyme from the following group: protease, amylase or lipase.

The water in the wash tank is agitated and the enzymes attack the adhesive and organic layers on the plastic base, in order to loosen and remove these materials. The film is then placed in a rinse tank containing either a weak aqueous glycol solution or a weak caustic solution which is agitated. Finally the film is removed from the rinse tank and dried. Liquid is removed from the wash tank and treated to remove silver therefrom.

Such a process is shown in Figure 144. As shown, the numeral **10** designates a wash tank which may have a capacity of about 1,000 gallons, and which contains hot water at a temperature between 30° and 60°C, and an enzyme or combination of enzymes. A pump **12** is responsible for continuously recirculating and agitating the water in the wash tank.

The used or unused photographic film is placed in a basket (not shown) and lowered into the wash tank **10** where it remains for a matter of minutes. The length of time will depend on how much film is placed in the tank at any given time, on the amount of enzyme present, and on the degree of agitation of the water.

During the dwell time within the wash tank **10**, the enzymes in the hot water contained in the wash tank attack the protein, gelatin and adhesive on the plastic base of the film, and loosen these materials in such a manner that the agitated water within the tank can dislodge them. Naturally, a method of heating the water in the tank **10** and controlling the range of its temperature would also be provided, although this has not been specifically illustrated.

Figure 144: Keltek Processing, Inc. Process for Stripping Photographic Film for Silver and Film Base Recovery

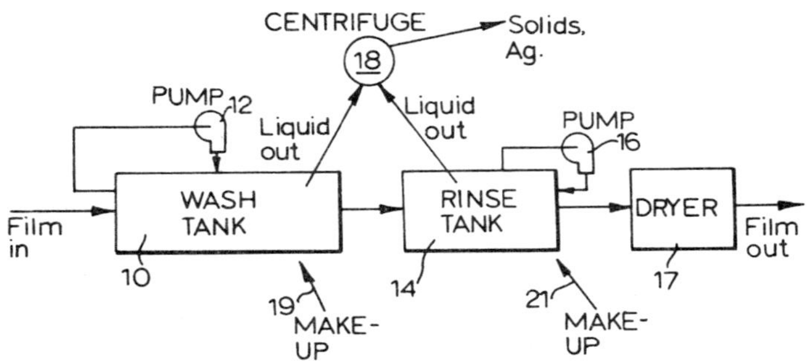

Source: U.S. Patent 4,150,977

After the protein, gelatin and adhesive have been loosened from the plastic base of the film, the basket in which the film resides is removed and is immersed in a rinse tank **14** which may contain either a weak aqueous glycol solution (about 10% glycol) or a very weak caustic solution (about 1% caustic). Each of these materials is particularly suited to removing any remaining adhesive.

A further pump **16** recirculates the water in the rinse tank in order to allow agitation to remove any bits of the protein, gelatin or adhesive which may have remained on the film base during transport from the wash tank to the rinse tank. After the last of these materials has been removed through agitation from the film base, the film is removed and passed through an air dryer **17**, from which it emerges in a condition ready to be recycled.

It will be appreciated that the liquid contained in the wash tank, in the absence of any other provision, would gradually accumulate more and more of the removed protein, gelatin and adhesive from the film which is being processed. Also, the enzymes would tend to become used up after a period of time. For this reason, provision is made to remove the liquid periodically or at a slow continuous rate to a centrifuge **18** which is able to separate from the liquid the solids and the solid silver in particular which are contained in suspension in the liquid. A make-up provision **19** is also provided in order to maintain the quantity of water and the concentration of enzymes in the wash tank.

The rinse tank also will gradually accumulate amounts of protein, gelatin and adhesive which are removed from the film, although at a slower rate than takes place in the wash tank. For this reason, provision is made for periodically removing the liquid from the rinse tank and passing it through the centrifuge.

The same centrifuge can be used for both tanks. Again, any loss of liquid would have to be replenished from time to time, through a make-up provision **21**.

It will be understood that the outlets from the two pumps **12** and **16** will be located in the respective tanks in such a manner as to promote most efficiently the removal of the loosened materials from the film base.

In regard to the enzymes used in the wash tank **10**, it has been found that the family of enzymes known as protease is able to carry out the necessary function quite satisfactorily, so long as the pH is maintained within the limits of 5 to 9, and preferably with the limits of 6.5 to 7.5. It has been found that the closer the solution can be maintained to exactly neutral, the better are the results obtained. For this reason, a particularly preferred range for the pH is 6.8 to 7.2.

In addition to the protease family of enzymes, it is considered that the amylase and lipase families would also function satisfactorily. It is considered that these three enzyme families could be employed separately or together in any combination.

SILVER FROM SCRAP SILVER

Indicative of the importance of scrap to the silver industry was the fact that the quantity of silver recovered from old scrap was 25% greater than domestic mine production in 1977. The increased secondary recovery in recent years was due primarily to higher market prices, which resulted in improved collection methods and recovery techniques.

A major quantity of domestic refinery silver production is recovered from old scrap; from 1973-77, old scrap averaged about 44% of total refinery production. This recovery was from used products, such as photographic materials, plating solutions, silverware, jewelry, batteries, coins and electronic scrap. Secondary silver recovered from both old and new scrap by refineries totaled 99.6 million ounces in 1977 with 47.9 million ounces attributed to old scrap.

SILVER FROM SILVER-BRAZED STAINLESS STEEL SCRAP

It is necessary in supersonic military aircraft that certain parts including flaps, ailerons and engine supporting struts are made of stainless steel rather than aluminum and the other metals and alloys used in the rest of the aircraft because of the high temperatures and stresses to which they are subjected. The stainless steel sections comprise contoured outer sheets with inner support fabricated as honeycomb sections. The honeycomb sections are brazed to the contoured outer sheets with a silver-based brazing compound.

When these military aircraft are declared obsolete and discarded for scrap, the silver-brazed stainless steel sections are separated from the rest of the aircraft for separate recovery of the metals therein or other scrapping procedures. These silver-brazed stainless steel sections contain approximately 5 to 30% by weight silver and 2 to 15% by weight copper with the balance stainless steel.

Previous methods used for electrolytic recovery of silver have not been able to be used to recover silver from metals containing low percentages of silver. For example, the Moebius cells needs a silver concentration in the anodes of greater than 95% and the Thum or Balbach systems can accept any refinable metal whose silver content exceeds 75%.

The Thum and Balbach systems further suffer from the disadvantage that the anode baskets need to be lined with fabric. This fabric traps the anode mud but also increases the cell voltage so that part of the stainless steel would dissolve as well as the silver.

Since some obsolete supersonic military aircraft contain about 300 pounds of silver, an improved and economical process is needed for its recovery. Subsonic military aircraft previously used do not generally contain sections of similar metal content. An exception is the flame cone assembly found on some jet engines.

Therefore, a silver recovery process has been developed by *B.W. Dunning, Jr. and D.H. Chambers; U.S. Patent 4,090,935; May 23, 1978; assigned to U.S. Secretary of the Interior* which is an electrolytic refining process for recovering silver, copper and stainless steel from stainless steel honeycomb sections from military aircraft brazed with a silver-based brazing compound, the brazed sections comprising 5 to 30% by weight silver and 2 to 15% by weight copper, the balance comprising stainless steel. Such a process is shown in Figure 145.

The following is one specific example of the conduct of the process. Silver-brazed, stainless steel aileron sections from a supersonic military aircraft were sheared to make pieces of 12" maximum dimension, and then shredded to make pieces of 1.5" maximum dimension. The pieces were screened on a 30 mesh per inch screen, and the anode basket from the electrolytic refining cell was filled with pieces from those left on the screen.

The anode basket was made from 30 mesh per inch stainless steel and was of cylindrical configuration. The electrolytic refining cell used contained this anode basket, which was encircled by a cylindrical sheet stainless steel cathode spaced 1" away from the wall of the anode basket. The electrolytic refining cell contained as electrolyte a solution of copper nitrate of concentration 20 g/ℓ of copper and silver nitrate of concentration 30 g/ℓ of silver.

The refining was carried out by passing a direct current of 30 A/ft^2 of cathode and 0.1 to 5 V through the cell. The anode basket was rotated at 0.5 to 5 rpm to circulate the electrolyte. A plastic coated wooden scraper was attached to the anode basket and as the anode basket rotated, the scraper scraped silver crystals from the cathode, which dropped into a basket lying at the bottom of the refining cell. These silver crystals were washed and dried and magnetically separated from any stainless steel fines which may have dropped into the basket. These silver crystals of 95 to 99% purity were further purified by conventional methods to a 99.9% purity. The stainless steel in the anode basket was washed and dried.

Copper scraps were added to the spent electrolyte from the electrolytic refining cell to precipitate the silver, leaving a copper nitrate solution. This copper nitrate solution was used as the electrolyte in a refining cell and copper was deposited at the cathode.

The anode mud, which collected on the bottom of the refining cell beneath the basket in which the silver was collected, was purified to recover palladium, silver and stainless steel fines.

Figure 145: U.S. Dept. of the Interior Process for Recovering Silver, Copper and Stainless Steel from Silver-Brazed Stainless Steel Sections

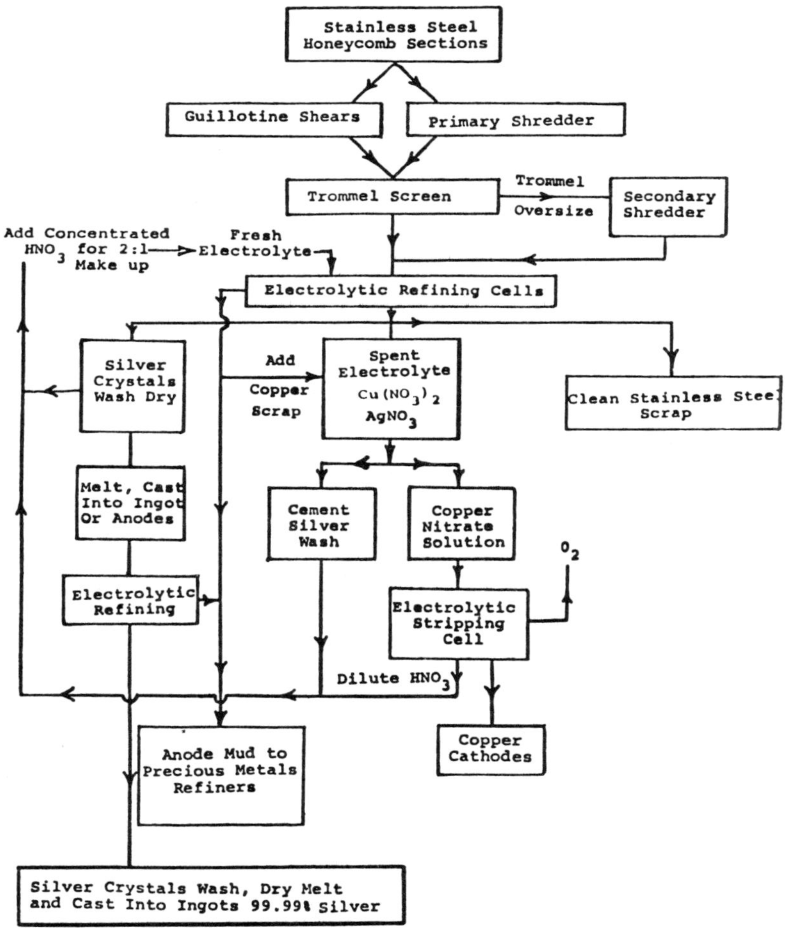

Source: U.S. Patent 4,090,935

SILVER FROM SPENT PHOTOPROCESSING SOLUTIONS

By way of background, in the fixing of photographic film, sodium thiosulfate or analogous agents, such as potassium thiosulfate and ammonium thiosulfate, or combinations of the foregoing are used to remove the unreduced silver from the film, and this silver concentrates in the fixing bath and is also carried over to the water which is subsequently used to wash the film.

The recovery of silver from used photographic solutions is very desirable for a number of reasons. First of all, the recovered silver is not only economically profitable, but it is also essential to conserving a natural resource wherein only about 65% of the silver which is used annually in industrial products is obtained from mining while the remainder must be obtained from other sources including the reclamation from industrial wastes.

The photographic industry is estimated to consume approximately one-third of the total silver used by industry and, therefore, silver recovery from the wastes of the photographic industry comprises a significant means for conserving silver. In this respect, it is estimated that the photofinishing, x-ray and graphic arts industries in the United States alone utilize about 2,000 tons of silver per year, and on the assumption that half of this amount remains on the photographic film or paper, then about 1,000 tons or 2,000,000 pounds of silver are dissolved by the processing solutions, and are subject to reclamation.

A second extremely significant reason for recovering silver from photographic wastes is because silver is classified as a water pollutant and its salts have an extremely deleterious physiological effect. In this respect, *Lange's Handbook of Chemistry* specifies that silver is as toxic as chromium or arsenic or lead and the concentration of any of these elements in excess of 0.05 part per million results in causing water to be unsuitable for drinking purposes.

Various processes were considered for use in reclaiming the silver from spent photographic fixing solutions. Certain of these processes were not practical from a commercial viewpoint. Still other processes were only able to recover up to 90% of the silver from the spent photographic fixing solution, thereby permitting an estimated 100 tons, or 200,000 pounds of silver to be dumped in the United States with its attendant economic loss and polluting effect, considering that 0.05 part per million causes water to be unsuitable for drinking purposes. Other processes were economically undesirable because of the initial cost of the equipment or the cost of maintenance, or the cost of operation, or because of the fact that their productivity was not sufficiently high to warrant commercial usage.

In addition, these methods for commercial recovery of silver do not adequately break down the sulfur compounds which are present in the fixer solution, these compounds being objectionable from a pollution viewpoint because they create an oxygen demand by their slow oxidation and under certain circumstances also react to form foul smelling and dangerous compounds. Therefore, considering that at least 2,000,000 pounds of silver are present in fixer solutions, as discussed above, there must also be an equivalent amount of sulfur compounds which are not treated but are disposed of to produce the undesirable oxygen demand noted above.

An apparatus developed by *J.S. Bentley; U.S. Patent 3,901,777; August 26, 1975; assigned to Photographic Silver Recovery Limited, England* is an intermittently-operating electrolytic cell for the recovery of silver from used photographic bleach-fix solution.

The cathode assembly and the solution are automatically separated when the plating current is switched off. A buffer tank upstream of the cell, from which solution is delivered to the cell by a metering pump during plating, has a bottom level switch that switches off the cell and feed and switches on an emptying pump for pumping the cell solution back to the buffer tank. The emptying pump is stopped by a cell bottom level switch; and a filling pump is started by a top level switch of the buffer tank, which also starts the plating cell and the metered feed, the filling pump being stopped by a top level switch of the cell. The cathode assembly is self-draining.

A process developed by *K.G. Burgess; U.S. Patent 3,926,768; December 16, 1975; assigned to Hydrospace Industries, Inc.* is one in which silver is recovered from spent photographic solutions by feeding the spent solutions into a precollecting vessel where the solution collects until a predetermined volume has been received, at which time it is automatically dispensed, by a self-triggering syphon, into an electrolysis chamber, and direct current is automatically turned on for a predetermined period of time between an anode and a cathode in the chamber to plate out the silver onto the cathode. A magnetic agitator is automatically energized to keep the silver suspended.

A similar process has been described by *K.G. Burgess; U.S. Patent 3,959,110; May 25, 1976; assigned to Hydrospace Industries, Inc.* Silver is electrolytically recovered from photographic solutions, such as spent hypo, by feeding such a solution to a precollecting or measuring vessel where the solution collects until a predetermined volume has been received, at which time it is automatically dispensed, by a self-triggering siphon, into an electrolysis chamber of larger volume than the precollecting vessel and direct current is automatically turned on for a predetermined flow between an anode and a cathode in the chamber to plate out the silver on the cathode.

An agitator, preferably a magnetic agitator, is also automatically energized to keep the solution homogeneously mixed during electrolysis, for maximum silver recovery efficiency.

A process developed by *M. Korosi; U.S. Patent 3,982,932; September 28, 1976; assigned to Eastman Kodak Company* is one in which silver is recovered from photographic waste liquids containing emulsified silver, silver compounds and gelatin by reacting therewith a proteolytic enzyme while the reactants are maintained in an alkaline condition. Then the reactants are acidified to a pH of 4.2 or less by introducing an acid such as HCl or H_2SO_4.

Precipitation of gelatin-bound silver and silver compounds occurs upon acidification, and the precipitate is allowed to settle out in a settling tank. Supernatant liquid is removed from the top, neutralized and sent to sewer. Settled sludge is removed from the bottom and silver recovered, as by incineration. The operation can be conducted either batchwise or continuously.

Figure 146 shows a batch version of this process. The waste washing water from photographic emulsion manufacture, or photographic gelatin emulsion removed from scrap film, is treated with proteolytic enzyme material in a reactor tank **11** containing 2,000 gallons of the liquid. Steam is injected into the liquid until the temperature reaches 50°C and the pH is then adjusted to about 8 by introducing an aqueous alkali such as NaOH or KOH. Approximately 5 ppm by weight of a proteolytic enzyme is added, and the mixture is allowed to digest for approximately one-half hour at 50°C. This treatment is of specific value when using an enzyme active under alkaline conditions such as Bioprase. Many proteolytic enzymes may be used. In each case the pH of the liquid is adjusted to the optimum range for the enzyme used.

Figure 146: Eastman Kodak Company Process for Recovery of Silver from Gelatinous Photographic Wastes

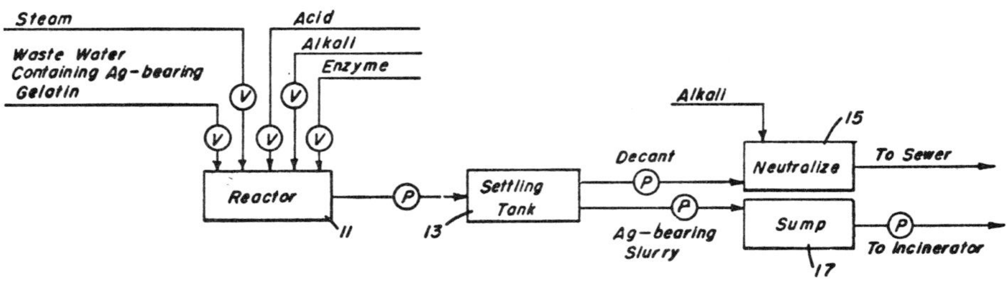

Source: U.S. Patent 3,982,932

With the reaction mixture containing Bioprase still being in the reactor tank **11**, the pH is lowered to a value below 4.2, such as between 4.2 and 2.5, preferably 3.5, by acid addition such as 98% undiluted sulfuric acid of which ½ to 1 liter may be added per 2,000 gallons in the reactor.

During the acid addition the contents of the reactor are continuously stirred by a suitable pump. A fine precipitation is observed within 1 minute, and coalescing gradually increases the size of the flakes. The time allowed for this operation is about 10 minutes, after which the contents of the tank are pumped into a 10,000 gallon settling tank **13**. The complete cycle for the reactor treatment occupies about 1½ hours.

Several reactor tanks and settling tanks can be used alternately and successively. The sedimentation achieved by this process is found to be quite rapid. After 4 hours of settling the residual silver content in the supernatant liquid seldom exceeds 4 ppm, and after overnight settling it is generally less than 2 ppm and is often small enough to escape detection.

The supernatant liquid is therefore ready for decantation and discharge into the sewer, although before doing so its pH may have to be raised to about 7 by introducing aqueous alkali such as NaOH or KOH in a drip-feed neutralizer **15** to comply with sewerage regulations. In a typical case the discharge to sewer contains about 0.2% gelatin, 0.1% salt and less than 1 ppm of silver.

The wet precipitate slurry is collected in a sump **17** from the bottom of the settling tank **13** (and also from reactor **11** when present therein) after a sufficient quantity has accumulated. The wet precipitate is then injected into an incinerator as one way of recovering its silver content. The solids content of the precipitate before incineration consists of approximately 33% by weight of silver in metal or halide form, and about 60% of gelatin.

Emulsion waste which has been removed from scrap film may be mixed into waste washing waters in the reactor tank **11** in any desired proportions, e.g., 100 gallons emulsion mixed into the 2,000 gallons in the reactor.

The initial pH adjustment in the reactor **11** may be effected by a 46% aqueous sodium hydroxide solution. The sewer neutralizer added at **15** may be 35% aqueous sodium hydroxide solution diluted with 3 parts H_2O per 1 part of solution. Approximately 1 to 2 liters are used for each 2,000 gallon batch in the reactor **11**, and about 3 liters per 10,000 gallons of discharge into the sewer.

Regarding the enzyme, about 5 to 10 ppm may be used for waste washing waters, and 10 to 50 ppm when 1 part of emulsion has been mixed in 20 parts of waste washing waters. Preferably the temperature is 48° to 52°C and the pH 7.0 to 10.6 when the enzyme is added. The volume of precipitate sludge may be $1/200$ of the volume of the original waste liquor.

The following is one specific example of the conduct of this process. Reactor tank **11** is filled with 2,000 gallons of waste wash waters from photographic emulsion manufacture containing 130 ppm of silver and steam is injected until the temperature is 50°C. Then 46% sodium hydroxide is added until the pH is raised to 8, at which point 5 ppm of Bioprase PN-10 is introduced in aqueous solution and digestion proceeds for 15 minutes.

Sulfuric acid (98%) is then added in the tank until pH is reduced to 3.5, the isoelectric point, and a fine precipitation begins within 1 minute. After 10 minutes the tank contents are pumped into a settling or sedimentation tank **13**. After settling for 24 hours the silver content of the supernatant liquid is less than 1 ppm. The supernatant liquid is pumped to tank **15** where it is neutralized with aqueous NaOH to pH 7 and then discarded.

The slurry in the bottom of tank **13** contains, on a dry weight basis, 23% silver (in metal and halide form) and 60% gelatin, and is pumped to sump **17**, and then injected into an incinerator where the water is vaporized and silver-bearing ash is formed.

A process developed by *K. Moeglich; U.S. Patent 4,021,319; May 3, 1977; assigned to Silrec Systems, Inc.* is an improved process which is capable of removing 100% of the silver from solutions associated with photographic fixing in a relatively simple manner and which also removes an amount of elemental sulfur equivalent to the amount of silver which has been removed, thereby not only reclaiming an amount of silver previously unreclaimable by the use of certain prior art processes with the attendant economic saving and lessening of pollution, but also removing elemental sulfur from the wastewater which would have in its compound form an undesirable oxygen demand if it were to remain in the disposed solution.

This improved process involves the steps of placing the photographic fixer solution in a cell having an anode chamber and a cathode chamber, with the chambers being separated by a diaphragm having a pore size from 1.0 to 500 microns and applying a direct current across the anode and cathode to create a current density of between 0.001 and 5 A/in^2 at the cathode to thereby cause argentic oxide to precipitate at the cathode and an equivalent amount of elemental sulfur to precipitate at the anode. The cell is preferably maintained at a temperature of from 20° to 60°C. The cell can be operated as a flow cell by maintaining a hydraulic head between the chambers to determine the flow across the diaphragm. When the cell is operated as a flow cell, a minimum specific amount of photographic fixer must be conveyed to the cell to prevent oxidation of sulfur and escape of SO_2 gas.

Figure 147 shows a suitable form of apparatus for conducting this process. It consists of a cell or tank **10** having a cathode compartment **11** separated from an anode compartment **12** by a permeable diaphragm **13**. Tank **10** is preferably fabricated from an inert insulator material, such as glass, plastic or ceramic material. A storage tank **14** contains used photographic fixing solution **16** and supplies it to cathode chamber **11** when valve **17** is opened. The used photographic fixing solution contains sodium ions, ammonium ions, thiosulfate ions, silver ions and bromine ions.

Figure 147: Silrec Systems, Inc. Electrolytic Process for Recovery of Silver from Photographic Fixer Solution

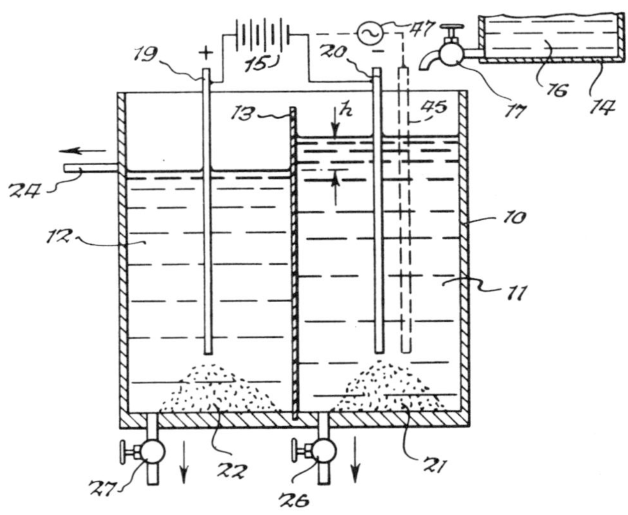

Source: U.S. Patent 4,021,319

Summarizing in advance, after the tank **10** has been filled with spent fixer solutions, such as sodium, ammonium, or potassium thiosulfate, or combinations thereof, argentic oxide **21** will precipitate in cathode compartment **11** and an equivalent amount of elemental sulfur **22** will precipitate in anode compartment **12** as a result of applying a direct current across anode **19** and cathode **20**.

With respect to the foregoing, when operating as a batch process, it is preferable to place the fixer in the cathode compartment and place water in the anode compartment. However, the batch process will also operate if fixer is placed in both compartments, but unwarranted by-products, such as SO_2 or Ag_2S, may be formed in the anode compartment.

After the desired amount of silver in the form of argentic oxide has been removed from solution, which is preferably 100%, valve **26** will be opened to drain the cathode compartment and valve **27** will be opened to drain the anode compartment. The precipitates, namely S and AgO, will be separated from solution by filtration or the like before the solution is disposed of. Thereafter, valve **17** will be opened to add a new supply of fixer solution from tank **14** to cell **10**. Thus, silver can be removed by a batch treatment process.

The foregoing description applies to a batch process with the understanding that no head **h** exists between the compartments and there is no outflow from conduit **24**. When operating as a batch process, there is the possibility that SO_2 gas may be given off from the solution if the solution is not sufficiently dilute, or other unwanted by-products may be formed.

Cell **10** is preferably operated as a continuous flow cell wherein the flow is at a rate which permits removal of 100% of the silver in the used or spent fixer as argentic oxide. When operating as a continuous flow cell, valve **17** will remain at a desired open setting to provide the required rate of flow to the inlet compartment, whether it be the anode or the cathode compartment, and the treated solution will be discharged from outlet spout such as **24** in the other compartment when there is a sufficient head on diaphragm **13** to cause the desired rate of liquid flow through the diaphragm.

While the figure shows the diaphragm cell with the inflow at the cathode compartment, which is preferable, it will be appreciated that it can be operated with the inflow at the anode compartment and the discharge at the cathode compartment, in which event the various conduits would be reversed. It is also to be noted that there is a head **h** in between the center compartment and the outer compartments.

The operation of the cell shown may be improved by inserting an additional electrode **45** in the cathode compartment and applying alternating current across electrodes **20** and **45** as shown. The foregoing results in increased removal of silver. As can be seen, the battery **15** provides the direct current to electrodes **19** and **20** and an alternating power source **47** applies alternating current to electrodes **45** and **20**.

The ac and dc power sources are electrically isolated from each other. The alternating current may have a frequency of between 1 and 800 Hz, preferably between 10 and 120 Hz, and most preferably about 60 Hz for practical reasons. Furthermore, the alternating current density may be between 0.1 and 20 A/in^2, preferably between 1 and 10 A/in^2, and most preferably between 3 and 7.5 A/in^2.

A somewhat similar cell and process are described by *K. Moeglich; U.S. Patent 4,021,320; May 3, 1977; assigned to Silrec Systems, Inc.* It involves electrochemically removing silver in the form of AgO from a photographic fixer solution containing silver ions without destroying the utility of the solution comprising the steps of applying alternating current having a frequency of between 0.5 and 800 Hz and a current density of between 0.1 and 20 A/in^2 across electrodes immersed in the solution to cause AgO to precipitate at the electrodes. The method can also be practiced above the range of about 20 A/in^2 at the cost of destroying the utility of the solution. The method can also be used to remove silver in the form of AgO from other solutions.

A process developed by *Y. Idota and H. Shibaoka; U.S. Patent 4,111,766; September 5, 1978; assigned to Fuji Photo Film Co., Ltd., Japan* is one for the recovery of silver from a fixing solution used in processing silver halide photographic light-sensitive materials wherein an electrolytic cell having a cathodic cell and an anodic cell separated by a diaphragm is used. Silver is recovered from the used fixing solution in the cathodic cell and the anodic cell is filled with a fixing solution which has been used or not used in photographic processing, or an electrolyzed fixing solution.

Figure 148 shows a suitable form of apparatus for conducting such a process. The electrolytic cell **1** comprises a cathodic cell **2**, an anodic cell **3**, cathode **4**, anode **5** and a diaphragm **6** separating the cathodic cell **2** and the anodic cell **3**.

Figure 148: Fuji Photo Film Co. Process for the Recovery of Silver from Fixing Solutions

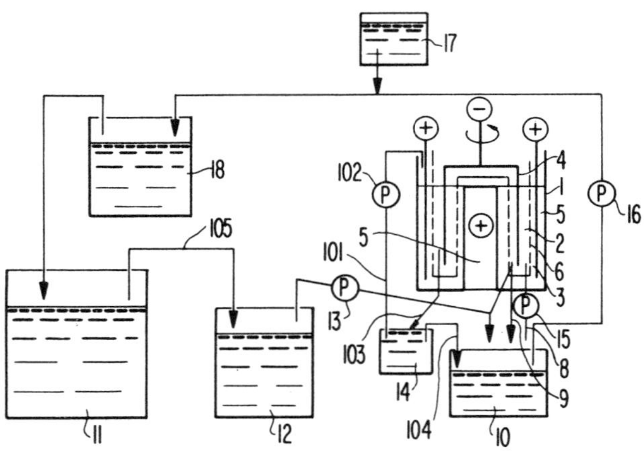

Source: U.S. Patent 4,111,766

The used fixing solution is pumped from overflow solution tank **12** into circulation tank **10** via pumps **13** (it will be noted that the line existing from pump **13** is later split into two lines; one line supplies solution to cell **2** and the other line supplies solution to tank **10**) and then into cathodic cell **2**. Used fixing solution is, of course, overflown from tank **11** to tank **12** via line **105**. Overflow solution tank **12** receives overflowing solution from the fixing tank **11** which is not per se a part of the developing apparatus.

Used fixing solution in circulation tank **10** is pumped into cathodic cell **2** via line **8** and pump **15**, and used fixing solution is returned from cathodic cell to circulation tank **10** via line **9**. Used fixing solution thus circulates between circulation tank **10** and cathodic cell **2**. The anodic cell **3** is previously filled with the fixing solution. When a definite amount of the used fixing solution has been introduced, a circulation pump **15** is activated and the preparation prior to electrolysis is completed.

As indicated previously, the purpose of circulation pump **15** is to circulate used fixing solution between cathodic cell **2** and circulation tank **10**. When the volume of cathodic cell **2** is smaller than the volume of the used fixing solution to be electrolyzed, pump **15** is necessary to control the amount of the solution being charged from circulation tank **10** to cathodic cell **2**. On the other hand, when the volume of cathodic cell **2** is greater than the volume of the fixing solution to be electrolyzed, pump **15** can essentially be omitted from the system.

Thereafter, when an electric current is passed through the system, desilverization and liquid-regeneration proceeds in the cathodic cell **2**. After the electrolysis is carried out for a definite period of time (preferably, the time for the amount of residual silver to reach 0.5 g/ℓ), the anodic solution is discharged from the system, essentially as waste. The cathodic solution is returned via line **103** to a constant quantity cup **14** and thence to the circulation tank **10** via line **104**.

The volume of the constant quantity cup **14** is determined by the amount to be discharged out of the system, the volume of the anodic cell, the concentration of the sulfurous acid ion in the fixing solution, the time of electrolysis, and the electric current density. It is to be noted, in this regard, that when the volume to be discharged from the anodic cell, the volume of the constant quantity cup **14**, the volume of the regeneration solution from tank **17** and the volume of the anolyte are all equal, the supply of constant supplemental solution is insured.

The fixing solution can be used as the anodic solution until the amount of the sulfite ions (stabilizer) reaches zero through electrolysis, etc., and the thiosulfate acid ions are converted into sulfur. The minimum volume of anodic cell is not less than 0.5 liter, and the maximum volume of the anodic cell is 50 volume percent of the used fixing solution to be electrolyzed. On a practical scale, 0.5 liter is the minimum volume of the electrolyte solution sent to waste, which is equal to the minimum volume of the anodic cell.

At a lesser volume, one is essentially dealing with an apparatus which is too small to be practically useful to process large amounts of fixing solution. The minimum volume of the anodic cell will also be equal to the minimum volume to be discharged from the system (i.e., the minimum volume sent to waste).

The fixing solution remaining in the constant quantity cup **14** is pumped into the anodic cell **3** and is made ready for the next electrolysis. When the fixing solution (anodic solution) has no sulfite ions, the thiosulfate ions are converted to sulfur. The life of the fixing solution being used as the anodic solution depends on the current density and the volume of the anodic solution, as one skilled in the art will appreciate. The fixing solution remaining in the constant quantity cup **14** is pumped into the anodic cell **3** by way of line **101** and pump **102**.

Where the anolyte is allowed to stand in the anode compartment for a long period of time, the overflow solution is again introduced. Since the anodic solution is the regenerated fixing solution, there is no danger of other ingredients being intermixed; thus, this method is advantageous for this reason.

The regenerated fixing solution is pumped via pump **16** from the circulation tank **10**. A regeneration agent is added thereto from a regeneration agent tank **17**, and then the regenerated fixing solution is conveyed into a replenisher tank **18** (which stores replenishing solution before supplying the same into fixing tank **11**) and then into a fixing tank **11**. In this way, the regeneration process is completed.

A process developed by *M. Kitajima and A. Abe; U.S. Patent 4,155,810; May 22, 1979; assigned to Fuji Photo Film Co., Ltd., Japan* is based on the discovery that some chemosynthetic sulfur bacteria, particularly those belonging to the *Thiobacillus* family, rapidly grow and propagate (or proliferate) under aerobic conditions utilizing reductive sulfur compounds as electron donors. It is considered that the bacteria efficiently take up the silver ions as a part of the donors present in the photographic waste and accumulate the silver in the cells thereof.

During growth and propagation (or proliferation), the bacteria take up and metabolize various additional components such as acetic acid, carbonate ion, inorganic ions such as ammonium ion, sodium ion or potassium ion as the sources of carbon, nitrogen and inorganic ions. Therefore, the utility of this method will further increase if an improvement is achieved in making use of the bacteria cells as fertilizers or feed stuffs.

SILVER FROM WASTEWATERS

See "Gold from Wastewaters."

SODIUM FERROCYANIDE FROM HALOGEN TIN ELECTROPLATING SLUDGES

See "Tin from Halogen Tin Electroplating Sludges."

SODIUM HYDROXIDE FROM SPENT PULPING LIQUOR

In the conventional Kraft process, raw cellulosic fibrous material, generally wood chips, is digested, by heating, in a pulping liquor, known as white liquor and containing sodium sulfide and sodium hydroxide as the active pulping chemicals, to provide a pulp and spent pulping liquor, known as black liquor. The black liquor is separated from the pulp by washing in a brown stock washer and the pulp is then passed to the bleach plant for brightening and purification operations.

The black liquor then is passed to the recovery and regeneration system in which the black liquor first is concentrated, usually by evaporation, and the concentrated black liquor is burned in a furnace to yield a smelt containing primarily sodium carbonate and sodium sulfide. A sodium- and sulfur-containing compound, generally sodium sulfate, is added to the black liquor generally prior to feed of the concentrated black liquor to the furnace, although such sodium- and sulfur-containing compound may be added at any other convenient point, such as to the white liquor prior to the digestion step, to make up sodium and sulfur values lost from the recovery system.

The smelt is dissolved in water to yield a raw green liquor which then is clarified to remove undissolved solids. The clarified green liquor is causticized with slaked lime whereby the sodium carbonate is converted to sodium hydroxide. The resulting liquor is white liquor which then may be recycled to the digestion step to provide at least part of the pulping liquor.

In addition, another pulping process to which this process may be applied is the soda process. In this process, the pulping liquor consists predominantly of aqueous sodium hydroxide solution. Spent pulping liquor is subjected to a recovery and regeneration procedure as in the Kraft process. The smelt which results from furnacing in the soda process contains primarily sodium carbonate which, after formation into an aqueous solution, is recausticized to regenerate sodium hydroxide solution for recycle to provide at least part of the pulping liquor.

The spent wash waters from bleach plants generally have been discharged to water bodies, such as streams, rivers, lakes and oceans, without any attempt to recover chemicals therefrom, although in some instances solid particle recovery operations have been made. One of the main reasons that no attempt has been made to recover these chemicals is because they are very dilute and of small value.

The bleach plant also produces spent bleaching liquor effluents and spent caustic extraction effluents. These effluents have objectionable color and are toxic and harmful to aquatic and marine biota and polluting since they contain fibers and materials consuming oxygen present in the water. It is desirable to avoid such environmental pollution, and hence avoid the discharge of these effluents from the mill.

In a process developed by *J.A. Lukes; U.S. Patent 3,909,344; September 30, 1975; assigned to Erco Envirotech Ltd., Canada* the bleach plant effluent preferably is added to the spent pulping liquor recovery and regeneration operation, and in this way this effluent is retained within the mill.

In this process the smelt from the spent pulping liquor recovery operations contains sodium carbonate and sodium chloride, and optionally other components, depending on the pulp mill system. The smelt is made up into a hot aqueous solution thereof, and is evaporatively cooled to crystallize hydrated sodium carbonate, while avoiding deposition of sodium chloride.

After separation of the crystallized sodium carbonate, and other crystallized salts, for example, sodium sulfide and sodium sulfate if a Kraft mill system is employed, the mother liquor is evaporated with heating to deposit sodium chloride in a substantially pure form. The sodium carbonate may be converted into an active pulping chemical for recycle, typically into sodium hydroxide by recausticization.

Such an overall process is shown in Figure 149. As shown there, wood chips, or other raw cellulosic fibrous materials, are fed by line **10** to a digester **12** wherein the wood chips are digested with a pulping liquor fed by line **14** and containing sodium hydroxide as the active pulping chemical in the soda process.

The resulting pulp and spent pulping liquor are separated and the pulp is washed in a brown stock washer **16**. The pulp is washed, in the embodiment illustrated, with aqueous bleach plant effluent fed by line **18**. Alternatively, the pulp may be washed with water or contaminated condensate, and the bleach plant effluent may be utilized elsewhere in the system, as described in more detail below, or discharged as desired.

The washed and unbleached pulp is fed by line **20** to a bleach plant **22** wherein the pulp is subjected to a series of bleaching and purification processes involving the use of one or more chlorine-containing bleaching agents. Generally, the bleaching and purification processes involve bleaching with chlorine, chlorine dioxide or mixtures thereof fed by line **24** and purification by caustic extraction, using aqueous sodium hydroxide solution fed by line **26**, typically in a CEDED (chlorine/extraction/chlorine dioxide/extraction/chlorine dioxide) sequence.

The pulp is washed during the bleach plant operations, typically after each bleaching or caustic extraction operation, by water fed by line **28**. The spent wash water from the bleach plant washing operations together with the spent chemicals from the bleaching and caustic extraction steps provide the bleach plant effluent in line **18**.

Preferably, the washing operations involve countercurrent flow of pulp and wash water through the bleach plant. Preferably, the quantity of sodium hydroxide used in the caustic extraction of the pulp is slightly in excess of the stoichiometric requirement of one sodium atom for each chlorine atom in the bleaching chemicals, whereby the bleach plant effluent in line **18** has an alkaline pH. The bleached and purified pulp of required brightness is recovered from the bleach plant **22** by line **30** and is sold as such, or passed to paper making procedures.

If desired, the bleach plant effluent in line **18** may be added directly to the spent pulping liquor in line **32**, although this procedure is less preferred, since the overall water requirement thereby is increased.

The bleach plant effluent in line **18** contains considerable quantities of sodium chloride which are transferred to the spent pulping liquor in line **32**. The spent pulping liquor is evaporated in an evaporator **34** prior to passage by line **36** to a furnace **38** of any convenient construction. The water recovered from the evaporator **34** by line **40** may be used to provide at least part of the water requirement of the system, for example, as at least part of the water fed to the bleach plant in line **28**, after suitable cleaning, if required.

Figure 149: Erco Envirotech Ltd. Process for Sodium Hydroxide Recovery as Pulping Liquor from Spent Pulping Liquor and Bleach Plant Effluent

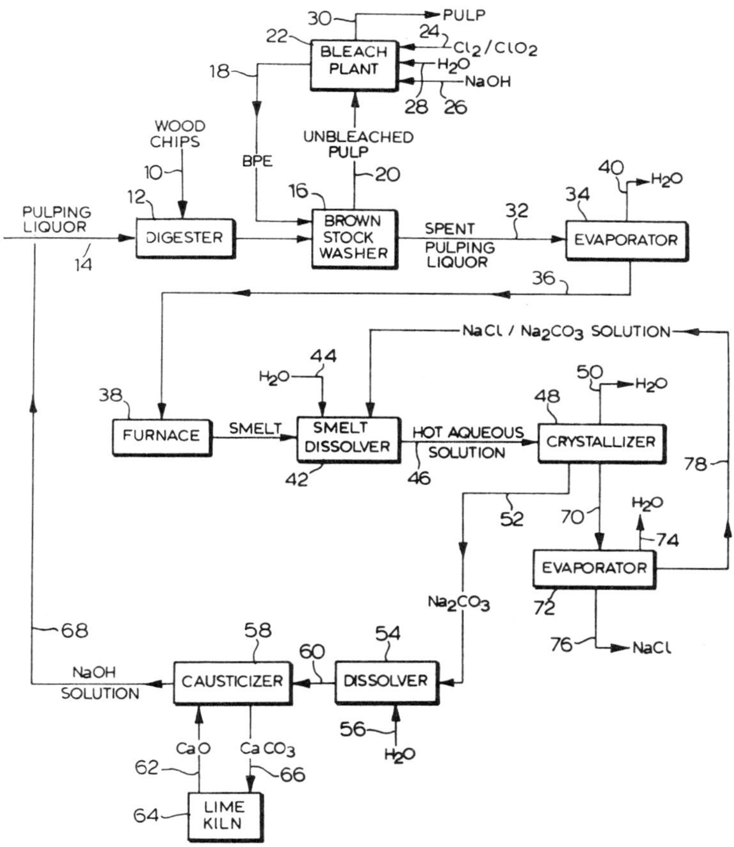

Source: U.S. Patent 3,909,344

The spent pulping liquor forms in the furnace **38** a smelt containing sodium carbonate, and additionally unreacted components, consisting substantially of sodium chloride.

Thus, there is obtained from the furnacing operation a smelt containing sodium carbonate and sodium chloride. The smelt is dissolved in a smelt dissolver **42** in water fed by line **44**. The water may be constituted by BPE (bleach plant effluent) evaporator condensate or weak wash water, if available after dissolving hydrated sodium carbonate in **54**. The water passed by line **44** to the smelt dissolver **42** may be partially bleach plant effluent from line **18**, especially in the case where fresh water or contaminated condensate is fed to the brown stock washer **16**.

The resulting hot aqueous solution in line **46** also contains recycled chemicals as described in more detail below.

If desired, all the smelt need not be dissolved in the aqueous material fed by line **44**, but instead the smelt may be leached with the aqueous material fed by line **44** to dissolve substantially all the sodium chloride values and part of the sodium carbonate values of the smelt, leaving substantially pure sodium carbonate. Operation in this manner reduces the quantity of water which requires evaporation. The substantially pure sodium carbonate may be used to form pulping liquor along with the sodium carbonate later separated from the hot aqueous solution resulting from the leaching.

The hot aqueous solution is cooled in a crystallizer **48** to crystallize hydrated sodium carbonate, preferably sodium carbonate decahydrate, while avoiding substantially the crystallization of sodium chloride. The cooling preferably is achieved by evaporation of water from the hot solution under vacuum, the evaporated water being removed by line **50**.

The hot aqueous solution in line **46**, which generally has a very high temperature, typically about 180° to 210°F, may be subjected to an initial cooling, with or without evaporation of water, to reduce the cooling load to be borne by the sodium carbonate crystallizer, prior to further evaporative cooling to cause crystallization of the sodium carbonate. More than one such initial cooling step may be used, if desired. The initial cooling may be omitted, if desired.

In a typical operation, the hot aqueous solution at an initial temperature of about 200°F is cooled to about 120°F with some evaporation of water and then the solution is evaporatively cooled to below about 70°F, preferably to 60°F, causing crystallization of sodium carbonate decahydrate. Since the sodium carbonate is crystallized in the form of the decahydrate, more water is removed from the solution.

The evaporative cooling and crystallization of sodium carbonate preferably are continued until the mother liquor is substantially saturated with respect to sodium chloride. The resulting slurry preferably is diluted with recycled mother liquor to improve the workability thereof and sodium carbonate, substantially free from sodium chloride, is separated from the mother liquor by line **52**. While the crystallization of sodium carbonate and its separation from the mother liquor are illustrated as occurring in the same vessel, this is for ease of illustration and normally separate vessels are used.

The sodium carbonate is passed by line **52** to a dissolving tank **54** wherein it is dissolved in water, or any other suitable aqueous medium, preferably weak wash water, fed by line **56** to form an aqueous sodium carbonate solution which is passed to a causticizer **58** by line **60**. The sodium carbonate in the aqueous solution is converted substantially completely to sodium hydroxide by the action of lime fed by line **62** from a lime kiln **64**.

The calcium carbonate mud precipitated in the causticizer **58**, after suitable washing to remove entrained alkali values, is returned to the lime kiln **64** by line **66** for reconversion to lime. The weak wash water resulting from the calcium carbonate mud washing may, in part, be used as the aqueous material fed by line **44** to dissolve the smelt.

The aqueous sodium hydroxide solution resulting from the recausticization process is recycled by line **68** to provide at least part of the pulping liquor fed by line **14** to the digester **12**.

The mother liquor resulting from separation of the sodium carbonate, containing dissolved quantities of sodium chloride and sodium carbonate, and saturated at least with respect to sodium carbonate, is forwarded by line **70** to an evaporator **72** wherein the mother liquor is heated and water is evaporated therefrom to deposit sodium chloride, while avoiding substantially the crystallization of sodium carbonate.

Generally, the evaporation of water in the evaporator **72** is achieved by boiling the solution under a reduced pressure, preferably until the solution is substantially saturated with respect to sodium carbonate. The water evaporated in the evaporator **72** is removed by line **74**.

The temperature to which the mother liquor is heated to achieve the desired evaporation and crystallization of sodium chloride depends on the temperature at which the sodium carbonate crystallization occurred in the crystallizer **48** and the relative concentrations of sodium carbonate and sodium chloride in the mother liquor. The temperature must be higher than a temperature at which sodium carbonate will crystallize, if substantially pure sodium chloride is to be recovered, and temperatures as high as 220°F may be used. At these temperatures, the temperature to which the hot aqueous solution is cooled to crystallize hydrated sodium carbonate must be quite low, typically about 32° to 50°F. Temperatures in the range of 100° to 140°F have proved to be most desirable.

In the typical procedure outlined above wherein the hot aqueous solution was cooled to about 60°F to achieve crystallization of hydrated sodium carbonate, the mother liquor may be heated to about 120°F to achieve crystallization of sodium chloride.

The sodium chloride which is crystallized is removed by line **76** in substantially pure form. Preferably, the quantity of sodium chloride removed in this manner is substantially the same as the quantity of sodium chloride present in the hot aqueous solution which is subsequently cooled to crystallize hydrated sodium carbonate, less any sodium chloride present in the recycle stream in line **78**.

The sodium chloride removed by line **76** may be put to a variety of uses. Typically, the sodium chloride is utilized to regenerate bleach plant chemicals. For example, the sodium chloride may be used to generate sodium hydroxide by electrolysis of an aqueous solution thereof, the sodium hydroxide being used in the bleach plant in line **26**. Alternatively, the sodium chloride may be used to generate chlorine dioxide and chlorine by reaction with sodium chlorate and sulfuric acid, the chlorine dioxide and chlorine being fed to the bleach plant by line **24**. Further, the sodium chloride may be electrolyzed as an aqueous solution to sodium chlorate for use in a chlorine dioxide producing reaction involving reduction of the sodium chlorate in an acid medium.

The evaporation of the mother liquor from the separation of sodium carbonate and the separation of the sodium chloride crystallized therefrom is illustrated as occurring in a single vessel **72**. In practice, usually two separate vessels are used, the slurry resulting from the evaporation in one vessel being passed to a second vessel for separation of the solid sodium chloride.

The mother liquor resulting from the sodium chloride separation will usually still contain quantities of sodium chloride and sodium carbonate and is recycled by line **78** to the smelt dissolver **42** along with the aqueous material in line **44**, or, alternatively, is recycled to hot aqueous solution in line **46**.

The mother liquor in line **78** may contain small quantities of sodium hydroxide, which would not be removed from the system by the illustrated procedure. The sodium hydroxide may arise from weak wash water from the washing of the calcium carbonate mud where such water is used to provide at least part of the aqueous material fed by line **44** to the smelt dissolver **42**. The sodium hydroxide also may arise from small quantities of sodium oxide values in the smelt.

The mother liquor in line **78** may be carbonated using carbon dioxide, or in any other convenient manner, to convert such sodium hydroxide values to sodium carbonate, prior to passage of the mother liquor in line **78** to the hot aqueous solution in line **46**. In this way, any buildup of sodium hydroxide values in the recycled mother liquor in line **78** is avoided.

If desired, following the separation of the sodium chloride therefrom, the resulting mother liquor may be cooled again to precipitate further quantities of hydrated sodium carbonate.

It will be apparent, therefore, that in the process illustrated, the sodium carbonate and sodium chloride in the smelt are separated one from another into substantially pure product streams which contain quantities of the hydrated sodium carbonate and sodium chloride which are substantially equal to the quantities of these materials present in the smelt.

SODIUM NITRATE FROM HEAT TREATING OPERATIONS

Commercial hardening techniques for ferrous metal components typically involve the steps of heating the components in a furnace or a bath of molten salt, such as a chloride, quenching the components in a bath of another molten salt such as a nitrate, nitrite, or mixture thereof, and thereafter causing the parts to be passed through one or more rinse baths to clean off salts which cling to the parts as a result of the heating and quenching steps. A certain amount of each of the molten salts is transferred by drag-out to the rinse bath, thus requiring occasional replenishment of the molten salts, as well as replacement of the rinse water.

It is desirable to recover the salts from the rinse water and return them to the heat treating baths. However, a number of problems have heretofore prevented such recycling on a straightforward and economical basis. First, it is difficult to justify economically the installation of recovery apparatus unless the quantities of reclaimed salts are very high, and the recovery process can be carried out without shut-down of the heat treating system.

Secondly, it has been found that the impurities which typically find their way into heat treating baths tend to build up to unacceptable levels particularly fast when the salts are recovered from the rinse and recycled into the heat treating bath. For example, chloride salts from a heating bath are dragged into the quench bath and thence into the rinse where recycling tends to return them to the quench bath. Recycled chloride impurities quickly reach insolubility levels in the quench, and crystallize out as a sludge which must be removed from time to time. In addition, chloride tends to substantially reduce the quench efficiency of a nitrate/nitrite bath and, thus, removing the chlorides is highly desirable.

Another impurity which will build up quickly as a result of salt recycling is carbonate, a reaction product of decomposition of the salt itself. For example, a sodium nitrate/nitrite salt tends to form Na_2CO_3, the carbon being taken out of the CO_2 present in the air. The sodium carbonate is quite soluble in the rinse water but only sparingly soluble in the nitrate/nitrite quench bath and, thus, forms a precipitate in the bath which impairs quenching efficiency and coats mechanical components in contact with the bath.

It should be understood that both the chloride and carbonate impurities, as well as many other impurities such as hard water ions, are typically present in the heat treating salt baths of prior art systems. However, the levels of concentration of such impurities are kept reasonably and acceptably low for longer periods of use by reason of the fact that impurities as well as desirable salts are continuously lost by drag-out. Lost salt, of course, is replaced by fresh salt which does not contain the impurities. Where salt is recycled, however, undesirable impurities dragged into the rinse and which are soluble in the rinse water are retained in the recycled materials and, after a short time, reach unacceptable concentrations.

A process developed by *R.W. Foreman and G.P. Dubal; U.S. Patents 4,149,702; April 17, 1979; and 4,158,579; June 19, 1979; both assigned to Park Chemical Company* provides for recycling water-soluble heat treating salts, typically by reclaiming such salts from a rinse bath and returning them to the heat treating bath in such form as to prevent the build-up of undesirable impurities such as chlorides and reaction products from decomposition of the original heat treating salts.

Figure 150 illustrates such a process in some detail. The figure shows a typical heat treating arrangement for hardening ferrous metal parts, and the recovery and recycling of nitrate/nitrite salts from a quench bath. The parts are heated by immersion in a molten chloride contained in a gas-fired refractory furnace **10**. It will be understood that the furnace could also be an atmospheric furnace as far as the effect of heat treatment is concerned.

After heating to approximately 1500°F in the chloride bath, the parts are transferred to a refractory quench furnace **12** containing a bath of molten salt such as a sodium nitrate/nitrite mixture maintained in a molten condition at a temperature of about 475° to 700°F. Furnace **12** is also typically gas-fired. The metallurgical result of heating and quenching ferrous metal parts is well known and will not be described herein.

From the quench bath **12** the parts are removed to an open rinse tank **14** containing water at approximately 150°F for the purpose of rinsing off salts which are dragged from the quench bath **12** by the parts under treatment as well as the carrier apparatus for the parts.

Figure 150: Park Chemical Company Process for Recycling Heat Treating Salts

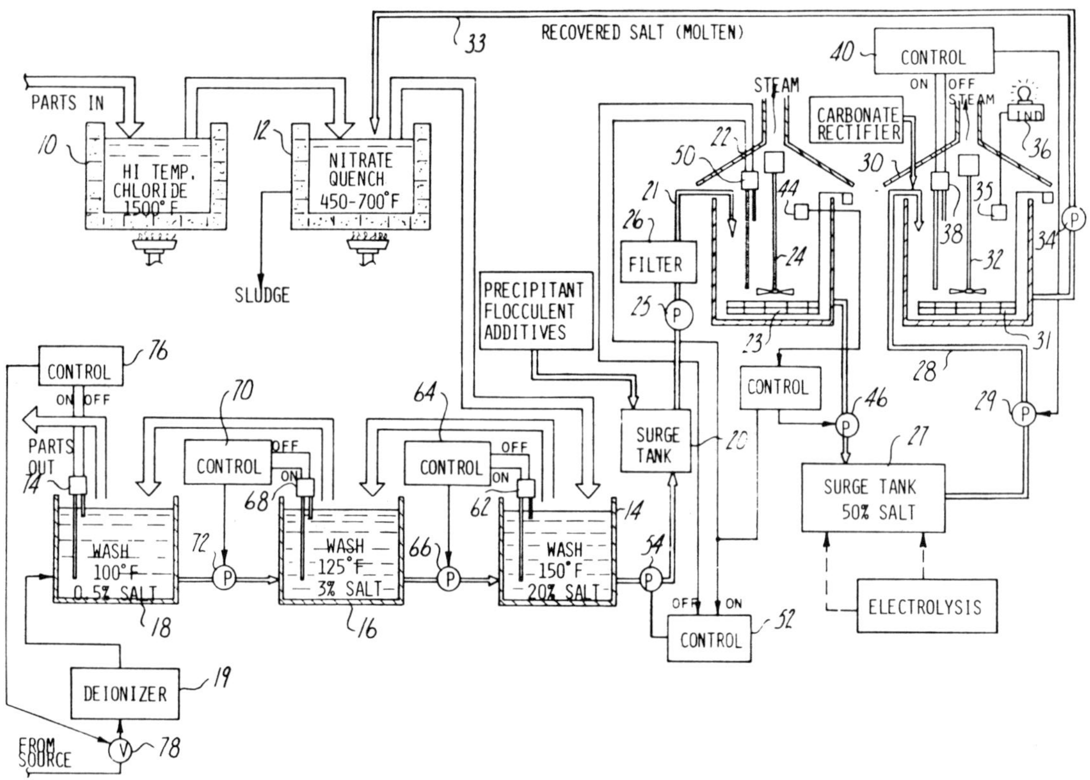

Source: U.S. Patent 4,149,702

Typically, the quench bath and the rinse tank are located in close physical proximity to one another such that a substantial portion of the molten salt dragged out of the bath **12** on the parts finds its way to the rinse tank **14**. The parts thereafter may go through a sequence of additional interconnected open rinse tanks **16** and **18** containing water at temperatures of 125°F and 100°F, respectively, where substantially all the remaining salts clinging to the parts and carriers are washed away. The sequential rinse results in gradually decreasing salt content in the rinse water of baths **14, 16** and **18** as shown.

It can be seen that some quantity of chloride salt from the heating furnace **10** is dragged into the nitrate/nitrite salt quench bath where it is regarded as an impurity. The chloride, and other impurities, are removed to the rinse tank **14** by drag-out and are dissolved in the rinse water along with the nitrate/nitrite salts which are to be recovered.

The first step in the recovery process involves concentrating the rinse water salt impurities solution. This is accomplished in two ways. First, a countercurrent flow system is connected between rinse tanks **14, 16** and **18** such that the highest concentration of recoverable salt exists in tank **14**. Fresh water enters tank **18** through a deionizer **19**. Tank **18** overflows into tank **16** which, in turn, overflows into tank **14**. It will be understood that the term overflows is not necessarily to be construed literally for, although a system of weirs and gravity flow channels can be used, the illustrated embodiment employs electronic level detectors and controlled pumps as hereinafter described.

After achieving a recoverable salt concentration in tank **14** of about 20%, the solution is transferred from tank **14** to a surge tank **20** which is preferably closed and of a capacity at least as great as that of the rinse tank **14**.

From surge tank **20**, the rinse water salt/impurity solution is transferred through a fluid circuit **21** to a hooded concentrator **22** for further concentration by evaporation of the rinse water. A concentrate of approximately 50% salt can be achieved by maintaining a uniform temperature in the solution of from 240° to 250°F. An over-the-side electric immersion heater **23** and an air-motor driven agitator **24** are employed for this purpose.

Fluid circuit **21** comprises, in addition to conduit of a suitable noncorrosive material such as Inconel, a pump **25** and a filter **26** which removes foreign matter such as scale.

After achieving the 50% concentration in unit **22**, the solution is transferred into a second, closed surge tank **27** of a capacity about three to four times that of the unit **22**. From surge tank **27**, the 50% solution is transferred through fluid circuit **28** containing pump **29** to a hooded melter unit **30** which is similar to unit **22** but which operates at a higher temperature for reasons to be described. Melter unit **30** comprises an electric immersion heater **31** and an air-motor driven agitator **32** to maintain uniform temperature throughout the contents thereof.

Melter unit **30** is connected via a controlled fluid circuit **33** to the nitrate/nitrite bath **12** for the return of molten salt thereto. The transfer of molten salt from melter **30** to quench bath **12** is preferably carried out by manual operation of pump **34** in fluid circuit **33**.

Assuming nitric acid rectification, this transfer is accomplished after the rectification process is completed and the temperature of the molten salt is adjusted to the temperature of the bath by suitable operation of heater **31**. The readiness of the molten salt for both transfer and rectification may be indicated by a temperature sensor **35** of the immersible type which is electrically interconnected to an alarm or other suitable indicator **36**.

Once transfer from melter **30** to bath **12** begins, the level of molten salt in the melter begins to drop. Molten salt level is detected and monitored by immersible liquid level detector **38** which is electrically connected to a control circuit **40**. When the level reaches the bottom of the longest immersible sensor, control **40** turns on pump **29** in fluid circuit **28** to transfer the 50% salt solution from the surge tank **27** to the melter **30**. The liquid level detector **38** is preferably mounted in the melter **30** so as to prevent the level from dropping to the point where the heater element **31** is exposed. Once the upper level has been achieved, liquid level detector **38** operates through control **40** to turn off pump **29**.

To transfer the 50% solution from concentrator **22** to surge tank **27** a temperature detector **44** is electrically interconnected to pump **46** to start the pump. It will be understood that where gravity flow is possible pumps may be replaced with simple valves to accomplish the same end result. Pump **46** may be shut off manually when concentrator **22** is nearly empty.

As the level of solution in concentrator **22** drops level detector **50** operates through control **52** to turn pump **54** on and off to maintain the proper level of solution in the concentrator **22**. It will be understood that as the 20% salt solution is transferred from the surge tank **20** to the concentrator **22** the temperature drops toward the 150°F level. Heater **23** slowly raises the temperature of the solution while agitator **24** circulates the solution to maintain temperature uniformity.

As the water is evaporated off the temperature slowly rises toward the 240° to 250°F range which signifies the achievement of a 50% salt solution. When this is reached, the temperature sensor **44** operates the pump **46** in the manner previously described to transfer the contents of concentrator **22** into the surge tank **27**.

To maintain the operability of the counterflow rinse system, the rinse tanks **14, 16** and **18** are provided with liquid level detectors, control circuits and pumps in the interconnecting lines between rinse tanks. Specifically, rinse tank **14** is provided with liquid detector **62** operating through circuit **64** to control pump **66**. Tank **16** is provided with liquid detector **68** operating through circuit **70** to control pump **72**. Finally, tank **18** is provided with liquid level detector **74** operating through circuit **76** to control a valve **78** between a pressure water source and the tank **18**.

The result is that as each tank drops in level to a minimum the in-coming fluid circuit is actuated to restore the fluid level to the predetermined maximum. As level changes and fluid transfers are reflected back down the line the next control circuit comes into operation as will be apparent from the foregoing description.

SODIUM NITRITE FROM ABSORBER LIQUOR

It is well known that if nitrous gases, i.e., gases containing nitric oxide and nitrogen dioxide, are brought into contact with alkaline aqueous solutions, the nitric oxide and nitrogen dioxide will be absorbed with the formation of nitrites and nitrates. Such gases are typically obtained from a wide variety of sources, as for example, by the arc process of nitrogen fixation, by the air-oxidation of ammonia or as waste gases from an absorption process of which nitric acid is the primary product. By controlling the ratio of nitric oxide and nitrogen dioxide so that more nitric oxide than nitrogen dioxide is present in the gas feed, the amount of nitrates that is formed can be minimized, thereby allowing maximum conversion of the nitrogen oxide to the nitrites.

A process developed by *C.F. Hecklinger and D.E. Crook; U.S. Patent 3,965,247; June 22, 1976; assigned to Allied Chemical Corporation* is one in which sodium nitrite is recovered from an aqueous solution containing dissolved sodium nitrite and sodium chloride.

The process comprises (1) subjecting the solution to crystallization in a first crystallization zone under conditions sufficient to effect crystallization of sodium chloride, including operating at temperatures of from 110° to 130°C, to produce a first slurry containing sodium chloride crystals; (2) separating the crystals from the first slurry to produce a solution containing dissolved sodium nitrite; (3) subjecting the sodium nitrite solution to crystallization in a second crystallization zone under conditions sufficient to effect crystallization of sodium nitrite from the solution, including operating at temperatures of from 10° to 60°C, to form a second slurry containing sodium nitrite crystals; and (4) recovering the sodium nitrite crystals from the second slurry.

SODIUM SULFITE FROM SPENT PULPING LIQUOR

The waste liquor from sodium-based sulfite cooking contains remarkable amounts of sodium chemicals, the recovery of which is profitable. For this purpose the waste liquor has been concentrated by evaporating and burning in a recovery boiler. From the burning there have been obtained flue gases containing sulfur dioxide and carbon dioxide, and a smelt which is dissolved into water to form so-called green liquor which contains sodium sulfide. Sulfur dioxide has been washed out from the flue gases in a washer with a sodium carbonate solution made of green liquor. The sodium sulfite obtained can be used as such in sodium sulfite pulping.

Due to the high capital investment requirements for present methods of recovery of chemicals in sodium sulfite and sodium bisulfite pulping and the difficulties encountered in side reactions producing sodium thiosulfate that is detrimental to the pulp digesting operation, there is need for improved, lower cost methods of recovery of the sodium and sulfur components for reuse in the cooking operation.

U.S. Patent 3,787,283 discloses a process and apparatus for recovering the sodium from sodium-based spent pulping liquors in which the concentrated spent liquor is combusted in the form of solid pellets rather than as a sticky molten smelt. This recovery process has numerous advantages over the prior conventional recovery processes. For example, the process can be carried out with relatively inexpensive furnace equipment such as a rotary kiln rather than the more expensive Kraft or glass furnaces which are required to handle a molten smelt.

Further, the sodium is recovered directly as sodium sulfite, whereas in most of the prior practices employing Kraft type furnaces, the smelt obtained therefrom is a mixture of principally sodium carbonate and sodium sulfide which must thereafter be processed by several means to recover the sodium as sodium sulfite.

A process developed by *D.R. Sheeley and J.H. Rion; U.S. Patent 4,035,228; July 12, 1977; assigned to Sonoco Products Company* provides a further improvement in such a technique.

More specifically, this process is one in which concentrated alkali metal-containing waste liquor is mixed with recycled alumina and formed into solid pellets by mixing the concentrated liquor-alumina mixture with recycled alkali metal aluminate furnace ash. The solid pellets are fed through a furnace operating at a temperature below the fusion temperature of the aluminate for combusting the organic portion of the pellets and reacting the alkali metal content thereof with the alumina to form additional alkali metal aluminate ash.

A portion of the furnace ash is recycled for reuse in the formation of pellets, while another portion of the ash is dissolved in water to form a solution of alkali metal aluminate. The alkali metal aluminate solution is mixed with slurry recycled from later in the process to initiate limited precipitation of alumina from the alkali metal aluminate. The mixture is then reacted with sulfur dioxide gas to effect additional precipitation of alumina and produce a slurry of the alumina in a solution containing the alkali metal. The alumina is separated for reuse in the treating process and the alkali metal-containing solution is recovered.

A process developed by *S. Mizuguchi, I. Nomura, T. Naito, M. Onodera and K. Saito; U.S. Patent 4,141,785; February 27, 1979; assigned to Ebara Manufacturing Co., Ltd., Japan* is one for recovery of chemicals from a sodium sulfite pulping waste liquor in which a smelt obtained from the waste liquor is introduced into an aqueous slurry containing solidified smelt while make-up water and a weak aqueous slurry are supplied to effect incomplete dissolution of the smelt into the aqueous slurry to maintain the content of total solid and the temperature of the slurry at constant levels.

The resulting aqueous slurry is subjected to a solid-liquid separation to obtain a wet cake having the molar ratio of S/Na_2O substantially equal to that of the smelt. The wet cake is mixed with hot particles of sodium carbonate and sodium sulfite while hot air is supplied to effect oxidation of sodium sulfide in the wet cake to sodium sulfite and then the oxidized product is dissolved in aqueous medium.

Sulfur dioxide-containing gas, preferably the exhaust gas from the recovery boiler, is contacted with the resulting aqueous solution to convert sodium carbonate into sodium sulfite, whereby the overall process is carried out in a closed system and the sulfur component and the sodium component present in the waste cooking liquor are recovered and regenerated into a cooking liquor.

Such a process is illustrated in Figure 151. As shown, concentrated black liquor is burned in recovery boiler **1** to form a smelt which is supplied continuously in the form of stream **2** to smelt hopper **3**. Steam or high pressure air stream **4** is directed to the smelt stream to effect cooling and dividing out of the smelt into fine particles. An aqueous slurry recycled from circulating tank **6** is flowed downwardly on the inner surface of the hopper to prevent accumulation of smelt thereon. The rate of the slurry to be supplied to the hopper is 100 times that of the smelt, by weight. The smelt particles drop on conveying screen **5** by which fine particles and large particles are separated, the former being supplied to the slurry circulating tank **6** and the latter to green liquor tank **7**.

Figure 151: Ebara Manufacturing Co., Ltd. Process for Recovery of Chemicals in Sodium Sulfite Pulping Liquor

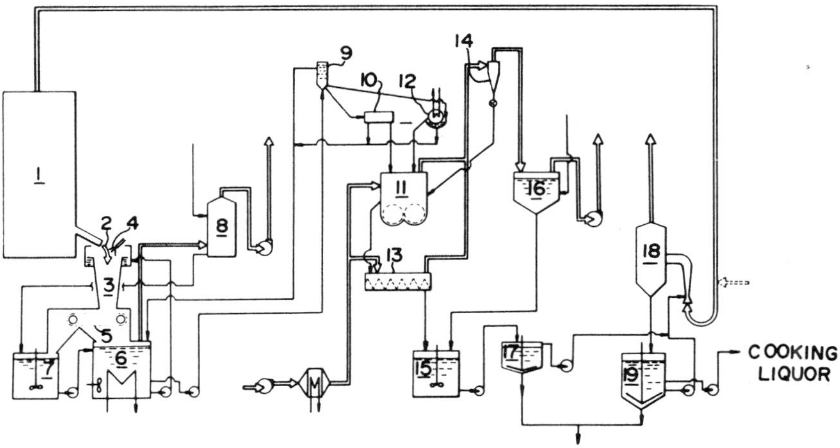

Source: U.S. Patent 4,141,785

In the green liquor tank, the large particles are dissolved under agitation in water, which is supplied from scrubber **8** for washing exhaust gas from the hopper, to form a weak green liquor which is pumped to the slurry circulating tank.

Though the concentration of the green liquor produced varies depending upon the amount of water to be supplied, which is determined taking in account the total water balance throughout the overall operation, the concentration is a factor in determining the concentration of slurry to be formed in the slurry circulating tank and is usually maintained at about 10% by weight.

Under steady operation conditions, a well established balance of water between (1) the sum of the supply to oxidation step and the water discharged together with air from the smelt hopper and (2) water supplied from the green liquor tank, is maintained to hold the solid content in the green liquor tank at a constant level.

The green liquor is mixed in the slurry circulating tank with the aqueous slurry, the solidified smelt particles and a weak aqueous slurry recycled from the second step under agitation, to form a slurry comprising a multicomponent aqueous phase containing mainly sodium sulfide as well as sodium hydroxide, sodium polysulfides and other sodium salts such as sodium thiosulfate, sodium sulfate, sodium carbonate and sodium chloride and a solid phase containing mainly sodium carbonate and other undissolved components above and various derivatives therefrom.

The molar ratio of S/Na_2O of total solid in the slurry is considerably higher than that of the smelt, for example, the ratio of smelt being about 0.5 the ratio of slurry being from 0.8 to 1.0. The temperature of the slurry is maintained at from 55° to 90°C, preferably 75° to 83°C, for example, by means of cooling water passing through a cooling coil or jacket and maintained at a temperature from 5° to 40°C, preferably 10° to 20°C, lower than that of the slurry.

The use of cooling water of too low temperature results in the precipitation of solid on the cooling surface to impede cooling efficiency. If desired, cooling by passing cold air through the slurry may be used in addition to such water cooling. In this case, some of the sodium sulfide is oxidized.

The total content of the chemicals in the slurry is maintained within a range of from 35 to 70%, preferably 45 to 60% by weight. At a concentration below 35%, though cooling efficiency is improved, the wet cake obtained in the second step contains more sodium carbonate than required for maintaining the desired molar ratio of S/Na_2O. Further, if the slurry or the filtrate is cooled on a drum flaker, there is insufficient solidification, or the resultant flakes contain too much water which lowers the oxidation temperature in the oxidation reactor to which the flakes are supplied.

On the other hand, at a concentration above 70%, there are disadvantages in that it becomes more difficult to maintain the slurry temperature below 90°C and there is clogging of the pipe lines.

The slurry is pumped via head tank **9** to screw decantor **10**. With the provision of the head tank, the slurry can be supplied continuously at a constant pressure to the decantor. By the decantor, a part of the solid phase is separated and removed as a wet cake from the aqueous slurry and the wet cake is supplied to oxidation reactor **11** while the remainder is recycled to the slurry circulating tank as a weak aqueous slurry.

In the case where the smelt has a proportion of sodium sulfide less than 40%, especially less than 30%, by weight, it is easy to adjust the molar ratio of S/Na_2O of the wet cake to that of the smelt.

On the other hand, if the proportion of sodium sulfide in the smelt increases to more than 30%, especially more than 40%, by weight, it is naturally necessary to increase the proportion of sulfur component in the aqueous slurry in order to adjust the S/Na_2O ratio of the wet cake to that of the smelt. If the concentration of sodium sulfide in the aqueous slurry increases, the cooling efficiency of the slurry is reduced due to the precipitation of sodium sulfide on the cooling surface, then, it is difficult to lower the temperature of the aqueous slurry to the required level. In such a case, part of the slurry discharged from the head tank is divided out and directed to drum flaker **12**; alternatively a part of the filtrate from the decantor is directed to the drum flaker.

By the drum flaker, the chemicals present in the slurry or the filtrate are solidified on the drum which is usually cooled to a temperature below 70°C, preferably 30° to 50°C, and the flakes formed are removed and supplied to the main oxidation reactor.

The flakes have a composition similar to that of the aqueous slurry supplied. For easy operation of the drum flaker, it is preferred to solidify only a part of the aqueous slurry and to recycle the remaining aqueous slurry to the circulating tank. Thus, by controlling the amount of flakes supplied to the reactor, the molar ratio of S/Na_2O of the combined filter cake and flake is readily adjusted to that of the smelt.

The main oxidation reactor is a double-shaft Z type kneader, the lower portion of which contains a large amount of solid particles of the oxidized product containing a small amount of water of, for example, a few percents and up to 5% by weight. As the wet cake and the flake are supplied to the reactor, they are immediately mixed with the oxidized product particles, while hot air (100° to 200°C, preferably 150° to 185°C) is simultaneously supplied.

Since the exhaust gas from the reactor contains a considerable amount of moisture, if too small an amount of air is supplied, the exhaust gas will have a higher moisture content which results in water droplets condensing on the surface of various parts, for example, cyclone separator and duct. Such droplets catch fine particles entrained in the exhaust gas and cause clogging. The supplied air also facilitates agitation and mixing of the reaction mixture in the reactor; therefore, an adequate air supply is desirable.

However, too much air requires more energy for heating and supplying air, and in addition, causes the escape of a large amount of particles from the reactor. Thus, the amount of air to be supplied is from 2 to 20 times, preferably 10 times, that required to completely oxidize the total sodium sulfide present in the smelt.

The exhaust gas from the reactor is fed to cyclone **14** in which particles entrained are recovered and recycled to the reactor. The oxidized product which contains unoxidized chemicals and oxidation intermediates is discharged from the reactor via the overflow chute and is supplied to suboxidation reactor **13**.

The purpose of the suboxidation reactor is to effect as complete an oxidation reaction as possible. If such unoxidized chemicals and intermediates are introduced in the subsequent step, they react with sulfur dioxide to form hydrogen sulfide which is a pollutant gas and sodium thiosulfate which is undesirable for pulp making.

To the suboxidation reactor, air is introduced to effect additional oxidation at a temperature of from 100° to 300°C, preferably 150° to 250°C. The water, up to about 5% by weight, present in the solid particles discharged from the main reactor is enough for performing the conversion of residual sodium sulfide into sodium sulfite. However, if less amount of water is present, additional water may be fed in order to facilitate the oxidation reaction. The exhaust gas from the suboxidation reactor is supplied to the cyclone **14** in which entrained solid particles are recovered.

If the entrained particles are not completely separated in the cyclone, then the exhaust gas is further treated in scrubber **16** to which fresh water or a dilute aqueous chemical solution is supplied to dissolve solid particles as completely as possible.

The oxidized product is discharged from the suboxidation reactor and supplied to dissolving tank **15** to which the aqueous solution discharged from the scrubber **16** is supplied to form an aqueous solution containing sodium carbonate and sodium sulfite.

The amount of aqueous solution to be supplied to the dissolving tank is controlled so that the concentration of the resulting aqueous solution is about 20% by weight. At a concentration above 20%, there is encountered difficulty in treating the aqueous solution with sulfur dioxide for producing a pulp cooking liquor; on the other hand, too dilute an aqueous solution cannot give a liquor having a concentration suitable for cooking.

The aqueous solution is supplied to sediment tank **17** in which insoluble material is separated as sludge, and the clarified liquor is supplied to absorber **18**. In the absorber, the clarified liquor and the exhaust gas discharged from the recovery boiler are intimately contacted to effect conversion of sodium carbonate into sodium sulfite to the extent required for the desired cooking liquor composition. The exhaust gas from the absorber does not contain sulfur dioxide and is vented to atmosphere.

The cooking liquor thus produced is clarified in thickener **19** and used for pulp making. The sludge recovered from the thickener is combined with the sludge from the sediment tank, washed with water, filtered and removed from the processing system. The washing and the filtrate are recycled to the scrubber and/or the dissolving tank.

An improved method for recovering spent sodium and sulfur components in sodium sulfite and sodium bisulfite pulping operations has been described by *W.G. Farin; U.S. Patent 4,148,684; April 10, 1979; assigned to MEI Systems Inc.* The method involves reduction burning of spent liquor to convert it to smelt and the treatment of this smelt in a sequence of operations that accelerate the reaction steps used to produce sodium hydrosulfide, hydrogen sulfide, sulfur dioxide, sodium carbonate, sodium bicarbonate and to recover sodium sulfite, sodium bisulfite, sodium carbonate and/or sodium hydroxide while preventing formation of sodium thiosulfate.

The method utilizes a sufficient quantity of recycled sodium bicarbonate addition to precarbonated liquor to maintain the partial pressure of carbon dioxide created by the solution above that provided by the carbon dioxide in the gas used for stripping out the hydrogen sulfide, recovering the carbon dioxide and water vapor for stripping and precarbonation by decomposition of the sodium bicarbonate and recovering carbon dioxide by pressure decarbonation and pressure neutralization, for the high pressure carbonation to produce the sodium bicarbonate.

An apparatus developed by *P.K. Rimpi; U.S. Patent 4,187,279; February 5, 1980; assigned to OY Tampella AB, Finland* consists of a precarbonation reactor, a H_2S-stripping column, a carbonation reactor, a washing tower for flue gases and a cooling tower for washed flue gases.

The carbonation reactor, the cooling tower and the washing tower are arranged on top of each other in the same column which is provided with trays so that means for feeding flue gases from the washing tower into the cooling tower and further into the carbonation reactor are openings made in the trays to allow the washed flue gases to rise from the washing tower into the cooling tower above it and further into the carbonation reactor above the cooling tower.

Common blower means are provided for feeding flue gases into the washing tower for blowing flue gases through the washing tower, the cooling tower, the carbonation reactor and the precarbonation reactor. Means for discharging a solution containing sodium carbonate and sodium bicarbonate from the stripping column are connected with means for feeding this solution directly into the carbonation reactor and the washing tower.

Such an apparatus is shown in Figure 152. When thick liquor is burnt in the recovery boiler, a smelt is obtained which is dissolved in water into green liquor. Dry solids are separated from the green liquor by clarifying. The clarified green liquor **10** is pumped into the uppermost part **1** of the column where it is treated with flue gases **11** according to the principle of countercurrent flow.

The precarbonation stage **1** can be realized either according to the spraying, packing piece or tray reactor principle. The sodium sulfide of the green liquor reacts with the carbon dioxide of the flue gases and is partly or completely precarbonated into sodium hydrogen sulfide. The flue gas **12** of the precarbonation stage leaves the column from the upper end.

The partly or completely precarbonated green liquor **13** is led into the next unit **2** of the column where it is treated with a large extra amount (2 to 4 times the equivalent amount of the reaction) of sodium bicarbonate. The liberated hydrogen sulfide **15** is transferreo from the liquid phase to the gas phase and out from the column by a transfer medium **16** being either steam, flue gas or some other inert gas.

The liquid **17** flowing out from the separation stage **2** contains mainly sodium carbonate and sodium bicarbonate and a minor amount of sodium sulfate and sodium thiosulfate. The sodium sulfide content is either nought or so small that it does not disturb further use of the solution in the process. The separated hydrogen sulfide gas **15** is transferred into further treatment, for example, to be burnt in a sulfur burner.

The sodium carbonate-sodium bicarbonate flow **17** is divided into two flows **18** and **19**. The flow **18** is led into the next column unit **3** where it is treated with flue gases **20** which have been cooled down to 20° to 40°C. The sodium carbonate reacts with the carbon dioxide of the flue gases and is converted into sodium bicarbonate. The mixture **14** of sodium bicarbonate and sodium carbonate is returned to the H_2S-separation stage **2**. The flue gas **21** goes directly out of the column and the process. The carbonation stage **3** can be realized for example on reactor trays according to the principle of countercurrent flow.

The sodium carbonate-bicarbonate flow **19** is further divided into two flows **22** and **23**. The flow **22** goes out of the column and the process, but the flow **23** is led into the flue gas washing stage **4**.

Figure 152: OY Tampella AB Apparatus for Recovering Sodium Chemicals from Green Liquor and Flue Gases

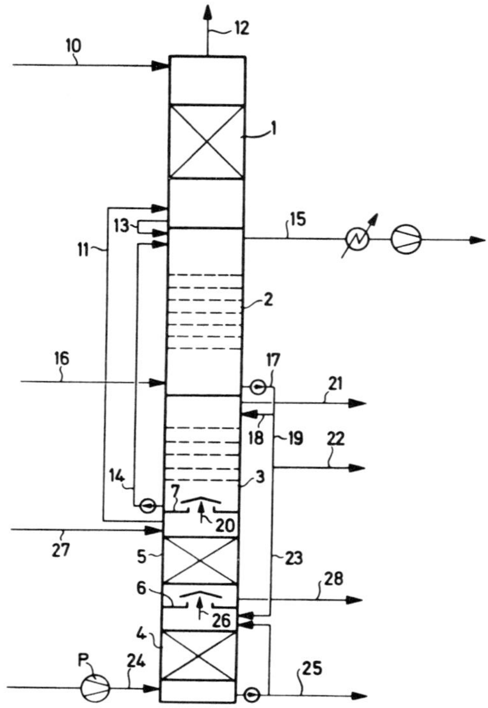

Source: U.S. Patent 4,187,279

The carbon dioxide and/or sulfur dioxide containing gases **24** are led into the lowermost section of the column. The lowermost section of the column serves as flue gas washing stage **4**, where mainly sulfur dioxide, sometimes also hydrogen sulfide, HCl and solid sodium sulfate are washed from the vent gases of the recovery boiler or other equipment. The washing in the column can take place in one or several phases either according to the principle of spraying, tray reactor or packing piece.

The liquid flow **25** leaving the washing stage **4** of the column is sodium sulfite solution which can be used either as such for sodium-based sulfite pulping or for other purposes.

The gases **26** leaving the washing stage are led into the cooling stage **5** where they are cooled with water **27** according to the principle of countercurrent flow either by spraying or in a packing piece layer. The warm water **28** leaving stage **5** is led out of the process and it can be used for different purposes. The cooled flue gases are led in flows **11** and **20** further into precarbonation **1** and carbonation **3**.

SODIUM SULFITE FROM TNT PROCESS WASTES

The red liquor waste stream from TNT manufacturing contains recoverable sodium sulfite. In a process developed by *W.R. Cook, D.R. Sheeley and J.A. Tyner; U.S. Patent 4,154,643; May 15, 1979; assigned to Sonoco Products Company* such an alkali metal-containing spent liquor may be treated to recover the alkali metal content for reuse without the need for recycling furnace ash and forming solid pellets therefrom.

A mixture of aluminum hydrate and alkali metal-containing waste liquor is directed through a heated furnace and dried to form granules of the aluminum hydrate and waste liquor solids, and the organic content of the granules are combusted and reacted to form granular alkali metal aluminate furnace ash. The SO_2-containing flue gases are reacted with the sodium aluminate to give sodium sulfite and aluminum hydroxide, which is recycled to the process.

In treatment of many waste liquors, as for example, spent pulping liquors, the organic content of the liquor is sufficient to supply the needed carbon to support combustion and provide a reducing atmosphere in the furnace for reduction of the alkali metal compounds.

However, in the treatment of some waste streams, as for example, the sodium and sulfur-containing red waste liquor stream from a TNT manufacturing operation, additional carbon must be provided and this may be supplied by mixing powdered coal, fuel oil, or any other convenient carbon-containing material with the waste liquor stream prior to introduction into the furnace. The amount of carbon to be added varies depending upon the carbon content of the liquor stream and upon other factors, and this can be readily ascertained by routine testing.

The following is a specific example of the conduct of such a process. 10,000 pounds per hour of a waste stream from a trinitrotoluene (TNT) manufacturing plant is concentrated from 10% solids to 2,000 pounds per hour of 50% solids red liquor. This liquor contains 320 pounds of sodium and 220 pounds of sulfur per hour. This concentrated liquor is mixed with 700 pounds per hour of powdered coal and the mixture is fed onto the bed of the second hearth of a multiple hearth furnace where it mixes with 1,085 pounds per hour of aluminum hydroxide which has been dried on the first hearth. Drying is completed on the second hearth and combustion takes place on the remaining hearths which reach temperatures of 1700°F or higher.

1,200 pounds per hour of ash is discharged from the bottom of the furnace where it is cooled to 400°F in a cooler by countercurrent air. The ash is then ground, stored in a surge bin, and finally goes to a dissolver tank where it is added to about 400 gallons of water per hour to produce a 25% solution of sodium aluminate. The hot flue gases from the furnace pass through a cyclone, are cooled to 160°F and are then directed to the absorber. The solution of sodium aluminate is contacted with the cooled flue gases in the absorber and aluminum hydroxide is precipitated and sodium sulfite is formed.

The discharge from the absorber is composed of a slurry of aluminum hydroxide in a solution of sodium sulfite at a pH of about 8.5. The slurry is carried to a belt filter washer where the filter cake is removed and conveyed back to the top hearth of the furnace for drying. The filtrate contains approximately 850 pounds of sodium sulfite which passes through a clarifier and finally to product storage.

SODIUM THIOCYANATE FROM USED ACRYLIC FIBER SPINNING SOLUTIONS

In the acrylic fiber production process including polymerization, dissolving and spinning steps, various sorts of organic and inorganic compounds are generally used as polymerization catalysts, anticoloring agents, dyeability improving agents, pigments, dyes, metal chelating agents, dispersing agents, plasticizers, ultraviolet ray absorbing agents, chain transfer agents, flame retardants, flame retardant synergists, antistatic agents, softening agents, deterioration preventing agents, etc. All or part of these compounds, as they are or after having been decomposed or changed into other compounds by reacting with coexisting compounds, are dissolved or come off into the fiber solvent, thus contaminating the solvent.

Further, among the fiber-forming polymers, low molecular weight polymers form a cause of the contamination of the solvent. In order that the fiber solvent thus contaminated by various impurities should be reused in the fiber production process, it has been necessary to subject the fiber solvent containing impurities to a suitable purification process after its recovery.

Among the acrylic fiber solvents, inorganic solvents cannot be purified by distillation as opposed to organic solvents, so that complicated purification processes such as adsorption filtration, precipitation filtration, recrystallization, etc., are employed. By the adsorption filtration process using activated carbon, ionizable impurities cannot be removed. Also, the precipitation filtration process utilizing a difference in solubility can remove only impurities having a low degree of solubility. Further, the recrystallization process suffers various defects such that the removal of impurities is incomplete, a large amount of inorganic solvent remains in the mother liquor and the recovery ratio is low, and these defects render this process impracticable.

In the purification of organic solvents, distillation is usually employed. In this process, however, a loss of solvent can be caused by the decomposition of the solvent per se. In addition, the process possesses a defect that a large amount of energy may be consumed because of a high boiling point of the solvent or contamination of azeotropic compounds. Therefore, a less expensive and more efficient purification process has been demanded.

Therefore, an improved process has been developed by *M. Ozaki and S. Nakajima; U.S. Patent 4,079,000; May 14, 1978; assigned to Japan Exlan Company Limited, Japan.* This process comprises (1) feeding an acrylic fiber solvent containing impurities or an aqueous solution thereof to a packed layer of a polymer consisting of one or more kinds of ester type monomers containing at least one polymerizable double bond and at least one ethylene oxide unit per molecule or a polymer consisting of such monomer(s) and one or more kinds of monomers containing at least one polymerizable double bond for one molecule, to cause the layer to catch the impurities and fiber solvent, (2) passing an eluent through the packed layer thus eluting the impurities together with the eluent, and (3) passing a further amount of the eluent through the packed layer thereby separating an eluate containing the fiber solvent.

Using this process, the following solutions have been rid of specific impurities as shown in the following tabulation:

Fiber Solvent Solution	Impurities
5% aqueous sodium thiocyanate solution	Polyethylene glycol 600, polyethylene glycol 6000 and ethylene glycol, in concentration of 0.1%, respectively
1 N aqueous nitric acid solution	Sodium sulfate and sodium β-sulfopropionitrile, in concentration of 1%, respectively
5% aqueous zinc chloride solution	Sodium sulfate in concentration of 1%
40% aqueous dimethylformamide solution	Sodium sulfate in concentration of 1%
43% aqueous dimethyl sulfoxide solution	Sodium sulfite in concentration of 1%

SOIL MODIFIERS FROM TITANIUM DIOXIDE PROCESS WASTE

It is known that in the manufacture of titanium dioxide from ilmenites or ilmenite slags by the sulfate method, one of the more serious problems is constituted indeed by the waste materials.

In fact, owing to the scant possibility of using these waste materials, considering also their high impurity content, and since their direct disposal causes considerable pollution, as their reducing properties, their toxicity and the large amounts involved lead to considerable environmental changes, the need is evident to convert them suitably into products devoid of the drawbacks described above. Obviously the best solution would be to convert the waste products of the manufacture of titanium dioxide into materials which can be advantageously used.

That is the aim of a process developed by *L. Piccolo and A. Paolinelli; U.S. Patent 4,077,794; March 7, 1978; assigned to Societa Italiana Resine S.I.R. SpA, Italy.* In this process, a soil modifier comprising from 80 to 95% by weight of a mixture of ferric oxide and ferric sulfate in an oxide/sulfate weight ratio of from 0.25:1 to 1:1, is prepared by calcining at 500° to 600°C in an oxidizing atmosphere a blend of by-product obtained in the manufacture of titanium oxide by the sulfate process, with at least 20% by weight with respect to the by-product, of a material based on ferric sulfate and/or ferric oxide. Calcining is then continued under the same conditions, replacing the material in the blend continuously fed in by a recycle fraction of the product continuously discharged.

STAINLESS STEEL FROM STAINLESS STEEL SCRAP

See "Silver from Silver-Brazed Stainless Steel Scrap."

STEEL FROM MACHINING SCRAP

A process developed by *G. Mathern, J. Lefevre, R. Tricot and A. Guessier; U.S. Patent 4,077,108; March 7, 1978; assigned to Ugine Aciers, France* is a process for upgrading metallurgical products in the divided state which are very often regarded as waste, such as scrap from sheet metal, turnings, cuttings, chips and granules.

This process yields steels having improved machinability. This process provides for the upgrading of divided metallurgical products, which are consolidated without fusion, with incorporation of a dispersed phase. The process is used in particular for scrap consisting of sheet metal or turnings, which is mixed in the desired proportion with an additive such as, for example, sulfur or lead. After it has been compressed the mixture is extruded in a press in the form of bars and the like.

The process is particularly suitable for producing stainless steels containing sulfur or lead. The lead stainless steels thus produced have an excellent corrosion resistance, in addition to their machinability characteristics.

STEEL FROM SCRAP AUTOMOBILES

Steel mills require high density high grade scrap for use as charges for steel furnaces in order to produce the requirements of industry. Unfortunately, it is extremely difficult to obtain sufficient quantities of high grade steel scrap because the sources of such scrap are heavily contaminated with materials adversely affecting the physical and chemical characteristics of any steel produced from such scrap.

One of the major sources of low grade scrap steel is discarded or junked automobile bodies which have been mechanically densified in various ways. It is, for example, common practice to first dismantle the bodies by physically removing the doors, trunk lids and engine hoods, and then cut up the remainder of the body and/or frame to provide relatively large panels or sections which may then be mechanically shredded with mangling devices to produce smaller metal fragments. These fragments are then sometimes processed through a magnetic separating apparatus to roughly sort the basically ferrous from the nonferrous metals.

The shredded materials, whether separated or not, may then be compressed by presses or drop hammers into sheets of material which subsequently may be sliced into strips and then baled. Sometimes the shredded material is baled directly, and in some cases the large sections are compressed into a bale without having been first shredded. In any event, all of these methods result in relatively low density, highly contaminated scrap material.

A number of processing plants for processing vehicle bodies into scrap or for converting vehicle bodies into reusable steel or high grade pig ore are well known in the art. However, the processing plants suffer from a number of deficiencies. For example, the source of fuel utilized in the plants is frequently very expensive and, furthermore, effective use is not made of the heat produced in the melting furnaces. In addition, because of the emphasis being placed upon the elimination of air pollution, it is essential that vehicle processing plants provide an effective means for removing the various pollutants from the end of the flue disposed downstream of the furnace.

In some cases, attempts to reduce the contaminants by burning of the discarded automobile bodies have been attempted, but the results have not been satisfactory because of the relatively low temperatures achievable by open dump type burning. Moreover, most communities prohibit such burning procedures under the air pollution statutes and ordinances. The low grade of scrap which consequently results from the abovedescribed treatments of discarded automobile bodies is not readily salable so that such old and discarded automobile bodies are not in demand for scrap salvage purposes.

As a result, more and more communities have large numbers of old abandoned cars accumulating on their streets and roadways. Salvage yards have found it more and more difficult to dispose of such car bodies when they have been stripped of whatever salable parts were salvageable.

For additional detail on material recovery from scrap automobiles, the reader is referred to the earlier volume by this author (3) and to references cited therein including:

> "Automobile Scrapping Process and Needs for Maryland," by Management Technology, Inc., Report SW-10d, Washington, D.C., U.S. Public Health Service, 1970.
>
> "Automotive Scrap Recycling–Processes, Prices and Prospects," by J.W. Sawyer, Jr., Baltimore, Md., Johns Hopkins University Press, 1974.
>
> "The Automobile Cycle–An Environmental and Resource Reclamation Problem," Washington, D.C., U.S. Environmental Protection Agency, 1972.

No really new technology has been developed since the previous review (3); hence the brief coverage here.

A process developed by *P.E. Rhinehart; U.S. Patent 3,917,239; November 4, 1975* provides a vehicle recycling plant wherein scrap vehicles are at least partially disassembled, processed and converted into suitable products. More particularly, the vehicle recycling plant is used for processing automobile bodies and automobile parts wherein the conversion of the vehicle body and vehicle parts into other products can be accomplished utilizing an inexpensive source of fuel and wherein the residual heat produced during the burning process can be effectively utilized to provide a number of functions.

The vehicle processing plant makes maximum use of the heat generated during the melting down of the vehicle without measurably contributing to air pollution.

The arrangement of equipment which may be used in the conduct of such a process is shown in Figure 153. As shown, the vehicle recycling plant comprises a furnace **1** into which scrap tires **2** are introduced through opening **3** as a source of fuel therefor. A conveyor belt **4** can be conveniently used to introduce the tires into the furnace. The furnace is provided with a plurality of burners **5** which are not only utilized to initiate the combustion of the tires in the furnace but also function as a supplemental heat source. The fuel utilized to supply the burners can be any conventional type of fuel known in the art.

The bottom of the furnace is provided with sloping surfaces **6** containing a plurality of apertures **7** for the removal of ash from the furnace. The sloping surfaces of the bottom of the furnace not only facilitate the removal of the ash from the system but also the removal of steel beads collected from the burning of steel belted tires, the steel beads being removed through a central aperture **8** disposed at the apex of the sloping surfaces. Advantageously, the bottom of the furnace can be provided with a vibrating means **6'** and as such, would function as a vibrating hamper for enhancing the removal of the ash and the steel beads from the furnace.

A container **9** is disposed beneath the aperture **8** for collecting the steel beads recovered from the burned tires. Of course, the steel recovered from the burning of steel belted tires can be subsequently processed and used to produce other elements. As stated above, a conveyor belt is provided at the inlet to the furnace for introducing rubber tires thereto as the source of fuel.

Immediately above the furnace and disposed at a slight angle with respect thereto, is located another conveyor system **10** for introducing scrap vehicles **11** into the vehicle recycling plant of this process. The conveyor system is disposed in close proximity to the furnace so that the heat radiating from the furnace can be effectively utilized to burn undesirable materials and initiate the melting down process of the scrap vehicles.

Advantageously, the conveyor system and the vehicles introduced into the system above the furnace are housed in a duct **12** so that the radiation heat which is conducted through the duct to the scrap vehicles is not lost to the environment. Although the figure shows the vehicles being conveyed on a conveyor **10** substantially parallel to the top of the furnace, the conveyor can be slightly inclined toward the top of the furnace.

The furnace is provided with a heated gas outlet **13** which communicates with a cupola at approximately the position where the scrap vehicles are introduced into the cupola. Accordingly, the scrap vehicles are first exposed to a substantial amount of heat as they are conveyed above the furnace and furthermore, the extremely high temperature gas which is exhausted from the furnace also contacts the vehicle as it is discharged into the cupola for purposes of being melted down to the basic metal.

Because the heated gas exhausted from the furnace communicates with the cupola, the gas functions as one of the main heat sources for the cupola. However, since additional heat may frequently be needed to complete the melting down process, carbon electrodes **14** capable of producing, for example, 3,200 kVA current are disposed in the cupola and are adapted to be lowered above the molten steel to completely convert it to a liquid phase.

Figure 153: Rhinehart Process for Vehicle Recycling for Steel Recovery Featuring Energy Economy

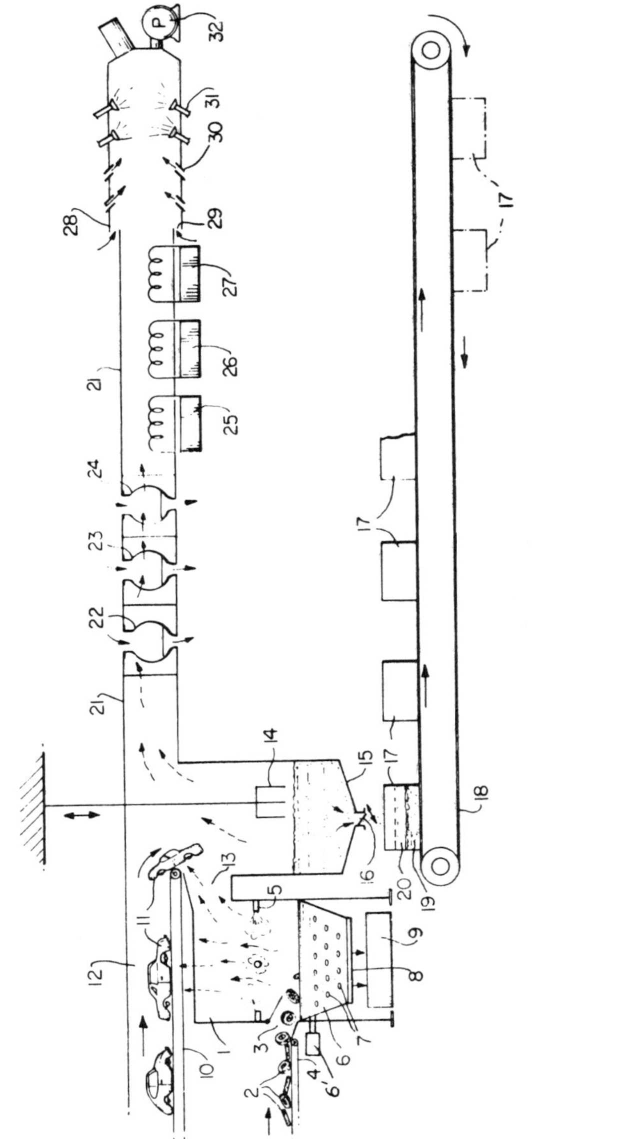

Source: U.S. Patent 3,917,239

The carbon electrode heating source is capable of increasing the temperature in the cupola up to about 3000°F and even higher. The carbon electrodes are automatically adjustable in the vertical position so that a specific distance can be automatically maintained between the surface of the molten metal and the electrodes. Once the steel disposed in the cupola has been finished off, that is, completely converted to a molten stage, the carbon electrodes are automatically withdrawn from the cupola. The bottom of the cupola **15** is sloped to facilitate the removal of the molten steel therefrom.

When the steel in the cupola has been completely converted to the molten state, the steel is removed from the cupola through an aperture **16** provided in the bottom of the cupola and introduced into a holding tank **17** disposed upon a conveyor belt **18**. In the holding tank, the molten steel separates into two levels, the heavier steel level **19** disposed in the bottom of the tank and the lighter slag level **20** disposed in a layer above the steel layer.

The holding tank is then conveyed to any desired location and after a suitable period of time when the slag has sufficiently separated from the steel, the steel is drained from the holding tank through an aperture provided therein and further processed in, for example, a rolling mill to manufacture light structural materials, such as, for example, reinforcing bars, fence posts, and the like. The remaining slag contained in the holding tanks may be disposed of in any suitable manner as is well known in the art.

As shown, the holding tanks can be suitably attached to the conveying means so that after the contents contained therein have been discharged, the containers can be returned to the cupola for reuse.

As will be noted, since the exhaust from the furnace communicates with the cupola, the heat produced in the furnace as well as the heat produced by the carbon electrodes combine to produce the molten slurry in the cupola. The heated gas from the furnace and the cupola is continuously removed from the system through a substantially horizontal flue **21** which contains a plurality of devices for making effective use of the amount of heat contained in the exhaust gas. For example, the flue **21** is provided with a number of smelters **22, 23** and **24** where various component parts of the vehicle which were previously separated from the vehicle shell can be melted down and made into suitable products.

Of course, the products which require the greatest amount of heat to melt them down are introduced into the smelter closest to the cupola. For example, in copper smelter **22**, alternators, radiators, generators, starters, voltage regulators, and the like are introduced thereto. Another smelter (not shown) can be utilized to process the motor block of the vehicle. Thus, after the motor has been disassembled to remove the steel portions, the cast iron portions of the motor are introduced into a smelter where they are melted down for resale to foundries.

A zinc smelter **23** can be utilized for handling transmissions, carburetors, and the like, and a glass smelter **24** can be utilized to melt down all of the glass contained in the vehicle. The horizontal flue stack **21** can be provided with other smelters as well as a number of other heat recovery elements such as, for example, elements **25, 26** and **27** which can be used to produce steam for running generators, etc.

Following the metal processing units and heat recovering units disposed in the horizontal flue, the hot gas traveling through the flue must be further cooled down and cleaned up before it is exhausted to the atmosphere. Accordingly, the flue **21** is provided with an enlarged section **28** which is sufficiently large to provide a gap **29** between the inside diameter of the enlarged section and the outside diameter of the flue. Thus, cool ambient air is drawn into the system through this air gap **29**.

The enlarged section is also provided with a plurality of air vents **30** which are angled in the direction of the flow of the exhaust gas so that the air introduced therethrough thoroughly mixes with the hotter gas flowing therein thereby producing a uniformly cooler exhaust gas. Downstream of the air vents, the enlarged section of the horizontal stack is further provided with a plurality of spray nozzles **31** which are provided for removing any existing contamination in the exhaust gas. Through the use of the vacuum produced by the pump **32** disposed in the end of the horizontal flue, the gas is finally discharged to the atmosphere at a substantially reduced temperature and substantially free from contamination.

In reviewing the overall system as described above, it is clear that the vehicle recycling plant makes maximum use of the heat produced in the recycling plant for providing a number of different functions and through the effective use of the heat also substantially converts all of the vehicle components into reusable products, including, for example, the use of tires as an inexpensive source of fuel. Tires are particularly desirable because their sulfur content produces a very high temperature in the furnace.

The same technical ground is covered by *P.E. Rhinehart; U.S. Patent 4,014,681; March 29, 1977.* That disclosure includes a block flow diagram of the process as shown in Figure 154.

STEEL FROM SCRAP STEEL

In various industries, such as the foundry industry and the production of cast iron and steel, it is customary to charge a furnace with scrap metal. In foundries the scrap metal is almost entirely the source of raw material employed to obtain the casting metal. In the basic iron and steel industry, great quantities of scrap are employed in the furnaces in combination with iron ore.

The type of scrap material used may take various forms. One source of scrap is such items as sheet shearings, punchings, turnings, chips, gates, risers and scraped castings. Other sources of scrap, which have become of increasing use, are scrap automotive parts such as shattered engine blocks, transmissions, and differentials.

One of the problems in using scrap is the presence of various contaminants such as oil and grease, especially in the automotive scrap. Where the scrap has been stored outdoors, there will be entrapped moisture present in the scrap. It is undesirable to charge such contaminated scrap into a furnace since the entrapped moisture and other readily vaporizable material, such as grease, will expand rapidly in the furnace with the effect of an explosion.

Figure 154: Block Flow Diagram of Rhinehart Process for Recovering Steel and Other Components from Scrap Automobiles

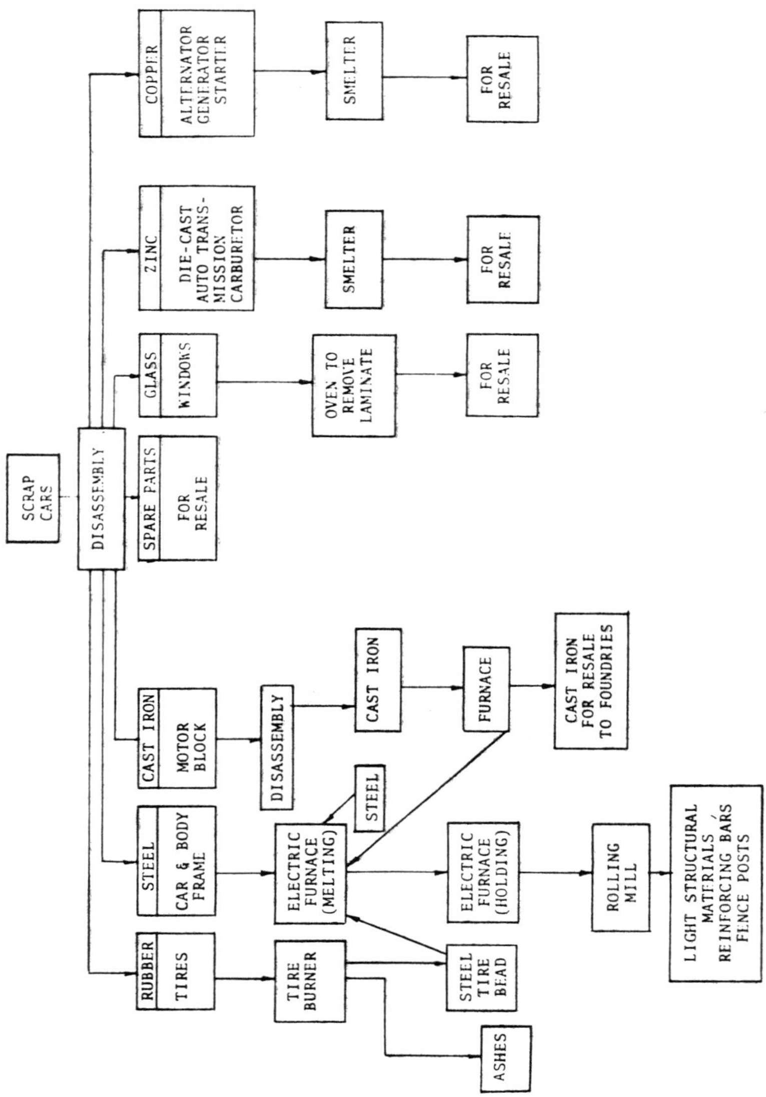

Source: U.S. Patent 4,014,681

In addition, grease and oil which does not vaporize immediately can result in contamination of the molten metal.

Charging of cold scrap into a furnace will increase the time for the batch to reach pouring temperature as well as the load on the furnace as compared to preheated scrap. Thus, many foundries desire preheating scrap to improve furnace output and efficiency.

For the foregoing reasons, various types of equipment have been developed for the purpose of preheating scrap and removing moisture and grease or oil from the scrap. One such type of equipment is a preheater which employs an elongate conveyor which passes scrap under a hood. The hood includes a plurality of burners that direct flames upon the scrap metal on the conveyor to burn off the moisture and grease and preheat the scrap.

One of the problems encountered in metal preheating systems of this type is the pollution in the form of smoke and unburned hydrocarbons which results from operation of the system. One way in which to reduce the amount of unburned hydrocarbons is to permit an excess of air under the hood beyond that needed for the burner flame to aid in complete burning of the grease and oil on the scrap. This could be accomplished by operating the burners with an excess of air or by permitting outside air to enter under the hood. However, the presence of excess air under the hood results in oxidation of metal which is undesirable.

As a result of the oxidation problem, preheating systems of the hood type are generally operated with a reducing atmosphere. The resultant heavy concentration of unburned hydrocarbons is permitted to escape to the atmosphere in some cases. In other cases, they are vented from the furnace and afterburners are used to clean up the pollutants in the exhaust gases.

Another problem encountered in the conveyor-hood type preheating system is the uneven preheating of the scrap. The thickness of the bed of scrap can vary along the conveyor. Additionally, the spacing of the burners, by necessity, must be spread out along the hood. As a result, the flames from the burners will impinge more directly upon the scrap directly beneath the burners than on the scrap between adjacent burners.

Further yet, the flames from the burners only strike the top surfaces of the scrap. The sides and underneath portions of the scrap as well as that positioned deeply in the bed are not engaged as directly by the flames from the burners resulting in less uniform heating and contaminants thereon being removed to a lesser degree.

Therefore, an improved scrap preheating system has been developed by *J.R. Fellnor and T.J. Dzakowic; U.S. Patent 4,083,675; April 11, 1978; assigned to Melting Systems, Inc.* It incorporates a scrap preheating system of the type employing a conveyor upon which the scrap is placed, an elongate burner hood surrounding the conveyor and burners within the hood which direct flames upon the scrap.

The side walls of the burner hood include an elongate exhaust chamber and side wall ports between the exhaust chamber and the interior of the hood. An exhaust fan induces a negative pressure under the hood through the exhaust chambers which pulls the unburned hydrocarbons from under the hood through the side wall ports into the exhaust chamber. Additionally, air is drawn upwardly between the conveyor and side walls of the hood and through the side wall ports into the exhaust chambers wherein combustion is completed and pollution reduced. The length of hood and burner array exceeds the length of scrap material on the conveyor.

The conveyor is periodically jogged during the preheating cycle to slightly advance and mix the scrap and only those burners positioned above the scrap are operated as it advances. The burners are initially operated with a slightly oxidizing flame until hydrocarbons are burned from the scrap and thereafter operated with a slightly reducing flame to prevent significant oxidation of the metal.

Electrothermal smelting of steel from scrap has been reserved mainly for electric arc furnaces. Other smelting equipment, such as induction furnaces, find their use in the steel industry limited by the special prerequisites of charge materials, as well as by the limitations on practicable slag metallurgy.

The advantages of electric arc furnaces, such as high smelting capacity due to concentration of great quantities of energy, considerable independence from the physical quality of the scrap (however, the latter having an essential influence on energy consumption), as well as the possibility of a controlled slag metallurgy, are accompanied by disadvantages. They are noise pollution, particularly during the smelting period, strong fluctuations in power requirements, particularly during the smelting periods and consequently great demand on the power lines, and high combustion losses due to the direct effect of the arc on the scrap during the smelting period, and on the steel bath during the refining period. This is accompanied by great iron losses due to evaporation, as well as by the accumulation of large quantities of fumes.

Other disadvantages include increased iron losses due to working with slag of relatively high iron content with relatively high specific slag quantity, increased energy losses due to constant working with an arc, which as an unfavorable thermal efficiency factor resulting from radiation losses, and working with relatively great specific slag quantities so that the kinetically unfavorable border area reactions between the steel bath and the slag take place within a reasonable period of time.

Still further disadvantages include absorption of nitrogen from the furnace atmosphere by ionization in the arc, and the consequent influence on the bare steel bath at the focal point of the electrodes, utilization of expensive graphite electrodes, and the limitation of furnace capacity resulting from limited diameter of graphite electrodes.

Also, there is great wear on the refractory lining material especially caused by radiation heat from the arc, by the additional mechanical and thermal stresses of the charge and tilt operations, as well as by the additional chemical attack on masonry and particularly the cap by the iron contained in the flue gas, burning in the arc and oxidizing in the furnace atmosphere, and the intense heat, especially due to the charge-type feeding of scrap and, during refining periods, operating with partially opened furnace doors.

It is the purpose of a process developed by *G. Rath; U.S. Patent 4,119,454; October 10, 1978; assigned to Demag AG, Germany*, while maintaining the advantages of the electric arc furnace, to eliminate the disadvantages noted above with respect to pollution of the environment, to improve the specific consumption and operating data, as well as to obtain optimum thermal and metallurgical performance in the entire process. The disadvantages are eliminated by having the scrap continuously smelted in a bath constantly covered by a layer of slag, whereby heat is supplied by means of electrodes immersed in the slag layer.

A suitable form of apparatus for use in such a process is shown in Figure 155. A stationary electric arc reduction furnace is heated by means of electrodes **1**, preferably preannealed carbon electrodes, such as Soderberg electrodes. The furnace cap **2** accommodates at **3** a device for continuous feeding of reduced scrap. At **4**, a charge or feed device is provided, by means of which scrap quantities which are connected, by welding, for example, are supplied.

Figure 155: Demag AG Process for Improved Electric Furnace Smelting of Steel Scrap

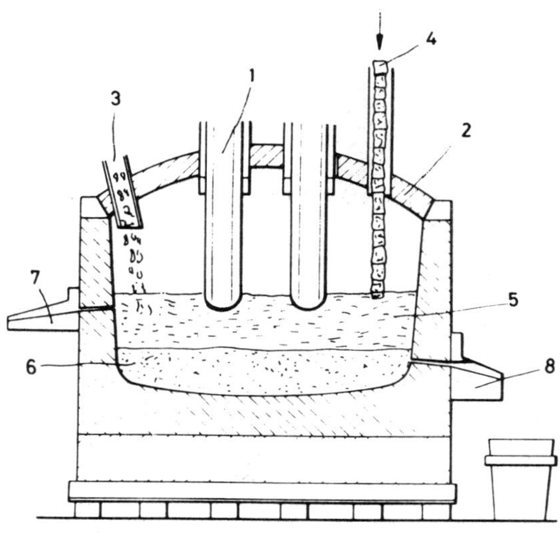

Source: U.S. Patent 4,119,454

Within the furnace, a thick slag layer **5** is maintained constantly covering the metal bath **6**, with the electrodes being immersed in the slag layer **5**. Slag tapping is done at **7**. The metal tap **8** is provided below tap **7** at the distance of the desired minimum height of the slag layer.

The method includes maintaining a continuous slag layer in the furnace vessel of a specific thickness and temperature, maintaining the furnace electrodes immersed in the slag layer, preheating the charge material to a specific range, and relating the charging speed to the processing speed of the furnace under prevailing operating parameters.

By immersing the electrodes in the slag layer and avoiding direct contact thereof with the preheated charge material, heat transfer takes place between the electrodes and the slag, causing the slag layer to apply heat directly and continuously to the charge as it passes through the slag layer, thus avoiding violent reactions at the charge-slag interface, avoiding heat loss by radiation and causing continuous decarbonizing, desulfurizing and dephosphorizing of the charge throughout the slag layer.

STEEL FROM WHOLE SCRAP TIRES

See "Reclaimed Rubber from Whole Scrap Tires" in Reference (1) as well as "Blast Furnace Fuel from Scrap Tires" in Reference (1).

STRONTIUM NITRATE FROM TRACER AMMUNITION

See "Magnesium from Tracer Ammunition."

SULFATE COOKING LIQUOR FROM SPENT PULPING LIQUOR

In a modern Kraft Pulp Mill, the chemical recovery boiler represents the single most expensive item of capital investment. Black liquor from pulping and brown stock washing is evaporated to 55 to 65% solids and sprayed in the boiler. Here it is burned to recover heat and pulping chemicals.

A process developed by *K.O. Henricson; U.S. Patent 4,093,508; June 6, 1978; assigned to A. Ahlstrom Osakeyhtio, Finland* is one for the recovery of chemicals from the waste liquors of sulfate cellulose digestion simultaneously with those from the wastewaters of bleaching.

A melt from a soda-ash roaster, mainly containing sodium sulfide and sodium carbonate, but also sodium chloride, is dissolved and clarified to produce green liquor containing carbonate, sulfide and chlorides. The carbonate is separated from the sulfide and the chlorides, at least part of the sodium chloride is separated from the sulfide in solution, at least part of the carbonate is causticized into hydroxide, and the hydroxide and sulfide in solutions are mixed together in a ratio suitable to form a digestion solution with a desired sulfide content.

The essentials of such a process are shown in the block flow diagram in Figure 156. As shown, the waste liquor is burned in the soda-ash roaster **1** and the obtained melt, which mainly consists of sodium carbonate and sodium sulfide, is dissolved in the dissolving unit **2**, thereby producing green liquor. The wastewater flow **20** from the bleaching unit is fed into the dissolving unit. The green liquor is cleared in the settler **3**, whereafter it is fed into the sodium carbonate separator **4**. The sodium carbonate is separated by crystallization from the green liquor **10** fed into the separator; the crystallization is performed in such a manner that the chlorides remain in the mother liquor. The crystallization can best be performed by cooling, since the carbonate is then precipitated in the form of a decahydrate, which reduces the evaporation requirement.

Figure 156: A. Ahlstrom Osakeyhtio Process for Recovering Chemicals from the Waste Liquors of Sulfate Cellulose Digestion and the Wastewaters of Bleaching

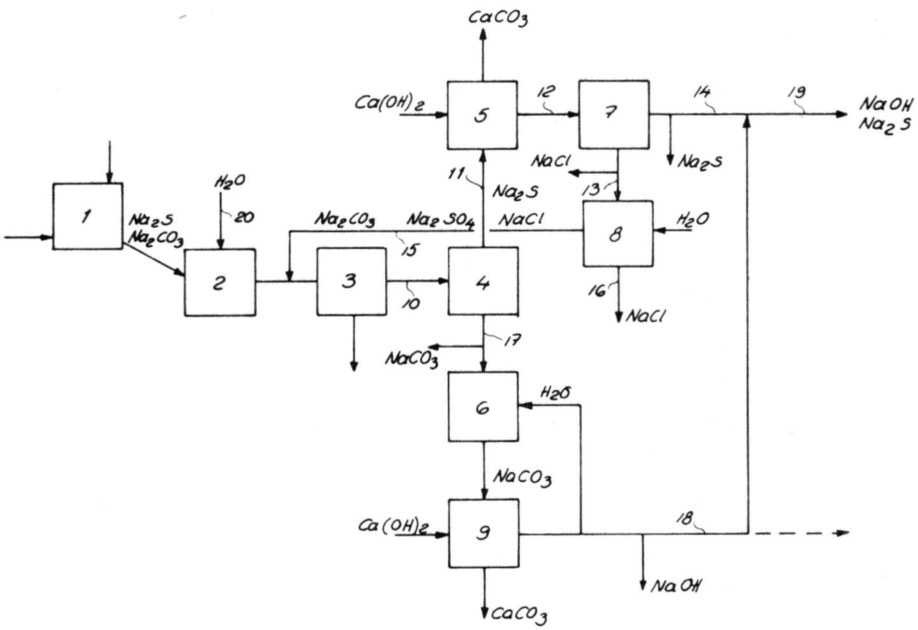

Source: U.S. Patent 4,093,508

The flow **11** from the soda separator, containing a great quantity of sulfide, is fed into the causticization apparatus **5** for the removal of the remaining carbonate, although this is not always necessary. The causticization can best be performed by slaking, in a sulfide solution, the entire quantity of lime required in the liquor system, whereby water is combined with the lime, which reduces the evaporation requirement in the crystallization apparatus **7**. Simultaneously, the large quantity of lime produces a rapid, high-degree causticization.

The crystalline sodium carbonate **17** obtained from the soda separator **4** is dissolved in the dissolving apparatus **6** and fed into the causticization apparatus **9**. Thereby a solution **18** is obtained which mainly contains sodium hydroxide, although certain impurities may be present. The sodium hydroxide solution is used for preparing the digestion liquor, but it can also be used in any unit processes in which sodium hydroxide is required.

The flow **11** consists of a sulfide solution in which certain impurities are present; above all sodium chloride is significant among the impurities. The flow **11** contains only a small part of the sodium carbonate because a great part of it has been separated in the separator **4**. If the flow **11** is causticized, most of the sodium carbonate present in it is converted into sodium hydroxide, in which case it can be assumed that the flow **12** leaving the causticization apparatus contains very little carbonate.

The sulfide flow **12** is directed into the evaporation crystallization apparatus **7**, in which the sodium chloride is precipitated in a crystalline form. It is possible to obtain a sodium chloride concentration of less than 4% by weight in the sulfide solution **14** leaving the crystallization apparatus, while the sodium concentration may rise to more than 25% by weight without the sodium sulfide precipitating in a crystalline form.

The crystalline material **13** contains certain impurities, above all sodium carbonate and sodium sulfate. Their quantities are, however, so small that they can usually be removed from the process together with the sodium chloride.

If the losses incurred during the process grow too great, the sodium carbonate and sodium sulfate can be recovered by leaching the crystalline material in the apparatus **8**. The recovered chemicals **15** can be returned to some stage of the process, prior to the soda separator **4**.

The digestion solution is prepared by mixing the flows **14** and **18** in a suitable ratio so that the correct sulfide content of the digestion liquor is obtained. Excess sodium present in this system can be easily removed in the form of crystalline sodium carbonate from the process flow **17**. The excess sulfur can be removed in the form of a concentrated sodium sulfide solution from the process flow **14**. The excess chlorine is removed in the form of crystalline sodium chloride from the process flows **13** and **16**.

A process developed by *G.C. Dehaas; U.S. Patent 4,135,968; January 23, 1979; assigned to Weyerhaeuser Company* is one for significantly increasing the capacity of and decreasing the pollution from an existing recovery boiler.

The spent liquor is concentrated to 55 to 65% by weight solids and divided into two portions. One portion, containing 10 to 65% by weight of the solids is pyrolyzed, reducing the original fuel value of that portion by 25 to 70%. The remaining carboniferous char and inorganic material is carried to the recovery furnace. The other portion, containing the remainder of the solids, is carried directly to the furnace. The two portions may be combined prior to entering the furnace. In the latter case, the solids content of the combined portion should not be greater than 80% by weight.

Such a process is shown in Figure 157. As shown, the material **10** from the pulp digestion system enters the separation system **11** in which the pulp **12** is separated from the spent liquor **13**. Spent liquor enters the evaporation system **14**, normally multieffect evaporation, in which water **15** is removed from the liquor and the solid content of the liquor is raised to between 40 and 65% by weight. The evaporators may be followed by a concentrator, not shown, as is typical in the industry, to raise the solids content to 55 to 65% by weight.

Figure 157: Weyerhaeuser Company Process for Spent Sulfate Cooking Liquor Recovery Featuring Increased Capacity and Decreased Pollution

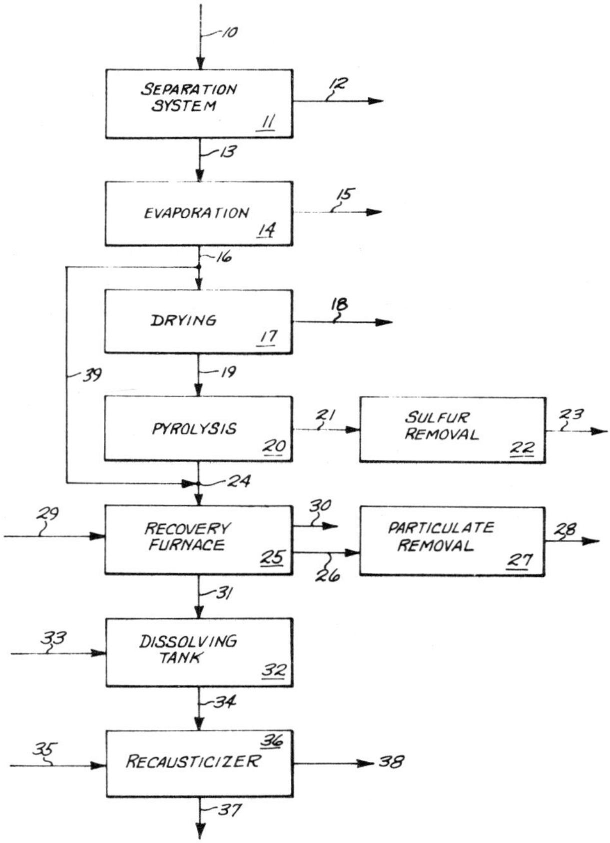

Source: U.S. Patent 4,135,968

A portion of the concentrated liquor may enter a drying system **17** to remove additional water **18**. A number of systems of drying known in the prior art are suitable for this step. Typical of these would be flash drying or spray drying. A modified spray drying system in which the liquor is atomized into superheated steam has been found to be quite effective. The saturated exhaust steam from this dryer may be used in numerous plant locations, such as the first effect of the evaporator. The remainder of the concentrated waste liquor stream **39** goes directly to the recovery boiler **25**.

The dried spent liquor **19**, preferably about 90 to 100% solids, enters pyrolysis unit **20** in which some of the organic material in the liquor is converted to fuel gas **21**. This gas may be sent through a sulfur removal unit **22** and the cleaned fuel gas **23** used for its energy values.

The residue **24** from the pyrolysis unit, containing both organic and inorganic material, enters a standard recovery boiler **25**. If it is not combined with stream **39** but remains in a solid particulate form and not the usual liquid form, it may be introduced into the recovery boiler in a variety of locations. This could be either the oxidizing, transition or reducing zones. The char could also advantageously be introduced directly onto the smelt bed. In this way, it is possible for the material to enter the boiler in a zone of relatively lower temperature and gas velocity.

Thus, there is less chance for the solids to become entrained as particulate matter or to be volatilized in the hotter oxidizing zone and carried out with the flue gas **26** into the boiler superheater section and out into the environment. The flue gas, however, does pass through an existing particulate removal system **27** before exiting to the atmosphere as clean gas **28**. The heat generated in the recovery furnace is recovered by converting incoming water **29** to high pressure steam **30** in the boiler and superheater section of the furnace.

The inorganic material **31** from the recovery furnace enters dissolving tank **32** in which it is contacted with water **33** to form green liquor **34**. The green liquor is contacted with burned lime **35** in recausticizer **36**. Here the sodium carbonate is reconverted to sodium hydroxide. The recausticized liquor **37** is recycled as cooking liquor. Calcium carbonate **38**, which forms during recausticization, is sent to the lime kiln for burning and reuse.

SULFUR FROM HYDROCRACKING PROCESS EFFLUENTS

See "Zinc Chloride from Spent Catalysts."

SULFUR DIOXIDE FROM SPENT SULFURIC ACID

In several industrial processes there are formed large quantities of by-product or spent sulfuric acids of varying H_2SO_4 content, which contain varying amounts of impurities, of organic and inorganic nature. Spent or waste acids containing organic impurities may, for example, be derived from the sulfuric acid used in petroleum refining operations including the alkylation process, and from sulfuric acid used in sulfation, sulfonation and nitration processes.

Spent sulfuric acids containing inorganic impurities may, for example, be obtained from hydrometallurgical leaching operations, metal pickling operations, or in the process of making titanium dioxide pigment, and the like. Large amounts of such spent acids are produced daily and pose a serious disposal problem. One avenue to resource recovery from such spent acids is recovery in the form of sulfur dioxide.

A process developed by *T.S. Harrer, B.I. Karsay and R.L. Sturtevant; U.S. Patent 4,053,573; October 11, 1977; assigned to Allied Chemical Corporation* is one for recovering sulfur values in the form of sulfur dioxide from spent sulfuric acid by decomposing the spent sulfuric acid at elevated temperature in the presence of elemental sulfur to generate a sulfur dioxide-containing gas stream.

This process involves (a) introducing the spent sulfuric acid into a pool of molten sulfur maintained at a temperature of at least 250°C to thereby generate a gaseous stream comprising sulfur dioxide, vaporous elemental sulfur, and water; (b) cooling the gaseous stream to a temperature above the melting point of the sulfur but below 160°C to condense elemental sulfur therefrom, separating the condensed elemental sulfur and returning it to the pool of molten sulfur; followed by (c) further cooling the gaseous stream from which elemental sulfur has been condensed to condense water therefrom, and separating the condensed water from the gaseous stream.

Such a process is shown in Figure 158. The equipment employed includes reactor **2**, sulfur condenser **5**, sulfur separator **6**, water condenser **9**, water separator **10**, pressure control valves **12** and **12a**, drying tower **13**, sulfur dioxide stripping column **16** with associated reflux condenser **19**, post reactor **21**, circulating pump **22**, sulfur heater **24**, and sulfur burner **28**, together with associated lines and ducts and other auxiliary equipment. Equipment employed is constructed of corrosion resistant material.

Figure 158: Allied Chemical Corporation Process for Recovery of Sulfur Values from Spent Sulfuric Acid as Sulfur Dioxide

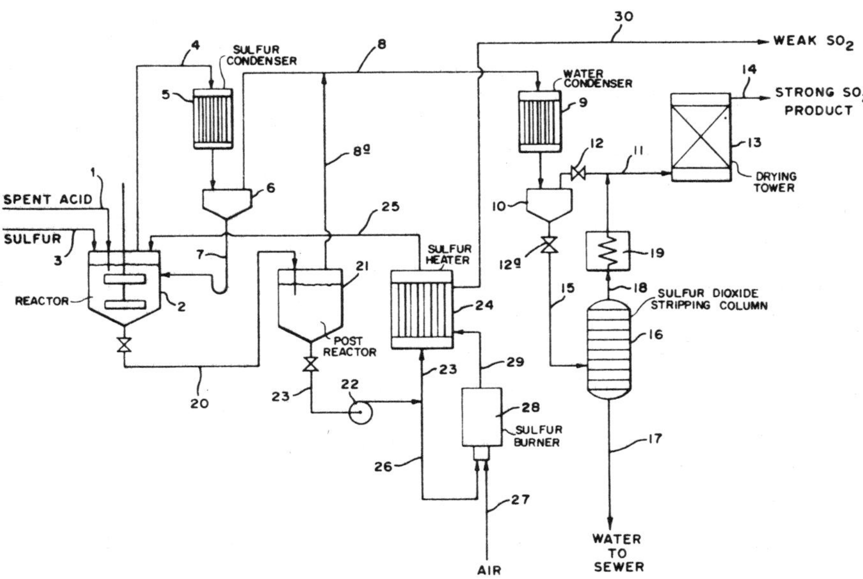

Source: U.S. Patent 4,053,573

Reactor **2** is a closed pressure reaction vessel, agitator-equipped and internally baffled. Reactor **2** contains 30,000 parts by weight of molten sulfur maintained at a temperature of about 370°C. Spent alkylation acid is introduced into reactor **2** through spent acid feed line **1** at a rate of 9,840 parts per hour, representing 9,212 parts per hour of sulfuric acid, 314 parts per hour of organic matter, and 314 parts per hour of water.

Concurrently therewith, molten sulfur is introduced into reactor **2** through sulfur feed line **3** at the rate of 4,480 parts per hour. Reactor **2** and equipment associated therewith is maintained under superatmospheric pressure of about 7 atmospheres. Spent acid fed to reactor **2** is decomposed in the liquid sulfur contained therein to form 8,350 parts per hour of sulfur dioxide, 2,360 parts per hour of water, and 269 parts per hour of carbon.

Gaseous products leaving reactor **2** are passed via line **4** through sulfur condenser **5**, which is a shell-and-tube type heat exchanger wherein the gas stream is cooled to about 140°C to condense 2,668 parts of sulfur per hour. The gaseous stream from which sulfur has been condensed and the condensed sulfur are permitted to separate in sulfur separator **6**, and the condensed sulfur is returned to reactor **2** via sulfur return line **7** which forms a lute. Sulfur generator **6** is an empty vessel with a bottom outlet for liquid sulfur and a top outlet for the gas stream.

The sulfur dioxide-containing gas stream from which sulfur has been condensed is then routed via line **8** to water condenser **9**, a shell-and-tube type heat exchanger, wherein it is cooled to a temperature of about 35°C. In water condenser **9** water is condensed at the rate of about 2,019 parts per hour. Together with the condensed water there is removed from water condenser **9** dissolved sulfur dioxide at the rate of about 220 parts per hour.

Water and gas are routed from water condenser **9** to water separator **10** wherein aqueous and gaseous phases are separated. The gaseous phase is withdrawn from water separator **10** through pressure control valve **12** as a sulfur dioxide-containing gas stream containing between 90 to 100% SO_2, the balance being carbon dioxide, carbon monoxide and some water vapor. Optionally, the sulfur dioxide-containing gas stream released through pressure control valve **12** and line **11** may be further dried, as by contacting it with strong sulfuric acid in drying tower **13** before removal at **14**.

Condensed water separated from the gaseous stream in water separator **10** is discharged therefrom through pressure control valve **12a** and is fed through line **15** into sulfur dioxide stripping column **16** connected by line **18** with associated reflux condenser **19**, wherein it is heated and refluxed to strip the sulfur dioxide. Water from which sulfur dioxide has been stripped is discharged from sulfur dioxide stripping column **16** through water line **17** to the sewer. Sulfur dioxide which has been stripped in stripping column **16** and is passed through reflux condenser **19** may be combined with the sulfur dioxide-containing gas stream passed through pressure control valve **12**.

Molten sulfur is constantly withdrawn from sulfur reactor **2** via line **20** at the rate of about 290,400 parts per hour and passed through post reactor **21**, wherein there is maintained a pool of molten sulfur representing about 290,400 parts of sulfur, providing retention time of about 1 hour. Sulfur dioxide generated within post reactor **21** is vented via line **8a** into line **8** to be combined with the main stream of sulfur dioxide-containing gas generated in reactor **2** for introduction into water condenser **9**.

From post reactor **21** sulfur is withdrawn by means of circulating pump **22**. The stream of sulfur discharged from circulating pump **22** is split into a first and a second stream. The first stream, representing 287,108 parts per hour of sulfur is routed via line **23** through sulfur heater **24**, which is a shell-and-tube type heat exchanger, to be heated therein with hot gases of combustion generated in sulfur burner **28** to a temperature of about 425°C, then to be returned to reactor **2** via line **25**.

The second stream representing 3,300 parts of sulfur per hour is routed through line **26** to sulfur burner **28**, wherein it is burned in contact with air supplied through air feed line **27**. Products of combustion generated in sulfur burner **28** containing about 13.4% by volume of sulfur dioxide are brought into direct heat exchange contact with the sulfur circulated through sulfur heater **24** via line **29**, thereafter to be discharged through line **30** for optional use as sulfur dioxide feed stream for a contact sulfur acid plant.

Combustion on sulfur withdrawn from reactor **2** in sulfur burner **28** effects purge of carbonaceous residue accumulated in the pool of molten sulfur wherein reaction is conducted in reactor **2**. The process upstream of pressure control valves **12** and **12a** is conducted under pressure of about 7 atmospheres, absolute.

SULFURIC ACID FROM SPENT PICKLE LIQUOR

When iron and steel products such as sheets, bands, or wires are descaled or pickled with sulfuric acid, iron dissolves in the sulfuric acid pickle liquor to form ferrous sulfate solution. With continued use, the free sulfuric acid content of the pickle liquor diminishes and the concentration of ferrous ion increases to a point where pickling no longer proceeds at a practical rate, at which point the pickle liquor is considered spent.

The spent pickle liquor may contain from 0.5 to 10% by weight sulfuric acid and up to 10% by weight or more of iron as ferrous sulfate. Practical methods for recovering both the sulfuric acid and the iron values of such spent pickle liquors are becoming increasingly desirable both for environmental and for economic reasons.

A number of processes are known for treating spent pickle liquor. One group of processes involves crystallization of ferrous sulfate heptahydrate from the solution by cooling and/or evaporative concentration of spent pickle liquor. Such processes consume large quantities of energy and yield the iron values in an undesirable form.

A process developed by *J.C. Peterson; U.S. Patent 3,900,955; August 26, 1975* is a simple crystallization process for recovering sulfuric acid and ferrous sulfate crystals from a spent aqueous sulfuric acid pickling solution. The apparatus includes a treating tank, settling tank and crystal strainer. The tanks utilize sloped bottoms to aid in crystal collection, and the settling tank includes baffle plates to promote settling of crystals. The crystal strainer is of a continuously operable type, and the system is automatically controlled.

A process developed by *R.N. Smith; U.S. Patent 4,128,465; December 5, 1978; assigned to Sybron Corporation* is an electrodialytic (ED) process for converting soluble metal sulfates from a pickle liquor into insoluble metal oxides. The pickle liquor stream entering the pickle liquor chamber of ED apparatus under turbulent flow conditions is excited by acoustic energy causing the insoluble oxide to remain in suspension thus preventing the depositing of precipitate on the membrane and environs.

Such a process is shown schematically in Figure 159. There is shown a conventional acid and alkali resistant electrodialytic apparatus **2** divided into chambers **14, 16, 18, 20** and **22**, the chambers being separated from each other by ion permeable membranes **4, 6, 8** and **10**, respectively. Suitable frames and other materials of construction (not shown) well known in the art for defining chambers and assembling electrodialytic apparatuses may be used.

Figure 159: Sybron Corporation Process for H_2SO_4 Recovery by Electrodialysis of Pickle Liquor

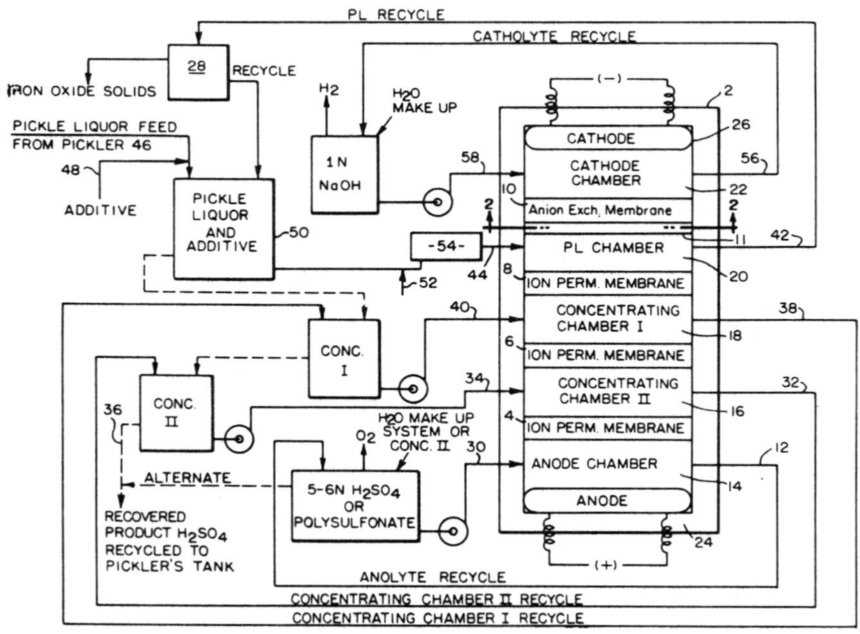

Source: U.S. Patent 4,128,465

Between anion exchange membrane **10** and the cathode is positioned an open mesh plastic screen which serves several functions, i.e., its open mesh forces flow of cathode rinse to fully fill the cathode chamber via a sheet-flow characteristic while permitting ready movement of the hydroxyl ions formed electrically at the cathode to move through the alkaline rinse thus carrying the dc current to the surface of the first anion exchange membrane **10**, therethrough into the pickle liquor chamber stream. Additionally, the screen mesh serves to push the membrane against the pickle liquor (PL) tortuous path frame to force that fluid to flow along the defined tortuous path and thereby minimizing the possibility of the PL stream bypassing portions of the defined flow path allowing portions of the path to be deficient in a vigorously flowing rinse stream.

Similarly, the other face of the tortuous path PL frame has the ion permeable membrane, on that side pushed oppositely against the PL frame to keep the PL rinse fluid flowing exclusively through the tortuous path assuming full flow conditions therein. The other membrane defined chambers containing screen mesh separation from the anode electrode to the PL chamber all serve to force the ion selective membrane **8** against the PL frame to form the second hydraulic face for the tortuous path in the PL chamber.

In the process shown, ion permeable membranes **4, 6, 8** and **10** are anion permeable membranes. Chamber **14** is an anode chamber by virtue of anode positioned therein. Chamber **22** is a cathode chamber by virtue of cathode **26** positioned therein. Pickle liquor chamber **20** is the chamber adjacent cathode chamber **22** and separated from cathode chamber **22** by anion permselective membrane **10**.

Two concentrating chambers designated as concentrating chamber **I** or first concentrating chamber and represented by numeral **18** and concentrating chamber **II** or second concentrating chamber designated by numeral **16** are disposed between anode chamber **14** and pickle liquor chamber **20**. Concentrating chamber **I** is defined by ion permeable membrane **8** and ion permeable membrane **6**. Concentrating chamber **II** is defined by ion permeable membrane **6** and ion permeable membrane **4**.

As with the case of membrane **8** and membrane **4**, membrane **6** may be any membrane or barrier which will permit migration of anions therethrough. Anode **24** and cathode **26** are connected to a suitable source of direct current (not shown). A solids removal device designated by numeral **28** is schematically shown.

Each chamber is provided with inlet and outlet means. Anode chamber **14** is provided with outlet **12** and inlet **30** for circulating anolyte within the chamber. Outlet **32** and inlet **34** connected by a suitable conduit are used for the circulation of electrolyte with concentrating chamber **16**. Conduit **36** is provided for the removal of recovered H_2SO_4 product. Outlet **38** and inlet **40** connected by a suitable conduit are used for the circulation of electrolyte within concentrating chamber **18**. Outlet **42** and inlet **44** are used for the circulation of liquor within spent liquor chamber **20** and the lines are operatively connected with solids removal device **28** by means of suitable conduits.

Conduit **46** provides spent pickle liquor feed to the system and conduit **48** provides chelating agent to maintain a concentration in the pickle liquor feed when needed. **50** is a surge tank for pickle liquor chelating agent mix. Conduit **52** provides gas such as air to the pickle liquor feed and **54** is the means to pulsate or vibrate the feed before entering the pickle liquor chamber **20** via conduit **44**.

Outlet **56** and inlet **58** are provided for the circulation of catholyte within catholyte chamber **22**.

In the operation of the abovedescribed apparatus for the regeneration of spent pickle liquor, the liquor along with a chelating agent is circulated in chamber **20** while an aqueous solution of sulfuric acid enriched concentration chamber electrolyte from lean pickle liquor is circulated within anode chamber **14**; an aqueous catholyte solution, preferably an alkali metal hydroxide is circulated within cathode chamber **22**; and an electrolyte, preferably an aqueous solution of sulfuric acid or lean pickle liquor is circulated within first concentrating chamber **18**.

Lean pickle liquor as used herein is spent pickle liquor from which the free acid and the bulk of the iron has been removed by previous treatment, such as, the effluent liquor from spent liquor chamber **20** of the operating electrodialytic apparatus described herein. The effluent from first concentrating chamber **18** flows from outlet **38** into inlet **34** where it becomes the influent of second concentrating chamber **16**.

A closed loop system of a five-chambered cell is used for concentrating, regenerating and recovering sulfuric acid from a pickle liquor stream. A sulfuric acid pickle liquor stream used to clean iron and steel products and containing metal ions is circulated in chamber **20**. The catholyte in chamber **22** is preferably an aqueous sodium hydroxide solution (1 N) and the anoylte in chamber **14** is preferably an aqueous sulfuric acid solution.

Various streams circulate in their respective chambers via pump and conduits. A direct current of electricity is passed through the cell by means of anode **24** and cathode **26** thus causing cations and anions within the chambers to migrate toward the cathode and anode respectively. Current flow causes two directions of ions flow to occur in solution. Sulfate and hydroxyl anions move toward anode and hydrogen and iron cations in PL chamber move toward cathode. The electrical neutrality is maintained at all times.

Within anode chamber **14** which contains dilute sulfuric acid, hydrogen ions are formed,

$$H_2O \rightarrow 2H^+ + ½O_2 + 2e$$

and migrate toward cathode **26** until they meet anion permeable membrane **4** which normally prevents the migration of cations therethrough. However, a substantial portion of the hydrogen ions formed at anode **24** and migrating toward cathode **26** leak through the anion permeable membrane **4**. These migrating hydrogen ions have a high mobility and enter second concentrating chamber **16** where they are incorporated into the acid electrolyte solution therein and enrich the electrolyte with hydrogen ions.

Similarly, some hydrogen ions in second concentrating chamber **16** will also migrate toward the cathode and enter first concentrating chamber **I** where they are incorporated into the acid electrolyte therein and enrich the electrolyte with hydrogen ions. Likewise, some hydrogen ions from the first concentrating chamber tend to leak slowly into spent PL chamber **20** which contains spent pickle liquor comprising principally iron cations, hydrogen cations and sulfate anions.

Within spent liquor chamber **20** which contains spent pickle liquor comprising principally iron cations, hydrogen cations and sulfate anions, the sulfate anions SO_4^{-2} migrate toward anode **24** and since membrane **8** is anion permeable the sulfate anions pass through the membrane barrier into first concentrating chamber **18** thereby increasing the overall concentration of sulfate ions within the aqueous acid electrolyte solution circulating in first concentrating chamber **18**. The hydrogen ions present in the chamber and those leaking into it from second concentrating chamber with the migrating sulfate anions in chamber **18** form sulfuric acid.

The electrolyte in the second concentrating chamber similarly becomes enriched with sulfuric acid. Except for the highly mobile hydrogen ions, some of which may pass through anion permeable membrane **10** cations formed within spent liquor chamber **20** are met by anion permeable membrane **10** in their attempt to migrate toward cathode **26** and are also repelled thereby and substantially confined to chamber **20**.

Within cathode chamber **22** containing aqueous alkaline catholyte, preferably an aqueous alkali metal hydroxide such as sodium hydroxide or potassium hydroxide, hydroxyl anions are formed at the cathode **26** and migrate toward anode **24** passing through anion permeable membrane **10** into the tortuous pickle liquor chamber **20**. The following reaction is believed to take place at the cathode to produce the hydroxyl ions:

$$2H_2O + 2e \rightarrow 2OH^- + H_2$$

After the free acid has been essentially removed from the PL, the net effect of the operative natural ion migration coupled with the anion permselective barrier between chambers **22** and **20** is the meeting of the iron cations and hydroxyl anions within chamber **20**, whereupon iron hydroxides and/or oxides are formed and are precipitated within chamber **20**. The precipitated solids along with circulating solutions are then carried from outlet **42** to solids removal device **28** and removed as high purity magnetite which is easily removed.

Sulfate anions which migrate from first concentrating chamber migrate through membrane **6** reach second concentrating chamber **16** and which remain unassociated with cations in the electrolyte solution contained in second concentrating chamber **16** pass through ion permeable membrane **4** and enter anode chamber **14** where they can associate with hydrogen ions formed at anode **24** and migrating in the direction of cathode **26**.

In this manner, sulfuric acid is also formed in anode chamber **14** and thereby becomes a part of the anolyte solution. Thus, sulfuric acid is formed in concentrating chambers **18, 16** and the respective electrolyte and anolyte therein becomes enriched with the sulfuric acid.

In the circulation scheme described according to this process, the liquor effluent from spent liquor chamber **20** which passes through conduit **42** to solids removal device **28** and exits therefrom depleted in acid and iron value and depleted in sulfate content becomes enriched with sulfuric acid in first concentrating chamber **18**, in second concentrating chamber **16** and finally in anode chamber **14** where it reaches its maximum sulfuric acid concentration.

Figure 160 is a sectional view of the frame in Figure 159 indicated by the line **2-2**.

Figure 160: Tortuous Flow Path Frame Used in Sybron Corporation Apparatus of Figure 159

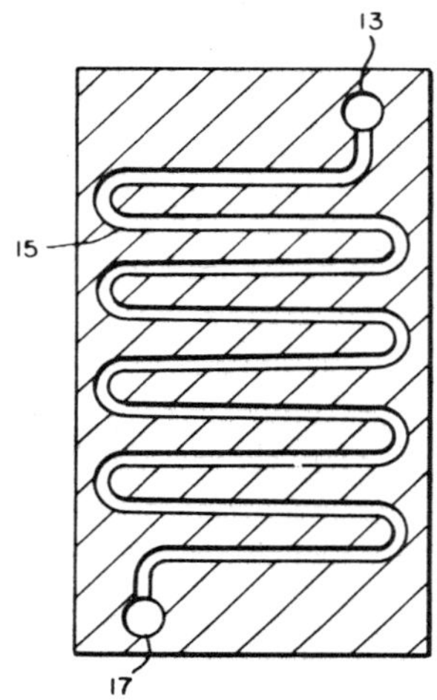

Source: U.S. Patent 4,128,465

A process developed by *Z.L. Burke; U.S. Patent 4,149,946; April 17, 1979; assigned to Davis Walker Corporation* is one in which spent sulfuric acid pickle liquor containing ferrous ion is regenerated and iron metal is recovered therefrom by electrolysis in a cell having an anode in an anode chamber and a cathode in a cathode chamber separated by a cation-selective membrane.

The spent pickle liquor is placed in the cathode chamber, aqueous ammonium sulfate solution is placed in the anode chamber, and electrical current is passed through the cell. Iron plates out on the cathode. Ammonium ion passes through the membrane into the cathode chamber to form aqueous ammonium sulfate solution therein, and the ammonium sulfate anolyte is converted to sulfuric acid solution useful for pickling. Ammonium sulfate solution from the cathode chamber is recycled to the anode chamber and the process is repeated with fresh spent pickle liquor.

A variety of cell designs may be used. Any acid-resistant material of construction, such as plastic or fiber glass, can be used. The simplest cell includes an anode chamber and a cathode chamber separated by a cation-selective membrane.

Another design is shown in Figure 161 wherein an electrolytic cell **10** has a central cathode chamber **7** having a cathode **16** between two anode chambers **3** and **5** having anodes **12** and **14** respectively. Cation-selective membranes **20** and **22** separate the anode chambers from the cathode chamber.

Figure 161: David Walker Corporation Process for Recovery of Spent Pickle Liquor and Iron Metal

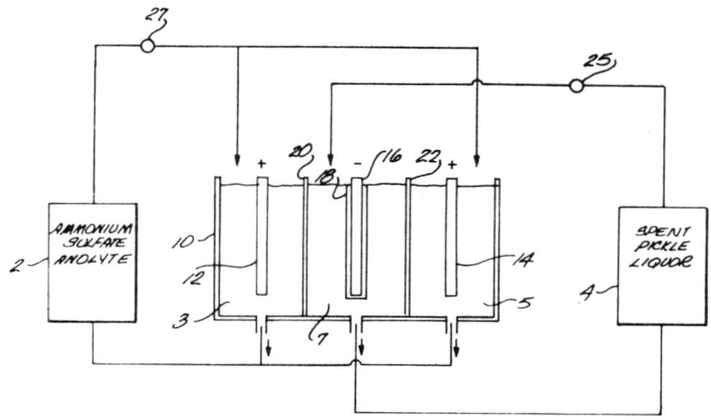

Source: U.S. Patent 4,149,946

A catholyte reservoir **4** contains spent pickle liquor which is circulated with a pump **25** through the cathode chamber. An anolyte reservoir **2** contains ammonium sulfate anolyte which is circulated through the two anode chambers with a pump **27**. A coating of deposited iron **18** is shown on the cathode.

A number of such cells can be connected in series or in parallel both electrically and hydraulically. This type of cell makes use of both sides of the cathode. It can be modified to include a plurality of anode chambers and a plurality of cathode chambers arranged in alternating succession, each pair of adjacent chambers being separated by a cation-selective membrane. In such a cell both sides of each electrode except the end electrodes are used, and the number of anodes required is smaller than the number required in a series of three-chambered cells having the same number of cathodes.

A plurality of cells are used. Spent pickle liquor containing dissolved ferrous sulfate and dissolved ammonium sulfate is continuously introduced into the cathode chamber of the first cell and is passed in series through the cathode chambers to the cathode chamber of the last cell. Catholyte from the last cell is continuously transferred to the anode chamber of the first cell and is passed in series through the anode chambers to the last cell. Regenerated sulfuric acid solution is continuously withdrawn from the anode chamber of the last cell.

A process developed by *R. Gerken, H. Guth, C. Mucke, W. Potes and H. Wieschen; U.S. Patent 4,153,628; May 8, 1979; assigned to Bayer AG and Metallgesellschaft AG, both of Germany* is one in which spent sulfuric acid, either alongside or mixed with contaminated metal salts such as iron sulfate heptahydrate, is regenerated to fresh acid by countercurrent contact with gases from cleavage of the sulfate salts, such gases containing SO_3 which effects build up of the acid concentration.

The built up acid is evaporated in another stage to obtain pure acid. The metal sulfate solids are subjected to high temperature cleavage, generating SO_3 which effects the previous concentration. Cleavage also produces some SO_2 which is converted to SO_3 by wet catalysis.

SULFURIC ACID FROM SPENT SULFURIC ACID

A number of processes have been used successfully in industry for the technically important working up of diluted sulfuric acid. Especially for concentrating higher percent acids the concentration process in outside heated vessels with a mounted dephlegmator according to Pauling or the concentration by direct contact with hot gases has been used.

Difficulties may, however, appear in these processes, if the sulfuric acid contains besides water and gaseous components, further impurities. Organic substances among the impurities such as, for example, up to 2% by weight aromatic compounds resulting from the nitration of benzene or naphthaline derivatives are generally eliminated in the regeneration process by distillation or oxidation.

Mineral salts such as, for example, copper, iron, ammonium, alkaline earth metals and sodium sulfates are, however, removed only partially in the process. When reusing the concentrated acid certain mineral salts may accumulate in the acid rendering impossible a repeated use of the regenerated acid for the same purpose. It was therefore desirable to transform a water-containing sulfuric acid having a content of mineral salts and/or organic impurities into a pure and concentrated sulfuric acid.

A process developed by *J. Alagy, C. Busson and M. Cessou; U.S. Patent 3,935,100; January 27, 1976; assigned to Institut Francais du Petrole, France* is one for treating a wastewater containing sulfuric acid and/or ammonium sulfate comprising (1) reacting the water with barium sulfide to form barium sulfate; (2) reacting the latter with carbon containing inorganic impurities, at high temperature, thereby forming barium sulfide, at least a major portion of which is recycled to step (1), and a gas which is separated; (3) dissolving in water a minor portion of the barium sulfide produced in step (2), separating the insoluble impurities therefrom and recycling at least one portion of the resulting purified solution to step (1).

A process developed by *H. von Plessen and S. Schiessler; U.S. Patent 3,972,987; August 3, 1976; assigned to Hoechst AG, Germany* is a two-step process for regenerating dilute sulfuric acid which may contain organic and inorganic impurities such as heavy metal salts. In the first step the impure acid is distilled in vacuo and in the second step concentrated in a Pauling apparatus.

A process developed by *K. Bodenbenner, G. Müller and H. Müller; U.S. Patent 4,157,381; June 5, 1979; assigned to Hoechst AG, Germany* is one for the regeneration of dilute sulfuric acid containing organic impurities and possibly also inorganic salt in several steps.

In the first step, the dilute acid is concentrated by indirect heat exchange with the vapors of the second step, and in the second step it is concentrated by direct heat exchange with a countercurrent mixture of steam and acid vapors coming from a third step. In this third step, the acid discharged from the second step is mixed in direct heat exchange with superheated steam. The regenerated sulfuric acid is separated from this mixture and the remaining steam/vapor mixture is forwarded to the second step. In the case where the dilute acid contains organic substances, an oxidant is added in at least one of these process steps.

Such a process is illustrated in Figure 162. A 20% sulfuric acid containing salt and organic impurities and having a temperature of 20°C is fed to the heat exchanger **1**, where it is preheated by means of the condensate of the circulation evaporator **2**, before it is forwarded to the latter one.

Figure 162: Hoechst AG Process for Regeneration of Sulfuric Acid

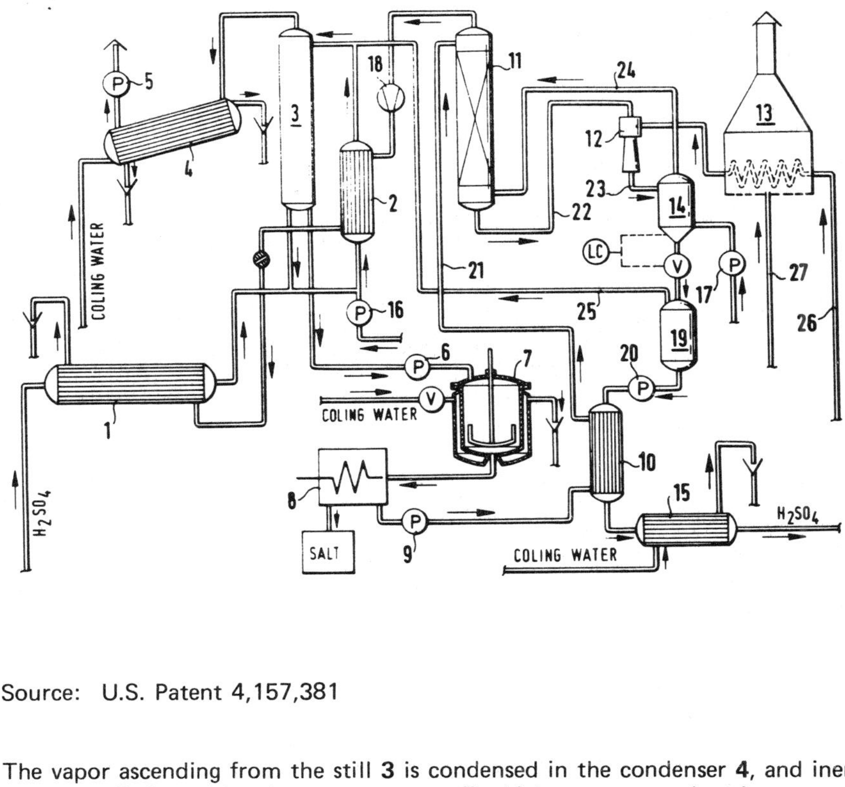

Source: U.S. Patent 4,157,381

The vapor ascending from the still **3** is condensed in the condenser **4**, and inert gases are eliminated by the vacuum pump **5** which ensures a reduced pressure during the evaporation. Part of the acid leaving the still is mixed with the acid leaving the heat exchanger and fed to the circulation evaporator. The remaining part of the acid leaving the still (which corresponds to the amount forwarded from heat exchanger **1** to circulation evaporator **2**), in the case where it contains inorganic salts, is forwarded by means of pump **6** to the agitation vessel **7** provided with a cooling jacket. Precipitated inorganic salt is separated in the centrifuge **8**. The acid leaving the latter centrifuge is forwarded to the heat exchanger **10** by means of pump **9**, and from there to the second step, that is, the column **11**, via duct **21**.

The acid fed in at the top of the column **11** is treated with a countercurrent flow of a mixture of steam and acid vapor. The acid leaving the sump of the column **11** is forwarded via duct **22** to the third step, that is, the jet mixer **12**, where it is thoroughly intermixed with superheated steam being heated to a temperature of from 600° to 700°C in heater **13**, thus forming a two-phase mixture having a temporary equilibrium temperature of about 290° to 300°C.

This two-phase mixture is forwarded via duct **23** to the separator **14**, where the acid is isolated. The mixture of steam and acid vapor leaving the separator **14** at the top is fed to the bottom of column **11** via duct **24**. The vapor leaving the column **11** at its top is fed via the blower **18** to the circulation evaporator **2** in order to serve as heating means. The acid leaving the separator **14** may be depressurized in a suitable depressurizing device, for example, the flash chamber **19** which, at its vapor end, is connected to the outlet of the circulation evaporator **2** via duct **25**.

The steam obtained in the depressurizing operation is utilized in the first step as direct heat conductor. The liquid phase is passed through the heat exchanger **10** by means of pump **20**, where it yields a great part of its heat, while the remaining heat is dissipated via the heat exchanger **15**. Depressurizing in the flash chamber **19** enables one to obtain a final concentration of 97% of sulfuric acid. Oxidants are fed to the individual steps by means of pumps **16** and **17**. Duct **26** serves for the feed of low pressure steam and duct **27** for the feed of fuel.

Example 1: 422 kg/hr of 23.4% sulfuric acid having a black color and a content of 0.6% carbon and containing organic impurities is fed at 20°C to an apparatus as described above, but without agitation cooler **7**, centrifuge **8** and flash chamber **19**. In the circulation evaporator **2**, operating under a pressure of 43 mm Hg, 255 kg/hr of water are distilled off.

The sump of the evaporator having a temperature of 83°C and containing 59% H_2SO_4 is heated to from 110° to 120°C by heat exchange with the hot 94% sulfuric acid in heat exchanger **10** and subsequently fed to the top of the column **11** from where the vapors escape at a temperature of 135°C. The acid leaving the sump of the column **11** is mixed in the jet mixer **12** with 263 kg/hr of superheated steam having a temperature of 620°C.

In the separator **14**, a temperature of 290°C results. 4.6 kg/hr of 100% HNO_3 are fed to the separator as oxidant, that is, 4.4% relative to the concentrated acid, and the carbon content of the 94% acid amounts then to 0.05%. The acid is practically colorless. When the flash chamber **19** is used, the concentration of the acid discharged increases to 96%, and simultaneously, the temperature drops to 205°C.

Example 2: A dilute sulfuric acid containing $FeSO_4$ is composed as follows: 31% H_2SO_4, 65.4% H_2O and 3.6% $FeSO_4$. After evaporation according to Example 1 (the apparatus containing heat exchanger **7** and centrifuge **8** but not flash chamber **19**), a hot acid is formed in this first step which contains 70%

H_2SO_4, 22.7% H_2O and 7.3% salt. After cooling and separating the precipitated iron sulfate, the acid is dehydrated by means of hot steam to have a final concentration of 94%.

SULFURIC ACID FROM TITANIUM DIOXIDE PROCESS WASTE

A process developed by *K.H. Dörr, G. Daradimos, H. Grimm, G. Schmidt, R. Gerken, C. Mücke and H. Wieschen; U.S. Patent 4,163,047; July 31, 1979; assigned to Metallgesellschaft AG and Bayer AG, both of Germany* is for recovery of sulfuric acid from waste sulfuric acid containing iron sulfate and from solid iron sulfate of high water content of crystallization which consists essentially of:

A. concentrating waste sulfuric acid to an acid concentration of 25 to 55% by weight, based upon the suspension, by removing water therefrom;

B. mixing the concentrated acid of Step A with recycled concentrated sulfuric acid obtained from Step E to form a resultant acid mixture of acid concentration of 30 to 65% by weight, based upon the suspension;

C. adding the solid iron sulfate of high water content of crystallization to the acid mixture of Step B thereby obtaining iron sulfate of low water content of crystallization;

D. separating the iron sulfate of low water content from the resultant sulfuric acid solution of Step C;

E. concentrating the separated sulfuric acid solution of Step D to an acid concentration of 45 to 70% by weight, based on salt-free acid, and recycling at least a portion thereof to Step B; and

F. thermally decomposing the iron sulfate from Step D to produce iron oxide and sulfur dioxide.

Such a process is shown in Figure 163, wherein waste acid from holding tank **2** is fed via line **2a** to a concentrator **3** wherein water is removed. The concentrated waste sulfuric acid is removed from the concentrator via line **3a** to an acid mixing tank **4**.

Iron sulfate or green salt is removed from the holding vessel **1** and passed through line **1a** to a dehydration zone **5** into which is passed an acid mixture from tank **4** via line **4a**. This acid mixture comprises concentrated waste acid from vessel **3** as well as concentrated liquid filtrate acid from vessel **8**.

In dehydration zone **5**, the iron sulfate of higher water content is converted to iron sulfate of lower water content by virtue of the acid treatment. The mixture is passed to a filtration zone **6** via line **5a**. In filtration zone **6**, the ferrous sulfate of lower water content is separated off and passed via line **6b** to a thermal dissociation stage **7**.

Liquid filtrate comprising sulfuric acid is removed from the filtration zone **6** via line **6a** and introduced into a concentrator **8** wherein water is removed to provide a more concentrated sulfuric acid containing filtrate. A portion of this more concentrated acid is introduced via line **8a** to the acid mixing vessel **4**, while another portion is withdrawn via line **8b** where it can be employed anew in a TiO_2 solubilizing stage **9**.

Figure 163: Metallgesellschaft AG and Bayer AG Process for Producing Sulfuric Acid from Waste Acid and Iron Sulfate

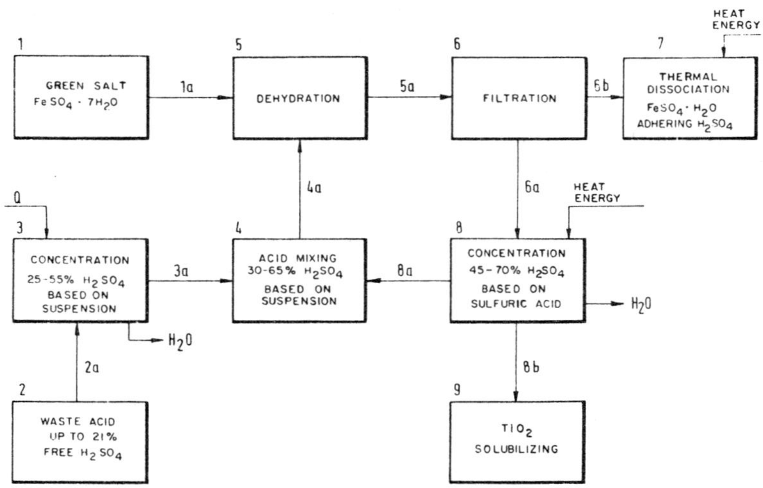

Source: U.S. Patent 4,163,047

The following is a specific example of the conduct of this process. As a result of the solubilizing treatment of Norwegian ilmenite with sulfuric acid, 4 tons of green salt and 8 tons of waste sulfuric acid become available per ton of TiO_2.

The green salt contains 90% by weight of $FeSO_4 \cdot 7H_2O$ and 5% by weight adhering moisture, balance metal sulfate hydrates. The waste sulfuric acid has approximately the following chemical composition: 15% by weight $MeSO_4$; 21% by weight H_2SO_4; and 64% by weight H_2O.

At a rate of 8,000 kg/hr, the waste sulfuric acid is fed from supply tank **2** through conduit **2a** to the strengthening stage **3**, in which heat **Q**, is supplied to the waste sulfuric acid to increase its concentration to 30.4% by weight H_2SO_4, based on the suspension. At a rate of 5,520 kg/hr, the suspension is fed through conduit **3a** to an acid-mixing stage **4**. At a rate of 4,480 kg/hr, additional waste acid containing 65% by weight H_2SO_4 is fed from a strengthening stage **8** to the acid-mixing stage.

As a result, sulfuric acid having a concentration of 45.9% by weight is fed at a rate of 10,000 kg/hr from the acid-mixing stage through conduit **4a** to the dehydrating stage **5**, which is also fed from a dump **1** through conduit **1a** with green salt at a rate of 4,000 kg/hr. The dehydrating stage **5** is thus fed at a total rate of 14,000 kg/hr and contains acid mixture and green salt in a ratio of 2.5:1. The resulting suspension has a sulfuric acid concentration of 32.8% by weight, based on the suspension, and contains about 230 g $MeSO_4$ per kg.

The suspension is filtered in a filtering stage **6**. The resulting filter cake becoming available at a rate of 5,200 kg/hr contains $MeSO_4 \cdot H_2O$ as well as 30% by weight adhering sulfuric acid having an H_2SO_4 concentration of 44.3% and is fed to a dissociating stage **7**. The filtrate which becomes available at a rate of about 8,800 kg/hr is strengthened in the strengthening stage **8** from 44.3 to 65% by weight H_2SO_4 by a supply of heat.

Of the total rate of about 6,000 kg/hr, part of the filtrate, at a rate of 4,480 kg/hr, is fed through conduit **8a** to the acid-mixing stage. The remainder, amounting to about 1,520 kg/hr and containing sulfuric acid at a rate of about 990 kg/hr, may be used to solubilize the TiO_2. This rate corresponds to about 50% of the sulfuric acid content of the free sulfuric acid which is contained in the waste sulfuric acid. The temperature in the filtering stage amounted to 40° to 50°C.

T

TANTALUM FROM SPENT CATALYSTS

A process developed by *R. Hulme; U.S. Patent 4,065,405; December 27, 1977; assigned to Exxon Research and Engineering Company* is one in which tantalum and/or niobium pentafluorides may be recovered from a deactivated or partially deactivated hydrocarbon conversion catalyst comprising: (a) a metal pentafluoride selected from the group consisting of tantalum pentafluoride, niobium pentafluoride and mixtures thereof; and (b) hydrogen fluoride, by distilling the catalyst in the presence of a Lewis acid containing neither of these Group V metals, thereby displacing a pentahalide of tantalum and/or niobium into the vapor phase from which it can be condensed. Addition of hydrogen fluoride then converts the pentahalide to the pentafluoride.

The added Lewis acid serves to displace the metal pentafluoride component from its poisoned complexes. When the added Lewis acid is not itself a fluoride, then, in general, halogen exchange will occur and tantalum and/or niobium pentahalides will be recovered overhead. In a preferred embodiment, hydrogen fluoride is removed from the deactivated catalyst prior to the abovedescribed distillation, i.e., heating and condensation, of the catalyst with the Lewis acid. Uncomplexed metal pentafluoride may also be removed from the catalyst prior to the distillation.

TELLURIUM FROM COPPER REFINERY ANODE SLIMES

As described by A. Petrick, Jr., H.J. Bennett, K.E. Starch and R.C. Weisner (17), the by-product tellurium subsystem duplicates a part of the selenium subsystem during the early stages of processing. This is due to the fact that the soda-nitre slag from the Cupel furnace carries all of the recoverable tellurium and about one-half of the recoverable selenium.

The two are separated at the neutralization tank where tellurium dioxide is precipitated and the neutralized selenium solution goes on to further processing. Therefore, the tellurium subsystem can be confined to the metallurgical flowsheet concerned with further processing of the tellurium dioxide pulp precipitated in the neutralization tank. Such a subsystem is illustrated in Figure 164.

Assuming 90% recovery of tellurium in the Cupel furnace and 90% recovery in the balance of the subsystem shown, the 400 lb of tellurium in the anode mud and slimes result in 324 lb of tellurium metal.

See also "Selenium from Copper Refinery Anode Slimes" and especially U.S. Patent 4,163,046.

Figure 164: By-Product Tellurium Subsystem

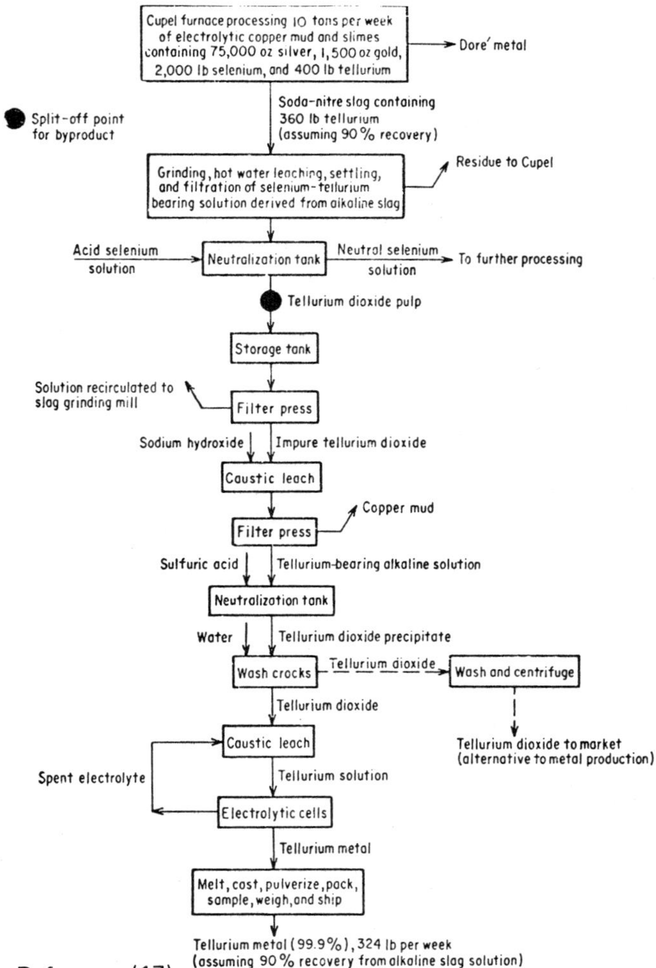

Source: Reference (17)

THALLIUM FROM ZINC SMELTER FLUE DUSTS

The U.S. supply of thallium is derived principally from the processing of zinc concentrates. The American Smelting and Refining Company at Denver, CO, has been the sole domestic producer of thallium for many years. Commercial thallium is derived generally from flue dusts generated in the roasting of sulfides or from cadmium-rich material derived from zinc ores. It is commonly associated with the cadmium by-product system. The first steps of thallium recovery are part of the cadmium subsystem. The final steps of thallium recovery involve precipitation with zinc, purification by reprecipitation and subsequent electrolytic recovery of the metal.

TIN FROM DROSS FROM HOT DIP COATING

In manufacturing facilities where copper products may be tin-plated as by the hot dip method, such plating processes generate tin dross having high tin count and the tin thus represented must be recovered for obvious reasons of economy. In practically any tinning operation, the residue formed contains a quantity of tin and this is what is commonly termed tin dross.

Over the years, tin platers have experimented with a variety of methods for the avowed purpose of holding down tin consumption, methods which for the most part have only demonstrated impracticality and inefficiency. Such methods as have been generally known and accepted in the trade have been identified as creating more problems than they solved and as causing more expense in the recovery program than could be represented in the dollar value of any recovered tin.

Therefore, an improved technique has been developed by *R.M. Gancarz, Sr.; U.S. Patent 4,119,136; October 10, 1978* for recovering tin from tin dross consisting of positioning a barrel having an apertured bottom and charged with a supply of the tin dross upon a partitioned collecting tin mold, with the components thus arranged being placed into a furnace for subjecting the components and tin dross charge to a temperature approximating 700°F for a time duration approximating 5 to 6 hours and allowing the developed molten free tin to flow downwardly of the barrel and through the apertures thereof and into and between the partitions of the collecting tin mold, and withdrawing the collecting tin mold for cooling before removing the cast tin bars therefrom.

TIN FROM HALOGEN TIN ELECTROPLATING WASTES

Among the commercial processes used to produce tin-plated steel strip is the so-called halogen tin electro-deposition process. In this process, pickled and washed steel strip is passed into a bath containing a complex stannous fluoride anion, thought to be SnF_6^{4-} [hexafluorostannate(II)], sodium bifluoride, hydrochloric acid and minor amounts of various addition agents such as grain refiners, all well-known to those skilled in the art.

The strip is made cathodic as it passes into the bath and a metallic tin anode is immersed in the bath. In operation, an electric current is applied causing the stannous fluoride complex ion to approach the cathodic steel strip surface where it is reduced to metallic tin, resulting in the deposition of a layer of tin on the steel strip.

One of the disadvantages of this halogen tin process is that the bath tends to accumulate an ever-increasing amount of sludge during operation, the major component of which sludge is sodium hexafluorostannate(IV), Na_2SnF_6, together with some hydrated stannic oxide, $SnO_2 \cdot xH_2O$. The hexafluorostannate(IV) complex is formed by air oxidation of the hexafluorostannate(II) complex. This oxidation of stannate(II) to stannate(IV) results from the entrapment of air in the plating solution due to vigorous agitation of the plating bath caused by the high line speed of the strip. Although the oxidation is kinetically slow, the rate may be catalytically increased in the presence of dissolved iron, which is present in the plating bath due, at least in part, to incomplete washing of the strip following the acid pickle.

In order to minimize the amount of ferrous ion present in the electroplating bath and thereby prevent catalysis of the hexafluorostannate(II)/hexafluorostannate(IV) oxidation, sodium ferrocyanide, $Na_4[Fe(CN)_6]\cdot 10H_2O$, is periodically added to the bath. The ferrocyanide ion, $Fe(CN)_6^{4-}$, has great affinity for ferrous ion, and will readily combine therewith to form a so-called ferro-ferrocyanide complex ion, which will precipitate from the bath solution as $Fe_2Fe(CN)_6$. This salt is, in turn, slowly oxidized by the air which is drawn into the bath, to ferriferrocyanide, $Fe_4[Fe(CN)_6]_3$, which along with the Na_2SnF_6 and $SnO_2\cdot xH_2O$ then forms a portion of the bath sludge which must be periodically removed. Sodium ferrocyanide additions are then made periodically in order to replenish the ferrocyanide available to remove any ferrous ion present.

The bath sludge is of commercial value due to the presence of the high-grade tin source compound, Na_2SnF_6, and a well-known process is customarily used for its recovery. This process involves treating the sludge with a sufficient volume of hot water to leach the majority of the Na_2SnF_6 from the sludge. The insoluble matter remaining, which is hereafter called secondary sludge, is removed from the hot water solution and, depending on the tin content, discarded or sold to a smelter. The clear solution remaining is then further processed to obtain a tin-bearing compound or metallic tin.

A process developed by *R.E. Horn and D.W. Grenda; U.S. Patent 3,926,759; December 16, 1975; assigned to Pitt Metals and Chemicals, Inc.* is one in which the waste rinse water from a halogen tin-plating process is first filtered to remove insoluble materials such as iron ferrocyanides. The filtrate is then treated to remove residual organic compounds. The treated solution is then rectified in an electrodialytic cell where the solution is introduced into a depletion compartment which is separated from the anode compartment by one or more anionic membranes with compartments therebetween and by a cationic membrane from the cathode compartment.

The anode compartment and the compartments between the anode compartment and the depletion compartment contain an acid solution, preferably hydrofluoric acid. The cathode compartment contains a caustic solution, preferably a sodium hydroxide solution. An insoluble inert anode is positioned in the anode compartment and the current is impressed across the electrodialytic cell. The halogen ions migrate through the anionic membranes from the depletion compartment to the intermediate or receiving compartment and increase the concentration of the halogen acid therein. The sodium ions migrate through the cationic membrane to increase the concentration of sodium hydroxide therein.

The rectified rinse water, depleted of excess halogens, is withdrawn from the depletion compartment and concentrated by a suitable process, such as reverse osmosis, to provide a solution having the desired concentration as a solution for use in the plating bath. The caustic from the cathode compartment and the acid from the intermediate compartment are withdrawn therefrom and reacted to form sodium bifluoride for reuse in the rinse water of the plating process. Such a process is shown in Figure 165.

Figure 165: Pitt Metals and Chemicals, Inc. Process for Recovering Tin Salts from the Waste Rinse Water of a Halogen Tin Plating Process

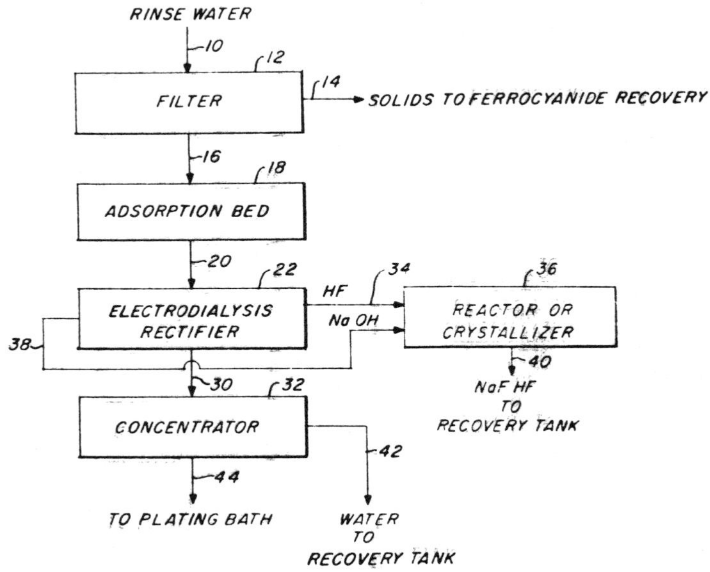

Source: U.S. Patent 3,926,759

Referring to the figure, the waste rinse water from the recovery section combined with any leakage or drippings from the pumps, cells, valves and the like, is conveyed from the recovery section through conduit **10** to a conventional filter **12**. The insoluble materials, particularly iron ferrocyanides, which have been generated by the precipitation of dragged in iron by sodium ferrocyanide, are filtered by the filter **12**. The solids are removed by a suitable conveying device **14** and are conveyed to a suitable device for recovering the ferrocyanide. The filtrate solution is then conveyed through conduit **16** to a tank or column **18** containing a bed of activated carbon. The solution passes through the bed of activated carbon where residual organic compounds are removed therefrom. The treated solution is then conveyed through conduit **20** to an electrodialysis rectifier **22**.

The rinse water flowing through conduit **20** is introduced into the depletion or rectifying compartment and additional water is introduced into the cathode compartment and the intermediate or receiving compartment of the electrodialytic cell **22**. In the cell, where the rinse water is introduced into the depletion or rectification compartment, the ratio of fluoride to tin units in the rinse water is reduced from its original mol ratio of fluoride to tin of between 20 and 25 to 1, to a fluoride to tin mol ratio of between 6 and 7 to 1. The rectified rinse water is withdrawn from the electrodialytic cell **22** through conduit **30** and is introduced into a suitable concentrator **32**, which may be a reverse osmosis cell or an electrodialytic cell.

The fluoride units in the form of HF are removed from the electrodialytic cell **22** through conduit **34** and conveyed to a tank **36**. Sodium hydroxide is also withdrawn from the cell **22** through conduit **38** and introduced into tank **36** where it is reacted with the HF to form the sodium bifluoride. The sodium bifluoride is withdrawn from tank **36** through conduit **40** and may be used as an additive to the rinse water to prevent the hydrolysis of the plating solution in the rinse tank or may be crystallized. Where desired, a plurality of intermediate compartments in series with suitable recycle may be used in the electrodialytic cell **22** to form a tin-free fluoride solution.

Water is removed from the concentrator through conduit **42** and may be recycled to the rinse tank or discarded. The concentrated solution of the rectified rinse water contains sodium and fluoride units in the desired mol ratio for use in the plating bath and is conveyed through conduit **44** from the concentrator **32** to the plating tanks.

A process developed by *T.W. Fisher, S.E. Rauch, Jr. and R.N. Steinbicker; U.S. Patent 4,006,213; February 1, 1977; assigned to Bethlehem Steel Corporation* is one for the extraction and recovery of hydrated stannic oxide and alkali metal ferrocyanide from the sludge formed in a halogen tin electrodeposition bath. The process comprises dissolving the ferrocyanide compounds and the tetravalent tin compounds present in the sludge in an alkaline medium, removing any insoluble materials, precipitating hydrated stannic oxide from the solution by neutralization of the solution, and separating the hydrated stannic oxide from the solution which then contains ferrocyanide ion together with acid and base counterions. Such a process is shown in Figure 166.

Figure 166: Bethlehem Steel Corp. Halogen Tin Electrodeposition Bath Sludge Treatment Process

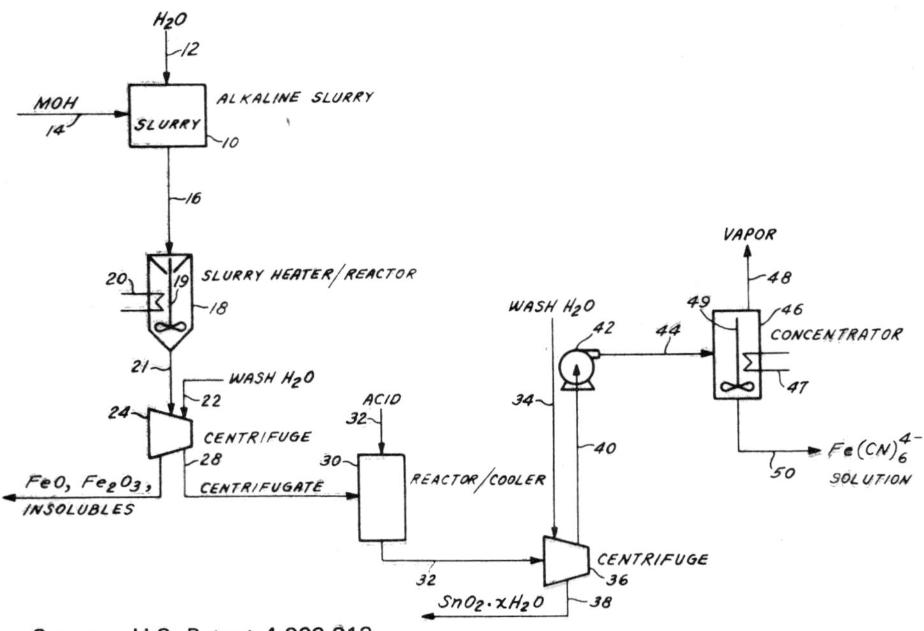

Source: U.S. Patent 4,006,213

Referring to the figure, the halogen tin bath sludge, either initial or secondary, containing Na_2SnF_6, SnO_2, $Fe_2[Fe(CN)_6]$ and $Fe_4[Fe(CN)_6]_3$, is placed in a slurry chamber **10**. Water and alkali metal hydroxide solution are introduced into slurry chamber **10** via lines **12** and **14**, respectively. The resulting slurry is fed, via line **16**, into reactor **18**, which is equipped with stirring and heating means, shown respectively as a stirrer **19** and a heating coil **20**.

After reaction is complete, the slurry is transferred via line **21** to centrifuge **24** where it is centrifuged to separate the ferrous and ferric oxides, and other insoluble material, from the solution containing the tin and ferrocyanide ions. The centrifugate or separated liquid is transferred through line **28** to a reactor **30**. The separated solid is then washed while still in centrifuge **24** with water from line **22**, centrifuged again, and the centrifugate transferred to reactor **30** via line **28**. The solid material is then removed from the reactor **24** and either discarded or used as a high grade iron ore.

The combined solution in reactor **30** is allowed to cool and an HF or HCl solution is introduced via line **32**. Addition of acid is continued until the solution pH is between about 6.5 and 7.5.

The resultant solution containing the $SnO_2 \cdot xH_2O$ is transferred via line **32** to centrifuge **36** where it is centrifuged. The clear centrifugate is drawn, through line **40**, by pump **42** and directed through line **44** to evaporator **46**. The hydrated stannic oxide remaining in centrifuge **36** is washed with water from line **34** and the washings are also pumped to the evaporator **46**. The washed hydrated stannic oxide is then removed at **38** for use as a high grade tin ore.

The combined centrifugate/wash solution in evaporator **46** is then concentrated by evaporation induced by heating coil **47** with simultaneous stirring by stirrer **49** to a ferrocyanide concentration of the equivalent of not less than 10 g/ℓ of $Na_4[Fe(CN)_6] \cdot 10H_2O$. The concentrated ferrocyanide solution removed from evaporator **46** via line **50** may, alternatively, either be used as a source of ferrocyanide for direct halogen tin bath addition or may be treated by any well-known means, such as further evaporation, fractional crystallization or ion exchange separation to isolate pure alkali metal ferrocyanide. Water vapor removed by evaporation is removed via line **48** to suitable condensation means not shown.

TIN FROM SCRAP TIN

Of the tin recycled in 1974, 86% was as an alloy constituent of bronzes, brasses, solders, and bearing and type metals. A small amount also remained in chemical compounds. Only 11% of the recycled tin, mostly from new tinplate scrap, found its way to the market as metal. Although small quantities of tin have been recovered at detinning plants from used tin containers, such reclamation has usually been attractive only under emergency conditions because of the high cost of collecting, transporting, and cleaning the cans. High prices for steel scrap have served as an incentive to increase the reclamation of tin from used containers.

Almost all primary tin is imported to the U.S. and secondary tin constitutes about one-third of total U.S. production so the scrap business is very important in the tin industry.

TIN FROM SLAGS

A process developed by *J.M. Floyd; U.S. Patent 3,905,807; September 16, 1975; assigned to Commonwealth Scientific and Industrial Research Organization, Australia* is one for producing tin from a low grade tin concentrate comprising the following steps:

(1) Melting and partial reduction of the mixture of concentrate and flux, to form a liquid concentrate slag (a) having a high tin content;

(2) Mixing and reacting the liquid concentrate slag (a) with a liquid tin-iron alloy (hard-head) (b) having a high iron content to yield a liquid crude tin metal product (c) of low iron content and a liquid intermediate slag (d) of high tin content; and

(3) Reduction of the tin in the liquid intermediate slag (d) resulting from step (2) to produce liquid tin-iron alloy (b) for recycle to step (2) and a liquid discard slag (e) of low tin content.

The process may be operated either continuously or batchwise.

TIN/ALUMINUM ALLOY FROM MULTILAYER BEARINGS

Multilayer bearing material may consist of a strong backing of steel, which is relatively inexpensive and a bearing lining of some more expensive material, for example, a tin/aluminum alloy or a material having platinum as a component. Such multilayer materials are frequently formed by bonding the layers together rather than by sintering or alloying, but even so, it is only economical to recover the lining if it can be separated from the backing reasonably easily and with something approaching recovery of 100% of the material separated.

A process developed by *A.D. Michael, J.A. White and G.L. Griffin; U.S. Patent 4,086,083; April 25, 1978; assigned to The Glacier Metal Company, Limited, England* is one for recovering an expensive bearing lining alloy from a multilayer bearing material comprising the alloy bonded to a steel backing, possibly through a pure aluminum foil. The multilayer scrap material is heated sufficiently for severance of the bond between the components to be started, and then is given repeated mechanical shocks which complete the severance of the components which can then be separated from one another by a magnetic method, or possibly by a gravitational method.

TITANIUM FROM SCRAP TITANIUM

Use of titanium raw materials, pigment, welding fluxes, ferroalloys, cutting tools, and most other product forms is dissipative and recycling does not take place. Only in the production of the ingot, mill products and fabricated metal parts does recycling of titanium scrap play an important part.

Scrap titanium generally accounts for one-third of ingot production. Only uncontaminated and well-segregated scrap, much of it generated by producers of ingot and mill products, is blended with sponge metal and alloys and remelted to ingot. A large quantity of scrap is generated by consumers of mill products in fabricating finished components. About one-third of such scrap returns to the ingot melt cycle. An estimated 75 to 80% of ingot metal becomes scrap while being processed to finished parts. Selected chips and turnings as well as massive croppings and other forms are recycled.

Most of the light scrap, however, consists of mixed alloys and contaminated material. Some of this is consumed in making ferrotitanium; some is used by the steel industry as a deoxidizing and grain-refining ladle additive; and some is used in aluminum smelting. Several thousand tons is exported annually, although this is partly offset by imports. The remainder is discarded or accumulated pending development of reclamation technology. Titanium grindings and fine turnings may catch fire; thus, handling and storing this type of scrap may create a safety hazard.

A process developed by *J. Dickson; U.S. Patent 3,933,473; January 20, 1976; assigned to Airco, Inc.* is one for recovering a usable alloy consisting of 2.5 to 3.5% aluminum, 1.7 to 2.50% tin and balance titanium from titanium scrap. The method comprises cleaning the scrap, charging the cleaned scrap into an electron beam furnace and refining the charge under a hard vacuum so as to attain a predetermined composition.

TRISODIUM PHOSPHATE FROM ALUMINUM BRIGHT DIPPING EFFLUENT

Bright dipping is a commonly used process in industry for chemical polishing of aluminum, and commercial bright dip operations utilize a solution mainly based on phosphoric acid. In the bright dip process, cleaned aluminum parts are immersed for a few minutes in an accurately proportioned hot aqueous solution of phosphoric acid, aluminum phosphate and nitric acid. This hot solution attacks aluminum vigorously with the release of nitrogen and nitrogen oxides and produces a bright, highly reflective finish. For high quality work at minimum chemical cost, bright dip baths are normally operated in the following range on a weight percent basis.

Phosphoric acid (H_3PO_4)	64-70
Aluminum phosphate ($AlPO_4$)	10-14
Nitric acid (HNO_3)	2.8-3.2
Water (H_2O)	17-23
Copper (Cu)	0.01-0.02

Other brightening baths based principally on phosphoric and nitric acids are also used.

In bright dipping operations, control of bath composition is critical. Variance of concentration of any component outside an acceptable range will directly affect the quality of work. With a bright dip bath, the bath constituents adhere to the parts being polished and are constantly being transferred to the subsequent rinse tank. This process is commonly referred to as drag-out. Hence, by normal drag-out and frequent addition of fresh acids, the aluminum concentration in the bright dip bath is controlled at equilibrium so that the aluminum dissolved from the parts being polished is balanced by the aluminum dragged out of the bath. While this ensures that the bath can be worked continuously, it is a wasteful procedure. In practice, this means that only 10 to 15% of the purchased phosphoric acid added to the bright dip bath is actually consumed in chemically polishing the surface. The remainder is lost to the subsequent rinses.

In addition to being an expensive operation, commercial bright dip lines also create a serious problem in the handling and disposal of the waste liquor because phosphate is one of the more objectionable materials from the standpoint of stream pollution. Chemical destruction methods for the neutralization and treatment of the rinse acid have been developed, but none of these has achieved widespread acceptance because of the cumbersome equipment required and high initial cost thereof and the large quantities of treating chemicals consumed. However, one preliminary recovery technique widely used is that of countercurrent rinsing wherein sufficient rinse tanks are operated in such a way that water-flow is countercurrent to the flow of parts to control the first rinse at acid concentrations in the range of 30 to 40%. Rinse acid over 30% H_3PO_4 is sold directly to liquid fertilizer manufacturers as a raw material, but the price is generally low and storage facilities are required due to the seasonal nature of the fertilizer business.

A refinement of this recovery for the fertilizer industry is evaporative recovery, where the rinse acid is concentrated by evaporation to higher strength for resale, for example, 75% H_3PO_4. Although 75% acid requires less storage space and costs less to ship than 30% rinse acid, recovery of 75% acid requires additional capital and operating costs.

Various other techniques have been proposed to recover the rinse acid, but these systems have not been totally satisfactory, being either impractical in operation or economically infeasible.

A process developed by *A.Y.-H. Fang; U.S. Patent 4,044,106; August 23, 1977* is one in which rinse acid collected from an aluminum bright dipping operation is reacted with soda alkalies at 80°C or higher to yield a solution of trisodium phosphate, Na_3PO_4; sodium aluminate, $NaAlO_2$; sodium nitrate, $NaNO_3$; and trace insolubles. The hot solution is quickly filtered to remove all insoluble impurities and then cooled so that the trisodium phosphate crystallizes out, which, after filtering, leaves a mother liquor containing some remaining trisodium phosphate and soluble impurities, namely, sodium aluminate and sodium nitrate. The mother liquor is then recycled repeatedly with further batches of rinse acid to maximize the phosphate recovery. Such a process is illustrated in Figure 167.

In a typical commercial bright dip and anodizing line, aluminum parts are first cleaned in a hot mild alkali cleaner **10**, rinsed in cold water as indicated at **12**, cleaned in a hot acid cleaner **14** and then rinsed in one or more stages as indicated at **16** before being chemically polished in a bright dip bath **18**. The constituents of the bright dip bath may be phosphoric acid, nitric acid and water, together with dissolved aluminum (in the form of aluminum phosphate) and trace quantities of aluminum alloying constituents, as well as other bath additives such as copper.

Figure 167: Fang Process for Reclamation of Sodium Phosphate from Bright Dip Drag-Out

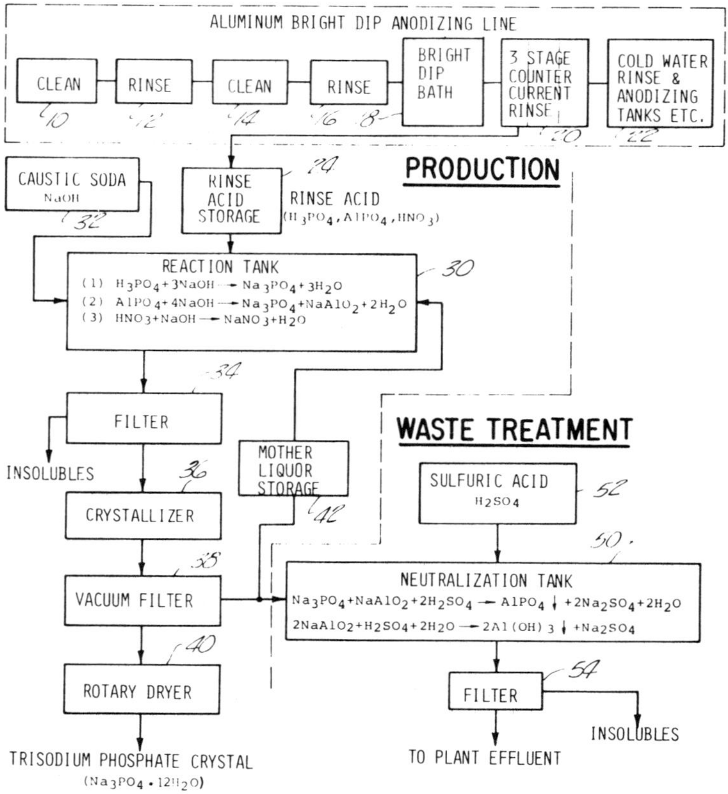

Source: U.S. Patent 4,044,106

After chemical polishing in the bright dip bath **18**, the parts are then transferred to a three-stage countercurrent rinse stage **20** and, after rinsing, into a typical anodizing tank and then to additional sealing, rinsing and drying operations as indicated generally at **22**. As indicated earlier, there is a continual transfer from the bright dip bath **18** into the rinse **20** of the constituents in the bright dip bath, commonly referred to as drag-out; namely, a transfer of phosphoric acid, nitric acid, water and dissolved aluminum. In a typical three-stage countercurrent rinse, for example, the first stage is what is known as a dead rinse and is maintained at a maximum concentration of 35% phosphoric acid, with the second and third stages being maintained at 10 and 1% phosphoric acid, respectively.

As the concentration exceeds 35% in the first stage, the rinse acid is removed and replaced by acid from the 10% concentration rinse acid in the second stage which, in turn, is replaced by the 1% concentration rinse acid in the third stage.

Although typically the first stage in the three-stage rinse **20** is maintained at a maximum of 35% concentration of phosphoric acid, the rinse can go to about a 40% phosphoric acid content without any substantial adverse effect on the aluminum part, provided that the rinse is maintained at a sufficiently low temperature and the part is not subjected to the high concentration rinse for an excessively long period of time. In some bright dip lines, the rinse acid is controlled at concentrations as low as approximately 30% phosphoric acid. It is noted that lower rinse acid concentrations reduce the loss to subsequent rinses. Hence, the rinses are generally operated in the range of 30 to 40% concentration and, preferably, at about 35% concentration.

This type of rinse acid, generally in the range of from 30 to 40% concentration (preferably at about 35% concentration, but commercially practical down to about 10% concentration), can be easily converted into trisodium phosphate and the trisodium phosphate can be readily and effectively separated from the solution, even though the rinse acid has a high aluminum content which, heretofore, has been considered a most costly impurity to remove. Hence, the rinse acid is periodically removed from rinse **20** and transferred to storage at **24** for subsequent transfer to a reaction tank **30**, where it is alkalized by the addition of caustic soda (NaOH) indicated at **32** to produce trisodium phosphate, sodium aluminate and sodium nitrate as shown by the reactions indicated in the figure.

During the reaction, aluminum could be precipitated in the pH ranges from 3 to 10, but a large quantity of valuable phosphate would be lost with the precipitated aluminum as aluminum phosphate. The pH at tank **30** is above 13 and the aluminum is totally converted into soluble sodium aluminate, $NaAlO_2$, by reaction with sufficient sodium hydroxide. Depending on the concentration of the rinse acid used and the amount of recycled mother liquor, as will be described hereinafter, it may be necessary to add more or less water with the caustic soda.

The reaction between the rinse acid and caustic soda is exothermic so that the alkalized reaction is at a temperature substantially above the room temperature. The hot alkalized solution from the reaction tank **30** is then filtered as indicated at **34** to remove insolubles such as heavy metal hydroxides including hydroxides formed from alloying elements in the aluminum being processed. The filtered alkalized solution is then processed at a low temperature crystallizer **36**.

It is important that the alkalized solution be filtered while hot so that the trisodium phosphate stays in solution, permitting insoluble impurities to be removed without removing trisodium phosphate. This results in a purer trisodium phosphate end product. It should be noted that the exothermic reaction can conveniently be used to heat the alkalized solution prior to filtering. However, depending on the specific variables in the process, in order to keep the trisodium phosphate in solution during filtering, it may be necessary to heat the solution prior to or during filtering.

Another important aspect of the process is based on the fact that sodium aluminate is very soluble in both hot and cold water, but trisodium phosphate is much less soluble in water at low temperatures than it is at elevated temperatures, and sodium nitrate is much more soluble than trisodium phosphate.

The crystallized trisodium phosphate can then be removed from the cooled alkalized solution by a vacuum filter or other suitable techniques indicated at **38** and dried as by a rotary dryer **40** to produce trisodium phosphate crystals. The mother liquor from filter **38** goes to storage at **42** from which it can be recycled at the reaction tank **30** with another batch of rinse acid and caustic soda.

The mother liquor generally includes, in solution, sodium nitrate, sodium aluminate, some uncrystallized trisodium phosphate and some excess caustic soda. Although cooling to about room temperature (~25°C) is preferred for economic reasons, it should be understood that the process can operate effectively over a wide range of cooling temperatures. Of course, the initial phosphate recovery increases with decreasing temperatures, and the particular cooling temperature selected will determine the economics of recycling the mother liquor. Although crystallization by cooling is preferred, crystallization by other techniques might be used, for example, by evaporation or by precipitation.

It may be required to reduce phosphorus content to 1 mg P/ℓ before spent mother liquor from filter **38** can be discharged as a plant effluent. The spent mother liquor from the above reclamation process contains 2 to 5% Na_3PO_4, depending on the final crystallization temperature. However, the phosphorus can be easily and efficiently removed from this waste solution since sodium aluminate, which is highly present in the solution, is a most efficient coagulant for phosphorus precipitation. Therefore, the spent mother liquor from filter **38** can be periodically fed to a neutralization tank **50** to which sulfuric acid **52** is added to lower the pH in the range of 6 to 9.

During this neutralization, phosphate is precipitated in the form of aluminum hydroxyphosphate or other complex. The chemical reactions may be represented as follows:

$$Na_3PO_4 + NaAlO_2 + 2H_2SO_4 \rightarrow AlPO_4\downarrow + 2Na_2SO_4 + 2H_2O$$

$$2NaAlO_2 + H_2SO_4 + 2H_2O \rightarrow 2Al(OH)_3\downarrow + Na_2SO_4$$

Insolubles are removed by filtration at **54** so that an effluent with a minimum phosphorus content is obtained.

TUNGSTEN FROM SINTERED METAL CARBIDES

In a method used to date to recover metal from sintered metal carbide scrap, the sintered metal carbide scrap including, for example, WC and Co, is roasted at a temperature more than 1700°C, which is a higher temperature than an ordinary sintering temperature of 1400 to 1450°C, and then quenched. The cemented body becomes brittle and is crushed into powder. In this method, WC and Co cannot be decomposed and recovered free of each other without using very complicated after-processes.

Hence, an improved process has been developed by *Y. Kobayakawa; U.S. Patent 4,140,597; February 20, 1979; assigned to Toho Kinzoli Co., Ltd., Japan,* which provides a simple and economical method for recovering the component metals from a sintered metal carbide. The sintered metal carbide is used as an anode and treated by electrolysis in an acid solution. The metal carbide, e.g., WC, dissolves and precipitates as a metal hydroxide. At the same time, the metal, e.g., Co, which is a sintering binder, is electrodeposited on the surface of a cathode. The metal hydroxide may be washed, calcined and reduced in hydrogen gas to obtain a pure metal powder. Figure 168 shows a suitable form of apparatus for the conduct of this process.

Figure 168: Toho Kinzoli Co., Ltd. Apparatus for Recovering the Component Metals from Sintered Metal Carbides

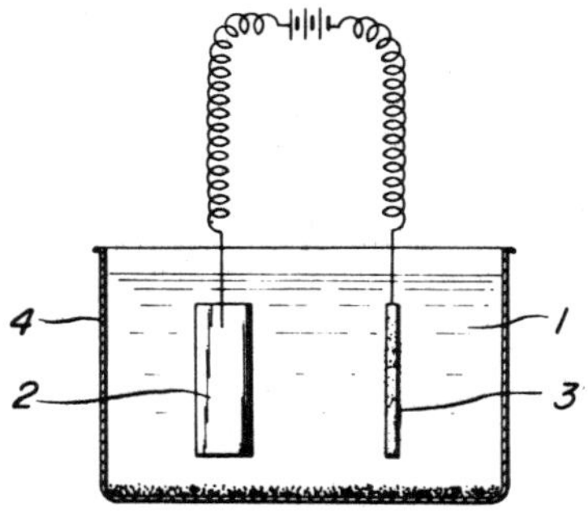

Source: U.S. Patent 4,140,597

A sintered metal carbide scrap comprised of metal carbide, e.g., WC, and binding metal, e.g., Co, is dipped into acid electrolyte **1**, preferably HNO_3 solution, as an anode **2**; at the same time an insoluble pole, e.g., W-bar, **3** is also dipped into the electrolyte as a cathode **3**; then electrolysis is performed. During this electrolysis, the WC is dissolved and becomes H_2WO_4; then, the H_2WO_4 is precipitated on the bottom of vessel **4** according to the following chemical reaction:

$$WC + 6H_2O \rightarrow H_2WO_4 + CO_2 + 5H_2 \text{ (10\% } HNO_3 \text{ solution)}$$

$$WC + 6H_2O \rightarrow WO_3{\cdot}H_2O \text{ (or } WO_3{\cdot}2H_2O) + CO_2 + 5H_2 \text{ (generally).}$$

The precipitated H_2WO_4 or $WO_3{\cdot}2H_2O$ is collected, washed with water and calcined; the pure W powder is then recovered by H_2-reduction. The Co is electrodeposited on the surface of the cathode; the deposited metal Co is stripped, washed and easily recovered.

TUNGSTEN FROM SPENT CATALYSTS

It is known to epoxidize or hydroxylate water-soluble olefinic compounds, e.g., allyl alcohol or crotyl alcohol, with hydrogen peroxide using compounds of transition metals as catalysts. Thus, compounds of vanadium, molybdenum and tungsten are suited for this purpose. Especially preferred for the epoxidation are tungsten compounds. Since the catalysts are a major factor in the economics of the process because they are expensive, there are a number of known processes for the recovery of tungsten-containing catalysts.

Thus there result in the epoxidation or hydroxylation of allyl alcohol to glycide or glycerine, aqueous reaction solutions from which the catalyst must be recovered.

One recovery process, as described by *H. Friedrich, W. Heim, A. Kleemann, H. Kolb and G. Schreyer; U.S. Patent 4,197,161; April 8, 1980; assigned to Deutsche Gold- und Silber-Scheideanstalt vormals Roessler, Germany,* is one in which such catalysts are recovered by separating the catalyst as a distillation residue. The distillation residue is introduced into an inert material which is in whirling movement, e.g., as a fluidized bed. The organic compounds contained in the distillation residue are burned with oxygen or oxygen-containing gases. The catalyst compound separated on the inert whirling material, together with the portion of the catalyst compound which accumulates in the solids separator connected at the outlet of the reactor, is dissolved in water and the solution obtained again used for epoxidation or hydroxylation. The process is shown in Figure 169.

Figure 169: Process for the Recovery of Tungsten Catalysts Used in the Epoxidation and Hydroxylation of Olefinic Compounds

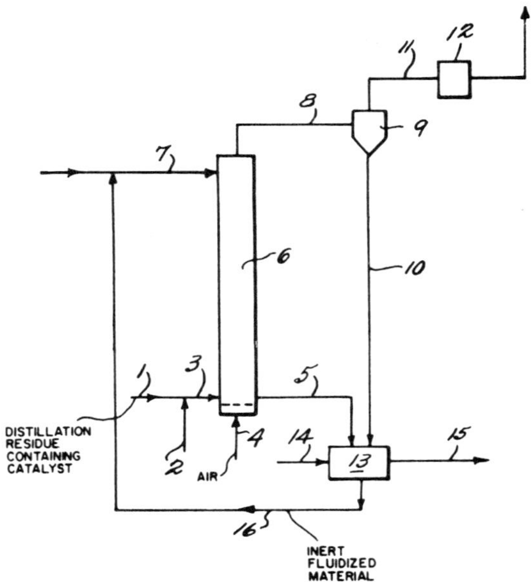

Source: U.S. Patent 4,197,161

The catalyst-containing distillation residue is, in case it is not pumpable, brought into a pumpable condition either by heating or by dilution with water or the washing solution from the wet wash.

The catalyst-containing distillation residue is then pumped via line **1** into the lower part of reactor **6**. Simultaneously, air is blown in through line **2** in order to avoid eventual clogging. The mixture is dosed into the reactor via line **3**. The main amount of the air necessary for combustion and fluidization is supplied via line **4** and is optionally preheated.

The reactor is heated to the desired temperature (at least 500°C). The heating can take place with the help of a starting heater, while the fluidized reactor is brought to the necessary temperature for starting the combustion reaction either directly through the hot burned gas or through heated air in the starting heater by heat exchange. After the beginning of the combustion reaction in the fluidized reactor, the excess heat of reaction is led off through a cooling system.

Preferably, for heating up the fluidized reactor there is used a salt bath which, for example, can be the known nitrite-nitrate melt (sodium nitrite-sodium nitrate). The salt bath in this case serves, after the starting of the reactor, as coolant, because it carries away the heat set free and thus the temperature is held at a constant level.

The addition of the material to be fluidized takes place via line **7**; the withdrawl, together with the catalyst, takes place via line **5**. The catalyst-containing fluidized material goes into a container **13**, where it is washed out with water (supplied via line **14**). The aqueous catalyst solution is separated from the fluidized material and can then be led directly via line **15** to the epoxidation or hydroxylation reaction. The dried fluidized material is led again via line **16** to the fluidized reactor.

The waste gas which, for environmental reasons should contain little carbon monoxide, goes via line **8** into the cyclone **9** where entrained solids are separated. These separated solids are led via line **10**, likewise, to the wash container **13** and freed of catalyst. The waste gas, after a further purification (for example, through a filter or wash **12**), is withdrawn.

TUNGSTEN FROM TUNGSTEN SCRAP

Tungsten obtained from scrap is believed to be a significant part of total supply. However, data on tungsten recovery from scrap are incomplete. Most tungsten-bearing scrap is recycled within the individual generating company. Tungsten superalloy scrap is reused without separation of the constituent elements. Scrap tungsten metal is saved at producing wireworks and either sold or returned to chemical processing units. Substantial quantities of tungsten carbide in sludges from metal-cutting operations are not recovered because of high chemical processing costs and low tungsten content. Metallurgical International, Inc. reportedly produced more than 800,000 lb of tungsten carbide powder in 1974, exclusively from scrap purchased on the open market.

A study conducted by Shwayder Chemical Metallurgy Corp. indicated that several thousand pounds of tungsten are discarded each month in the form of 1 to 2% thoriated tungsten welding rod ends.

The high volume of throwaway carbide tool bits used in metal-cutting operations prompted the investigation and development of methods to economically recover tungsten carbide. The comparative cost ratio of tungsten carbide made from virgin tungsten metal powder to that recovered from scrap is roughly 4 to 1.

High-purity tungsten wire end croppings represented about 75% of the feed material for crushed and cast tungsten carbide. In addition, some high-purity tungsten sludge was obtained in the cooling systems of abrasive cutting wheels during the preparation of tungsten contact points. After being dried, these sludges were added directly to steel melts (14).

TUNGSTEN CARBIDE FROM SCRAP MINING BITS

Mining bits for rock drilling and the like utilize tungsten carbide as the abrading medium and the tungsten carbide is in the form of short cylinders or buttons, known as inserts, embedded in and projecting from sockets in a steel body. In use, the tungsten carbide inserts become ineffective as a result of wear and the bit thereby is rendered useless.

Worn mining bits still contain considerable quantities of tungsten carbide, however. In view of the raw material cost of tungsten carbide, it is economically attractive to recover the worn inserts from the scrap mining bits for recycle to produce new bits.

Hence, a process has been developed by *F.J. Piche; U.S. Patent 4,170,513; October 9, 1979; assigned to Fernand Piche Enterprises Limited, Canada,* in which tungsten carbide inserts are recovered from scrap mining bits by treatment in a sulfuric acid bath to eat away part of the steel surrounding the inserts and then heating the bit and vibrating the heated bit to knock out the inserts. In a modification used on cross-type bits, heat and vibration alone are used.

U

URANIUM FROM UO_2 PROCESS WASTEWATERS

In preparing fuel elements for nuclear reactors it is desirable to employ UO_2 for this purpose. A widely used method is to process UF_6 to prepare ammonium diuranate (ADU) which is then calcined and reduced to UO_2. In preparing ADU, enriched UF_6 gas is reacted with water to produce an aqueous uranyl fluoride (UO_2F_2) solution which is then treated with an excess of ammonium hydroxide to cause a precipitate of ADU which is recovered as an aqueous slurry. The precipitate can be filtered out or centrifuged from the aqueous phase of the slurry. The resulting wastewater contains a high concentration of fluorides, primarily as NH_4F, with excess ammonia, and small quantities of dissolved uranium.

The disposal of this wastewater entails considerable problems and difficulty because of the toxicity of the fluorides as well as the excess ammonia and the small but significant quantities of radioactive uranium. On both ecological and health accounts it would be highly desirable to eliminate or greatly reduce this disposal problem, as well as to recover from the wastewater substantially all the traces of enriched uranium, which latter is a very costly material as well as being a health hazard.

A process developed by *T.J. Crossley; U.S. Patent 3,961,027; June 1, 1976; assigned to Westinghouse Electric Corporation* is one in which wastewater containing large amounts of fluorides and ammonia and trace amounts of uranium, as is produced during the hydrolysis and ammonium hydroxide treatment of uranium hexafluoride (UF_6) to produce ammonium diuranate (ADU) therefrom, is rendered suitable for cyclic reuse by initially treating the wastewater with sufficient lime to precipitate substantially all of the fluorides present in the wastewater to a relatively insoluble CaF_2 precipitate.

The treated solution is subjected to distillation to drive off ammonia for reuse in the ADU precipitation. The CaF_2 precipitate is separated from the aqueous distilland leaving water with dissolved calcium. The distilland is treated by a cationic ion-exchange material to remove substantially all of the calcium and other cationic metal impurities and the resulting water containing small amounts of uranium, fluoride and ammonia is recycled to react with UF_6 or to be combined with the ammonium hydroxide distillate and then treated with additional concentrated ammonium hydroxide to form a solution of the desired NH_3 content for use in precipitating ADU. This recycling is most important for ecological reasons, and for cost improvement, as well as health and safety purposes.

Such a process is shown in Figure 170. Referring to the drawing, which appears on the following page, from a tank **10** containing hydrated lime [$Ca(OH)_2$], measured quantities of lime are conveyed by a screw conveyor line **12**, to a lime slurry tank **14**. If desired, the lime is weighed out in batches and then conveyed to tank **14.** Sufficient water to produce an aqueous slurry is added from pipe **16** by suitable control valve operation. Though hydrated lime is easier to handle and use, it will be understood that calcium oxide or other reactive alkali earth metal compounds can be employed.

Figure 170: Westinghouse Electric Corp. Cyclic Process for Reuse of Wastewater Generated During the Production of UO_2

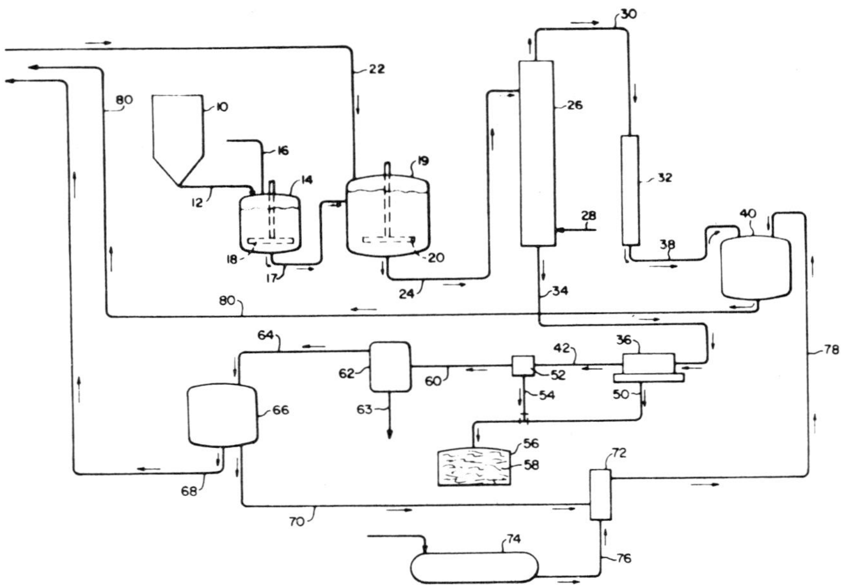

Source: U.S. Patent 3,961,027

Since commercial hydrated lime contains various impurities, such as magnesium compounds, aluminum compounds, silica, and some iron compounds, it is desirable to employ the purer available limes containing less than 5% of impurities. Magnesium fluoride is only slightly more soluble than calcium fluoride and it can be tolerated better than other impurities. There are readily available hydrated limes having a combined total of over 98% calcium and magnesium hydroxide. A uniform slurry of from 2 to 20 pounds of lime per 100 pounds of water may be prepared in the tank by action of the stirrer **18** as weighed amounts of lime and water are added.

Into a tank **19** provided with a stirrer **20,** there is conveyed via pipe **22** wastewater from the ADU centrifuge. After a weighed amount of the wastewater, at a pH of about 9.5, has been added, a predetermined amount of the lime slurry in tank **14** is pumped by line **17** into vessel **19** so as to provide a slight excess of calcium (and magnesium) hydroxide over that required to react with all of the fluoride in the wastewater brought by the pipe from the ADU centrifuge.

For every 74 pounds of NH_4F in, for example, 3,900 pounds of wastewater, there is introduced slightly over 74 pounds of lime in the slurry. For a slurry of 15 pounds of lime in 100 pounds of water, some 575 pounds of slurry, plus a slight excess, is required to react with and precipitate all of the fluoride.

It is not necessary to add the stoichiometric amount or slightly more of lime to react with all of the fluorides in the wastewater. An amount of lime slightly below the stoichiometric quantity, for example, 1 to 5% less, is acceptable since a small portion of fluoride can be present in the treated water when recycled without harm.

After a thorough mixing by operation of stirrer **20** so as to cause a thorough and complete reaction of the lime slurry with the fluorides in the wastewater, and so that large crystals of calcium fluoride are formed, the slurry in vessel **19** is pumped via pipe **24** to a distillation column **26** wherein live steam from a conduit **28** is introduced to cause ammonium hydroxide plus steam to vaporize and pass by pipe **30** to a condenser **32** where it is cooled and condensed to, for example, 25% NH_4OH solution, which solution is conveyed by pipe **38** to an ammonia storage tank **40.**

From the bottom of still **26,** the still residue comprising calcium fluoride precipitate along with some undissolved lime and impurity particles, all suspended in water, is conveyed to a centrifuge **36** wherein one stream output is a sludge or paste of calcium fluoride plus some calcium hydroxide and impurity solids with small amounts of water and is carried by line **50** as solids waste **56** to a waste receptacle **58.**

It is feasible to remove the calcium fluoride and other residues from the wastewater before distillation, since some ammonia is lost, and it is preferable to distill off the ammonia before centrifuging off the precipitates.

The liquid stream from separator **36** is conveyed by a pipe **42** to a polishing filter **52** in order to remove the last traces of any fine suspensions therefrom. The small volume of solids from filter **52** is also conveyed by a line **54** to the solids waste receptacle **58.** Since practically no uranium is present in the waste solids it is suitable for use in applications calling for calcium fluoride, for example, in glass making.

The clarified water effluent from the filter comprises some 200 mg/ℓ of free ammonia, not over 50 mg/ℓ of fluoride, calcium and magnesium and other soluble impurities, as well as practically all of the traces of dissolved uranium originally brought in by pipe **22.** This water is carried by a line **60** to a cation ion-exchange column **62** which absorbs the calcium, magnesium, iron and other metal impurities and preferably exchanges hydrogen for these metals. If desired, a second ion-exchange unit may be used to remove anions, particularly chlorides. Calcium is particularly undesirable in producing ADU. The ion-exchange resin can be regenerated and the cations are removed via pipe **63** to an acid-treatment tank for regeneration and reuse. Suitable cationic exchange resins are the well-known sulfonic acid treated divinyl benzene-styrene copolymers.

The relatively pure water with a small portion of ammonia, produced by the ion-exchange column **62** is conveyed by pipe **64** to a water storage vessel **66** where it may be piped by a line **68** to the UF_6 dissolving vessel, or to the lime slurry tank **14**, or by a pipe **70** to an ammonia make-up tank **72** wherein concentrated ammonia shipped in and stored in a tank **74**, is also conveyed by pipe **76** and the two are combined in suitable proportions to produce the proper concentration of 24 to 29% NH_3 to use in the ADU precipitation step. Any small amount of fluorine in the water in tank **66** is not objectionable. The minute traces of uranium in this water in fact are desirably present so that the uranium is not wasted.

Thus, it will be seen that, outside of calcium fluoride solids which are commercially soluble and usable, the wastewater from the ADU process is all recovered and recycled whereby to avoid difficult disposal problems.

Utilizing the wastewater treatment procedures of this process wherein treated wastewater is recycled for use in converting more UF_6 into ADU, no water need be added, or at most only small make-up additions need be made to the system. The only water losses from the system are in the ADU sludge going to the calciners and in the CaF_2 sludge, both taking relatively small quantities of water. This last amount of water loss can be very nearly compensated for by the water in the additional aqueous ammonia brought in to replace the ammonia present in the ADU. Consequently, no wastewater need be disposed of or discarded.

URANIUM OXIDE FROM NUCLEAR FUEL PROCESS SCRAP

The residue generally called "scrap" or "scrap materials" from many nuclear fuel processing operations can be in various forms and contain a sufficiently high concentration of the nuclear fuel to justify its recovery. Generally, one of the preliminary steps in a nuclear fuel recovery process comprises calcining scrap materials to remove the volatile matter and then dissolving the soluble portion of the resulting calcine in an acid. Such a process has been conducted using a calciner and a slab-shaped leaching apparatus.

The prior art also includes dry processes for nuclear fuel recovery from scrap. One such process for recovering hard scrap nuclear reactor fuel material is one in which the scrap is first oxidized in a fluidized bed and then comminuted, after which the oxidized material is reduced back to the original chemical form prior to the oxidation step with improved sintering characteristics.

The foregoing processes are not suitable for the efficient recovery of enriched nuclear fuel, such as uranium compounds, from filter media used to filter the air drawn from nuclear fuel processing operations. Typically, these filters (e.g., High Efficiency Particulate Air or HEPA filters) are made of fiber glass or include some fiber glass in the filter media. When incinerated, the fiber glass forms a clinker that can encapsulate the uranium and is insoluble in most commonly used acids.

In the past, ventilation filters have simply been disposed of by burial, or have been acid leached as a whole unit in a large tank. The leaching of filter media precludes complete recovery of the nuclear fuel and is, in addition, a manpower intensive operation that utilizes large volumes of acid.

Typically, the materials other than filter media containing small amounts of enriched nuclear fuel result from the processing of the enriched nuclear fuel into a form, commonly oxide or carbide, suitable for utilization in nuclear reactors. In greater detail, such materials are rags, uniforms, gloves, plastic, oils, etc. It has been found that when such materials are incinerated, they form chunks of various sizes that need to be comminuted or ground to a smaller uniform size in order to be efficiently leached in an acid solution. Accordingly, it has remained desirable to have a system for the efficient recovery of nuclear fuel materials from filter media, incinerator ash and other particulate materials.

A process described by *L.A. Divins and L.E. Short; U.S. Patent 4,177,241; Dec. 4, 1979; assigned to General Electric Company* is particularly useful when the nuclear fuel is an oxide of uranium or mixture of uranium and plutonium oxides.

The process yields an acid solution with the nuclear fuel dissolved therein. Except for spent filter media, the scrap materials are calcined to produce an oxidized material, which is reduced to a particulate material and leached in an acid bath that contains some recycled acid to yield an acid solution of the nuclear fuel material. The scrap materials comprised of spent filter media are mulched and contacted with the acid solution. The insoluble materials are separated from the acid solution, and at least a portion of the acid solution is recycled to the leaching step, while the remainder of the acid solution is collected for subsequent treatment to recover the dissolved nuclear fuel material.

Figure 171 is a block flow diagram of such a process. The first step of the process is calcining the scrap material other than filter media at a temperature in the range of 750° to 1000°C to yield an oxidized scrap material or ash containing the nuclear fuel in readily soluble form.

Since the oxidized material has a wide range of particle sizes, the material is screened in an optional step in order to separate any particles greater than about 250 μ in size. These oversized particles are comminuted, preferably by hammer milling or jaw crushing, to yield a particulate material of about 10 to about 250 μ in size. When the screening step is not used, all the material is subjected to a grinding step.

The calcine, now more conveniently referred to as particulate material, is contacted with an acid in a slab-shaped leaching zone of nuclear-safe geometry, preferably in a nitric acid bath maintained at a temperature in the range of 60° to 100°C with an acid concentration of about 3 to about 8 normal. The particles are agitated by mechanical agitation and are maintained in contact with the acid for a residence time of about 15 to 30 minutes to ensure dissolution of the nuclear fuel compounds, thus forming an acid solution containing a concentration of about 30 to 100 g of nuclear fuel/liter.

While it is preferred to use nitric acid, other acids can be used, such as sulfuric acid. The controlled feeding of material to the leaching zone limits the reaction rate of acid with the nuclear fuel compound and the evolution of gas from the reaction and yields an acid solution containing dissolved nuclear fuel.

Spent filter media containing nuclear fuel material are separated from their filter frames and then mulched so that the media are converted to a particulate material. The particles of filter media are contacted with the acid solution to dissolve the nuclear fuel material held by the media.

Figure 171: General Electric Co. Process for Recovering Nuclear Fuel from Scrap Material

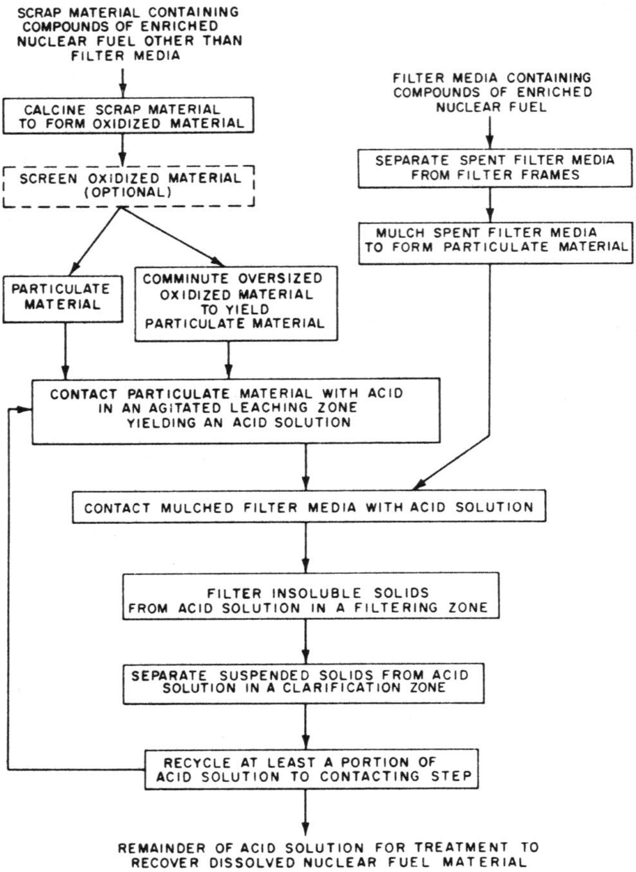

Source: U.S. Patent 4,177,241

The nonsuspended insoluble solids and filter media are separated from the acid solution by filtering in a filtering zone and suspended solids are separated from the acid solution in a clarification zone. At least a portion of the acid solution is recycled to the slab leaching zone, with the remainder being collected for subsequent treatment for recovery of the dissolved nuclear fuel.

The portion of the acid solution from the clarification zone not recycled can be emptied into a storage zone for accumulation of a sufficient quantity, at which time the nuclear fuel solution is contacted with an organic solvent to extract the nuclear fuel material in a liquid-liquid extraction zone. Also, the nuclear fuel may be precipitated from the acid solution by using hydrogen peroxide or alkaline materials.

V

VANADIA-TITANIA CATALYST FROM SPENT DENITRATION CATALYST

A process developed by *H. Kohama and S. Mizukami; U.S. Patent 4,142,989; March 6, 1979; assigned to Kobe Steel, Ltd., Japan* is directed at the improvement of methods for removing effectively and economically nitrogen oxides contained in smokes and exhaust gases, which are considered to be one of the causes of photochemical smogs and the like. As one of such methods, there can be mentioned a catalytic reduction method using ammonia as a reducing agent. In order to practice this catalytic reduction method to economic advantage it is necessary to regenerate a catalyst which has been used for the reaction and degraded.

It is generally considered that this regeneration can be accomplished effectively according to the heating regeneration technique. According to the conventional heating regeneration method customarily used in various denitration systems, a catalyst is withdrawn from a denitrating apparatus and a large quantity of the degraded catalyst is regenerated at one time in a heating furnace of the batch type. This conventional method, however, involves various defects such as mentioned below:

(1) When the degraded catalyst is withdrawn from the denitration apparatus, the operation of the denitration apparatus must be stopped.

(2) When it is desired to operate the denitration apparatus during the regeneration operation, it is necessary to arrange and use a spare catalyst.

(3) The operation of delivering the degraded catalyst to the regeneration apparatus is indispensable.

(4) An additional space is necessary for the regeneration apparatus.

Thus, an improved process has been proposed which features a continuous type catalyst regeneration process for continuously heating and regenerating a degraded catalyst from a denitration apparatus and recycling the regenerated catalyst to the denitration apparatus for repeated use thereof. In accordance with the process, a degraded catalyst is intermittently and by predetermined amounts charged into a heating furnace through a first damper mechanism in a gas-sealed state while the surface level of a catalyst held in the heating furnace is maintained lower than the outlet of catalyst regenerating gas and is therein subjected to heat-treatment at a temperature of 400° to 650°C with catalyst-regenerating gas introduced into the heating furnace.

Thus, regenerated catalyst is intermittently and by predetermined amounts discharged from the heating furnace toward a second damper mechanism through a catalyst delivery mechanism and recycled to the denitration apparatus.

VANADIUM FROM SPENT CATALYSTS

See also "Cobalt from Spent Catalysts," especially U.S. Patent 4,145,397.

It is known that complex catalysts containing vanadium oxide, which are used, for example, for the production of maleic anhydride by catalytic oxidation of butane with molecular oxygen, become deactivated with time. A process for reactivating these deactivated catalysts, by treatment with a reducing agent, is also known. However, the recovered activity and selectivity of the catalyst does not last long, so that the reductive treatment must be repeated after only a relatively short time.

It is also known how to convert the vanadium pentoxide of the spent catalysts into volatile vanadium oxychloride by treatment with thionyl chloride or phosgene, and to convert this compound into vanadium pentoxide in the usual manner. In an even better way, vanadium pentoxide may be recovered from catalysts which contain vanadium pentoxide by treating the catalysts with carbon tetrachloride at not less than 100°C and reacting the volatile reaction mixture with ammonia to give ammonium vanadate.

In these processes, the catalyst is not actually regenerated and instead only vanadium is selectively recovered, and at the same time auxiliary materials are consumed and problems of recovery or disposal both of the by-products resulting therefrom, and of the residual constituents of the catalyst, arise.

A process developed by *K. Rohbock, R. Wiemers, P. Woditsch and R. Thiel; U.S. Patent 4,111,832; September 5, 1978; assigned to Bayer AG, Germany* is a method for regenerating a deactivated oxidation catalyst, especially one containing vanadium oxide, wherein the deactivated catalyst is treated with an aqueous solution of ammonia and/or an amine at an elevated temperature so as to dissolve at least a portion of the catalyst and the resultant solution or suspension is treated in a known manner to obtain the reactivated catalyst.

The following is a specific example of the conduct of such a process.

A catalyst of which the composition by weight is $V_2O_5:P_2O_5:CuO$ = 1:1.5:0.3 was prepared in accordance with the process of U.S. Patent 3,625,863 from a solution of 1,250 g of 85% strength by weight phosphoric acid (remainder water) in 2,100 ml of water, 686 g of ammonium vanadate and 800 g of copper acetate in 600 ml of water, and 3,000 ml of 25% strength ammonia, and was employed for the oxidation of a butane-butene mixture with atmospheric oxygen.

The single reaction tube of the reactor used had a diameter of 25 mm and a length of 3 m and was surrounded by an electrically heated salt melt. The tube was filled with the loose catalyst to a height of 260 cm.

Per liter of catalyst, 3.6 Nm^3/hour of air, charged with 38 g/Nm^3 of a C_4 mixture which consisted to the extent of about 80% of n-butenes, were passed through. The reaction temperature was 420°C.

85% by weight, relative to the butene employed, of maleic anhydride was thus obtained.

After continuous operation for several years, the yield had fallen to 71% by weight of maleic anhydride, relative to butene employed.

1,000 g of this deactivated catalyst were ground and treated with 10 ℓ of 27% strength ammonia solution in a 20 ℓ stirred autoclave for 5 hours at about 150°C and a pressure of 25 bars. This resulted in 95% solution of the deactivated catalyst (degree of solution 95%).

The suspension thus obtained was subsequently evaporated at 150°C in a rotary evaporator until a firm paste formed. This paste was subsequently dried for 4 hours at about 210°C. The dried product was then heated from 200° to 450°C, raising the temperature about 10°C per hour, in the presence of oxygen, and the temperature of about 450°C was then maintained for a further 16 hours.

The material was then ground to a particle size of 40 to 500 μ and pressed, with addition of 2% by weight of aluminum stearate as a binder, to give 6 mm spheres.

The regenerated catalyst thus obtained was then used in the apparatus described above for the oxidation of the C_4 mixture. The yield of maleic anhydride, relative to the butene employed, was again 85% by weight. Even after 1,500 operating hours, the average yield was still 84% by weight of maleic anhydride, relative to the butene employed, and showed no greater decrease than is also observed with a catalyst which has directly been prepared new.

A process developed by *R.Z. Pyrih and R.S. Rickard; U.S. Patent 4,115,110; September 19, 1978; assigned to Earth Sciences, Inc.* is one for recovering vanadium values from carbonaceous type vanadium ores, and vanadium scrap, such as vanadium contaminated spent catalyst. This process comprises: (1) roasting the vanadium containing material in air at a temperature less than about 600°C to produce a material substantially devoid of organic matter; (2) subjecting the roasted material to a further oxidizing roast in an oxygen atmosphere at a temperature of at least about 800°C for a period sufficient to convert substantially all of the vanadium to the soluble form; (3) leaching the calcine with a suitable dilute mineral acid or water at a pH of neutral to about 2 to recover vanadium values; (4) precipitating vanadium values as iron vanadate from the leach solution with a soluble iron compound at a pH from neutral to about 1; and (5) recovering ferrovanadium from the iron vanadate by a reduction vacuum smelting operation.

The conversion of vanadium to the soluble form by the oxidizing roast is accomplished without the addition of an alkaline salt during calcining.

A process developed by *G. Gravey, J. Le Goff and C. Gonin; U.S. Patent 4,182,747; January 8, 1980; assigned to Metaux Speciaux SA, France* is one for recovering metallic constituents of spent or waste catalysts containing metallic compounds fixed on an aluminous support. The process is particularly applicable for the recovery of metals such as Al, Mo, V, Ni and Co, contained in spent catalyst. The process comprises transforming the metals to be recovered into volatile chlorides by carbochlorination and then separating the chlorides obtained by dry means and fixing them successively in a selective manner. The separation of $AlCl_3$ is accomplished by passage through granules of anhydrous NaCl, and the separation of $MoCl_5$ by passage through granules of crystallized KCl. The process is particularly suitable for the treatment of waste catalyst from the catalytic hydrocracking or hydrodesulfurization of oils.

Because of the presence of carbon in the pores of the catalyst, a reduction of the oxides is produced simultaneously, and particularly of the alumina, with chlorination of the reduced metal. The sulfur which is also present in the catalyst, principally in the form of metallic sulfides, like the carbon, probably also acts as an oxide reducer. However, when there is a great excess of carbon, i.e., a stoichiometric excess, relative to the quantities of oxides to be reduced, it is noted that the major part of the sulfur, and probably almost all of it, is found combined in the form of sulfur chloride in the gases which are discharged.

Tests have shown that there is a very high yield or recovery of chlorides of aluminum, molybdenum and vanadium from the spent catalysts treated, which recovery is generally above 90%. The chloride yields depend upon operating conditions and they are often a bit lower because an excess of chlorine must be present to optimize the process, but this is not a major problem, because this excess can be easily recycled.

The following is a specific example of the conduct of such a process.

A sample of 1 kg of a waste catalyst with an active alumina support and containing V, Mo, Co and Ni compounds was used. This sample was first treated for two hours at 400°C in a nitrogen current, to eliminate the hydrocarbons and the water. The treatment was carried out in a tubular enclosure rotating on its horizontal axis and of 140 mm diameter and 600 mm length, heated from the outside and having a flow of nitrogen running through at a rate of 20 ℓ/hr. After 2 hours of treatment, it was cooled. 825 g of catalyst with volatile materials removed was recovered, and for the most part the volatiles were recovered by condensation. These volatiles had approximately ⅓ water and ⅔ a mixture of hydrocarbonated compounds. After this treatment the principal constituent contents, by weight, were as follows:

	Percent		Percent
C	21	Ni	2.4
S	8.4	Co	1.7
V	9.4	Si	0.14
Mo	5.6	Al	27

This catalyst was then chlorinated in a vertical nickel tube of 80 mm diameter and 500 mm height, heated from the outside to a temperature between 500° and 600°C. A chlorine flow was injected from bottom to top at approximately 80 ℓ/hr for 10 hours. A fixed residue of 245 g was recovered containing by weight:

	Percent		Percent
C	36	Ni	8.1
S	4.8	Al	2.8
Co	5.6	Si	0.6

This residue contained only traces of Mo and V, which shows that these two metals had been entirely volatilized in the form of chlorides. These metals were then separated from an aqueous solution thereof, for example, in the form of hydrates or carbonates by known methods.

The gaseous phase from the chlorination step was introduced to the bottom of a column filled with granules of 1 to 2 cm of sodium chloride heated to 350°C. At the base of the column was observed the outflow of approximately 1 kg of a compound, the composition of which corresponded essentially to $AlCl_3 \cdot NaCl$. The other constituents of the gaseous mixture not retained by the salt were discharged at the top of the column, particularly the molybdenum and vanadium chlorides.

These gases were then injected at the base of a vertical column of approximately 60 mm diameter and 40 cm height, filled with potassium chloride in crystallized state obtained by crystallization from an aqueous solution and comprising grains of approximately 1 to 2 mm and at a temperature of 400°C. In this column, the molybdenum chloride is fixed to the surface of the KCl grains.

After passing through the bed of KCl, the gases were discharged from the column at the top and then passed through a condenser at a temperature of 60°C, in which the vanadium chloride, VCl_4, is retained.

Finally, the residual gases were passed over absorbers of known type intended to retain the silicon chloride and sulfur chloride.

Tests have shown that the quantity of molybdenum retained in the form of chloride by the column containing the potassium chloride exceeds 90% of the molybdenum present in the waste catalyst. The quantity of vanadium recovered on the condenser is also more than 90% of the vanadium present in the waste catalyst.

VANADIUM CATALYST FROM ADIPIC ACID PROCESS WASTES

In the process for nitric acid oxidation of cyclohexanol and/or cyclohexanone, significant amounts of succinic acid and glutaric acid are formed as by-products in admixture with the adipic acid. Various well-known schemes of crystallization, concentration and further crystallization have been used heretofore in industry to attempt separation thereof and usually a substantial amount of the adipic acid can be removed by these procedures. Ultimately, however, a mother liquor is obtained from these operations which contains succinic acid, glutaric acid, and a small amount of adipic acid in such proportions that further concentration and crystallization steps will yield only mixtures of these dibasic acids.

In addition, as a result of the removal of the adipic acid by crystallization and removal of the water and nitric acid by volatilization, the concentration of the metal catalyst in this final mother liquor becomes relatively high. Hence, loss of these materials, particularly the catalytic components, provides a distinct economic disadvantage as the relatively high proportion of valuable and reusable products is lost.

It is clear, therefore, that a distinct need remains in the art for processes by which this nitric acid oxidation mother liquor may be processed so as to recover the valuable components contained therein in an economic and efficient manner.

Therefore a process has been developed by *S.S. Mims; U.S. Patent 4,076,948; February 28, 1978; assigned to El Paso Products Company* which involves treatment of aqueous nitric acid solutions resulting from the nitric acid oxidation of cyclohexanol and/or cyclohexanone in the presence of metal compound oxidation catalysts for recovery of the valuable components contained therein. Such a process comprises adding an alcohol to the aqueous nitric acid solution to esterify the acids present, contacting the resultant alcoholic mixture with a water-immiscible solvent, separating layers into an aqueous phase and an organic phase, treating the aqueous phase to recover the nitric acid and catalyst components and treating the organic phase to recover the acid derivatives.

The process is shown schematically in Figure 172. In the drawing, it will be seen that a series of extractors and settlers are employed in combination with aqueous feed tanks so as to provide a system for the continuous extraction of the esterified purge stream.

In the illustration, the first and second extractors are stirred vessels or pots in order to obtain good admixture and thus good ester extraction. The third extractor is a packed column operated so that the aqueous phase is maintained as a continuous phase. Hold-up times of the aqueous phase should be maintained in this extractor to provide for complete reaction. Hold-up times of about one-quarter to one hour, preferably one-half hour are satisfactory. The aqueous feed tanks are of a size such that there is also sufficient hold-up time, i.e., about one hour, in each tank in order to obtain good admixture and complete reaction. The extractors and feed tanks are generally maintained at an elevated temperature as described hereinabove and preferably at about 55°C or above.

Referring now to the drawing specifically, a purge stream **1** from an adipic acid mother liquor source is fed into the first aqueous feed tank **3** together with alcohol by line **2**. The alcohol and the purge stream are preferably introduced into the first aqueous feed tank in about equal volumes. This solution is then fed continuously to the series of extractors and settlers and feed vessels while an appropriate volume of benzene or other water-immiscible solvent is fed by line **18** to the opposite end of the series of extractors, settlers and feed vessels so that the entire system operates in a countercurrent fashion.

Thus, in the first aqueous feed tank, the alcohol and purge stream components react to esterify the acids and the mixture is then passed by line **4** to the first extractor **5** where it is contacted with extractant coming from second settler **13** by line **21**. This mixture passes to first settler **7** where layers are separated and crude extract removed continuously via line **22**.

In the meantime, the aqueous phase from first settler **7** is passed by line **8** to second feed tank **9**, then by line **10** to second extractor **11** where it is mixed with further extractant coming from the third extractor **17** by line **20**. The mixture is then passed via line **12** to second settler **13** for layer separation, the extract being removed by line **21** and sent to first extractor **5**, and the aqueous phase passing by line **14** to the third feed tank **15** and then by line **16** to the third extractor. Fresh water-immiscible solvent such as benzene is added to the third extractor by line **18** so as to maintain sufficient extractant present to remove all the esters and this overhead is forwarded to second extractor **11** by line **20**. The aqueous phase is removed by line **19** for further processing as desired.

Figure 172: El Paso Products Co. Process for Treating Adipic Acid Mother Liquor to Recover Metal Catalysts and By-Product Organic Acids

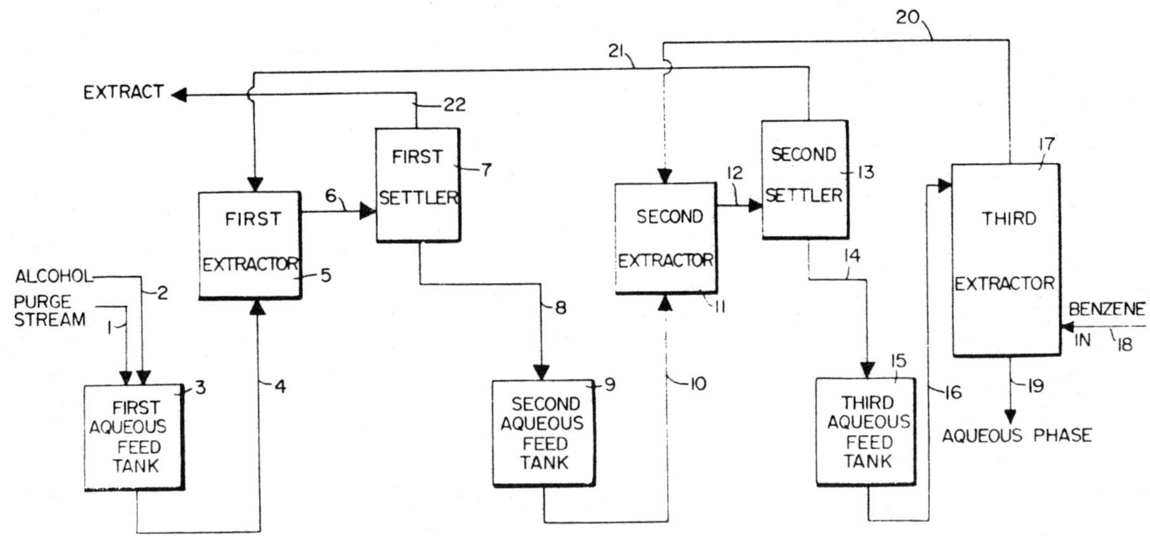

Source: U.S. Patent 4,076,948

Thus, by utilizing this method in a continuous manner, the purge stream can be continuously processed to continuously recover a crude extract from line **22** containing the esters formed and the aqueous phase from line **19** containing the alcohol excess, water, nitric acid and catalyst components. Then these streams may be further processed as desired.

This process thus provides a number of advantages over processes of the art which have been used in an attempt to recover these valuable components. Hence, in this process, the removal of nitric acid and water from the mixture is not required as, for example, by volatilization, as this is a hazardous step when carried to dryness. Further, distillation of the high boiling dibasic acids is not required; there is no necessity for the addition of nonvolatile acids or other inorganic materials which would accumulate in recycle stream; costly crystallization and filtration steps are eliminated and the aqueous and organic phases are obtained in uncontaminated form.

As opposed to prior art techniques, this process is simple and economical but still affords the simultaneous recovery of dibasic acids as well as of the metallic catalysts contained in the mixture. Moreover, the materials are recovered in highly useful form, the dibasic acids being recovered as diester derivatives useful as a mixture or easily processed by further means to yield individual compounds such as the individual esters, by hydrolysis to recover the acids, and by transesterification to produce more useful, particularly high molecular weight, esters. The metal catalysts are recovered as a concentrated solution in nitric acid and thus, in an ideal form for reuse in the cyclohexanol/cyclohexanone nitric acid oxidation reaction.

Very similar technology is employed by *C.A. Newton; U.S. Patent 4,105,856; August 8, 1978; assigned to El Paso Products Company.* By-product or waste streams from adipic acid plants which contain adipic acid, glutaric acid, succinic acid, nitric acid and catalyst values of copper and vanadium are treated to substantially recover all the components thereof by esterifying the dibasic acids with an alcohol, such as normal butanol, which creates esters that are substantially immiscible in the aqueous solution; allowing the organic esters to form a separate layer from the aqueous phase in which the metal values are retained; recovering the organic layer containing the esterified dibasic acids; and thereafter reusing the aqueous solution containing the catalyst values in the adipic acid process.

W

WATER TREATING AGENT FROM SPENT CRACKING CATALYST

The consumption of silica-alumina catalyst has shown a yearly increase with the recent growth of the petroleum industry, and it now is consumed in an amount of approximately 100,000 tons a year. This means that the same amount of waste catalyst has been discharged annually, while as for the means of disposal of this enormous volume of waste catalyst, the method of burying it in the ground has so far been exclusively adopted. However, a waste catalyst of this kind is composed essentially of ingredients similar to zeolite and is in the form of a firm sinter as a result of repeated exposure to high temperatures within the cracking apparatus. Therefore, subject to imparting cation-exchange property similar to that of the abovementioned natural zeolite by treating the waste catalyst in some way or other, not only an effective utilization of waste catalyst is feasible but also there can be produced a sewer wastewater treating agent useful for the removal of ammoniacal nitrogen and/or heavy metal ions and free of disintegration even when brought in contact with water.

Thus, in a process developed by *S.Sanga and Y. Nishimura; U.S. Patent 3,960,760; June 1, 1976; assigned to Catalysts and Chemicals Industries Co., Ltd., Japan* a sewer wastewater treating agent of superior quality is produced by alkali treatment of waste catalyst grains arising from the fluid-bed type catalytic cracking process, the sewer wastewater treating agent being capable of facilitating solid-liquid separation.

A process developed by *A.W. Liles and R.D. Schwartz; U.S. Patent 3,968,036; July 6, 1976; assigned to Exxon Research and Engineering Company* is a process for the treatment of wastewater from municipal and industrial sources by contacting the wastewater with an activated sludge at conditions at which biological oxidation takes place. In this process various water-insoluble inorganic oxides selected from the group consisting of silica, alumina, and silica-alumina are combined with the sludge whereby increased rates of biological oxidation are obtained and the sludge shows increased settleability.

In a most preferred embodiment of this method, the water-insoluble inorganic oxide is a spent cracking catalyst, that is, a zeolite which has been used in fluid cracking of hydrocarbon feeds and thus contains vanadium, iron, nickel, copper, and/or carbon. This catalyst may be recovered from aqueous scrubber solutions which are utilized to reduce stack losses in fluidized cracking processes and which present a solids waste disposal problem.

WOOL SCOURING LIQUOR FROM SPENT SCOURING LIQUOR

In ordinary wool scouring, impurities are removed from the wool by washing it in several bowls containing detergent solutions. In the best practice, wool passes from bowl to bowl in one direction, and liquor is pumped in the opposite direction. The raw wool enters the first bowl, and becomes cleaner as it moves towards the last. Clean water enters the last bowl and becomes dirtier as it moves towards the first. Detergent chemicals are added continuously to the first few bowls in an amount proportional to the weight of wool passing through the machine.

The dirty liquor in the first bowl is pumped continuously through a centrifuge and back to the bowl at a rate sufficient for the whole contents of the bowl to pass through the centrifuge in one to two hours. The centrifuge separates grease and dirt from the liquor, both products being accompanied by quantities of water. The grease product is usually washed and dewatered in other centrifuges, and the dirt in the form of a thin slurry is discharged, sometimes accompanied by additional centrifuged liquor. The quantity discharged in this way is commonly between 10 to 20 liters for each kilogram of greasy wool passing through the machine.

In this method of scouring, the centrifuge removes only a part of the dirt and grease, so the concentration of these contaminants in the first bowl gradually increases. Eventually the liquor becomes too dirty for acceptable scouring, and the bowl has to be discharged and cleaned. Thus, in addition to the centrifuge discharge described above, an additional quantity of highly polluted liquor is discharged whenever the bowls need cleaning.

The total waste liquor typically contains 7,000 mg of Biological Oxygen Demand (BOD), 6,000 mg of suspended solids and 4,000 mg of grease per liter, and is very difficult to purify by known methods. One method of purification involves removing the suspended dirt and grease particles from the liquor and then subjecting it to a biological digestion to remove the dissolved impurities.

A process has, therefore, been developed by *G.F. Wood; U.S. Patent 4,168,143; September 18, 1979; assigned to Commonwealth Scientific and Industrial Research Corporation, Australia* for reducing the pollutant concentration of effluent liquor from wool-scouring operations. The suint concentration of scouring liquor is allowed to build up to a point where suspended dirt and grease particles are destabilized and hence more efficiently removed by centrifugation. The centrifuged liquor is then recycled for further wool-scouring.

Y

YTTRIUM OXYSULFIDE FROM COLOR TV TUBE MANUFACTURE

The high costs of rare earth phosphors such as yttrium oxysulfide has motivated phosphor makers and phosphor users to develop processes for reclaiming these phosphor materials from contaminated screening wastes collected during color cathode ray tube screening operations. A number of phosphor reclaim processes have been developed and placed into commercial use which are directed to purifying the reclaimed phosphor to a state of virgin quality. It has been found, however, that it is not sufficient merely to restore the purity of a used phosphor material by removing screening contaminants. The phosphor material must also be given good screening properties. The properties that cause a phosphor material to screen satisfactorily may be stated as follows:

(1) The phosphor material must adhere to the surface of a cathode ray tube faceplate (glass). In "black surround" tubes, the phosphor material must also adhere to a layer used to form the black surround (typically graphite); and

(2) The phosphor material must have a high resistance to cross-contamination of earlier-deposited phosphor materials.

Cross-contamination is herein intended to mean the adherence of particles of one phosphor material upon previously deposited phosphor elements having another color emission characteristic. For example, when the blue-emitting phosphor material, typically deposited subsequent to the deposition of the green phosphor material, is contaminated by particles of later-deposited red-emitting phosphor material, upon energization of the blue-emitting phosphor, its emission spectrum will be altered due to the presence of the red phosphor particles.

The property of a reclaimed rare earth oxysulfide phosphor material to adhere to its substrate during rescreening has been addressed previously. It is common practice to subject such reclaimed phosphor materials during a reclaim process to a wash in nitric acid in order to attack and cleanse the outer surface of the phosphor particles. The use of nitric acid to enhance the adherability of a reclaimed phosphor material has proven successful; however, this process has a number of drawbacks:

(1) Nitric acid is undesirably costly;

(2) The high corrosiveness of nitric acid results in a relatively high yield loss in the reclaimed phosphor, for example, 3 to 4%, or more;

(3) The use of nitric acid requires extensive afterwashing in order to adequately cleanse the reclaimed phosphor material; and

(4) Nitric acid causes the reclaimed phosphor material to clump, requiring an undesirable amount of blending and sifting and thus undesirable abrasion of the phosphor particles, in order to restore the phosphor material to a loose, powdery state.

The problems associated with the tendency of a reclaimed phosphor material to cross-contaminate have not been overcome by prior art processes. It is known that certain reclaim processes involve subjecting the reclaimed phosphor material to a wash in ammonium carbonate in order to improve its resistance to cross-contamination. It has been found, however, that such prior art processes do not significantly affect the tendency of a reclaimed phosphor material to cross-contaminate.

Prior art phosphor reclaim processes have also been found to be deficient in their effectiveness in separating graphite contaminants from the phosphor material being reclaimed.

Thus, an improved process has been developed by *H. Forest and J. Harkin; U.S. Patent 3,954,657; May 4, 1976; assigned to Zenith Radio Corporation*. It is a process for reclaiming from contaminated color cathode ray tube screening waste rare earth oxysulfide phosphors, especially the red-emitting phosphor yttrium oxysulfide. The screening waste from which the phosphor material is to be reclaimed may contain, in addition to the phosphor material contaminants such as hardened photoresist material, blue-emitting and green-emitting zinc sulfide type phosphor materials and graphite.

The phosphor reclaim process disclosed is characterized by having steps including subjecting the partially purified oxysulfide phosphor material to a wash in dilute nitric, acetic, hydrochloric or citric acid, preferably citric acid, in order to enhance the adherability of the reclaimed phosphor material during subsequent rescreening thereof. The phosphor material is subjected to a wash with sodium tripolyphosphate to minimize during rescreening of the phosphor material cross-contamination of underlying blue-emitting or green-emitting phosphor materials.

In accordance with another aspect of the disclosed reclaim process, graphite contaminants are separated from the phosphor material being reclaimed by suspending the graphite-contaminated phosphor material in an aqueous bath, adding tripolyphosphate to the bath to cause the phosphor material to settle and then decanting the still suspended graphite.

A process developed by *U. Viohl and R. Zondler; U.S. Patent 4,159,931; July 3, 1979; assigned to International Standard Electric Corporation* is one for separating a red rare earth phosphor of a color television picture tube from the sediment in the drain channels of a screen-coating room, which sediment contains, in addition to various impurities, zinc-sulfide-base and zinc-cadmium-base green and blue phosphors.

Figure 173 shows a suitable form of apparatus for the conduct of such a process. To remove any coarse foreign matter, the sediment from the drain channels of a screen-coating room is first passed through a sieve whose apertures are approximately 100 μm in diameter. Residues of ammonium bichromate, polyvinyl alcohol and other substances soluble in water are washed out, and the settling solid constituents are heated to approximately 450°C in order to burn any organic constituents that may be left. After cooling, pulverization takes place, and the powder is mixed with an aqueous ammonium halide solution. For example, 1 kg ammonium chloride (NH_4Cl) is dissolved in 13 ℓ of demineralized water, and 1 kg of the powder is mixed with this solution in a mixing vessel **1** by means of a stirrer **2**.

Figure 173: International Standard Electric Corp. Apparatus for Treating Color Television Screen-Coating Room Waste to Recover Red Rare Earth Phosphors

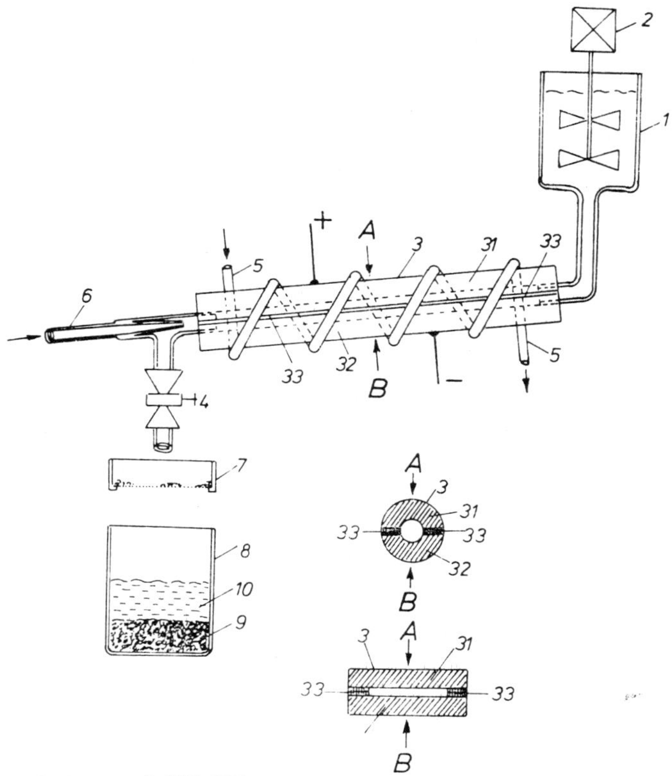

Source: U.S. Patent 4,159,931

This mixture then flows through an electrolysis apparatus. The chief component of this apparatus is a longitudinally divided graphite tube **3** whose halves **31** and **32**, which are mechanically joined together again by means of insulating layers but are electrically isolated from each other, serve as electrodes. Instead of using a round, longitudinally divided graphite tube **3** of circular section, two graphite plates **31** and **32** may be joined together by means of insulating parts **33** to form a tube of rectangular section, as shown in the detail views in the figure. A valve **4**, e.g., a hose clamp, disposed behind the tube **3** allows the rate of flow through the tube to be regulated. This flow rate is adjusted so that the above mixture flows through the electrolytic arrangement in about 4.5 hours at an electrode current of 9 to 14 A, for example.

The graphite tube is surrounded by a cooling facility **5** which may be a cooling coil as shown in the figure, but also a cooling jacket, and is traversed by the cooling liquid, e.g., cooling water, countercurrently to the mixture in the tube **3**.

By means of a blow-in pipe **6** air or nitrogen is blown into the tube **3**, also countercurrently to the mixture, to prevent the tube halves **31** and **32**, which serve as electrodes, from becoming inactivated due to deposition of phosphor particles, sulfur, and cathode deposit.

After flowing through the valve **4**, the electrolytic products pass through a sieve **7** whose apertures are about 30 μm in diameter and which separates the sulfur and the cathode deposit from the electrolytic products. The residue is collected in a collecting tank **8**, the solid constituents **9** sinking to the bottom and being separable by decanting the supernatant liquid. The liquid **10** can be used for one or two additional, like electrolytic processes, while the solid constituents are washed, reheated to about 450°C, cooled down, and pulverized.

The powder is then mixed again with an aqueous ammonium halide solution according to the same recipe and subjected to an electrolytic process of the same kind. Either the same electrolysis arrangement may be passed through again, or a second, like arrangement may follow the first. After decanting, the ammonium halide solution used in the second electrolysis may be reused, too.

The powder obtained after washing and drying is an almost pure red rare earth phosphor whose impurity level is not higher than that of the red phosphor to be regenerated by known regenerating methods, so that it can be processed by such a method and reused in the manufacture of screens, this being possible with and without addition of phosphor straight from the factory, without any deterioration in the quality of the screens. According to the method of this process, approximately 95% of the red rare earth phosphor contained in the sediment of the drain channels of a screen-coating room can be recovered.

Although the method according to this process is much more expensive than the prior art methods of regenerating only slightly contaminated rare earth phosphors, in view of the high price of such phosphors, their recovery from the sediment in the drain channels of a screen-coating room is economically worthwhile.

Z

ZINC FROM BRASS MILL DUSTS

See "Copper from Brass Mill Dusts."

ZINC FROM BRASS SKIMMINGS

Whenever molten brass comes in contact with air (or other oxygen-containing gases), the zinc content thereof rapidly forms a surface layer of zinc oxide as an impurity. Thus in the manufacture of virgin brass and in the subsequent foundry melting of the billets, a significant amount of the zinc content of the brass is lost as zinc oxide which forms as a floating dross and additional significant amounts of the brass alloy itself is lost in skimming off the floating impurities and fluxes prior to pouring the molten brass. As a result, it has been estimated that 5% of all melted brass is lost to such skimmings. The typical metallic values in such skimming are in excess of 85% (primarily copper and zinc in alloyed form and zinc oxide).

It can be appreciated that such losses represent a significant economic consideration, which the industry since its inception has long sought to minimize by various methods for recovering the metallic content of such skimmings. Two major brass mills alone each generate an average of 400 to 750 tons of brass skimmings a month respectively.

Over the years several different methods have been attempted to recover these losses. While some of these have succeeded to some degree in reducing the losses, they all still involve substantial loss of the metal content of the brass skimmings and are additionally quite costly. There have been three principal prior art methods. Some secondary manufacturers treat the brass skimmings pyrometallurgically, for example, in a rotary furnace, tying up the zinc oxide in fluxes, thus losing 30 to 40% of the zinc content and 10 to 20% of the copper content and additionally charging about 1 to 3% of the market value of the brass nuggets returned for the processing.

Recovery has also been achieved by mechanical means by grinding the brass skimmings, such as by ball milling, and separating the metals from the impurities by hydraulic classification. This recovers up to 90% of the material in the brass skimmings present in metallic form with the washed out fines being discarded as waste constituting essentially all of the zinc oxide mixed with other debris (including carbon and silica) and as much as 10 to 14% by weight of copper.

In a third method, the brass skimmings are smelted and the zinc oxide layer is blown off as waste with a 1 to 3% copper loss. Most of the zinc content of the alloy is converted to zinc oxide during this process and is also lost during the blowing process. The brass is then cast into an anode with the copper content thereof purified and recovered separately by electrolytic deposition.

Thus a process has been developed by *H. Bzura; U.S. Patent 3,905,808; Sept. 16, 1975; assigned to Madison Industries, Inc.* which involves recovering metals and a zinc salt by-product from brass skimmings by chemical methods after crushing to loosen and expose zinc oxide from the metallics thereof. The process then involves optionally screening out the larger particles (which are mainly metallic) for direct recycling, adding the remainder to a reaction vessel together with concentrated hydrochloric or sulfuric acid with agitation at a rate and ratio to react the zinc oxide but not the metallics (preferably accomplished by controlling such relative additions so as to maintain the reaction zone at a pH between 0.1 and 1.5). Finally, the process involves rapidly settling and removing for recycling the metallics from the vessel and overflowing the zinc salt reaction product in solution (as a useful by-product).

Figure 174 is a schematic of the apparatus used and its place in the process. As shown there, the reaction vessel **10** is charged by the feed device **12** receiving the minus No. 4 mesh material from the screen **14** which has already been crushed by the hammer mill **16**. Acid is added to the reaction vessel **10** through inlet **18** at a controlled rate. If required, water is added to the reaction vessel through pipe **20**. The reaction mass in the vessel is continuously mixed and agitated by agitator **22** driven by motor **24**.

Depending upon the by-product desired, either concentrated hydrochloric acid or concentrated sulfuric acid is added through inlet **18**. The exothermic reaction of the acid with the powdered zinc oxide present in the brass skimmings is very rapid and therefore requires a reaction time of less than a minute with no heat nor pressure being required. The reaction time of the acid in attacking the metals present in the brass skimmings is comparatively slow, particularly since the relative surface area is so small due to the presence of the majority of the metal in the form of nuggets, shot, and other relatively large particles (even after the aforesaid screening).

In this foregoing example of the preferred embodiment, the brass skimmings are added to the reaction vessel **10** at the rate of 5 tons/hr. The pH of the reaction vessel is monitored by sensing device **26** which in turn controls valve **28** thus metering in the acid through inlet **18** so as to keep the reaction mass within a specific pH range.

If the reaction mass is kept within the preferred pH range, then the reaction mass is acid enough to rapidly attack effectively all of the zinc oxide and yet is not so acid as unnecessarily to attack the metallics present in the brass skimmings. For hydrochloric acid, the desirable pH range is from 0.1 to little more than 1.0 with the optimum range being from 0.3 to 0.5. For sulfuric acid, the desired range is from 0.1 to 1.5 with the optimum range being from 0.1 to 0.8. Although this process can be practiced outside these ranges, it is generally undesirable. With too low a pH, there will be an excess of acid, requiring the addition of zinc dust (or its equivalent) in order to neutralize the excess acid overflowing the weir **30** from the reaction vessel **10** and to convert the copper salts (resulting from acid attack of the brass metal) to the desired zinc salts and to precipitate out metallic copper.

Figure 174: Madison Industries, Inc. Process for the Recovery of Metallics from Brass Skimmings

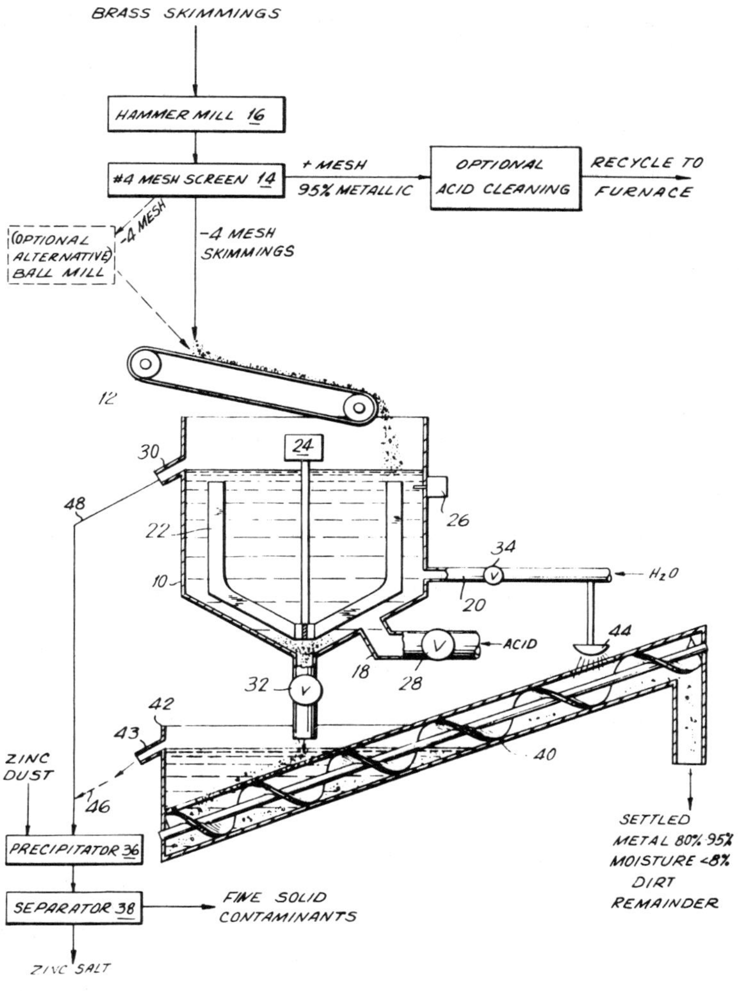

Source: U.S. Patent 3,905,808

On the other hand, with too high a pH there will be incompletely reacted zinc oxide contaminating the reaction products exiting from the weir **30**. As the pH of 3.5 is approached, the reaction becomes too slow for practical purposes, even if the residence time in the vessel **10** is greatly increased.

It is possible to control the reaction by monitoring other parameters or even by metered addition or preanalyzed reactants; none of which, however, are as simple and convenient as the pH control method.

The reaction product in this preferred example exits from the reaction vessel **10** by a weir overflow **30**. The unreacted heavier particles, mainly consisting of solid metallics, settle to the bottom of the vessel and exit through valve mechanism **32**. This can be a simple diaphragm valve, a rotary valve, or even a double valve for batch removal of solids.

The reaction within vessel **10** can be subject to foaming. However, this is normally not too serious a problem which can be generally controlled, if necessary, by proper agitation or other known methods including additive foam suppressants.

The acid and crushed brass skimmings are added in a conventional manner sufficiently remote from the weir overflow **30** to insure thorough mixing and sufficient residence time of the reactants in the reaction zone of the vessel to assure that the zinc oxide is converted to the zinc salt before it can reach the weir overflow. The addition of water from supply pipe **20** to the reaction vessel controlled by valve **34** is intended mainly to keep the reaction mass sufficiently dilute to insure that the zinc salt reaction product remains in solution. This is mainly required for zinc sulfate.

The liquid overflow from the weir **30** is mainly a zinc salt solution; however, it may contain minor amounts of very fine metallics, other contaminants, and perhaps some other metal salts, normally the copper salt. Consequently, (if necessary), this overflow can be subsequently treated by the addition of zinc dust to convert the other more electropositive metals to their metallic form thus creating additional zinc salts. This step has been illustrated in the form of a precipitator **36**. Thereafter the liquid mass is filtered or continuously centrifuged, for example, in a separator **38**, to remove the solid fines. These solid contaminants will usually contain a large mass of dirt with <0.5% copper (by weight of the original charge). This copper content is usually too small to be of concern, or even of value for reclamation.

The separated solids exiting through valve mechanism **32** of reaction vessel **10** are preferably subjected to a gentle washing. In the figure is illustrated an inclined screw conveyor **40** rising from an overflow wash tank **42** against a countercurrent of wash water from spray device **44**. If desired, the soluble salts washed from the solid metallics can be reclaimed by combining the overflow from the tank **42** via the conduit **46** (indicated by dots as being an optional alternative) to feed into conduit **48**. In this case the wash water **44** is minimized to avoid unnecessary dilution of the zinc salt solution. Conduit **48** leads from overflow weir **30** to the precipitator **36**.

Depending to a large extent upon the quality of the brass skimmings received, the material recovered from the screw conveyor **40** will contain 80 to 95% metal, up to 8% moisture, and the remainder will be dirt. Further purification has been found to be unnecessary. The recovered material can be recycled directly to the melting furnaces.

Typically 95 to 99% or more of the zinc oxide is solubilized and separated from the metallics by conversion to a useful zinc salt by-product.

The contaminants in the recovered metals are significantly reduced with any remaining amounts being well within acceptable limits for recharging to the brass furnaces. The zinc salt solution by-product is ammonia-free and has <10 ppm of copper. Since the zinc salts result almost exclusively from the zinc oxide and not from the zinc in the metallic alloy, the alloy composition in the brass skimmings remains substantially unchanged and can usually be recycled directly, with little or no necessity for additional speltering.

ZINC FROM ELECTROPLATING WASTES

The electrolytic deposition of zinc, which is a well-known industrial process, encounters certain difficulties, one of which is raised by the necessity of eliminating the magnesium sulfate which accumulates in the solution to be electrolyzed when the sulfuric liquors issuing from the electrolyzer are recycled during the treatment of the zinc ore with a view to extracting new amounts of zinc sulfate used for electrolytic deposition. This elimination is necessary since the presence of magnesium salts has unfavorable effects above a certain threshold of concentration on the formation of electrolytic zinc deposits, from a quantitative as well as from a qualitative point of view.

It is thus necessary to purify the residual electrolytic solution as soon as the magnesium content reaches a certain limit value, the latter being generally between 15 and 20 g/ℓ, lest the energetic efficiency of the electrolytic process be considerably decreased; the starting ore may contain 0.4 to 4% magnesium. Furthermore, traces of other impurities which are deleterious to the electrolysis of the zinc ore, such as Fe, Co, Ni, Mn, etc., are present in the solution. Since the solutions thus withdrawn from the cycle still contain Zn, there is a loss of zinc when these solutions are discarded. The problem is increased by the necessity of treating the effluents with a view to diminishing the polluting effect of the effluents prior to discarding the same by rejecting them into the natural environment. The neutralization of the sulfuric acid present requires the use of considerable amounts of lime or any other convenient neutralizing agent.

It is possible to consider the application of known processes for recovering the zinc from the solutions obtained after purification: the purification solution is treated with a view to eliminating the zinc therefrom by means of a cascade of special electrolytic vats in such a manner that a maximum amount of zinc is eliminated from the treated solution, the remaining zinc then being precipitated and the sulfuric acid being neutralized prior to rejecting the residual solution into the natural environment. However, this known method is not satisfactory from a technical point of view, nor from an economical point of view.

Therefore, an improved process has been developed by *J.-P. Bernat and J. Sors; U.S. Patent 4,108,744; August 22, 1978; assigned to Societe Nationale Elf Aquitaine, France* for the recovery of the zinc contained in the residual solutions obtained after electrolytic deposition by fixing the zinc (none of the other metals present in the solution being included in the complex), in the form of a complex by adding of an alkali metal halide (sodium chloride preferred) to the residual solution adjusted to a low acidity, for example, by dialysis. The process permits the almost total recovery of zinc, and a large part of sulfuric acid.

The flow diagram of such a process is given in Figure 175. As shown in the drawing, the effluent solution **1** passes through a dialyzer **2**, flows upwards in compartment **2a** along a diaphragm **2c** which separates compartment **2a** from an inner compartment **2b**. The latter defines a flow path for demineralized water which enters compartment **2b** by conduit **3** and flows downwardly, as shown in the drawing. With a view to simplifying the drawing, only one dialytic element is shown, it being well understood that in industrial practice a whole series of such elements are generally used.

Figure 175: Societe Nationale Elf Aquitaine Process for the Recovery of the Zinc Contained in the Residual Solutions Obtained After Electrolytic Deposition

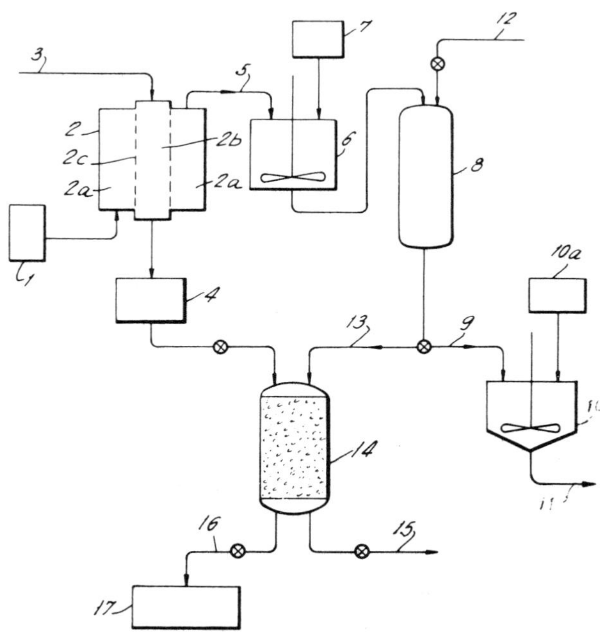

Source: U.S. Patent 4,108,744

The residual solution stemming from the electrolytic deposition process, which solution constitutes the effluent solution **1**, has a high sulfuric acid content; this acid thus will diffuse through diaphragm **2c** into compartment **2b**; consequently the solution issuing from compartment **2a** through conduit **5** has a lower sulfuric acid content, but it still contains the entire amount of zinc which was initially present in the solution, as well as the entire initial amount of magnesium and other impurities.

The residual solution, which thus has a decreased sulfuric acid content, flows into a reactor vat **6** wherein the convenient complexing agent is added to the solution, this complexing agent being advantageously constituted by NaCl fed from a tank **7** or a similar source. Consequently the zinc sulfate is converted into the anionic zinc complex, particularly Na_2ZnCl_4. The solution of this complex is passed into the upper end of column **8** which contains an anion exchanger advantageously constituted by a synthetic resin.

The halogenide concentration (more particularly the NaCl concentration) in vat **6** being conveniently adjusted, only the zinc is incorporated in the complex, whereas the other metals present, including Na, remain in the solution. Consequently only the complex zinc anion is fixed to the exchanger in column **8**. The remaining liquor leaves this column by flowing through conduit **9**, and is neutralized in a container **10**, generally by means of lime issuing from **10a**. The impurities, which are thus rendered nonpolluting, are finally evacuated at **11**.

The extraction of the zinc, retained by the anion exchanger in column **8**, is then performed by elution by means of demineralized water issuing from **12**. The eluate which is then withdrawn from column **8** through conduit **13** is a hydrochloric solution of $ZnCl_2$; if there is no possibility of using this solution directly the same is introduced into the upper part of a column **14** loaded with a cation exchanger. This cation exchanger may be a commercially available synthetic resin.

During this operation, the zinc in the form of cation is, in this case, fixed on the exchanger of column **14**, while the liquor which has flowed through the column is directed through **15** toward a neutralization station and then evacuated. The thus fixed zinc is then recovered in the form of a sulfate solution, by eluting with the weak sulfuric acid solution issuing from **4**, which has been obtained by the dialysis at **2**, as described above. The zinc sulfate solution thus recovered flows through conduit **16** toward the storing station **17** from which it can be taken up again so as to be reintroduced into the electrolytic cycle.

The following example illustrates the process in a more detailed manner. A residual solution of an installation for electrolytic deposition of zinc has the following composition (expressed in g/ℓ): Zn, 40; Mg, 15; Fe + Cu + Ni + Co + Mn, 2; and H_2SO_4, 120.

This solution is dialyzed with 1.1 m^3 demineralized water per m^3 of solution. 1 m^3 sulfuric acid at 100 g/ℓ is then obtained, while the residual solution only contains 20 g/ℓ H_2SO_4. 65 kg NaCl is added to this solution in the reaction vat **6**, with a view to forming the anionic zinc complex. The solution of the complex then flows into column **8** loaded with IRA 400 resin. The liquor which has flown through the resin bed is evacuated through **9** into container **10** and is then treated with lime and evacuated, as described above.

The resin is then eluted with 3 m^3 demineralized water issuing from **12**, and the solution thus obtained is introduced through **13** into column **14** filled with IR 120 resin. The liquor issuing from the lower end of the filling material **14** is withdrawn through **15** and neutralized, and then evacuated. When the exchange in column **14** is finished, the resin is eluted with 1 m^3 diluted sulfuric acid issuing from **4**; an amount of 2 m^3 $ZnSO_4$ solution containing 20 g/ℓ zinc is then obtained, this solution being sufficiently pure and substantially free of magnesium so as to be adapted to be recycled in the lixiviation treatment of the zinc ore.

ZINC FROM GALVANIZING WASTES

Galvanizers' dross is a semisolid pasty mixture that forms in the bottom of the galvanizers' kettle. It is created by the attack of molten zinc on iron salts that coat the surface of the piece to be galvanized. These iron salts are formed by the the customary pickling and pregalvanizing treatments. Lesser quantities of dross are formed by the solvent action of zinc on the galvanizing pot and the steel or iron piece being coated. Accumulation of this dross eventually interferes with the operation of the galvanizing bath so that periodic removal is necessary. Dross, as it is first formed, is believed to correspond to the formula $FeZn_7$ (10.9% Fe).

However, the average iron content of the final dross is considerably below this value because the $FeZn_7$ gradually absorbs zinc without loss of its crystalline character until it exists as a solid solution. Dross, as it is removed from the galvanizing bath, usually contains from 2 to 4% iron and approximately 1% lead.

A process developed by *J.B.S. Menendez, P.M. Sagrado and A.S. Alonso; U.S. Patent 3,902,894; September 2, 1975* is one for the recovery of zinc from galvanization mattes, scrap and residues of this metal which comprises essentially the following steps:

(a) Melting the mattes, scrap or residues of zinc until a liquid phase is obtained;

(b) Adding to the molten mass a metal selected from among the Groups II-A and III-A and/or an alloy thereof; and

(c) Treating the mass in the reactor to cause the disintegration thereof, favoring the formation and separation of oxides and aluminides, by means of the turboagitator contained in the reactor which impels the mass toward the bottom thereof hitting it and the walls of the reactor, thereby obtaining a closed cycle flow in vertical direction from top to bottom through the center of the mass and from bottom to top along the periphery, cut or interrupted by the blades of the turboagitator.

The apparatus permits obtaining the maximum yield in recovering up to 98% of the base metal of the scrap or galvanization matte, with the product obtained having a minimum of iron as impurities.

A process developed by *G. Thorsen; U.S. Patent 3,958,986; May 25, 1976* is one for the recovery of zinc and other metal values from zinc waste or other metal waste containing chlorine and/or other anions forming undesirable metal salts. The waste is leached with a liquid organic phase containing a cation exchanger. The impurities are removed from the organic phase with water, and the zinc and other metal values are extracted from the organic phase with mineral acid. Thereafter the zinc and other metal values can be recovered with known methods.

A process developed by *K. Kuwano and Y. Kanbara; U.S. Patent 4,003,559; January 18, 1977; assigned to Mitsui Mining & Smelting Co., Ltd., Japan* utilizes a compact apparatus for recovery of metallic zinc from the dross in which the filling ratio of dross is increased and metallic zinc is effectively squeezed out from the dross without external supply of heat and use of a flux.

More specifically, there is provided an apparatus for recovering metallic zinc in dross discharged from a zinc melting furnace, which comprises a saucer-like container for receiving dross. The container has a bottom portion of an inverse conical form inclined downwardly, and a plurality of agitating scraper plates supported around a vertical rotary shaft in the container to squeeze out metallic zinc from dross. It has discharge means for metallic zinc and discharge means for secondary dross consisting mainly of oxides; these discharge means are mounted on an opening formed on the bottom of the container.

A process developed by *R.W. Leonard; U.S. Patent 4,075,008; February 21, 1978; assigned to United States Steel Corporation* is one in which zinc is removed in a substantially pure form from the dross skimmings of galvanizing baths. The dross is poured into a mold and solidified. The solidified mold is then placed on a permeable base and in the substantial absence of agitation, reheated to a temperature of 420° to 700°C to melt out the zinc in an uncontaminated form.

ZINC FROM NONFERROUS SLUDGES

See also "Metals from Mixed Metal Slimes."

ZINC FROM SCRAP AUTOMOBILES

See "Steel from Scrap Automobiles."

ZINC FROM STEEL MILL FURNACE DUSTS

See also "Iron from Steel Mill Furnace Dusts."

It is known that the dusts entrained with the gases of a blast furnace and from a steel-making furnace contain a large part of the zinc introduced into the bath, e.g., in the form of galvanized scrap. The dusts are particularly rich in zinc when the melt in either furnace contains large proportions of galvanized-iron or galvanized-steel scrap.

It has already been proposed to recover the zinc in the form of its oxide from such dusts by extraction of the dusts from the gas by wet or dry separation. None of these techniques have proved to be commercially successful since the zinc content of the product is insufficiently high and the recovery itself is not satisfactory.

When the dusts are returned to the furnace they have a tendency to increase the zinc content of the melt and to cause problems in the iron or steel making process.

It is known, for example, that the presence of zinc in the fusion bed of a blast furnace causes difficulties in the operation thereof, not only because of an increased consumption of coke but also because of dangers to the personnel of the plant. It is possible to reduce the problem by reducing the input of zinc to the furnace but this requires special methods in selecting the scrap or utilizing ores which are free from zinc or contain only minor quantities thereof.

A process developed by *A. Wagner, R. Pixius and F. Schon; U.S. Patent 4,069,315; January 17, 1978; assigned to Acieries Reunies de Burbach-Eich-Dudelange SA (ARBED), Luxembourg* is one in which the zinc-containing dust from an iron-making (blast) furnace or a steel-refining (e.g., converter) furnace is recovered and extracted with water saturated with carbon dioxide to solubilize the zinc. The zinc solubilized in the CO_2-saturated water is then precipitated by the elimination of carbon dioxide therefrom. Such a process is illustrated in the flow diagram in Figure 176.

The process is applicable either to a blast furnace **10** or a steel-making converter **17** or both in a conventional ferrous metallurgy plant. The blast furnace **10** is provided with the usual gas recovery hood **11** which can be connected to a cyclone **12** from which the dust can be removed from the gas. The zinc-containing dust is led at **13** to a pelletizing stage **14** and the pellets are then introduced into an autoclave **15** which has previously been charged with water saturated with carbon dioxide and possibly containing sulfur dioxide.

Alternatively, or in addition, the gases from the converter **17** are recovered by a hood **18** and led by a duct **19** to a cyclone **20** in which the dust is separated from the gas, the latter being drawn through an outlet **21** by a blower **22**.

The zinc-containing dust is fed at **23** directly to the autoclave **15** or to the latter via a pelletizing stage as shown at **14**. The gases, freed from the dust, are led via ducts **16** and **21** to an inlet **24** of a washer or scrubber **25** which is supplied with water at **26**. Since these gases contain large quantities of carbon dioxide and, possibly, sulfur dioxide, the water emerging at **27** is saturated in carbon dioxide. It has been found to be advantageous to carry out the washing in the tower **25** at the same pressure as is maintained in the autoclave **15**.

The CO_2-saturated water is supplied to the autoclave as represented by the line **27**. The autoclave **15** can be agitated, e.g., rotated, and is maintained at the elevated pressure selected for recovery of the zinc.

The solution is decanted at **28** and after filtration or centrifugation, is delivered to a zinc oxide precipitating stage **29** in which it can be aerated as represented by arrow **31**, can be heated as represented at **32** and can be evacuated as shown at **30**, in combination or individually, to drive off the carbon dioxide. The resulting zinc oxide precipitate, recovered from the solution by filtration, centrifugation or decantation, as represented at **33**, can be introduced into a zinc producing plant **34**. The liquid phase, consisting primarily of water and any acids which may be added, are returned to the washer **25** as represented by line **35** as part of the water used for washing the gas.

The solids from which the extracting solution has been decanted at **28** are recovered at **36** and may either be discarded as shown at **37** or recycled as indicated at **38** to the blast furnace or, if desired, to the steel-making furnace.

Figure 176: ARBED SA Process for the Recovery of Zinc from Iron-Making or Steel-Refining Furnace Dust

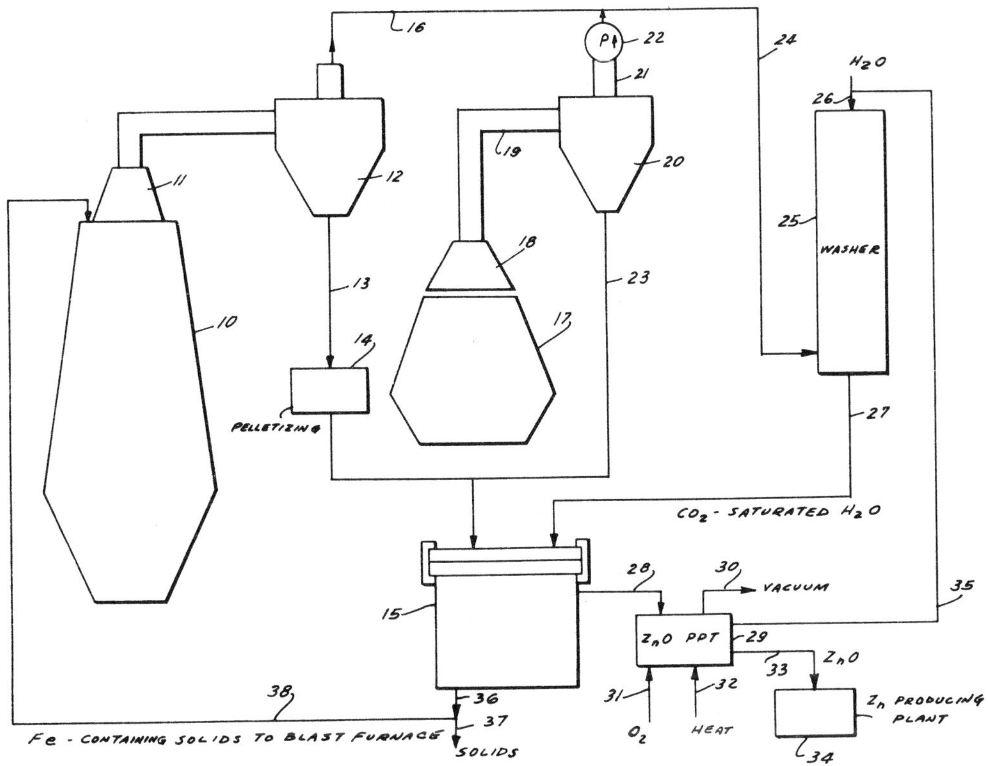

Source: U.S. Patent 4,069,315

A process developed by *M.A. Peters; U.S. Patent 4,071,357; January 31, 1978; assigned to Hazen Research, Inc.* is one for recovering a substantially impurity-free zinc oxide product from steel-making flue dust or a similar material which comprises leaching the flue dust with concentrated ammonia and carbon dioxide to dissolve zinc and unwanted impurities, cementing the leach filtrate with zinc to remove copper, cadmium and lead impurities, conducting a steam distillation on the cementation filtrate to precipitate basic zinc carbonate, removing the ammonia, carbon dioxide and iron impurities, and filtering to provide a residue of essentially basic zinc carbonate, sulfur, and chromium.

The residue is then washed to remove soluble sulfates, dried and calcined at high temperatures to break down the basic zinc carbonate into zinc oxide, water washed to remove chromium, and the residue of the water wash dried to produce the desired impurity-free zinc oxide product. The two water washes may be combined into one step performed after the calcining step to remove both sulfur and chromium in one wash step. Such a process is shown in the flow diagram in Figure 177.

Figure 177: Hazen Research Inc. Process for Recovering Zinc from Steel-Making Flue Dust

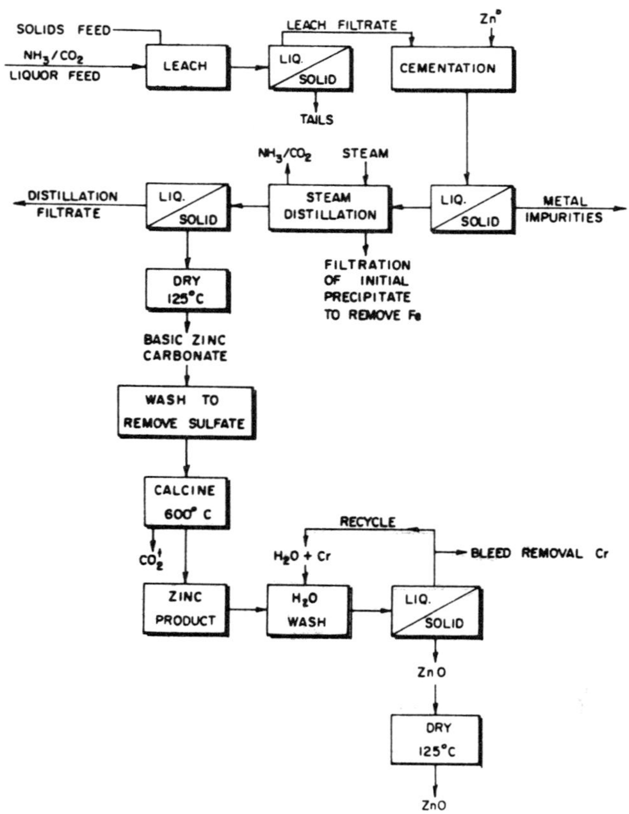

Source: U.S. Patent 4,071,357

A process developed by *H. Reinhardt, H.D. Ottertun and S.O.S. Andersson; U.S. Patent 4,124,462; November 7, 1978; assigned to MX-Processer Reinhardt & Co. AB, Sweden* is one in which zinc is recovered from a material containing zinc and iron. The material is leached with sulfuric acid to produce a final pH of 3 to 5.5. Zinc powder is added to the leaching solution to precipitate polluting metals. Zinc is then extracted with a solution of a dialkyl phosphoric acid in an organic solvent. Zinc is extracted from the organic solution with sulfuric acid and is recovered from the sulfuric acid solution as metal or as zinc sulfate. Such a process is shown in Figure 178.

Figure 178: MX-Processer Reinhardt & Co. AB Process for Zinc Recovery from Flue Dusts

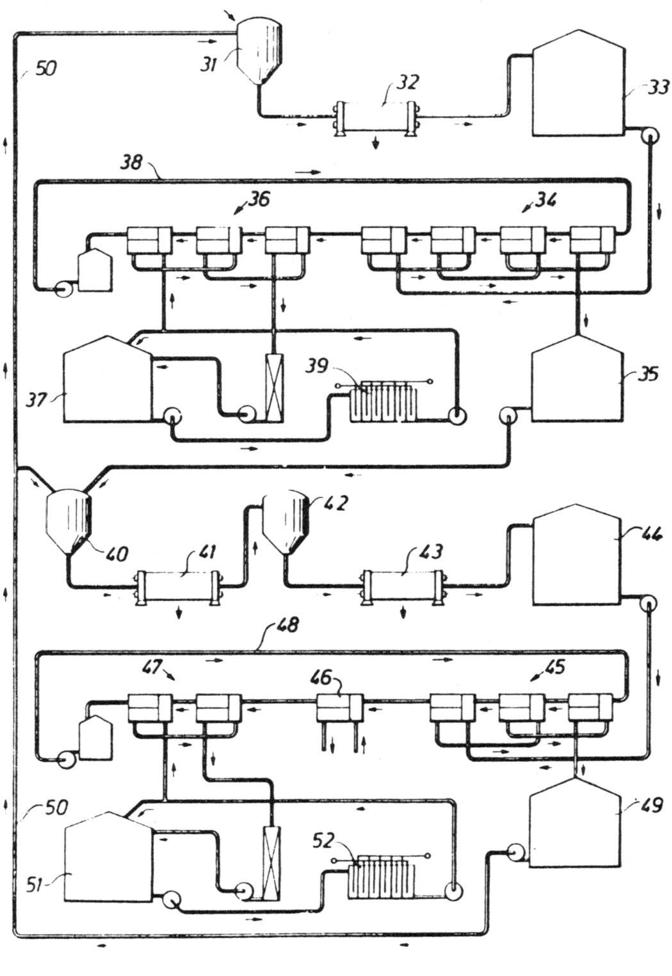

Source: U.S. Patent 4,124,462

The sulfuric acid and flue-gas dust are supplied to a leaching vessel **31**. The leaching solution is filtered in a filter **32**, stored in a tank **33** and conveyed to a liquid-liquid extraction apparatus **34**, comprising four mixer-settler units. Here the leaching solution is extracted in countercurrent flow with a copper reactant in an organic solvent. The raffinate is conveyed to a vessel **35** and the organic solution is led to an extraction apparatus **36** comprising three extraction units where copper is reextracted by means of a sulfuric acid solution in countercurrent flow. The organic solution is returned through a pipe **38** for renewed extraction. The sulfuric acid solution is conveyed via a storage tank **37** to an electrolyzer **39** where metallic copper is recovered.

From the storage tank **35** the raffinate, now almost free of copper, is conducted to a mixing tank **40** where the pH is adjusted to from about 3 to about 5.5. This is done either by adding an alkaline material, or by contacting the raffinate with a copper-free material in another leaching operation. After filtering in a filter **41**, zinc powder is added to the solution in a mixing tank **42**. Metal particles precipitated by means of cementation and any remaining zinc particles are separated in a filter **43**.

The liquid is led by way of a storage tank **44** to a series **45** of three mixer-settler units where the liquid is extracted in countercurrent flow with an organic solution containing a zinc reactant in the form of a dialkyl phosphoric acid. The raffinate is returned via a storage tank **49** and pipe **50**, some to the leaching or neutralizing tank **40** and some to the leaching tank **31**. The zinc-bearing organic solution is washed in a mixer-settler **46**, after which zinc is reextracted with a sulfuric acid solution in a set **47** of two mixer-settler units. The organic solution returns through a pipe **48** to the extraction units **45**. The zinc-containing sulfuric acid solution is conducted via a storage tank **51** to an electrolyzer **52** where zinc is obtained in metallic form.

ZINC FROM ZINC SCRAP

Recovery of zinc from old scrap, mainly in the form of diecastings, engravers' plate, and brass and bronze, represents less than 5% of the total supply. New scrap comes principally from zinc- and copper-base alloys from manufacturing operations and drosses from molten galvanizing and diecasting pots. New scrap is either sold to secondary smelters or processed in captive operations as run-around scrap. The large use of zinc in galvanizing and in compounds in which the zinc is lost is a major obstacle to increased recycling. The zinc-base alloys are recovered by remelting and redistillation to commercial zinc products, whereas the copper-base alloys are remelted and the contained zinc is reused in brass and bronze ingot.

The area of major potential in recovery of old scrap is in the salvage of diecastings from discarded automobiles and appliances. There are 100 to 150 automobile shredders in the United States with an annual capacity of 6 to 8 million tons of scrap, of which an estimated 200,000 tons is nonferrous material. The nonferrous fraction contains about 57% zinc, 33% aluminum, 8% copper, and 2% other metallics.

During the past decade, zinc recovered form old zinc-base scrap has amounted to 8% of the zinc consumed in zinc-base alloys, and zinc recovered from old copper-base scrap amounted to 21% of the zinc consumed in copper-base alloys (14).

ZINC CHLORIDE FROM SPENT CATALYSTS

There is a process for the hydrocracking of predominantly polynuclear aromatic hydrocarbons such as coal extracts in the presence of molten zinc halide catalysts to yield gasoline boiling range hydrocarbons.

A process for the regeneration of the zinc halide catalyst has been described by *E. Gorin; U.S. Patent 4,081,400; March 28, 1978; assigned to Continental Oil Company*.

The regeneration process comprises four steps. In the first step, the spent zinc halide melt from the hydrocracking zone is subjected to vapor phase oxidative treatment by combustion of the carbon and sulfur components, as well as any ammonia that may be present. In this step, the temperature is maintained at least high enough to assure vaporization of the zinc halide. The effluent vapors contain solids composed of zinc oxide and zinc oxide complexes, along with inorganic ash that may be in the spent melt, as well as residual organic residue.

The second step of the process consists of the separation of the zinc halide vapors from the solids by any conventional means, generally a cyclone. The third step consists of the treatment of the separated solids with hydrogen halide gas under conditions favoring the reaction of zinc oxide and the hydrogen halide at a temperature sufficiently high to form zinc halide in the vapor state. In the fourth step, vaporous zinc halide is separated from the ash and condensed to the molten state for recycle to the hydrocracking zone.

A similar process has been described by *C.W. Zielke; U.S. Patent 4,136,056; January 23, 1979; assigned to Continental Oil Company and The U.S. Department of Energy.* It is based on the observation that the rate of recovery of water-insoluble zinc values from the separated solids from the second step of the process is relatively slow. An improvement consists in treating the separated solids with both hydrogen chloride and calcium chloride, whereby the rate of recovery of the water-insoluble zinc values is improved. This behavior may be explained by the following equation:

$$ZnO{\cdot}SiO_2 + CaCl_2 \rightarrow CaO{\cdot}SiO_2 + ZnCl_2 \uparrow$$

The improved process is shown in the flow diagram in Figure 179. Referring to the drawing, spent zinc chloride melt is fed to a combustor **10** through a conduit **12**. The spent zinc chloride melt resulting from the hydrocracking of coal or coal derived products contains, in addition to zinc chloride, carbonaceous residue as well as ash, sulfur, and nitrogen components. The carbon, sulfur and nitrogen components in the spent catalyst are oxidized by air fed to the combustor via a conduit **14**. The air preferably contains a low concentration of HCl to aid in preventing hydrolysis of zinc chloride by steam.

Figure 179: Continental Oil Co. Process for Regeneration of Zinc Chloride Hydrocracking Catalyst

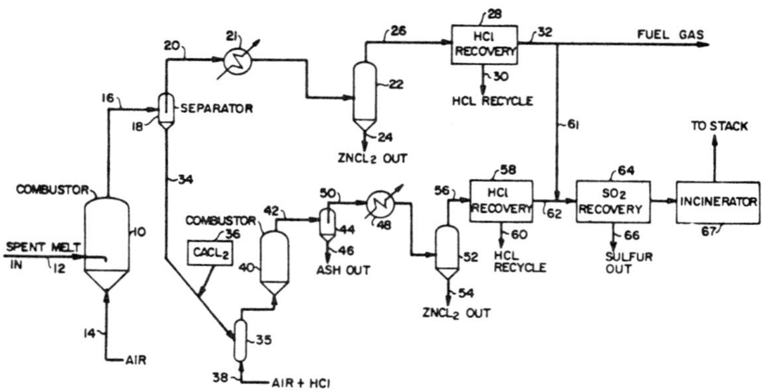

Source: U.S. Patent 4,136,056

The combustion conducted in the combustor may be either with less than or greater than, the stoichiometric amount of air. In the former case, fuel gas will be produced; in the latter case, flue gas. Heat is generated in the combustor by oxidation of the carbon contained in the spent melt and is sufficient to maintain the temperature in the combustor above the vaporization temperature of zinc chloride. The latter, in vapor form, together with steam, noncondensable gases, N_2, CO and CO_2, unreacted HCl, and entrained solids, is withdrawn from the combustor **10** through a conduit **16** to a hot separator **18** which is any cyclone commonly used for the separation of gases and solids at high temperature. In this instance, the temperature of the cyclone is not permitted to drop below the condensation point of zinc chloride.

Zinc chloride and water vapors and noncondensable gases CO, CO_2, N_2 and HCl are withdrawn from the separator **18** through a conduit **20** to a condenser **22** for selective condensation after passing through a precooler **21**. Zinc chloride condensate in molten form is discharged from the condenser through a conduit **24** for return to the hydrocracking zone (not shown). The noncondensable gases are withdrawn from the condenser through conduit **26** to a suitable HCl recovery zone **28** from which HCl is recovered through conduit **30** for reuse in the process. A gas is withdrawn from the HCl recovery zone through conduit **32** which is a low sulfur fuel gas whenever substoichiometric quantities of air are used in combustor **10**. A typical analysis of the separated solids from separator **18** is as follows:

Constituent	Percent by Weight
Zn*	2.94
Ins. Zn**	4.54
Cl	4.92
H	0.08

(continued)

C	0.09
S	0.37
Na_2O	0.10
K_2O	0.03
CaO	11.40
MgO	4.45
Fe_2O_3	3.90
TiO_2	0.94
SiO_2	42.86
Al_2O_3	18.41
Total	95.03***

*In the form of water-soluble zinc compounds
**In the form of water-insoluble zinc compounds
***Other compounds were obviously present that could not be readily determined

The solids discharged from the hot separator **18** are conducted by a conduit **34** to a premixer and preheater **35.** $CaCl_2$ in finely divided form is introduced into conduit **34** from a hopper **36.** The amount is preferably twice the amount needed to react with the water-insoluble zinc compounds, as determined by analysis. The solids are mixed with a stream of HCl-containing gas introduced through a conduit **38.** The gas may conveniently be air or combustion gases.

The mixture of gases and entrained solids is transferred to a second combustor **40** where the zinc oxide and zinc oxide complexes are converted to zinc chloride by reaction with HCl and $CaCl_2$. The temperature of the combustor is maintained above the vaporization temperature of zinc chloride, for example, 1700°F, by the oxidation of carbon in the case where oxygen is contained in the HCl-containing gas. The carbon is that contained in the unconsumed organic residue carried over from the first combustor or is added as required to the second combustor expressly for that purpose. Such added or supplemental carbon may be in the form of coal.

Zinc chloride vapors, ash and noncondensable gases are withdrawn from the combustor **40** through a conduit **42** to a hot cyclone separator **44** where the ash containing added calcium is rejected through a line **46.** The vapors and gases pass through a precooler **48** in a conduit **50** to a condenser **52** adapted to selectively condense zinc chloride in molten form. The molten zinc chloride is withdrawn through line **54** for recycle. The noncondensable gas is withdrawn through a conduit **56** to a suitable HCl recovery zone **58** for recovery of HCl through a conduit **60.**

The HCl-free gas, together with any sulfur-rich gas withdrawn from conduit **32** through a conduit **61,** is conducted through a conduit **62** to an SO_2-reduction zone **64** for conversion of SO_2 in the gas to elemental sulfur. The amount of SO_2 in conduit **32** and in conduit **62** is a function of the amount of air used in combustor **10,** the larger the amount of air, the greater the amount of SO_2 in conduit **32,** and conversely the smaller the amount, the larger the amount in conduit **62.** In either case, it is desirable to recover such SO_2 as sulfur. The CO and H_2 contained in the gas provides a suitable source of reducing agent for SO_2 reduction. The sulfur product is recovered through a conduit **66.** The remaining gas may be incinerated in an incinerator **67** to remove residual sulfur compounds.

ZINC CHLORIDE FROM USED ACRYLIC FIBER SPINNING SOLUTIONS

See "Sodium Thiocyanate from Used Acrylic Fiber Spinning Solutions."

ZIRCONIUM FROM SPENT PICKLE LIQUOR

Basically what occurs in the process of pickling of zirconium and zirconium metal alloys is as follows: the pickle acid used for pickling zirconium and its alloys comprises HF and HNO_3. Of the two types of acid present in the pickle acid solution, only hydrofluoric acid reacts with the zirconium metal. Among other functions, the nitric acid primarily reacts with tin which is a common component in the zirconium alloy and also is a source of hydrogen ion which accelerates the HF attack. As the zirconium is eaten away by the HF, the tin is simultaneously removed by the HNO_3. If the HNO_3 was not present, a layer of tin would be left behind on the surface of the zirconium metal and the pickling reaction would be suppressed.

Since tin is present in only very small amounts in zirconium alloys, the amount of HNO_3 consumed during the course of pickling is very low compared to the amount of HF consumed.

In terms of a chemical reaction, the following occurs during pickling: 4 HF molecules react with 1 Zr metal atom to form 1 ZrF_4 (zirconium fluoride) molecule and 2 hydrogen gas molecules.

At the time that a tank of pickle acid is spent and ready to be dumped, the following substances are present in the acid:

(1) Residual HF which has not been consumed by pickling.

(2) Nitric acid, a small fraction of which has been consumed.

(3) Dissolved ZrF_4 (zirconium fluoride).

(4) Other impurities such as dissolved tin which reacted with the nitric acid.

This spent pickle acid must be disposed of in some way since as it is it is unusable. This presents a pollution problem since it cannot be just dumped. Presently, the spent pickle acid is neutralized with lime. This requires a large expenditure for lime along with there being no saving of any acid.

A process developed by *J.A. Megy and R.L. Propst; U.S. Patent 4,105,469; August 8, 1978; assigned to Teledyne Industries, Inc.* is one in which a HF-HNO_3 pickle acid bath for pickling zirconium, zirconium alloys, hafnium, and hafnium alloys is regenerated by adding to the spent pickle acid a sufficient amount of NaF to precipitate out, in the case of zirconium, ZrF_4 as Na_2ZrF_6. HF and HNO_3 are added to the bath to make up losses and the pickle acid is recycled for use.

A suitable arrangement of equipment for the conduct of such a process is shown in Figure 180.

Figure 180: Teledyne Industries Inc. Process for Regenerating a HF-HNO_3 Zirconium Pickle Acid Bath and Recovering Zirconium Therefrom

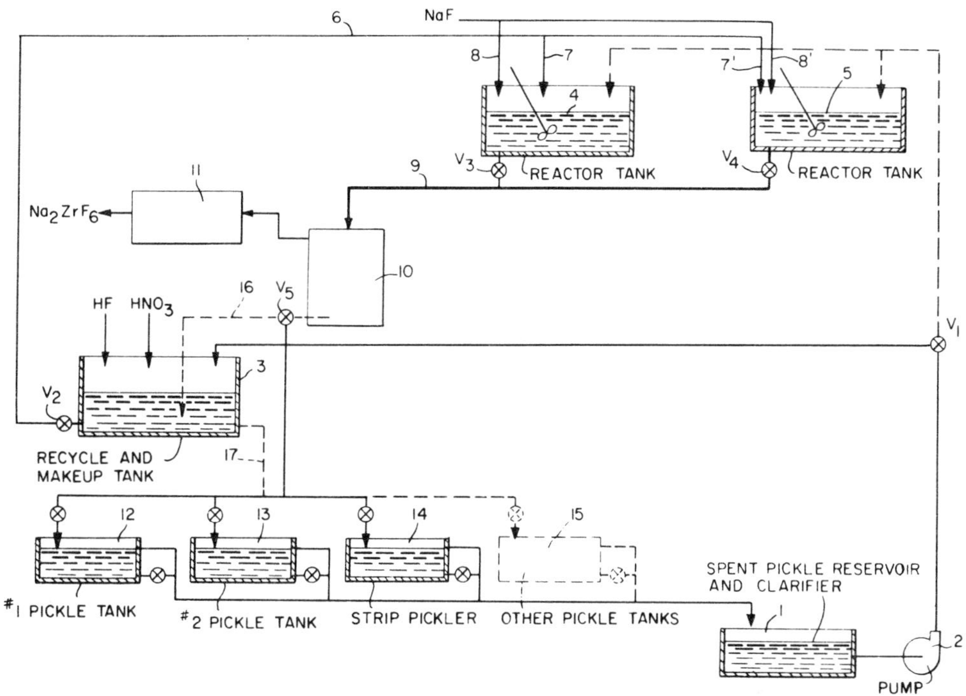

Source: U.S. Patent 4,105,469

As shown in the drawing, the spent pickle acid is stored and clarified in tank **1.** It is then pumped by pump **2** to make-up tank **3** by way of valve **V1.** Here HF is added to make up for that amount of acid used in pickling the zirconium and its alloys and, when necessary, HNO_3 is added to bring the solution up to the pickling concentration. From here the solution is pumped by way of valve **V2** to reactors **4** and **5** by way of lines **6** and **7** and **7'** where NaF is added by lines **8** and **8'** to precipitate ZrF_4 as Na_2ZrF_6 in accordance with the following equation:

$$ZrF_4 + 2NaF \rightarrow Na_2ZrF_6$$

The solution containing the precipitate is then pumped through line **9** by way of valves **V3** and **V4** to filter **10** where the solution is separated from the precipitate. The precipitate is then dried in drier **11** and the solution is pumped by way of valve **V5** to the pickle acid tanks **12, 13, 14** and **15.**

The spent pickle acid can be bulked to pickle concentration either before or after the precipitation of the Na_2ZrF_6. In the latter case, the spent pickle acid from tank **1** is pumped directly to reactor tanks **4** and **5** where NaF is added to precipitate ZrF_4 as Na_2ZrF_6. The solution containing the precipitate is then run through line **9** by way of valves **V3** and **V4** to filter **10.** The precipitate is then dried in drier **11** and the solution of pickle acid is pumped by way of valve **V5** and line **16** to make-up tank **3** where HF and HNO_3, when necessary, are added to bring the pickle acid to pickling concentration. The pickle acid is then pumped to pickle tanks **12, 13, 14** and **15** by way of line **17.**

Organic and Polymer Waste Reclaiming

The contents of a companion volume, *Organic and Polymer Waste Reclaiming Encyclopedia* by the same author, and also available from Noyes Data Corporation are given below.

ABS Plastics from Plated ABS Scrap
ABS Plastics from Storage Battery Scrap
Acetal Polymers from Acetal Resin Scrap
Acetic Acid from Furfural Manufacture
Acrylic Resins from PVC/Acrylic Resin Mixtures
Acrylonitrile from Acrylonitrile Process Effluents
Acrylonitrile Polymers from Plastic Bottle Waste
Acrylonitrile/Styrene Polymers from Storage Battery Scrap
Activated Carbon from Rubber Tire Discards
Activated Carbon from Sawdust
Activated Carbon from Sewage Sludge
Activated Carbon from Spent Activated Carbon
Activated Carbon from Spent Pulping Liquor
Adhesives from Polyolefin Resin Wastes
Adipic Acid from Nylon Waste
Alcohols from Lube Oil Additive Manufacturing Residues
Animal Feed from Cottonseed Hulls
Animal Feed from Fermentation Residues
Animal Feed from Pear Waste
Animal Feed from Wastepaper
Animal Feed Supplement from Sewage Sludge
Anthraquinone from Tall Oil
Aromatic Amines from Spent Color Photography Solutions
Asphalt from Oil Spills
Asphaltic Concrete from Old Pavement

Barrier Materials from Plastic Bottle Waste
Bis(Hydroxyethyl) Terephthalate Oligomers from Polyethylene Terephthalate Scrap
Blast Furnace Fuel from Scrap Tires
Building Blocks from Sawdust
Butyric Acid from Oxo Residues

Caprolactam from Caprolactam Still Bottoms
Caprolactam from Nylon Waste
Carbon Black from Rubber Tire Discards
Carbon Monoxide from Gas Mixtures
Cellophane from Waste Cellophane
Cellulose from Coated Cellophane Film Scrap
Cellulose Fibers from Milk Cartons
Cellulose Fibers from Municipal Waste
Cellulose Fibers from Packaging Scrap
Cellulose Fibers from Paper/Foil Laminates
Cellulose Fibers from Wastepaper
Cellulose Fibers from Wastewaters
Charcoal from Sawdust
Chipboard from Sawdust
Chlorinated Isocyanurates from Wastewaters
Chlorinated Solvents from Dry Cleaning Operations
Chlorinated Solvents from Photoresist Manufacture
Chlorinated Solvents from Waste Gases

Coating Resins from Crosslinked Polyethylene Scrap
Color Developing Agents from Spent Color Photography Solutions
Cotton Fibers from Cotton/Synthetic Fiber Blends
Cotton Linters from Cottonseed Hulls
Cyclododecanol/one from Oxidation Process Wastewaters

Deinked Cellulose Pulp from Printed Wastepaper
Developing Agents from Spent Photographic Solutions
Dewaxing Aids from Dewaxing Processes
Diamines from Urethane Resin Scrap
1,2-Dichloroethane from Vinyl Chloride Production Effluents
Diethylene Glycol from Polyester Process Wastes
Dimethyl Formamide from Cellulose Spin Baths
Dimethyl Formamide from PVC Reactor Cleaning Solvents
Dimethyl Formamide from Used Acrylic Fiber Spinning Solutions
Dimethyl Sulfoxide from Used Acrylic Fiber Spinning Solutions
Dimethyl Terephthalate from Oxidation Process Residues
Dimethyl Terephthalate from Polyethylene Terephthalate Waste
Dyes from Textile Dyeing Wastewaters

Ebonite from Storage Battery Scrap
Emulsion Coolant from Can Manufacturing Process
Ethyl Acetate from Sulfite Pulping Wastes
Ethylene from Polyethylene Scrap
Ethylene Dichloride from Process Off-Gases
Ethylene Glycol from Polyester Process Wastes
Ethylene Glycol from Polyester Still Bottoms
Ethylene Glycol from Polyethylene Terephthalate Waste
Ethylene Glycol from Wastewaters
Ethylene Glycol Dinitrate from Wastewaters
Ethylene Oxide from Sterilizing Process

Fats from Damaged Butter and Oleo Packages
Fats from Edible Oil Refinery Effluents
Fertilizer from Citrus Pulp
Fertilizer from Fermentation Residues
Fertilizer from Sawdust
Fertilizer from Sewage Sludge
Fertilizer from Spent Sulfite Liquor
Fiberboard from Barking Wastes
Fiberboard from Municipal Wastes
Foam Polyurethanes from Urethane Foam Waste
Formaldehyde from Wastewaters
Formic Acid from Furfural Manufacture
Fuel from Rubber Tire Discards
Fuel from Waste Lubricating Oils
Fuel Oil from Rubber and Plastic Wastes
Furfural from Oat Hulls
Furfural from Sulfite Pulping Wastes
Furfural from Wood Chips

Gasoline from Fuel Transfer Operations
Gasoline from Organic Wastes
Gasoline from Rubber and Plastic Wastes
Gelatin from Chrome Tanning Operations

Hexamethylene Diamine from Nylon Waste
Hexamethylene Tetramine from Formaldehyde Wastes
HMX (Cyclotrimethylene Nitramine) from Solid Propellant Waste
Hydraulic Fluids from Used Hydraulic Fluids
Hydrocarbons from Refinery Waste Gases
Hydrogen Cyanide from Ammonoxidation Process Wastes

Industrial Oils from Waste Industrial Oils

Latex-Base Paint from Wastewater Paint Solids
Lead Alkyl Compounds from Wastewaters
Lubricating Oil from Used Crankcase Oil

Textile Sizes from Spent Size Solutions
Thermosetting Resins from Scrap Polymer
Tire Cord from Rubber Tire Discards
Toluene from PVC Reactor Cleaning Solvents
Toluenediamine from Urethane Resin Scrap
Triarylboranes from Spent Catalysts
Tributyl Phosphate from Nuclear Fuel Processing
Trichloroethane from Degreasing
Trichloroethane from Photoresist Manufacture
Trichloroethylene from Waste Gases

Urea from Phenolic Solutions of Urea

Vanillin from Pulping Waste Liquors
Vinyl Chloride Monomer from Polyvinyl Chloride Latex
Vinyl Coated Fabric from Vinyl Coated Fabric Scrap
Vinyl Flooring from Scrap Floor Covering Material

Waxes from Refinery Sludges
Wood Fluff Fibers from Disposable Diaper Rejects

Xylitol from Cotton Seed Hulls

Yeasts from Baking Industry Wastes

Sources of Waste Feed Materials

Listed below in alphabetical order are the sources of waste feed materials for the various reclaiming processes and products described in this book. Each entry is followed by the material(s) which may be reclaimed from this waste feed source.

Caprolactam manufacture
  Ammonium sulfate

Caustic/chlorine process sludges
  Mercury

Caustic sludges
  Caustic

Cell sludges
  Magnesium

Cellulose spin bath
  Nitrogen tetroxide

Cement kiln dust
  Cement

Chlorination of organics
  Hydrogen chloride

Chromic anhydride production
  Chromium

Coal ash slag
  Portland cement

Color TV tube manufacture
  Yttrium oxysulfide

Cooling tower blowdown water
  Chromium

Copper refinery anode slimes
  Copper
  Selenium
  Silver
  Tellurium

Copper smelter flue dusts
  Arsenic

Copper sulfide ores
  Molybdenum

Cyclohexane oxidation wastes
  Boric acid

Debismuthizing dross
  Bismuth

Dehydrogenation process wastes
  Iodine

Dilute HCl solutions
  Hydrogen chloride

Drilling muds
  Barite

Dross from hot dip coating
  Tin

Electrical conductor scrap
  Aluminum

Electrical scrap
  Copper

Electroplating rinse waters
  Chromium

Electroplating wastes
  Silver
  Zinc

Flue dusts
  Lead oxide

Flue gas scrubber sludges
  Fertilizer

Fly ash
  Aluminum
  Paving material
  Porous ceramics
  Pozzolan

Furnace pot linings
  Hydrogen fluoride

Furnace skim or dross
  Aluminum

Galvanizing wastes
  Zinc

Glass container waste
  Glass

Glass fiber scrap
  Glass fibers

Gun club soils
  Lead

Halogen tin electroplating sludges
  Sodium ferrocyanide

Halogen tin electroplating wastes
  Tin

Heat treating operations
  Sodium nitrate

Hydrocracking process effluents
  Sulfur

Ready-mix concrete residues
  Cement

Red mud from Bayer process
  Alumina
  Building bricks
  Flocculating agent
  Iron

Rolling mill scale
  Iron

Rolling mill sludge
  Iron oxide

Scarfer spittings
  Abrasive material

Scrap aluminum
  Aluminum

Scrap antimony
  Antimony

Scrap automobiles
  Cast iron
  Copper
  Glass
  Nonferrous metals
  Steel
  Zinc

Scrap beryllium
  Beryllium

Scrap chromium
  Chromium

Scrap cobalt
  Cobalt

Scrap columbium
  Columbium

Scrap copper
  Copper

Scrap electrodes
  Ruthenium

Scrap furnace lining brick
  Magnesia

Scrap gold
  Gold

Scrap iron
  Iron

Scrap magnesium
  Magnesium

Scrap mercury
  Mercury

Scrap mining bits
  Tungsten carbide

Scrap molybdenum
  Molybdenum

Scrap nickel
  Nickel

Scrap photographic film
  Silver

Scrap silver
  Silver

Scrap steel
  Steel

Scrap tin
  Tin

Scrap tires
  Iron

Scrap titanium
  Titanium

Sewage sludge incineration residues
  Lime

Silver-brazed stainless steel scrap
  Silver

Silver refinery muds
  Gold

Sintered metal carbides
  Cobalt
  Tungsten

Slags
  Tin

Smelter flue dust
  Copper
  Molybdenum

Smelter slags
  Copper

Solder skimmings
  Lead

Tungsten scrap
  Tungsten

TV picture tubes
  Glass

UO_2 process wastewaters
  Calcium fluoride
  Uranium

Uranium process sludges
  Scandium

Urea plant wastewater
  Ammonia

Urethane reaction products
  Selenium

Used acrylic fiber spinning solutions
  Nitric acid
  Sodium thiocyanate
  Zinc chloride

Used foundry sand
  Foundry sand

Waste asbestos slurries
  Asbestos

Waste chromic acid solutions
  Chromic acid

Wastewaters
  Chlorine
  Gold
  Rhenium
  Silver

Water treatment sludge
  Paving material

Whole scrap tires
  Steel

Xerographic equipment discards
  Selenium

Zinc refinery flue dusts
  Cadmium
  Lead

Zinc scrap
  Zinc

Zinc smelter flue dusts
  Germanium
  Thallium

References

(1) Sittig, M., *Organic and Polymer Waste Reclaiming Encyclopedia,* Noyes Data Corp. (In Press), Park Ridge, NJ.

(2) U.S. Dept. of the Interior, Bureau of Mines, "Minerals and Materials–A Monthly Survey," Washington, D.C.

(3) Sittig, M., *Resource Recovery and Recycling Handbook of Industrial Wastes,* Noyes Data Corp., Park Ridge, NJ (1975).

(4) Barton, A.F.M., *Resource Recovery and Recycling,* John Wiley and Sons, New York, NY (1979).

(5) Nuss, G.R., Franklin, W.E., Hahlin, D., Park, W., Urie, M. and Cross, J. (Midwest Research Institute), *Base Line Forecasts of Resource Recovery, 1972 to 1990,* Report No. EPA/530/SW107c, U.S. Environmental Protection Agency, Washington, D.C. (March 1975).

(6) Kusik, C.L. and Kenahan, C.B. (Arthur D. Little, Inc.), *Energy Use Patterns for Metal Recycling,* Information Circular No. 8781, U.S. Bureau of Mines, Washington, D.C. (1978).

(7) Office of Technology Assessment, *Technical Options for Conservation of Metals,* Washington, D.C. (September 1979).

(8) National Technical Information Service, *Solid Waste Reclamation and Recycling, Part 1–Packaging and Containers,* Publ. No. NTIS/PS-79/0826, National Technical Information Service, Springfield, VA (1979).

(9) National Technical Information Service, *Solid Waste Reclamation and Recycling, Part 2–Plastics,* Publ. No. NTIS/PS-79/0827, National Technical Information Service, Springfield, VA (1979).

(10) National Technical Information Service, *Solid Waste Reclamation and Recycling, Part 3–Metals,* Publ. No. NTIS/PS-79/0828, National Technical Information Service, Springfield, VA (1979).

(11) National Technical Information Service, *Solid Waste Reclamation and Recycling, Part 4–Glass,* Publ. No. NTIS/PS-79/0829, National Technical Information Service, Springfield, VA (1979).

(12) National Technical Information Service, *Solid Waste Reclamation and Recycling, Part 5–Paper,* Publ. No. NTIS/PS-79/0830, National Technical Information Service, Springfield, VA (1979).

(13) National Technical Information Service, *Solid Waste Reclamation and Recycling, Part 6–Tires,* Publ. No. NTIS/PS-79/0831, National Technical Information Service, Springfield, VA (1979).

(14) U.S. Bureau of Mines, *Mineral Facts and Problems,* Bulletin 667, Washington, D.C. (1975).

(15) Nack, H., Murthy, K., Stambaugh, E., Carlton, H. and Smithson, G.R., Jr. (Battelle Columbus Laboratories), *Development of an Approach to Identification of Emer-*

gency Technology and Demonstration Opportunities: Environmental Protection Technology, U.S. Environmental Protection Agency, Washington, D.C. (1974).

(16) U.S. Dept. of the Interior, Bureau of Mines, *Mineral Commodity Profile Series,* Washington, D.C. (1978 and 1979).

(17) U.S. Bureau of Mines, *The Economics of By-Product Metals,* Information Circular No. 8569 and 8570, Washington, D.C. (1973).

BIODEGRADATION TECHNIQUES FOR INDUSTRIAL ORGANIC WASTES 1980

Edited by D.J. De Renzo

Pollution Technology Review No. 65
Chemical Technology Review No. 158

This book provides engineering and waste management personnel with data on the biological treatment of hazardous and toxic organic compounds in problematic industrial waste effluents. Selected biological treatment operations which can successfully handle those wastes not readily treatable by conventional processes are identified and characterized.

Various methodologies are assessed to determine those treatment techniques applicable to the broadest range of waste classes. Design, performance and economic comparisons of the technologies are presented. The results of in-depth studies carried out at several biological treatment facilities are also included. A condensed table of contents appears below:

ISBN-08155-0800-X **358 pages**

BUILDING MATERIALS FROM SOLID WASTES 1979

Edited by O.S. Renfroe

Pollution Technology Review No. 61

This book examines the feasibility of utilizing products available from industrial, mining and municipal wastes for residential and commercial building construction. Utilization of waste products for these applications is a worthwhile disposal technique, often resulting in a higher quality, better reinforced building structure at less cost.

For example: The addition of surplus magnesium oxychloride during the manufacture of plastic laminated panels increased the tensile resistance to strain and in addition to strength, the laminate would neither burn nor smoke.

Sheet rock, dry wall or gypsum boards are highly susceptible to damage from impact or water. Comparative boards made from rice hulls or peanut shells proved superior in hardness and screw-retention qualities.

Other lumber substitutes of superior quality can be made from inorganic and organic by-products as demonstrated in this book. Some of the experimental processes are backed up by encouraging cost figures.

The partial table of contents below lists chapter headings and **examples of some** subtitles.

ISBN 0-8155-0771-2 **275 pages**

PRIORITY TOXIC POLLUTANTS
Health Effects and Allowable Limits 1980

Edited by Marshall Sittig

Enivronmental Health Review No. 1

This book is a practical manual with its contents arranged alphabetically in encyclopedic form. It is intended to provide specific information on the 65 priority toxic pollutants (actually reflecting 129 individual compounds), their derivatives or degradation products, or intermedia transfers.

Compelled by a consent decree obtained in a federal court by public interest groups, the Environmental Protection Agency has promulgated criteria on allowable limits and profuse guidelines on how to interpret them and also how to determine further such criteria.

Here, in ready reference form, are the essentials of the criteria and guidelines which so far have been published for the control of priority toxic pollutants. Each typical pollutant is characterized with descriptions of its:

Occurrence
Physical Properties
Chemical Properties
Uses
Toxic Effects
Current Levels of Exposure
Special Groups at Risk
Existing Guidelines & Standards
Summary of Proposed Criteria
Bases for the Human Health Criteria
Pertinent References

A list of the 65 original priority pollutants follows here, however all of the 129 compounds are covered in this book.

1. Acenaphthene
2. Acrolein
3. Acrylonitrile
4. Aldrin/Dieldrin
5. Antimony & Compounds
6. Arsenic & Compounds
7. Asbestos
8. Benzene
9. Benzidine
10. Beryllium & Compounds
11. Cadmium & Compounds
12. Carbon Tetrachloride
13. Chlordane
14. Chlorobenzenes
15. Chlorinated Ethanes
16. Chloroalkyl Ethers
17. Chlorinated Naphthalenes
18. Chlorinated Phenols
19. Chloroform
20. 2-Chlorophenol
21. Chromium & Compounds
22. Copper & Compounds
23. Cyanides
24. DDT & Metabolites
25. Dichlorobenzenes
26. Dichlorobenzidine
27. Dichloroethylene
28. 2,4-Dichlorophenol
29. Dichloropropane/propene
30. 2,4-Dimethylphenol
31. Dinitrotoluenes
32. Diphenylhydrazines
33. Endosulfan
34. Endrin
35. Ethylbenzene
36. Fluoranthene
37. Haloethers
38. Halomethanes
39. Heptachlor
40. Hexachlorobutadiene
41. Hexachlorocyclohexane
42. Hexachlorocyclopentadiene
43. Isophorone
44. Lead & Compounds
45. Mercury & Compounds
46. Naphthalene
47. Nickel & Compounds
48. Nitrobenzene
49. Nitrophenols
50. Nitrosamines
51. Pentachlorophenol
52. Phenol
53. Phthalate Esters
54. Polychlorinated Biphenyls
55. Polynuclear Aromatics
56. Selenium & Compounds
57. Silver & Compounds
58. Tetrachlorodibenzo-p-dioxin
59. Tetrachloroethylene
60. Thallium & Compounds
61. Toluene
62. Toxaphene
63. Trichloroethylene
64. Vinyl Chloride
65. Zinc & Compounds

ISBN 0-8155-0797-6 **370 pages**

HAZARDOUS AND TOXIC EFFECTS OF INDUSTRIAL CHEMICALS 1979

by Marshall Sittig

There is a continuing need to assess the status of potentially dangerous substances including those now available and those that may reach commercial availability in the future. This should be done with a predictive view to avoid or at least to ameliorate catastrophic episodes similar to those that have occurred with methyl mercury, polychlorinated biphenyls, vinyl chloride monomer, dioxin and a number of pesticides.

This handbook is intended to be a working guide for the industrial hygienist and other interested persons. Most of the individual chemical listings contain a number of literature references.

This book is intended to be a guide to these hazardous chemicals, to give first warning signals, and to serve as an introduction to the published literature and to government agencies offering detailed guidance.

Information (where available) under each chemical listing includes:

Description. Derivation, chemical structure and systematic chemical name.
Synonyms. Other chemical names, generic names and trademarks.
Potential Occupational Exposures.
Permissible Exposure Limits.
Routes of Entry.
Harmful Effects. Local, systemic.
Medical Surveillance.
Special Tests.
Personal Protective Methods.
Bibliography. References to articles in journals and government publications.

The book consists of about 250 individual monographs arranged alphabetically by the common name of each substance described. It is impossible to indicate here all of the chemical entries in this book. In order to give an indication of the vast scope of information dealt with here, the names of most entries in the beginning of the book are reproduced here.

Acetaldehyde
Acetates
Acetic Acid
Acetic Anhydride
Acetone (See Ketones)
Acetonitrile
Acetylaminofluorene
Acetylene
Acridine
Acrolein
Acrylamide
Acrylonitrile
Alkanes
Allyl Alcohol
Allyl Chloride
Aluminum & Compounds
Aminodiphenyl
Ammonia
Amyl Alcohol
Aniline
Antimony & Compounds
Arsenic
Arsine
Asbestos
Asphalt Fumes
Barium & Compounds
Benzene
Benzidine & Salts
Benzoyl Peroxide
Benzyl Chloride
Beryllium & Compounds
Bis(Chloromethyl) Ether
Bismuth & Compounds
Boron & Compounds
Boron Hydrides
Boron Trifluoride
Brass
Bromine
1,3-Butadiene
Butyl Alcohol
Butylamine
Cadmium & Compounds
Calcium Cyanamide
Calcium Oxide
Carbaryl
Carbon Dioxide
Carbon Disulfide
Carbon Monoxide
Carbon Tetrachloride
Carbonyls
Cement Dust
Cerium & Compounds
Chlorinated Benzenes
Chlorinated Lime
Chlorinated Naphthalenes
Chlorine
o-Chlorobenzylidene Malonitrile
Chlorodiphenyls & Derivs.
Chloroform
Chloromethyl Methyl Ether
Chloroprene
Chromium & Compounds
Coal Tar Products
Cobalt & Compounds
Copper & Compounds
Cotton Dust
Creosote
Cresol
Cyclohexanone
Diacetone Alcohol
2,4-Diaminoanisole
4,4'-Diaminodiphenylmethane
Dibromochloropropane
1,2-Dibromoethane
3,3'-Dichlorobenzidine
1,2-Dichloroethane
1,2-Dichloroethylene
Dichloroethyl Ether
Diisobutyl Ketone
N,N-Dimethylacetamide
4-Dimethylaminoazobenzene
N,N-Dimethylformamide
Dimethyl Sulfate
Dinitrobenzene
Dinitro-o-Cresol
Dinitrophenol
Dinitrotoluene
Dioxane
Diphenyl
Epichlorohydrin
Ethanolamines
Ethyl Alcohol
Ethylbenzene
plus 160 other chemicals

ISBN 0-8155-0731-3 **$42** **460 pages**